The Complete PCI Express† Reference

Design Implications for Hardware and Software Developers

Edward Solari
Brad Congdon

Copyright © 2003 Intel Corporation. All rights reserved.

ISBN 0-9717861-9-4

No part of this publication may be reproduced, stored in a retrieval system or transmitted in any form or by any means, electronic, mechanical, photocopying, recording, scanning or otherwise, except as permitted under Sections 107 or 108 of the 1976 United States Copyright Act, without either the prior written permission of the Publisher, or authorization through payment of the appropriate per-copy fee to the Copyright Clearance Center, 222 Rosewood Drive, Danvers, MA 01923, (978) 750-8400, fax (978) 750-4744. Requests to the Publisher for permission should be addressed to the Publisher, Intel Press, Intel Corporation, 2111 NE 25th Avenue, JF3-330, Hillsboro, OR 97124-5961. E-Mail: intelpress@intel.com.

This publication is designed to provide accurate and authoritative information in regard to the subject matter covered. It is sold with the understanding that the publisher is not engaged in professional services. If professional advice or other expert assistance is required, the services of a competent professional person should be sought.

Intel Corporation may have patents or pending patent applications, trademarks, copyrights, or other intellectual property rights that relate to the presented subject matter. The furnishing of documents and other materials and information does not provide any license, express or implied, by estoppel or otherwise, to any such patents, trademarks, copyrights, or other intellectual property rights.

Intel may make changes to specifications, product descriptions, and plans at any time, without notice.

Fictitious names of companies, products, people, characters, and/or data mentioned herein are not intended to represent any real individual, company, product, or event.

Intel products are not intended for use in medical, life saving, life sustaining, critical control or safety systems, or in nuclear facility applications.

Intel and the Intel logo are trademarks or registered trademarks of Intel Corporation.

† Other names and brands may be claimed as the property of others.

"The information in this book comes from many sources and will be implemented in many ways. Any errors in the information, omissions in the information, typographical errors, or printing errors shall not imply and shall not incur directly or indirectly any liability by the authors and Research Tech Inc. The authors and Research Tech Inc. make no warranty for the correctness in this book of the information or technical explanations, and assume no direct or indirect liability for the interpretation of the information or technical explanations. The authors and Research Tech Inc. reserve the right to make changes to the book without any notice, without incurred liability, and without warranty."

This book is printed on acid-free paper.

Publisher: Richard Bowles
Editor: David J. Clark
Managing Editor: David B. Spencer
Content Manager: Matthew Wangler
Text Design & Composition: Wasser Studios
Graphic Art: Donna Lawless and Rich Jevons (illustrations), Ted Cyrek (cover)

Library of Congress Cataloging in Publication Data:

Printed in the United States of America

    10 9 8 7 6 5 4 3 2 1

Because of the close relation of PCI Express to PCI, reference books that describe PCI software, especially the different editions of *PCI Hardware and Software: Architecture and Design* by Edward Solari and George Willse, are still valuable and were helpful in creating this book. In particular, Ed was a great motivator and mentor to me from the first day of writing this book. I couldn't have done it without him.

Saving the best for last, I couldn't have written this book without the wholehearted support of my lovely wife Sarah, who makes me a better man.

Brad

Writing a book of this type is a very time-consuming effort and I am very dependent on my co-author to complete his portion of the book independently. Brad has been super in undertaking this task and making sure all of the material was first rate.

I would like to dedicate this book to my father, mother, and of course my very understanding wife. Additionally I would like to thank my wife for letting me take time away from other things to work on this book.

Edward

Contents at a Glance

Chapter 1	Architecture Overview	1
Chapter 2	PCI Express Architecture Overview	41
Chapter 3	Implementation	93
Chapter 4	Addressing and Routing	121
Chapter 5	Transaction Packets	171
Chapter 6	Transaction Layer	215
Chapter 7	Data Link Layer and Packets	315
Chapter 8	Physical Layer and Packets	359
Chapter 9	Errors	421
Chapter 10	Transaction Ordering	501
Chapter 11	Flow Control Protocol: Part 1	519
Chapter 12	Flow Control Protocol: Part 2	581
Chapter 13	PME Overview, Wake Events, and Reset	637
Chapter 14	Link States	671
Chapter 15	Power Management	731
Chapter 16	Hot Plug Protocol	775
Chapter 17	Slot Power Protocol	797
Chapter 18	Advanced Switching	807
Chapter 19	Interrupts	813
Chapter 20	Lock Protocol	833
Chapter 21	Mechanical and Electrical Overview	841
Chapter 22	Configuration Overview	865
Chapter 23	Configuration Registers	893
Chapter 24	Configuration Capabilities	937
Chapter 25	Real-Time Analysis	977
Appendix A	Capability IDs	1001
Glossary		1005
Index		1019

For information on The Complete PCI Express Reference Tutorial, turn the page.

The Complete PCI Express Reference Tutorial

> The specification provides the outline and this book provides the detailed implementation information for PCI Express. PCI Express defines a wide range of interactions among different elements. Many of these elements are new concepts for bus protocols implemented by PCs and servers. In order to help designers to quickly understand and implement all PCI Express elements, The Complete PCI Express Reference Tutorial was developed.
>
> The Complete PCI Express Reference Tutorial is a self-paced tutorial tailored to this book. The tutorial consists of special instructional text, illustrations, and flow charts that directly reference more detailed text, illustrations, and flow charts in this book. Additionally, the tutorial includes a smart index that alerts the designer to locations of relevant information in this book. The tutorial is a PowerPoint file downloadable for FREE from
>
> **www.intel.com/intelpress/pciexpresscomplete**

Contents

Acknowledgements   xxi

Chapter 1   Architecture Overview   1
  The Original PCI Bus   2
  PCI Platforms   3
    Early Architecture   3
    PCI Hub-Based Architecture   5
    PCI 64 Data Bits and 66.6 Megahertz Emerge   6
    PCI Protocol Basics   7
    Limitations of the PCI Bus Architecture   14
    PCI Bus Feature Summary   15
  PCI-X Platforms   16
    Split Transaction Protocol   17
    Improved Performance of PCI-X 1.0 over PCI   17
    PCI-X 2.0 Increases Performance Further   19
    PCI-X 1066   20
    PCI-X Bus Feature Summary   21
  PCI Express   23
    PCI Express Improves Upon PCI and PCI-X   23
    Platform Implementations with PCI Express   26
    PC and Server Platforms   27
    PCI Express Specific Performance   31
    Other PCI Express Performance and Features   33

Chapter 2 PCI Express Architecture Overview 41

    Generic PCI Express Platform Architecture 41
        Hierarchy and Hierarchy Domain 44
        Other Interconnects between PCI Express Devices 44
    Generic Architecture of PCI Express Devices and Flow
        Control Considerations 48
        Root Complex 48
        Switches 52
        Endpoints 54
        Bridges 54
    Overview of Transaction Flow and Participants 57
        Restrictions of Root Complex as Requester/Completer 57
        Restrictions on Switch TLP Porting 58
        Endpoints 59
        Restrictions on Bridges TLP Porting 61
        In-band and Advanced Peer-to-Peer Communications 62
    Transactions and Packets 62
        Requester/Completer Protocol 63
        PCI Express Layers and Packets Overview 65
        Transmitting and Receiving Packets: An Example 69
    Flow Control Protocol 74
        Quality of Service 75
        Traffic Classes 76
        Virtual Channels 78
        Link Flow Control 78
    Other PCI Express Compatibilities and Differences
        with PCI and PCI-X 78
        Address Spaces 79
        Messages 84
        Interrupts 84
        Lock Function 85
        Power Management 85
        Hot Plug 86
    Mechanical and Electrical Overview 86
        Mechanical 86
        Electrical 88

Chapter 3 Implementation 93

    Generic Root Complex and Endpoints 95
    Generic Switch 98

Generic Bridge   100
Other Implementations   101
   Other Implementations of the Root Complex   102
   Other Implementations of Switches   106
   Other Implementations of Bridges   109
Other Bridge Considerations   112
   PCI Express/PCI Bridge and PCI Express/PCI-X Bridge Considerations   112

Chapter 4 Addressing and Routing   121

Address Spaces Defined in PCI Express   123
   Memory Address Space   123
   I/O Address Space   126
   Configuration Address Space   128
   Message Address Space   140
Participants of Transactions   143
   Participants of Memory, I/O and Configuration Transactions   143
   Message Transaction Participants   146
Addressing and Routing   150
   Addressing of Memory and I/O Transactions   151
   Configuration and Completer Transaction ID Routing   152
   Routing Message Transaction Packets   154
Associated Addressing Protocols   161
   Type 0 and Type 1 Configuration Requester Transactions   161
   TAG# and Transaction ID   168

Chapter 5 Transaction Packets   171

Header Field Information   174
Memory Transaction Packets and Execution Protocol   175
   Requester ID, Completer ID, and Tag#   176
   Memory Read Transaction Packets   178
   Memory Read Transaction Execution   180
   Treatment of Other Fields in the Memory Read Transaction Packets   188
   Memory Write Transaction Packets   191
   Memory Write Transaction Execution   192
   Treatment of Other Fields in Memory Write Transaction Packets   192
I/O Transaction Packets and Execution Protocol   193
   I/O Transaction Packets   194

I/O Read and Write Execution   196
Treatment of Other Fields in the I/O Transaction Packets   196
Configuration Transaction Packets Protocol   197
Configuration Transaction Packets   198
Configuration Read and Write Execution   200
Treatment of Other Fields in the Configuration Transaction Packets   200
Message Transaction Packets Protocol   202
Message Baseline Transaction Packets   203
Message Baseline Transaction Execution   204
Treatment of Other Fields in Message Baseline Transaction Packets   208
Message Vendor-Defined Transaction Packets   209
Message Vendor-Defined Transaction Execution Packets   211
Treatment of Other Fields in Message Vendor-Defined Transaction Packets   212
Message Advanced Switching Requester Transaction Packets   213

Chapter 6   Transaction Layer   215

Transaction Layer   217
The Control Element   218
The Address Element   218
The Flow Control Protocol Element   219
The Repeater/Completer Protocol Element   219
The Data Element   220
The Integrity Element   220
The Message Element   221
Transaction Layer Implementation   221
Header Field Formats   222
Header Field Formats for Memory Read Transactions   224
MEMORY READ REQUESTER TRANSACTION   227
MEMORY READ COMPLETER TRANSACTION   232
Header Field Formats for Memory Write Transactions   238
MEMORY WRITE REQUESTER TRANSACTION   240
Header Field Formats for I/O Read Transactions   245
I/O READ REQUESTER TRANSACTION   247
I/O READ COMPLETER TRANSACTION   250
Header Field Formats for I/O Write Transactions   255
I/O WRITE REQUESTER TRANSACTION   256

I/O WRITE COMPLETER TRANSACTION   260
Header Field Formats Configuration Read Transactions   263
    CONFIGURATION READ REQUESTER TRANSACTION   265
    CONFIGURATION READ COMPLETER TRANSACTION   269
Header Field Formats for Configuration Write Transactions   274
    CONFIGURATION WRITE REQUESTER TRANSACTION   275
    CONFIGURATION WRITE COMPLETER TRANSACTION   279
Header Field Formats for Message Baseline Transactions   282
    MESSAGE BASELINE REQUESTER TRANSACTION   284
Header Field Formats for Message Vendor-defined Transactions   293
    MESSAGE VENDOR-DEFINED REQUESTER TRANSACTION   295
Header Field Formats for Message Advanced Switching Transactions   300
    MESSAGE ADVANCED SWITCHING REQUESTER TRANSACTION   302
Data Field and Digest Field Formats   305
    Data Field   305
    Digest Field   312

Chapter 7   Data Link Layer and Packets   315

Overview of Data Link Layer Packets   317
    DLLP   317
    LLTP   317
    Other Considerations   318
Link Activity States   318
    Link_UP and Link_DOWN   319
    Link Activity States Application   319
    Details of Link Activity States   322
Flow Control Initialization   324
    Set of Buffers   325
    Flow Control Initialization Protocol   329
    Flow Control Protocol Update of Available Buffer Space   340
DLLP Formats   342
    Reserved   343
    Flow Control DLLPs   343
    Acknowledgement DLLPs   348
    Power Management (PM) DLLPs   350
    Vendor-Specific DLLPs   353

LLTP Format   355
    Reserved   355
    LLTPs   355

Chapter 8   Physical Layer and Packets   359
Parallel LLTPs and DLLPs versus Serial Physical Packets   362
    Step 1: Conversion of LLTP and DLLP Portion of
        Physical Packet   365
    Step 2: Addition of FRAMING to Complete the
        Physical Packet   370
    Step 3: Encoding and Decoding Protocols and
        Reference Clock   372
    Physical Packet and Other Symbol Transmission   381
Lane Parsing   383
    Parsing Example with Four Lanes   383
    K and D Symbols' Application and Lane Parsing   385
    Application of K and D Symbols Relative to Physical Packets and
        Actual Lane Parsing   387
    Receiver Error   395
Application of Ordered Sets   395
    Skip OS   396
    Fast Training Sequence OS   397
    Electrical Idle OS   398
    Training Sequence 1 and 2 OSs
        (also Known as TS1 and TS2 OSs)   401
    Inclusion of Ordered Sets into Lane Parsing   403
    Receiver Error   404
Special Transmitting and Receiving Considerations   405
    Scrambling   405
    Lane Polarity Inversion   407
    Disparity   407
Link Training and Link Configuration   408
    Link Configuration Protocol Iterative Process Discussion   409

Chapter 9   Errors   421
Relationship to PCI   422
Error Protocol Specific to PCI Express   423
Error Reporting   424
Basic Transaction Layer Errors   425
    Errors Related to TLPs and Checked at Requester and
        Completer Destinations   427

Special Error Considerations Specific to Message Requester
        Transaction Packets   438
    Special TLP Error Considerations for Switches and Bridges
        (Excluding Error Poisoning)   439
    Errors Related to Transaction Layer: Error Forwarding
        (also Known as Data Poisoning)   442
Errors Related to Transaction Layer: Completion Timeout Error   444
Errors Related to Transaction Layer: Receiver Overflow Error   445
Errors Related to Transaction Layer:
    Flow Control Protocol Error (FCPE)   446
Transaction Layer: TLP Fields Checked for Errors   447
    MEMORY and I/O REQUESTER TRANSACTIONs contain TYPE,
        ATTRIBUTE, REQUESTER, and TAG HDWs: 448
    CONFIGURATION REQUESTER TRANSACTIONs contain TYPE,
        ATTRIBUTE, REQUESTER, and TAG HDWs: 452
    COMPLETER TRANSACTIONs contain TYPE, ATTRIBUTE,
        COMPLETER, STATUS, REQUESTER, and TAG HDWs: 455
    MESSAGE BASELINE REQUESTER TRANSACTIONs contain TYPE,
        ATTRIBUTE, REQUESTER, and TAG HDWs   460
    MESSAGE VENDOR-DEFINED REQUESTER
        TRANSACTIONs contain TYPE, ATTRIBUTE, REQUESTER,
        and TAG HDWs   464
    MESSAGE ADVANCED SWITCHING REQUESTER
        TRANSACTIONs contain TYPE, ATTRIBUTE, REQUESTER,
        and TAG HDWs   467
Data Link Layer Errors   469
    Special Error Considerations for Switches   471
    Errors Related to CRC, LCRC, and Nullified TLP Packets   471
    Errors Related to Retry_Timer Timeout and Retry_Num# Counter
        Rollover   473
    Data Link Layer Protocol Errors (DLLPEs)   473
Physical Layer Errors   474
    Special Error Considerations for Switches   475
    Receiver Errors   475
    Training Errors   477
CRC, LCRC, ECRC, and Digest Field   478
    CRC and LCRC   478
    ECRC   479
    Protocols for Calculating and Checking CRC, LCRC
        and ECRC   480

Error Reporting and Logging   484
    General Error Considerations   485
    Message Error Requester Transaction Packets   485
    Minimum Error Reporting Protocol   488
    Advance Error Reporting Protocol   492

Chapter 10 Transaction Ordering   501

Transactions Begin and End   501
    Simplest Transaction Execution   501
    More Complex Transaction Execution   502
Buffers in the PCI Express Fabric   503
    Buffers versus FIFOs   504
    Solution to Livelock and Deadlock Situations   505
PCI Express Transaction Ordering Protocol   505
    Inclusion of Endpoints   505
    Attribute   506
    Traffic Class   506
Transaction Ordering Protocol Implementation   507
    Specifics   512
    Special Considerations   516

Chapter 11 Flow Control Protocol: Part 1   519

Introduction to the Flow Control Protocol, Part 1   519
    Flow Control Protocol Elements Overview   520
    Elements of the Flow Control Protocol   526
Flow Control Protocol: Virtual Channels and Traffic Classes   528
    Flow of Transaction Packets and Definition of *Packet* in
        This Chapter   528
    Virtual Channels   531
    Additional Considerations   539
    Traffic Classes   541
Flow Control Protocol: Buffer Management   553
    FCC and Set of Buffers   556
    FCC   557
    Flow Control Protocol Execution   559

Chapter 12 Flow Control Protocol: Part 2   581

Introduction to the Flow Control Protocol: Part 2   581
    Receiving Ports versus Transmitting Ports   582
    VC0 versus VCx   582

LLTP Tracking 582
    LLTP Tracking at Transmitting and Receiving Ports 583
PCI Express Arbitration 605
    Other Arbitration Considerations 607
Transaction Layer Packet (TLP) Arbitration 607
    Switches with Upstream TLP Flow 608
    Port Arbitration 611
    VC Arbitration 613
    Switches with Downstream Flowing and Peer-to-Peer TLP 616
    Non-Switch Port Arbitration, VC Arbitration, and Buffer
        Considerations 617
    Priority Schemes 620
Quality of Service 624
    Latency 625
    Effective Bandwidth 625
    Bandwidth Consistency 625
    Backpressure 625
    Total Latency 626
    Isochronous Transfers 627

Chapter 13 PME Overview, Wake Events, and Reset 637

Introduction 637
    Other Background Information 638
Power Management Protocol Overview 644
    Transition from Sleep (L2) or Powered Off (L3) to
        Powered On (L0) 644
    Lowered Power 646
    Placement into Sleep or Powered off Link State 648
Power Cycling 650
    Main Power and Vaux 651
    Powered On to Powered Off or Sleep 651
    Powered Off to Powered On 653
    Sleep to Powered On 654
Wake Events: Wakeup Protocol and Wakeup Event 655
    The Other Wake Event 656
    PCI Legacy PME# Signal Line 656
    Wakeup Protocol 657
Resets 662
    PERST# Signal Line Considerations 663
    Fundamental Reset Protocol 664
    Hot Reset Protocol 666

Chapter 14 Link States 671

Introduction to Link States 671
    Other Considerations 672
Normal Operation Link State 677
    L0 Link State 677
Power Management Category of Link States 682
    L0s Link State 682
    L1 Link State 686
    L2/3 Ready Link State 688
    L2 Link State 689
    L3 Link State 692
Link Training Link States 692
    Detect Link State 693
    Polling Link State 696
    Configuration Link State 700
Other Link States 714
    Recovery Link State 714
    Disabled Link State 720
    Hot Reset 723
Testing and Compliance 724
    Loopback Link State 724
    Polling.Compliance Link Sub-state 729

Chapter 15 Power Management 731

Introduction to Power Management 731
    Support Requirements 732
    Powered-on, Powered-off, and Sleep 733
L0 and Power Management Link States 734
    L0 Link State 734
    L0s Link State 735
    L1 Link State 736
    L2/3 Ready Link State 738
    L2 Link State 739
    L3 Link State 740
PCI-Power Management Software Protocol 740
    "D" States for PCI-PM Software Protocol 741
    TLPs for PCI-PM Software Protocol 744
    DLLPs for PCI-PM Software Protocol 753
    PCI-PM Software Protocol Application of "D" States, TLPs, and DLLPs 753

Active-State Power Management Protocol 758
    TLP for Active-State PM Protocol 759
    DLLPs for Active-State PM Protocol 759
    Active-State PM Protocol Application of the TLP and DLLPs 760

Chapter 16 Hot Plug Protocol 775

Introduction to the Hot Plug Protocol 775
Hardware Components for the HP Protocol 776
    Power Indicator 777
    MRL and MRL Sensor 777
    Attention Indicator 778
    Attention Button 779
    Power Controller (Fault) 780
    Add-in Card Present 780
Non-Hardware Components for the HP Protocol 781
    Message Baseline Requester Transactions 782
    Configuration Address Space Registers for
        Hot Plug Protocol 786
Hot Plug Protocol 791
    HP Protocol with Normal Power 792
    HP Protocol with Sleep or Lower Power State 794
Other Considerations for HP Protocol 795

Chapter 17 Slot Power Protocol 797

Introduction to Slot Power 797
Slot Power (SP) Protocol 798
    Power Limits 798
    Power Associated Message Transactions and Configuration
        Registers 799
    Slot Power Limit Protocol Implementation 803
    Message Baseline Requester Transaction Packets: Slot Power
        Messages 804
Associated Slot Power Items 805

Chapter 18 Advanced Switching 807

Introduction to Advanced Switching 807
Advanced Peer-to-Peer Links 810
Advanced Peer-to-Peer Communications 811

Chapter 19 Interrupts   813
Introduction to Interrrupts   813
Legacy Interrupt Protocol Implementation   814
    Message Interrupt Requester Transaction Packets   815
    Message Interrupt Requester Transaction Packet Flow   816
    Legacy Interrupt Protocol Implementation by Legacy Endpoints, Bridges, and Switches   817
    Legacy Interrupt Protocol Implementation by the Root Complex   821
    Boundary Conditions for Disabling INTx   821
    Unique Considerations for the Legacy Interrupt Protocol for Re-Mapping through Each Virtual PCI/PCI Bridge   823
Message Signaled Interrupt Protocol   832

Chapter 20 Lock Protocol   833
Introduction to the Lock Protocol   833
    EHA Participants   834
Lock Function (LF) Protocol   835
    Lock Transaction Packets   835
    LF Protocol to Establish EHA   835
    LF Protocol after EHA Established   836
    Restriction of Accesses to Locked Device   837
    LF Protocol to Terminate LF between Root Complex and Locked Device   838
MWr Issue   839

Chapter 21 Mechanical and Electrical Overview   841
Introduction   841
Mechanical   841
    Non-Mobile Add-in Cards   842
    Mobile Add-in Cards: PCI Express Mini Cards   850
    Cable Modules   853
    Signal Line Definitions   854
Electrical   857
    PHY   858
    Waveform   863

Chapter 22 Configuration Overview 865

Introduction   865
Features of Configuration Space   866
The Target of a Configuration Address   867
Configuration Space Size   867
Configuration Space Hierarchy   868
Creating a Configuration Address   872
Legacy PCI Express Configuration Mechanism   872
   Config_Address Register   872
   Config_Data Register   873
   Config_Address and Config_Data Example   873
   Limitation of Config_Address and Config_Data   875
Native PCI Express Configuration Mechanism   875
How Configuration Addresses Reach Their Destination   876
Type 0 and Type 1 Configuration Accesses   876
   Configuration Access Example 1   877
   Configuration Access Example 2   878
   Configuration Access Example 3   879
Access Rules for Configuration Space   880
   Access Rules for Configuration Space Reads   880
   Access Rules for Configuration Space Writes   880
How to Detect PCI Express Devices   881
PCI Express Bus Enumeration   882
PCI Express Bridge Architecture   884
   Two Rules for Connecting PCI Express Bridges   885
   Other Rules for PCI Express Bridges   891

Chapter 23 Configuration Registers 893

Introduction to Configuration Registers   893
Basic Function of Configuration Space Registers   894
Configuration Space Architecture   895
Configuration Space Layout   896
Root Complex Register Block   898
Configuration Header Register Block   899
Common Header Region   899
Type 0 Header Region   908
Type 1 Header Region   919

Chapter 24 Configuration Capabilities 937

Introduction to Configuration Capabilities   937
   Defined Capabilities   938

Required Capabilities 938
Extended Capabilities 938
Configuration Capability Architecture 939
Device-independent Region 940
Header Type Region 940
Device-Dependent Region 940
Extended Configuration Region 940
Shared Registers 941
Capabilities List Bit 941
How to Locate the Capabilities List 941
How to Scan the Capabilities List 943
Capability Structure Layout 943
Capability Linked List 945
Required Capabilities Overview 946
PCI Express Capability Structure 946
PCI Power Management Capability Structure 947
Message Signaled Interrupts Capability Structure 948
PCI Express Capability Structure 948

Chapter 25 Real-Time Analysis 977

Configurations 977
X1 and X4 Configuration Test Setup 977
X8 Configuration Test Setup 978
Mid-Bus Probe: 979
PCI Express Transaction Flow Overview 979
Transaction Layer 981
TLP Fields 981
TLP Routing 982
TLP Request/Completion 983
Data Link Layer 986
Physical Layer 987
Configuration Space 992
Traffic Class and Virtual Channels 999

Appendix A Capability IDs 1001

Glossary 1005

Index 1019

Acknowledgements

A book of this size requires an extensive staff at Intel to publish. Edward and Brad would like to thank Richard Bowles and David Spencer for supporting this project, Matt Wangler for managing it, and Matt and David for their direct contributions. We also would like to thank those people on the front lines of the day to day book development: David Clark, Donna Lawless, and Rich Jevons. David did the primary editing of the text and Donna and Rich prepared all the figures for printing.

The material content of a book of this type is critical. Edward and Brad would like to thank Fred Rastgar and Mark Abbas of Computer Access Technology Corporation (CATC) for providing all of the material in Chapter 25. We would also like to thank the following people for reviewing the draft of the book and providing critical feedback: Marc Pyne, Peter Teng, Robert Birch, Dave Coleman, and Adam Wilen. Furthermore, the Intel architects involved in developing the PCI Express Specification did a superb job creating the best written PCI specification to date. We wish to thank them for their help, and in particular Sridhar Muthrasanallur for his consistent cheerful willingness to answer our questions.

We would also like to acknowledge Ted Cyrek for the book cover graphics.

Chapter 1

Architecture Overview

It is useful to begin a discussion of PCI Express with a review of the PCI and PCI-X bus architectures. An understanding of the limitations of the PCI and PCI-X buses make it much easier to understand the approach taken by the architects in developing PCI Express. Another reason for reviewing PCI is that PCI Express retained the basic configuration, software models, and power management protocols from PCI. This chapter also identifies the limitations of PCI and PCI-X and explains how PCI Express solves these problems.

Please note that the discussions of PCI and PCI-X are not intended to detail the operation of these buses, but rather to provide insight into the limitations of these buses that ultimately led to the development of PCI Express.

The first two specifications related to PCI Express are *PCI Express Base Specification revision 1.0* and *PCI Express Card Electromechanical Specification revision 1.0*, both released July 23, 2002. The *PCI Express Base Specification revision 1.0a* and *PCI Express Card Electromechanical Specification revision 1.0a* were released on April 15, 2003. The *PCI Express to PCI/PCI-X Bridge Specification revision 1.0* was released February 18, 2003 and *PCI Express Mini Card Electromechanical Specification revision 1.0* was released June 2, 2003. This book includes information on all of these specifications and associated errata. When further errata are released by the PCI-SIG, updates to this book will be available at www.intel.com/intelpress/pciexpresscomplete. Information about any new related specifications will also be available at this website.

At the print time of this book, the Advanced Switching specification had not been released. The advanced switching information provided in this book is included to provide the interconnection between PCI Express and advanced switching. However, the final Advanced Switching specification may change the information in this book. When further advanced switching information related to advanced switching is released by the PCI-SIG, updates to this book will be available at www.intel.com.

Additional information is available in the following books: *PCI and PCI-X Hardware and Software* by Edward Solari and George Willse (ISBN 0-929392-63-9), *PCI Power Management* by George Willse, Edward Solari, Jim Erwertz (ISBN 0-929392-62-0), and *PCI Hot-Plug Application and Design* by Alan Goodwin (ISBN 0-929392-60-4).

The Original PCI Bus

The original concept of PCI was chip-to-chip interconnection developed by Intel Corporation. Indeed, PCI is an acronym for *peripheral component interconnect*. Prior to the public release of the PCI bus specification, it was modified to include add-in card slots. With the addition of add-in card slots Intel made a decision in the early 1990s to focus on PCI, to depart from the bus standards of their MultiBus I and II buses, and the other industrial bus standard VESA. As the PCI bus protocol was included in platforms it eventually replaced the ISA and EISA buses in personal computers and servers.

The implementation of the PCI bus in both high-end personal computers and servers placed demands on the PCI bus to provide higher levels of performance; consequently, the PCI bus evolved from 32 to 64 data bits and its clock reference from 33.3 megahertz to 66.6 megahertz. The final specification specific to PCI is *PCI Local Bus Specification revision 2.3* released March 29, 2002. This is the core specification to which all other specifications refer. In the case of PCI Express, the basic configuration and software models for PCI are implemented. PCI Express also retains compatibility with the PCI power management protocol with the implementation of "D" states within the functions of PCI Express devices. Consequently, the other PCI-related specifications applicable to both PCI and PCI Express are:

- *PCI-to-PCI Bridge Architecture revision 1.1*
- *PCI Bus Power Management Interface revision 1.1*
- *Advance Configuration and Power Interface Specification revision 2.0*

- *PCI Hot Plug revision 1.1*
- *PCI Standard Hot Plug Controller and Subsystem revision 1.1*
- *PCI BIOS revision 2.1*

The PCI, PCI-X and PCI Express specifications listed above are owned and supported technically by an industrial sponsored group: the PCI Special Interest Group (commonly referred to as PCI-SIG), whose website is www.pcisig.com.

PCI Platforms

The PCI architecture has changed over the years to solve performance problems and accommodate other changes. The following sections describe the major changes in the evolution of PCI system architecture.

Early Architecture

The PCI bus was first implemented in a personal computer with the legacy ISA bus illustrated in Figure 1.1 with North and South Bridges. The implementation of a server is similar, but with some of the PC-compatible functions removed. The shared architectural concepts between personal computers and servers are as follows:

- The *FSB (front-side bus)* isolated from the PCI bus segments by a *North Bridge* (also known as the *Host/PCI Bridge*). The isolation of the two entities permits any processor to implement a PCI platform, and for PCI to operate and evolve independently of the operation and evolution of the CPUs. In a similar fashion, the ISA bus segment is isolated from the PCI bus segments by a *South Bridge* (also known as the *PCI/ISA Bridge*). The South Bridge permits PCI to operate and evolve independently of the operation of the ISA bus segment.

- Platform memory is the primary high performance memory and is closely associated with CPU and caches. It is also isolated from PCI and ISA bus segments by the North Bridge for performance reasons.

- Certain functions need very high performance that would use up all of the available bandwidth of PCI bus segments. In this example, the video function (AGP) is directly connected to the North Bridge.

- The PCI-centric portion of the personal computer or server is downstream of the North Bridge and consists of a hierarchy of bus segments, as illustrated in Figure 1.2. The PCI devices are either a PCI bus master and target or both.

- PCI bus masters can access targets on the same bus segments with arbitration for ownership of the one bus segment. PCI bus masters accessing targets on other bus segments require interconnecting bridges to arbitrate on behalf of the PCI bus master for every bus segment between the PCI bus master and the target.

- All PCI devices as well as devices residing within or downstream of the South Bridge that wish to access main memory must all send transactions across the PCI bus 0. This approach can lead to the PCI bus 0 becoming a bottleneck for PCI devices attempting to access memory.

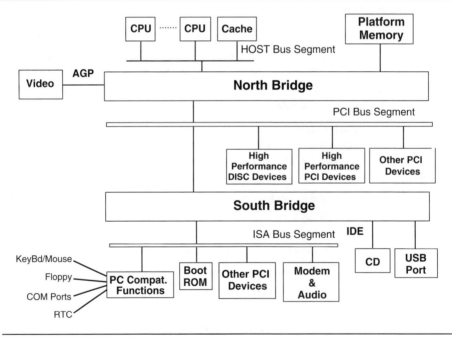

Figure 1.1   Overview of PCI and ISA Platform

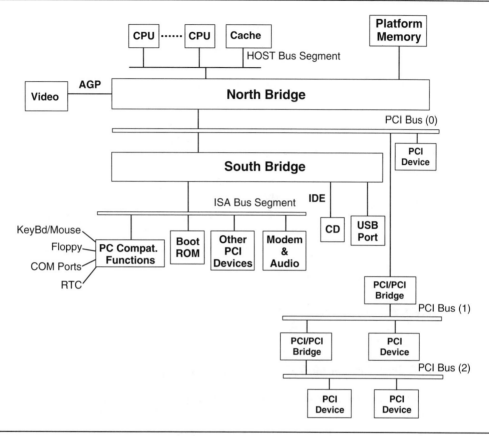

Figure 1.2   Hierarchical PCI Bus Segments

PCI Hub-Based Architecture

The introduction of the PCI bus specification quickly eliminated all ISA buses from personal computers and all EISA buses from servers. At the same time that the ISA and EISA bus segments were being eliminated, other ports and buses emerged and USB expanded in importance. As illustrated by Figure 1.3, the focus moves from a North and South Bridge to a hub focus. The North Bridge is replaced with a Memory Controller Hub and the South Bridge is replaced by an I/O Controller Hub. This approach carries transactions between memory and a wide variety of devices connected to the I/O Controller Hub (including the PCI devices). The hub architecture effectively eliminated the performance bottleneck to main memory caused by all the devices sending transactions across PCI bus 0.

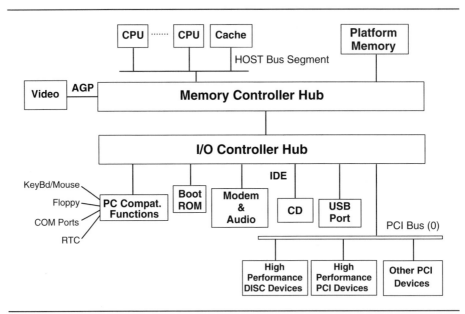

Figure 1.3 Overview of PCI Only Platform: Improved Performance

PCI 64 Data Bits and 66.6 Megahertz Emerge

The original concept of PCI was a minimal number of signal lines supporting 32 data bits with a 33.3-megahertz clock reference. It became clear in early PCI platforms that simple tasks such as updating the video display or accessing the hard disk caused the overall platform bandwidth to be totally consumed. A wider data bus (64 bits) and faster clock rate (66.6 megahertz) improved overall performance, but at a relatively high cost. The increased pin count significantly increases the cost of implementation, and the higher clock frequency reduces the number of PCI devices that can be connected to each bus segment.

In order to improve platform performance by implementing higher performance PCI bus segments, these PCI bus segments must be connected to the higher performance Memory Controller Hub. As illustrated in Figure 1.4, the PCI 64-bit/66.6-megahertz bus segments are connected directly to the Memory Controller Hub, retaining the same architecture as the I/O Controller Hub. Note that the increased clock frequency of PCI 64 bits/66.6 megahertz has a tradeoff between increased performance and the limited number of components or add-in card slots per PCI bus segment.

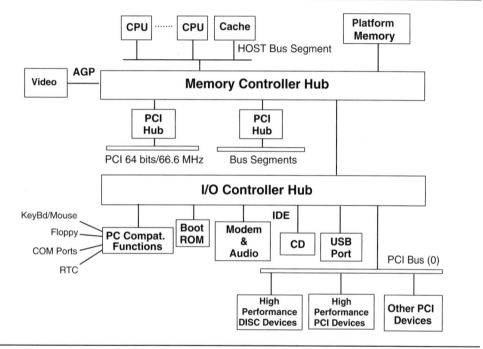

Figure 1.4 Overview of Hub Architecture with 66.6-Megahertz Bus Segments

PCI Protocol Basics

The previous discussions have focused on the overall PCI bus architecture from a system perspective. The following discussion focuses on the key elements and characteristics of the PCI bus.

Multi-drop Parallel Bus

PCI is designed to support a variety of devices residing on each bus segment. This design approach has several advantages for desktop systems such as:

- Traces on the motherboard reduced to a single set because of shared buses.

- A single protocol (PCI) used for all embedded devices, add-in cards, and chipset components connected to the bus segment.

The fact that multiple PCI devices including connectors connect to a shared bus means that trace lengths and electrical complexity limit the maximum usable clock speed. For example, a generic PCI bus has a maximum clock speed of 33.3 megahertz; the PCI specification permits increasing the clock speed to 66.6 megahertz, but the number of devices/connectors on the bus is very limited.

Synchronous Bus Protocol and Burst Transfers

The PCI protocol permits data transfers during each cycle of the PCI reference clock. Thus, depending upon clock speed and data bus width, fairly high bandwidth can be achieved. However, the protocol also allows up to eight PCI clock periods for each data transfer, which potentially drops performance significantly. Each PCI clock period is the inverse of the frequency of the CLK signal line.

The Effects of PCI's Reflected-Wave Switching

The PCI bus uses reflected-wave switching for signaling. Each signal that is driven on the rising edge of a reference clock creates a wave front that eventually reaches the ends of the trace. However, the amplitude of the incident wave front is not sufficient to cause a transition of the logic level. When the waveforms reach the ends of the trace that is not terminated, the reflection of the standing wave causes the amplitude of the wave front to be approximately doubled. This reflected wave causes the logic of the input circuitry to transition as it reaches PCI devices on its journey back to the originating PCI device. The propagation of the wave front and settling time must occur within a single clock period. Reflected-wave switching severely limits bus length, clock frequency, and bus loading.

The electrical load restriction for 33.3 megahertz reference clock is nominally ten integrated-chip type loads for any PCI device. The implementation of add-in card slots requires two loads per slot; consequently, this limits the number of chips and add-in card slots (PCI devices) possible for a specific PCI bus segment. Increasing the reference clock to 66.6 megahertz further restricts the load capacity.

The PCI bus implements a clock signal line as a clock reference to indicate when the other signal lines are valid. Transmission line effects of each signal line include propagation delay and wave reflection, which are affected by the bus capacitance and inductance. The transition from one voltage level (logical level) to another of a signal line creates a wave front that is reflected from the ends of the bus segment. Every signal line transition received by a component is a combination of the initial transition (incident wave) by the sending PCI device and the resulting reflective wave front. This combination results in a *settling time* parameter, which is the time between the voltage level transition at the transmitting PCI device to the detection of the reflected transition by another PCI device on the bus segment. Because the PCI reference clock signal line is required to validate other signal lines, both signal lines must consider the settling time. The settling time increases with the length of the bus segment and the number of PCI devices connected to the bus segment. In this way, settling time limits the bus segment performance. Consequently, the highest possible performance is achieved with the shortest bus segment and smallest load (that is, a point-to-point interconnect between two devices).

Table 1.1 lists the data bus widths, number of signal lines, reference clock frequencies, maximum bandwidth, and the maximum number of connectors supported by PCI.

Table 1.1 PCI Bus Bandwidth and Add-in Card Slot Limitation

Data Bus Width	Signal Lines (excluding power, arbitration, and test)	Clock Signal Line Frequency (MHz)	Maximum Bandwidth (Megabytes/second)	Maximum Add-in Card Slots
PCI 32 bits	49	33.3	133.2	4 (1)
PCI 32 bits	49	66.6	266.4	2 (1)
PCI 64 bits	81	33.3	266.4	2 (1)
PCI 64 bits	81	66.6	532.8	2 (1)

Table Notes:

1. Other platform components in addition to the bridge or add-in slot are possible, but the entry is the limit of add-in cards with one platform bridge and one other platform component.

PCI Arbitration

PCI bus transactions require that the PCI bus master originating the bus transaction or the bridge forwarding the bus transaction, arbitrate to gain ownership of the each PCI bus segment between the PCI bus master and the target. Consequently, PCI bus masters and PCI/PCI Bridges (collectively referred to as bridges) may be stalled for relatively long periods of time while waiting to take ownership of each bus segment. Even with just two PCI devices on a bus segment, arbitration must occur for bus segment ownership.

Only the PCI device winning bus arbitration can perform a bus transaction across a PCI bus segment as the PCI bus master or bridge acting on behalf of a PCI bus master. PCI permits one bus transaction on a bus segment in one direction at a given time. This has the effect of impeding overall transaction flow when bus transactions need to flow in opposite directions at the same time.

When a hierarchy of PCI bus segments is implemented as illustrated in Figure 1.5, a bus transaction may only proceed to the next bus segment when each of the intervening bridges has gained ownership of the next bus segment. Figure 1.5 depicts PCI bus master on PCI bus segment (3) performing a memory write transaction to the platform memory. The PCI bus master must arbitrate for ownership of PCI bus segment (3) prior to executing the memory write transaction. In turn, each intervening bridge must arbitrate for ownership of the next PCI bus segment until the bus transaction reaches PCI bus segment (0). Once the bus transaction arrives on PCI bus segment (0) the data can be written to the platform memory via the HOST/PCI Bridge. The ability of the PCI bus master to write data to platform memory is straightforward because the address and data can be posted at each bridge, and each bridge can forward the memory write transaction towards platform memory. The posted address and data of the bus transaction is held in the bridge until bus segment ownership can be achieved and only then can the bus transaction be forwarded.

The performance of PCI devices on a shared bus segment may vary widely in terms of required bandwidth, tolerance for bus access latency, typical data transfer size, and so on. All of this complicates arbitration on the PCI bus segment when multiple PCI bus masters wish to initiate bus transactions.

Delayed Transaction Protocol

The flow of the memory write transactions as discussed in the preceding section illustrates the advantage of posting the address and data at each intervening bridge and at the final HOST/PCI Bridge. If the bus transaction is a memory read, I/O read, or configuration read, the bus transaction is not complete until read data is returned. Similarly, I/O write and configuration write transactions are not complete until completion status of data write is returned. For none of these bus transactions can the address and data be posted as done for the memory write transaction. The PCI platform implements the Delayed Transaction protocol for all transactions that are not memory write transactions.

12 ■ The Complete PCI Express Reference

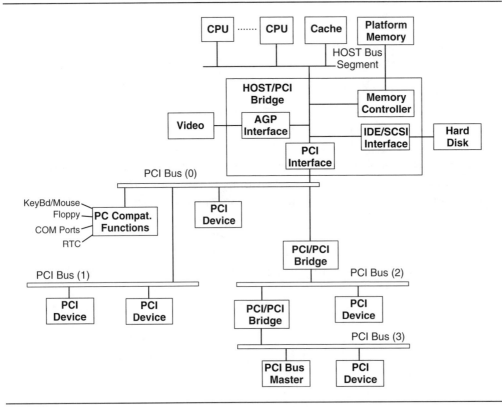

Figure 1.5 PCI Platform with Hierarchical Bus Segments

Consider Figure 1.5 with PCI bus master on PCI bus segment (3) performing a memory read transaction instead of a memory write transaction. The memory write transaction execution was straightforward because the address and data could be posted at each bridge. In the case of a memory read transaction, only the address can be posted. The bus transaction is not complete until the data is read and returned to the PCI bus master. PCI/PCI Bridges C and B must each arbitrate for ownership of their respective bus segments and forward the bus transaction to main memory. If the PCI bus master and these bridges start the transaction and wait for the data to be returned, all three bus segments (0, 2, and 3) will be occupied for the entire duration of the bus transaction. Consequently, all other PCI devices residing on these bus segments, and any other PCI bus master attempting to access main memory, will likely be stalled for an extended period of time. To improve platform performance in situations like this, PCI implemented the *Delayed Transaction* protocol. This protocol defines the following rules and procedures:

- A target device (platform memory or bridge) must end the first data transfer within 16 PCI CLK signal line periods.

- If a target device is unable to complete the first data phase within 16 PCI CLK signal line periods, it must signal to the PCI bus master (original or bridge) to terminate the transaction and retry it later.

- PCI devices implementing the Delayed Transaction protocol signal retry, latch the bus transaction, and start the bus transaction on the bus segment as quickly as possible.

- When the bus transaction is completed, the PCI device implementing the Delayed Transaction protocol waits for the PCI bus master to retry the bus transaction again and then completes the bus transaction.

- The PCI bus master must retry bus transactions repeatedly until the target is able to complete the transaction.

- Each bridge acts like a PCI bus master on behalf of the original PCI bus master on the bus segment on the side of the bridge opposite the original PCI bus master. The bridge acts like a target on the bus segment on the same side as the original PCI bus master.

For example, to execute a memory read transaction in Figure 1.5, original PCI bus master (PCI/PCI Bridges C and B), and the HOST/PCI Bridge implement the Delayed Transaction protocol. The original PCI bus master begins execution of the memory read transaction after gaining bus segment ownership of PCI bus segment (3). The bridge between PCI bus segments (3) and (2) acts as the target on PCI bus segment (3), posts the address, and acts like a PCI bus master on PCI bus segment (2). The bridge between PCI bus segments (2) and (0) acts as the target on PCI bus segment (2), posts the address, and acts like a PCI bus master on PCI bus segment (0). The address for the memory read transaction is delivered from PCI bus segment (0) to platform memory via the HOST/PCI Bridge. As the memory read transaction flows from the original PCI bus master to the platform memory, the posting of the address at each bridge permits the associated bus segment to be made available for bus segment ownership by other PCI devices. The original PCI bus master and bridges retries the memory read transaction until data begins following from the platform memory to the original PCI bus master. Between the retries of the memory read transaction the bus segments between the platform memory and the platform memory are available to other PCI devices.

Limitations of the PCI Bus Architecture

As popular as it has been, PCI presents additional problems that contribute to performance limitations:

- Bus transaction size is not limited and is not known, which makes it difficult to size buffers and causes frequent disconnect signaling by targets. PCI devices are also allowed to insert numerous wait states during each data phase, which contributes to this problem.

- All PCI bus transactions by PCI devices that target platform memory generally require a snoop cycle by CPUs to assure coherency with internal caches. This impacts both CPU and PCI bus performance.

- PCI's data width for each bus transaction very limited (32/64 data bits).

- Because of the PCI electrical specification (low-power, reflected wave signals), each PCI bus segment is physically limited in the ratio of ICs and connectors to CLK signal line frequency.

- PCI bus segment arbitration is vaguely specified. Access latencies can be long and difficult to quantify. If a second PCI bus segment is added using a PCI/PCI Bridge, arbitration for the secondary bus segment typically resides in the new bridge. This further complicates PCI arbitration for traffic moving vertically to memory.

- Only one bus transaction can be performed at one time and only in one direction on a given PCI bus segment.

- PCI masters have no idea whether the target device can accept the bus transaction that they send. Consequently, the bus transaction may need to be retried many times before it completes.

PCI Bus Feature Summary

A variety of PCI features in addition to those discussed previously are listed in Table 1.2.

Table 1.2 Major PCI Features

Feature	Description
Processor independence	Components designed for the PCI bus are PCI-specific, not processor specific, thereby isolating device design from the processor upgrade treadmill.
Limited number of devices on single bus segment	A typical PCI bus implementation supports approximately ten electrical loads, and each device presents a load to the bus. Each connector represents two loads for a maximum of approximately four to five connectors per bus segment.
Multiple functions within each device	Each device on a PCI bus segment may contain up to eight individual PCI functions.
Additional PCI bus segments can be Added	Because the number of electrical loads that can be supported on a single bus segment is limited, the specification supports PCI/PCI Bridges that can provide additional hierarchical bus segments. The maximum number of PCI bus segments that can be supported is 256.
Low-power consumption	A major design goal and subsequent accomplishment of the PCI specification was the creation of a system design that would draw as little current as possible.
Bursts can be performed on all read and write transfers	The 32 data bit PCI bus supports the transfer of 4 bytes per clock cycle; the 64 data bit PCI bus segment supports transfer of 8 bytes per clock cycle.
Bus speed	Revision 2.0 specification supported PCI bus speeds up to 33.3 megahertz. Revision 2.1 adds support for 66.6 megahertz bus operation.
Concurrent bus operation	Bridges support full bus concurrency with the processor bus, PCI bus segments, and the expansion bus simultaneously in use.
Bus master support	PCI bus masters allow peer-to-peer PCI bus access, as well as access to main memory and expansion bus devices through PCI/PCI and expansion bus bridges.
Arbitration	Due to shared bus segments, masters must arbitrate for ownership of bus segment before starting a bus transaction.
Pin count	Parallel bus signal lines require a minimum of 47 pins for PCI targets and 49 pins for masters.

Table 1.2 Major PCI Features *(continued)*

Feature	Description
Bus transaction integrity	Parity is checked on the address, command, and data signal lines only. The integrity of the control signal lines is not checked.
Three address spaces	PCI supports 32- or 64-bit Memory addressing, 32-bit I/O addressing, and 256 bytes of Configuration address space per function.
Auto-configuration	PCI supports full bit-level specification of the configuration registers necessary to support automatic device detection and configuration.
Add-in cards	The PCI specification includes a definition of PCI connectors and add-in cards.
Add-in card size	The PCI specification defines three card sizes: long, short and variable-height short cards, and low-profile add-in cards.

PCI-X Platforms

As personal computer and server applications required ever increasing performance, another evolution of a variant of the PCI bus was developed to address the demands of these applications occurred called the PCI-X bus. The PCI-X bus retains much of the basic bus protocol as the PCI bus, but increases the frequency of the reference clock (CLK signal line) and in later implementations includes a strobe reference. PCI-X also solves many of the problems associated with PCI.

Over the years there have been three levels of PCI-X bus specifications to improve bus performance. The different levels of PCI-X reflect the effective clock frequency or strobe rate. (*66* stands for 66.6 megahertz clock reference, *133* stands for 133.2 megahertz, and so on.) The three levels are:

- *PCI-X Addendum Specification revision 1.0* defines the PCI-X 66 and PCI-X 133 protocols.

- *PCI-X Addendum Specification revision 2.0* defines the PCI-X 66, PCI-X 133, PCI-X 266, and PCI-X 533 protocols.

- *PCI-X Local Bus Specification revision 3.0* defines PCI-X 1066. This latest specification was introduced as a "to be done" specification on November 13, 2002. No release date for the actual specification has been announced.

The following sections discuss the performance features introduced by each revision of PCI-X bus specification and the associated limitations.

Split Transaction Protocol

One of the major differences between PCI and PCI-X is the replacement of the Delayed Transaction protocol with the Split Transaction protocol. As previously discussed, the Delayed Transaction protocol required the PCI bus master and bridges to retry a bus transaction until data or completion status was ready to be returned. A Split Transaction protocol does not retry the bus transactions. The address and data (if applicable) are posted from PCI –X to PCI-X device. The PCI-X device that is the target will source an associated completion transaction back to the official PCI-X bus master when the data and completion status is ready. The Split Completion protocol eliminates the waste of retrying bus transactions and thereby using up platform bandwidth.

Improved Performance of PCI-X 1.0 over PCI

PCI-X was developed to address the need to increase performance over PCI for use in the server environment. It does so while maintaining backward compatibility with PCI. The same signals and connectors are used and all PCI-X devices are required to operate in PCI mode. PCI-X made many performance improvements:

- The maximum CLK signal line frequency increased to 133 megahertz.
- Split Transaction protocol replaced Delayed Transaction protocol, thereby eliminating multiple retries of a bus transaction.
- PCI-X added an attribute phase to the bus transaction protocol that carries a variety of information to improved performance.
- Data transfer size is specified to improve efficiency of split transactions.
- No snoop bit reduces platform memory access time.
- Relaxed ordering bit can improve performance of some memory read transactions

- PCI-X uses the reflected-wave switching like PCI, but includes modified electrical characteristics that permit greater loads at the higher frequencies. The protocol also requires registered signaling versus the signal sampling supported by PCI.
- The PCI 64-bit/66.6 megahertz bus segments as illustrated in Figure 1.4 are typically replaced by the PCI-X bus segments.

Figure 1.6 shows a PCI-X based system with four 64-bit PCI-X bus segments connected to a Memory Controller Hub. In addition to the performance improvements listed here for PCI-X, some of higher performance functions such as Ethernet may move from the PCI bus segment to be directly connected to the I/O Hub Controller. It should be noted that Figure 1.6 illustrates a server platform; consequently, the inclusion of the PC-compatible functions is optional.

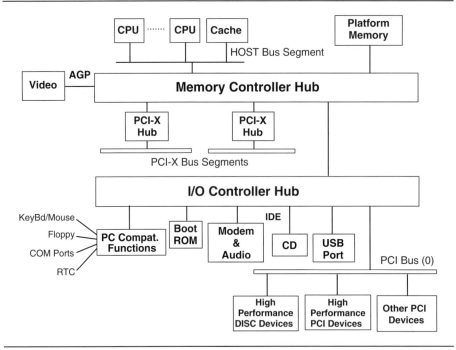

Figure 1.6 Overview of PCI and PCI-X Only Platform: Higher Performance

PCI-X 2.0 Increases Performance Further

To further improve platform performance, the clock reference of the PCI-X bus segment needed to increase. However, the increase of the clock reference beyond 133 megahertz was not possible with a PCI-like CLK signal line. PCI-X addressed this issue by defining source-synchronous clock references for PCI-X 266 and PCI-X 533. PCI-X 266 and PCI-X 533 use the CLK signal line as a clock reference for the Address/Data and Attribute signal lines during the address phase. During the data phase the C/BE# signal lines are used as source-synchronous strobes (clock references) for the Address/Data signal lines. The resulting effect is to provide two strobe points (DDR) and four strobe points (QDR) per CLK signal line period for PCI-X 266 and PCI-X 533, respectively. To achieve the maximum performance provided by PCI-X 266 or 533 bus segments, the PCI-X devices must be connected with a point-to-point interconnect. As illustrated in Figure 1.7, the point-to-point implementation requires the number of PCI-X/PCI-X Bridges to greatly increase. Also, to achieve high performance PCI-X/PCI-X Bridges must be tightly coupled with the processor and platform memory.

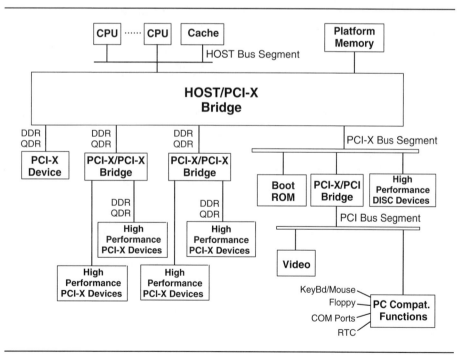

Figure 1.7 Overview of PCI-X DDR/QDR Platform: Highest Performance

Figure 1.7 also emphasizes two other considerations compared with the lower performance platform of Figure 1.6. First, the focus shifts from two hub controllers to a single HOST/PCI-X Bridge. The need to maintain two chips is removed. Second, the some of the devices located on the I/O Controller Hub and PCI bus segment in Figure 1.6 move to a lower performance PCI-X bus segment or even lower performance PCI bus segment. The location of all of these devices is determined by the individual implementation. The key concept to remember is that very high performance devices are connected to PCI-X DDR/QDR bus segments. The PCI-X DDR/QDR bus segments must be connected directly to the HOST/PCI-X Bridge for performance and the bus segments are actually point-to-point interconnections. The point-to-point interconnections require many more PCI-X/PCI-X Bridges. All of the other devices in the platform share lower performance PCI-X or PCI bus segments.

To further understand the impact of point-to-point implementation and the requirement for more PCI-X/PCI-X Bridges, see Figure 1.8. The point-to-point implementation requires the number of PCI-X/PCI-X Bridges to greatly increase and requires non-bridge functions ("+PCI-X" in the figure) to make the implementation practical. Also, the designer may want to provide a local version of the PCI-X bus segment (0) and a buffered version with higher bandwidth to the first downstream PCI-X/PCI-X Bridge. The interconnection of PCI-X/PCI-X Bridges begins to resemble the PCI Express topology. However, the 82 signal lines exclusive of power, arbitration, and test required for PCI-X bus segments make point-to-point interconnection impractical.

Figure 1.8 also introduces the stick frame model that will be used in this book to provide the internal details to key components. The peripherals of SCSI, Ethernet, Boot ROM, and so on are de-emphasized in order to emphasize the architecture of the many PCI-X/PCI-X Bridges needed to interconnect multiple PCI-X DDR/QDR devices.

PCI-X 1066

PCI-X 1066 had just been announced at the time this book was going to press. Consult the PCI-SIG website www.pcisig.com for more information.

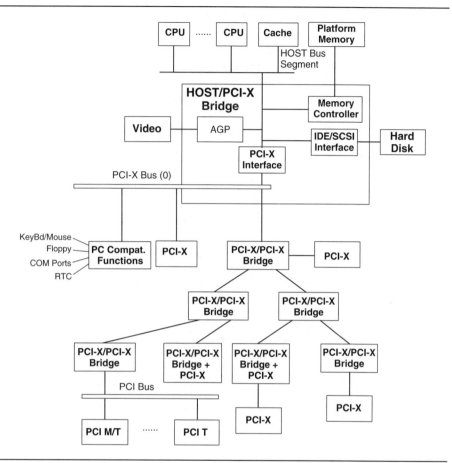

Figure 1.8 Generic PCI-X DDR/QDR Platform

PCI-X Bus Feature Summary

The fundamental limitations of PCI bus architecture also apply to the PCI-X bus architecture to a large degree. Also, the features of the PCI bus summarized in Table 1.2 are essentially the same features of the PCI-X bus. The four major differences between PCI and PCI-X are the implementation of the Spit Transaction Protocol, the increase in the CLK signal line frequency, the introduction of strobes for a faster clock references, and the reduction in the number of PCI-X devices that can reside on a bus segment.

The key distinction in implementation of the three levels of PCI-X is that of reference clock frequencies. The clock reference as implemented by the CLK signal line has increased to 133.2 megahertz. The clock references are extended with the inclusion of strobes. The implementation of the higher frequency CLK signal line and strobes severely limit the number of PCI-X devices that can be connected to each bus segment. This limitation effectively forces point-to-point interconnection of PCI-X devices.

Table 1.1 summarized the data bus widths, number of signal lines, clock frequencies, maximum bandwidth, and the maximum number of connectors supported by PCI. Table 1.3 summarizes these considerations for PCI-X.

Table 1.3 PCI-X Bus Bandwidth and Add-in Card Slot Limitation

Data Bus Width	Signal Lines (excluding power, arbitration, and test)	Clock Signal Line Frequency (MHz)	Maximum Bandwidth Megabytes/ Second	Maximum Add-in Card Slots
PCI-X 1.0 64 bits	82 lines	66.6	532.8	4 (1)
PCI-X 1.0 64 bits	82 lines	100	800	2 (1)
PCI-X 1.0 64 bits	82 lines	133.2	1065.6	1 (1)
PCI-X 2.0 64 bits	82 lines	Source synchronous DDR (3)	2131.2	1 (2)
PCI-X 2.0 64 bits	82 lines	Source synchronous QDR (3)	4262.4	1 (2)
PCI-X 3.0 64 bits	82 lines	Source synchronous (3)	8524.8	1 (2)

Table Notes:

1. Other platform components in addition to the bridge or add-in slot are possible, but the entry is the limit of add-in cards with one platform bridge and one other platform component.
2. Essentially, the add-in card is the only item connected to the platform bridge of this associated bus segment.
3. PCI-X 266 and PCI-X 533 implement source synchronous for data via dual use of C/BE# signal lines. Implementation per function for PCI-X 1066 is pending.

PCI Express

In 2001, engineers at Intel decided that a fresh approach to personal computer and server buses was required. Intel concluded that parallel bus segments used in PCI and PCI-X buses could not support ever increasing performance requirements of applications.

PCI Express Improves Upon PCI and PCI-X

Several important steps were taken to evolve PCI and PCI-X platform performance:

- Dedicated buses (some proprietary) were incorporated into the HOST/PCI Bridge (video and hard disk) to remove the associated bandwidth burden on the PCI and PCI-X bus segments.

- Higher clock reference frequencies were defined for the CLK signal line and source synchronous strobes were introduced.

- Data width was increased to 64 bits.

All of the above steps were not sufficient, however, to address the ever increasing demands for bandwidth on PCI and PCI-X. Once the performance limitations were understood, the interest in developing a new personal computer and server platform grew. Consequently, work began in defining a new platform under the code name *Third Generation Input and Output* or *3GIO*. Prior to public release of the specification, the name was changed to PCI Express.

The experience of implementing PCI and PCI-X platforms led to several key conclusions that were incorporated into the design and definition of the PCI Express platform:

- Point-to-point interconnections between components permit higher transmission rates and therefore greater bandwidth.

- Wide parallel bus implementations are too costly when using point-to-point connections.

- Reference CLK signal lines used to validate other signal lines slow transmission rates due to settling time delays.

- Arbitration for bus segment ownership with the restriction of one bus transaction in one direction at a time on each bus segment limits overall platform performance.

- The Delayed Transaction protocol employed by PCI resulted in poor bus efficiency. The Split Transaction protocol used by PCI-X was much better and is improved upon slightly by the Requester/Completer protocol of PCI Express.

- PCI Express retains key elements of PCI and PCI-X to be compatible with PCI and PCI-X software models, specifically configuration and power management.

Each of these considerations is discussed in the following sections.

Point-to-Point Interconnections Support Higher Transmission Rates

The advantages of the PCI and PCI-X shared parallel bus implementations are diminished as the reference clock frequency is increased to improve performance. The implementation of higher bandwidth PCI and PCI-X bus segments restricts the number of loads on each bus segment to essentially a point-to-point interconnection. The PCI Express design recognizes this limitation and has defined point-to-point interconnections called *links* to replace the parallel bus segment. The PCI Express link minimizes the transmission line effects and promotes maximum transmission performance.

Parallel Buses Too Costly for Point-to-Point Connections

When multiple PCI or PCI-X devices can be connected to the same (shared) bus segment, the number of signal lines is not a concern. As the number of PCI or PCI-X devices per bus segment approaches the one-to-one interconnection, the number of signal lines becomes critical. That is, the many signal lines connected to a bridge to support a bus segment with only one PCI or PCI-X device and possibly one bridge is not practical on the basis of power and chip package considerations. The PCI and PCI-X devices that implement the highest reference clock frequency need to be directly connected to the HOST/PCI Bridge in a one-to-one interconnection. The signal line count for a one-to-one interconnection to the HOST/PCI Bridge makes this chip package impossibly large.

Furthermore, PCI Express's implementation of point-to-point links is not an improvement over PCI or PCI-X bus segments if the signal line count is not reduced. Consequently, PCI Express replaces the parallel structure with a pair of signal lines per direction. This set of four signal lines is termed a lane and multiple lanes can be defined for each link. In this way each lane provides incremental bandwidth to the link.

Reference Clock Encoding Improves Performance

The implementation of a short bus segment with two devices and a CLK signal line used to validate other signal lines as a reference clock does not permit the highest performance. One solution is to provide a strobe signal line as the effective reference clock running in parallel to the specific signal lines it is referenced to. This concept minimizes the settling time caused by a single clock signal reference, and is the basis of the strobe reference clock in PCI-X 2.0 and 3.0. Another solution is to integrate the reference clock with the information and data being transmitted onto the same signal line, thus eliminating the difference in settling time between the reference clock and other signal lines. Integrating these elements onto the same differentially driven signal line pair is achieved by encoding a reference clock in the information and data bit streams. PCI Express implements such an approach with the encoding of a reference clock with the associated information and data into a single pair of signal lines for each direction with 8/10b encoding. This approach permits a very high bit stream rate between the two devices on a link.

Dual Simplex Signaling Yields Greater Performance

PCI and PCI-X permit one bus transaction on a bus segment in one direction at a specific time. Even with just two components on a PCI or PCI-X bus, arbitration must occur for bus ownership because they are designed as shared buses. Because PCI Express reduces pin count to a single pair, one pair can be implemented in each direction between the two PCI Express devices. The bit stream direction of each pair of signal lines is fixed and permits bit streams to flow simultaneously in both directions. This capability is known as *dual simplex* signaling. The two pairs of signal lines that provide bi-directional bit streams are called a *lane*. PCI Express allows for the definition of a multiple number of lanes on a link between the two PCI Express devices. Collectively these lanes are called a *link*. PCI Express also allows for the definition of a multiple number of links between the two PCI Express devices. The implementation of each lane further improves the performance of the link while maintaining a signal line count of the PCI Express link less than that of a PCI or PCI-X bus segment.

Requester/Completer Protocol Improves Efficiency

PCI and PCI-X implement different bus transaction protocols, called Delayed Transactions and Split Transactions, respectively. PCI Express uses a form of Split Transaction protocol called the Requester/Completer protocol. Each of these protocols is summarized below:

- The Delayed Transaction protocol requires the PCI bus master and bridges to continuously repeat a bus transaction until the target is ready to complete the bus transaction. The repetition of bus transitions uses bus segment bandwidth needlessly.

- The Split transaction protocol requires the PCI-X bus master to initiate the bus transaction only once. The target responds whenever it is ready to complete the bus transaction. The Split Transaction protocol eliminates the repetition of bus transitions required by the Delayed Transaction protocol.

- The Requester/Completer protocol is essentially the same as Split Transaction protocol except that it is required for all transactions at all times. (The Split Transaction protocol required the target to request the protocol or immediately finish the bus transaction.)

PCI Express Retains Software Investment

A major investment for the personal computer and server market is the software. Associated with the application software is the automated convenience of configuration software. A platform based on PCI Express is designed to appear as a PCI or PCI-X platform to the software. Consequently, the implementation of a PCI Express platform viewed at the hardware design level is that of a virtual PCI platform. Also, the elements of power management as they relate to PCI software were retained in the PCI Express power management protocol. PCI Express has made other power management improvements, but these operate at the hardware level to preserve software compatibility.

Platform Implementations with PCI Express

The basic PCI Express topology and platform interconnection is a logical evolution of the PCI and PCI-X platforms. PCI and PCI-X platforms have five types of devices: HOST/PCI Bridge, PCI bus master, PCI bus masters / targets, targets, and bridges. PCI Express platforms have four devices: Root Complex, switches, endpoints, and bridges. The Root Complex is essentially a HOST/PCI Express Bridge similar to the HOST/PCI Bridge. The switches are essentially the PCI/PCI Bridges. The endpoints are collectively the PCI bus master, PCI bus masters/targets, and targets. The bridge in a PCI Express platform is similar to the PCI/PCI Bridge except that it is a PCI Express/PCI Bridge. Similar comparisons can be made between PCI-X and PCI Express platforms.

PC and Server Platforms

PCI Express platform designs may utilize existing PCI platform approaches as illustrated by Figure 1.9. Comparing Figure 1.9 to Figure 1.3, the emphasis of this PCI Express platform is to connect many PCI Express links to the I/O Controller Hub.

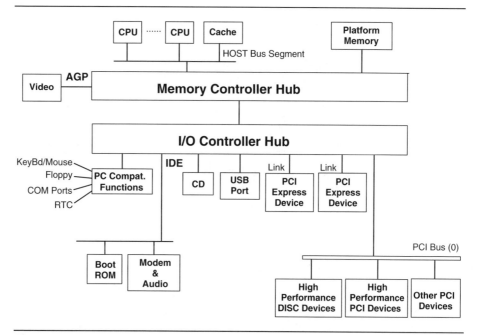

Figure 1.9 Overview of Minimal PCI Express Platform

Figure 1.10 shows a more server-centric implementation. Here the emphasis is on supporting a large number of peripheral devices, which requires the implementation of switches. The switches act like PCI-X / PCI-X Bridges. Also illustrated in Figure 1.10 is the combining of the Memory Controller Hub and I/O Controller Hub into a single entity called the *Root Complex*.

In Figure 1.10 and other figures, *PCI EXP* is an abbreviation for PCI Express.

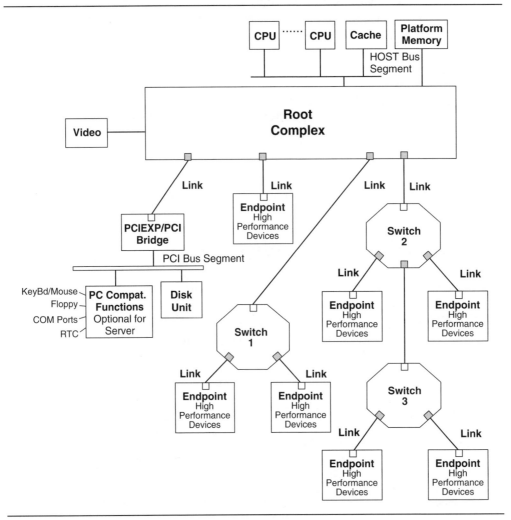

Figure 1.10   Overview of Typical PCI Express Platform

Figure 1.11 continues stick frame model that is used in this book to provide the internal details to key components. This particular diagram provides additional details regarding the internals of the Root Complex. Notice that the HOST/PCI and PCI/PCI Bridges provide the elements of compatibility with PCI software. The other devices such as PC-compatible functions, video interface, hard disk interface, and so on, are not included for purposes of clarity.

Figure 1.11 also illustrates several terms used in this book. These and other key terms can also be found in the Glossary at the back of the book:

- **Root Complex**: Collectively the interface between the HOST bus segment, platform memory, and all of the PCI Express devices downstream of its ports.

- **Switch**: The circuitry that interconnects the link connected to its upstream port to the links connected to multiple downstream ports.

- **Endpoints**: The implementation-specific circuitry that is on the downstream side of a link.

- **PCI Express/PCI Bridge**: The PCI Express device that translates the PCI Express transaction protocol to a PCI bus transaction protocol to allow connection of PCI bus segments to the PCI Express fabric. The PCI Express/PCI-X Bridge is for the support of downstream PCI-X bus segments and devices. Collectively called *bridges*.

- **Link**: The interconnection between any two PCI Express devices.

- **Upstream port:** Interconnection point between a PCI Express device and a link with the port facing towards the Root Complex.

- **Downstream port:** Interconnection point between a PCI Express device and a link with the port facing away from the Root Complex.

- **Upstream flow:** Transaction flow that traverses a link to arrive at a downstream port.

- **Downstream flow:** Transaction flow that traverses a link to arrive at an upstream port.

- **PCI Express devices:** This is a collective term for the Root Complex, switches, endpoints, and bridges.

- **PCI Express fabric:** The collection of all PCI Express devices connected to a single Root Complex and the associated links. It also includes a single Root Complex. This book defines that the PCI and PCI Express bus segments and devices below a bridge are not part of the PCI Express fabric. Note: The PCI Express fabric as defined by this book includes only one Root Complex. In Figure 1.12 the two Root Complexes "a" and "b" are defined as two

PCI Express fabrics. One PCI Express fabric contains all PCI Express devices with the "a" suffix. The other PCI Express fabric contains all PCI Express devices with the "b" suffix.

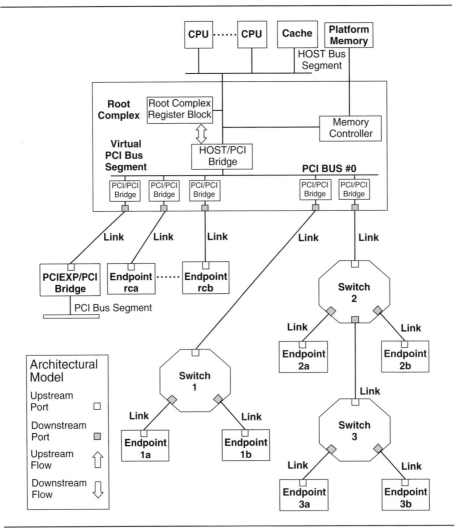

Figure 1.11 Generic PCI Express Platform

PCI Express Specific Performance

As previously discussed, PCI Express only interconnects two components on a link and a link consists of one or more lanes. Each lane contains a set of two differentially driven pairs of signal wires. The set provides bi-directional transmission of bit streams. Multiple lanes can be implemented by the link (interconnect) between two PCI Express devices. Table 1.4 summarizes the higher bandwidths possible on a point-to-point link. Part of this increase in bandwidth is due to the minimization of loads and minimization of transmission line effects, but the majority of bandwidth improvement is to due to the PCI Express reference clock. PCI Express's reference clock is done with the integration of a clock into the bit stream of information and data on each lane of the link. The protocol is called 8/10b encoding. It effectively encodes 8-bit information and data into a 10-bit pattern that provides transition edges for a receiver's phase lock loop to generate a reference clock. This approach permits the transition points of the bit stream to switch at higher frequencies and thus provides a higher bandwidth. Any limitation of a signal line switching to remain synchronized to a clock signal line has been removed.

Table 1.4 PCI Express Signal lines and Link Bandwidth

PCI Express Bit Stream Rate versus Bandwidth							
Default 2.5Gb/sec per Differential Pair (Each Direction)							
Lanes Per Link	x1	x2	x4	x8	x12	x16	x32
Signal Lines Each direction	2	4	8	16	24	32	64
Raw Bit Stream per second	2.5 Gb/s	5.0 Gb/s	10.0 Gb/s	20.0 Gb/s	30.0 Gb/s	40.0 Gb/s	80.0 Gb/s
Bandwidth Bytes per second	250 MB/s	500 MB/s	1000 MB/s	2000 MB/s	3000 MB/s	4000 MB/s	8000 MB/s

Two of the consequences of nonparallel signal lines are the requirement that all of the information (address and control) and data are transmitted as a sequential bit stream. These consequences, plus the integration of the clock reference places some overhead on the transmission of the data payload (that is, useful data per bus transaction). The high bandwidth available in the PCI Express link provides more than sufficient effective data bandwidth per lane.

Table 1.5 summarizes efficiency of a PCI Express memory write transaction transmission on a link with a single lane (x1) versus a maximum number of lanes (x32) under worst case conditions. The efficiency reflects the ratio of actual data bits transferred versus the total number of bits transferred per transaction. Table 1.5 also summarizes the effect on data bandwidth of a memory transaction. The overall link's effective bandwidth increases linearly with the inclusion of each additional lane. The "Data Bytes Payload per Packet" columns simply reflect the number of data bytes per transaction.

Table 1.5 PCI Express Effective Bandwidth

PCI Express Effective Bandwidth of Link Layer Transaction Packets per a Link 2.5Gb/sec (x1) with Raw 250 MB/s versus 80 Gb/sec (x32) with Raw 8000MB/s

Per Differential Driven Pair in Each Direction not including bandwidth used by lane fillers, Ordered Sets, and Data Link Layer Packets

Data Bytes Payload per Packet	128	256	512	1024	2048	4096
Efficiency						
32-Bit Address	84.21%	91.42%	95.52%	98.84%	98.84%	99.42%
64-Bit Address	83.12%	90.78%	95.17%	97.52%	98.75%	99.37%
Effective Data Bandwidth per Second for 32-Bit Address	210.5 MB/s versus 6736.8 MB/s	228.6 MB/s versus 7315.2 MB/s	237.5 MB/s versus 7600 MB/s	244.2 MB/s versus 7817.6 MB/s	247.1 MB/s versus 7907.2 MB/s	248.6 MB/s versus 7955.2 MB/s
Effective Data Bandwidth per Second for 64-Bit Address	207.5 MB/s versus 6640 MB/s	227.0 MB/s versus 7264 MB/s	237.9 MB/s versus 7613.6 MB/s	243.8 MB/s versus 7801.6 MB/s	246.9 MB/s versus 7900 MB/s	247.5 MB/s versus 7920 MB/s

Other PCI Express Performance and Features

The previous discussion has focused on improved platform performance through increasing the reference clock and thus the associated PCI or PCI-X bus segment or PCI Express link bandwidth. Improved platform performance by increasing available bandwidth is of course an important feature, but PCI Express implements additional methods to improve platform performance: the Flow Control protocol, Quality of Service (QoS), robust messaging, and the Requester/Completer protocol.

Flow Control

The Flow Control protocol provides software the opportunity to assign priority-related information to each bus transaction. As the bus transactions flow through the PCI Express platform between PCI Express devices via links, link bandwidth must be considered. Of particular note are the switches. The transactions (in the form of transaction packets) from different sources may merge in the switch's buffer and vie for link bandwidth on the next link they traverse. Balancing the bandwidth needs of the different links versus the bandwidth needs of the different transactions in the different switch buffers requires a priority protocol for transmission onto the link. The Flow Control protocol provides a mechanism for each transaction to be assigned priority information by its source that can be compared at the switches with the priority information of other transactions from other sources. The protocol permits each transaction packet's transmission to be prioritized relative to other transaction packets at each switch output connected to each link.

Quality of Service (QoS)

The execution of some transactions is sensitive to the time it takes to begin the execution (latency), the speed at which the data can be transferred (bandwidth), and the time that passes before the execution is completed. These are the basic issues of QoS. The Flow Control protocol provides the mechanism to fine-tune the paths and relative transmission priorities of transactions and thus defines a QoS for each transaction. One of the key applications of QoS is the isochronous transfer. Isochronous transfers rely on a known latency and bandwidth to ensure proper operation. For example, a video stream relies on continuous transmission of transactions. The time it takes a specific transaction to begin (latency) and how well the data is transferred (bandwidth) will determine whether the video display is smooth or choppy in appearance.

PCI Express Messages

Messages are not a new concept, but they were never used in PCI or PCI-X 66 and PCI-X 133 platforms. PCI-X 266 and PCI-X 533 do implement a minimal message protocol. In a PCI Express platform messages are simply bursts of information for control or data transfer that are posted from buffer to buffer between PCI Express devices. PCI Express implements messages on three levels.

First, messages in the form of message bus transactions replace all of the discrete "wires" used for interrupt, Lock function, power management, and so on. Once the related message is transmitted no further bandwidth is required to support these "wires."

Second, the types of different information and data that can be transmitted in a PCI Express platform have expanded over the types for PCI or PCI-X platforms. For example, error-related information is transmitted from any PCI Express device to the Root Complex to alert the software and provide error logging. Also, vendor-specific information can be transmitted in a PCI Express platform that goes beyond the confines of the PCI Express specification.

Third, the application of messages also supports information and data that transfers between multiple PCI Express fabrics through the use of advanced peer-to-peer links, as illustrated in Figure 1.12. The messages that traverse advance peer-to-peer links permit the encapsulation of other packets specific to other technologies not related to PCI, PCI-X, or PCI Express.

> The inclusion of the advanced peer-to-peer link is to reflect future enhancements to the PCI Express specification. The PCI Express specification was first released with the advanced peer-to-peer link and message advanced switching requester transactions defined. The errata available at the time this book was printed removed most references to these entities, but retained others. See Chapter 18 for more information on the advanced peer-to-peer link and message advanced switching requester transactions.

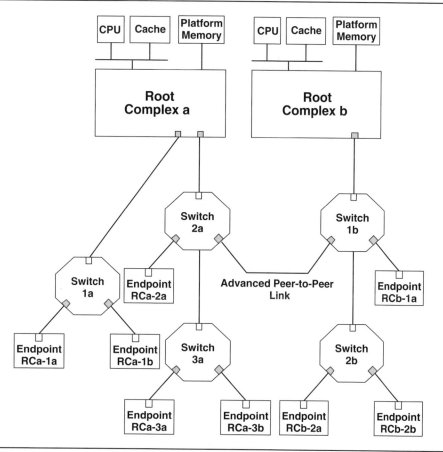

Figure 1.12 Overview of PCI Express Platform

Requester/Completer Protocol

As previously discussed PCI platforms implemented the Delayed Transaction protocol, which used a lot of platform bandwidth executing bus transactions over and over (retry) as a prompt to the PCI device being accessed. The PCI device responds to the prompt when it is ready to complete the bus transaction. A PCI-X devices use much less bandwidth by not repeating bus transactions over and over. It is the responsibility of the PCI-X device being accessed to complete the transaction when it can without any prompting, using the Split Transaction protocol. The associated PCI and PCI-X protocols were Delayed Transaction and Split Completion, respectively. PCI Express implements the Requester/Completer protocol, which is very similar to the Split Transaction protocol. The key concept is that a transaction consists of requester and completer transactions. One PCI Express device sources the requester transaction once. The PCI Express device being accessed sources the completer transaction(s) when ready. More detailed information about requester transactions and completer transactions is provided in later chapters. The example discussed in the following section simply provides an introduction the concepts of the Requester/Complete protocol and an introduction to the basic transaction flow throughout the PCI Express fabric.

Requester/Completer Protocol Example Transactions are between two PCI Express devices called *participants*. All bus transactions except for memory write and message consist of requester transactions and completer transactions. Memory write and message transactions only consist of requester transactions. The execution of a transaction begins with the transmission of the requester transaction from a PCI Express device (requester source) to access another PCI Express device (requester destination). In response to the requester transaction, the requester destination responds (except for memory write or message) with a completer transaction. With this response the requester destination becomes the completer source and the requester source becomes the completer destination.

Figures 1.13 and 1.14 illustrate the requester transaction and completer transaction sequence of the Requester/Completer protocol for some possible transactions. Figure 1.13 shows the requester sources and destinations for the four examples. Figure 1.14 shows the completer sources and destinations for the four examples. As will be discussed in later chapters, not all transactions can be executed between all PCI Express devices.

- **Transaction Example 1:** A memory requester transaction is transmitted from endpoint rcb to the Root Complex. The associated completer transaction is transmitted from Root Complex to the endpoint rcb. The two participants are endpoint rcb and the Root Complex. If the actual access is to the platform memory, the Root Complex is still defined as the participant as viewed by the PCI Express fabric.

- **Transaction Example 2:** An I/O requester transaction is transmitted from the Root Complex to endpoint 1b. The associated completer transaction is transmitted from endpoint 1b to the Root Complex. The two participants are the endpoint rcb and the Root Complex. Switch 1 ports the associated transactions between the downstream link and the upstream link but is not defined as a participant. If the associated requester transaction was a memory write or a message, no completer transaction is transmitted.

- **Transaction Example 3:** A configuration requester transaction is transmitted from the Root Complex to switch 2. The associated completer transaction is transmitted from switch 2 to the Root Complex. The two participants are switch 2 and the Root Complex. Switch 2 is a participant for the purposes of the Root Complex accessing the configuration register block within the switch.

- **Transaction Example 4:** A message requester transaction is transmitted from endpoint 3a to switch 3. No completer transaction is transmitted from switch 3 to endpoint 3a in response to the associated message requester transaction transmitted by endpoint 3a. The two participants are endpoint 3a and switch 3. Switch 3 is a participant for the purposes of the endpoint transmitting a message transaction.

38 ■ The Complete PCI Express Reference

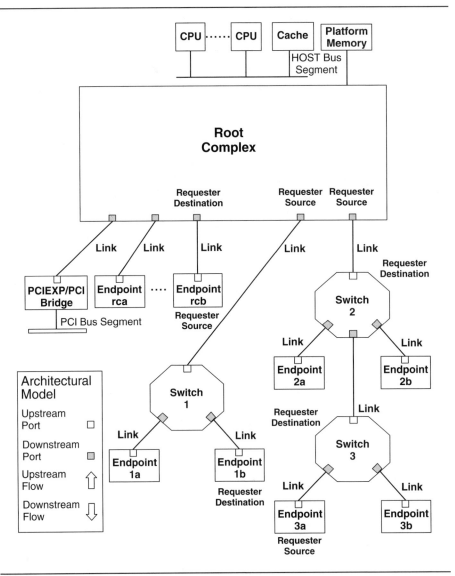

Figure 1.13   Requester Transactions

Chapter 1: Architecture Overview ■ **39**

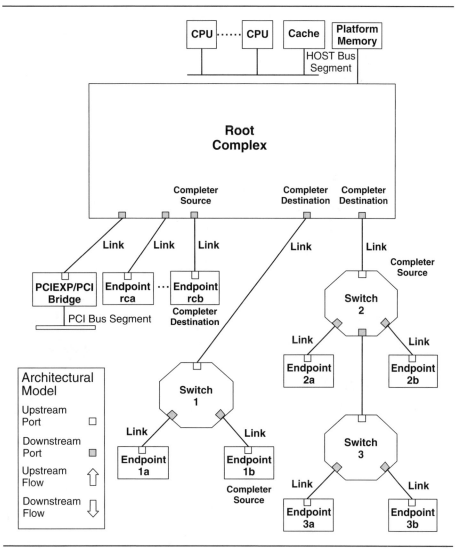

Figure 1.14 Completer Transactions

Chapter 2

PCI Express Architecture Overview

The purpose of this chapter is to provide the next level of detail to permit easier reading of the more detailed chapters later.

In the previous chapter, the generic PCI Express platform was introduced and restated here in Figure 2.1, also in the previous chapter was the introduction to the Requester/Completer protocol for transactions flowing throughout the PCI Express fabric. This chapter provides more detailed information about the PCI Express devices and associated transaction flows contained within the PCIEX fabric.

This chapter will first provide incremental details of the generic PCI Express platform architecture and the architecture of the PCI Express devices. It then introduces more information about the possible PCI Express devices that can participant in a common transaction as well as more details of the Requester/Completer protocol. The chapter will then detail how the PCIEX Transaction, Data Link, and Physical Layers operate to support the transmission and reception of transactions via different packets. The final part of the chapters discusses the details of each of the address space, PCI related issues, and a brief mechanical-electrical overview.

Generic PCI Express Platform Architecture

The PCI Express fabric consists of multiple PCI Express devices interconnected with links, as illustrated in Figure 2.1. Chapter 1 introduced the PCI Express devices: the Root Complex, switches, endpoints, and bridges (PCI Express/PCI Bridge or PCI Express/PCI Bridge). Within a PCI Express fabric all interconnections of PCI Express devices are possible if these four requirements are met:

- Only one Root Complex is implemented and it is the only interface between the other PCI Express devices, the HOST bus segment, and platform memory.
- Each switch can have only one upstream port. Any number of downstream ports is possible (up to 256).
- Each endpoint and bridge can have only a single upstream port.
- Each link can interconnect only one downstream port and one upstream port. The exception is advanced peer-to-peer discussed below.

These four requirements do not restrict the combination of PCI Express devices that are interconnected. For example, the PCI Express platform in Figure 2.1 has only one bridge connected to the Root Complex. It is possible to have multiple bridges connected to the Root Complex or the downstream ports of switches. Figure 2.1 also shows a switch connected to the Root Complex and a switch connected to another switch. All switches may be connected to the Root Complex. A PCI Express platform may also have no switches at all. With no switches, all endpoints and bridges must be connected directly to the Root Complex. PCI Express devices may also be interconnected in such a way that only endpoints exist and no bridges, or vice versa.

An important part of the interconnection of PCI Express devices is the link. In the PCI Express fabric each link has precisely two PCI Express devices connected to it and each link supports simultaneous bi-directional transactions; consequently, PCI Express devices do not arbitrate for the link. PCI Express transactions may traverse several links between the source and destination PCI Express devices of the transaction. For example, PCI Express endpoint 3a may need to access platform memory. This requires the transaction to pass through switch 3 and then switch 2 prior to reaching the Root Complex. At the same time endpoint 3b may also be accessing platform memory via the same switches and associated links. These transactions, contained in Transaction Layer Packets (TLPs), are buffered at the upstream ports of the switches 2 and 3 as they flow upstream. At the upstream ports of the switches these buffered TLPs vie for bandwidth of the connected upstream link. At the same time other PCI Express devices have TLPs vying for link bandwidth. The allocation of link bandwidth among the vying TLPs is managed by the PCI

Express Flow Control protocol. The important concept to grasp is that the arbitration is not for link ownership but for the appropriation of link bandwidth among TLPs to be transmitted from the same upstream port. This concept also applies to TLPs flowing in the downstream direction that are transmitted by a downstream port.

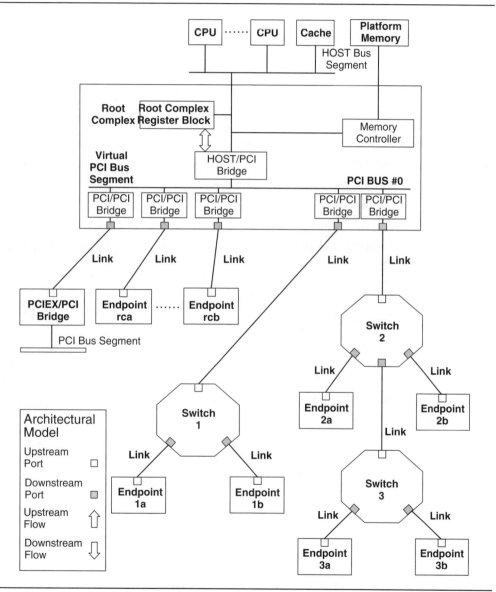

Figure 2.1 Generic PCI Express PC Platform Model

Hierarchy and Hierarchy Domain

The PCIEX specification defines a hierarchy and hierarchy domain. A hierarchy is the collection of all PCI Express devices connected to a specific Root Complex and the Root Complex itself. Effectively the PCIEX fabric for a specific Root Complex is the hierarchy.

Each link connected to the Root Complex defines a connection to a hierarchy domain. All of the PCIEX devices downstream of that connection point are part of that specific hierarchy domain. As exemplified in Figure 2.2, a Root Complex with only 5 downstream ports can only define 5 hierarchy domains.

As illustrated in Figure 2-2, the possible hierarchy domains are as follows:

- Endpoint rca, endpoint rcb, or the bridge are each a hierarchy domain connected to the Root Complex.

- For switches 1 and 2 each is the upstream interconnect point for an individual hierarchy domain.

- Switch 3 and the endpoints 1a, 1b, 2a, 2b, 3a, and 3b are connected to switches (not the Root Complex). Consequently, individual hierarchy domains are not defined at their respective upstream.

Other Interconnects between PCI Express Devices

The interaction between PCI Express devices is typically between the Root Complex and another downstream PCI Express device. The PCI Express specification also defines interactions between two PCI Express devices not including the Root Complex. These other interactions implement either peer-to-peer transactions or advance peer-to-peer links.

Chapter 2: PCI Express Architecture Overview

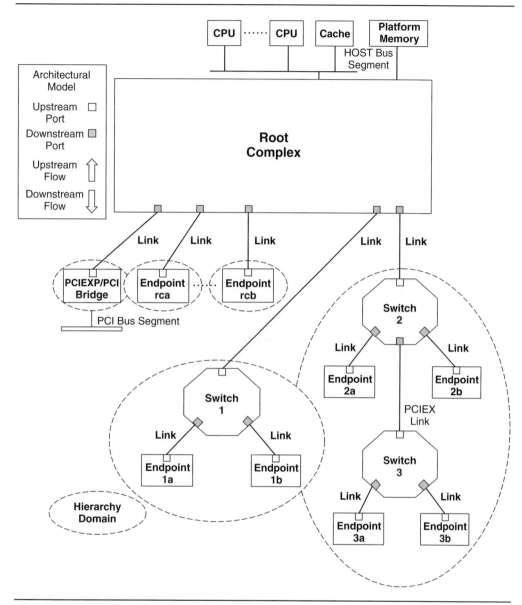

Figure 2.2 Generic PCIEX PC Platform Model with Hierarchy

Peer-to-Peer Transactions

In addition to the upstream and downstream of TLPs discussed above, TLPs flowing through a switch can port from one downstream port to another downstream port. These transactions between two downstream ports of a switch are defined as *peer-to-peer transactions*. Peer-to-peer transactions are only defined between two endpoints, two bridges, or a combination of bridges and endpoints as the participants. A unique characteristic of a peer-to-peer transaction is that it moves in both upstream and downstream directions within the execution of a single specific transaction. The inflection point is the switch where the TLPs of the upstream flow become the TLPs of the downstream flow. The participants of a peer-to-peer transaction may or may not be connected to the same switch. For example, in Figure 2.1 some of the possible peer-to peer transitions are between endpoints 2a and 2b, endpoints 2a and 3a, endpoints 3a and 3b, endpoints 3a and2b, and so forth.

The Root Complex typically does not support peer-to-peer transactions. If a Root Complex does support transaction flowing from one downstream port to another, it is because a virtual switch resides inside the Root Complex, which is a very atypical implementation. Consequently, a switch may provide the only inflection point for peer-to-peer transactions. A Root Complex that does not support peer-to-peer transactions restricts the participants to endpoints and bridges that reside downstream of a common switch. As shown earlier, endpoints or bridges are the only possible participants in peer-to-peer transactions and both must exist downstream of a switch that can operate as the inflection point.

Peer-to-peer transactions can be memory, I/O, and message transactions, but not configuration transactions.

Advanced Peer-to-Peer Link

A specific PCI Express fabric consists of switches, endpoints, and bridges interconnected with links to a single Root Complex. Only one Root Complex is defined per specific PCI Express fabric. It is possible for a system to have multiple Root Complexes and thus multiple PCI Express fabrics. The PCI Express specification defines the specific link that interconnects between two PCI Express fabrics as the advanced peer-to-peer link. As illustrated in Figure 2.3, the advanced peer-to-peer link is only connected between two downstream ports of two switches, each switch part of a different PCI Express fabric. The PCI Express specification does not clarify whether multiple advance peer-to-peer links can be defined between two specific PCI Express fabrics. See Chapter 18 for more information.

Chapter 2: PCI Express Architecture Overview ■ **47**

The transactions that traverse advanced peer-to-peer links are comprised only of message advanced switching requester transactions. None of the other types of transactions can traverse advanced peer-to-peer links.

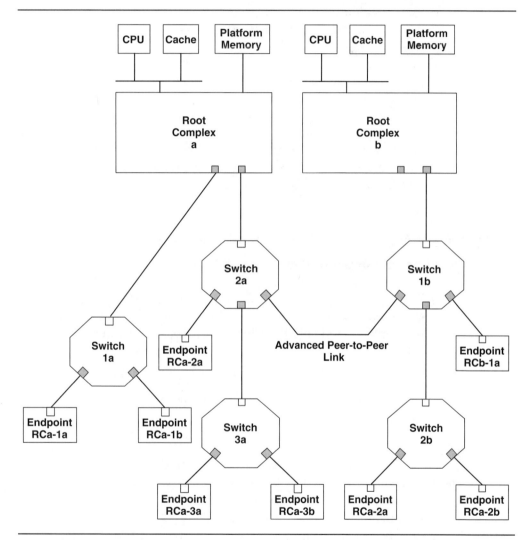

Figure 2.3   Advanced Peer-to-Peer Transaction Architectural Model

> **Important Note**
>
> The inclusion of the advanced peer-to-peer link reflects planned future enhancements to the PCI Express specification. The PCI Express specification was first released with the advanced peer-to-peer link and message advanced switching requester transactions defined. The errata available at the time this book was printed removed most references to these entities, but retained others. Please check `www.PCISIG.com` and `www.Intel.com` for more information.
>
> The inclusion of advanced peer-to-peer link and message advanced switching requester transactions in this book serves two purposes. First, it provides reference points in the book compatible to the final specification of these entities. Second, some of the configuration registers still reference the advanced peer-to-peer link.
>
> See Chapter 18 for more additional information.

Generic Architecture of PCI Express Devices and Flow Control Considerations

The following architectural models of PCI Express devices are the generic ones used throughout the book. Other possible architectural models can be implemented, and they are summarized in Chapter 3.

The key architectural requirements are discussed for each PCI Express device. Except for endpoints, a primary function of the PCI Express devices is to support the flow of transitions throughout the PCI Express fabric. Consequently, where applicable, the Flow Control consideration is introduced.

Root Complex

The HOST/PCI Bridge in PCI based platforms has been replaced by the Root Complex and the Root Complex is designed to support the following platform connections, as illustrated in Figure 2.1:

- Interconnecting the HOST bus segment with platform memory.
- Interconnecting the HOST bus segment with the PCI Express portion of the platform via downstream ports and links.
- Interconnecting the platform memory with the PCI Express portion of the platform via downstream ports and links.

The Root Complex also contains a variety of PCI Express and platform resources. These resources can be configured by accessing their registers within the Root Complex via the memory or I/O address space of the HOST bus segment. These resources are:

- The Interrupt controller for handling PCI Express interrupts. See Chapter 19 for more detailed information on this controller.

- Initialization, Power Management, and Hot Plug controllers. See Chapters 13, 15, and 16 respectively for more detailed information on each of these controllers.

Figure 2.4 provides the next level of detail for the Root Complex and is the basic architectural model to be used throughout this book. As previously discussed, all PCI Express devices reside downstream of the Root Complex. The platform CPUs only reside on the HOST bus segment upstream of the Root Complex. Also, the platform memory only resides upstream of the Root Complex. The PCI Express specification does not provide a specific implementation of the Root Complex, but Figure 2.4 represents the simplest and most straightforward. This Root Complex architectural model reflects the requirement that PCI Express hardware must be compatible with PCI and PCI-X configuration and general application software.

The key elements of the Root Complex are as follows:

- **The Virtual HOST/PCI Bridge** operates as the central interface point for the CPUs, platform memory, and all downstream PCI Express devices.

- **The Root Complex Register Block (RCRB)** is the set of configuration registers used to support the HOST/PCI Bridge.

- **The Virtual PCI bus segment** is simply a PCI-like structure to interconnect the virtual HOST/PCI Bridge and multiple virtual PCI/PCI Bridges. The IDSEL signal lines are also included as part of this structure to provide a PCI-like implementation. The IDSEL signal lines are effectively chip selects for the configuration register blocks in the virtual PCI/PCI Bridges.

- **Virtual PCI/PCI Bridges** only reside at the downstream ports. They fan out transactions sourced from the CPUs to the PCI Express devices or to fan in transactions to main memory from downstream PCI Express devices.

Of particular note is the interface at the Root Complex's downstream ports. With the use of virtual PCI/PCI Bridges at each downstream port and without any IDSEL-like signal lines external to the Root Complex, each connected link appears to the architectural model as a single PCI bus segment of a specific bus number. Without the implementation IDSEL-like signal line external to the Root Complex, all PCI Express devices on the other end of these links can only be defined as PCI devices with a DEVICE# 0.

All of the PCI-centric elements in the Root Complex are effectively virtual implementations. The form, fit, and function of these elements must be compatible with their actual PCI counterparts, but their implementation may be radically different.

Flow Control Features of the Root Complex

The Root Complex has the following features related to Physical Packet flow:

- **Peer-to-Peer Support:** A minimal Root Complex does not support peer-to-peer transactions between its downstream ports. Peer-to-peer transactions may be supported between the Root Complex's downstream ports if a virtual switch is internally constructed between two or more downstream ports. Only downstream ports connected to the virtual switch can support peer-to-peer transactions through the Root Complex. For example, in Figure 2.1, assume the Root Complex's downstream ports connected to endpoints rca and rcb are internally connected to the downstream ports of a virtual switch. No other Root Complex's downstream ports are connected to the virtual switch. In this example, peer-to-peer transactions occur between endpoints rca and rcb via the virtual internal switch. The virtual peer-to-peer transactions are actually a form of peer-to-peer transactions, which are only defined for switches.

- **Virtual PCI Bridges Manage Flow Through Root Complex:** The links in the PCI Express fabric are connected to the HOST bus segment and platform memory via virtual HOST/PCI Bridge and virtual PCI/PCI Bridges within the Root Complex. The downstream ports also consist of virtual PCI/PCI Express interfaces. The software model for configuration and programming of the Root Complex are illustrated in Figure 2.4. In this implementation IDSEL signal lines common to the PCI configuration mechanism represent an internal chip select mechanism. The IDSEL signal lines represent an internal decoding mechanism to select the appropriate virtual PCI/PCI Bridges. Actual hardware to implement the virtual HOST/PCI Bridge, PCI/PCI Bridge, or virtual PCI/PCI Express interfaces is implementation-specific.

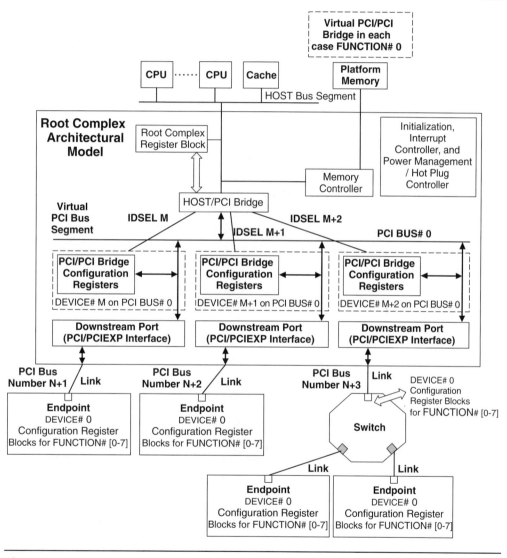

Figure 2.4 Root Complex Architectural Model

Switches

The switches and links provide the tree-like effect of the overall PCI Express fabric. Links are connected to the downstream ports of the Root Complex or other switches to interconnect with other switches, bridges, and endpoints in the PCI Express fabric.

As illustrated in Figure 2.5, the upstream port of a switch appears as a virtual PCI/PCI Bridge designated as DEVICE# 0 and FUNCTION # 0. The DEVICE# 0 designation follows the convention that a link appears to configuration software as a PCI bus segment with only one PCI device connected. According to the PCI specification, multiple functions can be defined for this virtual PCI/PCI Bridge, but as illustrated in Figure 2.5 only FUNCTION# 0 is defined. Chapter 3 discusses other switch implementations that result in the upstream port of the switch appearing as a collection of virtual PCI/PCI Bridges on the downstream ports, each as a function in DEVICE# 0. For discussion purposes, this book assumes no memory or I/O addresses are implemented by the switches.

An additional requirement of the PCI Express specification is that switches must merge message interrupt requester transactions received on the downstream ports and transmit upstream a representative message interrupt requester transaction. See Chapter 19 for more information on message interrupt requester transactions.

Flow Control Features of Switches

The primary function of a switch is to transport TLPs moving both upstream and downstream to the port leading to the destination. Each switch is designed with the following characteristics:

- Switches' upstream ports consist of virtual PCI/PCI Bridges defined as DEVICE# 0 or a collection of virtual PCI/PCI Bridges on the downstream ports as functions in DEVICE# 0.

- The links in the PCI Express fabric are interconnected by a virtual PCI bus segment and virtual PCI/PCI Bridges within the switches. The architectural software model of the upstream and downstream ports also consists of virtual PCI Express/PCI and virtual PCI/PCI Express interfaces, respectively. Actual hardware to implement the virtual PCI/PCI Bridges or virtual PCI Express/PCI and PCI/PCI Express interfaces will vary greatly. In this implementation IDSEL signal lines are used to exemplify an internal chip select mechanism. The IDSEL signal lines are representative of an internal decoding mechanism to select the appropriate virtual PCI/PCI Bridges.

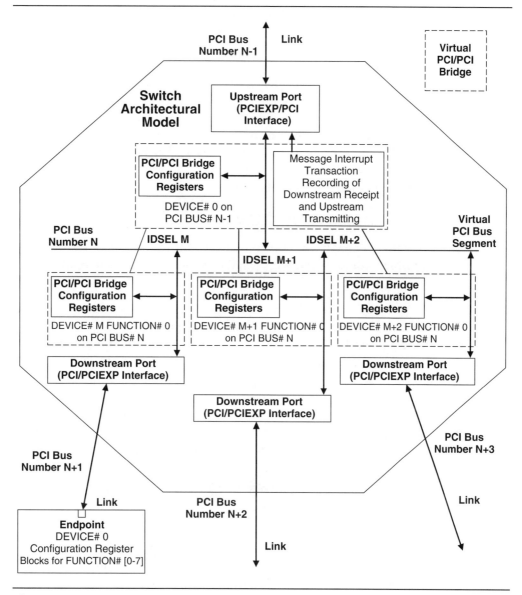

Figure 2.5   Switch Architectural Model

Endpoints

Endpoints are PCI Express devices that replace the PCI bus master and target resources. Consequently, each endpoint relative to the link appears as a PCI-like DEVICE # 0 with one to eight functions. Each function contains a configuration register block and responds to Type 0 configuration requester transactions. The endpoint is effectively any implementation-specific PCI device.

As discussed later in this chapter and Chapter 4, there are two types of endpoints: legacy and PCI Express. The major distinction is that the legacy endpoint can source requestor transactions and the PCI Express endpoint cannot. The transaction flow control requirements for endpoints are discussed later in the chapter. Endpoints have no specific Flow Control features other than the buffers implemented for TLPs that are received or are waiting for transmission.

Flow Control Features of Endpoints

There are no specific Flow Control features for endpoints other than the buffers implemented for TLPs received or waiting for transmission.

Bridges

As previously discussed the one or more PCI Express/PCI or PCI Express / PCI-X Bridges can be connected to one or more links. The primary purpose of a bridge is to translate PCI Express transactions to PCI or PCI-X bus transactions.

PCI Express supports bridges to permit the connection of other bus architectures. Discussions in this book are limited to interconnecting PCI and PCI-X bus segments to the PCI Express fabric, herein collectively called *bridges*. The connections are as follows:

- Figure 2.1 depicts a bridge connected to the Root Complex. A single bridge can be connected to any downstream port of the Root Complex or any switch. Multiple bridges can also be connected to the Root Complex and/or switch(es).

- A bridge's upstream port consists of a virtual PCI/PCI Bridge defined as DEVICE# 0 or a collection of virtual PCI/PCI Bridges on the downstream ports as functions in DEVICE# 0.

- For discussion purposes, this book assumes no memory or I/O addressable devices exist within the bridges. That is, the bridges only contain virtual PCI/PCI Bridges for the purpose of transferring transactions between PCI Express links and PCI or PCI-X bus segments.

As illustrated in Figure 2.6, the upstream port of a bridge appears as a virtual PCI/PCI Bridge designated as DEVICE# 0 and FUNCTION # 0. The DEVICE# 0 designation follows the convention that a link appears to configuration software as a PCI bus segment with only one PCI device connected. The PCI specification allows multiple functions to be defined for this virtual PCI/PCI Bridge, but as illustrated in Figure 2.6 only FUNCTION# 0 is defined. Chapter 3 discusses other bridge implementations that result in the upstream port of the bridge appearing as a collection of virtual PCI/PCI Bridges on the downstream ports as functions in DEVICE# 0. For discussion purposes, this book assumes no memory or I/O addresses are implemented by bridges.

All PCI or PCI-X bus segments must reside downstream of the bridge. The operation of the bus segments are defined by the PCI or PCI-X specification. To retain compatibility with existing hardware the bridge must appear to PCI or PCI-X devices on the bus segments as any other PCI/PCI Bridge or PCI-X/PCI-X Bridges on the related platforms. As illustrated in Figure 2.6 the downstream side of the bridge is a series of virtual PCI/PCI Bridges.

Additionally, that bridge is required to translate the interrupt signal lines defined downstream of the bridge on the PCI and PCI-X bus segment to message interrupt requester transactions. See Chapter 19 for more information on message interrupt requester transactions.

Flow Control Features of Bridges

The upstream port must be compliant with PCI Express protocol and resemble the upstream ports of the aforementioned switches.

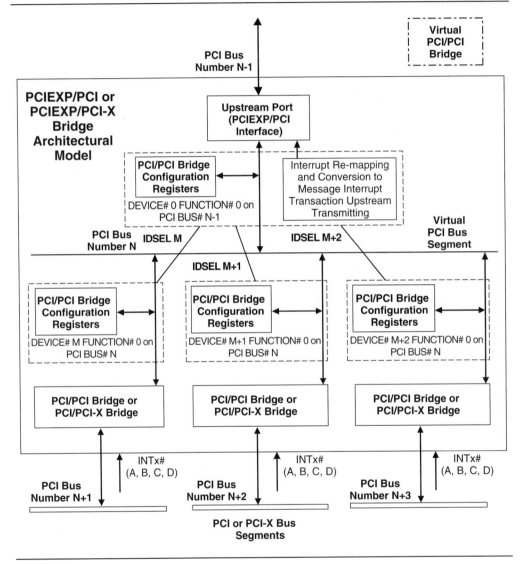

Figure 2.6 Bridge Architectural Model

Overview of Transaction Flow and Participants

As discussed above, transactions are between two PCI Express devices called "participants." As the transactions flow through the PCI Express fabric contained in TLPs, address or routing bits in the Header fields of the TLPs are used for directing the TLPs from the source to the destination.

Independent of the addressing or routing bits used for a specific type transaction, there are restrictions relative to the participants of a specific transaction. Some of these restrictions are by convention of PCI Express specification, while others are due to the architecture of the PCI Express fabric. The following discussions focus each of the PCI Express device types, including the restrictions each has relative to their role as participants of transactions and associated flow across the PCI Express fabric. More detailed information about participants, addressing bits, and routing bits is in Chapter 4.

> This book uses the following terms:
>
> - *Requester source* and *requester destination:* The two participants in the transmitting and receiving of the requester transaction packets.
>
> - *Completer source* and *completer destination:* The two participants in the transmitting and receiving of the completer transaction packets.

Restrictions of Root Complex as Requester/Completer

The Root Complex is optionally a requester source on behalf of the CPUs on the HOST bus segment or the requester destination and completer source for all PCI Express transactions with the following qualifications:

- The Root Complex must be requester source and completer destination for configuration transactions. The Root Complex cannot be requester destination and completer source for configuration transactions.

- The Root Complex cannot be a requester transaction destination and completer transaction source for the Lock function, but it can optionally be a Lock function requester transaction source and completer transaction destination. This option supports the

Lock function for legacy endpoints and PCI or PCI-X devices downstream of a bridge. The Root Complex can also optionally support the Lock function between the HOST bus segment and the platform memory. There is no support of the Lock function flowing upstream. There is no support of the Lock function by PCI Express endpoints.

Restrictions on Switch TLP Porting

Switches must port all transactions upstream, downstream, and peer-to-peer transactions with the following qualifications:

- Switches must support porting of TLPs for requester transactions from the upstream ports to downstream ports for the Lock function support in legacy endpoints and PCI or PCI-X devices downstream of a bridge. Switches must support porting of TLPs for completer transactions from downstream ports to the upstream ports for the Lock function support in legacy endpoints and PCI or PCI-X devices downstream of a bridge. Switches do not support porting of TLPs for requester transactions from the downstream ports to the upstream ports for the Lock function for legacy endpoints and PCI or PCI-X devices downstream of a bridge. Switches do not support porting of TLPs for completer transactions from the upstream ports to the downstream ports for the Lock function for legacy endpoints and PCI or PCI-X devices downstream of a bridge. Switches do not support porting of TLPs for requester or completer transactions between downstream ports for the Lock function; that is, no support of peer-to-peer transactions.

- As illustrated in Figure 2.5 (an example of switch 2 in Figure 2.1), the upstream port is connected to all downstream ports to support upstream, downstream, and peer-to-peer transaction flow via distribution on the internal virtual PCI bus segment.

- TLPs for downstream transaction flow can only be ported from the upstream port to a single downstream port; no snooping of the TLPs occurs for another downstream port. The one exception is the TLPs for message baseline requester transactions with routing number 011b in the TYPE HDW. This is used as a broadcast mechanism from the Root Complex to all PCI Express devices downstream.

- TLPs for upstream transaction flow can only be ported from the one to more downstream ports to the upstream port; no snooping of the TLPs occurs for another downstream port. That is, a TLP for a transaction flowing between the upstream port and a downstream port cannot be detected, monitored, or changed by other downstream ports.

- TLPs for peer-to-peer transactions can only be ported between two downstream ports of a switch at a time. A third downstream port and the upstream port cannot snoop the TLPs for peer-to-peer transactions porting between two other downstream ports. That is, a TLP for a transaction flowing between the two downstream ports port cannot be detected, monitored, or changed by other downstream ports or the upstream port.

- Switches must support porting of TLPs for configuration requester transactions from the upstream ports to downstream ports. Switches must support porting of TLPs for configuration completer transactions from downstream ports to the upstream ports. Switches do not support porting of TLPs for configuration requester transactions from the downstream ports to the upstream ports. Switches do not support porting of TLPs for completer transactions from the upstream ports to the downstream ports. Switches do not support porting of TLPs for configuration transactions between downstream ports (no peer-to-peer transactions support).

- Physical Packets containing bus transitions cannot be split into smaller Physical Packets by a switch.

Endpoints

Endpoints must be defined as requester destinations and completer sources of Type 0 configuration transactions. The endpoints also represent other non-PCI Express, non-PCI, and non-PCI Express buses like USB (that is, USB controller). The two types of endpoints, legacy endpoints and PCI Express endpoints, are discussed below.

Legacy Endpoints and Restrictions

The structure of legacy endpoints reflects the need to be compatible with existing PCI and PCI-X software. Of particular note is the support of the Lock function. Legacy endpoints must support as requester source and destination and a completer source and destination of all PCI Express transactions with the following qualifications:

- Legacy endpoints must be requester destinations and completer sources for Type 0 configuration transactions. Legacy endpoints cannot be requester source and completer destination for Type 1 configuration transactions.

- Legacy endpoints can optionally be requester destinations and completer sources for memory, I/O, and message transactions. Legacy endpoints can optionally be requester sources and completer destinations for memory, I/O, and message transactions.

- Legacy endpoints must be requester destinations and completer sources for transactions that support the Lock function. Legacy endpoints cannot be requester source and completer destination for transactions that support the Lock function.

- Legacy endpoints do not support the Legacy Interrupt protocol but do support Message Signaled Interrupts (MSI) via memory write transactions.

PCI Express Endpoints and Restrictions

PCI Express endpoints must support as requester source and destination and a completer source and destination of all PCI Express transactions with the following qualifications:

- PCI Express endpoints must be requester destinations and completer sources for Type 0 configuration transactions (that is, configuration transactions targeting this endpoint). Legacy endpoints cannot be requester source and completer destination for Type 1 configuration transactions (that is, configuration transactions targeting a PCI Express endpoint on another link).

- PCI Express endpoints can optionally be requester destinations and completer sources for memory, I/O, and message transactions. PCI Express endpoints can optionally be requester sources for message transactions. PCI Express endpoints can optionally be requester sources and completer destinations for message transactions. PCI Express endpoints cannot be requester sources and completer destinations for I/O transactions.

- PCI Express endpoints cannot be requester destinations and completer sources for transactions that support the Lock function. PCI Express endpoints cannot be requester sources and completer destinations for transactions that support the Lock function.

- PCI Express endpoints do not support the Legacy Interrupt protocol but do support Message Signaled Interrupts (MSI) via memory write transactions.

Restrictions on Bridges TLP Porting

Bridges must port all transactions upstream, downstream, and peer-to-peer (downstream bus segment to downstream bus segment) with the following qualifications:

- Bridges must support porting of requester transactions from the upstream ports to downstream ports for Lock function support in PCI or PCI-X devices downstream of a bridge. Bridges must support porting the equivalent of completer transactions from PCI or PCI-X devices downstream of a bridge to the upstream ports for the Lock function. Bridges do not support porting of equivalent requester transactions from the PCI or PCI-X devices downstream of a bridge to the upstream ports for the Lock function. Bridges do not support porting of equivalent completer transactions from the upstream ports to the downstream ports for the Lock function for PCI or PCI-X devices downstream of a bridge. Bridges do not support porting of requester or completer transactions between downstream ports for the Lock function (no peer-to-peer transactions support).

- Bridges must support porting of configuration requester transactions from the upstream ports to downstream ports. Bridges must support porting of equivalent configuration completer transactions from downstream ports to the upstream ports. Bridges do not support porting of equivalent configuration requester transactions from the downstream ports to the upstream ports. Switches do not support porting completer transactions from the upstream ports to the downstream ports. Bridges do not support porting of configuration requester or configuration completer transactions between downstream ports (no peer-to-peer transactions support).

- The balance of the bridge design must be compatible to Revision 2.3 or later of the PCI local bus specification. See the book and specification references in Chapter 1.

In-band and Advanced Peer-to-Peer Communications

The transaction packets that flow within a specific PCI Express fabric are defined as accessing memory, I/O, and configuration address spaces. The aforementioned peer-to-peer transactions are defined as part of the PCI Express fabric of a specific Root Complex, and consist of transactions accessing memory and I/O address spaces. In addition to transactions accessing memory, I/O, and configuration address spaces, transactions are defined for in-band communications and advanced peer-to-peer communications.

In-band communications comprise message baseline requester transactions and message vendor-defined requester transactions defined to flow within a specific PCI Express fabric. The participants of in-band communications are any of the PCI Express devices within the PCI Express fabric as part of a specific Root Complex.

PCI Express also defines message advanced switching requester transactions between PCI Express fabrics of two or more different Root Complexes, as illustrated in Figure 2.4. These transactions are defined as part of the advanced peer-to-peer communication and only traverse the advanced peer-to-peer link. See Chapter 18 for more information.

Transactions and Packets

In the traditional PC platform model (like PCI), a bus interconnects a master and target. The simplest execution of a write transaction is for a master to send an address/send data to a target, the target acknowledges receipt of the address/data and requests completion of transaction, and the master completes the transaction. The simplest execution of a read transaction is for a bus master to send an address to a target, the target acknowledges by returning data and requests completion of the transaction, and the master completes the transaction. In both transactions an implied assumption is made that both transaction participants are on the same bus, and that the master can wait on the bus until the target requests transaction completion. Extending this general discussion to a PCI Express platform, the Root Complex (master) and an endpoint (target) may not be on the same link (bus). The transaction may have to pass through multiple switches interconnected with links. The simplest execution of a transaction is not possible unless all intervening switches and links are dedicated to the transaction's execution. Complete dedication of the intervening switches and links is not possible in order to provide platform-wide bandwidth; consequently, PCI Express implements the Requester/Completer protocol.

As discussed in Chapter 1, the transactions are executed on links consisting of 1 to 32 lanes. Each lane is consists of two signal line pairs, one pair for each direction. Each signal line pair supports a bit stream that contains an integrated reference clock, destination information (address or routing), and sometimes data of the transaction. Transactions must be converted into bit stream packets called Physical Packets. The Physical Packets are transmitted onto the link by a PCI Express device (source) and the PCI Express device (destination) at the other end of the link extracts the transaction. Thus, transactions traverse the links contained within Physical Packets. The complete process of converting a transaction into a Physical Packet and vice versa involves several steps that include two other types of packets. In addition, if intervening switches are located between a transaction's source and destination, the switch extracts the transaction from a Physical Packet at the receiving port and then converts it to a Physical Packet at the transmission port.

This section will first present a more detailed discussion of the Requester/ Completer Transaction protocol that was introduced in Chapter 1. This section will then introduce how the transactions are contained with three types of packets, and an overview of their purposes and those of their associated PCI Express layers. The second part of this section will discuss the Requester/Completer as an introduction to transaction flow between PCI Express devices in the next section.

Requester/Completer Protocol

PCI Express adopted an approach similar to the PCI-X Split Transition protocol called the Requester/Completer protocol. The Requester/ Completer protocol requires that all transactions from the device core begin as requester transactions; some are completed with the receipt of completer transactions. The requester and completer transactions are contained in TLPs that can be posted in the PCI Express fabric. The requester transactions from the original source are assembled into TLPs that traverse the links in Link Layer Transaction Packets (LLTPs) contained within Physical Packets. The possible original sources are the Root Complex, bridges, and endpoints. At each switch the Physical Packets from the link are received and the TLPs are extracted, posted, and merged with other TLPs moving in the same direction in the switch. The collection of TLPs ready to transmit from a specific switch port are assembled into LLTPs to be contained in Physical Packets traversing the next link.

Once the Physical Packets arrive at the final destinations the destinations respond with completer transactions if applicable. The possible destinations are the Root Complex, bridges, and endpoints. The completer transactions are assembled into TLPs that traverse the links in LLTPs contained within Physical Packets and post in switches and the final destination as described for requester transactions. The final destinations of completer transactions are the original sources of their associated requester transactions.

Two key concepts of the Requester/Completer protocol are reflected in the above consideration. First, the TLPs can be posted as they flow through the PCI Express fabric. Second, for those requester transactions that are associated with completer transactions, the transaction is not complete until the completer transaction is received. Otherwise, the transaction is complete with the assembly of the requester transaction into the TLP.

To simplify the discussion and use of the Requester/Completer terms, the Requester/Completer protocol also extends to PCI Express memory write and message requester transactions. PCI Express memory write and message transactions require requester transactions but do *not* require completer transactions. This is unlike Delayed Transaction and Split Transaction protocols where PCI or PCI-X memory write transactions are not part of the Delayed Transaction or Split Transaction protocols, respectively.

The Requester/Completer protocol applies to all PCI Express transactions. The different types of transactions and the application of the Requester/Completer are summarized in Table 2.1. The possible participants (original source and final destination) of the above consideration are the Root Complex, bridges, and endpoints. For specific conditions related to configuration transactions, switches can be a participant and not simply a PCI Express device that ports TLPs from one port to another port.

Table 2.1 Packet Flow in PCI Express Fabric

Type of Transaction	Completer transaction associated with requester transaction?	Location of Participants	
		Same Link	Separated by Switch(s)
Memory write and message requester	No	TLPs are posted	TLPs are posted
Memory read, I/O read and write, config. write and read requester	Yes	TLPs are posted Completer transaction required	TLPs are posted Completer transaction required
Completer	NA	TLPs are posted	TLPs are posted

PCI Express Layers and Packets Overview

PCI Express defines three layers that detail the process of converting and extracting transactions, which can be sent via one or more links to a destination. These layers apply to both transmitting and receiving Physical Packets over the link as represented by Figure 2.7. The PCI Express layers are defined as follows.

- **The Transaction Layer:** This layer builds the transaction packet based on the interaction with the PCI Express device core. The interaction with the PCI Express device core will cause either a requester transaction packet or a completer transaction packet. The requester and completer transaction packets are part of the Requester/Completer protocol discussed below. The collective name for these two types of packets is Transaction Layer Packet (TLP). As discussed in Chapter 6, the TLP consists of a Header field, Data field, and a Digest field.
    - The Header Field contains the type of transaction, attributes and address or routing information.
    - The Data field is included if applicable. As the name implies it contains the data associated with the address and routing information in the Header field.
    - The Digest is optional and simply contains CRC information over the Header and Data fields. This specific CRC information is called ECRC to distinguish if from the LCRC and CRC associated with the packets of the Data Link Layer.

- **The Data Link Layer:** The primary responsibility of the Data Link Layer is to ensure that the TLP packet is delivered successfully across the link. To accomplish this task it implements the Link Layer Transaction Packet (LLTP). As discussed in Chapter 7, a sequence number and a LCRC ("long" CRC) are attached to each TLP to form the LLTP. The sequence number ensures that all TLPs transverse the link in a strict order and the LCRC information ensures the integrity of the LLTP. In addition, the Data Link Layer is responsible for overall link management. As discussed in Chapter 7, there are certain activities the Data Link Layer must perform to ensure the link is functioning correctly. The link management information is contained within the Data Link Layer Packet (DLLP). The DLLP consists of the link management information and CRC information to ensure the integrity of the DLLP.

- **The Physical Layer:** This layer converts the DLLP or LLTP into a serial bit stream defined as the Physical Packet. The conversion includes the integration of a reference clock into the bit stream, and framing information. Framing simply adds unique bits that denote the beginning and end of a Physical Packet. As discussed in Chapter 8, Physical Packets are transmitted as bit streams across the link over a pair of differentially driven signal lines.

The specification does not define the upper layers illustrated in Figure 2.7. However, the PCI Express device core represents the functions associated within a given device or the host processor, and may consist of only hardware or a combination of hardware and software. The core interfaces to the PCI Express layers via an interface that is also implementation-specific. This section focuses on defining the functionality of each layer and the basic contents of the associated packets.

Overview of Services Provided by Each Layer

The following is brief summary of the primary functions and services provided by each layer as represented in Figure 2.7. More detailed information for the Transaction, Data Link, and Physical Layers are provided in Chapters 6, 7, and 8 respectively.

Chapter 2: PCI Express Architecture Overview ■ 67

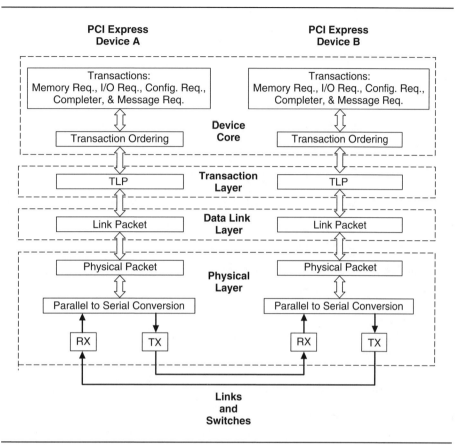

Figure 2.7   PCI Express Layers

PCI Express-Core Interface

The PCIEX-Core interface is implementation specific and not a layer defined by the PCIEX specification. Its basic activities:

- ■ Translate requests originating in the device core and translate requester and completer transactions outbound to the PCI Express layers. This may include data transfers, link management, etc.

- ■ Translate requester and completer transactions inbound from PCI Express layers to the format required by the device core.

Transaction Layer

The upper layer of the PCI Express protocol is the Transaction Layer. Services include:

- Assembly and disassembly of TLPs.
- Storage of negotiated link capability and other configuration information.
- Power management control/status.
- Credit-based link flow control.
- Application of Quality of Service (QoS) policies through use of Virtual Channel and Traffic Class numbers.
- Enforcement of TLP ordering rules.
- Management of TLP's integrity, including generation/checking of ECRC and support of data poisoning.

Data Link Layer

The second layer of the PCI Express protocol is the Data Link Layer. Services include:

- Assembly and disassembly of DLLPs and LLTPs.
- Initialization and Power Management, which involves porting requests between the Transaction Layer and the PCI Express device core and the Physical Layer.
- Conveyance of actual link state information from the Physical Layer back to the PCI Express device core via the Transaction Layer.
- Management of DLLPs and LLTPs integrity. This includes generation/checking of LCRC and CRC, ACK and retry of messages, packet retry on error, and so on.

Physical Layer

The third layer of the PCI Express protocol is the Physical Layer. Services include:

- Assembly and disassembly of Physical Packets.

- Link configuration, control, and status by link state transitions of the Link Training and Status State Machine (LTSSM). Link configuration includes detection and configuration of lanes to define a configured link, and establishment of link width. The activities of this layer and the LTSSM are detailed in Chapter 8 and include:
  - 8/10b encoding and 10/8b decoding, integration of a clock reference, and related symbol and bit synchronization.
  - Link transmitter(s) and Link receiver(s) implemented by the PCI Express device's ports.
  - FRAMING, scrambling, descrambling, and lane skew control.

- Design for test (DFT) and compliance features.

Transmitting and Receiving Packets: An Example

The following simple example in Figure 2.8 illustrates the PCI Express layers. The example introduces the bits fields associated with the assembling an outbound TLP and LLTP (requester or completer source), followed by disassembling an inbound TLP and LLTP (requester or completer destination) by the receiver on the other side of a link. Later chapters describe the low-level details of the three protocol layers. Refer to Figure 2.8 during this discussion.

This example does not discuss the DLLPs exchanged between the Data Link Layers of the PCI Express devices for link management. Link management operates in parallel with the transmitting and receiving of Physical Packets that contain LLTPs, which in turn contain TLPs.

Physical Packet Transmission

Each transaction originates within the Device Core. Requests for a PCI Express data transfer or control/status event are conveyed from the core to the Transaction Layer via the PCI Express-Core interface. Each PCI Express layer then performs the required operations.

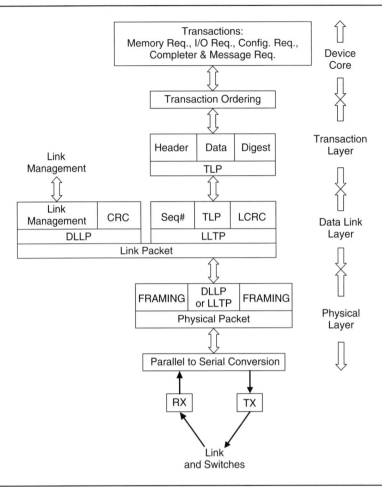

Figure 2.8   Overview of PCI Express Packets

Transaction Layer Based on the Device Core request, the Transaction Layer assembles the outbound TLP. The major components of a TLP include the Header, Data, and Digest fields as depicted in Figure 2.8, and include the following characteristics and information:

- The **TLP Header Field** consists of 3 or 4 Double Words (DWORDs). The format varies, but in each case the TLP Header Field defines the type of transaction and associated attributes. Key individual bit fields that may be included in the Header field are listed here.
    - **Type** identifies whether the transaction is for memory, I/O, configuration, message, or completer.
    - **TC (Traffic Class)** provides a value to identify priority of this TLP relative to other TLPs vying for link bandwidth. The Transaction Ordering protocol applies to all bus transactions with the same TC number. The TC number assigned to a TLP determines its Virtual Channel. Virtual Channel arbitration determines the transmission priority among Virtual Channels.
    - **Attributes** define the transaction ordering to be used and whether snooping is required.
    - **Address** identifies the memory address, I/O address, or Configuration ID.
    - **Length#** is number of DWORDs to be read or written.
    - **Requester ID#** contains the BUS#, DEVICE#, and FUNCTION# to identify the source of the requester transaction. The associated completer transactions use this Requester ID#.
    - **Tag#** identifies which requester transaction the subsequent completer transaction is to be aligned to.
    - Error and Poisoning: Identifies whether bus transaction packet integrity has been affected in transfer between the two participants.

- The **TLP Data Field** exists only in transactions that have an immediate data payload (for example, memory or I/O write). The data payload immediately follows the Header field. The data payload size varies according to the transaction and is defined by the LENGTH field in the Header field.

- The **TLP Digest (optional)** supports end-to-end CRC generation and checking (ECRC). If ECRC is used, it is inserted in the Digest position of the TLP. The Digest may also be used to indicate a data error occurred at the target and data is "poisoned" (invalid).

Data Link Layer When the TLP arrives, the Data Link Layer attaches strict ordering and error checking elements to it to create the LLTP:

- **Sequence Number** is a 12-bit number attached to the front of the TLP.

- **LCRC** is a 32-bit CRC (LCRC) on the Sequence Number and the TLP. It is appended to the end of the TLP.

Once the Sequence Number and LCRC have been added to the TLP it is referred to as a *Link Layer Transaction Packet (LLTP)* in this book. The PCI Express specification does not provide a label for this packet.

The Sequence Number and LCRC are used in conjunction with the subsequent Ack/Nak packets to verify that each TLP packet has been transferred across this link successfully. The Ack/Nak packets originate at the Data Link Layer of the receiving port, and returned to the transmitting port to report the result of the transmission. A copy of each TLP is kept in the Data Link Layer's retry buffer in the event a Nak is returned. Failed transfers may be retried by the sourcing device.

Physical Layer The Physical Layer prepares the LLTP for transmission across the link. The packet is encoded and driven onto the link by the Physical Layer. This process involves the following key steps:

- **Framing the packet:** This involves adding symbols to the beginning and end of the packet.

- **Encoding the packet:** 8/10-bit encoding is used to ensure sufficient transitions in the bit stream to permit the receiver to maintain synchronization with the transmission bit rate (2.5 gigabits per second). The encoding provides a stream of 10-bit symbols.
- **Converting parallel to serial:** This permits transmission of the transaction using fewer signal lines.
- **Parsing of symbols across the lanes:** When more than one lane is used, the symbols are parsed across all lanes to speed the transfer.
- **Differential signaling:** This reduces noise associated with symbol transmission.

The packets sent across the link are referred to as the *Physical Packet* in this book. The PCI Express specification does not define a term for the framed LLTP or DLLP.

Physical Packet Reception

When the Physical Packet is received by the port at the opposite end of the link, it must process the packet, including verification that the packet was received without errors. The PCI Express layers perform the following actions when receiving a packet:

Physical Layer This layer performs the inverse of the operation performed by the transmitting port. The Physical Layer:

- Receives differential bit stream.
- Performs serial to parallel conversion of the incoming packet.
- Reconstructs the information and data sequence from each lane (if multiple lanes are used).
- Verifies correct framing information.
- Performs the 10/ 8-bit decoding and recovers the LLTPs and DLLPs.

If the Physical Layer detects a framing error, this information is passed to the Data Link Layer so it can generate a Nak response (instead of Ack).

Data Link Layer When the LLTP arrives, the Data Link Layer performs the following error checks:

- It checks for Physical Layer framing errors.
- It verifies that the Sequence Number is correct.
- It calculates a CRC value for the received LLTP and compares it with the LCRC value associated with the packet.

Once the error checks have been made, one of two possible actions is taken:

- If no errors are detected, the Data Link Layer is permitted to forward the packet to the Transaction Layer and to send an Ack DLLP back to the sending port. An Ack DLLP notifies the sending port that it can flush the LLTP from its retry buffer.
- If an error is detected, the Data Link Layer discards the packet and sends a Nak packet back to the sending port.

Transaction Layer The TLP is received from the Data Link Layer and the Transaction Layer passes key information extracted from the inbound TLP up to the PCI Express device Core interface. The information may be data or completion status information.

Flow Control Protocol

Flow Control protocol in the context of this discussion is a global concept that relates to how transactions flow between participants across the PCI Express fabric. Flow Control protocol considers that TLPs may traverse one or more links on between their source and destination, and that many TLPs may be flowing in each direction at the same time. Several concepts and mechanisms are involved in managing TLP flow including:

- Quality of Service (QoS)
- Traffic Class
- Virtual Channel
- Link Flow Control

The role of each of these topics relative to Flow Control protocol is described in the following sections. This is an introductory discussion of key elements related to Flow Control protocol. See Chapters 11 for a discussion of foundation elements of Traffic Class, Virtual Channel, and Buffer Management. Chapter 12 focuses on the application of foundation elements.

Quality of Service

Quality of Service (QoS) is a generic term that normally refers to the ability of a network or other entity (such as PCI Express) to provide predictable and differentiated results. QoS is of particular interest when applications require continuous platform bandwidth, such as streaming video. In this respect, PCI Express defines isochronous transactions that require a high degree of QoS. However, QoS can apply to any transaction or series of transactions that flow through the PCI Express fabric.

QoS can involve many elements of performance including:

- Transmission rate
- Effective Bandwidth
- Latency
- Error rates
- Other parameters that affect performance

Many aspects of the PCI Express architecture lead to improvements over the PCI and PCI-X platforms that improve QoS. However, of special interest is the concept and implementation of traffic classes, which help predict transaction performance. Consider that a specific high-bandwidth PCI Express transaction associated with a specific application may be required to traverse a series of links interconnected by switches. Many other transactions may be traversing the same links and switches; consequently, achieving consistent results for a specific high-bandwidth PCI Express transaction is difficult. It is desirable to allocate greater bandwidth for the transactions of applications that need it and less bandwidth to those applications that need less bandwidth. This bandwidth allocation is managed through the assignment of different traffic class numbers, which permits bandwidth differentiation between the transactions.

Traffic Classes

Traffic classes give the application or PCI Express device the ability to determine the relative priority of transactions it sources versus others. The Traffic Class (TC) number is carried in the TLP and may include up to eight different numbers to be assigned. These numbers permit a higher degree of differentiation between different streams of TLPs flowing in the PCI Express fabric with varying requirements for PCI Express platform bandwidth.

PCI Express switches perform a pivotal role in managing the priority of TLPs. Switches map TC numbers into Virtual Channel (VC) numbers that are used to determine the priority of a given TLP at the transmission port. The process of prioritizing TLP streams involves allocating different amounts of bandwidth to the TLPs that are being ported to the same output port. The process of arbitrating between different TLP streams also requires the use of buffers at each port as illustrated in Figure 2.9.

The PCI Express implementation of Traffic Classes (TC) and Virtual Channels (VC) provides an opportunity for the transmission of specific transactions (via TLPs) and the associated path of switches and links they flow through to be "tuned" by system-level software. Recall that each PCI Express device (or associated application) assigns a specific TC number to each TLP, but the way in which different TLP streams are managed is based on a TC number mapping to a specific VC number for each output of each switch. The VC number is determined by system software. The assignment of a specific TC number to a specific VC number on a switches' output port determines the grouping of the TLPs that vie for the bandwidth of a specific link.

Chapter 2: PCI Express Architecture Overview ■ **77**

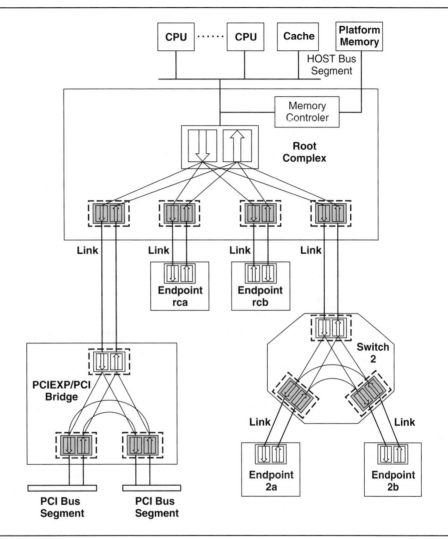

Figure 2.9   Overview of PCI Express Buffers

Virtual Channels

Switches use the concept of *Virtual Channels* (VCs) to determine transmission priority from the transmitting port. As the name implies, VCs are not actual physical channels. VCs are a way to provide transmission priority that will best use the link bandwidth. For example, the transmitting port may have different groups of TLPs mapped to different VCs (based on TC number). The transmission priority for allocation link bandwidth between the groups TLPs mapped to a specific VC number is called Port Arbitration. The transmission priority for link bandwidth between different VC numbers of a specific output port is called VC Arbitration.

Once the priority of transmission is established by the switch, TLPs are placed into buffers pending transfer to the port at the other end of the link.

Link Flow Control

All transmitting ports in the PCI Express fabric must implement link flow control, which is managed individually for each link that a TLP traverses. Before a TLP can be transmitted to the receiving port, the link flow control portion of the Flow Control protocol is checked to determine whether the TLPs receiving port's buffers can hold the TLP. If the receiving buffer is full, the switch moves on to other TLPs, knowing that this TLP would be rejected if sent. This mechanism avoids the retries that waste bandwidth in the PCI and PCI-X environments.

Link flow control uses a credit-based mechanism that allows the transmitting port to maintain a count of the credits (buffer space) available at the receiving port. The receiving port updates the transmitting port regularly by transmitting the number of credits that have been freed up, using the UpdateFC DLLP.

Other PCI Express Compatibilities and Differences with PCI and PCI-X

PCI and PCI-X platforms both implement what are commonly called *address spaces* as a protocol to access different elements in the PCI and PCI-X devices. These address spaces are also implemented by PCI Express platforms in addition to the message address space defined only by the PCI Express specification.

Address Spaces

The PCI, PCI-X, and PCI Express platforms all implement the memory, I/O and configuration address spaces.

Memory and I/O Address Space

The memory and I/O address spaces illustrated in Figure 2.10 for PCI, PCI-X, and PCI-Express are all implemented as "flat" address space across the platform. That is, only one address value is assigned to each memory or I/O byte throughout a platform. In order to ensure backward compatibility with PCI and PCI-X software the same memory address space is defined for PCI Express platforms. The memory address space is a maximum of 4 gigabytes for 32 address bits or 16 exabytes for 64 address bits.

The I/O address space defined for PCI, PCI-X, and PCI Express are all 4 gigabytes for 32 address bits.

Typically a platform has one PCI Express fabric, but multiple PCI Express fabrics are possible. The memory and I/O address spaces of each PCI Express fabric are independent of other memory and I/O address spaces, respectively. As previously discussed, the only communication over an advanced peer-to-peer link is via message advanced switching requester transactions. See Chapter 18 for more information on message advanced switching requester transactions.

Configuration Address Space

The PCI Express configuration address space is backward compatible to PCI and PCI-X. The basic configuration address space for PCI, PCI-X 66, and PCI-X 133 is 256 bytes per function in each device, as illustrated in Figure 2.11. In order to support the expanded features of PCI-X 266, PCI-X 533, PCI-X 1066, and PCI Express; the configuration address space has been expanded to 4096 bytes per function in each device. In the case of PCI Express some of the expanded features for example are advanced error reporting.

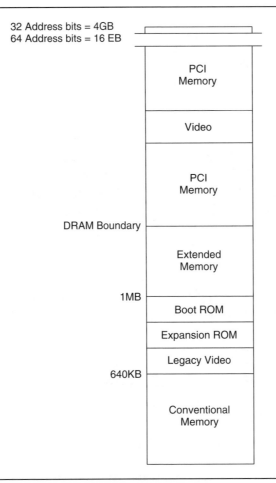

Figure 2.10 Overview of PCI Express Memory Map

The configuration address space allocated per function is called a configuration register block. The major consequence of the configuration register block of PCI Express being 4096 bytes versus PCI being 256 bytes is that the overall configuration address space is larger. PCI Express is a total of 256 megabytes (268.44 million bytes to be exact) versus a PCI total of 16 megabytes (16.777 million bytes). The only PCI Express device that can access the configuration address space is a CPU on the HOST bus segment through the Root Complex. The access to the configuration register blocks throughout the PCI Express platform is implemented by two methods: legacy configuration and PCI Express configuration.

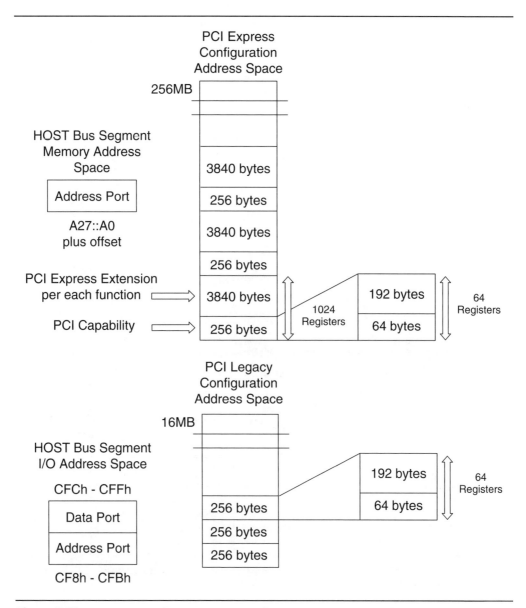

Figure 2.11 Overview of PCI Express Configuration Register Block

As illustrated in Figure 2.12, legacy configuration address space is accessed by the HOST bus segment CPUs via the I/O address space (CF8h to CFFh).

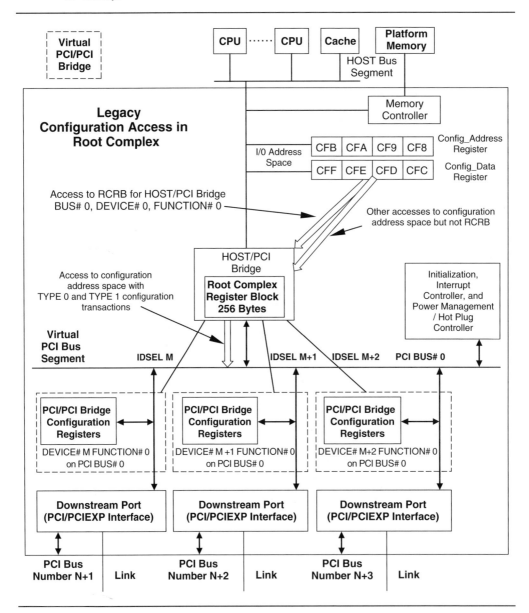

Figure 2.12 Overview of Legacy Configuration in Root Complex

Chapter 2: PCI Express Architecture Overview ■ 83

As illustrated in Figure 2.13, PCI Express configuration address space is accessed by the HOST bus segment CPUs via the memory address space via ADD [27::00].

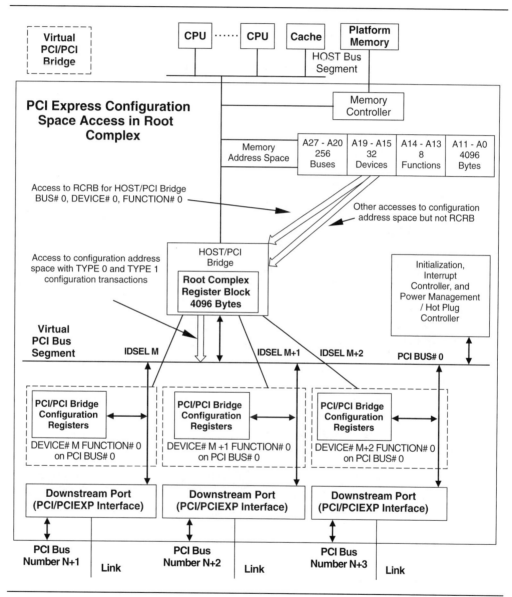

Figure 2.13 Overview of PCI Express Configuration in Root Complex

Messages

Some of the PCI and PCI-X hardware functions such as INT#x signal lines, Lock# signal line, and so on are supported in PCI Express by message transactions. This type of message is used as a virtual wire. Message transactions also provide additional features specific to PCI Express. The message address space is associated with message transactions. The message address space implements Implied Routing with two exceptions.

- The message vendor-defined requester transaction can also use ID Routing.

- The message advanced switching requester transaction only uses RID Routing. RID is Route Identifier, which is a 15-bit global.

Implied Routing encodes a binary pattern into the r field (contained within the Header field of a message requester transaction packet) that defines which PCI Express device is the destination. ID Routing uses the BUS#, DEVICE#, and FUNCTION# of the Vendor-defined ID HDW contained within of the Header field of a message vendor-defined requester transaction packets. The ID Routing defines which PCI Express device is the destination. RID Routing is unique to message advanced switching requester transactions. See Chapter 18 for more information.

Interrupts

Interrupts can be implemented in two ways, with MSI (Memory Signaled Interrupt) transactions, and with Message Interrupt Requester transactions. Each is described briefly here. You can find more information on interrupts in Chapter 19.

MSI

The preferred method of signaling interrupts is MSI. This implementation does not involve PCI Express message transactions as its name might imply. Instead, MSI is simply a memory write requester transaction. This is the native mechanism used by PCI Express devices to signal interrupts to the host.

Message Interrupt Requests

The message interrupt requester transaction implementation is only for supporting legacy endpoints and bridges. The message interrupt requester transactions consist of message Assert_INTx and Deassert_INTx requester transactions. These message requester transactions emulate the assertion and deassertion of INTx# signal lines found on PCI and PCI-X bus segments. The message interrupt requester transaction cannot be implemented by PCI Express endpoints.

Lock Function

The Lock function can be implemented in two ways: Exclusive Hardware Access (EHA) and Exclusive Software Access (ESA). The purpose of the Lock function is to permit interaction between two PCI and PCI-X devices without interference by a third PCI or PCI-X device. For example, a semaphore operation must be completed without risk of the third PCI or PCI-X device changing the values read and written.

The ESA implementation relies only on coordination between application software. The Lock function implemented by ESA is independent of the PCI, PCI-X, or PCI Express platform. The EHA implementation for the Lock function on a PCI Express platform is only defined between the Root Complex and a legacy endpoint or bridge. No EHA Lock function is defined between the Root Complex and a PCI Express endpoint, between any two endpoints, between any two bridges, or between an endpoint and a bridge.

The EHA implementation requires specific hardware. On the PCI and PCI-X bus segments a LOCK# signal line is defined. PCI Express links has no signal lines except for those associated with the lanes of the link. Consequently, PCI Express implements two entities to emulate the LOCK# signal line: unique transactions and a unique message transaction. The unique transactions to begin the Lock function are memory read lock requester transactions and completer lock transactions. The unique message transaction to end the Lock function is the message unlock requester transaction. For more information on the LOCK function, see Chapter 20.

Power Management

For backward compatibility to PCI and PCI-X, PCI Express has maintained compatible power management with support of D0 to D3 states within each function of a device. The D0–D3 states are programmed by the power management software. In some cases, PCI Express implements message requester transactions and link management packets to support power management between the two PCI Express devices on a link. PCI Express has also added Active-State power management, which operates invisibly to PCI, PCI-X, or PCI Express application software.

Hot Plug

PCI Express also supports the hot plug capability of add-in cards. It implements the PRSNT1#, PRSNT2#, WAKE# signal lines and 3.3 Vaux in the same functional way as PCI and PCI-X. PCI Express does complement hot plug functionality with message requester transactions to operate indicators, attention buttons, power supplied to the slot, and mechanical safety latches.

Mechanical and Electrical Overview

The section provides an overview of the mechanical and electrical definitions of the PCI Express specification.

Mechanical

In the same manner that PCI Express protects the software investment by using the PCI configuration mechanism, the mechanical definition of PCI Express protects the mechanical platform investment made for PCI and PCI-X by using the same add-in card types. The only departure is in the connector and the power rails supports. See Chapter 21 for more information on these mechanical differences. The summary of platform mechanical requirements is as follows:

- **Add-in Card Sizes**
    - **PCI-Compatible Card Sizes:** The three basic add-in card sizes defined by PCI and PCI-X are retained by PCI Express: low profile, short length (variable height), and standard length. These basic add-in card sizes are typically implemented in desktop PCs and server platforms.
    - **PCI Express Mini Card:** This add-in card is smaller in size than the PCI-compatible add-in card sizes and is typically defined for mobile platforms.
    - **Specialty Add-in Cards:** At the print time of this book there are four specialty add-in cards defined—NEWCARD, Server I/O Modules, Advance TCAs, and Cable Modules. As of the print date of this book, the final specifications were not available.

Chapter 2: PCI Express Architecture Overview

- **Signal Lines in addition to the PCI Express Link**
  - **REFCLK**: Reference Clock (Required): Similar to the CLK signal line of PCI and PCI-X except that it is 100 megahertz and used for the clock reference to transmit the symbol bit stream per 8/10-bit encoding (2.5 gigabits per second). REFCLK consists of two differentially driven clocks: REFCLK+ and REFCLK-.
  - **CLKREQ#** (Output from add-in card; mobile add-in cards only): For mobile add-in cards that implement PCI Express, the assertion (active low) of these signal lines enables the REFCLK to be sourced from the platform to the add-in card. This is an open collector signal line.
  - **PERST#** (Input to add-in card): Typically the PERST# signal line is used as a reference point for Cold Reset as an indicator for the power level of main power. It is also used as a reference point for a Warm Reset. See Chapter 13 for more information.
  - **PWRGD**: Power good connection (Required). PWRGD provides a common interface to a power supply for PCI, PCI-X, and PCI Express add-in cards.
  - **JTAG** (optional): Five connections used for platform production testing.
    - **TRST#**- Test reset
    - **TCK**- Test clock
    - **TDI**- Test data in
    - **TDO** - Test data out
    - **TMS**- Test mode select
  - **SMBus** (Optional): SMBus is a system management bus that can be implemented between PCI, PCI-X, and PCI Express components and add-in cards. SMBus is a low power and lower bandwidth serial bus. The bus consists of SMBCLK and SMDAT. It is a general purpose bus for purposes that range from platform management to inputs for non-bus equipment.
  - **PRSNT1# and PRSNT2#** connections (Required): Used by PCI, PCI-X, and PCI Express for hot-plug /removal support.

- **WAKE#** signal line (Required for add-in cards and optional for other PCI Express devices) and 3.3 Vaux signal line (Optional): WAKE# provides backward compatibility with PCI and PCI-X sleep mode.
- **Power Rails:** PCI, PCI-X, and PCI Express all support +12, +3.3 and 3.3 Vax volts. PCI and PCI-X also support -12 volts. Earlier revisions of he PCI and PCI-X defined a + 5 volt power rail as well.

The connectors defined for PCI Express are different in pin count than PCI or PCI-X, but of the same connector design and type as PCI and PCI-X. The possible lanes per link are x1, x2, x4, x8, x12, x16, and x32. Currently x2 and x32 connector sizes have not been defined; consequently x2 and x32 links cannot yet be implemented via an add-in card slot. Future revisions of the PCI Express specification may include x2 or x32.

The different connector sizes simply define connectors with different pin counts as follows: x1 = 36 pins, x4 = 64 pins, x8 = 98 pins, and x16 = 164 pins. For comparison: PCI and PCI-X require 124 pins and 188 pins for 32 data bits and 64 data bits, respectively.

A link with lane x1 can be implemented on any connector and x2 can be implemented in the connectors. A link with lane x2, x4, and x8 can be implemented on any connector x2, x4, x8 respectively. Links with lane x12 or x16 can only be implemented on connector x16.

Electrical

The electrical can be viewed in three parts: PHY, wave form, and timing. This is a brief overview to introduce the reader to some key concepts. See Chapter 8 for more information about PHY topics. See Chapter 21 for more information on wave-form and timing.

The PHY defines the protocol for transmitting and receiving the Physical Packet between ports on a link. The PHY transmitting protocol is illustrated in Figure 2.14. The addition of FRAMING bytes identifies and demarcates the link packets of the LLTP and the DLLP encapsulated in the Physical Packet. Physical Packets of parallel orientation are converted to a serial byte stream. As previously discussed, each link consists of one or more lanes. Each lane consists of two signal line pairs, one pair for each direction. Parsing distributes the byte stream across the link width to provide greater transmission bandwidth to the Physical Packet. If only one lane is available, no parsing occurs. As illustrated in Figure 2.15, two lanes are defined for the link. The byte stream of 8-bit bytes is parsed to the two lanes. Beginning with the FRAMING byte as the first byte; the first byte, third byte, fifth byte and so on of the Physical Packet are transmitted onto LANE# 0. The second byte, fourth byte, sixth byte, and so on of the Physical Packet are transmitted on to LANE# 0.

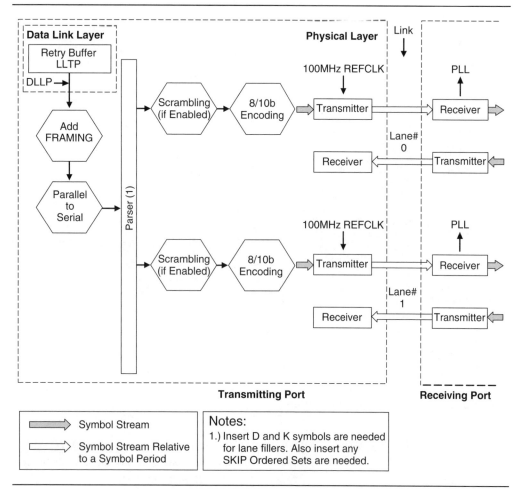

Figure 2.14 Physical Packet Transmission

Each 8-bit byte of the Physical Packet is encoded to a 10 bit *sbytes* (8/10-bit encoding). The *sbyte* is a term adopted by this book to identify this unique 10-bit byte. The 10-bit sbytes are also called symbols; they ensure that sufficient transitions of the bits in the symbol stream allow the receiving port to maintain clock synchronization (lock).

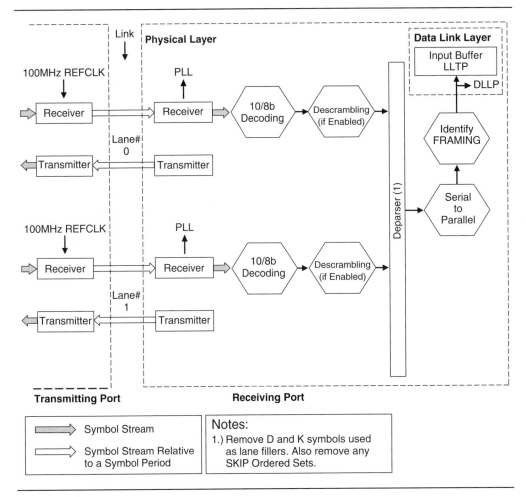

Figure 2.15 Physical Packet Reception

The last part of the PHY transmitting protocol provides a bit time (period) for each 10-bit sbyte symbol transmitted. That is, the differentially driven waveform representing the bit stream must have transition points. These transition points are defined by a bit period. The bit period is generated from a 100 megahertz clock to provide a 2.5 gigabit per second bit rate (the only bit rate per revision 1.0 of PCI Express specification). Ten-bit periods define a common symbol period among the lanes of a specific link.

The PHY receiving protocol is illustrated in Figure 2.15. The reception of the symbol stream requires the extraction of the reference clock. The 8/10-bit encoding provides a reference clock to which a PLL (phase lock loop) in the receiver can synchronize. This permits the receiver to determine the valid portion of each bit period. On each lane the symbol stream is decoded into a stream of 8-bit bytes with 10/8b decoding. The resulting byte streams from the two lanes are deparsed into a single byte stream. The FRAMING bytes demarcate the link packet and identify a DLLP versus LLTP. The DLLP executes link management and the LLTP is placed in an input buffer.

Chapter 3

Implementation

The PCI Express specification does not provide implementation examples. Indeed, it implies certain PCI elements should be used in the design PCI Express devices, but does not make them requirements. For example, the PCI Express specification requires that the Root Complex and switches have multiple "PCI/PCI-like" bridges and "PCI-like" internal bus segments, but does not define the exact implementation. The PCI Express specification does have strong requirements for configuration address space and software application compatibility with PCI.

The major chip set suppliers will develop the Root Complex, switches, and bridges. Individual hardware designers will focus on endpoints and in some cases portions of bridges to front end existing PCI designs. All will rely on experience and availability of existing PCI elements. Additionally, the interface between a link and the circuitry in the endpoint will need to be PCI-compatible in two ways: First, the configuration address space must be compatible with PCI software. Second, in order to avoid deadlock and livelock situations, the basic bus transaction ordering of the endpoint's interface buffer must follow PCI protocol.

The support of existing PCI- and PCI-X-compatible designs and add-in cards will dictate that the initial PCI Express platforms include PCI and PCI-X bus segments. Individual software developers will focus on PCI-like programming models, both from experience and the requirements of PCI Express. The implementation of PCI Express is essentially that of a point-to-point PCI platform. No matter how the hardware designers or software developers implement PCI Express, they both must have a point-to-point PCI implementation as the central focus.

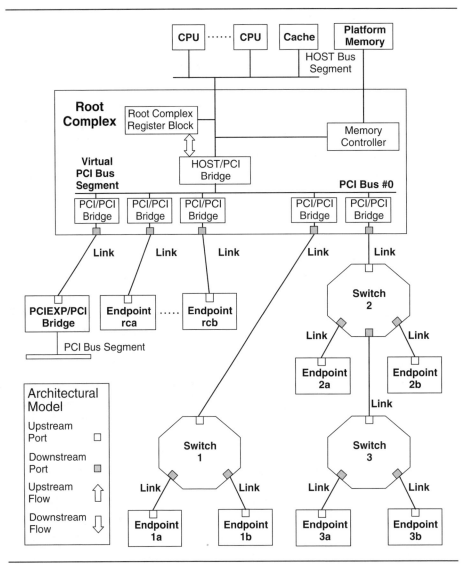

Figure 3.1   Generic Implementation of a PCI Express Platform

Given the above considerations, this book references a generic PCI Express platform and PCI Express devices based on PCI elements as an example implementation of the PCI Express specification. The PCI elements are portions of present components used in PCI platforms that can be adapted to the design of components compatible with PCI Express. Figure 3.1 is the overall generic implementation of a PCI Express platform. Different versions of this platform appear throughout the book. It includes endpoints connected to the Root Complex and switches, switches connected to switches, and a bridge connected to the Root Complex. More bridges can be added to the Root Complex or switches, but the concepts remain the same.

The generic PCI Express platform consists of four major PCI Express devices: Root Complex, switch, endpoint, and bridge. Each of these PCI Express devices can be constructed in different ways. As an application example, this book focuses on the generic PCI Express devices constructed with the maximum number of PCI elements. At the end of this chapter the other possible ways to construct the PCI Express devices are reviewed.

Generic Root Complex and Endpoints

Figure 3.2 represents a generic implementation of the Root Complex and illustrates the following key design considerations:

- The HOST bus segment and platform, cache, and platform memory are Root Complex implementation-specific. All of these platform resources are upstream of the HOST/PCI Bridge.

- The primary interface between the resources and the upstream portion of the PCI Express fabric is the HOST/PCI Bridge. The PCI-like configuration address space is the Root Complex Register Block.

- The interface to the PCI fabric external to the Root Complex is a series of virtual PCI/PCI Bridges connected to an internal virtual PCI bus segment (BUS# 0). Each bridge is modeled after a generic PCI/PCI Bridge whose configuration register blocks are selected by IDSEL signal lines.

- Endpoints are also included in Figure 3.2. They are highly specific to the implementation. As discussed in Chapter 2, the two types of endpoints are legacy and PCI Express. All endpoints must appear as DEVICE# 0 with FUNCTION# 0 (minimum) on the link. Up to seven other functions can be assigned to DEVICE# 0. If it is legacy endpoint, the PCI-compatible configuration address space (256 bytes per function) is supported. If it is PCI Express endpoint, the configuration address space compatible with

PCI Express (4096 bytes per function) is supported. There are other attributes unique to legacy endpoints versus PCI Express endpoints discussed in Chapter 2.

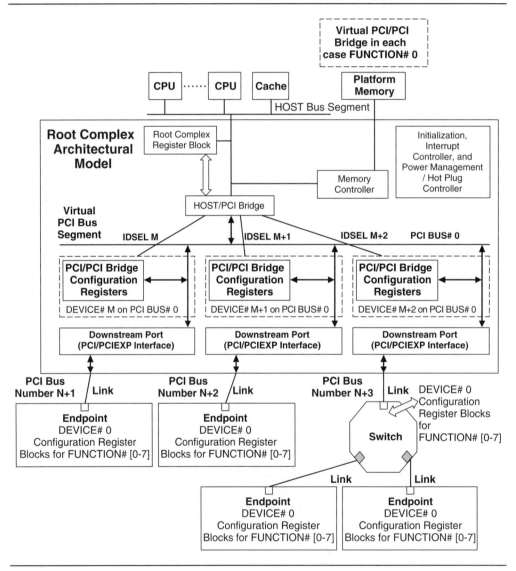

Figure 3.2 Generic Implementation of Root Complex and Endpoint

Chapter 3: Implementation ■ 97

The implementation of the generic Root Complex has two versions dependent on software compliance. For platforms designed only for PCI-compatible software, the access to the PCI Express configuration address space is via the I/O address space of the HOST bus segment, as illustrated in Figure 3.3. For platforms designed for software compatible with PCI Express, the access to the PCI Express configuration address space is via the memory address space of the HOST bus segment, illustrated in Figure 3.4.

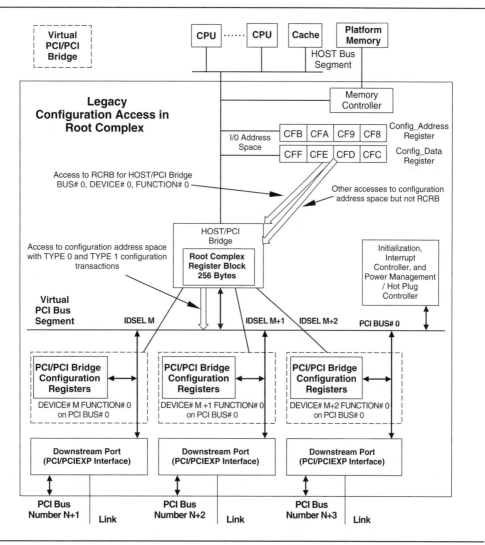

Figure 3.3 Generic Implementation of Root Complex for PCI Software

98 ■ The Complete PCI Express Reference

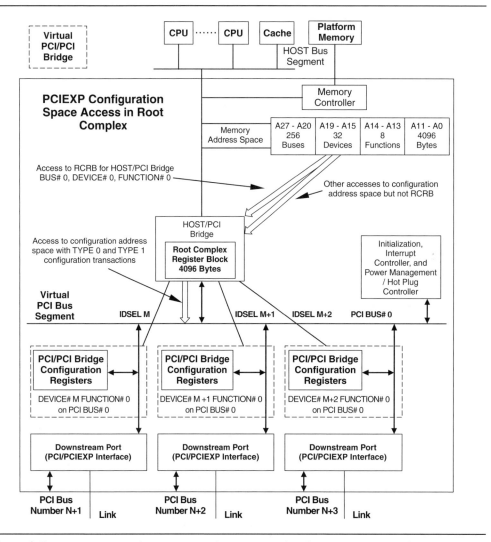

Figure 3.4 Generic Implementation of Root Complex for PCI Express Software

Generic Switch

Figure 3.5 represents a generic implementation of a switch and illustrates the following key design elements:

- The upstream port consists of a virtual PCI/PCI Bridge with additional logic to support legacy interrupts. See Chapter 19 for more information on interrupts.

- The downstream ports that interface with the downstream links consist of virtual PCI/PCI Bridges connected to an internal virtual PCI bus segment (BUS# N). Each virtual PCI/PCI Bridge is modeled after a generic PCI/PCI Bridge whose configuration register blocks are selected by IDSEL signal lines.
- Each downstream virtual PCI/PCI Bridge is a device on the virtual PCI bus segment; up to 32 of these devices are possible. Each device can have up to eight functions.

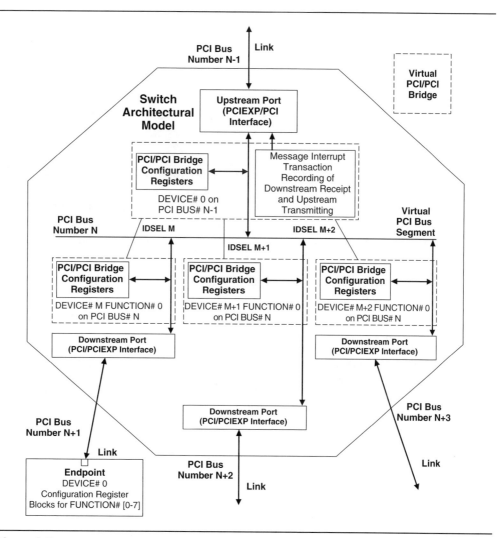

Figure 3.5 Generic Implementation of a PCI Express Switch

Generic Bridge

Throughout this book there are discussions about PCI/PCI, PCI-X/PCI-X, PCI Express/PCI, and PCI Express/PCI-X Bridges. PCI/PCI and PCI-X/PCI-X Bridges interconnect PCI and PCI-X bus segments, respectively. For more detailed information about PCI/PCI Bridges, please see the specifications and books referenced in Chapter 1.

> This book uses the term *bridge* to collectively refer to both PCI Express/PCI and PCI Express/PCI-X Bridges. The term does not refer to PCI/PCI or PCI-X/PCI-X Bridges. The book distinguishes between a PCI Express/PCI Bridge and a PCI Express/PCI-X Bridge only when PCI or PCI-X differ in their behavior relative to PCI Express.

Figure 3.6 represents a generic implementation of a PCI Express/PCI or PCI Express/PCI-X Bridge and illustrates the following key design considerations:

- The upstream port consists of a virtual PCI/PCI Bridge with additional logic to support legacy interrupts. The support of the legacy interrupts includes discrete open collector interrupt signal lines connected to the downstream side of the bridge. See Chapter 19 for more information on interrupts.

- The interface to the downstream bus segments is a series of PCI/PCI Bridges connected to an internal virtual PCI bus segment (BUS# N). Each bridge is modeled after a generic PCI/PCI Bridge whose configuration register blocks are selected by IDSEL signal lines.

- Each downstream virtual PCI/PCI Bridge is a device on the virtual PCI bus segment; up to 32 of these devices are possible. Each device can have up to eight functions.

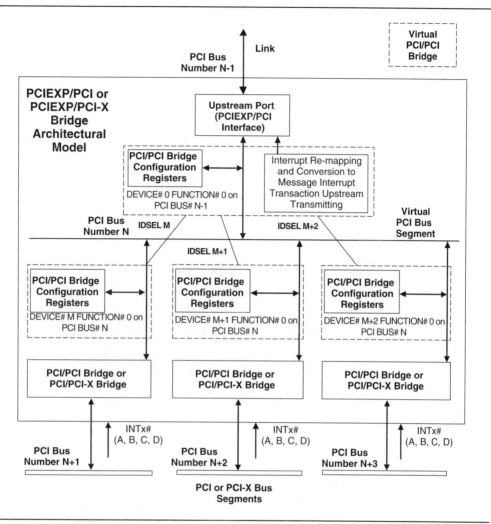

Figure 3.6 Generic Implementation of a Bridge

Other Implementations

As previously discussed, the PCI Express specification does not provide implementation examples. The generic Root Complex, switch, endpoint, and bridges discussed in the previous sections are implementations used for discussions in this book. However, it is possible for the Root Complex, switches, bridges, and endpoints to be implemented slightly differently by reducing the number of devices and virtual PCI/PCI Bridges. The implementations discussed in previous sections and those that follow can be intermixed within a PCI Express platform.

Other Implementations of the Root Complex

As illustrated in Figures 3.7 and 3.8 it is possible to make of the downstream virtual PCI/PCI Bridges of the Root Complex of different functions of one device number.

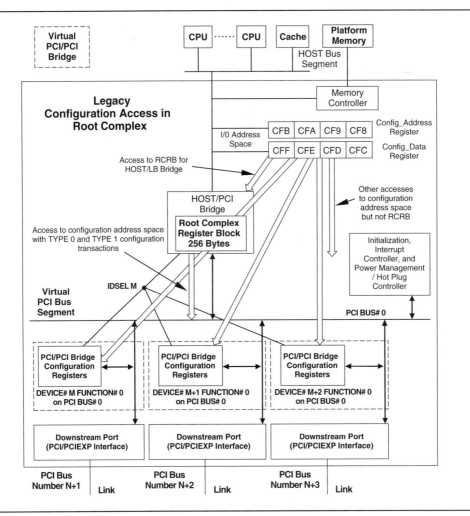

Figure 3.7 Other Implementation of Legacy Root Complex with Single Internal Device

Chapter 3: Implementation ■ 103

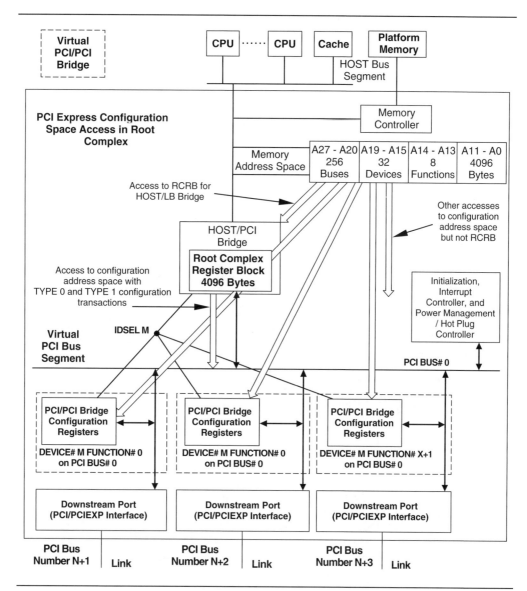

Figure 3.8 Other Implementation of PCI Express Root Complex with Single Internal Device

104 ■ The Complete PCI Express Reference

As illustrated in Figures 3.9 and 3.10, it is possible to make of the downstream virtual PCI/PCI Bridges of the Root Complex of different functions have different device numbers.

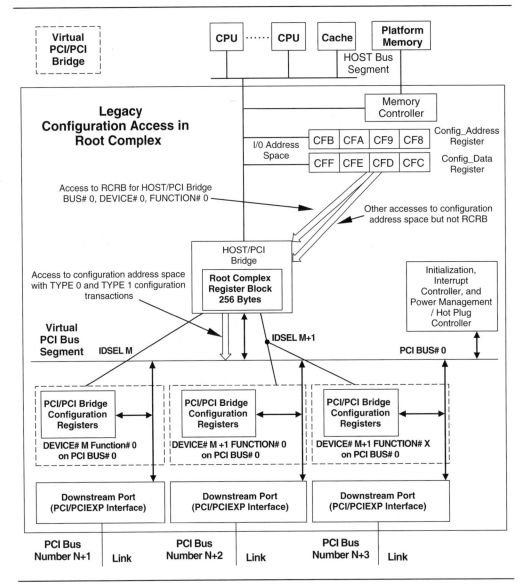

Figure 3.9 Other Implementation of Legacy Root Complex with Internal Device and Function

Chapter 3: Implementation ■ **105**

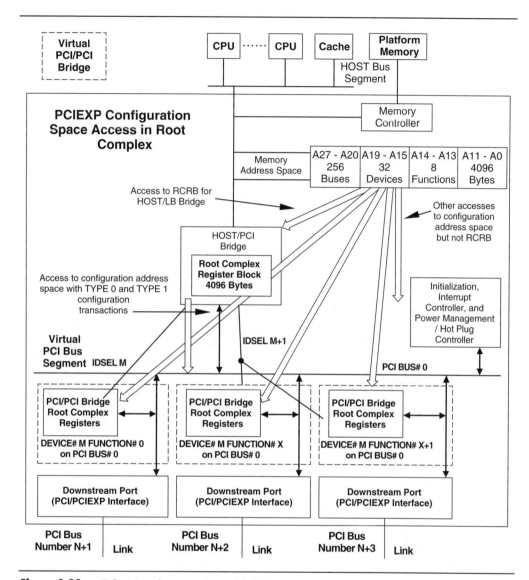

Figure 3.10 Other Implementation of PCI Express Root Complex with Internal Devices and Functions

Other Implementations of Switches

As illustrated in Figure 3.11, a switch can be implemented without an upstream virtual PCI/PCI Bridge. The switch appears to the upstream link as a single DEVICE# 0 with multiple functions. The virtual PCI bus segment is replaced by a local bus (LB). The LB can be of any design. The virtual PCI/PCI bridges are replaced with virtual LB/PCI Bridges. The other key design considerations are:

- The upstream interface to the link is not a virtual PCI/PCI Bridge; consequently, all of the downstream virtual LB/PCI Bridges collectively are DEVICE# 0 on BUS# N of the upstream link.

- The interface to the downstream links is a series of virtual LB/PCI Bridges collectively part of DEVICE# 0; consequently, each virtual LB/PCI Bridge is a function.

- Only eight functions can be assigned to DEVICE# 0; consequently, a maximum of eight downstream links is possible.

- As illustrated in Figures 3.12, a switch can be implemented that is a combination of the one depicted in Figures 3.5 and 3.11. One of the drawbacks of the implementation in Figure 3.11 is that the maximum number of eight functions is available for both internal use and downstream links. The implementation in Figure 3.5 provides up to 32 devices (with FUNCTION# 0 each) for internal use and downstream links. In a similar fashion, it would be possible to implement up to eight functions per device in the implementation in Figure 3.5.

Chapter 3: Implementation ■ 107

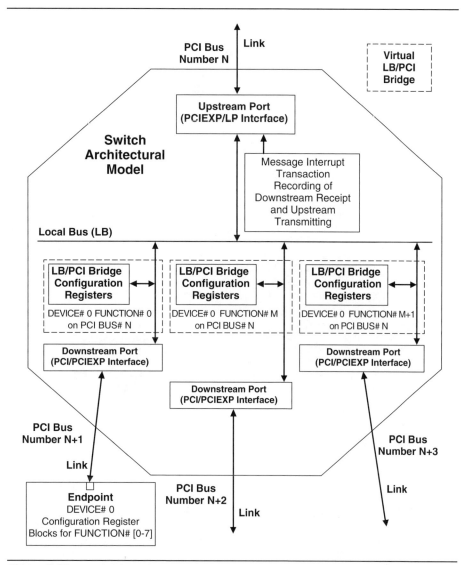

Figure 3.11 Other Implementation of Switches with Single Internal Device

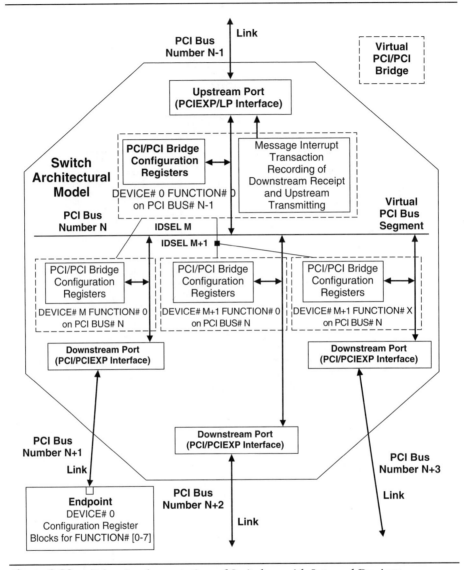

Figure 3.12 Other Implementation of Switches with Internal Devices and Functions

Other Implementations of Bridges

As illustrated in Figure 3.13, a bridge can be implemented without an upstream virtual PCI/PCI Bridge. The bridge appears to the upstream link as a single DEVICE# 0 with multiple functions. The virtual PCI bus segment is replaced by a local bus. The local bus can be of any design. The virtual PCI/PCI bridges are replaced with virtual LB/PCI Bridges. The key design elements are:

- The upstream interface to the link is not a virtual PCI/PCI Bridge; consequently, all of the downstream virtual LB/PCI or LB/PCI-X Bridges collectively are DEVICE# 0 on BUS# N of the upstream link.

- The interface to the downstream bus segments is a series of virtual LB/PCI or LB/PCI-X Bridges collectively part of DEVICE# 0; consequently, each virtual LB/PCI Bridge is a function.

- Only eight functions can be assigned to DEVICE# 0; consequently, a maximum of eight downstream bus segments is possible.

- As illustrated in Figures 3.14 it is possible to implement a bridge that is a combination of the ones depicted in Figures 3.6 and 3.13. One of the drawbacks of the implementation in Figure 3.13 is that the maximum number of eight functions is available for both internal use and downstream bus segments. The implementation in Figure 3.6 provides up to 32 devices (with FUNCTION# 0 for each) for internal use and downstream bus segments. In a similar fashion, it would be possible to implement up to eight functions per each device in the implementation in Figure 3.6.

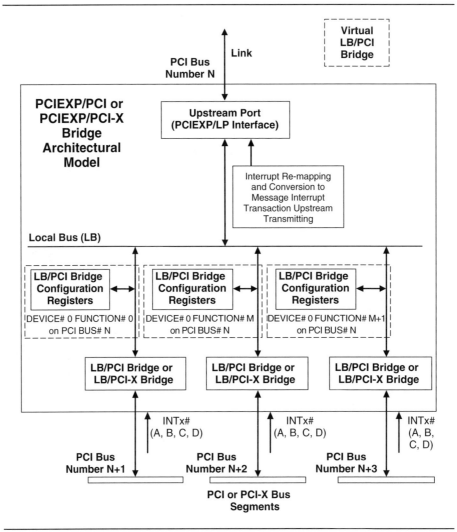

Figure 3.13 Other Implementation of Bridges with Single Internal Device

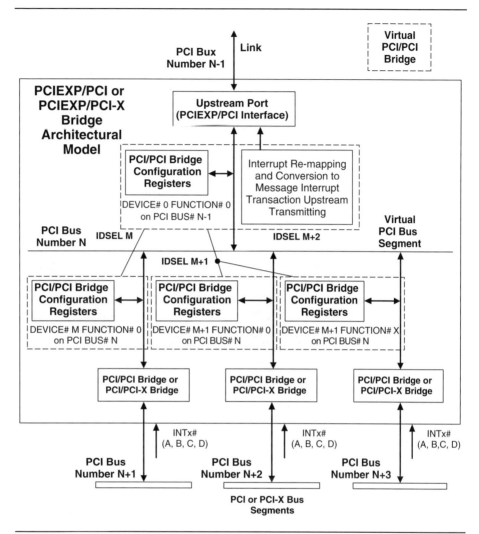

Figure 3.14 Other Implementation of Bridges with Internal Devices and Functions

Other Bridge Considerations

The PCI Express specification defines four types of bridges: PCI Express to PCI Bridge, PCI Express to PCI-X, PCI to PCI Express Bridge, and PCI-X to PCI Express Bridge. The first two types that were discussed previously are collectively called *bridges* and their abbreviated terms are PCI Express/PCI Bridge and PCI Express/PCI-X. The later two types of bridges are collectively called *reverse bridges* and their abbreviated terms are PCI/PCI Express Bridge and PCI-X/PCI Express Bridge. The implementation of bridges defines the PCI Express link to be connected to the upstream port of the bridge and the PCI or PCI-X bus segments to be connected downstream of the bridge. The implementation of reverse bridges defines the PCI Express link to be connected to the downstream port(s) of the bridge and the PCI or PCI-X bus segment to be connected upstream of the bridge. Bridges connected as illustrated in Figure 3.1 are the typical implementation and are the only type of bridge discussed in this book. Reverse bridges will rarely if ever be designed; consequently, reverse bridges are not a focus of this book. See Appendix A of the *PCI Express to PCI/PCI-X Bridge Specification* for more information on reverse bridges.

This primary focus of this book is PCI Express, so bridges are only described in sufficient detail to provide an understanding of their overall interaction with the PCI Express fabric. Detailed information for bridge component design and programming is beyond the scope of this book. For more information on this subject, see the *PCI Express to PCI/PCI-X Bridge Specification*.

PCI Express/PCI Bridge and PCI Express/PCI-X Bridge Considerations

The basic architecture of a bridge illustrated in Figures 3.6, 3.13, and 3.14 consists of an upstream port that is a PCI Express compliant link compliant link and one or more downstream ports. Each downstream port is compliant with either a PCI or a PCI-X bus segment. The activity of the upstream and downstream ports must be in compliance with the PCI Express and PCI or PCI-X specifications, respectively. Of note are PCI and PCI-X requirements that have no counterpart in PCI Express. All of these requirements must operate independent the PCI Express fabric on the PCI or PCI-X bus segments. For example, the arbitration circuitry for bus segment ownership must be entirely implemented on the PCI and PCI-X bus segments. Other examples are fulfilling bus segment parking requirements, PCI-X Initialization Pattern generation, IDSEL lines, and open collector interrupt signal lines.

The interaction of PCI Express transactions and PCI or PCI-X bus transactions across a bridge places unique requirements on the bridge. The following discussion covers the essential unique requirements. See the *PCI Express to PCI/PCI-X Bridge Specification* for the full and detailed discussion.

Internal Functions

If the access is from the upstream side of the bridge (PCI Express link) to a function or functions internal to the bridge, the internal device(s) containing the function(s) must be compliant with PCI Express and must operate as PCI Express endpoint(s). If the access is from the downstream side of the bridge (PCI or PCI-X bus segment) to a function or functions internal to the bridge, the internal device(s) containing the function(s) must be PCI or PCI-X compliant and operate as a PCI or PCI-X target(s).

PCI-X Bus Segments and PCIXCAP

The bridge interacts with the PCI and PCI-X bus segments compliant with the PCI and PCI-X bus specifications. If a PCI-X bus segment is implemented and a PCI device is installed in the bus segment per the PCIXCAP signal line, the downstream port of the bridge operates in compliance with the PCI specification.

Memory and I/O Address Spaces

The memory address space is defined on both sides of the bridge. The access to the memory address space can be ported in the upstream or downstream direction by the bridge. Any access to the memory address space upstream of the bridge from a PCI bus segment must be posted (if a write) or executed according to the Delayed Transaction protocol (if a read). Any access to the memory address space upstream of the bridge from a PCI-X bus segment must be posted (if a write) or executed according to the Split Transaction protocol (if a read). Any access to the memory address space downstream of the bridge from a PCI Express link must follow the Requester/Completer protocol. On behalf of the PCI Express link, the bridge executes the associated PCI or PCI-X bus transactions with posted writes, and Delayed Transaction protocol for PCI or Split Transaction protocol for PCI-X to execute reads.

The full 64 address bit decoding for memory address space must be implemented by all ports of the PCI Express/PCI-X Bridge. The full 64 address bit decoding for memory address space can optionally be implemented by all ports of the PCI Express/PCI Bridge. If the full 64 address is implemented, it is via the Dual Address Cycle of the PCI and PCI-X bus segments.

Access to memory mapped I/O must be supported by the bridge with the implementation of Memory Base and Memory Limit registers in the configuration register block.

The access to memory address space with Exclusive Hardware Access is not required to be supported. If supported, the Exclusive Hardware Access can only be ported in the downstream direction by the bridge.

The support of VGA addressing, expansion ROM, and prefetchable address space within the memory address space of the PCI and PCI-X is not required.

Accesses to the I/O address space upstream of the bridge are required. Accesses to the I/O address space downstream of the bridge are optional. Any access to the I/O address space upstream of the bridge from a PCI bus segment must be executed using the Delayed Transaction protocol. Any access to the I/O address space upstream of the bridge from a PCI-X bus segment must executed using the Split Transaction protocol. On behalf the PCI Express link the bridge executes the PCI or PCI-X bus transaction with Delayed Transaction protocol for PCI or Split Transaction protocol for PCI-X.

Configuration Address Space

The configuration address space is defined on both sides of the bridge. Access to the configuration address space on the PCI and PCI-X bus segments is required and can only be ported in the downstream direction by the bridge. If the extended configuration address space is accessed, the access must be ported through the bridge for PCI-X compliant bus segments or ported to the internal functions that must be contained in PCI Express compliant devices. Type 0 configuration transactions from the PCI Express link are only to functions within the bridge. Type 1 configuration transactions from the PCI Express link are for accessing configuration address space of functions internal to the bridge and devices on the PCI or PCI-X bus segments. Type 1 configuration transactions from the PCI Express link are for also for executing special bus transactions on the

PCI and PCI-X bus transactions. On behalf the PCI Express link, the bridge executes the PCI or PCI-X bus transaction with Delayed Transaction protocol for PCI or Split Transaction protocol for PCI-X.

Message Address Space

The message address space and associated message transactions are only defined on the upstream port compatible with PCI Express. The source or destination of message transactions can be either the upstream port of the bridge acting on behalf of the PCI and PCI-X bus segments or internal functions contained in PCI Express compliant devices in the bridge. Any access to the message address space upstream of the bridge from a PCI-X bus segment must be executed using the Split Transaction protocol. On behalf of the PCI Express transactions, the bridge executes the PCI bus transactions with Delayed Transaction protocol or PCI-X bus transactions with Split Transaction protocol.

Interrupt Acknowledge

Interrupt acknowledge bus transactions are not executed downstream of the bridge on the PCI and PCI-X bus segments.

General Items of Compliance Optional for Bridges

General items of compliance that are optional for bridges to implement are as follows:

Digest Field The bridge may optionally implement the generating and checking of the Digest field associated with Transaction Layer Packets. However, if the bridge is checking the Digest field for CRC errors, it must implement the Advance Error Reporting protocol.

Hot Plug Protocol The upstream port of the bridge can be directly connected to the edge fingers of an add-in card and may optionally implement the Hot Plug protocol.

Data Widths The bridge is required to support 32 data bit PCI and PCI-X bus segments. The bridge may optionally support 64 data bit PCI and PCI-X bus segments.

Frequency Support The bridge is required to support the applicable frequencies for the PCI and PCI-X specification for which the bus segment is compliant. The bridge may optionally support the 66.6 megahertz implementation of PCI on a PCI-compliant bus segment. The bridge is required to support the applicable source strobes of the revision of the PCI-X specification for which the bus segment is compliant.

PCI-X PCI-X Mode 1 and Mode 2 are supported by the bridge. The Device ID Message of PCI-X translation to and from PCI Express message transactions (vendor-defined) is optional. See the *PCI Express to PCI/PCI-X Bridge Specification* for the full and detailed discussion.

Translation between Transactions and Bus Transactions

Table 3.1 summarizes translation of transactions on a PCI Express link into the equivalent bus transactions on PCI and PCI-X bus segments with the bridge as Master on the bus segments. If the bridge is a Target on the bus segments, it will translate the bus transactions on the PCI and PCI-X bus segments into the equivalent transactions on PCI Express link. Any PCI Express transaction, PCI bus transaction, or PCI-X bus transaction not listed in Table 3.1 is not translated by a bridge.

Table 3.1 Translation between Transactions and Bus Transactions

Transaction Layer Packet on PCI Express Link		PCI on Downstream Ports (2)		PCI-X on Downstream Ports (2)	
Bridge As Transmitter on Upstream Port (1)	Bridge As Receiver on Upstream Port (1)	Bridge As Master	Bridge As Target	Bridge As Master	Bridge As Target
Req — Memory Read Requester	Req — Memory Read Requester	MR Req	MR Req	MR DW Req	MR DWB Req
		MRL, MRM Opt	MRL, MRM Req	MRB Req	MRB, Alias MRB Req
	Req (3) — Memory Read with Lock Requester	MR Req (5)		MR Req (5)	
		MRL, MRM Opt (5)		MR, MRL, MRM Opt (5)	
Req — Memory Write Requester	Req — Memory Write Requester	MW Req	MW Req	MW Req	MW Req
		MWI Opt (6)	MWI Req (6)	MWB Req (6)	MWB, Alias MWB Req (6)
Req — I/O Read Requester	Opt — I/O Read Requester	IOR Opt	IOR Req	IOR Opt	IOR Req
Req — I/O Write Requester	Opt — I/O Write Requester	IOW Opt	IOW Req	IOW Opt	IOW Req
	Req — Config. Read Requester Type 0				
	Req — Config. Write Requester Type 0				
	Req — Config. Read Requester Type 1	Config. Read Type 0,1 Req		Config. Read Type 0,1 Req	

Table 3.1 Translation between Transactions and Bus Transactions *(continued)*

Transaction Layer Packet on PCI Express Link		PCI on Downstream Ports (2)		PCI-X on Downstream Ports (2)	
Bridge As Transmitter on Upstream Port (1)	**Bridge As Receiver on Upstream Port (1)**	**Bridge As Master**	**Bridge As Target**	**Bridge As Master**	**Bridge As Target**
	Req	Config. Write Requester Type 1	Config. Write Type 0,1; Special Req		Config. Write Type 0,1; Special Req
Req	Message Baseline without Data (7)	Req	Message Baseline without Data (7)		
Req	Message Baseline Requester with Data (7)	Req	Message Baseline Requester with Data (7)		
Opt (4)	Message Vender-Defined with Data	Opt (4)	Message Vender-Defined with Data	Device ID Messages Opt	Device ID Messages Opt
Req	Completer without Data	Req	Completer without Data		
Req	Completer with Data	Req	Completer with Data		
Opt	Completer Lock without Data				
Opt	Completer Lock with Data				

Table Notes:

1. Also known as primary side of bridge.
2. Also known as secondary side of bridge.
3. The propagation to the PCI and PCI-X bus segments is optional.
4. The message vendor-defined requester with data may optionally be supported for Device ID Messages defined for the PCI-X bus segments. The message vendor-defined requester with data is not supported for the PCI bus segment.
5. In conjunction with assertion of LOCK# signal line.
6. MW must be used instead if the byte enables are not contiguous.
7. The message baseline transactions are for internal functions on the bridge itself. These are not translated to the PCI or PCI-X Bridge.

Chapter 4

Addressing and Routing

This chapter discusses the address and routing protocols for the different PCI Express transactions. Several types of transactions are executed in the PCI Express fabric and some of these are associated with memory, I/O and configuration address spaces. These PCI Express transactions are similar to those in PCI and PCI-X, as illustrated in Figure 4.1. Transactions to memory, I/O, and configuration address spaces require the familiar address bits that provide a specific address to identify the PCI Express device being accessed. For the memory and I/O transactions these address bits are defined as ADDRESS [31::00] or ADDRESS [63::00] (excluding I/O). For the configuration transaction these address bits are defined as BUS#, DEVICE#, and FUNCTION#. These items represent a unique ID for each PCI Express device.

As discussed in previous chapters, the Requester/Completer protocol applies to these transactions and each consists of a requester transaction and a completer transaction (except memory write). Memory write transactions only implement requester transactions. Completer transactions are not directly associated with address spaces, but are indirectly given their association with requester transactions. Completer transactions do not use the familiar address bits similar to PCI and PCI-X, but instead employ the requester transaction's Requester ID for routing the completer transaction to the requester source. This is called *ID Routing*.

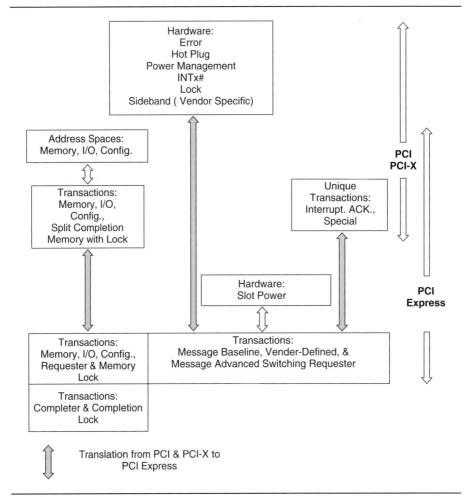

Figure 4.1   Transaction and Hardware Translation

PCI Express also defines message transactions that are not associated with the memory, I/O, and configuration address spaces. These message transactions are associated with the following:

- The replacement of PCI and PCI-X hardware, as illustrated in Figure 4.1
- The replacement of unique PCI transactions
- PCI Express in-band communication
- PCI Express advanced peer-to-peer communications

Message transactions comply with the posting of the requester transaction portion of the Requester/Completer protocol; there are no associated completer transactions. Message requester transactions do not employ the address bits used by other types of requester transactions to determine their destination. Instead message requester transactions determine their destination based on message type, which implies the destination. This is called *Implied Routing*. In the case of a message advanced switching requester transaction, the destination RID (Route Identifier) Routing applies.

Three primary elements are involved in accessing the proper PCI Express devices: the address spaces definition, the participants, and the address or routing protocol. This chapter covers each of these elements in detail.

Associated with these primary elements are additional addressing protocols for configuration transactions and implementation of a tag. These protocols are discussed at the end of this chapter.

Address Spaces Defined in PCI Express

> This book uses the following terms:
>
> - *Requester source* and *requester destination:* The two participants in the transmitting and receiving of the requester transaction packets.
>
> - *Completer source* and *completer destination:* The two participants in the transmitting and receiving of the completer transaction packets.

Memory Address Space

Memory address space is the largest address supported with 64 address bits viewed as a flat address across the PCI Express fabric. The 64 address bits provide an address space of 16 exabytes. The minimal addressable size is 32 data bits addressed as four bytes with a minimal data access of one byte (via byte enables). Memory address space has several attributes:

- **Variable Transfer Size:** The memory address space can be accessed (data can be transferred) as a single 32 data bit access for every one address in one transaction or as multiple 32 data bits accesses (transfers) for one beginning address in one transaction.

- **Sequential Addresses:** The address increases monotonically.

- **Distributed Address Space:** The platform memory is typically the largest single portion of the memory address space, but does not contain all of the memory address space. The balance is distributed throughout the PCI Express fabric in PCI Express endpoints and PCI or PCI-X resources downstream of bridges. By definition, the platform memory's controller (memory controller) is integral to the Root Complex.

Implementation Considerations for PCI Express Memory Address Space

In implementing the PCI Express memory address space, consideration must be given to the support of cacheable memory, prefetching, memory write posting, as well as transaction posting.

Cacheable Memory Support The memory address space permits multiple copies of a portion of the address space contained in the platform memory within different platform resources. Typically, this is cache memory on the HOST bus segment. The cache contains copies of different portions of the platform memory and contains the most recently written data. Consequently, a PCI Express device reading platform memory may require the data to be copied from the cache to the platform memory before the transaction is completed, or the data may be read directly from the cache. Similarly, a PCI Express writing data to the platform memory may actually be written to cache memory or both cache memory and platform memory. All of these considerations are defined as *cache coherency*.

It is possible to locate memory in PCI Express devices that are also cached on the HOST bus segment. PCI originally supported this possibility, but it was never implemented in the market place. PCI Express does not permit non-platform memory address space to be cached. Consequently, PCI Express only defines cache on the HOST bus segment and it is only coherent with the platform memory. The Root Complex must manage this coherency and complete the PCI Express bus transition with the proper data porting to and from cache or platform memory.

Prefetching Support Another consideration when accessing memory address space is data *prefetching*. Portions of the memory address space can be defined as prefetchable address space (collectively prefetchable addresses). PCI Express devices reading from prefetchable addresses receive the same data (provided no write to this address space has occurred) for multiple readings. Multiple readings and the possible discarding of data from prefetchable addresses by PCI Express devices do not have negative side effects.

Memory Write Posting Support One of the most important features of memory address space is that the Physical Packets contain memory write requester transaction packets that can be posted in switches and requester destinations. Posting permits the requester sources to transmit memory write requester transaction packets and immediately assume that the associated data in the Physical Packets have been written (posted) at the requester destinations. The possible requester sources are the Root Complex, endpoints, or bridges. The requester destinations are the Root Complex, endpoints, or bridges. Consequently, the memory write transaction protocol does not include completer transactions. Effectively, requester sources "post and forget" the Physical Packets. In the case of multiple switches between the participants, posting permits memory write requester packets to move though the PCI Express fabric with the minimal amount of overhead and with the minimal use of bandwidth. Switches post and forget the packets in much the same way requester sources post and forget to switches or requester destinations. That is, switches simply post the packets to the next switch or to the requester destinations. The key components of the memory write requester transaction packets that are being posted are address and data.

Transaction Posting For memory read transactions, it is clear that data cannot be posted as part of Physical Packets that contain memory read requester transaction packets. Consequently, requester sources transmit memory read requester transaction packets and immediately assume that the associated Physical Packets have been written (posted) at the requester destinations, but the transaction is not complete. Effectively, requester sources post and forget the Physical Packets, but the transactions are not complete (data read) until the completer transactions are received from the completer source (requester destination). Similar with memory write requester transaction packets, the completer sources transmit completer transaction packets and immediately assume that the packets have been written (posted) at the completer destination (requester source). Effectively, the completer source posts and forgets the packets. In the case of multiple switches between the participants, posting permits packets to move though the PCI Express fabric with the minimal amount of overhead and with the minimal use of bandwidth. Switches post and forget the packets much the same way requester and completer sources post and forget to switches or requester and completer destinations. That is, switches simply post the packets to the next switch or to the requester and completer destinations.

> As discussed in Chapter 3, the RCRB within the Root Complex is defined as part of the memory or I/O address space of the HOST bus segment. The Root Complex is not required to support an access to the RCRB that crosses DWORD aligned boundaries or use a Lock Function protocol to access it.

I/O Address Space

I/O address space is supported with 32 address bits addressed as four bytes viewed as a flat address across the PCI Express fabric. The 32 address bits provide an address space of 4 gigabytes. The minimal addressable size is 32 data bits. The minimal addressable size is 32 data bits with a minimal data access of one byte as selected by the byte enables (BE). I/O address space is more restrictive than the memory address space. The attributes of I/O address space are:

- **Fixed Transfer Size:** The I/O address space can only be accessed as a single 32 data bit access for each single address in one transaction.

- **Distributed Address Space:** The platform I/O is typically distributed throughout the PCI Express fabric in PCI Express endpoints including the PCI resources downstream of bridges. Unlike platform memory, no I/O resources are defined internally to the Root Complex.

- **Non-Cacheable:** The I/O address space does not permit multiple copies of a portion of the address space within different platform resources. That is, unlike with memory address space, cache coherency is not considered for the I/O address space.

- **Non-Prefetchable:** I/O address space permits the data to be read only once and multiple reading of the same address may not provide the same data for each read. I/O address space also assumes that the data written to a specific address may not be read back. The assumption for I/O address space is that each address has access from the PCI Express fabric and another access from a non-PCI Express port. This other access may be hardware related.

Implementation Considerations for PCI Express I/O Address Space

Transaction execution to access the I/O address space in a PCI Express device is simpler than transaction execution to access the memory address space, because cache coherency is not supported and the I/O address space has no prefetchable address range. One other consideration applies to both the memory and I/O address space: transaction posting.

Transaction Posting A limitation of the I/O address space is that I/O write requester transaction data cannot be posted in the same manner as memory write transaction data. The Requester/Completer protocol requires the requester sources to execute I/O requester transactions and wait for completer transactions to indicate data as been written at the endpoints or bridges. The possible requester sources are the Root Complex, endpoints, and bridges. The requester destinations do not include the Root Complex, because it does not contain I/O resources. Consequently, requester sources transmit I/O write requester transaction packets and immediately assume that the packets have been written (posted) at the requester destinations, but the transaction is not complete. Effectively, requester sources post and forget the Physical Packets, but the transactions are not complete (that is, the data has not been written) until the completer transactions are received from the completer source (requester destination). Similar to the memory write requester transaction packets, the completer sources transmit completer transaction packets and immediately assume that the packets have been written (posted) at the completer destination (requester source). Effectively, the completer source posts and forgets the packets. In the case of multiple switches between the participants, posting permits packets to move though the PCI Express fabric with the minimal amount of overhead and with the minimal use of bandwidth. Switches post and forget the packets in much the same way requester and completer sources post and forget to switches or requester and completer destinations. That is, switches simply post the Physical Packets to the next switch or to the requester and completer destinations.

For I/O read transactions, the protocol is the same as memory read transactions.

Configuration Address Space

Unlike the memory or I/O address spaces, which are viewed as a flat address across the PCI Express fabric, configuration address space consists of groups of 64 or 1024 registers defined for each function within each PCI Express device. Each register is four bytes wide. The minimum addressable size is 32 data bits addressed as four bytes with a minimum data access of one byte selected by byte enables (BE). Configuration address space is more restrictive than memory address space and has some special requirements. The configuration address space attributes are:

- **Fixed Transfer Size:** The configuration address space can only be accessed (data can only be transferred) as a single 32 data bit access for one address in one transaction.

- **Non-Cacheable:** The configuration address space does not permit multiple copies of the configuration register blocks assigned to each device. The configuration address space is accessed by addressing the memory address space, but the actual data associated with each configuration register block is located at the associated device.

- **Non-Prefetchable:** Configuration address space permits the data in the configuration register blocks to be is read only once, and multiple readings of the same register may not in all cases provide the same data for each read. The registers that reflect resource setup have re-readable data. Other registers that contain resource information, such as errors, do not always read the same each time. The configuration address space does assume that the data written to a device setup register can be read back.

Implementation Considerations for PCI Express Configuration Address Space (General)

The configuration address space has two unique access considerations relative to the memory and I/O address spaces. One of these access considerations stems from the requirement that all PCI Express devices implement a configuration register block. The other consideration has to do with transaction posting (like I/O address space).

Configuration Register Block The PCI Express configuration address space and associated transactions are required to be supported by all PCI Express devices. The PCI Express devices are contained within the Root Complex, switches, endpoints, and bridges. The Root Complex contains PCI Express devices in the form of a virtual HOST/PCI Bridge and virtual PCI/PCI Bridges (downstream ports). Switches and bridges contain PCI Express devices in the form of virtual PCI/PCI Bridges (upstream [optional] and downstream ports). Each virtual PCI/PCI Bridge is defined as a device and can contain up to eight functions (numbered 0-7). Upstream ports of endpoints are defined as a single DEVICE# 0 on the downstream end of the links. Each specific device number contains up to eight functions (numbered 0-7). All devices are required to support at a minimum FUNCTION# 0. See Chapter 3 for more information.

Each function must support one group of the configuration address registers defined as the configuration register block. PCI Express has expanded the address range of each configuration register block beyond the PCI-defined total of 256 bytes (64 DWORD registers). PCI Express defines each configuration register block to be 4096 bytes (1024 DWORD registers) as illustrated in Figure 4.2. As discussed in Chapter 3, 256 bytes are implemented by the Root Complex and PCI Express devices compatible with legacy software. The 1024 bytes are implemented by the Root Complex and PCI Express devices compatible with PCI Express software. Each grouping of four bytes is called a register.

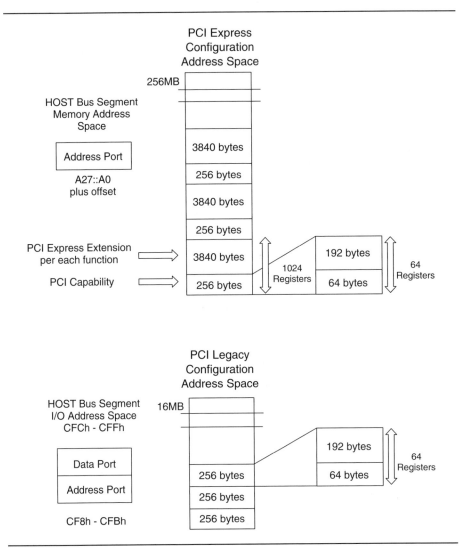

Figure 4.2 Configuration Address Space

Transaction Posting A limitation of the configuration address space is that configuration write requester transaction data cannot be posted in the same manner as memory write transaction data. The Requester/Completer protocol requires the requester source to execute configuration write requester transactions and wait for completer transactions to indicate data as been written at the requester destination. The only requester source defined is the Root Complex. The possible requester destinations are the switches, endpoints, or bridges. Consequently, requester sources transmit configuration write requester transaction packets and immediately assume that the associated Physical Packets have been written (posted) at the requester destinations, but the transaction is not complete. Effectively, requester sources post and forget the Physical Packets, but the transactions are not complete (that is, the data is not written) until the completer transactions are received from the completer source (requester destination). Similar to the memory write requester transaction packets, the completer sources transmit completer transaction packets and immediately assume that the associated packets have been written (posted) at the completer destination (requester source). Effectively, the completer source posts and forgets the Physical Packets. In the case of multiple switches between the participants, posting permits packets to move though the PCI Express fabric with the minimum amount of overhead and with the minimum use of bandwidth. Switches post and forget the packets in much the same way requester and completer sources post and forget to switches or requester and completer destinations. That is, switches simply post the packets to the next switch or to the requester and completer destinations.

For configuration read transactions, the protocol is the same as memory read transactions.

Implementation Considerations for PCI Express Configuration Address Space (Specific)

The PCI Express configuration address space is based on PCI to ensure software compatibility. However, the replacement of PCI bus segments with PCI Express links places restrictions on the configuration architecture of PCI Express devices. A BUS #, DEVICE #, and FUNCTION # is assigned to each configuration register block in each PCI Express device following protocols specific to the architecture of the device.

> As discussed in Chapter 3, the Root Complex, switches, and bridges can be structured in several ways relative to configuration address space. The discussion in this section uses the most PCI-like structures to provide the most detailed discussion. The implementations of the other structures relative configuration address space are a subset of this discussion.

Endpoints and Switches The assignment of BUS #, DEVICE #, and FUNCTION # to configuration register blocks for switches and endpoints is illustrated in Figure 4.3.

The specific implementation considerations for each endpoint are fairly simple. Each endpoint on each PCI Express link is defined a DEVICE# 0. Multiple PCI Express devices are not assigned to each link as would be done on a PCI bus segment; consequently DEVICE# [31::1] cannot be defined for an endpoint. Each endpoint must contain FUNCTION# 0 and its associated configuration register block. Optionally, the endpoint can contain up to eight functions, each function has a FUNCTION# and a configuration register block.

The specific implementation considerations for each switch are more complex than for the endpoint. Part of this complexity is due to the multiple PCI devices defined internally to each switch. The assignment protocol is as follows:

- **Overall Architecture of the Switch:** Each switch must contain an upstream port and a minimum of two downstream ports. The downstream ports appear to the configuration address space as virtual PCI/PCI Bridges with a PCI bus segment connected downstream to each. The connected PCI bus segment is actually the PCI Express link. Each link is assigned a BUS# just like a PCI segment (BUS# [N+1::N+3] in this example).

- **Downstream Ports Considerations:** The downstream ports of each switch appear as virtual PCI/PCI Bridges to the internal virtual PCI bus segment. The PCI/PCI Bridges represent all directly connected links (defined per BUS#s ... BUS# [N+1::N+3] in this example) and links connected via other downstream switches. The internal virtual PCI bus segment of each switch assigns a DEVICE# to each downstream port PCI/PCI Bridge (DEVICE# [M::M+2] in this example). The switch can internally implement virtual IDSEL signal lines or some similar device select mechanisms.

- **Upstream Port Considerations for BUS#:** The upstream port of each switch appears as a virtual PCI/PCI Bridge to the upstream link. In the same fashion as a link to an endpoint, this upstream link is assigned a BUS# (BUS# N-1 in this example). The switch's upstream port virtual PCI/PCI Bridge connects this upstream link with an internal virtual PCI bus segment (BUS# N in this example). The switch's internal hardware to interconnect the upstream and downstream ports may be designed in many different ways including an internal "micro" PCI bus segment. For the purposes of configuration address space the internal interconnect must appear as a PCI bus segment and thus this discussion assumes an internal virtual PCI bus segment.

- **Upstream Port Considerations for DEVICE# and FUNCTION#:** The upstream port of the switch is the only PCI Express device than can be connected to the link, unlike individual PCI bus segments that can have multiple PCI resources. Consequently, each virtual PCI/PCI Bridge of the switch's upstream port is assigned DEVICE# 0. Multiple PCI Express devices are not assigned to each link as would done on each PCI bus segment; consequently DEVICE# [31::1] cannot be defined for the upstream port's virtual PCI/PCI Bridge. The upstream port's virtual PCI/PCI Bridge must contain FUNCTION# 0 and its associated configuration register block. Optionally, the upstream port's virtual PCI/PCI Bridge can contain up to eight functions; each function has a FUNCTION# and a configuration register block. As discussed in Chapter 3, it is possible to construct a switch without a specific virtual PCI/PCI Bridge on the upstream port (Figures 3.11 and 3.12). A switch constructed in this manner can define multiple DEVICE numbers representing the virtual PCI/PCI Bridges of the switch's downstream ports.

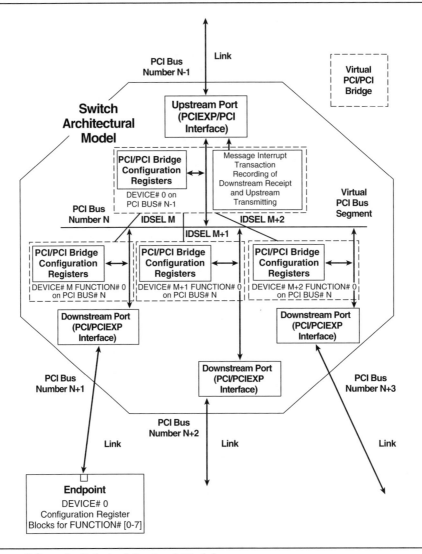

Figure 4.3   Switch Architectural Model

Bridges   The assignment of BUS #, DEVICE #, and FUNCTION # to configuration register blocks for bridges follows the same protocol as discussed for switches. Similarly, the implementation of virtual PCI bus segments and IDSEL signal lines is the same as with switches. The application is illustrated in Figure 4.4. As discussed in Chapter 3, it is possible to construct a bridge without a specific virtual PCI/PCI Bridge on the upstream port (Figures 3.13 and 3.14). A bridge constructed in this manner can define multiple DEVICE numbers representing the virtual PCI/PCI Bridges of the bridge's downstream ports.

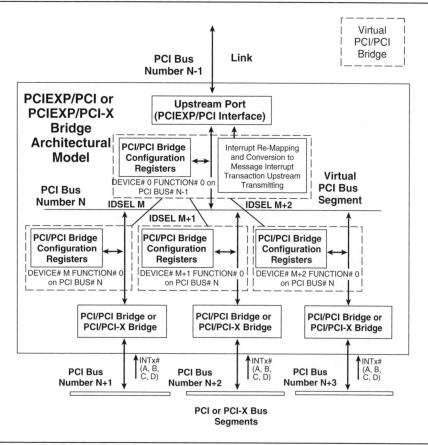

Figure 4.4 Bridge Architectural Model

Root Complex The assignment of BUS #, DEVICE #, and FUNCTION # to configuration register blocks for the Root Complex is illustrated in Figures 4.5, 4.6, and 4.7.

- **Overall Architecture of the Root Complex:** A Root Complex must contain a minimum of one downstream port. Each downstream port appears to the configuration address space as a virtual PCI/PCI Bridge with a single PCI bus segment connected downstream. The connected PCI bus segment is actually a PCI

Express link. Each PCI Express link is assigned a BUS# just like a PCI segment (BUS# [N+1::N+3] in this example). Unlike a PCI bus segment with multiple devices, only one downstream device assigned DEVICE# 0 is defined for an endpoint and typically for a switch or bridge. As discussed in Chapter 3, it is possible to construct a switch or a bridge without a specific virtual PCI/PCI Bridge on the upstream port (Figures 3.11 through 3.14). A switch or bridge constructed in this manner can define multiple DEVICE numbers representing the virtual PCI/PCI Bridges of the bridge's downstream ports.

- **Virtual Internal Components:** The downstream ports' virtual PCI/PCI Bridges are connected to an internal virtual PCI bus segment assigned BUS# 0. The Root Complex's internal hardware to interconnect the downstream ports, HOST bus segment, and platform memory may be designed in many different ways including an internal "micro" PCI bus segment and HOST/PCI Bridge. For the purposes of configuration address space discussion the internal interconnect must appear as a HOST/PCI Bridge and a PCI bus segment, thus this discussion assumes an internal virtual HOST/PCI Bridge and PCI bus segment.

- **Downstream Ports Considerations for BUS#:** The downstream ports of Root Complex appear as virtual PCI/PCI Bridges to the internal virtual PCI bus segment. The PCI/PCI Bridges represent all directly connected links (defined per BUS#s ... BUS# N+1::N+3 in this example) and links connected via other downstream switches.

- **Downstream Ports Considerations for DEVICE#:** The PCI/PCI Bridges in the downstream ports appear as devices with different DEVICE#s (DEVICE# [M::M+2] in this example). The Root Complex can internally implement virtual IDSEL signal lines or some similar device select mechanism.

Chapter 4: Addressing and Routing ■ 137

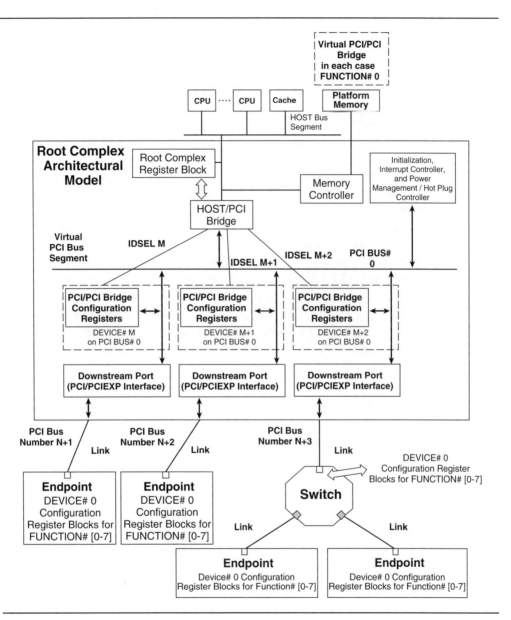

Figure 4.5   Root Complex Architectural Model

- **RCRB Considerations**: The configuration address space associated with the virtual HOST/PCI Bridge is the RCRB (Root Complex Register Block). The HOST CPU accesses to the RCRB and other configuration register blocks are implementation-specific as illustrated in Figures 4.6 and 4.7. For access compatible with legacy configuration, the Config_Address and Config_Data Registers are implemented in the HOST bus segment's I/O address space. For access compatible with PCI Express configuration, the reserved address range is implemented on the HOST bus segment's memory address range. The RCRB is accessed via the HOST bus segment's I/O or memory address spaces. The configuration register blocks associated with the virtual PCI/PCI Bridges on the downstream ports are accessed as Type 0 configuration transactions via HOST bus segment's I/O or memory address spaces. The Type 0 configuration transaction is executed in conjunction with the virtual IDSEL signal lines to select the correct configuration register block. Each of these virtual PCI/PCI Bridges must define FUNCTION# 0 with an associated configuration register block. Any of these virtual PCI/PCI Bridges may define up to eight FUNCTION#s, each with an associated configuration register block.

Chapter 4: Addressing and Routing ■ 139

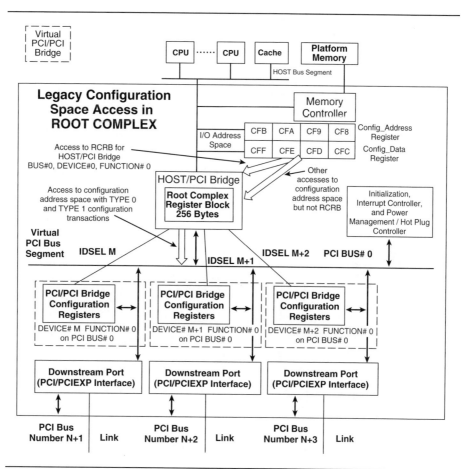

Figure 4.6 Legacy Root Complex Architectural Model

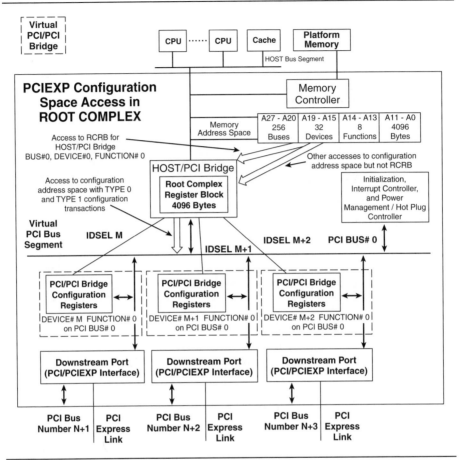

Figure 4.7 PCI Express Root Complex Architectural Model

Message Address Space

PCI, PCI-X, and PCI Express support the memory, I/O, and configuration address spaces with specific transactions, as illustrated earlier in Figure 4.1. Two elements that are not related to these address spaces are unique transactions and hardware. The unique PCI and PCI-X transactions for "interrupt acknowledge" and "special transactions" are not implemented by PCI Express. The PCI and PCI-X hardware not associated with an address space are the PERR# and SERR# signal lines, Hot Plug, the PME# signal line (power management), INTx# signal lines (discrete interrupt request to interrupt controller), the LOCK# signal line, and SIDEBAND

(SB) signal lines. These unique transactions and hardware are implemented in PCI Express via message transactions.

As previously discussed in this chapter, message transactions do define an address space. This is a convention used by the PCI Express specification. However, the message address space is not accessed in the same fashion as memory or I/O address spaces. The destination for message transactions is identified by ID and Implied Routing. These will be discussed later in this chapter.

Overview of Message Types

As outlined in Figure 4.1, message transactions fall into three categories: message baseline requester transactions, message vendor-defined requester transactions, and message advanced switching requester transactions. No completer transactions are associated with these message requester transactions. Each message category is discussed below.

Baseline Messages Group The baseline message group applies the in-band signaling between PCI Express devices residing in the same PCI Express fabric. These messages replace hardware signals used in PCI and PCI-X platforms, and can be thought of as virtual wires.

- **INTx Interrupt Signaling:** This message delivers interrupt requests from the PCI and PCI-X bus segments. Interrupt requests from PCI and PCI-X devices (in the form of INTx# signals) are routed to PCI/PCI-X-to-PCI Express bridges where they are converted into message baseline requester transaction packets.

- **Power Management:** This message replaces the Power Management Event signal (PME#) used by PCI and PCI Express devices. PCI/PCI-X-to-PCI Express bridges detect PME# signal assertion and convert the signal into message baseline requester transaction packets.

- **Error Signaling:** The PERR# and SERR# signals used in PCI and PCI-X to indicate errors are reported to the host via message baseline requester transaction packets.

- **Locked Transaction Support:** The lock signal in PCI and PCI-X is signaled via message baseline requester transaction packets.

- **Slot Power Limit Support:** Establishes the amount of power that can be supplied to an add-in card.

Vendor-Defined Messages These messages are part of the in-band communications discussed in Chapter 2. The participants are the PCI Express devices of a specific PCI Express fabric. The message vendor-defined requester transactions are unique to the vendor and are not defined in the PCI Express specification. This uniqueness means that interoperability is not guaranteed with all message vendor-defined transactions for PCI Express devices from different vendors

Advanced Switching Messages These messages are introduced in the PCI Express specification with minimal information. These are part of the advanced peer-to-peer communications that traverse the advanced peer-to-peer link discussed in Chapter 2. The participants are endpoints of two different PCI Express fabrics. See Chapter 18 for more information on advanced peer-to-peer links.

Message Transactions

For message transactions the minimum size is 32 data bits. Message transactions have the following attributes:

- **Variable Transfer Size:** Message baseline requester transactions can only be executed (that is, data can only be transferred) as a single 32 data bit access per transaction (if data is defined for the message transaction). Message vendor-defined and advanced switching requester transactions can be executed (that is, data can be transferred) as single or multiple 32 data bit accesses per transaction (if data is defined for the message transaction).

- **Non-Cacheable:** The message address does not permit multiple copies of a portion of the address space within different platform resources. That is, unlike with memory address space, cache coherency is not considered for the message address space.

- **Non-Prefetchable Is Not Applicable:** The message address space only supports writing of message data. Message transactions do not read data.

Implementation Considerations for PCI Express Messages

One of the most important features of message transactions is that the message requester transaction packets can be posted. Message requester transaction packets include the message to be written, in much the same way memory write transactions packets include data to be written. Posting permits requester sources (Root Complex, switch, endpoint, or bridge) to transmit message requester transaction packets and immediately assume that the associated packets have been written (posted) at the requester destinations (Root Complex, switch, endpoint, or bridge). Consequently, the message transaction protocol does not include completer transactions. Effectively, requester sources post and forget the packets. In the case of multiple switches between the participants, posting permits packets to move though the PCI Express fabric with the minimum amount of overhead and with the minimum use of bandwidth. Switches post and forget the packets in much the same way requester sources post and forget to switches or requester destinations. That is, switches simply post the packets to the next switch or to the requester destinations.

Participants of Transactions

Prior to explaining the protocols for implementing addressing versus protocols for routing, this chapter discusses in detail the possible participants of the different transactions. As introduced in Chapter 2, not all PCI Express devices can be participants of all transactions with all other possible PCI Express devices.

Participants of Memory, I/O and Configuration Transactions

Table 4.1 summarizes all the possible participants for the non-message transactions. The non-message transactions are memory, I/O, and configuration transactions. Each of these transactions except memory write consists of requester and completer transactions. Memory write transactions consist only of memory write requester transactions. As discussed in Chapter 2, the portion of peer-to-peer transactions between two endpoints consisting of memory and I/O transactions are included in this table.

Table 4.1 makes the following assumptions:

- The Root Complex represents the HOST bus segment CPUs and platform memory on the PCI Express fabric.
    - The platform memory is part of the memory address space accessible from the PCI Express fabric.
- The Root Complex on behalf of the HOST bus segment CPUs is the only source of configuration requester transactions. Access to the Root Complex Configuration Block (RCRB) contained within the I/O or memory address space is restricted to the HOST bus segment CPUs.
- The Root Complex does not contain any internal virtual switches. Consequently, peer-to-peer transactions between the Root Complex's downstream ports are not supported.
- The Root Complex contains no I/O address space relative to the PCI Express fabric. The I/O address space contained "within" the Root Complex is that of the HOST bus segment.
- Only the address space accessed internal to the switches is the configuration address space. This access is to the configuration register blocks internal to the switches and defines the switches as one of the participants.
    - Except for the accesses to configuration register blocks of switches, the switches are porting the transaction packets and are not one of the participants.
- The table only lists the requester transactions; it is implied that the completer transactions are between the two participants listed. For completer transactions, the requester source becomes the completer destination. Similarly, the requester destination becomes the completer source.
- If a participant combination is not listed in the table, then no memory, I/O, or configuration transactions occur between these participants.

Chapter 4: Addressing and Routing ■ 145

Table 4.1 Transactions Participants Exclusive of Message Transactions

Requester Source	Requester Destination	Memory Address Space	I/O Address Space	Config. Address Space
Root Complex	Switch	NA	NA	Required DS (1) (3)
Root Complex	PCI Express/ PCI Bridge or PCI Express/ PCI-X Bridge (7)	Opt. DS (2)	Opt. DS	Required DS
Root Complex	PCI Express endpoint	Opt. DS	Opt. DS	Required DS (3)
Root Complex	Legacy endpoint	Opt. DS (2)	Opt. DS	Required DS (3)
Root Complex	Root Complex	Required Internal (5)	NA	Required DS (4)
PCI Express/ PCI Bridge or PCI Express/ PCI-X Bridge (7)	Root Complex	Opt. US (6)	NA	NA
PCI Express endpoint	PCI Express endpoint	Opt PP	NA	NA
PCI Express endpoint	Legacy endpoint	NA	NA	NA
PCI Express endpoint	Root Complex	NA	NA	NA
Legacy endpoint	PCI Express endpoint	Opt PP	Opt. PP	NA
Legacy endpoint	PCI Express/ PCI Bridge or PCI Express/ PCI-X Bridge (7)	Required PP	Opt. PP	NA

Table 4.1 Transactions Participants Exclusive of Message Transactions *(continued)*

Requester Source	Requester Destination	Memory Address Space	I/O Address Space	Config. Address Space
PCI Express/ PCI Bridge or PCI Express/ PCI-X Bridge (7)	Legacy endpoint	Required PP	Required PP	NA
Legacy endpoint	Legacy endpoint	Opt PP	Opt PP	NA
Legacy endpoint	Root Complex	Opt US (6)	NA	NA

US = Upstream, DS =Downstream, PP = peer-to-peer transactions, NA= Not Applicable (Not defined as possible participants by the PCI Express specification)

Table Notes:

1. Only access is via the configuration address space.
2. Includes optional LOCK support.
3. Type 0 only.
4. Accessed to the RCRB is through the I/O or memory address space of the HOST bus segment.
5. HOST bus segment access to platform memory.
6. Platform memory is the destination and not RCRB in the memory address space.
7. Actually PCI or PCI-X bus segment. If there are any internal bridge functions being accessed these are treated as inputs.

Message Transaction Participants

Message requester transactions fall into three categories: message baseline, message vendor-defined, and message advance switching. Each category of message requester transaction has different types of PCI Express devices as participants.

Message Baseline Requester Transactions Participants

Tables 4.2 and 4.3 summarize the participants of the message baseline transaction given the restrictions of Implied Routing. The distinctions between address and routing mechanisms are discussed earlier in this chapter. The participants for message vendor-defined transactions and message advance switching transactions are discussed separately. The assumptions relative to Tables 4.2 and 4.3 are as follows:

- The Root Complex represents the HOST bus segment CPUs on the PCI Express fabric.
- The Root Complex does not contain any internal virtual switches. Consequently, peer-to-peer transactions between the Root Complex's downstream ports are not supported.
- The tables only list requester transactions; message baseline transactions do not have a completer transaction component.
- If a participant combination is not listed in the table, then no message baseline transactions occur between those participants.

Table 4.2 Message Baseline Transactions Only

Requester Source	Requester Destination	Message Baseline					
		Intx (3)	Unlock	Power Management			
				NAK	PME	OFF	ACK
Root Complex	Switch	NA	NA	DS	NA	NA	NA
Root Complex	PCI Express/PCI Bridge (4)	NA	DS	DS	NA	DS	NA
Root Complex	PCI Express/PCI –X Bridge (4)	NA	NA	DS	NA	DS	NA
Root Complex	PCI Express endpoint	NA	NA	DS	NA	DS	NA
Root Complex	Legacy endpoint	NA	DS	DS	NA	DS	NA
Switch	Switch	US	NA	DS	NA	NA	NA
Switch	PCI Express endpoint	NA	NA	DS	NA	NA	NA
Switch	Legacy endpoint	NA	NA	DS	NA	NA	NA
Switch	Root Complex	US	NA	US	US (1)	NA	US (2)
Switch	PCI Express/PCI Bridge (4)	NA	NA	DS	NA	NA	NA

Table 4.2   Message Baseline Transactions Only *(continued)*

Requester Source	Requester Destination	Message Baseline					
		Intx (3)	Unlock	Power Management			
				NAK	PME	OFF	ACK
Switch	PCI Express/PCI-X Bridge (4)	NA	NA	DS	NA	NA	NA
PCI Express/ PCI Bridge (4)	Root Complex	US	NA	NA	US	NA	US
PCI Express/PCI-X Bridge (4)	Root Complex	US	NA	NA	US	NA	US
PCI Express endpoint	Root Complex	NA	NA	NA	US	NA	US
Legacy endpoint	Root Complex	US	NA	NA	US	NA	US

DS=Downstream, US = Upstream, NA = Not Applicable (Not defined as possible participants by the PCI Express specification)

Table Notes:

1. Only as part of Hot Plug/Removal protocol, otherwise NA.
2. This is a special situation. The switch gathers all of the messages from downstream ports and transmits upstream a consolidation of these messages. See Chapter 15 for more information on this situation.
3. This is a special situation. The switch gathers all of the messages from downstream ports and transmits upstream a consolidation of these messages. See Chapter 19 for more information on this situation.
4. The message baseline transactions are interaction with functions within the bridge. The message baseline transactions are not being transferred to or from the PCI or PCI-X bus segments.

Table 4.3 Message Baseline Transactions Only

Requester Source	Requester Destination	Message Baseline		
		Error	Hot Plug / Removal	Slot Power
Root Complex	Switch	NA	DS	DS
Root Complex	PCI Express/PCI or PCI Express/PCI-X Bridge (1)	NA	DS	DS
Root Complex	PCI Express endpoint	NA	DS	DS
Root Complex	Legacy endpoint	NA	DS	DS
Switch	Switch	NA	DS	DS
Switch	PCI Express/PCI or PCI Express/PCI-X Bridge (1)	NA	DS	DS
Switch	PCI Express endpoint	NA	DS	DS
Switch	Legacy endpoint	NA	DS	DS
Switch	Root Complex	US	NA	US
PCI Express/PCI or PCI Express/ PCI-X Bridge (1)	Root Complex	US	NA	NA
PCI Express endpoint	Root Complex	US	NA	NA
Legacy endpoint	Root Complex	US	NA	NA

DS=Downstream, UP = Upstream, NA = Not Applicable (Not defined as possible participants by the PCI Express specification)

Table Notes:

1. The message baseline transactions are interactions with functions within the bridge. The message baseline transactions are not being transferred to or from the PCI or PCI-X bus segments.

Message Vendor-Defined Requester Transaction Participants

The two participants in the vendor-defined message are any PCI Express devices located in a specific PCI Express fabric (in-band). The exact participants, the inclusion of data, and the exact interpretation is vendor-specific. One vendor's implementation (data contained and purposes) of a message vendor-defined requester transaction may not be compatible with another vendor's.

The message vendor-defined requester transaction with data is defined for PCI Express /PCI-X Bridges. See the *PCI Express to PCI/PCI-X Bridge Specification* for more information.

Message Advanced Switching Requester Transaction Participants

As discussed earlier in this chapter, messages advance switching requester transactions are defined only for advanced peer-to-peer links. The advanced peer-to-peer links are connected to the downstream ports of switches contained in two different PCI Express fabrics. The actual participants behind these switches had not yet been defined by the revision of the PCI Express specification at the time this book was printed. Further information will be available in later revisions PCI Express specification. See Chapter 18 for more information.

Addressing and Routing

The different transactions use different addressing and routing mechanisms. All PCI Express devices that support a specific transaction must support the associated address or routing mechanism fully. That is, address decoding of some of the bits (aliasing) and partial decoding of the associated routing mechanisms is not permitted. In particular, switches must be able to direct transactions from port to port based on addresses and routing mechanisms. Similarly, the bridges must be able to port between the highest level PCI or PCI-X bus segment and the upstream PCI Express link transaction based on addresses and one of the routing mechanisms.

Transactions flow throughout the PCI Express fabric via packets. Transaction packets (TLPs) each contain a Header field (a topic covered in more detail in Chapter 6). Within the Header field of the different bus transition packets are a set of header words (HDWs). The exact set of HDWs and the fields within the HDWs are specific to the transaction. For addressing in memory and I/O requester transactions, ADDRESS HDWs are defined. For ID Routing in configuration requester transactions, CONFIGURATION and REGISTER HDWs are defined. For ID Routing in completer transactions, REQUESTER ID HDWs are defined. Three types of message transactions are used for routing: ID, Implied, and RID Routing. For Implied Routing in message transactions, TYPE ID HDWs are defined. For ID Routing in message transactions, VENDOR-DEFINED ID HDWs are defined. For RID Routing in message advanced switching transactions, RESERVED HDWs are defined. Each of these will be discussed in detail in this section.

Addressing of Memory and I/O Transactions

The protocol of transactions and their associated address and routing mechanisms are as follows:

- Memory and I/O requester transactions and 32 and 64 address bits, illustrated in Figure 4.8:
  - Both use two ADDRESS HDWs for 32 address bit addressing.
  - Only memory requester transactions also use four ADDRESS HDWs for 64 bit addressing.
  - The associated address spaces are flat across all PCI Express devices and all resources below the PCI Express/PCI Bridge.

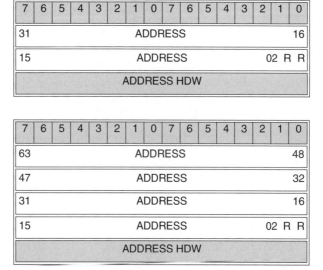

Figure 4.8 ADDRESS HDWs for Memory and I/O

> If the data to be transferred is in the first 4 gigabytes of the memory address space, the 32 address bit format *must* be used. If the 64 address bit format is used it is for accessing memory resources above the 4-gigabyte address boundary. In either situation, the full address must be provided.

Configuration and Completer Transaction ID Routing

ID Routing is a specific method to direct transaction packets to specific PCI Express devices without the use of the typical address bits.

Configuration Transactions

Configuration requester transactions implement ID Routing. The CONFIGURATION HDW contains the BUS#, DEVICE#, and FUNCTION# and the REGISTER HDW contains the register address the configuration register block illustrated in Figure 4.9. The BUS#, DEVICE#, and FUNCTION# portion of the configuration requester transactions uniquely identify the destination of configuration requester transactions. The possible destinations are any PCI Express devices and any PCI or PCI-X devices downstream of a bridge.

Additionally, each function in a PCI Express, PCI, or PCI Express device contains a configuration register block. The contents of the REGISTER HDW selects one of the 64 DWORD registers for PCI, PCI-X 66, or PCI-X 133 and one of the 1024 DWORD registers for PCI-X 266, PCI-X 533, and PCI Express.

See the section "Associated Addressing Protocols" at the end of the chapter for other addressing considerations in configuring requester transactions.

Completer Transactions

Completer transactions implement ID Routing. As discussed above, ID Routing is simply the use of the BUS#s, DEVICE#s, and FUNCTION#s to identify (ID) the destination of completer transactions. The REQUESTER ID HDW, illustrated in Figure 4.10, in the associated requester transaction contains the BUS#, DEVICE#, and FUNCTION# of the requester source. The completer source uses the REQUESTER ID HDW contained the read requester transaction to create its REQUESTER ID HDW in the completer transaction. It is the REQUESTER ID HDW in the completer transition that is used for the ID Routing.

The completer source also sources its BUS#, DEVICE#, and FUNCTION# in the COMPLETER ID HDW, illustrated in Figure 4.10, to identify to the requester source (completer destination) the source of the completer transaction packet. Another part of the completer transaction is the Transaction ID discussed in the section "Associated Addressing Protocols" at the end of the chapter.

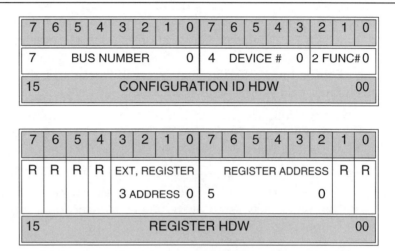

Figure 4.9   CONFIGURATION and REGISTER HDWs

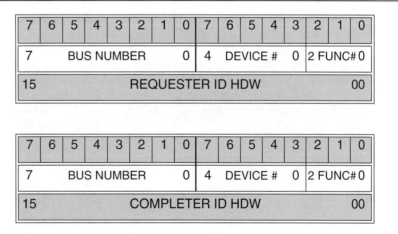

Figure 4.10   REQUESTER and COMPLETER HDWs

Routing Message Transaction Packets

Implied Routing is a specific method to direct transaction packets to specific PCI Express devices without the use of the typical address bits.

Message Baseline Transaction Implied Routing

Message baseline and vendor-defined requester transactions implement Implied Routing. Implied Routing is simply the use of the values of the r field which predefines the destination of message baseline and vendor-defined requester transactions. The routing information is contained in the r field in the TYPE HDW illustrated in Figure 4.11 within the Header field. The r field values are interpreted as follows:

- **r = 000b:** The destination of message baseline requester transaction packets is the Root Complex from any one of the sources defined in tables.

- **r = 011b:** The source of message baseline requester transaction packets is the Root Complex and the destination is all PCI Express devices listed in the tables (broadcast to multiple destinations). See the following shaded section.

- **r = 100b:** The source and destination are on the same link. The associated message baseline requester transaction packets are not ported through a switch.

- **r = 101b:** Gathered from several sources by switches and sent to Root Complex, this routing only applies to message PME_TO_ACK requester transaction. See the following shaded section.

- **r = 110b to 111b:** This range of values are RESERVED and terminated on the same link.

As discussed in the following shaded section, the Implied Routing values of r = 000b, 011b, 100b, and 101b are also implemented by message vendor-defined requester transactions.

> According to the present PCI Express specification, the r = 011b Implied Routing is only defined as a broadcast from the Root Complex and not from a downstream port of a switch. In the section of the PCI Express specification that discusses power management, the downstream port of a switch and Root Complex are defined as the two possible original sources of a message PME_Turn_Off requester transaction packet as it is defined in this book. Similarly, according to the current PCI Express specification, the r = 101b Implied Routing only defines the destination of the Root Complex. If the original source of a message PME_TO_Ack requester transaction is a downstream port of a switch, the destination of the associated message PME_TO_Ack requester transaction packets must also be the same downstream port. In the section of the PCI Express specification that discusses power management, the downstream port of a switch and Root Complex are defined as the two possible destinations a message PME_TO_Ack requester transaction packet can have as it is defined in this book. See Chapter 15 for more information.

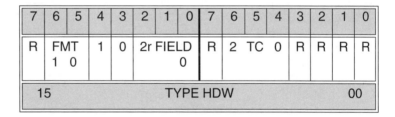

Figure 4.11   TYPE HDWs

In addition to Implied Routing, the message baseline transactions also implement addressing. When r = 001b, the destinations are selected by the ADDRESS HDWs in the Header field.

> No message baseline requester transaction defined in PCI Express revision 1.0 implements this Implied Routing value of r = 001b. The PCI Express specification defined this routing value as RESERVED. Also, this routing value is not defined for message vendor-defined requester transactions. Consequently, this book defines message baseline and vendor-defined requester transactions as not implementing any of the address bits similar to those used by memory or I/O transactions via the ADDRESS HDWs.

Tables 4.4 through 4.9 summarize the r values for Implied Routing and the Requester ID (BUS#, DEVICE#, and FUNCTION#) used in message baseline transaction packets. The Requester ID is included the message baseline transition packets to provide information at the destination of the source of the packets.

Table 4.4 Interrupt Related Messages

Name	In TYPE HDW r Field [2::0]	Requester ID Field Entries
Assert_INTx	r = 100b	BUS# = Actual DEVICE# = Reserved FUNC# = Reserved
Deassert_INTx	r = 100b	BUS# = Actual DEVICE#= Reserved FUNC# = Reserved

Table 4.5   Error Related Messages

Name	In TYPE HDW r Field [2::0]	Requester ID Field Entries
ERR_COR (Correctable)	r = 000b	BUS# = Actual DEVICE# = Actual FUNC# = Actual or Reserved (1)
ERR_NONFATAL (Uncorrectable)	r = 000b	BUS# = Actual DEVICE# = Actual FUNC# = Actual or Reserved (1)
ERR_FATAL (Fatal Error	r = 000b	BUS# = Actual DEVICE# = Actual FUNC# = Actual or Reserved (1)

Table Notes:

1. In some cases the actual FUNCTION# cannot be returned. For example, suppose the access is to a multiple FUNCTION device and the access is to a FUNCTION# not enabled (one that does not exist). In such a case the device returns a completer transaction packet with the Status field = unsuccessful and the FUNCTION# = 000b.

Table 4.6   Hot Plug Related Messages

Name	In TYPE DWORD r Field [2::0]	Requester ID Field Entries
Attention_Indicator_On	r = 100b	BUS# = Actual DEVICE# = Actual FUNC# = Actual
Attention_Indicator_Blink	r = 100b	BUS# = Actual DEVICE# = Actual FUNC# = Actual
Attention_Indicator_Off	r = 100b	BUS# = Actual DEVICE# = Actual FUNC# = Actual
Attention_Button_Pressed	r = 100b	BUS# = Actual DEVICE# = Actual FUNC# = Actual

Table 4.6   Hot Plug Related Messages *(continued)*

Name	In TYPE DWORD r Field [2::0]	Requester ID Field Entries
Power_Indicator_On	r = 100b	BUS# = Actual DEVICE# = Actual FUNC# = Actual
Power_Indicator_Blink	r = 100b	BUS# = Actual DEVICE# = Actual FUNC# = Actual
Power_Indicator_Off	r = 100b	BUS# = Actual DEVICE# = Actual FUNC# = Actual
PM_PME (Also defined for Power Management)	r = 000b	BUS# = Actual DEVICE# = Actual FUNC# = Actual

Table 4.7   Power Management Related Messages

Name	In TYPE DWORD r Field [2::0]	Requester ID Field Entries
PM_Active_State_Nak	r = 100b	BUS# = Actual DEVICE# = 00000b (Reserved) FUNC# = 000b (Reserved)
PM_PME (Also defined for Hot Plug / Removal)	r = 000b	BUS# = Actual DEVICE# = Actual FUNC# = Actual
PME_Turn_Off	r = 011b	BUS# = Actual DEVICE# = Actual FUNC# = Actual (1)
PME_To_Ack	r = 101b	BUS# = Actual DEVICE# = Actual FUNC# = Actual (1)

Table Notes:

1. Even though a specific function in a PCI Express device is defined, the entire PCI Express device is affected by the change in power provided.

Table 4.8   Lock Protocol Transaction Related Message

Name	In TYPE HDW r Field [2::0]	Requester ID Field Entries
Unlock	r = 011b	BUS# = Actual DEVICE# = Actual FUNC# = Reserved

Table 4.9   Slot Power Limit Message

Name	In TYPE HDW r Field [2::0]	Requester ID Field Entries
Set_Slot Power_Limit (Set power of upstream port)	r = 100b	BUS# = Actual DEVICE# = Actual FUNC# = Actual

Message Vendor-Defined Transaction ID Routing and Implied Routing

Message vendor-defined requester transactions implement ID Routing. As previously discussed, ID Routing is simply the use of the BUS#s, DEVICE#s, and FUNCTION#s to identify (ID) the destination of message vendor-defined requester transactions. As discussed above, ID Routing is also used for completer transactions.

For message vendor–defined requester transactions to use ID Routing, r = 010b in the TYPE HDW, as illustrated in Figure 4.11. The destinations are defined in the Vendor-defined ID HDW, illustrated in Figure 4.12, in the Header field. Vendor-defined ID HDW contains the actual BUS#, DEVICE#, and FUNCTION # of the transaction's destination.

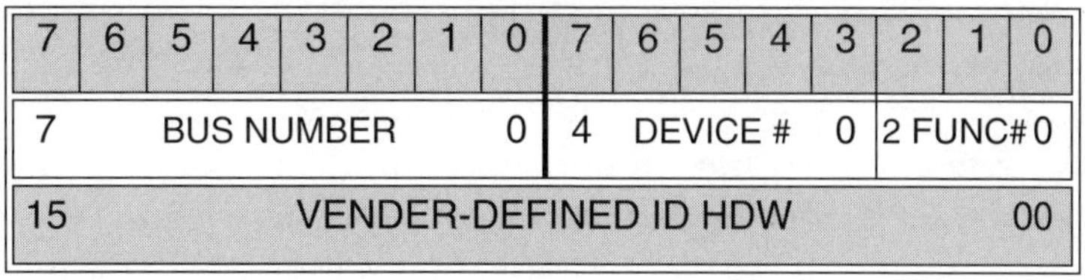

7	6	5	4	3	2	1	0	7	6	5	4	3	2	1	0
7			BUS NUMBER				0	4			DEVICE #		0	2 FUNC#	0
15							VENDER-DEFINED ID HDW								00

Figure 4.12   VENDOR-DEFINED HDWs

Message vendor-defined requester transactions also use Implied Routing. The implementation of Implied Routing by vendor-defined requester transactions is the same as that previously discussed for message baseline requester transactions.

Table 4.10 summarizes the r values for Implied Routing and the Requester ID (BUS#, DEVICE#, and FUNCTION#) used in message vendor-defined baseline transaction packets. The Requester ID is included in the message vendor-defined transition packets to provide information at the destination as to the source of the packets.

Table 4.10 Vendor-Defined Message

Name	In TYPE DWORD r Field [2::0]	Requester ID Field Entries
Vendor_Defined Type 0	r = 000 to 100b r =001b means ID Routing	BUS# = Implementation-Specific
		DEVICE# = Implementation-Specific
		FUNC# = Implementation-Specific
Vendor_Defined Type 1	r = 000 to 100b r =001b means ID Routing	BUS# = Implementation-Specific
		DEVICE# = Implementation-Specific
		FUNC# = Implementation-Specific

Message Advanced Switching Routing

The advance switching requester transaction packets implement RID (Route Identifier) Routing. Unlike the addressing schemes used by the message and vendor-defined requester transaction packets, the message advanced switching requester transaction packets use a 15-bit global address space. This was introduced in early versions of the PCI Express specification. Further information will be available in later revisions PCI Express specification. The c and n fields are defined as part of the RID. See Chapter 18 for more information on message advanced switching requester transactions.

Associated Addressing Protocols

Accessing the configuration address space consists of accessing the configuration register blocks with each function of each PCI Express device. The protocol to access to the configuration address space associated with configuration register blocks is more complex than the protocol to access the memory or I/O address spaces.

Type 0 and Type 1 Configuration Requester Transactions

Only the HOST bus segment CPU can access the configuration register blocks of all PCI Express devices external and internal to the Root Complex. Due to the PCI Express fabric, the flow of configuration requester transaction packets is in a downstream direction. Configuration requester transactions cannot be sourced from endpoints or bridges.

Configuration requester transactions are not defined for upstream flow, peer-to-peer transactions, or advanced peer-to-peer communications.

The definition of a configuration address space places the following unique protocol requirements on configuration transactions. The protocol of configuration transactions internal and external to the Root Complex is similar but different.

Execution of Configuration Transactions Internal to the Root Complex

As discussed earlier, the Root Complex contains a virtual HOST/PCI Bridge, virtual PCI/PCI Bridges in its downstream ports, and an internal virtual PCI bus segment (0). The HOST bus segment CPU accesses the virtual HOST/PCI Bridge configuration block registers by addressing its associated RCRB (via I/O or memory address space of the Host bus segment). If the configuration access is to the HOST/PCI Bridge, the I/O or memory address spaces need only be accessed, as illustrated in Figures 4.13 and 4.14. If the configuration register blocks are associated with the Root Complex's virtual PCI/PCI Bridges in the downstream ports, internal TYPE 0 configuration requester transactions (with the appropriate virtual IDSEL signal lines) are transmitted over the virtual PCI bus segment from the virtual HOST/PCI Bridge. If the configuration register blocks are associated with PCI Express devices external to the Root Complex, TYPE 1 configuration requester transactions are transmitted over the virtual PCI bus segment from the virtual HOST/PCI Bridge. The appropriate virtual PCI/PCI Bridge in one of the downstream ports will claim the Type 1 configuration requester transaction and transmit it on the link as a TYPE 0 or TYPE 1 configuration requester transaction. The Root Complex considers the access to a configuration register block complete with the receipt of the associated completer transaction.

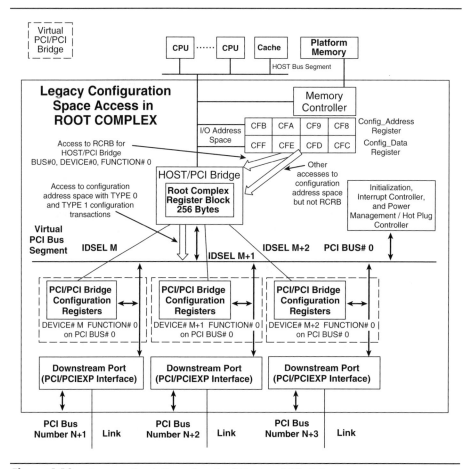

Figure 4.13 Legacy Root Complex Architectural Model

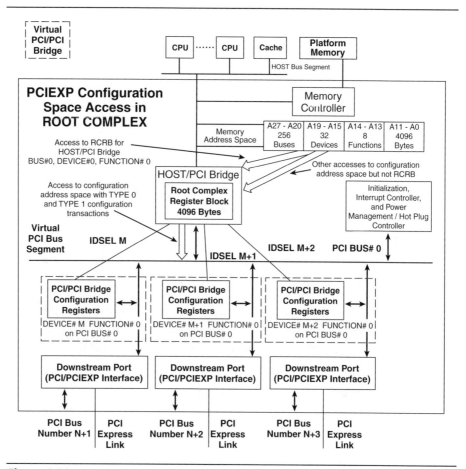

Figure 4.14 PCI Express Root Complex Architectural Model

Execution of Configuration Transactions External to the Root Complex

The following section describes the execution of configuration transactions external to the Root Complex.

Links Connected to the Root Complex Type 0 configuration requester transaction packets are used to access configuration register blocks of devices connected directly to the downstream end of the link. The virtual PCI/PCI Bridges of the Root Complex's downstream ports will convert the TYPE 1 configuration register transaction packets to a TYPE 0 if the BUS# if the connected downstream link matches the BUS#s in the Header field of the configuration requester transaction. The Root Complex's downstream port will subsequently transmit a Type 0 configuration requester transition packet to access the configuration register block of the device on the downstream end of the link. The device on the downstream end of the link is an endpoint, a switch, or a bridge. For the implementations illustrated in Figures 4.15 and 4.16, in the case of a switch or a bridge, its configuration register block is accessed in the upstream port's virtual PCI/PCI Bridge.

Otherwise the downstream port's virtual PCI/PCI Bridge will claim the Type 1 configuration requester packet if the BUS# in the Header field is within the range of BUS#s downstream of the port. The Root Complex's downstream port will subsequently transmit a Type 1 configuration requester transition to be processed by the switch or bridge on the downstream end of the link (as discussed in the next section).

Links Not Connected to the Root Complex Type 1 configuration requester transaction packets are used to access devices not directly connected to the link connected to the Root Complex. Devices that are not directly connected are the downstream ports' virtual PCI/PCI Bridges of switches, PCI or PCI-X resources on the downstream side of bridges, and PCI Express devices connected to the links on the downstream ports of switches. The Type 1 configuration requester transaction packets originate in the Root Complex. The upstream and downstream ports of the intervening switches route the Type 1 configuration requester transaction packets based on the BUS# in their Header fields.

At each switch a decision must be made whether to transmit the Type 1 configuration requester packet downstream, convert it to a Type 0 configuration requester packet and consume it internally to the switch, or convert it to a Type 0 configuration packet and transmit it downstream. Similarly, at each bridge the same decisions must be made. The protocol for these decisions is illustrated in Figure 4.15 for a switch and in Figure 4.16 for a bridge.

Chapter 4: Addressing and Routing ■ 165

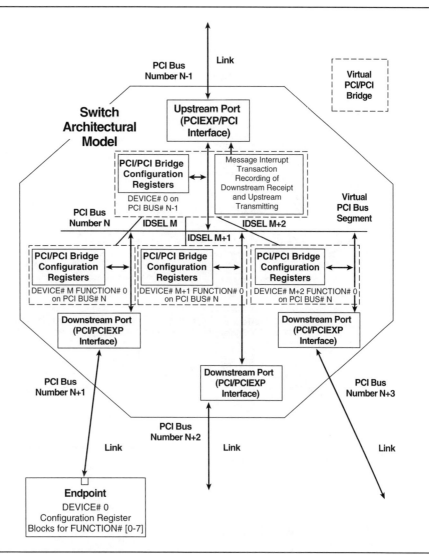

Figure 4.15 Switch Architectural Model

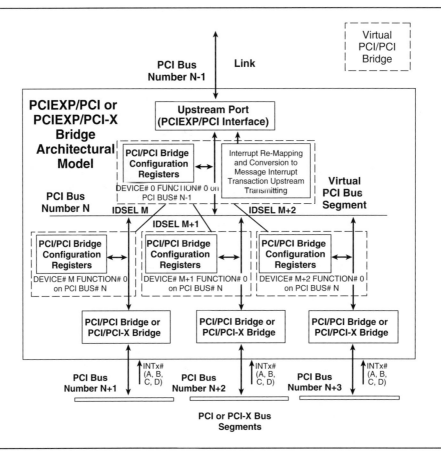

Figure 4.16 Bridge Architectural Model

The receipt of a Type 1 configuration requester transaction packet at the virtual PCI/PCI Bridge of the switch or bridge's upstream port either converts the Type 0 configuration requester transaction packet or passes through the switch as a Type 1 configuration requester packet depending on the BUS# as follows:

- **BUS# matches:** If the BUS# in the Header field of the Type 1 configuration requester transaction packet matches the number of the virtual PCI bus segment (Number N in this example) in the switch or bridge, the upstream port's virtual PCI/PCI Bridge converts it to a Type 0 configuration requester transaction packet. The Type 0 configuration requester transaction packet is ported to the correct downstream port's virtual PCI/PCI Bridge of the switch or bridge by asserting a virtual IDSEL signal line (IDSEL [M::M+2] in these examples), effectively operates as a chip select, based on the DEVICE# in the Header field. The receipt of a Type 0 configuration requester transaction packet by a downstream port's virtual PCI/PCI Bridge results in an access of its configuration register block.

- **BUS# does not match:** If the BUS# in the Header field of the Type 1 configuration requester transaction packet received by the upstream port's virtual PCI/PCI Bridge does not match the number of the virtual PCI bus segment (Number N in this example) in the switch or bridge, the upstream port's virtual PCI/PCI Bridge does not convert the Type 1 configuration requester transaction packet to a Type 0 configuration requester transaction packet. The upstream port's virtual PCI/PCI Bridge will transmit it to the internal virtual PCI bus segment to be claimed by one of the downstream ports' virtual PCI/PCI Bridges. A downstream port's virtual PCI/PCI Bridge will claim the Type 1 configuration requester packet if the BUS# in the Header field is within the range of BUS#s downstream of the port. If the BUS# of the connected downstream link matches the BUS# in the Header field, the TYPE 1 configuration requester transaction packet will be converted to Type 0 configuration requester transaction packet and transmitted onto the link. Otherwise, the downstream port's virtual PCI/PCI Bridge will transmit the Type 1 configuration requester transaction unchanged onto the link (for a switch) or a bus segment (for a bridge).

TAG# and Transaction ID

In addition to the address, ID Routing, and/or Implied Routing discussed in the previous section, some of the transactions implement the Tag# field in the TAG HDW within the Header field of the transaction packet. Each PCI Express device can source and have outstanding multiple requester transaction packets and each is waiting for the associated completer transaction packets. The receipt of the completer transaction packets may not be in the same order as the associated requester transaction packets. Consequently, the PCI Express requester source assigns a unique TAG# to each outstanding requester transaction. The requester source places the TAG# into the Tag# field of the requester transaction packet and this TAG# is returned in the Tag# field by the completer source in the associated completer transaction packet. The Requester ID field in conjunction with Tag# is defined as the Transaction ID.

When TAG# Not Is Used

The Tag# field is defined for all transaction packets except message advance switching transaction packets, which contain no Tag# field. Memory write requester, message baseline requester, and message vendor-defined requester transaction packets contain the Tag# field but it is unused. These transactions do not have an associated completer transaction; consequently, the Tag# field has no value. The Tag# field is not used by the requester destination and can be of any value.

When TAG# Is Used

The Tag# field is used by the requester destination for memory read requester, all I/O requester, and all configuration requester transactions. The completer source (requester destination) uses the TAG# received in the requester transaction packet. The Requester ID and TAG# in these requester transaction packets form a valid Transaction ID. When the Tag# is used the following applies:

- TAG# [4:0] is value is assigned by the requester source. The Tag# in conjunction with the Requester ID (requester source) is used by the completer source (requester destination) to provide the Transaction ID used in the completer transaction packet.

- TAG# [7::5] bits are not used and are defined as RESERVED (value = "000b") unless the Extended Tag Field bit is set to "1" in the configuration register block of the requester source. If the Extended Tag Field bit is set to "1", all 8 bits of the Tag# field are used.

- The completer source returns the TAG# in the completer transaction packet exactly as the TAG# was received in the associated requester transaction.

Other Considerations

The Transaction ID contains the both the Requester ID and the TAG#. The Requester ID includes the FUNCTION#; consequently, a function within a specific PCI Express device can assign a TAG# without regard to the TAG#s assigned by other functions within the same PCI Express device. The combination of the Requester ID and TAG# ensures Transaction ID's uniqueness. Similarly, the Requester ID includes the BUS# and DEVICE#; consequently, each PCI Express device with any BUS# can assign a TAG# without regard to the TAG#s assigned by other PCI Express devices in the platform. The combination of Requester ID and TAG# ensures the Transaction ID's uniqueness. Note: the Transaction ID does not distinguish what type of transaction it is connected to.

It is possible for a specific function within a PCI resource to implement unused FUNCTION#s in the PCI Express device. That is, two or more FUNCTION#s may be implemented by a specific function to attain a larger number of unique Transaction IDs. The completer transaction packet contains the Completer ID in the COMPLETER ID HDW in addition to the Transaction ID comprised of the components of the REQUESTER ID HDW and the TAG HDW. The Completer ID is the BUS#, DEVICE#, FUNCTION# of the completer source.

Chapter 5

Transaction Packets

As discussed in Chapter 2, the Transaction Layer is the interface between a PCI Express device's core and the Data Link Layer, illustrated in Figure 5.1. The PCI Express device's core has implementation-specific requirements for the different types of transaction to be executed in conjunction with other PCI Express devices. Typically it conveys read or write data or provides status information between the two PCI Express devices (participants). The transaction execution begins with one participant providing a requester transaction to the Transaction Layer to construct a Transaction Layer Packet (TLP), which in turn is transferred to the Data Link Layer to construct the Link Layer Transaction packet (LLTP). Finally, the LLTP is transferred to the Physical Layer to construct the Physical Packet. The Physical Packet traverses a combination of links and switches to the other participant. The other participant extracts the TLP from the LLTP and Physical Packet received. The other participant's transaction layer interfaces with its device core with the received requester transaction. Per the Requester/Completer protocol and depending on the type of requester transaction, the PCI Express device's core that received the requester transaction may have to execute a completer transaction. The PCI Express device's core that needs to execute a completer transaction interfaces with its Transaction Layer to contract a TLP, and so forth.

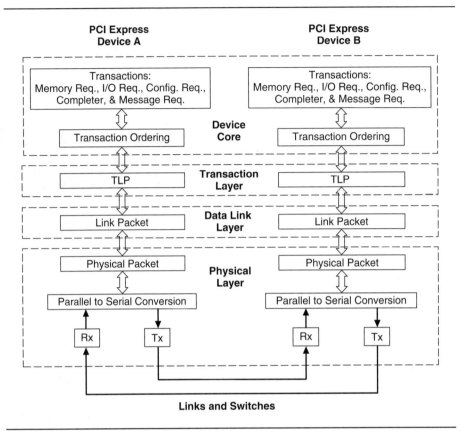

Figure 5.1   PCI Express Layers

As discussed in Chapter 2, there are several types of transactions and thus there is While several types of requester transactions exist, completer transactions have just two types: with and without data. Consequently, the TLPs are different for the different types of requester and completer transactions. Each TLP consists of a collection of bit fields contained within the major fields: Header, Data, and Digest, as illustrated in Figure 5.2. The differences in the TLPs are manifested in the collection of the individual bits fields within the Header field.

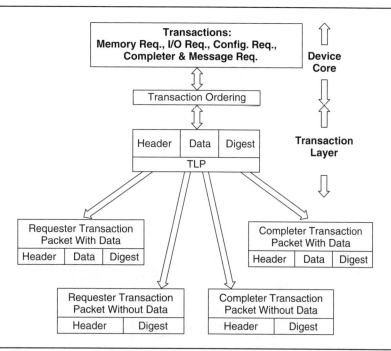

Figure 5.2   PCI Express Layers

Chapter 6 provides detailed information on the formats of the Header, Data, and Digest fields. Chapter 9 provides the detailed information about the Digest field. However, Chapter 6 will not provide all of the details for all the individual bit fields in the Header field. The composition of the individual bit fields in the Header field varies depending on the different types of transactions the TLP represents.

As will be discussed in Chapter 6, this book has adopted the convention that the Header field can be viewed as a collection of Header Words. The combination of different *Header Words* (HDWs) for a specific TLP reflects the type of transaction the TLP represents. The contents of the individual bit fields of the different HDWs are discussed in Chapter 6. Chapter 4 provides additional information about the address and routing bits found in the HDWs.

Header Field Information

Using the HDWs as a discussion vehicle and aligning the discussion with the different types of transactions, this chapter discusses in detail specific individual bit fields in the Header field not completely covered in the other chapters. In order to aid the reader in reviewing all the individual bit fields in the Header Field, Table 5.1 lists these fields and the associated chapters with relevant information.

Table 5.1 Header Field Reference Chapters

Individual Bit Fields in Header Field	Description	Chapters with Relevant Information
FMT	Format of Header Field	6
TYPE	Type of transaction packet	6
TD	Transaction Digest CRC of Header and Data Fields	6 and 9
EP	Error and Poisoning - data incorrect	6 and 9
ATTRI:	OR: ordering level and SN: snoop and caching level	5 and 10
rnc FIELD	Routing - "r" used for Implied Routing. " n" and "c " used for advanced switching protocol	4 and 18
EXT REGISTER ADDRESS	Configuration Register Address	4
REGISTER ADDRESS	Configuration Register Address:	4
BUS#, DEVICE#, FUNCTION#	Used for REQUESTER ID, COMPLETER ID, and CONFIGURATION HDW	4
TAG	Tag for transaction	4
MESSAGE CODE	Defines message type	5
TC	Traffic Class	5, 6, 10, and 11
LENGTH	Number DWORDs in Data field	5
FIRST BE	Byte enable information for starting address	5
LAST BE	Byte enable information for ending address	5

Table 5.1 Header Field Reference Chapters *(continued)*

Individual Bit Fields in Header Field	Description	Chapters with Relevant Information
BYTE COUNT	Remaining number of bytes to be provided by completer transaction(s)	5
LOWER ADDRESS	Reference address of data returned in completer transaction packet	5
BCM	Byte Count Modified – Special bit for PCI-X compatibility	5
STATUS	Status of completer transaction packet	6 and 9
ADDRESS	Address bits for memory and I/O address spaces	4
BUS#, DEVICE#, FUNCTION#	Used for VENDOR-DEFINED ID HDW	5 and 6
RESERVED	VENDOR ID and RESERVED HDWs	6

Memory Transaction Packets and Execution Protocol

A memory transaction protocol consists of a memory read or write requester transaction, and in the case of a memory read an associated completer transaction. The execution of a memory read transaction is not complete until the receipt of an associated completer transition at the requester transaction source. The execution of a memory write transaction is considered complete once it is transmitted over the link.

As discussed in Chapter 4, a memory requester transaction packet contains a Header field with either 32 or 64 address bits contained in the ADDRESS HDWs. In addition the requester source provides its Requester ID and a Tag# field that uniquely identifies this request and is termed the Transaction ID. The completer transition returns the Transaction ID and adds the Completer ID. Before discussing the details of the memory transaction protocol, here is a review of key elements of the Requester ID, Completer ID, and Tag# introduced in Chapter 4.

Requester ID, Completer ID, and Tag#

The execution of a memory read requester transaction is not complete until the receipt of an associated completer transaction by the requester source. The completer transaction packet routes to the requester source by the Requester ID field in the Header field. The inclusion of the Tag# field from the requester transaction with the Requester ID creates the Transaction ID in the Header field of the completer transaction packet. The Transaction ID provides the completer transaction packet's association with a memory read requester transaction packet by the requester source. The Requester ID in the completer transaction packet is also used to route the packet. The completer transaction packet also contains the Completer ID, which identifies to the requester source the completer source. The Requester ID in memory read requester packets and the Completer ID in the completer transaction packets provide useful for Advanced Error Reporting protocol. See Chapter 9 for more information on this protocol.

The generic format of the Requester ID, Completer ID, and Tag# fields are illustrated in Figure 5.3.

The execution of a memory write requester transaction is immediately considered complete when transmitted, and therefore has no associated completer transaction. Without any associated completer transaction the contents of the Tag# field can be any value. The Requester ID in memory write requester packets are only useful to provide as information for Advanced Error Reporting protocol. See Chapter 9 for more information.

In order to simplify the discussion, the memory transition discussions do not include any further reference to the Requester ID, Completer ID, Tag#, or Transaction ID.

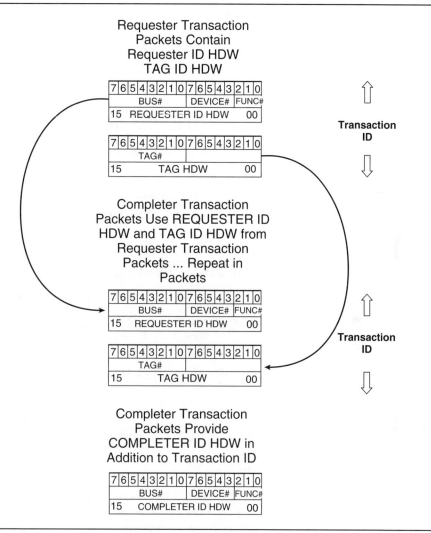

Figure 5.3 Requester and Transaction ID Field

Memory Read Transaction Packets

Memory read requester transaction packets contain the following bit fields that identify the source of the requested data to be returned in the completer transaction packets.

- **FIRST BE [3::0]**, **LAST BE [3::0]**, and **TAG#[7::0]** in the TAG HDW
- **LENGTH [9::0]** in the ATTRIBUTE HDW
- **ADDRESS [31::00]** or **ADDRESS [63::00]** in the ADDRESS HDW
- Requester ID in the REQUESTER ID HDW

The associated completer transaction packets contain the following components that collectively identify the data transferred from the completer source (requester destination) to the completer destination (requester source):

- **LENGTH [9::0]** in the ATTRIBUTE HDW
- **BYTE COUNT [11::0]** and **STATUS [2::0]** in the STATUS HDW
- **LOWER ADDRESS [6::0]** and **TAG# [7::0]** the TAG HDW
- Requester ID in the REQUESTER ID HDW
- Completer ID in the COMPLETER ID HDW

Bit Fields Defining Data to Access

FIRST BE [3::0] Bytes enables of the first DWORD (lowest address order) in the Data field. BE0 = byte 0 of the DWORD, BE1 = byte 1 of DWORD, and so on. The associated byte is read if BE = 1b.

LAST BE [3::0] Bytes enables of the last DWORD (highest address order) in the Data field BE0 = byte 0 of the DWORD, BE1 = byte 1 of DWORD, and so on. The associated byte is read if BE = 1b.

LENGTH [9::0] For a memory read requester transaction packet, **LENGTH [9::0]** is the amount of data that is requested to be read and returned in the associated completer transaction packet(s). For a completer transaction packet it is amount of data in the Data field. The Data field resolution is one DWORD. It must equal 0 0000 0000b when STATUS [2::0] does *not* equal 000b (unsuccessful read).

ADDRESS [31::00] or ADDRESS [63::00] For a single DWORD, reading or writing these bits in the requester transaction packet selects the DWORD in memory address space. For multiple DWORDs, reading or writing these bits in the requester transaction packet selects the beginning address in memory address space.

TAG# [7::0], Requester ID, and Completer ID Included in the requester transaction packet to provide a unique indicator for the requester transaction packet from the requester source identified by the Requester ID. The TAG# is used by the completer source to identify the requester transaction packet associated with the completer transaction packet(s). The Completer ID simply provides the completer destination (requester source) the source of the completer transaction packet(s). The Requester and Completer IDs contain BUS#, DEVICE#, and FUNCTION#.

BYTE COUNT [11::0] This is the number of actual bytes that must still be read including those in the present completer transaction packet if a part of a "set." If only one completer transaction packet is returned, the BYTE COUNT is the total valid bytes read and provided in the Data field. If multiple completer transaction packets are returned, the byte count represents the remaining bytes to be read in subsequent completer transaction packets plus the bytes in the present completer transaction packet. It must equal 0000 0000 0000b when STATUS [2::0] does *not* equal 000b (unsuccessful read). See the further discussion in the following section.

LOWER ADDRESS [6::0] It defines the lower seven address bits of start address of the associated data. LOWER ADDRESS [1::0] is the lowest order two bits of the lowest order byte in the Data field of the completer transaction packet. For the single completer transaction packet sourced (See Example 1 and 2a), the value of the two lowest order address bits (LOWER ADDRESS [1::0]) is linked to the contents of the FIRST BE field and is summarized in Tables 5.2 and 5.3. If FIRST BE [3::0] equals 0000b, the LOWER ADDRESS [1::0] equals 00b. Also, for the single completer transaction packet, LOWER ADDRESS [6::2] must equal the ADDRESS [6::2] of the associated memory read requester transaction packet for the single completer transaction packet. When a set of multiple completer transaction packets are sourced (Example 2b and 2c), the value of LOWER ADDRESS [6::0] may be different, as discussed in the following section. It must equal 000 0000b when STATUS [2::0] does *not* equal 000b (unsuccessful read).

Memory Read Transaction Execution

Memory read requester transaction packets are responded to by completer transaction packets with read data. The PCI Express device being read must make the determination whether all the data requested to be read can be returned in a single completer transaction packet or a set of multiple completer transaction packets (a *set* is defined as multiple completer transaction packets associated with a single requester transaction packet). If the PCI Express device decides to return the data in the set of multiple completer transaction packets, it must provide the data up to a specific address boundary called the RCB. Prior to discussing the protocols for single versus multiple completer transaction packets, the particulars of the RCB must first be defined.

Read Completion Boundary (RCB)

The Read Completion Boundary (RCB) is a naturally aligned address boundary of 64 or 128 bytes (selected in the Link Control Register in the configuration register block or hardwired). A PCI Express memory resource may have multiple RCBs (every address aligned with an integer multiple of the RCB value). The purpose of the RCB is to provide address termination boundaries for multiple completer transaction packets as discussed below.

Single versus Multiple Completions

Memory read requester transaction packets define the address range (ADDRESS plus LENGTH) of the DWORDs to be read. If the address range does not cross an RCB (ADDRESS field LENGTH [9:0]), and equals or is less than a RCB boundary defined for the completer source, it is required to source all of the data in a single completer transaction packet. If the address range does cross one or more of the RCBs defined for the completer source, it can optionally source the data in a single completer transaction packet or as a set of multiple completer transaction packets aligned to the RCBs. For accesses to the Root Complex the downstream port (completer source) is hardwired to support only an RCB equal to either 64 bytes or 128 bytes. For all other PCI Express devices the RCB value is either hardwired to 64 bytes or is programmable as 64 bytes or 128 bytes.

The implementations of the FIRST BE, LAST BE, LENGTH, ADDRESS, BYTE COUNT, and LOWER ADDRESS bits provided in the Header field are illustrated by two examples in the next section. Example 1 shows the return of all the data read in a single completer transaction packet. Example 2 shows the return of data read in one, two, or three completer transaction packets. As you can see from the two examples, the value of the individual field bits of the completer transaction packets vary depending on the address range of the data being read and the sequence of the completer transaction packet of a set. The values for each specific completer transaction packet are defined as follows.

- **Single Completer Transaction or First Completer Transaction Packet** For the single completer transaction packet or the first completer transaction packet of a "set" ("set" is defined as multiple completer transaction packets associated with a single requester transaction packet):
    - FIRST BE, LAST BE, and LENGTH values in the requester transaction packets in conjunction with Tables 5.2 and 5.3 determine the BYTE COUNT value in the completer transaction packet.
    - FIRST BE value in the requester transactions in conjunction with Tables 5.2 and 5.3 determines the LOWER ADDRESS [1::0] value.
    - The LOWER ADDRESS [6::2] of the completer transaction packet equals the ADDRESS [6::2] of the requester transaction packet.
    - The LENGTH [9::0] value in the completer transaction packet is the actual number of DWORDs in the packet's Data field.

- **Second or Subsequent Completer Transaction** (For the completer transaction packets of a set other than the first completer transaction packet of a set):
  - The BYTE COUNT value is the number of remaining bytes to be transferred including those contained in the current completer transaction packet's Data field. (See the discussion at the end of this section about unique interpretation of the BYTE COUNT if the BCM bit in the STATUS field is set to "1.")
  - The LOWER ADDRESS [6::0] must equal 000 0000b.
  - The LENGTH value in the completer transaction packet is the actual number of DWORDs in the packet's Data field.
- **Error Detected in Completer Transaction:** If the STATUS [2::0] field is not a successful completion ("Successful Completion" = 000b), additional considerations must be taken into account, and these can be found in the completer transaction protocol section at the end of this chapter. For the remainder of this discussion it is assumed that all of the completer transaction packets' STATUS [2::0] fields equal 000b.

> If a set of completer transaction packets is sourced, the associated address sequence must be maintained. For example, an individual completer transaction within the set cannot skip over the addresses associated with RCB n to RCB n+1.

BYTE COUNT and LOWER ADDRESS Tables The LAST BE, FIRST BE, and LENGTH bits provided in the memory read requester transaction packet are used to determine the BYTE COUNT and LOWER ADDRESS bytes in the completer transaction packet(s).

Table 5.2 summarizes the value of the BYTE COUNT and LOWER ADDRESS [1::0] in the completer transaction packets per the listed FIRST and LAST BEs when LAST BE is not equal to 0000b.

Table 5.3 summarizes the value of the BYTE COUNT and LOWER ADDRESS [1::0] in the completer transaction packets per the listed FIRST and LAST BEs when LAST BE equals 0000b. LAST BE equals 0000b only when LENGTH [9::0] equals 00 0000 0001b (1 DWORD).

Table 5.2 BYTE COUNT and LOWER ADDRESS [1::0] Value for LAST BE not equal to 0000b

LAST BE [3::0] of Requester	FIRST BE [3::0] of Requester	BYTE COUNT [11:00] of Completer	LOWER ADDRESS [1::0] of Completer
1xxx	xxx1	(LENGTH × 4) – 0	00
01xx	xxx1	(LENGTH × 4) –1	00
001x	xxx1	(LENGTH × 4) – 2	00
0001	xxx1	(LENGTH × 4) – 3	00
1xxx	xx10	(LENGTH × 4) – 1	01
01xx	xx10	(LENGTH × 4) – 2	01
001x	xx10	(LENGTH × 4) – 3	01
0001	xx10	(LENGTH × 4) – 4	01
1xxx	x100	(LENGTH × 4) – 2	10
01xx	x100	(LENGTH × 4) – 3	10
001x	x100	(LENGTH × 4) – 4	10
0001	x100	(LENGTH × 4) – 5	10
1xxx	1000	(LENGTH × 4) – 3	11
01xx	1000	(LENGTH × 4) – 4	11
001x	1000	(LENGTH × 4) – 5	11
0001	1000	(LENGTH × 4) – 6	11

Table 5.3 BYTE COUNT and LOWER ADDRESS [1::0] Value for LAST BE equals 0000b

LAST BE [3::0] of Requester	FIRST BE [3::0] of Requester	BYTE COUNT [11:00] of Completer	LOWER ADDRESS [1::0] of Completer
0000	1xx1	4	00
0000	01x1	3	00
0000	0011	2	00
0000	0001	1	00
0000	1x10	3	01
0000	0110	2	01
0000	1100	2	10
0000	0010	1	01
0000	1000	1	11
0000	0100	1	10

Example 1: Memory Read with Single Completer Transaction Packet

Example 1 Requester Assumption: The memory read requester packet contains

- ADDRESS [31::00] = 00 00 00 08H ... base DWORD address
- FIRST BE [3::0] = 0110b ... lowest order enabled byte of first DWORD (Start Address)
- LENGTH [9::0] = 00 0000 0100b ... last DWORD address is four DWORDs above the base address ... last address is 00 00 00 0BH
- LAST BE [3::0] = 0100b ... highest order enabled byte of last DWORDs (End Address)

Example 1 Memory Resource Assumption; The PCI Express memory resource being accessed in this example has a RCB of 64 bytes. The address range defined by the memory read requester transaction packet is less than 64 bytes and thus does not cross an RCB multiple. Consequently, a single completer transaction packet is required and contains all 64 bytes in a single transaction packet.

Example 1 with Single Completer Transaction Packet: In this response a single completer transaction packet is sourced that returns all of the requested data and contains

- LENGTH [9::0] = 00 0000 0100b
- BYTE COUNT [11::00] = 0000 0000 1110b
- LOWER ADDRESS [6::0] = 00 01001b

Notice that the BYTE COUNT = 14 in Example 1. The LENGTH of 4 DWORDs is 16 bytes, but FIRST BE does not enable the lowest order byte, and the LAST BE does not enable the highest order byte. Thus, the first and last bytes are *not* read. Consequently, the Data field in the completer transaction packet contains a value "X" for these bytes, the LENGTH defines DWORDs and is not concerned about the FIRST and LAST BE values read (thus LENGTH of completer = LENGTH of requester), and the BYTE COUNT does not include the first and last bytes. Also notice that FIRST BE [3] is not enabled and the LAST BE [1:0] are not enabled. Consequently, the Data field in the requester transaction packet contains a value "X" for these bytes. The LENGTH [9::0] is the number of DWORDs in the Data field including those DWORDs that contain data bytes defined as "X." The BYTE COUNT [11:0] does not include the data bytes defined as "X."

> The value "X" is undefined by the PCI Express specification revision 1.0. However, the CRC calculation must be done over the entire Header and Data fields including the bytes assigned the value "X."

> As discussed in Chapter 6, for a write requester transaction packet, the byte associated with a disabled BE ("0") is not written at the requester destination.
>
> For a read requester transaction packet, the byte associated with a disabled BE ("0") is not read at the completer source (requester destination) if within non-prefetchable address space. If within a prefetchable address space, the completer source can optionally read the data, but it will be discarded at the completer destination (requester source).

Example 2: Memory Read with One, Two, or Three Completer Transaction Packets

Example 2 Requester Assumption: The memory read requester transaction packet contains

- ADDRESS [31::00] = 00 00 00 08h ... base DWORD address
- FIRST BE [3::0] = 0010b ... lowest order enabled byte of first DWORD
- LENGTH [9::0] = 10 0000 0010b ... last DWORD address is 514 DWORDs above the base address
- LAST BE [3::0] = 0010b ... highest order enabled byte of last DWORD

Example 2 Memory Resource Assumption: The PCI Express memory resource being accessed has a RCB of 64 bytes and is *not* the ROOT COMPLEX. The address range defined by the memory read requester transaction packet greater than 64 bytes and does cross several RCB multiples. Consequently, the PCI Express completer source can respond (optionally) to the memory requester transaction packet in several ways, as described in Examples 2a, 2b, and 2c.

Example 2a with Single Completer Transaction Packet: In this response a single completer transaction packet is sourced that returns all of the requested data and contains

- LENGTH [9::0] = 10 0000 0010b (514 DWORDs)
- BYTE COUNT [11::00] = 1000 0000 0101b (2053 bytes)
- LOWER ADDRESS [6::0] = 00 010001b

Example 2b with Two Completer Transaction Packets: In this response a set of two completer transaction packets are sourced with the second beginning at the first RCB. Each completer transaction contains

- First completer transaction packet
  - LENGTH [9::0] = 00 0000 1000b This completer transaction stops at first 64 byte RCB (16th DWORD);The first 64 byte block = 0-63 with the next block starting at byte 64; The base ADDRESS = 00 00 00 08h with BE[0010] = start byte 33;Consequently, bytes 33–63, or a total of 31 bytes will be transferred
  - BYTE COUNT [11::00] = 1000 0000 0101b (2053 bytes)
  - LOWER ADDRESS [6::0] = 000 10001b
- Second completer transaction packet
  - LENGTH [9::0] = 01 1111 1010b (514 - 8 = 506 DWORDs)
  - BYTE COUNT [11::00] = 0111 1110 0110b (2053 - 31 = 2022 bytes)
  - LOWER ADDRESS [6::0] = 000 0000b

Notice that the starting LOWER ADDRESS [6::0] of the second completer transaction packet is by definition 000 0000b. The LOWER ADDRESS [6::0] value for the first completer transaction packet follows the same protocol as for the single completer transaction packet example. The LOWER ADDRESS [1::0] value is linked to the value of the FIRST BE and is summarized in Tables 5.1 and 5.2. If FIRST BE [3::0] = 0000b, the LOWER ADDRESS [1::0] = 00b. The LOWER ADDRESS [6::2] must equal the ADDRESS [6::2] of the associated memory read requester transaction packet.

The second completer transaction packet is required to begin at an RCB and in this example it is the first RCB (location 64). The second packet could have begun at any subsequent RCB. The LENGTH [9::0] value in the second packet is the number of DWORDs in the second packet's data field, which equals the balance of the data DWORDs to be transferred in this example. The BYTE COUNT [11::00] value in the second packet is the remaining number of bytes the data to be transferred. Note: the number of bytes transferred in the first completer transaction packet is 31. The eight DWORDs provide 32 bytes, but the FIRST BE specified that the lowest order byte is disabled.

In the two completer transaction examples 2a and 2b where a set of multiple completer transactions packets is sourced, it was assumed the source was not the Root Complex. If the source had been the Root Complex with RCB equal to 64 bytes, the value of LOWER ADDRESS [6] for the completer transaction packets (other than the first packet) would toggle.

Example 2c with Three Completer Transaction Packets: This response contains a set of three completer transaction packets. They are sourced with the second completer transaction beginning at the first RCB, the third beginning at the fourth RCB. Each completer transaction contains:

- First completer transaction packet
  - LENGTH [9::0] = 00 0000 1000b. The base ADDRESS = 00 00 00 08H and the first RCB is 64 bytes; consequently, the LENGTH value begins with the DWORD associated with ADDRESS value 00 00 00 08H up to DWORD containing byte 63 (16th DW), which is a LENGTH of eight DWORDs (64 bytes is RCB but 00 to 63 is a count of 64 bytes)
  - BYTE COUNT [11::00] = 1000 0000 0101b (2053 bytes)
  - LOWER ADDRESS [6::0] = 0001 0001b
- Second completer transaction packet
  - LENGTH [9::0] = 00 0011 0000b (first RCB +16 + 16 + 16 to the fourth RCB) = 48 DWORDs)
  - BYTE COUNT [11::00] = 0111 1110 0110b (2053 - 31 = 2022 bytes)
  - LOWER ADDRESS [6::0] = 000 0000b
- Third completer transaction packet
  - LENGTH [9::0] = 01 1100 1010b (514 - 48 - 8 = 458 DWORDs)
  - BYTE COUNT [11::00] = 0111 0010 0110b (2053 - 64 - 64 - 64 - 31 = 1830 bytes)
  - LOWER ADDRESS [6::0] = 000 0000b

Notice that the LOWER ADDRESS [6::0] value of the second and third completer transaction packets are by definition 000 0000b. The LOWER ADDRESS [6::0] value for the first completer packet follows the same protocol as for the first completer transaction packet of response of the above two completer transaction response.

The second completer transaction packet is required to begin at a RCB and in this example it is the first RCB. The third completer transaction packet is required to begin at a RCB and in this example it is the fourth RCB. The second or third packet could have begun at any RCB (provided the RCB of the third packet is greater address than the RCB of the second and first packets, and so forth).

The LENGTH [9::0] value in the second packet is the number of DWORDs in the second packet's data field, which equals the number data DWORDs to the fourth RCB transferred. The LENGTH [9::0] value in the third packet is the number of DWORDs to complete the transfer in this packet.

The BYTE COUNT [11::00] value in the second packet is the remaining number of bytes of data to be transferred. In response to the three completer transactions, the remaining number of bytes to transfer are done in two packets (second and third). The BYTE COUNT [11::00] value in the third packet is the remaining number of bytes of data to be transferred. In response 2c, the remaining number of bytes to transfer is done in this (third) packet.

Treatment of Other Fields in the Memory Read Transaction Packets

Within each memory read transaction packet there are additional fields that must be defined and implemented. These fields are beyond those specifically related to the size of the valid data read. However, these other bits are directly associated with the data read.

Transaction Completion Status

The previous examples assumed all of the completer transaction packets were successful (STATUS [2::0] = 000b). If a completer transaction packet is received by the completer destination that is *not* successful, the completer transaction packet changes to one without a Data field. See Chapter 9 for details associated with error reporting and handling.

Byte Count Modified (BCM) Status Bit

One additional factor must be considered relative to BCM (Byte Count Modified) in the STATUS HDW of completer transaction packets. The BCM bit is included in the STATUS HDW to support unique PCI-X protocol requirement. The BCM bit is implemented in the following way:

- **BCM set only to 0b:** The BCM is set to 0b for all completer transaction packets not associated with memory read requester transactions. The PCI Express completer destination (PCI Express requester source) ignores the BCM bit in the completer transaction packets associated with any I/O and any configuration requester transaction packets. The BCM is set to 0b for all completer transaction packets associated with a memory read requester transaction accessing the Root Complex, PCI Express/PCI Bridge, PCI legacy endpoint, or PCI Express endpoint.

- **BCM set to 0b or 1b:** The BCM is set to "0" or "1" for all completer transaction packets associated memory read requester transactions issued by a PCI Express/PCI-X Bridge or PCI-X legacy endpoint. The interpretation of the BYTE COUNT [11:00] value in the STATUS HDW in the completer transaction packet is:

    - If the BCM is set to 0b, BYTE COUNT [11:00] is the number of remaining bytes to be transferred including those contained in the current completer transaction packet's Data field.

    - If the BCM is set to 1b, BYTE COUNT [11:00] is the number of bytes in the current completer transaction packet's Data field, not the remaining bytes to be transferred in a set of multiple completer transaction set. Only the first completer transaction packet of a set of multiple completer transactions can have the BCM set to 1b. The subsequent completer transaction packets of the set of multiple completer transactions must have the BCM bit set to 0b. The BCM is set to 0b in the subsequent completer transaction packets to indicate that the BYTE COUNT [11:00] is the number of remaining bytes to be transferred, including those contained in the current completer transaction packet's data field. If the BCM bit is set 1b in a subsequent completer transaction packet of a set of multiple completer transaction packets, the completer transaction packet is defined as **malformed**. The malformed completer transaction packet is processed as discussed in Chapter 9.

Traffic Class Bits

Memory read requester transaction and associated completer transaction packets also contain the traffic class (TC) number. Application is as follows (See Chapters 10 and 11 for additional information):
For TC [2::0] in the TYPE HDW

- The requester source selects a TC number for each transaction. The completer source (as requester destination) simply uses the TC value received for its completer transaction packet.

- The TC assigned to a transaction packet establishes the other transactions that it must be ordered with, and the virtual channel it is assigned to.

- If the requester transaction is part of a LOCK function, the TC [2::0] must equal 000b.

Ordering and Cache Coherency Attribute Bits

Memory read requester transaction and associated completer transaction packets also contain the ordering (OR) and cache coherency (SN) attributes:
ATTRI[1:0] = [OR:SN] in the ATTRIBUTE HDW

- **OR = 1b**: PCI-X relaxed ordering is applied to the requester and completer transaction packet.

- **OR = 0b**: PCI-X relaxed ordering is applied to the requester and completer transaction packet.

- **SN = 1b**: Hardware cache snooping is not permitted.

- **SN = 0b**: Hardware cache snooping is permitted. The only entity that can be cached in a PCI Express platform is the platform memory and only the HOST bus segment CPU or downstream legacy or bridge can access it. The PCI Express bus specification makes no further clarification on what the implementation is relative to the bit.

Memory Write Transaction Packets

Memory write requester transaction packets contain the following components that collectively identify the data requested to be transferred (written to the PCI Express requester destination):

- **FIRST BE [3::0], LAST BE [3::0],** and **TAG# [7::0]** in the TAG HDW
- **LENGTH [9::0]** in the ATTRIBUTE HDW
- **ADDRESS [31::00]** or **ADDRESS [63::00]** in the ADDRESS HDW
- Requester ID in the REQUESTER ID HDW

Bit Fields Defining Data to Access

FIRST BE [3::0]   Bytes enables of the first DWORD (lowest address order) in the Data field. BE0= byte 0 of the DWORD, BE1= byte 1 of DWORD, and so on. The associated byte is written if BE = 1b.

LAST BE [3::0]   Bytes enables of the last DWORD (highest address order) in the Data field BE0= byte 0 of the DWORD, BE1= byte 1 of DWORD, and so on. The associated byte is written if BE = 1b.

LENGTH [9::0]   For a memory write requester transaction packet it is the amount of data in the Data field to be written. The Data field resolution is one DWORD.

ADDRESS [31::00] or ADDRESS [63::00]   For a single DWORD, reading or writing these bits in the requester transaction packet selects the DWORD in memory address space. For multiple DWORDs, reading or writing these bits in the requester transaction packet selects the beginning address in memory address space.

TAG# [7::0], Requester ID, and Completer ID Included in the requester transaction packet to provide a unique indicator for the requester transaction packet from the requester source identified by the Requester ID. The TAG# is used by the completer source to identify the requester transaction packet associated with the completer transaction packet. The TAG# is discarded by the resource destination because there are no associated completer transaction packets.

Memory Write Transaction Execution

Memory write requester transaction packets are not associated with completer transaction packets. Once the memory write requester transaction is transmitted it is considered completed. All of the data associated with a memory write requester transaction is written in a single requester transaction packet.

Address and Length Fields

Memory write requester transaction packets define the address range (ADDRESS HDW plus LENGTH) of the DWORDs to be written.

Treatment of Other Fields in Memory Write Transaction Packets

Within each memory write transaction packet there are other fields that must be defined and implemented. These fields are beyond those specifically related to the size of the valid data write. However, these other fields are directly associated with the data written.

Traffic Class Bits

Memory write requester transaction packets also contain the traffic class (TC) number. For TC [2::0] in the TYPE HDW, the application for requester packets is as follows (See Chapters 10 and 11 for additional information): TC [2::0] in the TYPE HDW.

- The requester source selects a TC value for each transaction.
- The TC assigned to a transaction packet establishes the other transactions that it must be ordered with, and the virtual channel it is assigned to.
- If the memory write requester transaction packet is used for the LOCK function, TC [2:0] must equal 000b.

Ordering and Cache Coherency Attributes Bits

Memory write requester transaction packets also contain the attributes for ordering (OR) and cache coherency (SN):

ATTRI[1:0] = [OR:SN] in the ATTRIBUTE HDW

- **[OR:SN] = 00b:** If the memory write requester transaction packet is used for a message signaled interrupt (MSI), ATTRI[1:0] must equal 00b.

- **OR = 1b:** PCI-X relaxed ordering is applied to the requester and completer transaction packet.

- **OR = 0b:** PCI-X relaxed ordering is applied to the requester and completer transaction packet.

- **SN = 1b:** Hardware cache snooping is not permitted.

- **SN = 0b:** Hardware cache snooping is permitted. The only entity that can be cached in a PCI Express platform is the platform memory and only the HOST bus segment CPU or downstream legacy or bridge can access it. The PCI Express bus specification makes no further clarification on what the implementation is relative to the bit.

I/O Transaction Packets and Execution Protocol

The I/O transaction protocol consists of I/O read and write requester transactions, and associated completer transactions. The execution of an I/O read or write transaction is not considered complete until the receipt of an associated completer transition at the requester transaction source.

As discussed in Chapter 4, a requester transaction contains a Header field with 32 address bits contained in the ADDRESS HDWs. In addition, the requester source provides its Requester ID and Tag# fields that uniquely identifies this request, and is termed the Transaction ID. The completer transition returns the Transaction ID and adds the Completer ID. The protocols of Requester ID, Completer ID, and Tag# discussed in the "Memory Transaction Packets and Execution Protocol" section at the beginning of this chapter also apply to I/O transactions. The only difference is that the I/O write requester transaction packet is associated with a completer transaction packet, which provides confirmation of the write. Consequently, the contents of the Tag# field in the I/O write requester packet has value.

The I/O transition discussions will not include any further reference to the Requester ID, Completer ID, Tag#, or Transaction ID to simplify these discussions.

I/O Transaction Packets

I/O read requester transaction packets contain the following bit fields that identify the source of the requested data to be returned in the completer transaction packets. Similarly, I/O write requester transaction packets contain the same bits fields that identify the destination of the data.

- **FIRST BE [3::0], LAST BE [3::0],** and **TAG# [7::0]** in the TAG HDW
- **LENGTH [9::0]** in the ATTRIBUTE HDW
- **ADDRESS [31::00]** in the ADDRESS HDW
- Requester ID in the REQUESTER ID HDW

The associated completer transaction packets contain the following bit fields that collectively identify the data transferred (read) or confirm write status (write) from the completer source (requester destination) to the completer destination (requester source) associated with I/O requester transactions.

- **LENGTH [9::0]** in the ATTRIBUTE HDW
- **BYTE COUNT [11::00]** and **STATUS [2::0]** in the STATUS HDW
- **LOWER ADDRESS [6::0]** and **TAG# [7::0]** the TAG HDW
- Requester ID in the REQUESTER ID HDW
- Completer ID in the COMPLETER ID HDW

Bit Fields Defining Data to Access

FIRST BE [3::0] Bytes enables of the one DWORD in the Data field. BE0 = byte 0 of the DWORD, BE1= byte 1 of DWORD, and so on. The associated byte is read or written if BE = 1b.

LAST BE [3::0] Defined as 0000b

LENGTH [9::0] For an I/O read requester transaction packet it is the amount of data that is requested to be read and returned in the associated completer transaction packet(s). For the associated completer transaction packet it is the amount of data in the Data field. In both cases it must equal 00 0000 0001b. In the associated completer transaction it must equal 00 0000 0000b when STATUS [2::0] does *not* equal 000b (unsuccessful read).

For an I/O write requester transaction packet it is the amount of data in the Data field to be written. In this case it must be equal to 00 0000 0001b. For the associated completer transaction packet it must be equal to 00 0000 0000b.

Data field resolution is one DWORD.

ADDRESS [31::00] For a single DWORD, reading or writing these bits in the requester transaction packet selects the DWORD in I/O address space.

TAG# [7::0], Requester ID, and Completer ID Included in the requester transaction packet to provide a unique indicator for the requester transaction packet from the requester source identified by the Requester ID. The TAG# is used by the completer source to identify the requester transaction packet associated with the completer transaction packet(s). The Completer ID simply provides the completer destination (requester source) the source of the completer transaction packet(s). The Requester and Completer IDs contain BUS#, DEVICE#, and FUNCTION#.

BYTE COUNT [11::0] For the completer transaction packets associated with I/O read requester transactions, the byte count must equal 0000 0000 0100b when STATUS [2::0] = 000b (successful). It must equal 0000 0000 0000b when STATUS [2::0] does *not* equal 000b (unsuccessful read).

For the completer transaction packets associated with I/O write requester transaction packets must equal 0000 0000 0000b.

LOWER ADDRESS [6::0] This is defined as 00 0000b.

I/O Read and Write Execution

Completer transaction packets respond to I/O read requester transaction packets primarily to provide read data. Completer transaction packets respond to I/O write requester transaction packets primarily to confirm successful writing of the data. For a specific I/O read or write transaction packet a single completer transaction packet is returned.

Treatment of Other Fields in the I/O Transaction Packets

Within each I/O transaction packet there are other fields that must be defined and implemented. These fields are beyond those specifically related to the size of the valid data transferred, which in the case of I/O transactions is fixed at 32 data bits. However, these other fields are directly associated with the data transfer.

Transaction Completion Status

The previous discussion assumed (except as noted) all of the completer transaction packets were successful (STATUS [2::0] = 000b). If a completer transaction packet is received by the completer destination that is *not* successful, the completer transaction packet associated with an I/O read requester transaction packet changes to one without a Data field.

Traffic Class Field

I/O requester transaction packets and the associated completer transaction packets also contain the traffic class (TC) number. The TC number is applied in the following manner:

TC [2::0] in the TYPE HDW

- The requester source selects a TC number for each transaction. The completer source (as requester destination) simply uses the TC value received for its completer transaction packet.

- The TC assigned to a transaction packet establishes the other transactions that it must be ordered with, and the virtual channel it is assigned to.

- TC [2:0] must equal 000b.

Ordering and Cache Coherency Attributes

I/O requester transaction packets and the associated completer transaction packets also contain the attributes for ordering (OR) and cache coherency (SN):

ATTRI[1:0] = [OR:SN] in the ATTRIBUTE HDW

- [OR:SN] = 00b.

- **OR = 0b**: PCI-X relaxed ordering is applied to the requester and completer transaction packet.

- **SN = 0b**: By definition, cache snooping does not apply to I/O requester transactions.

Configuration Transaction Packets Protocol

The configuration transaction protocol consists of configuration read and write requester transactions, and associated completer transactions. The execution of a configuration read or write transaction is not complete until the receipt of an associated completer transaction at the requester transaction source.

As discussed in the Chapter 4, the configuration requester transaction packet Header field format defines BUS#, DEVICE#, and FUNC# (CONFIGURATION HDW), and the 10 address bits (EXT REG. ADDRESS and REGISTER ADDRESS in the REGISTER HDW) for the configuration register block in each function. In addition, the requester source provides its Requester ID and Tag# fields that uniquely identify this request, and collectively they are termed the Transaction ID.

The completer transition returns the Transaction ID and adds the Completer ID. The protocols of Requester ID, Completer ID, and Tag# discussed in the "Memory Transaction Packets and Execution Protocol" section at the beginning of this chapter also apply to configuration transactions. The only difference is that the configuration write requester transaction packet is associated with a completer transaction packet, which provides confirmation of the write. Consequently, the contents of the Tag# field in the configuration write requester packet has value.

The configuration transition discussions do not include any further reference to the Requester ID, Completer ID, Tag#, or Transaction ID to simplify these discussions.

Configuration Transaction Packets

Completer transaction packets respond to configuration read requester transaction packets primarily to provide read data. Completer transaction packets respond to configuration write requester transaction packets primarily to confirm successful writing of the data. A single completer transaction packet is returned for a specific configuration read or write transaction packet.

Configuration read requester transaction packets contain the following bit fields that identify the source of the requested data to be returned in the completer transaction packets. Similarly, Configuration write requester transaction packets contain the same bits fields that identify the destination of the data.

- **FIRST BE [3::0]** and **TAG# [7::0]** in the TAG HDW.

- **BUS# [7::0]**, **DEVICE#[4::0]**, and **FUNCTION# [2::0]** in the CONFIGURATION HDW.

- **EXT REG. ADDRESS [3::0]** and **REGISTER ADDRESS [5::0]** in the REGISTER HDW.

- **LENGTH [9::0]** in the ATTRIBUTE HDW (Always = "1").

- Requester ID in the REQUESTER ID HDW.

The associated completer transaction packets contain the following bit fields that collectively identify the data transferred (read) or confirm write status (write) from the completer source (requester destination) to the completer destination (requester source) associated with configuration requester transactions.

- **LENGTH [9::0]** in the ATTRIBUTE HDW.

- **BYTE COUNT [11::00]** and **STATUS [2::0]** in the STATUS HDW.
- **LOWER ADDRESS [6::0]** and **TAG# [7::0]** the TAG HDW.
- Requester ID in the REQUESTER ID HDW.
- Completer ID in the COMPLETER ID HDW.

Bit Fields Defining Data to Access

FIRST BE [3::0]   Byte enables of the one DWORD in the Data field. BE0 = byte 0 of the DWORD, BE1 = byte 1 of DWORD, and so on. The associated byte is read or written if BE = 1b.

LAST BE [3::0]   This is defined as 0000b.

BUS# [7::0], DEVICE#[4::0], and FUNCTION# [2::0]   This bit combination in the CONFIGURATION HDW selects a specific function within a PCI Express device matching the DEVICE# located on a link with the matching BUS#. It may also select a virtual PCI device on a virtual PCI bus segment, or PCI or PCI-X device on a PCI or PCI-X bus segment with the matching BUS#.

EXT REG. ADDRESS [3::0] and REGISTER ADDRESS [5::0]   This bit combination in the REGISTER HDW selects a specific register (32 bit) in the configuration register block if a specific function within the selected PCI Express, virtual PCI, PCI, or PCI-X devices.

LENGTH [9::0]   For a configuration read requester transaction packet it is the amount of data that is requested to be read and returned in the associated completer transaction packet(s). For the associated completer transaction packet it is amount of data in the Data field. In both cases it must equal 00 0000 0001b. In the associated completer transaction it must equal 00 0000 0000b when STATUS [2::0] does *not* equal 000b (unsuccessful read).

For a configuration write requester transaction packet it is the amount of data in the Data field to be written. In this case it must be equal to 00 0000 0001b. For the associated completer transaction packet it must be equal to 00 0000 0000b.

Data field resolution is one DWORD.

TAG# [7::0], Requester ID, and Completer ID   Included in the requester transaction packet to provide a unique indicator for the requester transaction packet from the requester source identified by the Requester ID. The TAG# is used by the completer source to identify the requester transaction packet associated with the completer transaction packet(s). The Completer ID simply provides the completer destination (requester source) the source of the completer transaction packet(s). The Requester and Completer IDs contain BUS#, DEVICE#, and FUNCTION#.

BYTE COUNT [11::0]   For the completer transaction packets associated with configuration read requester transactions it must equal 0000 0000 0100b when STATUS [2::0] = 000b (successful). It must equal 0000 0000 0000b when STATUS [2::0] does *not* equal 000b (unsuccessful read).

For the completer transaction packets associated with configuration write requester transaction packets must equal 0000 0000 0000b.

LOWER ADDRESS [6::0]   This is defined as 00 0000b.

Configuration Read and Write Execution

Completer transaction packets respond to configuration read requester transaction packets primarily to provide read data. Completer transaction packets respond to configuration write requester transaction packets primarily to confirm successful writing of the data. For a specific configuration read or write transaction packet a single completer transaction packet is returned.

Treatment of Other Fields in the Configuration Transaction Packets

Within each configuration transaction packet there are other fields that must be defined and implemented. These fields are beyond those specifically related to the size of the valid data transferred, which in the case of configuration transactions is fixed at 32 data bits. However, these other fields are directly associated with the data transfer.

Transaction Completion Status

The previous discussion assumed (except as noted) all of the completer transaction packets were successful (STATUS [2::0] = 000b). If a completer transaction packet is received by the completer destination that is *not* successful, the completer transaction packet associated with configuration read requester transaction packet changes to one without a Data field. See Chapter 9 for details associated with error reporting and handling.

Traffic Class Field

Configuration requester transaction packets and the associated completer transaction packets also contain the traffic class (TC) number. The TC number is applied as follows (see Chapters 10 and 11 for additional information):

TC [2::0] in the TYPE HDW

- The requester source selects a TC number for each transaction. The completer source (as requester destination) simply uses the TC value received for its completer transaction packet.

- The TC assigned to a transaction packet establishes the other transactions that it must be ordered with, and the virtual channel it is assigned to.

- TC [2:0] must equal 000b.

Ordering and Cache Coherency Attributes

Configuration requester transaction packets and the associated completer transaction packets also contain the attributes for ordering (OR) and cache coherency (SN):

ATTRI[1:0] = [OR:SN] in the ATTRIBUTE HDW

- [OR:SN] = 00b.

- OR = 0b: PCI-X relaxed ordering is applied to the requester and completer transaction packet.

- SN = 0b: By definition, cache snooping does not apply to I/O requester transactions.

Message Transaction Packets Protocol

The message transaction protocol only defines requester transaction packets; it has no completer transaction packets. The execution of a message write transaction is considered complete once it is transmitted over the link.

As discussed in Chapter 4, the message baseline transaction packets are routed to the destination with Implied Routing. Message vendor-defined transaction packets are routed to the destination with Implied and ID Routing.

Implied Routing uses the values in the r field of the TYPE HDW to determine the destination of the message requester transaction packet. ID Routing is simply the use of the BUS#s, DEVICE#s, and FUNCTION#s to identify (ID) the destination of message vendor-defined requester transactions. For message vendor-defined requester transactions to use ID Routing, r = 010b in the TYPE HDW. The destinations are defined in the VENDER-DEFINED ID HDW in the Header field.

The message baseline and message vendor-defined requester transaction packets contain the Requester ID and Tag#. The implementation of these two fields is the same as for the memory write requester transaction packet. That is, the Requester ID is only useful to provide as information for the Advanced Error Reporting protocol and the contents of the Tag# field can be any value. The message transition discussions will not include any further reference to the Requester ID or Tag# to simplify these discussions.

The message advance switching requester transaction packets are routed to the destination by RID (Route Identifier) Routing. As previously discussed in this book inclusion of any information related to advance switching reflect future enhancements to the PCI Express specification. The PCI Express specification was first released with message advanced switching requester transactions defined. The errata available at the time this book was printed removed all details to these transactions. Consequently, this section will not provide any discussion of the message advance switching transaction protocol. See Chapter 18 for more information.

Message Baseline Transaction Packets

Message baseline transaction packets contain the following components that collectively identify the data requested to be transferred (written to the PCI Express requester destination).

- **r FIELD** in the TYPE HDW.
- **MESSAGE CODE [7::0]** and **TAG#[7::0]** in the TAG HDW.
- **LENGTH [9::0]** in the ATTRIBUTE HDW.
- Requester ID in the REQUESTER ID HDW.

Bit Field Defining Data to Access

r FIELD [2::0] This bit field implies the destination for the message baseline transaction packet (Implied Routing). Its value varies depending on which message baseline transaction is being executed. See Chapter 4 for the possible destinations referenced by these values. The possible values assigned are summarized in Tables 5.4 through 5.9.

MESSAGE CODE [7::0] The MESSAGE CODE [7::0] defines which message the message baseline requester transition packet represents. The values are summarized in Tables 5.4 through 5.9.

LENGTH [9::0] For a message baseline requester transaction packet, the LENGTH [9::0] may or may not contain a Data field. If a Data field is included in the packet, the LENGTH [9::0] defines the number of DWORDs in the Data field. The Data field resolution is one DWORD.

TAG# [7::0], Requester ID, and Completer ID Included in the requester transaction packet to provide a unique indicator for the requester transaction packet from the requester source identified by the Requester ID. The TAG# is used by the completer source to identify the requester transaction packet associated with the completer transaction packet. The TAG# is discarded by the resource destination because there are no associated completer transaction packets.

Message Baseline Transaction Execution

A message baseline transaction is executed with a single requester transaction packet. The information transferred to the destination is in two forms. First, for some message baseline transactions data is included in a Data field. For all message baseline requester tractions the MESSASE CODE [7::0] provides additional information.

The values of the MESSAGE CODE [7::0], TC [2::0], LENGTH [9::0], and whether a Data field is included are provided in Table 5.4 through 5.9. With each table is a brief description with reference to the appropriate chapter for the message baseline requester transaction packets identified in the table.

Message Baseline Transaction Packets: Interrupt Messages

Interrupt messages replace the individual PCI interrupt signal lines; consequently, they are implemented only by legacy endpoints and bridges. If possible, these PCI Express devices should use the MSI (message signal interrupt) protocol. The other PCI Express devices must use the MSI protocol. See Chapter 19 for a more information on the message signal interrupt protocol.

Table 5.4 Interrupt Related Message

Name	TAG HDW Message Code Field [7::0]	TYPE HDW TC [2::0]	Data Field? In ATTRI. HDW Length [9::0]
Assert _ INTx	A = 0010 0000b B = 0010 0001b C = 0010 0010b D = 0010 0011b	000b	No Data Field Length = 00 0000 0000b = Reserved
Deassert_ INTx	A = 0010 0100b B = 0010 0101b C = 0010 0110b D = 0010 0111b	000b	No Data Field Length = 00 0000 0000b = Reserved

Message Baseline Requester Transaction Packets: Error Messages

Error messages report to the Root Complex errors in the platform (general) and errors related to packets. See Chapter 9 for more information.

Table 5.5 Error Related Message

Name	TAG HDW Message Code Field [7::0]	TYPE HDW TC [2::0]	Data Field? In ATTRI. HDW Length [9::0]
ERR_COR (Correctable)	0011 0000b	000b	No Data Field Length = 00 0000 0000b = Reserved
ERR_NONFATAL (Uncorrectable)	0011 0001b	000b	No Data Field Length = 00 0000 0000b = Reserved
ERR_FATAL (Fatal Error)	0011 0011b	000b	No Data Field Length = 00 0000 0000b = Reserved

Message Baseline Requester Transaction Packets: Hot Plug Messages

Hot plug messages are for add-in card slots that support hot plug of add-in cards for the platform slots. See Chapter 16 for more information.

Table 5.6 Hot Plug Related Message

Name	TAG HDW Message Code Field [7::0]	TYPE HDW TC [2::0]	Data Field ? In ATTRI.HDW Length [9::0]
Attention_Indicator_On	0100 0001b	000b	No Data Field Length = 00 0000 0000b = Reserved
Attention_Indicator_Blink	0100 0011b	000b	No Data Field Length = 00 0000 0000b = Reserved
Attention_Indicator_Off	0100 0000b	000b	No Data Field Length = 00 0000 0000b = Reserved

Table 5.6 Hot Plug Related Message *(continued)*

Name	TAG HDW Message Code Field [7::0]	TYPE HDW TC [2::0]	Data Field ? In ATTRI.HDW Length [9::0]
Attention_Button_Pressed	0100 1000b	000b	No Data Field Length = 00 0000 0000b = Reserved
Power_Indicator_On	0100 0101b	000b	No Data Field Length = 00 0000 0000b = Reserved
Power_Indicator_Blink	0100 0111b	000b	No Data Field Length = 00 0000 0000b = Resreved
Power_Indicator_Off	0100 0100b	000b	No Data Field Length = 00 0000 0000b = N Data Field
PM_PME	0001 1000b	000b	No Data Field Length = 00 0000 0000b = Reserved

Message Baseline Requester Transaction Packets: Power Management

Power management messages control the power levels and requirements of PCI Express devices and associated links. See Chapter 15 for a more information.

Table 5.7 Power Management Related Message

Name	TAG HDW Message Code Field [7::0]	TYPE HDW TC [2::0]	Data Field ? In ATTRI. HDW Length [9::0]
PM_Active_State_Nak	0001 0100b	000b	No Data Field Length = 00 0000 0000b =Reserved
PM_PME	0001 1000b	000b	No Data Field Length = 00 0000 0000b = Reserved
PME_Turn_Off	0001 1001b	000b	No Data Field Length = 00 0000 0000b = Reserved
PME_To_Ack	0001 1011b	000b	No Data Field Length = 00 0000 0000b =Reserved

Message Baseline Requester Transaction Packets: Slot Power Messages

Slot power messages set the power limits for add-in cards connected to a slot downstream of a Root Complex or a switch. See Chapter 17 for more information.

Table 5.8 Slot Power Limit Message

Name	TAG HDW Message Code Field [7::0]	TYPE HDW TC [2::0]	Data Field? In ATTRI. HDW Length [9::0]
Set_Slot Power_Limit (Set power of upstream port)	0101 0000b	000b	Data Field Length = 00 0000 0001b

Message Baseline Requester Transaction Packets: Unlock Messages

The unlock message supports the LOCK function begun by the Root Complex. See Chapter 20 for more information.

Table 5.9 Lock Protocol Transaction Related Message

Name	TAG HDW Message Code Field [7::0]	TYPE HDW TC [2::0]	Data Field ? In ATTRI. HDW Length [9::0]
Unlock	0000 0000b	000b	No Data Field Length = 00 0000 0000b = Reserved

Treatment of Other Fields in Message Baseline Transaction Packets

Within each message baseline transaction packet there are other fields that must be defined and implemented. These fields are beyond those specifically related to the size of the valid data written, which in the case of message baseline transactions is either no data or a multiple of 32 data bits. However, these other fields are directly associated with the data written. Currently, only one message baseline transaction contains any data and that size is fixed at 32 data bits.

Traffic Class Bits

Message baseline requester transaction packets also contain the traffic class (TC) number. The application for requester packets follows (see Chapters 10 and 11 for additional information): TC [2::0] in the TYPE HDW.

- The requester source selects a TC value for each transaction.
- The TC assigned to a transaction packet establishes the other transactions that it must be ordered with, and the virtual channel it is assigned to.
- Per the message baseline transitions defined by the PCI Express specification as listed in the above tables, TC [2:0] = 000b.

Ordering and Cache Coherency Attributes Bits

Message baseline requester transaction packets also contain the attributes for ordering (OR) and cache coherency (SN):

ATTRI [1:0] = [OR:SN] in the ATTRIBUTE HDW

- **[OR:SN] = 00b** – For all message baseline requester transaction packets.

- **OR = 1b**: PCI-X relaxed ordering is applied to the requester and completer transaction packet.

- **OR = 0b**: PCI-X relaxed ordering is applied to the requester and completer transaction packet.

- **SN = 1b**: hardware cache snooping is not permitted.

- **SN = 0b**: hardware cache snooping is permitted. The only entity that can be cached in a PCI Express platform is the platform memory and only the HOST bus segment CPU or downstream legacy or bridge can access it. The PCI Express bus specification makes no further clarification on what the implementation is relative to the bit.

Message Vendor-Defined Transaction Packets

Message vendor-defined transaction packets contain the following components that collectively identify the data requested to be transferred (written to the PCI Express requester destination):

- **:r FIELD** in the TYPE HDW.

- Requester ID in the REQUESTER ID HDW.

- **BUS# [7::0], DEVICE# [4::0],** and **FUNCTION# [2::0]** in the VENDOR-DEFINED ID HDW.

- **MESSAGE CODE [7::0]** and **TAG# [7::0]** in the TAG HDW.

- **LENGTH [9::0]** in the ATTRIBUTE HDW.

Bit Fields Defining Data to Access

r FIELD [2::0] This bit field implies the destination for the message vendor-defined transaction packet (Implied Routing). Its value is implementation-specific. See Chapter 4 for the possible destinations referenced by these values. If the r FIELD [2::0] = 010b, the ID Routing is used. The ID Routing implements the VENDOR-DEFINED ID HDW. The possible values assigned are summarized in Table 5.10.

MESSAGE CODE [7::0] The MESSAGE CODE [7::0] defines which message type the message vendor-defined requester transition packet represents. Message vendor-defined requester transaction packets have two types, Type 0 and Type 1, which provide vendors with two levels of support. The receipt by the requester destination of a Type 0 message vendor-defined requester transaction packet when it is not supported results in an **unsupported request** error. See Chapter 9 for more information. The receipt by the requester destination of a Type 1 message vendor-defined requester transaction when it is not supported is simply discarded and no error is reported. This is the only distinction the PCI Express specification makes between these two types. The values are summarized in Table 5.10.

This bit combination in the VENDOR-DEFINED ID HDW selects a specific function within a PCI Express device matching the DEVICE# located on a link with the matching BUS#. It may also select a virtual PCI device on a virtual PCI bus segment, or PCI or PCI-X device on a PCI or PCI-X bus segment with the matching BUS#.

LENGTH [9::0] For a message baseline requester transaction packet, the LENGTH [9::0] may or may not contain a Data field. If a Data filed is included in the packet, the LENGTH[9::0] defines the number of DWORDs in the Data field. The Data field resolution is one DWORD.

TAG# [7::0], Requester ID, and Completer ID Included in the requester transaction packet to provide a unique indicator for the requester transaction packet from the requester source identified by the Requester ID. The TAG# is used by the completer source to identify the requester transaction packet associated with the completer transaction packet. The TAG# is discarded by the rcsource destination because there are no associated completer transaction packets.

Message Vendor-Defined Transaction Execution Packets

As discussed earlier in this chapter, message vendor–defined requester transactions implement both Implied Routing and ID Routing. The two participants in the vendor-defined message are any PCI Express devices located in a specific PCI Express fabric (in-band). The exact participants, the inclusion of data, and the exact interpretation are vendor-specific. One vendor's implementation (data contained and purposes) of a message vendor-defined requester transaction may not be compatible with another vendor's.

The execution of a message vendor-defined transaction is with a single requester transaction packet. The information transferred to the destination is in the optional Date field. If a Data field is not included, the information is implied by the receipt of the type (Type 0 or Type 1) of the message vendor-defined requester transaction packet.

The values of the TC [2::0], MESSAGE CODE [7::0], TC [2::0], LENGTH [9::0], and whether a Data field is included are provided in Table 5.10.

Table 5.10 Vendor-defined Message

Name	TAG HDW Message Code Field [7::0]	TYPE HDW TC [2::0]	Data Field? In ATTRI.HDW Length [9::0]
Vendor_Defined Type 0	0111 1110b	TC = 000b to 111b Implementation specific	Optional Data Length = Value
Vendor_Defined Type 1	0111 1111b	TC = 000b to 111b Implementation specific	Optional Data Length = Value

Treatment of Other Fields in Message Vendor-Defined Transaction Packets

Within each message vendor-defined transaction packet there are other fields that must be defined and implemented. These fields are beyond those specifically related to the size of the valid data written, which in the case of message vendor-defined transactions is either no data or a multiple of 32 data bits. However, these other fields are directly associated with the data written.

Traffic Class Bits

Message vendor-defined transaction packets also contain the traffic class (TC) number. The application for requester packets follows (see Chapters 10 and 11 for additional information): TC [2::0] in the TYPE HDW.

- The requester source selects a TC value for each transaction.
- The TC assigned to a transaction packet establishes the other transactions that it must be ordered with, and the virtual channel it is assigned to.
- Per the message baseline transactions, define the values of the TC [2:0] = 000b to 111b except 010b, which indicates ID Routing.

Ordering and Cache Coherency Attributes Bits

Message vendor-defined requester transaction packets also contain the attributes for ordering (OR) and cache coherency (SN):
ATTRI[1:0] = [OR:SN] in the ATTRIBUTE HDW

- **OR = 1b**: PCI-X relaxed ordering is applied to the requester transaction packet.
- **OR = 0b**: PCI-X relaxed ordering is applied to the requester transaction packet.
- **SN = 1b**: hardware cache snooping is not permitted.
- **SN = 0b**: hardware cache snooping is permitted. The only entity that can be cached in a PCI Express platform is the platform memory and only the HOST bus segment CPU or downstream legacy or bridge can access it. The PCI Express bus specification makes no further clarification on what the implementation is relative to the bit.

Message Advanced Switching Requester Transaction Packets

As discussed in earlier chapters, messages advanced switching requester transactions are defined only for advanced peer-to-peer links. As stated in the introduction of this section, the PCI Express specification was first released with message advanced switching requester transactions defined. The errata available at the time this book was printed removed all details to these transactions. Consequently, this book will not provide any discussion of the message advance switching transaction protocol. See Chapter 18 for more information.

Chapter 6

Transaction Layer

This is the first of three chapters that focus on the specific elements of each of the three layers introduced in Chapter 2: the Transaction Layer, Data Link Layer, and Physical Layer. As shown in Figure 6.1, the Transaction Layer interfaces with the device core and constructs the Transaction Layer Packets (TLPs) containing Header, Data, and Digest fields. The Header field contains control and address information. The Data field contains the data (if applicable). The Digest field is the cyclic redundancy check (called ECRC) computed over the Header and Data fields. The Data Link Layer converts the TLPs and any link management information into Link Layer Transaction Packets (LLTPs) and Data Link Layer Packets (DLLPs), respectively. The Physical Layer transmits onto the link LLTPs and DLLPs within Physical Packets and extracts them from the Physical Packets received from the link. Each type of packet is supported by an associated PCI Express layer. This chapter discusses the Transaction Layer and the associated TLPs. Chapters 7 and 8 discuss the Data Link Layer and Physical Layer with associated packets, respectively.

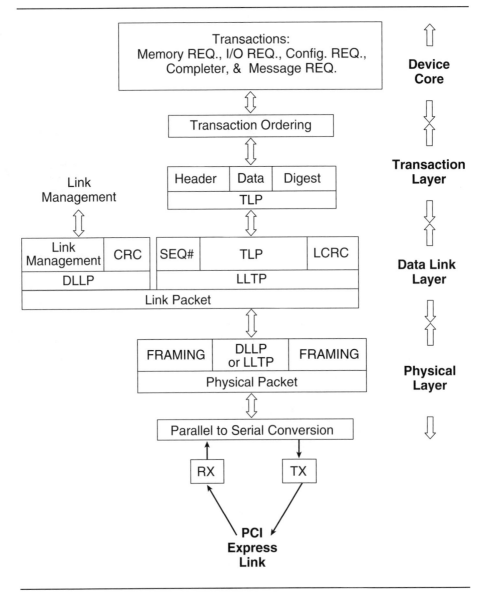

Figure 6.1 Overview of PCI Express Layers

Transaction Layer

The Transaction Layer is the only portion of the PCI Express fabric that interfaces with the implementation-specific transactions that are executed by the PCI Express devices' core. The transactions are defined (typically) by the implementation of the CPUs on the HOST bus segment, the CPUs and circuitry residing in the implementation-specific portion of endpoints, and PCI and PCI-X devices downstream of bridges. The structure of these transactions can vary widely; consequently, communication between platform resources supporting different transaction structures is impossible. The Transaction Layer of PCI Express provides a common structure in the TLPs. The Transaction Layer takes the control, address, and data elements of the different transaction structures and assembles them into similar elements in TLPs for transmission over the link. At the transactions' destinations, the Transaction Layer disassembles the TLPs into control, address, and data elements compatible with the destinations' device core.

The Transaction Layer must also consider that the TLPs flow within a PCI Express fabric or between PCI Express fabrics. The PCI Express fabric implements a Flow Control protocol and a Requester/Completer protocol that are unique to PCI Express. Thus, the Transaction Layer must provide elements in the TLP to support these protocols. In the case of the Flow Control protocol, the Transaction Layer assigns traffic class (TC) numbers to each TLP. In the case of the Requester/Completer protocol, two types of TLPs are executed by the Transaction Layer: requester transaction packets and completer transaction packets. The two exceptions are TLPs for memory write transactions and message transactions, which only consist of requester transaction packets. Each requester transaction packet contains information about source of the packet (Requester ID) and an association tag (TAG). The Transaction Layer at the source of the completer transaction packet uses the Requester ID to direct the packet to the associated requester transaction packet's source, and also supplies the TAG in the packet to provide association with the requester transaction packet.

Finally, PCI Express has two features not typically supported by CPUs on the HOST bus segment, the CPUs and circuitry residing in the implementation-specific portion of endpoints, or PCI and PCI-X devices downstream of bridges. These two features are transaction integrity and messaging. The Transaction Layer includes elements in the TLPs to support these two features.

The following sections summarize the Transaction Layer's structuring of the TLPs to group these elements. As previously discussed, the transaction packets are divided into Header, Data, and Digest fields.

The Control Element

The control elements of a TLP typically define the transaction type (memory, I/O, configuration, or message), read versus write, transaction execution order of device transactions, and cache coherency.

Header field of TLP contains:

FORMAT (FMT): Defines the size of the Header field.

TYPE: Defines transaction type and read versus write.

ATTRIBUTE (ATTRI): Defines the type of transaction ordering that applies and whether cache coherency is implemented.

STATUS: Provides information about the successful or unsuccessful execution of requester transaction packets.

The Address Element

The address elements of the TLP provide the address to select specific bytes within the memory and I/O address spaces. The address elements also provide the ID Routing and the register address to select the specific bytes of the configuration register block in the configuration address space. Finally, address elements also provide the ID and Implied Routing for the message address space.

Header field of TLP contains:

ADDRESS: The "typical" address bits for memory and I/O address space. The address can also be used in message vendor-defined transaction packets.

FIRST BE, LAST BE, and LOWER ADDRESS: Provides byte-level address resolution for the memory data being accessed.

FIRST BE: Provides byte-level address resolution for the I/O and configuration data being accessed.

BUS#, DEVICE#, and FUNCTION (FUNC)#: These fields are the components for ID Routing used by configuration transaction packets and by message vendor-defined transaction packets. They also provide the components for the ID Routing used by completer transactions. Collectively, the BUS#, DEVICE#, and FUNC# are the components of the Requester ID, Complete ID, and Vendor-defined ID contained in the HDWs of the same name.

r FIELD: This field is for Implied Routing used by message baseline and vendor-defined transaction packets.

REGISTER ADDRESS and EXTENDED REGISTER ADDRESS: These fields provide the addressing information for registers within the configuration register blocks of each PCI Express device.

The Flow Control Protocol Element

The Flow Control protocol element provides each TLP with information to permit switches to determine priority between different TLPs vying for link bandwidth.

Header field of TLP contains:

TRAFFIC CLASS (TC): Defines the traffic class number assigned to the TLP. The Flow Control protocol dictates that each TC number is mapped to a virtual channel as part of the arbitration for link bandwidth. The mapping of the TC number directly maps the associated TLP.

The Repeater/Completer Protocol Element

The Requester/Completer protocol assumes that multiple requester transaction packets will be transmitted before any of the associated completer transaction packets are returned to the source of the requester transaction packets. This element provides part of the association between the requester and completer transaction packets.

Header field of TLP contains:

TAG: The PCI Express device that begins the execution of the transaction (source of requester transaction packet) may have multiple transactions at different levels of completion. The completion of the transaction may require the receipt of a completer transaction packet. The TAG# is used in conjunction with the Requester ID to provide a unique label to link the requester transaction packet transmitted to the completer transaction packet received.

The Data Element

The data element of the transaction packet provides the actual data being accessed.

Header field of TLP contains:

LENGTH: The amount of data in the Data field of a specific requester or completer transaction packet.

BYTE COUNT: The remaining amount of data associated with a single memory read requester transaction packet to be returned in the present completer transaction packet or in a set of the completer transaction packets.

BCM: Defines the byte count modified flag for compatibility with PCI-X legacy endpoints (endpoints with a PCI-X device core) and PCI Express/PCI-X Bridges.

Data field of TLP contains:

The actual data being accessed. The Data field is only defined for write requester packets and successful completer transaction packets.

The Integrity Element

The transactions specific to the device core do not allow error checking. However the transaction is executed in the PCI Express fabric, the integrity of the associated transaction packets must be assured.

Header field of TLP contains:

Error and Poisoning (EP): Defines whether the Data field contains data with an error.

Transaction Digest (TD): Defines whether the Digest field is included in the packet.

Digest field of TLP contains:

Cyclic redundancy check (ECRC) over the Header and Data fields. For TLPs without Data fields, the ECRC is over the Header field. The Digest field is optional and included according to the value of the TD bit in the Header Field.

The Message Element

The implementation-specific transactions of the device core typically do not define messages. However, the messages are a useful feature within a PCI Express fabric and between PCI Express fabrics.

Header field of TLP contains:

MESSAGE CODE: Selects the type of message.

VENDOR ID — RESERVED: The VENDOR ID is defined by the vendor of the PCI Express device.

RESERVED: Other fields of the Header field that are presently reserved.

Transaction Layer Implementation

The exact protocol for how the Transaction Layer of a PCI Express device interfaces with each device core is implementation-specific. The structure of the TLPs assembled in a PCI Express device for transmission by the Transaction Layer is precisely defined by the PCI Express specification. The TLPs are constructed of precisely defined Header, Data (as needed), and Digest (optional) fields. The exact definition of each field varies depending on the type of transaction it represents. Once this TLP is assembled, it is transferred to the Data Link Layer by protocols discussed in Chapter 7. At the Transaction Layer of the receiving PCI Express device (the final destination), the precisely defined fields permit the disassembling of the TLPs into transactions compatible with the device core.

The remainder of this chapter focuses on the formats of the Header, Data, and Digest fields for the different types of transactions.

Header Field Formats

> This book uses the following terms:
> - *Requester source* and *requester destination:* The two participants in the transmitting and receiving of the requester transaction packets.
> - *Completer source* and *completer destination:* The two participants in the transmitting and receiving of the completer transaction packets.
> - Transactions that port through switches do not define these switches as participants. Switches are participants for accesses to their configuration register blocks and for some message transactions.

Header Words (HDWs)

This book adopts the convention that the Header field can be viewed as a collection of Header Words (HDWs). The combination of different HDWs and definition of their fields vary for the TLPs of the different transactions.

The values of the fields within the HDWs are either fixed or programmable, depending on the type of transaction being executed. These values are discussed in detail in the associated text. All other fixed and programmable field values are RESERVED.

RESERVED

A RESERVED field value is logical "0" and is marked with an R. The use of a RESERVED field (individual bits or collectively the entire field) for any purpose is not permitted. Unless otherwise specified, the PCI Express device creating the field must implement "0" in the RESERVED field. The PCI Express device that is processing the RESERVED field must ignore it except for cyclic redundancy check calculations.

The RESERVED field is included in the CRC, LCRC, and ECRC calculations with no preconceived notion by the receiving PCI Express device that the RESERVED field has a "0" value. This permits the same hardware to be used with future revisions of the PCI Express specification when some of the RESERVED bits or fields may be redefined. The CRC and LCRC are discussed in Chapter 7 as part of the Data Link Layer. The ECRC is discussed in the context of the Digest field in this chapter.

CROSS REFERENCES

This chapter discusses in detail specific individual bits fields in the Header field not completely covered in the other chapters. In order to aid the reader in reviewing all the individual bit fields in the Header Field, these fields and the associated chapters with relevant information are listed in Table 6.1.

Table 6.1 Header Field Reference Chapters

Individual Bit Fields in Header Field	Description	Chapters with Relevant Information
FMT	Format of Header Field	6
TYPE	Type of transaction packet	6
TD	Transaction Digest CRC of Header and Data Fields	6 and 9
EP	Error and Poisoning: data incorrect	6 and 9
ATTRI:	OR: ordering level and SN: snoop and caching level	5 and 10
rnc FIELD	Routing: "r" used for Implied Routing. "n" and "c" used for advanced switching protocol	4 and 18
EXT REGISTER ADDRESS	Configuration Register Address	4
REGISTER ADDRESS	Configuration Register Address	4
BUS#, DEVICE#, FUNCTION#	Used for REQUESTER ID, COMPLETER ID, and CONFIGURATION HDW	4
TAG	Tag for transaction	4
MESSAGE CODE	Defines message type	5
TC	Traffic Class	5, 6, 10, and 11
LENGTH	Number DWORDs in Data field	5
FIRST BE	Byte enable information for starting address	5

Table 6.1 Header Field Reference Chapters *(continued)*

Individual Bit Fields in Header Field	Description	Chapters with Relevant Information
LAST BE	Byte enable information for ending address	5
BYTE COUNT	Remaining number of bytes to be provided by completer transaction(s)	5
LOWER ADDRESS	Reference address of data returned in completer transaction packet	5
BCM	Byte Count Modified — Special bit for PCI-X compatibility	5
STATUS	Status of completer transaction packet	6 and 9
ADDRESS	Address bits for memory and I/O address spaces	4
BUS#, DEVICE#, FUNCTION#	Used for VENDOR-DEFINED ID HDW	5 and 6
RESERVED	VENDOR ID and RESERVED HDWs	6

Header Field Formats for Memory Read Transactions

Memory read requester transaction packets consists of Header and (optional) Digest fields and come in two formats: 32 address bits, shown in Figure 6.2, and 64 address bits, shown in Figure 6.3.

If the data to be transferred is in the first 4 gigabytes of the memory address space, the 32 address bits format *must* be used. If the 64 address bits format is used it is for accessing memory resources above the 4-gigabyte address boundary. In either situation, the full address must be provided.

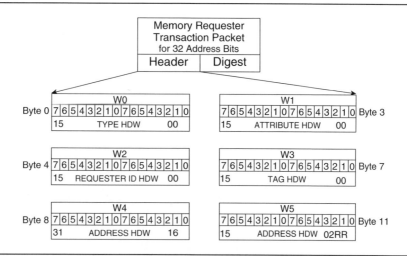

Figure 6.2   Memory Requester Transaction Packet for 32 Address Bits

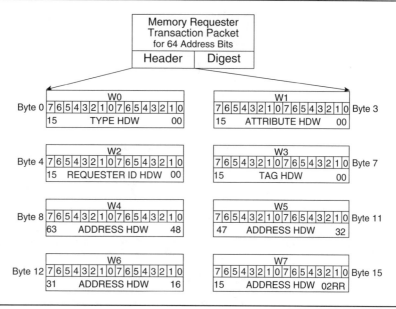

Figure 6.3   Memory Requester Transaction Packet for 64 Address Bits

The Repeater/Completer protocol dictates that the Root Complex, endpoint, or bridge that is being read responds with completer transactions. If the data read was successful (STATUS [2::0] = 000b), the completer transaction packet contains data (in the Data field) as shown in

Figure 6.4. If the data read is not successful (STATUS [2::0] not = 000b), no data is contained in the packet (no Data field), as illustrated in Figure 6.5.

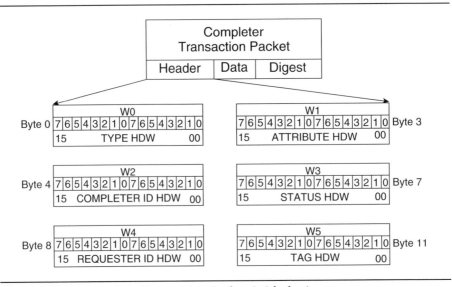

Figure 6.4 Completer Transaction Packet (with data)

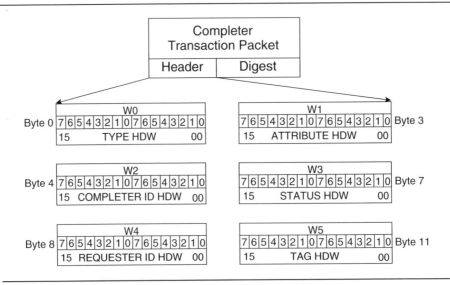

Figure 6.5 Completer Transaction Packet (with no data)

MEMORY READ REQUESTER TRANSACTION

7	6	5	4	3	2	1	0	7	6	5	4	3	2	1	0
R	\multicolumn{2}{l}{FMT 1 0}	\multicolumn{4}{l}{4 TYPE 1 L}		R	\multicolumn{3}{l}{2 TC 0}			R	R	R	R
TD	EP	\multicolumn{2}{l}{ATTRI 1 0}	R	R	9		\multicolumn{7}{c}{LENGTH}	0
7	\multicolumn{5}{c}{BUS NUMBER}		0	4	\multicolumn{3}{c}{DEVICE #}		0	\multicolumn{2}{l}{2 FUNC# 0}	
7	\multicolumn{5}{c}{TAG}		0	3	\multicolumn{3}{c}{LAST BE}		0	3	FIRST BE	0
31	\multicolumn{13}{c}{ADDRESS}			16
15	\multicolumn{11}{c}{ADDRESS}			02	R	R

64 Address bits

7	6	5	4	3	2	1	0	7	6	5	4	3	2	1	0
63	\multicolumn{14}{c}{ADDRESS}		48
47	\multicolumn{14}{c}{ADDRESS}		32
31	\multicolumn{14}{c}{ADDRESS}		16
15	\multicolumn{11}{c}{ADDRESS}			02	R	R

15	TYPE HDW	00
15	ATTRIBUTE HDW	00
15	REQUESTER ID HDW	00
15	TAG HDW	00
	ADDRESS HDW	

*MEMORY READ REQUESTER TRANSACTION HEADER FIELD
without LOCK: MRd ... with LOCK: MRdLk*

TYPE HDW

With 32 Address bits

FMT [1:0]: Format of Header field. FMT [0] = 0b for 3 DWORD versus FMT [0] = 1b for 4 DWORD in Header field. FMT [1] = 0b no Data field, in packet versus FMT [1] = 1b for Data field in packet.

- FMT [1::0] = 00b.

TYPE [4::0]: Type of transaction associated with packet.

- TYPE [0] = 1b LOCK function related transaction, 0b no LOCK function. Only defined for lock memory read requester or lock completer transaction packet. See Chapter 20 for more information on the LOCK function.
- TYPE [4::1] = 0000b.

TC [2::0]: Traffic Class

- TC [2::0] = 000b to 111b.
- TC [2::0] = 000b if lock memory read requester transaction packet.

With 64 Address bits

FMT [1:0]: Format of Header field. FMT [0] = 0b for 3 DWORD versus FMT [0] = 1b for 4 DWORD in Header field. FMT [1] = 0b no Data field, in packet versus FMT [1] = 1b for Data field in packet.

- FMT [1::0] = 01b.

TYPE [4::0]: Type of transaction associated with packet.

- TYPE [0] = 1b LOCK function related transaction, 0b no LOCK function. Only defined for lock memory read requester or lock completer transaction packet. See Chapter 20 for more information on the LOCK function.
- TYPE [4::1] = 0000b.

TC [2::0]: Traffic Class

- TC [2::0] = 000b to 111b.
- TC [2::0] = 000b if lock memory read requester packet.

ATTRIBUTE HDW

TD: Transaction Digest defines whether the transaction packet contains the Digest DWORD.

- TD = 0b No Digest DWORD is attached.
- TD = 1b Digest DWORD is attached.

EP: Error and Poisoning is optional and only defined for write requester transaction and read completer transaction packets. See Chapter 9 for more information on Error and Poisoning.

- EP = 0b.

ATTRI [1:0]: ATTRIBUTES field defines snoop and ordering information. For memory requester transaction packets with or without data ATTRI [1:0] = 00b to 11b.

- ATTRI [1] = (OR) 1b defines the PCI-X relaxed ordering for the packet.
- ATTRI [1] = (OR) 0b defines the PCI strong ordering for the packet. See Chapter 10 for more information on ordering.
- ATTRI [0] = (SN) 1b signifies that cache coherency is not supported.
- ATTRI [0] = (SN) 0b signifies that cache coherency is supported. See Chapter 5 for more information on cache coherency.

LENGTH [9::0]: LENGTH in DWORDs of Data to be read. The data is transferred in DWORD increments (4 bytes = 32 data bits total), naturally address aligned. For read requester transactions packets, the LENGTH [9::0] field is the number of DWORDs of data to be read and returned in completer transaction packet(s). Read requester transaction packets do not contain a Data field in the packet.

- For memory requester transaction packets, the data length is encoded into ten binary bits [9::0]. For example, 00 0000 0001b is one DWORD, 00 0000 0010 is two DWORDs, 11 1111 1111b is 1023 DWORDs, and so on.

- The maximum number is 1024 and the LENGTH [9::0] = 00 0000 0000b. The existence of the data packet is defined in the FMT [1] field; consequently the definition of NO data bytes and NO Data field in the packet is not from information of the LENGTH [9::0] field but from the information of the FMT [1] field.

- If the read requester transaction packet has LENGTH [9::0] = one DWORD and FIRST BE [3::0] = 0000b, the associated completer transaction packet must have a LENGTH = one DWORD and must provide one DWORD of data. This data has no meaning, the associated bytes were not read at the requester destination, and can be any value.

REQUESTER ID HDW

BUS# [7::0]: BUS NUMBER identifies the link or the virtual PCI bus segment in a Root Complex, endpoint, or bridge. (See Chapter 4.)

DEVICE# [4::0]: DEVICE NUMBER identifies the device on the link or on the virtual PCI bus segment in the Root Complex, endpoint, or bridge. (See Chapter 4.)

FUNC# [2:0]: FUNCTION NUMBER identifies one the eight functions within a device. (See Chapter 4.)

TAG HDW

TAG# [7::0]: A unique TAG# is assigned by the PCI Express requester source to each of its requester transaction packets. Each requester transaction packet is assigned a TAG# to permit the requester source to identify the associated completer transaction received. The completer source must retain and use the TAG# received as the requester destination in the completer transaction packets. More information on requester transaction packets is included in Chapter 4.

- If the Extended Tag Enable bit equals "0" (configuration address space) only TAG# [4::0] are defined and TAG# [7::5] are required to equal 000b.

- If the Extended Tag Enable bit equals "1" (configuration address space) all TAG# [7::0] bits are defined.

LAST BE [3::0]: LAST Byte Enables. LAST BE [3] refers to the highest order byte of the highest order DWORD (highest address, which is the last DWORD) of the Data field.

- Defines the enabled bytes of the last DWORD (highest order) in the Data field of the transaction packet. BE0 = byte 0 of the DWORD, BE1 = byte 1 of DWORD, and so on.

- The associated byte is read if BE = 1b. The associated byte is not read if BE = 0b.

- The BE pattern is not required to read contiguous bytes.

- Memory requester transaction packets with only one DWORD (defined by the LENGTH field) must use the FIRST BE field, and the LAST BE field must = 0000b.

- Memory requester transaction packets with two or more DWORDs (defined by the LENGTH field) require the LAST BE field to not equal 0000b.

- For memory read requester transactions the associated byte can optionally be pre-read by the completer sources if within a pre-fetchable address and if the associated BE=0b.

- For memory read requester transactions the associated byte must not be pre-read by the completer sources if not within a pre-fetchable address and if the associated BE = 0b.

FIRST BE [3::0]: FIRST Byte Enables. FIRST BE [3] refers to the highest order byte of the lowest order (lowest address, which is first DWORD) DWORD of Data field.

- Defines the enables bytes of the first DWORD (lowest address order) in the Data field of transaction packets. BE0= byte 0 of the DWORD, BE1= byte 1 of DWORD, and so on.

- The associated byte is read if BE = 1b. The associated byte is not read if BE = 0b.

- The BE pattern is not required to read contiguous bytes.

- Memory transaction packets with only one DWORD (defined by the LENGTH field) must use the FIRST BE.

- Memory requester transaction packets with two or more DWORDs (defined by the LENGTH field) require the FIRST BE not be equal to 0000b.

- If the memory read requester transaction packet has LENGTH = one DWORD and FIRST BE = 0000b, the associated completer transaction packet must have a LENGTH = one DWORD and provide one DWORD of data. This data has no meaning, the associated bytes were not read at the requester destination, and can be any value.
- For memory read requester transaction packets the associated byte can optionally be pre-read by the PCI Express completer source if within a prefetchable address and if the associated BE = 0b.
- For memory read requester transaction packets the associated byte must not be pre-read by the PCI Express completer source if not within a prefetchable address and if the associated BE=0.

ADDRESS HDW

ADDRESS [31::0]: Address to access a DWORD (32 data bit resolution) in memory address space. ADDRESS [1::0] are RESERVED.

ADDRESS [63::0]: Address to access a DWORD (32 data bit resolution) in memory address space. ADDRESS [1::0] are RESERVED.

MEMORY READ COMPLETER TRANSACTION

7	6	5	4	3	2	1	0	7	6	5	4	3	2	1	0
R	FMT 1 0		4 TYPE 1 T					R	2 TC 0			R	R	R	R
TD	EP	ATTRI 1 0		R	R	9		LENGTH							0
7	BUS NUMBER						0	4	DEVICE #				0	2 FUNC# 0	
STATUS 2 0		BCM	11					BYTE COUNT						00	
7	BUS NUMBER						0	4	DEVICE #				0	2 FUNC# 0	
7	TAG #						0	R	6	LOWER ADDRESS					0

15	TYPE HDW	00
15	ATTRIBUTE HDW	00
15	COMPLETER ID HDW	00
15	STATUS HDW	00
15	REQUESTER ID HDW	00
15	TAG HDW	00

MEMORY READ COMPLETER TRANSACTION HEADER FIELD with data without LOCK: CplD ... with LOCK:CplDLk without data and without LOCK: Cpl

TYPE HDW

Completer with data (successful read STATUS [2::0] = 000b)

FMT [1:0]: Format of Header field. FMT [0] = 0b for 3 DWORD versus FMT [0] = 1b for 4 DWORD in Header Field. FMT [1] = 0b no Data field in packet versus FMT [1] = 1b for Data field in packet.

- FMT [1::0] = 10b.

TYPE [4::0]: Type of transaction associated with packet and lock.

- TYPE [0] = 1b LOCK function related transaction, 0b no LOCK function. Only defined for lock memory read requester or lock completer transaction. See Chapter 13 for information on the LOCK function.
- TYPE [4::1] = 0101b.

TC [2::0]: Traffic Class. The TC [2::0] for completer transaction packets must be the same as the associated requester transaction packets.

- TC [2::0] = 000b to 111b.
- TC [2::0] = 000b if the completer transaction packet is for a lock memory read requester transaction packet.

Completer without data (successful read STATUS [2::0] not = 000b)

FMT [1:0]: Format of Header field. FMT [0] = 0b for 3 DWORD versus FMT [0] = 1b for 4 DWORD in Header Field. FMT [1] = 0b no Data field, no packet versus FMT [1] = 1b for Data field in packet.

- FMT [1::0] = 00b.

TYPE [4::0]: Type of transaction associated with packet and lock.

- TYPE [0] = 1b LOCK function related transaction, 0b no LOCK function. Only defined for lock memory read requester or lock completer transactions. For the other transactions this bit is ignored.
- TYPE [4::1] = 0101b.

TC [2::0]: Traffic Class. The TC [2::0] for completer transaction packets must be the same as the associated requester transaction packets.

- TC [2::0] = 000b to 111b.
- TC [2::0] = 000b if a completer transaction for a lock memory read requester.

ATTRIBUTE HDW

TD: Transaction Digest defines whether the transaction packet contains the Digest DWORD. If EP = 0b, the Digest DWORD contains a valid ECRC value. If EP = 1b, the packet is poisoned and ECRC is irrelevant.

- TD = 0b No Digest DWORD is attached.
- TD = 1b Digest DWORD is attached.

EP: Error and Poisoning is optional and only defined for write requester transaction and read completer transaction packets.

- If EP = 0b there is no error forwarding. (See Chapter 9.)
- If EP = 1b, there is error forwarding. (See Chapter 9.)

ATTRI [1:0]: ATTRIBUTES field defines snoop and ordering information. The ATTRI [1::0] for completer transaction packets must be as same as the associated requester transaction packets.

- ATTRI [1] = (OR) 1b defines the PCI-X relaxed ordering for the packet.

- ATTRI [1] = (OR) 0b defines the PCI strong ordering for the packet.

- ATTRI [0] = (SN) 1b specifies that cache coherency is not supported.

- ATTRI [0] = (SN) 0b specifies that cache coherency is supported.

LENGTH [9::0]: LENGTH in DWORDs of Data field included in packet. The data is transferred in DWORD increments (4 bytes = 32 data bits total) naturally address aligned. The LENGTH [9::0] field is the number of DWORDs in the Data field of the packet.

- For completer transaction packets, the Data field length is encoded into ten binary bits [9::0]. For example, LENGTH [9::0] = 00 0000 0001b is one DWORD, 00 0000 0010 is two DWORDs, 11 1111 1111b is 1023 DWORDs, and so on. For completer transaction packets with no Data field (STATUS [2::0] NOT = 000b), the LENGTH [9::0] = RESERVED and must equal 00 0000 0000b.

- The maximum number is 1024 and the LENGTH [9::0] = 00 0000 0000b. The existence of the Data field in the packet is defined in the FMT [1] field; consequently, the definition of NO data bytes and NO Data field in the packet is not from information of the LENGTH [9::0] field but from the information of the FMT [1] field.

- If the read requester transaction packet has LENGTH [9::0] = one DWORD and FIRST BE = 0000b, the associated completer transaction packet must have a LENGTH [9::0] = one DWORD and provide one DWORD of data. This data has no meaning, the associated bytes were not read at the requester destination, and can be any value.

COMPLETER ID HDW

BUS# [7::0]: BUS NUMBER identifies the link or the virtual PCI bus segment in a Root Complex, endpoint, or bridge. (See Chapter 4.)

DEVICE# [4::0]: DEVICE NUMBER identifies the device on the link or on the virtual PCI bus segment in the Root Complex, endpoint, or bridge. (See Chapter 4.)

FUNC# [2:0]: FUNCTION NUMBER identifies one the eight functions within a device. (See Chapter 4.)

STATUS HDW

STATUS [2::0]: Completer status defines the result of the associated requester transaction as follows. (See Chapter 9):

- 000b Successful Completion (SC)
- 001b Unsupported Request (UR)
- 010b Configuration Request Retry Status (CRS)
- 100b Completer Abort (CA)
- All other values are RESERVED

BCM: byte count modified. The BCM = 0b if the completer source is the Root Complex, a PCI Express/PCI Bridge, a PCI legacy endpoint (endpoint based on a PCI device core), or a PCI Express endpoint. BCM = 0b or 1b if the completer source is a PCI-X legacy endpoint (endpoint based on a PCI-X device core) or PCI Express/PCI-X bridge. See Chapter 5 for more information on requester transactions.

BYTE COUNT [11::00]

Each memory read requester transaction packet is responded to by a set of completer transaction packets (set = 1 or more). For each completer transaction packet received with STATUS [2::0] = 000b (successful) the BYTE COUNT [11::00] field is as follows (the bytes remaining include those bytes in the present completer transaction packet and subsequent complete transaction packets of the set):

Byte remaining = 1 byte, Byte Count [11::00] = 0000 0000 0001b.
Bytes remaining = 2 bytes, Byte Count [11::00] = 0000 0000 0010b.
And so forth...
Bytes remaining = 4095 bytes, Byte Count [11::00] = 1111 1111 1111b.
Bytes remaining = 4096 bytes, Byte Count [11::00] = 0000 0000 0000b.

If any completer transaction packet of the set of completer transaction packets returned for a single memory read requester transaction packet is not successful (STATUS [2::0] not = 000b), no further completer transaction packets of the set are transmitted. The unsuccessful completer transaction packet will have no Data field (Cpl or CplLK instead of CplD or CplDLK), *but* the BYTE COUNT [11::0] and LOWER ADDRESS [6::0] fields contain the same values as if the completer transaction packet were successful. See Chapter 5 for more information.

REQUESTER ID HDW

BUS# [7::0]: BUS NUMBER identifies the link or the virtual PCI bus segment in a Root Complex, endpoint, or bridge. (See Chapter 4.)

DEVICE# [4::0]: DEVICE NUMBER identifies the device on the link or on the virtual PCI bus segment in the Root Complex, endpoint, or bridge. (See Chapter 4.)

FUNC.# [2:0]: FUNCTION NUMBER identifies one the eight functions within a device bridge. (See Chapter 4.)

TAG HDW

TAG# [7::0]: A unique TAG# is assigned by the PCI Express requester source to each of its requester transaction packets. Each requester transaction packet is assigned a TAG# to permit the requester source to identify the associated completer transaction received. The completer source must retain and use the TAG# received as the requester destination in the completer transaction packets. See Chapter 5 for more information.

- If the Extended Tag Enable bit equals "0" (configuration address space) only TAG# [4::0] are defined and TAG# [7::5] are required to equal 000b.

- If the Extended Tag Enable bit equals "1" (configuration address space) all TAG# [7::0] bits are defined.

LOWER ADDRESS [6::0]: Defined for completer transaction packets associated with memory read requester transactions. It defines the byte address for the first byte returned (i.e. first enabled BE [3::0]) of the data in the Data field of the packet when STATUS [2::0] = 000b (successful). The LOWER ADDRESS [1::0] listed below for the first completer transaction packet becomes 00b for the subsequent completer transaction packet of the set of completer transaction packets returned for a single memory read requester transaction.

- LOWER ADDRESS [1::0] must equal 00b if FIRST BE equals 0000b or xxx1b.

- LOWER ADDRESS [1::0] must equal 01b if FIRST BE equals xx10b.

- LOWER ADDRESS [1::0] must equal 10b if FIRST BE equals x100xb.

- LOWER ADDRESS [1::0] must equal 11b if FIRST BE equals 1000b.

- LOWER ADDRESS [6::2] must equal the ADDRESS [6::2] of the associated memory read requester transaction for the first completer transaction packet. Otherwise LOWER ADDRESS [6::2] must equal 00000b.

- If the source of the completer transaction packet is the Root Complex with an RCB value of 64B, the LOWER ADDRESS field of the first completer transaction packet of the set of completer transaction packets is defined as above. For subsequent completer transaction packets of the set of completer transaction packets, the LOWER ADDRESS [5::0] = 000000b and LOWER ADDRESS [6] toggles according to the alignment of the 64B in the Data field.

- If any completer transaction packet of the set of completer transaction packets returned for a single memory read requester transaction is not successful (STATUS [2::0] not = 000b), no further completer transaction of the set are transmitted. The unsuccessful completer transaction packet will have no Data field (Cpl or CplLK instead of CplD or CplDLK), but the BYTE COUNT [11::0] and LOWER ADDRESS [6::0] fields contain the same values as if the completer transaction packet was successful. See Chapter 5 for more information.

> See Chapter 5 for more information on the relationship between LOWER ADDRESS [6::0], LENGTH [9::0] and BYTE COUNT [11::0].

Header Field Formats for Memory Write Transactions

Memory write requester transaction packets consist of Header, Data, and (optional) Digest fields and comes in two formats: 32 address bits shown in Figure 6.6 and 64 address bits shown in Figure 6.7. No completer transaction packet are associated with to these write requester transaction packets.

> If the data to be transferred is in the first 4 gigabytes of the memory address space, the 32 address bit format *must be* used. If the 64 address bit format is used, it is for accessing memory resources above the 4-gigabyte address boundary. In either situation, the full address must be provided.

Chapter 6: Transaction Layer ■ **239**

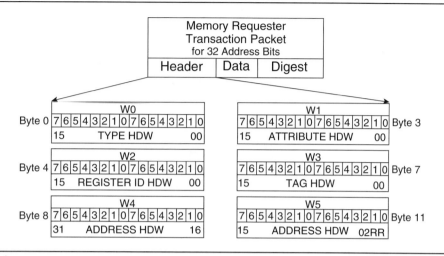

Figure 6.6 Memory Requester Transaction Packet for 32 Address Bits

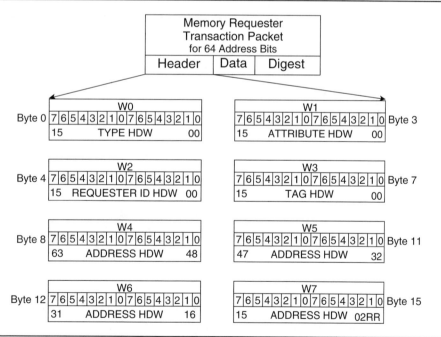

Figure 6.7 Memory Requester Transaction Packet for 64 Address Bits

MEMORY WRITE REQUESTER TRANSACTION

7	6	5	4	3	2	1	0	7	6	5	4	3	2	1	0
R	FMT 1 0		4 TYPE 1 L					R	2 TC 0			R	R	R	R
T D	E P	ATTRI 1 0		R	R	9		LENGTH							0
7			BUS NUMBER				0	4			DEVICE #		0	2 FUNC# 0	
7			TAG				0	3			LAST BE		0	3 FIRST BE 0	
31							ADDRESS								16
15							ADDRESS						02	R	R

64 Address bits

7	6	5	4	3	2	1	0	7	6	5	4	3	2	1	0
63							ADDRESS								48
47							ADDRESS								32
31							ADDRESS								16
15							ADDRESS						02	R	R

15	TYPE HDW	00
15	ATTRIBUTE HDW	00
15	REQUESTER ID HDW	00
15	TAG HDW	00
	ADDRESS HDW	

MEMORY WRITE REQUESTER TRANSACTION HEADER FIELD MWr

TYPE HDW

With 32 Address bits

FMT [1:0]: Format of Header field. FMT [0] = 0b for 3 DWORD versus FMT [0] = 1b for 4 DWORD in Header Field. FMT [1] = 0b no Data field in packet versus FMT [1] = 1b for Data field in packet.

- FMT [1::0] = 10b.

TYPE [4::0]: Type of transaction associated with packet.

- TYPE [0] = 0b no LOCK function. Only defined for lock memory read requester or lock completer transaction packet.
- TYPE [4::1] = 0000b.

TC [2::0]: Traffic Class

- TC [2::0] = 000b to 111b.
- If memory write requester associated with LOCK function TC [2::0] = 000b.
- Memory write requester transactions used for MSI does not require 000b. TC [2::0] can be any value.

With 64 Address bits

FMT [1:0]: Format of Header field. FMT [0] = 0b for 3 DWORD versus FMT [0] = 1b for 4 DWORD in Header Field. FMT [1] = 0b no Data field, no packet versus FMT [1] = 1b for Data field in packet.

- FMT [1::0] = 11b.

TYPE [4::0]: Type of transaction associated with packet and lock.

- TYPE [0] = 0b no LOCK function. Only defined for lock memory read requester or lock completer transaction.
- TYPE [4::1] = 0000b.

TC [2::0]: Traffic Class

- TC [2::0] = 000b to 111b.
- If memory write requester associated with LOCK function TC [2::0] = 000b.
- Memory write requester transactions used for MSI does not require 000b. TC [2::0] can be any value.

ATTRIBUTE HDW

TD: Transaction Digest defines whether the transaction packet contains the Digest DWORD. If EP = 0b, the Digest DWORD contains a valid ECRC value. If EP = 1b, the packet is poisoned and ECRC is irrelevant. (See Chapter 9.)

- TD = 0 No Digest DWORD is attached.
- TD = 1 Digest DWORD is attached.

EP: Error and Poisoning is optional and only defined for write requester transaction and read completer transaction packets.

- If EP = 0b there is no error forwarding. (See Chapter 9.)
- If EP = 1b, there is error forwarding. (See Chapter 9.)

ATTRI [1:0]: ATTRIBUTES field defines snoop and ordering information.

- ATTRI [1] = (OR) 1b defines the PCI-X relaxed ordering for the packet. (See Chapter 10.)
- ATTRI [1] = (OR) 0b defines the PCI strong ordering for the packet. (See Chapter 10.)
- ATTRI [0] = (SN) 1b indicates that cache coherency is not supported. (See Chapter 5.)
- ATTRI [0] = (SN) 0b indicates that cache coherency is supported. (See Chapter 5.)
- ATTRI [1:0] = 00b if the memory write requester transaction packet is used for a message signaled interrupt.

LENGTH [9::0]: LENGTH in DWORDs to be written. The data is transferred in DWORD increments (4 bytes = 32 data bits total) naturally address aligned. The LENGTH [9::0] field is the number of DWORDs in the Data field of the packet.

- For memory requester transaction packets, the data length is encoded into ten binary bits [9::0]. For example, 00 0000 0001b is one DWORD, 00 0000 0010 is two DWORDs, 11 1111 1111b) is 1023 DWORDs, and so on.

- The maximum number is 1024 and the LENGTH [9::0] = 00 0000 0000b. The existence of the Data field is defined in the FMT [1] field; consequently the definition of NO Data field in the packet is not from information of the LENGTH [9::0] field but from the information of the FMT [1] field.

REQUESTER HDW

BUS# [7::0]: BUS NUMBER identifies the link or the virtual PCI bus segment in a Root Complex, switch, or bridge. (See Chapter 4.)

DEVICE# [4::0]: DEVICE NUMBER identifies the device on the link or on the virtual PCI bus segment in the Root Complex, switch, or bridge. (See Chapter 4.)

FUNC# [2:0]: FUNCTION NUMBER identifies one the eight functions within a device. (See Chapter 4.)

TAG HDW

TAG# [7::0]: A unique TAG# is assigned by the PCI Express requester source to each of its requester transaction packets. Each requester transaction packet is assigned a TAG# to permit the requester source to identify the associated completer transaction received. The completer source must retain and use the TAG# received as the requester destination in the completer transaction packets. (See Chapter 4.)

- The TAG# [7::0] for a memory write requester transaction packets (i.e. no associated completer transaction packets) can be any value.

LAST BE [3::0]: LAST Byte Enables. LAST BE [3] refers to the highest order byte of the highest order (highest address that is last DWORD of Data field) DWORD.

- Defines the enabled bytes of the last DWORD (highest order) in the Data field of the transaction packet. BE0 = byte 0 of the DWORD, BE1 = byte 1 of DWORD, and so on.

- Associated byte is written BE = 1b. Associated byte is not written BE = 0b.

- The BE pattern is not required to write contiguous bytes.

- Memory write requester transaction packets with only one DWORD (defined by the LENGTH field) must use the FIRST BE field and the LAST BE field must equal 0000b.

- Memory write requester transaction packets with two or more DWORDs (defined by the LENGTH field) requires the LAST BE field to not = 0000b.

FIRST BE [3::0]: FIRST Byte Enables. FIRST BE[3] refers to the highest order byte of the lowest order lowest address, which is first the DWORD of data packet) DWORD.

- Defines the enables bytes of the first DWORD (lowest address order) in the data field of transaction packets. BE0= byte 0 of the DWORD, BE1= byte 1 of DWORD, and so on.

- Associated byte is written if BE = 1b.

- Associated byte is not written if BE = 0b.

- The BE pattern is not required to write contiguous bytes.

- Memory write requester transaction packets with only one DWORD (defined by the LENGTH field) must use the FIRST BE.

- Memory write requester transaction packets with two or more DWORDs (defined by the LENGTH field) require the FIRST BE not = 0000b.

- If the memory write requester transaction packet has LENGTH = one DWORD and FIRST BE = 0000b; the associated data has no meaning, can be any value, and is not written at the requester destination.

> See Chapter 5 for more information on the relationship between LAST BE [3::0], FIRST BE [3::0], and LENGTH [9::0].

ADDRESS HDW

ADDRESS [31::0]: Address to access a DWORD (32 data bit resolution) in memory address space. ADDRESS [1::0] are RESERVED.

ADDRESS [63::0]: Address to access a DWORD (32 data bit resolution) in memory address space. ADDRESS [1::0] are RESERVED.

Header Field Formats for I/O Read Transactions

I/O read requester transaction packets consist of Header and (optional) Digest fields and come in one format: 32 address bits, as shown in Figure 6.8.

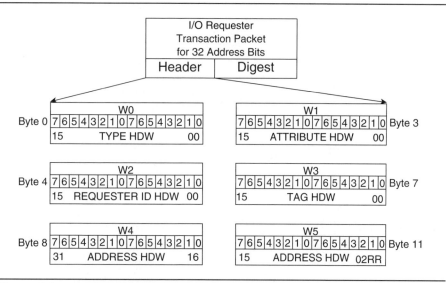

Figure. 6.8 I/O Read Requester Transaction Packet

The Repeater/Completer protocol dictates that the endpoint or bridge that is being read responds with completer transactions. If the data read was successful (STATUS [2::0] = 000b), the completer transaction packet contains data (in the Data field), as shown in Figure 6.9. If the data read is not successful (STATUS [2::0] not = 000b), no data is contained in the packet (no Data field), as shown in Figure 6.10.

246 ■ The Complete PCI Express Reference

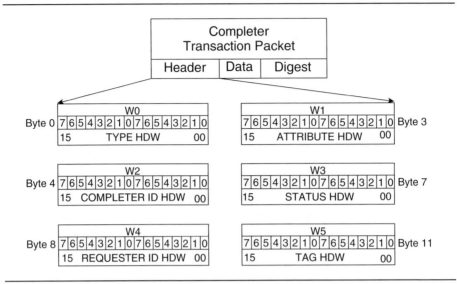

Figure 6.9 Completer Transaction Packet (with Data)

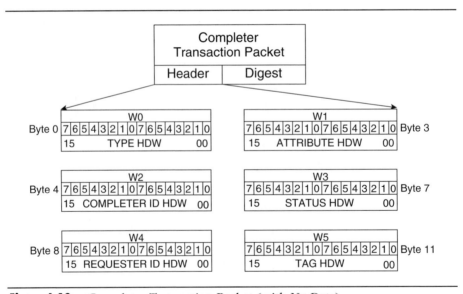

Figure 6.10 Completer Transaction Packet (with No Data)

I/O READ REQUESTER TRANSACTION

7	6	5	4	3	2	1	0	7	6	5	4	3	2	1	0
R	FMT 1 0		4 TYPE 1 L					R	2 TC 0			R	R	R	R
TD	EP	ATTRI 1 0		R	R	9		LENGTH							0
7	BUS NUMBER						0	4	DEVICE #			0	2 FUNC# 0		
	7 TAG 0							3	LAST BE			0	3	FIRST BE	0
31					ADDRESS										16
15					ADDRESS								02	R	R

15	TYPE HDW	00
15	ATTRIBUTE HDW	00
15	REQUESTER ID HDW	00
15	TAG HDW	00
31	ADDRESS HDW	16
15	ADDRESS HDW	02 R R

I/O READ REQUESTER TRANSACTION HEADER FIELD IORd

TYPE HDW

FMT [1:0]: Format of Header field. FMT [0] = 0b for 3 DWORD versus FMT [0] = 1b for 4 DWORD in Header Field. FMT [1] = 0b no Data field in packet versus FMT [1] = 1b for Data field in packet.

- FMT [1::0] = 00b.

TYPE [4::0]: Type of transaction associated with packet and lock.

- TYPE [0] = 0b no LOCK function. Only defined for lock memory read requester or lock completer transaction.
- TYPE [4::1] = 0001b.

TC [2::0]: Traffic Class

- TC [2::0] = 000b.

ATTRIBUTE HDW

TD: Transaction Digest defines whether the transaction packet contains the Digest DWORD. If EP = 0b, the Digest DWORD contains a valid ECRC value. If EP = 1b, the packet is poisoned and ECRC is irrelevant. (See Chapter 9.)

- TD = 0 No Digest DWORD is attached.
- TD = 1 Digest DWORD is attached.

EP: Error and Poisoning is optional and only defined for write requester transaction and read completer transaction packets. (See Chapter 9.)

- EP = 0b.

ATTRI [1:0]: ATTRIBUTES field defines snoop and ordering information. ATTRI [1:0] = 00b.

- ATTRI [1] = (OR) 0b defines the PCI strong ordering for the packet. (See Chapter 10.)
- ATTRI [0] = (SN) 0b indicates that cache coherency is supported. (See Chapter 5.)

LENGTH [9::0]: LENGTH in DWORDs to be read. The data is transferred in DWORD increments (4 bytes = 32 data bits total) naturally address aligned. For read requester transactions packets, the LENGTH [9::0] field is the number of DWORDs of data to be read. Read requester transaction packets do not contain a Data field in the packet.

- I/O read requester transaction packets are only defined to have a data length of 1 DWORD, the LENGTH [9::0] = 00 0000 0001b.

- If I/O read requester transaction packet has LENGTH = one DWORD and FIRST BE [3::0] = 0000b, the associated completer transaction packet must have a LENGTH = one DWORD and provide one DWORD of data. This data has no meaning, the associated bytes were not read at the requester destination, and can be any value.

REQUESTER HDW

BUS# [7::0]: BUS NUMBER identifies the link or the virtual PCI bus segment in a Root Complex, switch, or bridge. (See Chapter 4.)

DEVICE# [4::0]: DEVICE NUMBER identifies the device on the link or on the virtual PCI bus segment in the Root Complex, switch, or bridge. (See Chapter 4.)

FUNC# [2:0]: FUNCTION NUMBER identifies one the eight functions within a device. (See Chapter 4.)

TAG HDW

TAG# [7::0]: A unique TAG# is assigned by the PCI Express requester source to each of its requester transaction packets. Each requester transaction packet is assigned a TAG# to permit the requester source to identify the associated completer transaction received. The completer source must retain and use the TAG# received as the requester destination in the completer transaction packets. (See Chapter 4.)

- If the Extended Tag Enable bit equals "0" (configuration address space) only TAG# [4::0] are defined and TAG# [7::5] are required to equal 000b.
- If the Extended Tag Enable bit equals "1" (configuration address space) all TAG# [7::0] bits are defined.

LAST BE [3::0]: For I/O requester transaction packets LAST BE [3::0] = 0000b.

FIRST BE [3::0]: FIRST Byte Enables. FIRST BE [3] refers to the highest order byte of the lowest order address, (which is the first DWORD of data packet) DWORD.

- Defines the enables bytes of the first DWORD (lowest address order) in the data field of transaction packets. BE0= byte 0 of the DWORD, BE1= byte 1 of DWORD, and so on.

- Associated byte is read if BE = 1b. Associated byte is not read if BE = 0b.
- The BE pattern is not required to read contiguous bytes.
- I/O requester transaction packets only contain one DWORD (defined by the LENGTH field) and require the FIRST BE field to be used.
- If the I/O read requester transaction packet has LENGTH = one DWORD and FIRST BE = 0000b, the associated completer transaction packet must have a LENGTH = one DWORD and provide one DWORD of data. This data has no meaning, the associated bytes were not read at the requester destination, and can be any value.

See Chapter 5 for more information on the relationship between FIRST BE [3::0] and LENGTH [9::0].

ADDRESS HDW

ADDRESS [31::0]: Address to access a DWORD (32 data bit resolution) in memory address space. ADDRESS [1::0] are RESERVED.

I/O READ COMPLETER TRANSACTION

7	6	5	4	3	2	1	0	7	6	5	4	3	2	1	0
R	\multicolumn{2}{c}{FMT 1 0}		\multicolumn{4}{c}{4 TYPE 1 L}	R	\multicolumn{3}{c}{2 TC 0}	R	R	R	R
T D	E P	\multicolumn{2}{c}{ATTRI 1 0}	R	R	9		\multicolumn{8}{c}{LENGTH 0}
7	\multicolumn{6}{c}{BUS NUMBER}	0	4	\multicolumn{3}{c}{DEVICE #}	0	2	FUNC#	0
STATUS 2 0				B C M	11			\multicolumn{6}{c}{BYTE COUNT}		0 0
7	\multicolumn{6}{c}{BUS NUMBER}	0	4	\multicolumn{3}{c}{DEVICE #}	0	2	FUNC#	0
7	\multicolumn{6}{c}{TAG #}	0	R	6	\multicolumn{4}{c}{LOWER ADDRESS}			0

15	TYPE HDW	00
15	ATTRIBUTE HDW	00
15	COMPLETER ID HDW	00
15	STATUS HDW	00
15	REQUESTER ID HDW	00
15	TAG HDW	00

I/O READ COMPLETER TRANSACTION HEADER FIELD with data: ClpD ...without data: Cpl

TYPE HDW

Completer with data (successful read STATUS [2::0] = 000b)

FMT [1:0]: Format of Header field. FMT [0] = 0b for 3 DWORD versus FMT [0] = 1b for 4 DWORD in Header Field. FMT [1] = 0b no Data field, no packet versus FMT [1] = 1b for Data field in packet.

- FMT [1::0] = 10b.

TYPE [4::0]: Type of transaction associated with packet and lock.

- TYPE [0] = 0b no LOCK function.
- TYPE [4::1] = 0101b.

TC [2::0]: Traffic Class. The TC [2::0] for completer transaction packets must be the same as the associated requester transaction packets.

- TC [2::0] = 000b.

Completer without data (successful read STATUS [2::0] not = 000b)

FMT [1:0]: Format of Header field. FMT [0] = 0b for 3 DWORD versus FMT [0] = 1b for 4 DWORD in Header Field. FMT [1] = 0b no Data field, no packet versus FMT [1] = 1b for Data field in packet.

- FMT [1::0] = 00b.

TYPE [4::0]: Type of transaction associated with packet and lock.

- TYPE [0] = 0b no LOCK function.
- TYPE [4::1] = 0101b.

TC [2::0]: Traffic Class. The TC [2::0] for completer transaction packets must be the same as the associated requester transaction packets.

- TC [2::0] = 000b.

ATTRIBUTE HDW

TD: Transaction Digest defines whether the transaction packet contains the Digest DWORD. If EP = 0b, the Digest DWORD contains a valid ECRC value. If EP = 1b, the packet is poisoned and ECRC is irrelevant. (See Chapter 9.)

- TD = 0b No Digest DWORD is attached.
- TD = 1b Digest DWORD is attached.

EP: Error and Poisoning is optional and only defined for write requester transaction and read completer transaction packets.

- If EP = 0b there is no error forwarding. (See Chapter 9.)
- If EP = 1b, there is error forwarding. (See Chapter 9.)

ATTRI [1:0]: ATTRIBUTES field defines snoop and ordering information. The ATTRI [1::0] for completer transaction packets must be the same as the associated requester transaction packets, thus ATTRI [1:0] = 00b.

- ATTRI [1] = (OR) 0b defines the PCI strong ordering for the packet. (See Chapter 10.)
- ATTRI [0] = (SN) 0b indicates that cache coherency is supported. (See Chapter 5.)

LENGTH [9::0]: LENGTH in DWORDs of Data field included in packet. The data is transferred in DWORD increments (4 bytes = 32 data bits total) naturally address aligned.

- For completer transaction packets associated with I/O read transactions, the one DWORD is retuned LENGTH [9::0] = 00 0000 0001b (STATUS [2::0] = 000b). For completer transaction packets with no Data field (STATUS [2::0] NOT = 000b), the LENGTH [9::0] is RESERVED and must = 00 0000 0000b.

- The maximum number is 1024 and the LENGTH [9::0] = 00 0000 0000b. The existence of the Data field in the packet is defined in the FMT [1] field; consequently, the definition of NO data bytes and NO Data field in the packet is not from information of the LENGTH [9::0] field but from the information of the FMT [1] field.

- If the I/O read requester transaction packet has LENGTH [9::0] = one DWORD and FIRST BE [3::0] = 0000b, the associated completer transaction packet must have a LENGTH [9::0] = one DWORD and provide one DWORD of data. This data has no meaning, the associated bytes were not read at the requester destination, and can be any value.

COMPLETER ID HDW

BUS# [7::0]: BUS NUMBER identifies the link or the virtual PCI bus segment in a endpoint or bridge. (See Chapter 4.)

DEVICE# [4::0]: DEVICE NUMBER identifies the device on the link or on the virtual PCI bus segment in the endpoint or bridge. (See Chapter 4.)

FUNC# [2:0]: FUNCTION NUMBER identifies one the eight functions within a device.

STATUS HDW

STATUS [2::0]: Completer transaction status defines the result of the associated requester transaction as follows:

- 000b Successful Completion (SC)
- 001b Unsupported Request (UR)
- 010b Configuration Request Retry Status (CRS)
- 100b Completer Abort (CA)
- All other values are RESERVED

BCM: byte count modified. The BCM = 0b. Not defined for this type of completer transaction.

BYTE COUNT [11::00]

Each I/O read requester transaction packet is responded to with a single completer transaction packet. For I/O read requester transactions the associated completer transaction packets contain one DWORD in the Data field, the BYTE COUNT [11:00] must = 0000 0000 0100b when STATUS [2::0] = 000b (successful). The unsuccessful completer transaction packet will have no Data field (Cpl instead of CplD), but the BYTE COUNT [11::0] and LOWER ADDRESS [6::0] fields contain the same values as if the completer transaction packet was successful.

REQUESTER ID HDW

BUS# [7::0]: BUS NUMBER identifies the link or the virtual PCI bus segment in a Root Complex, endpoint, or bridge. (See Chapter 4.)

DEVICE# [4::0]: DEVICE NUMBER identifies the device on the link or on the virtual PCI bus segment in the Root Complex, endpoint, or bridge. (See Chapter 4.)

FUNC# [2:0]: FUNCTION NUMBER identifies one the eight functions within a device. (See Chapter 4.)

TAG HDW

TAG# [7::0]: A unique TAG# is assigned by the PCI Express requester source to each of its requester transaction packets. Each requester transaction packet is assigned a TAG# to permit the requester source to identify the associated completer transaction received. The completer source must retain and use the TAG# received as the requester destination in the completer transaction packets. (See Chapter 4.)

- If the Extended Tag Enable bit equals "0" (configuration address space) only TAG# [4::0] are defined and TAG# [7::5] are required to equal 000b.

- If the Extended Tag Enable bit equals "1" (configuration address space) all TAG# [7::0] bits are defined.

LOWER ADDRESS [6::0]: LOWER ADDRESS [6::0] must = 000 0000b for completer transaction packets not associated with memory read requester transactions. The unsuccessful completer transaction packet will have no Data field (Cpl instead of CplD), but the BYTE COUNT [11::0] and LOWER ADDRESS [6::0] fields contain the same values as if the completer transaction packet was successful.

> See Chapter 5 for more information on the relationship between LOWER ADDRESS [6::0], LENGTH [9::0] and BYTE COUNT [11::0].

Header Field Formats for I/O Write Transactions

I/O write requester transaction packets consists of Header, Data, and (optional) Digest fields and come in one format: 32 address bits, shown in Figure 6.11.

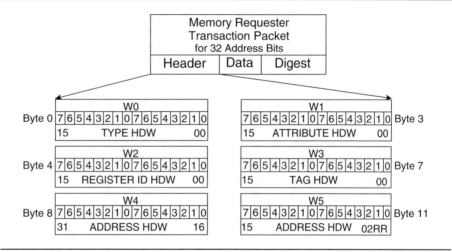

Figure 6.11   I/O Write Requester Transaction Packet

The Repeater/Completer protocol dictates that the endpoint or bridge that is being written to responds with completer transaction packets, as shown in Figure 6.12.

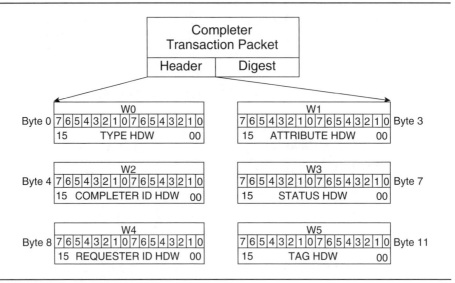

Figure 6.12 Completer Transaction Packet

I/O WRITE REQUESTER TRANSACTION

7	6	5	4	3	2	1	0	7	6	5	4	3	2	1	0
R	\multicolumn{2}{c}{FMT 1 0}	\multicolumn{5}{c}{4 TYPE 1 L}	R	\multicolumn{3}{c}{2 TC 0}	R	R	R	R
\multicolumn{2}{c}{TD}	\multicolumn{1}{c}{EP}	\multicolumn{2}{c}{ATTRI 1 0}	R	R	9	\multicolumn{7}{c}{LENGTH}	0
7	\multicolumn{6}{c}{BUS NUMBER}	0	4	\multicolumn{3}{c}{DEVICE #}	0	2	\multicolumn{2}{c}{FUNC#}	0
7	\multicolumn{6}{c}{TAG}	0	3	\multicolumn{2}{c}{LAST BE}	0	3	\multicolumn{2}{c}{FIRST BE}	0
31	\multicolumn{14}{c}{ADDRESS}	16
15	\multicolumn{12}{c}{ADDRESS}	02	R	R

15	TYPE HDW	00
15	ATTRIBUTE HDW	00
15	REQUESTER ID HDW	00
15	TAG HDW	00
31	ADDRESS HDW	16
15	ADDRESS HDW	02 R R

I/O WRITE REQUESTER TRANSACTION HEADER FIELD IOWr

TYPE HDW

FMT [1:0]: Format of Header field. FMT [0] = 0b for 3 DWORD versus FMT [0] = 1b for 4 DWORD in Header Field. FMT [1] = 0b no Data field in packet versus FMT [1] = 1b for Data field in packet.

- FMT [1::0] = 10b.

TYPE [4::0]: Type of transaction associated with packet and lock.

- TYPE [0] = 0b no LOCK function. Only defined for lock memory read requester or lock completer transaction.
- TYPE [4::1] = 0001b.

TC [2::0]: Traffic Class

- TC [2::0] = 000b.

ATTRIBUTE HDW

TD: Transaction Digest defines whether the transaction packet contains the Digest DWORD. If EP = 0b, the Digest DWORD contains a valid ECRC value. If EP = 1b, the packet is poisoned and ECRC is irrelevant. (See Chapter 9.)

- TD = 0 No Digest DWORD is attached.
- TD = 1 Digest DWORD is attached.

EP: Error and Poisoning is optional and only defined for write requester transaction and read completer transaction packets.

- If EP = 0b there is no error forwarding. (See Chapter 9.)
- If EP = 1b, there is error forwarding. (See Chapter 9.)

ATTRI [1:0]: ATTRIBUTES field defines snoop and ordering information. The ATTRI [1::0] = 00b.

- ATTRI [1] = (OR) 0b defines the PCI strong ordering for the packet. (See Chapter 10.)
- ATTRI [0] = (SN) 0b indicates that cache coherency is supported. (See Chapter 5.)

LENGTH [9::0]: LENGTH in DWORDs to be written. The data is transferred in DWORD increments (4 bytes = 32 data bits total) naturally address aligned. For write requester transaction packets, the LENGTH [9::0] field is the number of DWORDs in the Data field of the packet.

- I/O write requester transaction packets are only defined to have a data length of 1 DWORD, the LENGTH [9::0] = 00 0000 0001b.
- If I/O write requester transaction packet has LENGTH = one DWORD and FIRST BE [3::0] = 0000b; the Data field has no meaning, the associated bytes were not written at the requester destination, and can be any value.

REQUESTER ID HDW

BUS# [7::0]: BUS NUMBER identifies the link or the virtual PCI bus segment in a Root Complex, endpoint, or bridge. (See Chapter 4.)

DEVICE# [4::0]: DEVICE NUMBER identifies the device on the link or on the virtual PCI bus segment in the Root Complex, endpoint, or bridge. (See Chapter 4.)

FUNC# [2:0]: FUNCTION NUMBER identifies one the eight functions within a device. (See Chapter 4.)

TAG HDW

TAG# [7::0]: A unique TAG# is assigned by the PCI Express requester source to each of its requester transaction packets. Each requester transaction packet is assigned a TAG# to permit the requester source to identify the associated completer transaction received. The completer source must retain and use the TAG# received as the requester destination in the completer transaction packets. (See Chapter 4.)

- If the Extended Tag Enable bit equals "0" (configuration address space) only TAG# [4::0] are defined and TAG# [7::5] are required to equal 000b.

- If the Extended Tag Enable bit equals "1" (configuration address space) all TAG# [7::0] bits are defined.

LAST BE [3::0]: For I/O requester transaction packets this field is defined as LAST BE [3::0] = 0000b.

FIRST BE [3::0]: FIRST Byte Enables. FIRST BE [3] refers to the highest order byte of the lowest order lowest address, (which is the first DWORD of data packet) DWORD.

- Defines the enables bytes of the first DWORD (lowest address order) in the data field of transaction packets. BE0= byte 0 of the DWORD, BE1= byte 1 of DWORD, and so on.

- Associated byte is written if BE = 1b. Associated byte is not written if BE=0b.

- The BE pattern is not required to write contiguous bytes.

- I/O requester transaction packets only contain one DWORD (defined by the LENGTH field) and require the FIRST BE field to be used.

- If the write requester transaction packet has LENGTH = one DWORD and FIRST BE = 0000b, the associated data has no meaning, can be any value, and is not written at the requester destination.

> See Chapter 5 for more information on the relationship between FIRST BE [3::0] and LENGTH [9::0].

ADDRESS HDW

ADDRESS [31::0]: Address to access a DWORD (32 data bit resolution) in memory address space. ADDRESS [1::0] are RESERVED.

I/O WRITE COMPLETER TRANSACTION

7	6	5	4	3	2	1	0	7	6	5	4	3	2	1	0
R	FMT 1 0		4 TYPE 1 L					R	2 TC 0			R	R	R	R
TD	EP	ATTRI 1 0		R	R	9		LENGTH							0
7	BUS NUMBER						0	4	DEVICE #			0	2	FUNC#	0
STATUS 2 0		BCM	11					BYTE COUNT							00
7	BUS NUMBER						0	4	DEVICE #			0	2	FUNC#	0
7	TAG #						0	R	6	LOWER ADDRESS					0

15	TYPE HDW	00
15	ATTRIBUTE HDW	00
15	COMPLETER ID HDW	00
15	STATUS HDW	00
15	REQUESTER ID HDW	00
15	TAG HDW	00

I/O WRITE COMPLETER TRANSACTION HEADER FIELD: Clp

TYPE HDW

FMT [1:0]: Format of Header field. FMT [0] = 0b for 3 DWORD versus FMT [0] = 1b for 4 DWORD in Header Field. FMT [1] = 0b no Data field, no packet versus FMT [1] = 1b for Data filed in packet.

■ FMT [1::0] = 00b.

TYPE [4::0]: Type of transaction associated with packet and lock.

- TYPE [0] = 0b no LOCK function.
- TYPE [4::1] = 0101b.

TC [2::0]: Traffic Class: The TC [2::0] for completer transaction packets must be the same as the associated requester transaction packets.

- TC [2::0] = 000b.

ATTRIBUTE HDW

TD: Transaction Digest defines whether the transaction packet contains the Digest DWORD. If EP = 0b, the Digest DWORD contains a valid ECRC value. If EP = 1b, the packet is poisoned and ECRC is irrelevant. (See Chapter 9.)

- TD = 0b No Digest DWORD is attached.
- TD = 1b Digest DWORD is attached.

EP: Error and Poisoning. is optional and only defined for write requester transaction and read completer transaction packets. (See Chapter 9.)

- If EP = 0b.

ATTRI [1:0]: ATTRIBUTES field defines snoop and ordering information. The ATTRI [1::0] for completer transaction packets must be the same as the associated requester transaction packets, thus ATTRI [1:0] = 00b.

- ATTRI [1] = (OR) 0b defines the PCI strong ordering for the packet. (See Chapter 10.)
- ATTRI [0] = (SN) 0b indicates that cache coherency is supported. (See Chapter 5.)

LENGTH [9::0]: LENGTH in DWORDs of Data field included in packet. The data is transferred in DWORD increments (4 bytes = 32 data bits total) naturally address aligned, the LENGTH [9::0] field is the number of DWORDs in the Data field of the packet.

- For completer transaction packets associated with I/O write transactions, the data length is encoded into ten binary bits [9::0]. Completer transaction packets have no Data field, thus the LENGTH [9::0] is RESERVED and must = 00 0000 0000b.

COMPLETER ID HDW

BUS# [7::0]: BUS NUMBER identifies the link or the virtual PCI bus segment in an endpoint or bridge. (See Chapter 4.)

DEVICE# [4::0]: DEVICE NUMBER identifies the device on the link or on the virtual PCI bus segment in an endpoint or bridge. (See Chapter 4.)

FUNC# [2:0]: FUNCTION NUMBER identifies one the eight functions within a device. (See Chapter 4.)

STATUS HDW

STATUS [2::0]: Completer status defines the result of the associated requester transaction as follows:

- 000b Successful Completion (SC)
- 001b Unsupported Request (UR)
- 010b Configuration Request Retry Status (CRS)
- 100b Completer Abort (CA)
- All other values are RESERVED

BCM: byte count modified. The BCM = 0b. Not defined for this type of completer transaction.

BYTE COUNT [11::00]

For I/O write requester transactions the associated completer transaction packets contain no Data field, the BYTE COUNT [11:00] = 0000 0000 0000b, and LOWER ADDRESS [6::0] = 000 0000b.

REQUESTER ID HDW

BUS# [7::0]: BUS NUMBER identifies the link or the virtual PCI bus segment in a Root Complex, endpoint, or bridge. (See Chapter 4.)

DEVICE# [4::0]: DEVICE NUMBER identifies the device on the link or on the virtual PCI bus segment in the Root Complex, endpoint, or bridge. (See Chapter 4.)

FUNC# [2:0]: FUNCTION NUMBER identifies one the eight functions within a device. (See Chapter 4.)

TAG HDW

TAG# [7::0]: A unique TAG# is assigned by the PCI Express requester source to each of its requester transaction packets. Each requester transaction packet is assigned a TAG# to permit the requester source to identify the associated completer transaction received. The completer source must retain and use the TAG# received as the requester destination in the completer transaction packets. (See Chapter 5.)

- If the Extended Tag Enable bit equals 1b (configuration address space) only TAG# [4::0] are defined and TAG# [7::5] are required to equal 000b.

- If the Extended Tag Enable bit equals 1b (configuration address space) all TAG# [7::0] bits are defined.

LOWER ADDRESS [6::0]: LOWER ADDRESS [6::0] must = 000 0000b for completer transaction packets not associated with memory read requester transactions. For I/O write requester transactions, the associated completer transaction packets contain no Data field, the BYTE COUNT [11::00] = 0000 0000 0000b, and LOWER ADDRESS [6::0] = 000 0000b.

Header Field Formats Configuration Read Transactions

Configuration read requester transaction packets consist of Header and (optional) Digest fields, shown in Figure 6.13.

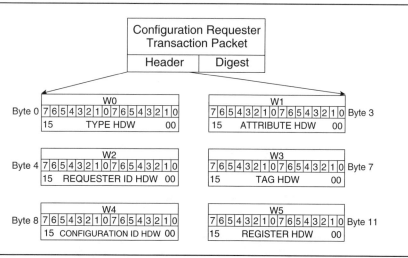

Figure. 6.13 Configuration Read Requester Transaction Packet

The Repeater/Completer protocol dictates that the endpoint, switch, or bridge that is being read responds with completer transactions. If the data read was successful (STATUS [2::0] = 000b), the completer transaction packet contains data (in the Data field), as shown in Figure 6.14. If the data read is not successful (STATUS [2::0] not = 000b), no data is contained in the packet (no Data field), as shown in Figure 6.15.

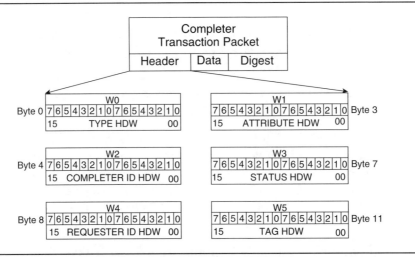

Figure 6.14   Completer Transaction Packet (with data)

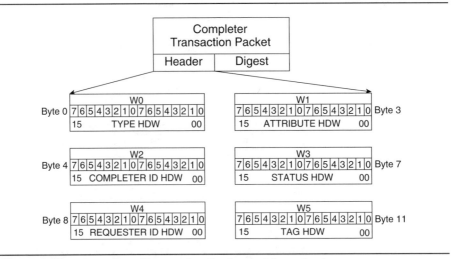

Figure 6.15   Completer Transaction Packet (with no data)

CONFIGURATION READ REQUESTER TRANSACTION

7	6	5	4	3	2	1	0	7	6	5	4	3	2	1	0
R	FMT 1 0		4 TYPE 1 T					R	2 TC 0			R	R	R	R
T D	E P	ATTRI 1 0		R	R	9		LENGTH							0
7	BUS NUMBER						0	4	DEVICE #			0	2 FUNC# 0		
7	TAG						0	3	LAST BE		0	3	FIRST BE		0
7	BUS NUMBER						0	4	DEVICE #			0	2 FUNC# 0		
R	R	R	R	EXT, REGISTER 3 ADDRESS 0				REGISTER ADDRESS 5 0						R	R

15	TYPE HDW	00
15	ATTRIBUTE HDW	00
15	REQUESTER ID HDW	00
15	TAG HDW	00
15	CONFIGURATION ID HDW	00
15	REGISTER HDW	00

CONFIGURATION READ REQUESTER TRANSACTION HEADER FIELD CfgRd

TYPE HDW

FMT [1:0]: Format of Header field. FMT [0] = 0b for 3 DWORD versus FMT [0] = 1b for 4 DWORD in Header Field. FMT [1] = 0b no Data field in packet versus FMT [1] = 1b for Data filed in packet.

- FMT [1::0] = 00b.

TYPE [4::0]: Type of transaction associated with packet and lock.

- TYPE [0] = 0b defines Type 0 versus TYPE [0] = 1b defines Type 1.
- TYPE [4::1] = 0010b.

TC [2::0]: Traffic Class

- TC [2::0] = 000b.

ATTRIBUTE HDW

TD: Transaction Digest defines whether the transaction packet contains the Digest DWORD. If EP = 0b, the Digest DWORD contains a valid ECRC value. If EP = 1b, the packet is poisoned and ECRC is irrelevant. (See Chapter 9.)

- TD = 0 No Digest DWORD is attached.
- TD = 1 Digest DWORD is attached.

EP: Error and Poisoning is optional and only defined for write requester transaction and read completer transaction packets.

- EP = 0b.

ATTRI [1:0]: ATTRIBUTES field defines snoop and ordering information. ATTRI [1:0] = 00b.

- ATTRI [1] = (OR) 0b defines the PCI strong ordering for the packet. (See Chapter 10.)
- ATTRI [0] = (SN) 0b indicates that cache coherency is supported. (See Chapter 5.)

LENGTH [9::0]: LENGTH in DWORDs to be read. The data is transferred in DWORD increments (4 bytes = 32 data bits total) naturally address aligned. For read requester transactions packets, the LENGTH [9::0] field is the number of DWORDs of data to be read. Read requester transaction packets do not contain a Data field in the packet.

- Configuration read requester transaction packets are only defined to have a data length of 1 DWORD, the LENGTH [9::0] = 00 0000 0001b.

- If the configuration read requester transaction packet has LENGTH = one DWORD and FIRST BE [3::0] = 0000b, the associated completer transaction packet must have a LENGTH = one DWORD and provide one DWORD of data. This data has no meaning, the associated bytes were not read at the requester destination, and can be any value.

REQUESTER ID HDW

BUS# [7::0]: BUS NUMBER identifies the link or the virtual PCI bus segment in the Root Complex. (See Chapter 4.)

DEVICE# [4::0]: DEVICE NUMBER identifies the device on the link or on the virtual PCI bus segment in the Root Complex. (See Chapter 4.)

FUNC# [2:0]: FUNCTION NUMBER identifies one the eight functions within a device. (See Chapter 4.)

TAG HDW

TAG# [7::0]: A unique TAG# is assigned by the PCI Express requester source to each of its requester transaction packets. Each requester transaction packet is assigned a TAG# to permit the requester source to identify the associated completer transaction received. The completer source must retain and use the TAG# received as the requester destination in the completer transaction packets. (See Chapter 4.)

- If the Extended Tag Enable bit equals "0" (configuration address space) only TAG# [4::0] are defined and TAG# [7::5] are required to equal 000b.
- If the Extended Tag Enable bit equals "1" (configuration address space) all TAG# [7::0] bits are defined.

LAST BE [3::0]: For configuration requester transaction packets LAST BE [3::0] = 0000b.

FIRST BE [3::0]: FIRST Byte Enables. FIRST BE [3] refers to the highest order byte of the lowest order lowest address, (which is the first DWORD of data packet) DWORD.

- Defines the enables bytes of the first DWORD (lowest address order) in the data field of transaction packets. BE0= byte 0 of the DWORD, BE1= byte 1 of DWORD, and so on.

- Associated byte is read if BE = 1b. Associated byte is not read if BE = 0b.
- The BE pattern is not required to read contiguous bytes.
- Configuration requester transaction packets only contain one DWORD (defined by the LENGTH field) and require the FIRST BE field to be used.
- If the configuration read requester transaction packet has LENGTH = one DWORD and FIRST BE = 0000b, the associated completer transaction packet must have a LENGTH = one DWORD and provide one DWORD of data. This data has no meaning, the associated bytes were not read at the requester destination, and can be any value.

> See Chapter 5 for more information on the relationship between FIRST BE [3::0] and LENGTH [9::0].

CONFIGURATION HDW

BUS# [7::0]: BUS NUMBER identifies the link or the virtual PCI bus segment. Virtual PCI bus segments exist in the Root Complex (BUS# 0), in switches, and in bridges. (See Chapter 4.)

DEVICE# [4::0]: DEVICE NUMBER identifies the device on the link or the virtual PCI bus segment. Virtual PCI bus segments exist in the Root Complex (BUS# 0), in switches, and in bridges. On the link the device is either the endpoint or switches upstream ports' virtual PCI/PCI Bridges. The link device is defined as only DEVICE# 0. The other devices as defined in the previous bullet can be any DEVICE NUMBER. (See Chapter 4.)

FUNC# [2:0]: FUNCTION NUMBER identifies one of the eight functions within a device. (See Chapter 4.)

REGISTER HDW

REGISTER ADDRESS [5::0]: Combined with the lowers order bits [1:0] of the EXTENDED REGISTER ADDRESS addresses the 256 DWORD registers defined by PCI for the configuration register block.

EXTENDED REGISTER ADDRESS [3::0]: Used with the REGISTER ADDRESS [5::0] in the previous bullet to address the 256 DWORD registers defined by PCI for the configuration register block. The upper two bits [3:2] address the extension of the configuration register block defined by PCI Express. The combination of REGISTER ADDRESS and EXTENDED REGISTER ADDRESS addresses 1024 DWORD registers.

See Chapter 4 for more information about BUS# [7::0], DEVICE# [4::0], FUNC# [2:0], REGISTER ADDRESS [5::0], and EXTENDED REGISTER ADDRESS [3::0].

CONFIGURATION READ COMPLETER TRANSACTION

7	6	5	4	3	2	1	0	7	6	5	4	3	2	1	0
R	FMT 1 0		4 TYPE 1 T					R	2 TC 0			R	R	R	R
TD	EP	ATTRI 1 0		R	R	9		LENGTH							0
7	BUS NUMBER						0	4	DEVICE #				0	2 FUNC# 0	
STATUS 2 0		BCM	11					BYTE COUNT							00
7	BUS NUMBER						0	4	DEVICE #				0	2 FUNC# 0	
7	TAG #						0	R	6 LOWER ADDRESS 0						

15	TYPE HDW	00
15	ATTRIBUTE HDW	00
15	COMPLETER ID HDW	00
15	STATUS HDW	00
15	REQUESTER ID HDW	00
15	TAG HDW	00

CONFIGURATION READ COMPLETER TRANSACTION HEADER FIELD with data: ClpD ... without data: Cpl

TYPE HDW

Completer with data (successful read STATUS [2::0] = 000b)

FMT [1:0]: Format of Header field. FMT [0] = 0b for 3 DWORD versus FMT [0] = 1b for 4 DWORD in Header Field. FMT [1] = 0b no Data field in packet versus .FMT [1] = 1b for Data field in packet.

- FMT [1::0] = 10b.

TYPE [4::0]: Type of transaction associated with packet and lock.

- TYPE [0] = 0b defines Type 0 versus Type [0] = 1b defines Type 1.
- TYPE [4::1] = 0101b.

TC [2::0]: Traffic Class: The TC [2::0] for completer transaction packets must be the same as the associated requester transaction packets.

- TC [2::0] = 000b.

Completer without data (successful read STATUS [2::0] not = 000b)

FMT [1:0]: Format of Header field. FMT [0] = 0b for 3 DWORD versus FMT [0] = 1b for 4 DWORD in Header Field. FMT [1] = 0b no Data field, no packet versus FMT [1] = 1b for Data filed in packet.

- FMT [1::0] = 00b.

TYPE [4::0]: Type of transaction associated with packet and lock.

- TYPE [0] = 0b defines Type 0 versus Type [1] = 1b defines Type 1.
- TYPE [4::1] = 0101b.

TC [2::0]: Traffic Class: The TC [2::0] for completer transaction packets must be the same as the associated requester transaction packets.

- TC [2::0] = 000b.

ATTRIBUTE HDW

TD: Transaction Digest defines whether the transaction packet contains the Digest DWORD. If EP = 0b, the Digest DWORD contains a valid ECRC value. If EP = 1b, the packet is poisoned and ECRC is irrelevant. (See Chapter 9.)

- TD = 0b No Digest DWORD is attached.
- TD = 1b Digest DWORD is attached.

EP: Error and Poisoning is optional and only defined for write requester transaction and read completer transaction packets.

- If EP = 0b there is no error forwarding. (See Chapter 9.)
- If EP = 1b, there is error forwarding. (See Chapter 9.)

ATTRI [1:0]: ATTRIBUTES field defines snoop and ordering information. The ATTRI [1::0] for completer transaction packets must be the same as the associated requester transaction packets, thus ATTRI [1:0] = 00b.

- ATTRI [1] = (OR) 0b defines the PCI strong ordering for the packet. (See Chapter 10.)
- ATTRI [0] = (SN) 0b indicates that cache coherency is supported. (See Chapter 5.)

LENGTH [9::0]: LENGTH in DWORDs of Data field included in packet. The data is transferred in DWORD increments (4 bytes = 32 data bits total) naturally address aligned.

- For completer transaction packets associated with configuration read transactions, the data length is encoded into ten binary bits [9::0]. When one DWORD is retuned LENGTH [9::0] = 00 0000 0001b (STATUS [2::0] = 000b). For completer transaction packets with no Data field (STATUS [2::0] NOT = 000b), the LENGTH [9::0] is RESERVED and must = 00 0000 0000b.
- The maximum number is 1024 and the LENGTH [9::0] = 00 0000 0000b. The existence of the Data field in the packet is defined in the FMT [1] field; consequently, the definition of NO data bytes and NO Data field in the packet is not from information of the LENGTH [9::0] field but from the information of the FMT [1] field.

- If the configuration read requester transaction packet has LENGTH [9::0] = one DWORD and FIRST BE [3::0] = 0000b, the associated completer transaction packet must have a LENGTH [9::0] = one DWORD and provide one DWORD of data. This data has no meaning, the associated bytes were not read at the requester destination, and can be any value.

COMPLETER ID HDW

BUS# [7::0]: BUS NUMBER identifies the link or the virtual PCI bus segment in an endpoint, switch, or bridge. (See Chapter 4.)

DEVICE# [4::0]: DEVICE NUMBER identifies the device on the link or on the virtual PCI bus segment in an endpoint, switch, or bridge. (See Chapter 4.)

FUNC# [2:0]: FUNCTION NUMBER identifies one the eight functions within a device. (See Chapter 4.)

STATUS HDW

STATUS [2::0]: Completer transaction status defines the result of the associated requester transaction as follows:

- 000b Successful Completion (SC)
- 001b Unsupported Request ((UR)
- 010b Configuration Request Retry Status (CRS)
- 100b Completer Abort (CA)
- All other values are RESERVED

BCM: byte count modified. The BCM = 0b. Not defined for this type of completer transaction.

BYTE COUNT [11::00]

Each configuration read requester transaction packet is responded to with a single completer transaction packet. For a configuration read requester transactions, the associated completer transaction packet contains one DWORD in the Data field, the BYTE COUNT [11:00] must = 0000 0000 0100b when STATUS [2::0] = 000b (successful). The unsuccessful completer transaction packet will have no Data field (Cpl instead of CplD), but the BYTE COUNT [11::0] and LOWER ADDRESS [6::0] fields contain the same values as if the completer transaction packet was successful.

REQUESTER ID HDW

BUS# [7::0]: BUS NUMBER identifies the link or the virtual PCI bus segment in the Root Complex. (See Chapter 4.)

DEVICE# [4::0]: DEVICE NUMBER identifies the device on the link or on the virtual PCI bus segment in the Root Complex. (See Chapter 4.)

FUNC# [2:0]: FUNCTION NUMBER identifies one the eight functions within a device. (See Chapter 4.)

TAG HDW

TAG# [7::0]: A unique TAG# is assigned by the PCI Express requester source to each of its requester transaction packets. Each requester transaction packet is assigned a TAG# to permit the requester source to identify the associated completer transaction received. The completer source must retain and use the TAG# received as the requester destination in the completer transaction packets. (See Chapter 4.)

- If the Extended Tag Enable bit equals "0" (configuration address space) only TAG# [4::0] are defined and TAG# [7::5] are required to equal 000b.

- If the Extended Tag Enable bit equals "1" (configuration address space) all TAG# [7::0] bits are defined.

LOWER ADDRESS [6::0]: LOWER ADDRESS [6::0] must equal 000 0000b for completer transaction packets not associated with memory read requester transactions. The unsuccessful completer transaction packet will have no Data field (Cpl instead of CplD), but the BYTE COUNT [11::0] and LOWER ADDRESS [6::0] fields contain the same values as if the completer transaction packet was successful. See Chapter 5 for more information.

> See Chapter 5 for more information on the relationship between LOWER ADDRESS [6::0], LENGTH [9::0] and BYTE COUNT [11::0].

Header Field Formats for Configuration Write Transactions

Configuration write requester transaction packets consist of Header, Data, and (optional) Digest fields, as shown in Figure 6.16.

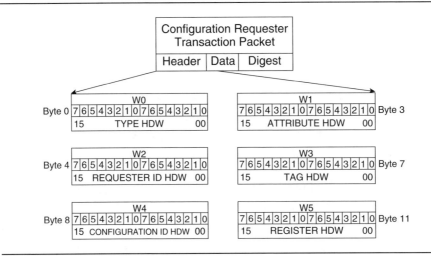

Figure 6.16   Configuration Write Requester Transaction Packet

The Repeater/Completer protocol dictates that the endpoint, switch, or bridge that is being written to responds with completer transactions, as shown in Figure 6.17.

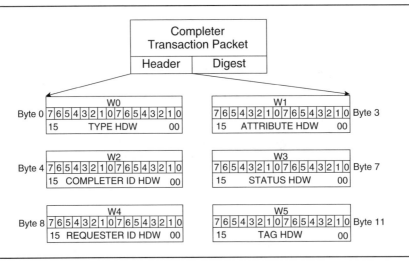

Figure 6.17   Completer Transaction Packet

CONFIGURATION WRITE REQUESTER TRANSACTION

7	6	5	4	3	2	1	0	7	6	5	4	3	2	1	0
R	FMT 1 0		4 TYPE 1 T					R	2 TC 0			R	R	R	R
TD	EP	ATTRI 1 0		R	R	9		LENGTH							0
7	BUS NUMBER						0	4	DEVICE #			0	2	FUNC#	0
7	TAG						0	3	LAST BE		0	3	FIRST BE		0
7	BUS NUMBER						0	4	DEVICE #			0	2	FUNC#	0
R	R	R	R	EXT, REGISTER 3 ADDRESS 0				REGISTER ADDRESS 5 0						R	R

15	TYPE HDW	00
15	ATTRIBUTE HDW	00
15	REQUESTER ID HDW	00
15	TAG HDW	00
15	CONFIGURATION ID HDW	00
15	REGISTER HDW	00

CONFIGURATION WRITE REQUESTER TRANSACTION HEADER FIELD CfgWr

TYPE HDW

FMT [1:0]: Format of Header field. FMT [0] = 0b for 3 DWORD versus FMT [0] = 1b for 4 DWORD in Header Field. FMT [1] = 0b no Data field in packet versus FMT [1] = 1b for Data field in packet.

- FMT [1::0] = 10b.

TYPE [4::0]: Type of transaction associated with packet and lock.

- TYPE [0] = 0b defines Type 0 versus Type [0] = 1b defines Type 1.
- TYPE [4::1] = 0010b.

TC [2::0]: Traffic Class

- TC [2::0] = 000b.

ATTRIBUTE HDW

TD: Transaction Digest defines whether the transaction packet contains the Digest DWORD. If EP = 0b, the Digest DWORD contains a valid ECRC value. If EP = 1b, the packet is poisoned and ECRC is irrelevant (Chapter 9).

- TD = 0 No Digest DWORD is attached.
- TD = 1 Digest DWORD is attached.

EP: Error and Poisoning. EP is optional and only defined for write requester transaction and read completer transaction packets.

- If EP = 0b there is no error forwarding. (See Chapter 9.)
- If EP = 1b, there is error forwarding. (See Chapter 9.)

ATTRI [1:0]: ATTRIBUTES field defines snoop and ordering information. The ATTRI [1::0] = 00b.

- ATTRI [1] = (OR) 0b defines the PCI strong ordering for the packet. (See Chapter 10.)
- ATTRI [0] = (SN) 0b indicates that cache coherency is supported. (See Chapter 5.)

LENGTH [9::0]: LENGTH in DWORDs to be written. The data is transferred in DWORD increments (4 bytes = 32 data bits total) naturally address aligned. For write requester transaction packets, the LENGTH [9::0] field is the number of DWORDs in the Data field of the packet.

- Configuration write requester transaction packets are only defined to have a data length of 1 DWORD, the LENGTH field must equal 00 0000 0001b.
- If the configuration write requester transaction packet has LENGTH = one DWORD and FIRST BE [3::0] = 0000b; the Data field has no meaning, the associated bytes were not written at the requester destination, and can be any value.

REQUESTER ID HDW

BUS# [7::0]: BUS NUMBER identifies the link or the virtual PCI bus segment in the Root Complex. (See Chapter 4.)

DEVICE# [4::0]: DEVICE NUMBER identifies the device on the link or on the virtual PCI bus segment in the Root Complex. (See Chapter 4.)

FUNC# [2:0]: FUNCTION NUMBER identifies one the eight functions within a device. (See Chapter 4.)

TAG HDW

TAG# [7::0]: A unique TAG# is assigned by the PCI Express requester source to each of its requester transaction packets. Each requester transaction packet is assigned a TAG# to permit the requester source to identify the associated completer transaction received. The completer source must retain and use the TAG# received as the requester destination in the completer transaction packets. (See Chapter 4.)

- If the Extended Tag Enable bit equals "0" (configuration address space) only TAG# [4::0] are defined and TAG# [7::5] are required to equal 000b.

- If the Extended Tag Enable bit equals "1" (configuration address space) all TAG# [7::0] bits are defined.

- **LAST BE [3::0]**: For configuration requester transaction packets this field is defined as LAST BE [3::0] = 0000b.

FIRST BE [3::0]: FIRST Byte Enables. FIRST BE [3] refers to the highest order byte of the lowest order lowest address, (which is the first DWORD of data packet) DWORD.

- Defines the enables bytes of the first DWORD (lowest address order) in the data field of transaction packets. BE0= byte 0 of the DWORD, BE1= byte 1 of DWORD, and so on.

- Associated byte is written if BE = 1b. Associated byte is not written if BE=0b.

- The BE pattern is not required to write contiguous bytes.

- Configuration requester transaction packets only contain one DWORD (defined by the LENGTH field) and require the FIRST BE field to be used.

- If the configuration write requester transaction packet has LENGTH = one DWORD and FIRST BE = 0000b; the associated data has no meaning, can be any value, and is not written at the requester destination.

> See Chapter 5 for more information on the relationship between FIRST BE [3::0] and LENGTH [9::0].

CONFIGURATION HDW

BUS# [7::0]: BUS NUMBER identifies the link or the virtual PCI bus segment. Virtual PCI bus segments exist in the Root Complex (BUS# 0), in switches, and in bridges. (See Chapter 4.)

DEVICE# [4::0]: DEVICE NUMBER identifies the device on the link or the virtual PCI bus segment. Virtual PCI bus segments exist in the Root Complex (BUS# 0), in switches, and in bridges. On the link the device is either the endpoint or switches upstream ports' virtual PCI/PCI Bridges. The link device is defined as only DEVICE# 0. The other devices as defined in the previous bullet can be any DEVICE NUMBER. (See Chapter 4.)

FUNC# [2:0]: FUNCTION NUMBER identifies one of the eight functions within a device. (See Chapter 4.)

REGISTER HDW

REGISTER ADDRESS [5::0]: Combined with the lowers order bits [1:0] of the EXTENDED REGISTER ADDRESS addresses the 256 DWORD registers defined by PCI for the configuration register block.

EXTENDED REGISTER ADDRESS [3::0: Used with the REGISTER ADDRESS [5::0] in the previous bullet to address the 256 DWORD registers defined by PCI for the configuration register block. The upper two bits [3:2] address the extension of the configuration register block defined by PCI Express. The combination of REGISTER ADDRESS and EXTENDED REGISTER ADDRESS addresses 1024 DWORD registers.

> See Chapter 4 for more information about **BUS# [7::0]**, **DEVICE# [4::0]**, **FUNC# [2:0]**, **REGISTER ADDRESS [5::0]**, and **EXTENDED REGISTER ADDRESS [3::0]**.

CONFIGURATION WRITE COMPLETER TRANSACTION

7	6	5	4	3	2	1	0	7	6	5	4	3	2	1	0
R	FMT 1 0		4 TYPE 1 T					R	2 TC 0			R	R	R	R
TD	EP	ATTRI 1 0		R	R	9		LENGTH							0
7	BUS NUMBER						0	4	DEVICE #			0	2	FUNC#	0
STATUS 2 0		BCM	11					BYTE COUNT							00
7	BUS NUMBER						0	4	DEVICE #			0	2	FUNC#	0
7	TAG #						0	R	6	LOWER ADDRESS					0

15	TYPE HDW	00
15	ATTRIBUTE HDW	00
15	COMPLETER ID HDW	00
15	STATUS HDW	00
15	REQUESTER ID HDW	00
15	TAG HDW	00

CONFIGURATION WRITE COMPLETER TRANSACTION HEADER FIELD: Clp

TYPE HDW

FMT [1:0]: Format of Header field. FMT [0] = 0b for 3 DWORD versus FMT [0] = 1b for 4 DWORD in Header Field. FMT [1] = 0b no Data field in packet versus FMT [1] = 1b for Data filed in packet.

- FMT [1::0] = 00b.

TYPE [4::0]: Type of transaction associated with packet and lock.

- TYPE [0] = 0b defines Type 0 versus Type [0] = 1b defines Type 1.
- TYPE [4::1] = 0101b.

TC [2::0]: Traffic Class: The TC [2::0] for completer transaction packets must be the same as the associated requester transaction packets.

- TC [2::0] = 000b.

ATTRIBUTE HDW

TD: Transaction Digest defines whether the transaction packet contains the Digest DWORD. If EP = 0b, the Digest DWORD contains a valid ECRC value. If EP = 1b, the packet is poisoned and ECRC is irrelevant. (See Chapter 9.)

- TD = 0b No Digest DWORD is attached.
- TD = 1b Digest DWORD is attached.

EP: Error and Poisoning is optional and only defined for write requester transaction and read completer transaction packets. (See Chapter 9.)

- If EP = 0b.

ATTRI [1:0]: ATTRIBUTES field defines snoop and ordering information. The ATTRI [1::0] for completer transaction packets must be the same as the associated requester transaction packets, thus ATTRI [1:0] = 00b.

- ATTRI [1] = (OR) 0b defines the PCI strong ordering for the packet. (See Chapter 10.)
- ATTRI [0] = (SN) 0b indicates that cache coherency is supported. (See Chapter 5.)

LENGTH [9::0]: LENGTH in DWORDs of Data field included in packet. The data is transferred in DWORD increments (4 bytes = 32 data bits total) naturally address aligned, the LENGTH [9::0] field is the number of DWORDs in the Data field of the packet.

- For completer transaction packets associated with configuration write transactions, the data length is encoded into ten binary bits [9::0]. Completer transaction packets have no Data field, thus the LENGTH [9::0] is RESERVED and must = 00 0000 0000b.

COMPLETER ID HDW

BUS# [7::0]: BUS NUMBER identifies the link or the virtual PCI bus segment in an endpoint, switch, or bridge. (See Chapter 4.)

DEVICE# [4::0]: DEVICE NUMBER identifies the device on the link or on the virtual PCI bus segment in the endpoint, switch, or bridge. (See Chapter 4.)

FUNC# [2:0]: FUNCTION NUMBER identifies one the eight functions within a device. (See Chapter 4.)

STATUS HDW

STATUS [2::0]: Completer status defines the result of the associated requester transaction as follows:

- 000b Successful Completion (SC)
- 001b Unsupported Request (UR)
- 010b Configuration Request Retry Status (CRS)
- 100b Completer Abort (CA)
- All other values are RESERVED

BCM: byte count modified. The BCM = 0b. Not defined for this type of completer transaction.

BYTE COUNT [11::00]

For configuration write requester the associated completer transaction packets contain no Data field, the BYTE COUNT [11:00] = 0000 0000 0000b, and LOWER ADDRESS [6::0] = 000 0000b.

REQUESTER ID HDW

BUS# [7::0]: BUS NUMBER identifies the link or the virtual PCI bus segment in the Root Complex. (See Chapter 4.)

DEVICE# [4::0]: DEVICE NUMBER identifies the device on the link or on the virtual PCI bus segment in the Root Complex. (See Chapter 4.)

FUNC# [2:0]: FUNCTION NUMBER identifies one the eight functions within a device. (See Chapter 4.)

TAG HDW

TAG# [7::0]: A unique TAG# is assigned by the PCI Express requester source to each of its requester transaction packets. Each requester transaction packet is assigned a TAG# to permit the requester source to identify the associated completer transaction received. The completer source must retain and use the TAG# received as the requester destination in the completer transaction packets. (See Chapter 5.)

- If the Extended Tag Enable bit equals "0" (configuration address space) only TAG# [4::0] are defined and TAG# [7::5] are required to equal 000b.

- If the Extended Tag Enable bit equals "1" (configuration address space) all TAG# [7::0] bits are defined.

LOWER ADDRESS [6::0]: LOWER ADDRESS [6::0] must = 000 0000b for completer transaction packets not associated with memory read requester transactions. For configuration write requester transactions, the associated completer transaction packets contain no Data field, the BYTE COUNT [11:00] = 0000 0000 0000b, and LOWER ADDRESS [6::0] = 000 0000b.

> See Chapter 5 for more information on the relationship between LOWER ADDRESS [6::0], LENGTH [9::0] and BYTE COUNT [11::0].

Header Field Formats for Message Baseline Transactions

Message baseline requester transaction packets consist of Header, Data (when applicable), and (optional) Digest fields, as shown in Figures 6.18 and 6.19. No completer transaction packets respond to message baseline requester transaction packets.

Chapter 6: Transaction Layer ■ **283**

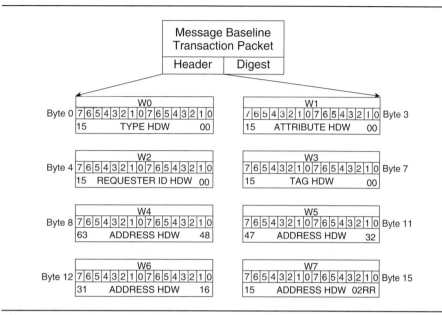

Figure 6.18 Message Baseline Requester Transaction Packet without Data

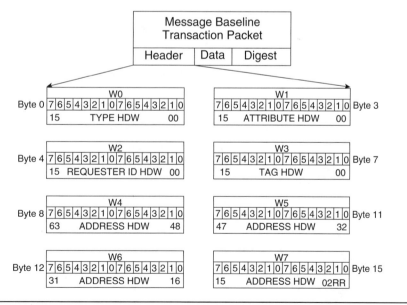

Figure 6.19 Message Baseline Requester Transaction Packet with Data

As discussed in Chapter 4, in addition to Implied Routing, message baseline transactions also implement addressing. When r = 001b, the destinations are selected by the ADDRESS HDWs in the Header field. No message baseline requester transaction defined in PCI Express revision 1.0 implements this Implied Routing value of r = 001b. The PCI Express specification defined this routing value as RESERVED. Consequently, this book defines message baseline requester transactions as not implementing any of the address bits similar to those used by memory or I/O transactions via the ADDRESS HDWs. The ADDRESS HDWs are defined as RESERVED.

MESSAGE BASELINE REQUESTER TRANSACTION

7	6	5	4	3	2	1	0	7	6	5	4	3	2	1	0
R	FMT 1 0			4 TYPE 0				R	2 TC 0			R	R	R	R
TD	EP	ATTRI 1 0		R	R	9		LENGTH							0
7	BUS NUMBER						0	4	DEVICE #			0	2 FUNC# 0		
7	TAG						0	7	MESSAGE CODE						0
63	ADDRESS HDW														48
47	ADDRESS HDW														32
31	ADDRESS HDW														16
15	ADDRESS HDW														00

15	TYPE HDW	00
15	ATTRIBUTE HDW	00
15	REQUESTER ID HDW	00
15	TAG HDW	00
	ADDRESS HDW Message baseline transactions defined in PCI Express Rev. 1.0 only implement Implied Routing, without using ADDRESS HDW. Presently this HDW is RESERVED.	

MESSAGE BASELINE REQUESTER TRANSACTION HEADER FIELD without data ... MsgB with data MsgBD

TYPE HDW

Without Data

FMT [1:0]: Format of Header field. FMT [0] = 0b for 3 DWORD versus FMT [0] = 1b for 4 DWORD in Header Field. FMT [1] = 0b no Data field in packet versus FMT [1] = 1b for Data field in packet.

- FMT [1::0] = 01b.

TYPE [4::0]: Type of transaction associated with packet.

- TYPE [4::0] = 10r2r1r0b.
- **r [2::0]:** Implied Routing information, discussed in detail in Chapter 4.
    - The "r" values are used for Implied Routing information of message baseline requester transactions. See Tables 6.2 through 6.7.

TC [2::0]: Traffic Class

- TC [2::0] = 000b to 111b.
- Actual value dependent on the specific message baseline transaction. See Tables 6.2 through 6.7.

With Data

FMT [1:0]: Format of Header field. FMT [0] = 0b for 3 DWORD versus FMT [0] = 1b for 4 DWORD in Header Field. FMT [1] = 0b no Data field, no packet versus FMT [1] = 1b for Data field in packet.

- FMT [1::0] = 11b.

TYPE [4::0]: Type of transaction associated with packet.

- TYPE [4::0] = 10r2r1r0b.
- r [2::0] – Implied Routing information, discussed in detail in Chapter 4.
  - The "r" values are used for Implied Routing information of message baseline requester transactions. See Tables 6.2 through 6.7.

TC [2::0]: Traffic Class

- TC [2::0] = 000b to 111b "possible."
- Actual value dependent on the specific message baseline transaction. See Tables 6.2 through 6.7.

ATTRIBUTE HDW

TD: Transaction Digest defines whether the transaction packet contains the Digest DWORD. If EP = 0b, the Digest DWORD contains a valid ECRC value. If EP = 1b, the packet is poisoned and ECRC is irrelevant. (See Chapter 9.)

- TD = 0 No Digest DWORD is attached.
- TD = 1 Digest DWORD is attached.

EP: Error and Poisoning is optional and only defined for requester write transaction and completer read transaction packets.

- If EP = 0b there is no error forwarding. (See Chapter 9.)
- If EP = 1b, there is error forwarding. (See Chapter 9.)

ATTRI [1:0]: ATTRIBUTES field defines snoop and ordering information. The ATTRI [1::0] = 00b.

- **ATTRI [1]** = (OR) 0b defines the PCI strong ordering for the packet. (See Chapter 10.)
- **ATTRI [0]** = (SN) 0b indicates that cache coherency is supported. (See Chapter 5.)

LENGTH [9::0]: LENGTH in DWORDs to be written. The data is transferred in DWORD increments (4 bytes = 32 data bits total) naturally address aligned. For message baseline requester transaction packets, the LENGTH [9::0] field is the number of DWORDs in the Data field of the packet.

- For message baseline requester transaction packets, the data length is encoded into ten binary bits [9::0]. For example, 00 0000 0001b is one DWORD, 00 0000 0010 is two DWORDs, 11 1111 1111b) is 1023 DWORDs, and so on.
- The maximum number is 1024 and the LENGTH [9::0] = 00 0000 0000b. The existence of the Data field is defined in the FMT [1] field; consequently the definition of NO Data field in the packet is not from information of the LENGTH [9::0] field but from the information of the FMT [1] field.
- For message baseline requester transaction packets with NO Data field the LENGTH [9::0] field is RESERVED and must = 00 0000 0000b.
- The values of LENGTH [9::0] supported by each specific message baseline transaction packet as listed in Tables 6.2 through 6.7.

REQUESTER HDW

BUS# [7::0]: BUS NUMBER identifies the link or the virtual PCI bus segment in the Root Complex, switch, endpoint, or bridge. (See Chapter 4.) Also, as listed in Tables 6.2 through 6.7, the value may the "actual" or RESERVED.

DEVICE# [4::0]: DEVICE NUMBER identifies the device on the link or on the virtual PCI bus segment in the Root Complex, switch, endpoint, or bridge. (See Chapter 4.) Also, as listed in Tables 6.2 through 6.7, the value may the "actual" or RESERVED.

FUNC# [2:0]: FUNCTION NUMBER identifies one the eight functions within a device. (See Chapter 4.) Also, as listed in Tables 6.2 through 6.7, the value may the "actual" or RESERVED.

TAG HDW

TAG# [7::0]: A unique TAG# is assigned by the PCI Express requester source to each of its requester transaction packets. Each requester transaction packet is assigned a TAG# to permit the requester source to identify the associated completer transaction received. The completer source must retain and use the TAG# received as the requester destination in the completer transaction packets. (See Chapter 5.)

- If the Extended Tag Enable bit equals "0" (configuration address space) only TAG# [4::0] are defined and TAG# [7::5] are required to equal 000b.

- If the Extended Tag Enable bit equals "1" (configuration address space) all TAG# [7::0] bits are defined.

MESSAGE CODE [7::0]: Defines which message baseline transaction the packet represents. See Tables 6.2 through 6.7.

ADDRESS HDW

ADDRESS [31::0]: RESERVED.

ADDRESS [63::0]: RESERVED.

Table 6.2 Interrupt Related Message

Name	TAG HDW Message Code Field [7::0]	TYPE HDW TC [2::0]	Data Field? In ATTRI. HDW Length [9::0]	Requester ID Field Entries
Assert _ INTx	A = 0010 0000b B = 0010 0001b C = 0010 0010b D = 0010 0011b	000b	No Data Field Length = 00 0000 0000b = Reserved	BUS# = Actual DEVICE#= Reserved FUNC# = Reserved
Deassert_ INTx	A = 0010 0100b B = 0010 0101b C = 0010 0110b D = 0010 0111b	000b	No Data Field Length = 00 0000 0000b = Reserved	BUS# = Actual DEVICE# = Reserved FUNC# = Reserved

Table 6.3 Error Related Message

Name	TAG HDW Message Code Field [7::0]	TYPE HDW TC [2::0]	Data Field? In ATTRI. HDW Length [9::0]	Requester ID Field Entries
ERR_COR (Correctable)	0011 0000b	000b	No Data Field Length = 00 0000 0000b = Reserved	BUS# = Actual DEVICE# =Actual FUNC# =Actual or Reserved (1)
ERR_NONFATAL (Uncorrectable)	0011 0001b	000b	No Data Field Length = 00 0000 0000b = Reserved	BUS# = Actual DEVICE# =Actual FUNC# =Actual or Reserved (1)

Table 6.3 Error Related Message *(continued)*

Name	TAG HDW Message Code Field [7::0]	TYPE HDW TC [2::0]	Data Field? In ATTRI. HDW Length [9::0]	Requester ID Field Entries
ERR_FATAL (Fatal Error)	0011 0011b	000b	No Data Field Length = 00 0000 0000b = Reserved	BUS# = Actual DEVICE# =Actual FUNC# =Actual or Reserved (1)

Table Notes:

1. In some cases the actual FUNCTION# cannot be returned. For example, the access is to a multiple FUNCTION device and the access is to a FUNCTION# not enabled (one that does not exist). The device returns a completer transaction packet with the STAUS field = unsuccessful and the FUNCTION# = 000b.

Table 6.4 Hot Plug Related Message

Name	TAG HDW Message Code Field [7::0]	TYPE HDW TC [2::0]	Data Field ? In ATTRI.HDW Length [9::0]	Requester ID Field Entries
Attention_ Indicator_On	0100 0001b	000b	No Data Field Length = 00 0000 0000b = Reserved	BUS# = Actual DEVICE# =Actual FUNC# =Actual
Attention_ Indicator_ Blink	0100 0011b	000b	No Data Field Length = 00 0000 0000b = Reserved	BUS# = Actual DEVICE# =Actual FUNC# =Actual

Table 6.4 Hot Plug Related Message *(continued)*

Name	TAG HDW Message Code Field [7::0]	TYPE HDW TC [2::0]	Data Field ? In ATTRI.HDW Length [9::0]	Requester ID Field Entries
Attention_ Indicator_Off	0100 0000b	000b	No Data Field Length = 00 0000 0000b = Reserved	BUS# = Actual DEVICE# =Actual FUNC# =Actual
Attention_ Button_ Pressed	0100 1000b	000b	No Data Field Length = 00 0000 0000b = Reserved	BUS# = Actual DEVICE# =Actual FUNC# =Actual
Power_ Indicator_ On	0100 0101b	000b	No Data Field Length = 00 0000 0000b = Reserved	BUS# = Actual DEVICE# =Actual FUNC# =Actual
Power_ Indicator_ Blink	0100 0111b	000b	No Data Field Length = 00 0000 0000b = Reserved	BUS# = Actual DEVICE# =Actual FUNC# =Actual
Power_ Indicator_ Off	0100 0100b	000b	No Data Field Length = 00 0000 0000b = N Data Field	BUS# = Actual DEVICE# =Actual FUNC# =Actual
PM_PME	0001 1000b	000b	No Data Field Length = 00 0000 0000b = Reserved	BUS# = Actual DEVICE# =Actual FUNC# =Actual

Table 6.5 Power Management Related Message

Name	TAG HDW Message Code Field [7::0]	TYPE HDW TC [2::0]	Data Field ? In ATTRI. HDW Length [9::0]	Requester ID Field Entries
PM_Active_State_Nak	0001 0100b	000b	No Data Field Length = 00 0000 0000b =Reserved	BUS# = Actual DEVICE# = 00000b =Reserved FUNC# = 000b =Reserved
PM_PME	0001 1000b	000b	No Data Field Length = 00 0000 0000b = Reserved	BUS# = Actual DEVICE# = Actual FUNC# = Actual
PME_Turn_Off	0001 1001b	000b	No Data Field Length = 00 0000 0000b = Reserved	BUS# = Actual DEVICE# = Actual FUNC# = Actual
PME_To_Ack	0001 1011b	000b	No Data Field Length = 00 0000 0000b =Reserved	BUS# = Actual DEVICE# = Actual FUNC# = Actual

Table 6.6 Slot Power Limit Message

Name	TAG HDW Message Code Field [7::0]	TYPE HDW TC [2::0]	Data Field? In ATTRI. HDW Length [9::0]	Requester ID Field Entries
Set_Slot Power_Limit (Set power of upstream port)	0101 0000b	000b	Data Field Length = 00 0000 0001b	BUS# = Actual DEVICE# = Actual FUNC# = Actual

Table 6.7 Lock Protocol Transaction Related Message

Name	TAG HDW Message Code Field [7::0]	TYPE HDW TC [2::0]	Data Field ? In ATTRI. HDW Length [9::0]	Requester ID Field Entries
Unlock	0000 0000b	000b	No Data Field Length = 00 0000 0000b = Reserved	BUS# = Actual DEVICE# = Actual FUNC# = 000b Reserved

Header Field Formats for Message Vendor-defined Transactions

Message vendor-defined requester transaction packets consist of Header, Data (when applicable), and (optional) Digest fields, as shown in Figures 6.20 and 6.21. No completer transaction packets are associated with message baseline requester transaction packets.

294 ■ The Complete PCI Express Reference

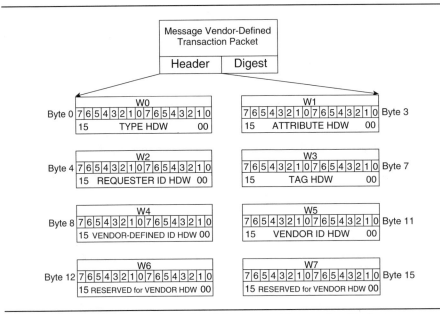

Figure 6.20 Message Vendor-defined Requester Transaction Packet Without Data

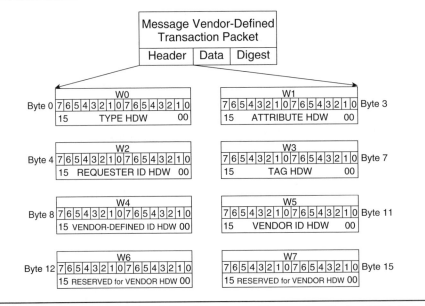

Figure 6.21 Message Vendor-defined Requester Transaction Packet With Data

As discussed in Chapter 4, message baseline requester transactions also implements addressing with ADDRESS HDW when r = 001b. PCI Express revision 1.0 also defines r = 001b for message vendor-defined requester transactions. If r is not equal to 001b, Implied Routing is used and the Vendor-defined ID HDW is RESERVED. If r = 001b, the Vendor-defined ID HDW defines BUS#, DEVICE#, and FUNCTION# for ID Routing.

MESSAGE VENDOR-DEFINED REQUESTER TRANSACTION

7	6	5	4	3	2	1	0	7	6	5	4	3	2	1	0
R	\multicolumn{3}{c}{FMT 1 0}	1	0	\multicolumn{3}{c}{2 r FIELD 0}	R	\multicolumn{3}{c}{2 TC 0}	R	R	R	R
TD	EP	\multicolumn{3}{c}{ATTRI 1 0}	R	R	9	\multicolumn{7}{c}{LENGTH}	0
7	\multicolumn{6}{c}{BUS NUMBER}	0	4	\multicolumn{3}{c}{DEVICE #}	0	2	\multicolumn{2}{c}{FUNC#}	0
7	\multicolumn{6}{c}{TAG}	0	7	\multicolumn{6}{c}{MESSAGE CODE}	0
7	\multicolumn{6}{c}{BUS NUMBER}	0	4	\multicolumn{3}{c}{DEVICE #}	0	2	\multicolumn{2}{c}{FUNC#}	0
15	\multicolumn{14}{c}{VENDOR ID HDW}	00
15	\multicolumn{14}{c}{RESERVED FOR VENDOR HDW}	00
15	\multicolumn{14}{c}{RESERVED FOR VENDOR HDW}	00

15	TYPE HDW	00
15	ATTRIBUTE HDW	00
15	REQUESTER ID HDW	00
15	TAG HDW	00
15	VENDOR-DEFINED ID HDW	00
15	VENDOR ID HDW	00
15	RESERVED FOR VENDOR HDW	00
15	RESERVED FOR VENDOR HDW	00

MESSAGE VENDOR-DEFINED REQUESTER TRANSACTION HEADER FIELD without data ... MsgV with data ... MsgVD

TYPE HDW

Without Data

FMT [1:0]: Format of Header field. FMT [0] = 0b for 3 DWORD versus FMT [0] = 1b for 4 DWORD in Header Field. FMT [1] = 0b no Data field in packet versus FMT [1] = 1b for Data field in packet.

- FMT [1::0] = 01b.

TYPE [4::0]: Type of transaction associated with packet.

- TYPE [4::1] = 1r2r1r0b.

- **r [2::0]** – Implied Routing information, discussed in detail in Chapter 4.
  - The "r" values are used for Implied Routing information of message vendor-defined requester transactions can be r = 000b, 010b, 011b, or 100b.
  - If r [2::0] = 010b ID Routing is used.

TC [2::0]: Traffic Class

- TC [2::0] = 000b to 111b.

With Data

FMT [1:0]: Format of Header field. FMT [0] = 0b for 3 DWORD versus FMT [0] = 1b for 4 DWORD in Header Field. FMT [1] = 0b no Data field, no packet versus FMT [1] = 1b for Data field in packet.

- FMT [1::0] = 11b.

TYPE [4::0]: Type of transaction associated with packet.

- TYPE [4::1] = 1r2r1r0b.

- **r [2::0]** – Implied Routing information, discussed in detail in Chapter 4.
  - The "r" values are used for Implied Routing information of message vendor-defined requester transactions can be r = 000b, 010b, 011b, or 100b.
  - If r [2::0] = 010b ID Routing is used.

TC [2::0]: Traffic Class

- TC [2::0] = 000b to 111b.

ATTRIBUTE HDW

TD: Transaction Digest defines whether the transaction packet contains the Digest DWORD. If EP = 0b, the Digest DWORD contains a valid ECRC value. If EP = 1b, the packet is poisoned and ECRC is irrelevant. (See Chapter 9.)

- TD = 0 No Digest DWORD is attached.
- TD = 1 Digest DWORD is attached.

EP: Error and Poisoning is optional and only defined for requester write transaction and completer read transaction packets. If not implemented for the transaction packet the, EP = 0b.

- If EP = 0b there is no error forwarding. (See Chapter 9.)
- If EP = 1b, there is error forwarding. (See Chapter 9.)

ATTRI [1:0]: ATTRIBUTES field defines snoop and ordering information. For message vendor-defined requester transaction packets with or without data ATTRI [1:0] = 00b to 11b.

- ATTRI [1] = (OR) 1b defines the PCI-X relaxed ordering for the packet. (See Chapter 10.)
- ATTRI [1] = (OR) 0b defines the PCI strong ordering for the packet. (See Chapter 10.)
- ATTRI [0] = (SN) 1b indicates that cache coherency is not supported. (See Chapter 5.)
- ATTRI [0] = (SN) 0b indicates that cache coherency is supported. (See Chapter 5.)

LENGTH [9::0]: LENGTH in DWORDs to be written. The data is transferred in DWORD increments (4 bytes = 32 data bits total) naturally address aligned. For message vendor-defined requester transaction packets, the LENGTH [9::0] field is the number of DWORDs in the Data field of the packet.

- For message vendor-defined requester transaction packets, the data length is encoded into ten binary bits [9::0]. For example, 00 0000 0001b is one DWORD, 00 0000 0010 is two DWORDs, 11 1111 1111b) is 1023 DWORDs, and so on.

- The maximum number is 1024 and the LENGTH [9::0] = 00 0000 0000b. The existence of the Data field is defined in the FMT [1] field; consequently the definition of NO Data field in the packet is not from information of the LENGTH [9::0] field but from the information of the FMT [1] field.
- For message vendor-defined requester transaction packets with NO Data field the LENGTH [9::0] field is RESERVED and must = 00 0000 0000b.

REQUESTER HDW

BUS# [7::0]: BUS NUMBER identifies the link or the virtual PCI bus segment in the Root Complex, switch, endpoint, or bridge. (See Chapter 4.) Also, as listed in Table 6.8, the value is implementation-specific.

DEVICE# [4::0]: DEVICE NUMBER identifies the device on the link or on the virtual PCI bus segment in the Root Complex, switch, or bridge. (See Chapter 4.) Also, as listed in Table 6.8, the value is implementation-specific.

FUNC# [2:0]: FUNCTION NUMBER identifies one the eight functions within a device. (See Chapter 4.) Also, as listed in Table 6.8, the value is implementation-specific.

TAG HDW

TAG# [7::0]: A unique TAG# is assigned by the PCI Express requester source to each of its requester transaction packets. Each requester transaction packet is assigned a TAG# to permit the requester source to identify the associated completer transaction received. The completer source must retain and use the TAG# received as the requester destination in the completer transaction packets. (See Chapter 5.)

- If the Extended Tag Enable bit equals "0" (configuration address space) only TAG# [4::0] are defined and TAG# [7::5] are required to equal 000b.
- If the Extended Tag Enable bit equals "1" (configuration address space) all TAG# [7::0] bits are defined.

MESSAGE CODE [7::0]: Defines which type of message vendor-defined transaction the packet represents. See Table 6.8.

VENDOR-DEFINED ID HDW

As previously discussed, this HDW is the destination of the message vendor-defined requester transaction when r = 001b (ID Routing). When r not = 001b (Implied Routing), this HDW is RESERVED.

BUS# [7::0]: BUS NUMBER identifies the link or the virtual PCI bus segment in the Root Complex, switch, endpoint, or bridge. (See Chapter 4.)

DEVICE# [4::0]: DEVICE NUMBER identifies the device on the link or on the virtual PCI bus segment in the Root Complex, switch, or bridge. (See Chapter 4.)

FUNC# [2:0]: FUNCTION NUMBER identifies one the eight functions within a device. (See Chapter 4.)

VENDOR ID

This is a unique number for each vendor and is provided by the PCI-SIG.

VENDOR-DEFINED HDW

This is RESERVED for use by the PCI Express device's vendor. No interoperability with PCI Express devices by other vendors is implied.

Table 6.8 Vendor-defined Message

Name	TAG HDW Message Code Field [7::0]	TYPE HDW TC [2::0]	Data Field? In ATTRI. HDW Length [9::0]	Requester ID Field Entries
Vendor_ Defined Type 0	0111 1110b	TC = 000b to 111b Implementation specific	Optional Data Length = Value	BUS# = Implementation Specific DEVICE# = Implementation Specific FUNC# = Implementation Specific
Vendor_ Defined Type 1	0111 1111b	TC = 000b to 111b Implementation specific	Optional Data Length = Value	BUS# = Implementation Specific DEVICE# = Implementation Specific FUNC# = Implementation Specific

Header Field Formats for Message Advanced Switching Transactions

Message advanced switching requester transaction packets consist of Header, Data (when applicable), and (optional) Digest fields, as shown in Figures 6.20 and 6.21. No completer transaction packets are associated with to message advanced switching requester transaction packets.

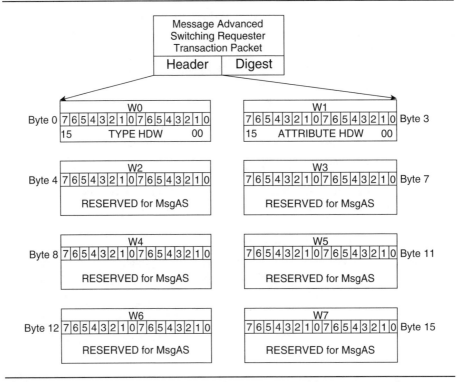

Figure 6.22 Message Advanced Switching Requester Transaction Packet Without Data

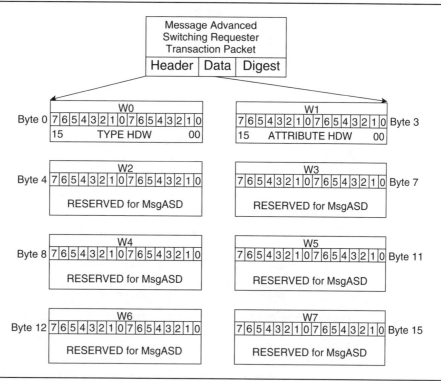

Figure 6.23 Message Advanced Switching Requester Transaction Packet With Data

MESSAGE ADVANCED SWITCHING REQUESTER TRANSACTION

7	6	5	4	3	2	1	0	7	6	5	4	3	2	1	0
R	FMT 1 0			1	0	2 n FIELD 0 2 c FIELD 0		R	2 TC 0			R	R	R	R
TD	EP	ATTRI 1 0		R	R	9		LENGTH							0
15					RESERVED HDW FOR MsgAS & MsgASD										00
15					RESERVED HDW FOR MsgAS & MsgASD										00
15					RESERVED HDW FOR MsgAS & MsgASD										00
15					RESERVED HDW FOR MsgAS & MsgASD										00
15					RESERVED HDW FOR MsgAS & MsgASD										00
15					RESERVED HDW FOR MsgAS & MsgASD										00

15	TYPE HDW	00
15	ATTRIBUTE HDW	00
15	RESERVED HDW FOR MsgAS & MsgASD	00
15	RESERVED HDW FOR MsgAS & MsgASD	00
15	RESERVED HDW FOR MsgAS & MsgASD	00
15	RESERVED HDW FOR MsgAS & MsgASD	00
15	RESERVED HDW FOR MsgAS & MsgASD	00
15	RESERVED HDW FOR MsgAS & MsgASD	00

MESSAGE ADVANCED SWITCHING REQUESTER TRANSACTION HEADER FIELD *without data ... MsgAS with data ... MsgASD*

> **Note:** The inclusion of the message advanced switching requester transaction information is to reflect future enhancements to the PCI Express specification. The PCI Express specification was first released with the message advanced switching requester transaction information. The errata available at the time this book was printed removed most reference to these entities, but retained others. The following information is included in order to tie in with future revisions of the PCI Express specification. See Chapter 18 for more information on message advanced switching requester transactions.

TYPE HDW

Without Data

FMT [1:0]: Format of Header field. FMT [0] = 0b for 3 DWORD versus FMT [0] = 1b for 4 DWORD in Header Field. FMT [1] = 0b no Data field in packet versus FMT [1] = 1b for Data field in packet.

- FMT [1::0] = 01b.

TYPE [4::0]: Type of transaction associated with packet.

- TYPE [4::0] = 11n2n1n0b.
- n [2::0] – RID Routing information.

TC [2::0]: Traffic Class

- TC [2::0] = 000b to 111b.

With Data

FMT [1:0]: Format of Header field. FMT [0] = 0b for 3 DWORD versus FMT [0] = 1b for 4 DWORD in Header Field. FMT [1] = 0b no Data field, no packet versus FMT [1] = 1b for Data field in packet.

- FMT [1::0] = 11b.

TYPE [4::0]: Type of transaction associated with packet.

- TYPE [4::0] = 11c2c1c0b.
- c [2::0] – RID Routing information. See Chapter 4 and 18.

TC [2::0]: Traffic Class

- TC[2::0] = 000b to 111b.

ATTRIBUTE HDW

TD: Transaction Digest defines whether the transaction packet contains the Digest DWORD. If EP = 0b, the Digest DWORD contains a valid ECRC value. If EP = 1b, the packet is poisoned and ECRC is irrelevant. (See Chapter 9.)

- TD=0 No Digest DWORD is attached.
- TD = 1 Digest DWORD is attached.

EP: Error and Poisoning is optional and only defined for requester write transaction and completer read transaction packets. If not implemented for the transaction packet the EP = 0b.

- If EP = 0b there is no error forwarding. (See Chapter 9.)
- If EP = 1b, there is error forwarding. (See Chapter 9.)

ATTRI [1:0]: ATTRIBUTES field defines snoop and ordering information. For message advanced switching requester transaction packets with or without data ATTRI [1:0] = 00b to 11b.

- ATTRI [1] = (OR) 1b defines the PCI-X relaxed ordering for the packet. (See Chapter 10.)
- ATTR I[1] = (OR) 0b defines the PCI strong ordering for the packet. (See Chapter 10.)
- ATTRI [0] = (SN) 1b indicates that cache coherency is not supported. (See Chapter 5.)
- ATTRI [0] = (SN) 0b indicates that cache coherency is supported. (See Chapter 5.)

LENGTH [9::0]: LENGTH in DWORDs to be written. The data is transferred in DWORD increments (4 bytes = 32 data bits total) naturally address aligned. For advanced switching requester transaction packets, the LENGTH [9::0] field is the number of DWORDs in the Data field of the packet.

- For message advanced switching requester transaction packets, the data length is encoded into ten binary bits [9::0]. For example, 00 0000 0001b is one DWORD, 00 0000 0010 is two DWORDs, 11 1111 1111b) is 1023 DWORDs, and so on.

- The maximum number is 1024 and the LENGTH [9::0] = 00 0000 0000b. The existence of the Data field is defined in the FMT [1] field; consequently the definition of NO Data field in the packet is not from information of the LENGTH [9::0] field but from the information of the FMT [1] field.

- For message advanced switching requester transaction packets with NO Data field the LENGTH [9::0] field is RESERVED and must = 00 0000 0000b.

RESERVED HDW FOR MsgAS & MsgASD

To be defined.

Data Field and Digest Field Formats

The following section describes the Data and Digest field formats.

Data Field

Transaction Packets with Data Field

The discussion to this point has been focused on the Header field contained within every transaction packet. With the Header field, the FMT field defines whether a Data field is included immediately following the Header field in the transaction packet. Also within the Header field is the LENGTH [9::0] field, which defines the number of DWORDS in the Data field. In the case of read requester transaction packets, the LENGTH [9::0] field is the number of DWORDs to be returned in the associated completer packets; but the read requestor transaction packets have no Data fields. All of these considerations are summarized in Table 6.9.

Table 6.9 Transaction Packets and the Data Field

Transaction Type	Data Field	Length [9::0] Field in Attribute HDW
Message Baseline (Except Slot Power) Requester MsgB	None	00 0000 0000b = RESERVED
Message Slot Power MsgBD	Yes	00 0000 0001b
Message Vendor-defined Requester MsgV	None	00 0000 0000b = RESERVED
Message Vendor-defined Requester MsgVD	Yes	Value
Message Advanced Switching Requester MsgAS	None	00 0000 0000b = RESERVED
Message Advanced Switching Requester MsgASD	Yes	Value
Memory Read Requester MRd	None	Value
Memory Write Requester MWr	Yes	Value
I/O Read Requester IORd	None	00 0000 0001b
I/O Write Requester IOWr	Yes	00 0000 0001b
Configuration Read Requester CfgRd	None	00 0000 0001b
Configuration Write Requester CfgWr	Yes	00 0000 0001b
Completer Clp	None	00 0000 0000b = RESERVED
Completer ClpD	Yes	Value

Payload

The Data field contains what is defined as the payload of the TLP. The payload is simply the number of bytes associated the Data field. For those TLPs that contain a Data field, only those associated with the memory transactions (requester and completer), message vendor-defined requester transaction, and message advanced switching requester can have a payload of one DWORD or greater. Note: The message baseline requester transaction may in the future have a larger payload than one DWORD, but according to *PCI Express Specification revision 1.0* only the message slot power requester transaction has data and that payload is one DWORD.

PCI Express defines a protocol for the actual payload transmitted in TLP and the payload size is limited for each PCI Express device. Within the configuration register blocks of each PCI Express device (specifically each function within a PCI Express devices) are the following bits associated with payload size:

- **Bits [2::0]** of Device Capabilities Register define the maximum payload a PCI Express device can support (maximum payload supported bits).

- **Bits [7::5]** of Device Control Register define the maximum payload a PCI Express device can support in any TLP (maximum payload size bits) except for memory read requester transaction.

- **Bits [14::12]** of Device Control Register define the maximum payload a PCI Express device can request in a memory read requester transaction (maximum read payload size bits).

These bits define a payload size of 128, 256, 512, 1024, 2048, or 4096 bytes. Each PCI Express device provides information about the maximum size it can accommodate in the Data fields of the TLPs it can transmit or receive (maximum payload supported bits). The configuration software takes the information in the maximum payload supported bits and determines the appropriate value to program into the maximum payload size bits and maximum read payload size bits. The maximum payload size bits define the maximum payload of the Data field of any TLP it can receive and any TLP it transmits. The maximum read payload size bits define maximum payload of the Data field associated with the associated completer transactions. That is, the maximum read payload size is manifested in each of the Data fields of the TLPs containing the completer transactions.

Transaction Packet Formats

Figures 6.24 through 6.31 illustrate Data fields. These figures show the orientation of the DWORDs of the Data field relative to the Header and Digest fields. Collectively, The Header, Data and Digest fields define the transaction packet TLP).

Figures 6.24 through 6.31 represent the general version of each type of transaction packet with the actual inclusion of the Data field as listed in Table 6.9. Chapter 8 provides more information about the order of the bytes transferred across the link.

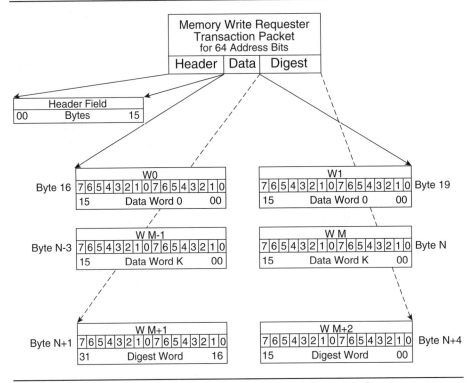

Figure 6.24 Memory Write Requester Transaction Packet for 64 Address Bits

Chapter 6: Transaction Layer **309**

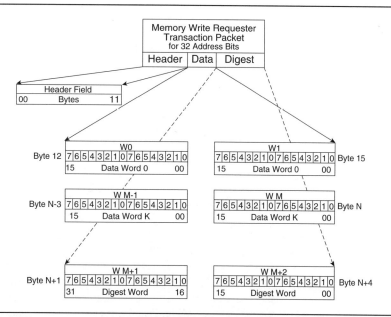

Figure 6.25   Memory Write Requester Transaction Packet for 32 Address Bits

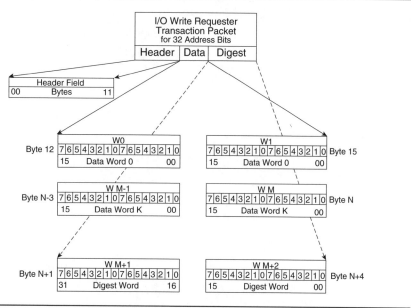

Figure 6.26   I/O Write Requester Transaction Packet for 32 Address Bits

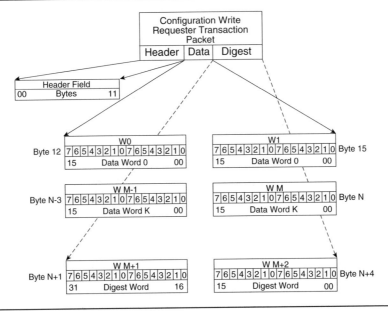

Figure 6.27 Configuration Write Requester Transaction Packet

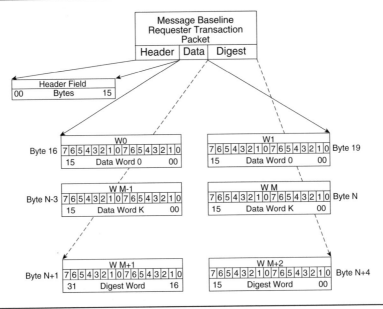

Figure 6.28 Message Baseline Requester Transaction Packet

Chapter 6: Transaction Layer ■ **311**

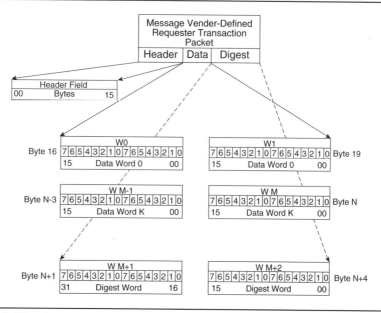

Figure 6.29 Message Vendor-Defined Requester Transaction Packet

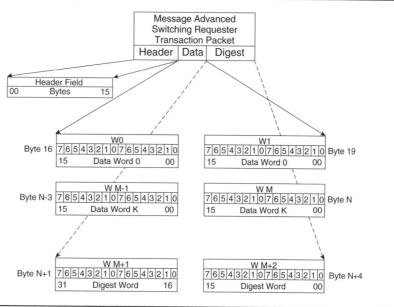

Figure 6.30 Message Advanced Switching Requester Transaction Packet

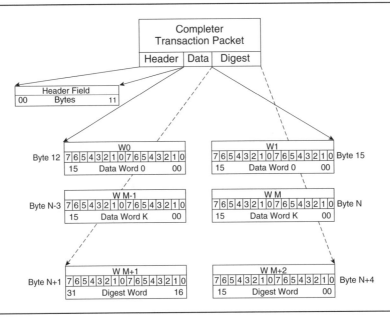

Figure 6.31   Completer Transaction Packet

Digest Field

The integrity of the link packets are protected as they traverse a specific link by the CRC and LCRC defined by the DDLP and LLTP, respectively. PCI Express defines an additional level of integrity for the transaction packets (TLP) in addition to link packets' error detection. This additional level of integrity for the transaction packets is within the Transaction Layer and is provided by the Digest field. The Digest field provides end-to-end coverage of transaction integrity. End-to-end coverage recognizes that the Physical, Data Link, and Transaction Layers have successfully transferred the associated packets between the PCI Express devices and possibly through switches. However, it is the requester source and requester destination (participants) or the completer source and the completer destination (participants) that are the actual PCI Express devices that use the contents of the TLPs.

The Digest field contains the ECRC value that provides optional cyclic redundancy checking over the Header and Data fields in the TLPs. If the Data field is not included in the TLP, the Digest field only applies to the Header field. The Digest field is required to be included in the transaction packet if the TD bit in the ATTRIBUTE DWORD of the Header field has "1" as its value. The generation of and checking for the ECRC is defined for a specific PCI Express device by the appropriate bits in the Advance Error Capabilities and Control Register in the configuration register block of the PCI Express device. The Advance Error Capabilities and Control Registers are only implemented as part of the Advance Error Reporting protocol. See Chapter 9 for more information on error reporting.

Figures 6.24 through 6.31 include the Digest field in the structure of the transaction packets. It is placed immediately following the Header and Data fields. If the transaction packet does not include a Data field the Digester field is placed immediately after the Header field.

Chapter 7

Data Link Layer and Packets

This is the second of three chapters that will focus on the specific elements of each of the three layers introduced in Chapter 2: Transaction Layer, Data Link Layer, and Physical Layer. As illustrated in Figure 7.1, the Transaction Layer interfaces with the PCI Express device core and provides the interface point between transactions of the core and Transaction Layer Packets (TLPs). The Data Link Layer converts the TLPs and any link management information into Link Layer Transaction Packets (LLTPs) and Data Link Layer Packets (DLLPs), respectively. The Physical Layer transmits LLTPs and DLLPs onto the link within physical packets and extracts them from the physical packets received from the link. Each type of packet is supported by an associated PCI Express layer. This chapter discusses the Data Link Layer and the associated LLTPs and DLLPs. Chapters 6 and 8 discuss the Transaction Layer and Physical Layer with associated packets, respectively.

The Data Link Layer transmits the transactions packets (TLPs) contained within link packets (LLTPs) across the link and manages the link to support this transmission with other link packets (DLLP). The key functions of the Data Link Layer are:

- Link activity states
- Flow Control Initialization
- Flow Control Update of Available Buffer Space
- DLLPs Formats
- LLTP Format

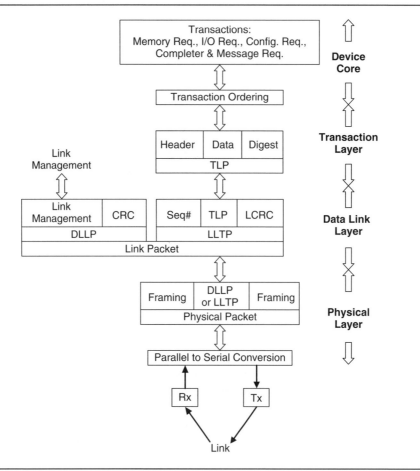

Figure 7.1   PCI Express Layers

> The terms *receiving port* and *transmitting port* are used in reference to the TLPs. *Receiving port* refers to the portion of the port that receives the TLPs. *Transmitting port* refers to the portion of the port that transmits the TLPs. Each port at each end of the link consists of both a receiving port and a transmitting port. As previously discussed, the TLPs are contained within link packets (LLTPs), which in turn are contained within the actual physical packets transmitted onto the link.

Overview of Data Link Layer Packets

The Data Link Layer defined two types of packets: the *Data Link Layer Packet* (DLLP) and the *Link Layer Transaction Packet* (LLTP). Both types of packets are transmitted across the link within Physical Packets of the Physical Link Layer. The DLLP contains link management information specific to the Data Link Layer. The LLTP contains the TLP of the Transaction Link Layer.

DLLP

As previously discussed, the Data Link Layer manages the link. Three activities comprise link management: Flow Control, Power Management, and Vendor-Specific activities. The Data Link Layer defines eight different DLLPs to support these activities. As illustrated in Figure 7.1, these activities are collectively defined as link management. To ensure integrity of the DLLP a CRC is included. Here is a brief overview of these DLLPs:

- The InitFC1, InitFC2, and UpdateFC DLLPs are part of the Flow Control protocol to track the available buffer space at the receiving port of a LLTP. Collectively these are called *Flow Control DLLPs* (FC DLLPs).

- The PM_Enter_L1, PM_Enter_L23, and PM_Active_State_Request_L1 DLLPs are transmitted by the port on the downstream side of the link to request the link to change power levels. The PM_Request_Ack DLLP is transmitted by the port on the upstream side of the link to acknowledge receipt of the request. Collectively these are called *Power Management DLLPs* (PM DLLPs).

- The Vendor-Specific DLLP has an implementation specific to the vendor of the PCI Express device. The contents of the DLLP are RESERVED for the vendor of the PCI Express device.

LLTP

The Data Link Layer is also responsible for transferring TLPs contained within the LLTPs across the link. The Data Link Layer supports this transfer by preserving strict ordering in TLPs transmission and by ensuring error-free transmission of the associated LLTPs.

As illustrated in Figure 7.1, the Data Link Layer attaches a SEQ# and a LCRC to each transaction packet transferred from the Transaction Layer to form each LLTP. The Data Link Layer subsequently transfers the LLTP to the Physical Layer of the transmitting port. The Physical Layer at the receiving port transfers the LLTP to the Data Link Layer upon reception. The Data Link Layer at the receiving port checks the SEQ# and recalculates the LCRC to check for errors. The SEQ# permits the Data Link Layer to maintain strong ordering among the TLPs contained in the LLTPs transmitted across the link. The LCRC provides cyclic redundancy checking for the SEQ# and the TLP of each LLTP. This is in addition to the ECRC contained within the TLP. Unlike the ECRC contained in the optional Digest field, the LCRC is required on all LLTPs.

The Ack and Nak DLLPs are associated with the Data Link Layers' transmission of LLTPs. The Ack and Nak DLLPs acknowledge successful and unsuccessful receipt of LLTPs; respectively. These DLLPs are used as part of the Flow Control protocol. See Chapter 11 for more information on the Flow Control protocol.

Other Considerations

The DLLPs are different than LLTPs in several ways:

- Link packets begin and end at the ports at each end of the link. The TLPs contained with the LLTPs are the only part of the LLTPs that pass through the ports.

- The DLLPs are transmitted across the link as needed and are immediately acted upon when received. Transmission priority must be considered when transmitting DLLPs as opposed to transmitting LLTPs. See Chapter 12 for more information on transmission priority.

- The concept of virtual channels (VCs) only applies to the transmission of LLTPs. VCs cannot transmit DLLPs. VCs have no actual hardware on the link. See Chapter 11 for more information in VCs.

Link Activity States

The PCI Express specification used the terms *link state* or *state of link* to refer to both the link activity state discussed here and the link states discussed in Chapters 14. This book uses the term *link activity state* to refer to the states of the Data Link Control and Management State Machine, listed in Table 7.1, and uses the term *link state* to refer to states of the link and the Link Training and Status State Machine discussed in Chapter 14.

In order for the link to support transmission of DDLPs and LLTPs contained in Physical Packets, two things must occur. First, the Physical Layer must successfully configure a link between two ports prior to transmitting DLLPs or LLTPs. Second, the Data Link Layer must establish the available buffer space at the receiving port of the LLTPs before transmitting LLTPs.

This chapter discusses in detail the link activity states associated with the Data Link Control and Management State Machine (DLCMSM) as executed in the Data Link Layer. The status of the DLCMSM in each PCI Express device is reported to the Transaction Layer as either Link_UP or Link_DOWN. See Chapter 13 for details on how the Transaction Layer reacts to the reports of Link_UP and Link_DOWN status from Data Link Layer.

Chapter 14 discusses in detail the link states associated with the link and the Link Training and Status State Machine (LTSSM) as executed by the Physical Layer. The status of the link and the LTSSM in each PCI Express device is reported to the Data Link Layer as Physical LinkUp = "0" or "1". When Physical LinkUp = "1" it is an indication from the Physical Layer to the Data Link Layer that the link has been configured and Physical Packets *can* be transmitted across the link. When Physical LinkUp = "0" it is an indication from Physical Layer to the Data Link Layer that the link has not been configured and Physical Packets *cannot* be transmitted onto the link. The Physical LinkUP status is generated by the Link Training and Status State Machine (LTSSM) and the link states.

Link_UP and Link_DOWN

Link_UP is a status indication from the Data Link Layer to the Transaction Layer that TLPs contained within LLTPs can be transmitted across the link. In order to transmit Physical Packets containing LLTPs across the link as indicated by Link_UP, the Physical LinkUP must = "1" and Flow Control Initialization must be completed on Virtual Channel 0 (VC0). Link_DOWN is an indication from the Data Link Layer to the Transaction Layer that TLPs contained within LLTPs cannot be transmitted across the link.

Link Activity States Application

The support of the Flow Control protocol by the Data Link Layer includes Flow Control Initialization. After a Hot, Cold, or Warm Reset the DLCMSM transitions to the DL_Inactive link activity state. The DLCMSM moves through the sequence of link activity states in parallel with the LTSSM moving through the sequence of link states used for Link Training. When Link Training is complete, the Physical Layer indicates with the Physical LinkUp = "1" status to the Data Link Layer that Physical Packets containing DLLPs and LLTPs can be transmitted across the link.

Before any LLTP containing TLPs can be transmitted across the link, the TLPs' receiving port must provide information on the available buffer space for at least VC0. The DLCMSM of each PCI Express device on the link sequences through the link activity states by the Data Link Layer to establish the available buffer space. The available buffers space is defined first for the VC0 for each Virtual Channel (VCs). The sequencing through the link activity states is part of the Flow Control Initialization for VC0. Flow Control Initialization is first applied to the VC0. Once the Flow Control Initialization is successfully completed for VC0, TLPs can be transferred from the Transaction Layer to the Data Link Layer to be contained in LLTPs to be transmitted across the link in Physical Packets. Table 7.1 summarizes the link activity states.

Table 7.1 Summary of Link Activity States

Present Link Activity State	Present Link Activity State Meaning	Next Link Activity State
DL_Inactive	Link is not operational, no communication between ports, or link not connected to port.	DL_Init
	Transition occurs when Physical LinkUp transitions from "0" to "1" status.	
	Link_DOWN is indicated to the Transaction Layer.	
DL_Init	Link is operational for initialization of VC0 (buffer space initialized) by transmitting DLLPs. LLTP transfers may occur on VC0 when one of the two ports enters the FC_Init2 sub-state of Flow Control Initialization.	DL_Active or DL_Inactive
	Transition to DL_Active occurs when Flow Control Initialization for VC0 is complete and Physical LinkUp = "1" value. Link_UP is indicated to the Transaction Layer.	
	Transition to DL_Inactive occurs when Flow Control Initialization for VC0 fails OR Physical LinkUp transitions from "1" to "0" status. Link_DOWN is indicated to the Transaction Layer.	

Table 7.1 Summary of Link Activity States *(continued)*

Present Link Activity State	Present Link Activity State Meaning	Next Link Activity State
DL_Active	Link is operational and VC0 Flow Control Initialization has been successfully completed and Flow Control Initialization is applied to the other enabled VCs on the link.	Remain in DL_Active or DL_Inactive
	Remains in the DL_Active if Physical LinkUp = "1". Link_UP is indicated to the Transaction Layer.	
	Transition to DL_Inactive occurs when Physical LinkUP transitions from "1" to "0" status. Link_DOWN is indicated to the Transaction Layer.	

Each receiving port buffer actually consists of a set of buffers assigned to each VC number. The set of buffers for a specific VC number consists of the different buffers assigned to different types of transactions. That is, those requester transactions of a specific VC number that post data use buffers PH and PD specific to that VC number. Those requester transactions of a specific VC number that do not post data use buffers NPH and NPD specific to that VC number. Finally, the completer transactions of a specific VC number use buffers CPLH and CPLD specific to that VC number. The "H" in the CPLH stands for the Header field of the TLP. The "D" in CPLD stands for the Data field of the TLP.

Flow Control Initialization of each VC number establishes the initial available space for the associated set of buffers. Flow Control Initialization is completed for a specific VC number when the available space is established for the associated set of buffers. The link activity states simply provide the Data Link Layer with a protocol to distinguish when VC0 has successfully completed Flow Control Initialization. VC0 is the minimum VC number that must be enabled and must be enabled first on the link. The successful completion of Flow Control Initialization permits LLTPs assigned to VC0 by the TC numbers in TLPs to be transmitted. After the successful Flow Control Initialization of VC0, the other VC numbers undergo Flow Control Initialization. As each of the other VC numbers complete Flow Control Initialization, it permits LLTPs assigned that VC number by the TC number in TLPs to be transmitted.

As part of the Flow Control protocol discussed in more detail in Chapter 11, the Gate circuitry in the Data Link Layer does not transfer a

TLP from the Transaction Layer to the Data Link Layer unless the receiving port has sufficient buffer space. The Flow Control Initialization provides the Gate circuitry with the initial value of available buffer space in the set of buffers for each VC number.

Details of Link Activity States

A portion of the link activity states are executed by the DLCMSMs of the Data Link Layer of each PCI Express device on the link prior to the link being able to transmit DLLPs and LLTPs. Physical LinkUp = "0" from the LTSSMs of the Physical Layer of each PCI Express device on the link indicates that the link cannot transmit DLLPs and LLTPs. The associated link and LTSSMs link state is not L0 link state. A portion of the link activity states is executed by the DLCMSMs of the Data Link Layer of each PCI Express device on the link after the link is able to transmit DLLPs and LLTPs. Physical LinkUp = "1" from the LTSSMs of the Physical Layer of each PCI Express device on the link indicates that the link can transmit DLLPs and LLTPs. The associated link and LTSSMs link state is L0 link state.

Here are the detailed attributes and activities of each link activity state as viewed by the Data Link Layer. Please refer to Figure 7.2 when reviewing these attributes and activities and the subsequent discussion of Flow Control Initialization and Updates. (See Chapter 13 for more information about the interaction between the link activities and Hot Reset, Cold Reset, and Warm Reset.)

DL_Inactive The DL_Inactive activity state is entered by the DLCMSM of the Data Link Layer in each PCI Express device after a Hot Reset, Cold Reset, or Warm Reset.

- The Data Link Layer indicates Link_DOWN status to Transaction Layer and Gate circuitry of the Data Link Layer. The Transaction Layer discards any outstanding transactions. That is, any requester transactions waiting for the associated completer transactions are discarded. The assumption of the downstream ports of the Root Complex and switches of Link_DOWN is equivalent to an **unexpected removal**. The assumption of the upstream ports of the switches, endpoints, and bridges of Link_DOWN is equivalent to a Hot Reset. See Chapter 13 for more information on Hot Resets.

- All link information is set to default and link retry buffer contents of the Data Link Layer are discarded. The Data Link Layer does not execute any DLLPs for link management.

- No DLLPs or LLTPs are transmitted or received as indicated by the Physical LinkUp = "0" status.

- All counters, flags, and timers used by Flow Control protocol are initialized but not enabled.

- Transition to DL_Init if Transactions Layer indicates a link is not disabled and the Physical Layer indicates the link is configured as indicated by the Physical LinkUp = "1" status.

DL_Init

- The Data Link Layer indicates Link _DOWN status to Transaction Layer and Gate circuitry of the Data Link Layer.

- Physical Layer indicates Physical LinkUp = "1" to show that DLLPs and LLTPs can be transmitted across the link.

- Initially, the Data Link Layer indicates Link _DOWN status to the Transaction Layer and Gate circuitry of the Data Link Layer while in sub-state FC_Init1 for VC0.

- The Flow Control Initialization begins with VC0, which by default is enabled.
  - The FI1 flag, FI2 flag, and resend timer are initialized and enabled when Flow Control Initialization is applied toVC0.
  - LLTPs' receiving port indicates to the LLTPs' transmitting port the available buffer space in the LLTPs' receiving ports for VC0 via Flow Control DLLPs.
  - The Data Link Layer indicates Link _UP status to Transaction Layer and Gate circuitry in the Data Link Layer while in sub-state FC_Init2 or when Flow Control Initialization is completed for VC0.
  - LLTPs are transferred on VC0 when a port has entered sub-state FC_Init2 for VC0; otherwise no LLTPs are transferred.

- The Transaction Layer in sub-state FC_Init1 must be able to receive and process DLLPs and LLTPs associated with VC0.

- Transition to DL_Active occurs once Flow Control Initialization is completed for VC0 (in both directions) and the Physical Layer indicates Physical LinkUp = "1" status.

- Transition to DL_Inactive occurs if Physical Layer indicates Physical LinkUp = "0" or if Flow Control Initialization for VC0 was not successful.

DL_Active

- The Data Link Layer indicates Link_UP status to Transaction Layer and Gate circuitry of the Data Link Layer.

- The Flow Control Initialization begins for the other enabled VC numbers (not VC0).

- LLTPs are transmitted on any VC number that has entered sub-state FC_Init2 or has completed Flow Control Initialization.

- Flow Control Initialization for enabled VC numbers executes indefinitely for those VC numbers that have not exited sub-state FC_Init2.

- Transition to DL_Inactive occurs if Physical Layer indicates Physical LinkUp ="0" status.

Flow Control Initialization

The following discussion of the Flow Control Initialization of a VC is representative of any VC number. As discussed in the previous section, VC0 must be the first VC number to undergo Flow Control Initialization within the DL_Init link activity state. Subsequently, the other VC numbers undergo Flow Control Initialization in the DL_Active link activity state. Consequently, LLTPs contained within Physical Packets are transmitted across the link via VC0 while other VC numbers are undergoing Flow Control Initialization. LLTP transmissions via other specific VC numbers begin as each specific VC number completes Flow Control Initialization.

Although the following discussion focuses on the Flow Control Initialization of VC0, the concepts discussed apply to the other VC numbers as well. The available buffer space defines a set of buffers for each VC number. For the purposes of this discussion assume the phrases *available buffer space* or *buffer space is available* refer to the entire set of buffers based on the specific VC number under consideration.

As discussed in Chapter 11, the transfer of a TLP to the Data Link Layer to be encapsulated in a subsequently transmitted LLTP by the Physical Layer occurs if buffer space is available at the receiving port. The Gate circuitry at the boundary of the Transaction Layer and Data Link Layer determines the buffer space needed for a TLP by its FCCs (Flow Control Credit) required for the Header field and the Data field of the TLP contained within the LLTP. The minimum initial FCC value that must be assigned to Header and Data fields of each type of TLP is summarized in Table 7.2. Each FCC equals 16 bytes. The values in Table 7.2 are the minimum values; if larger buffer space is available at the receiving port the initial values can be larger. It is the responsibility of the TLPs' receiving port on the link to supply this information to the TLPs' transmitting port on the link. The transmitting port makes no assumption of the initial buffer space available at the receiving port.

> The minimum FCC values for each type of buffer during Flow Control Initialization of a link are listed in Table 7.2. The transmitting port of the LLTP can optionally check to determine that the initial buffer space available is equal to or greater than the minimum; if not a **Flow Control Protocol Error (FCPE)** is optionally reported. See Chapter 9 for more information about **FCPE**.

Set of Buffers

A link is comprised of multiple virtual channels (VCs). Each VC is assigned a specific set of buffers. The set of buffers at each receiving port for each VC number is defined as follows:

- **Buffer type PH (posted header):** Buffer space for Header fields of memory write and message requester transaction packets.

- **Buffer type PD (posted data):** Buffer space for Data fields of memory write and message requester transaction packets.

- **Buffer type NPH (non-posted header):** Buffer space for Header fields of memory read, I/O read and write, and configuration read and write requester transaction packets.

- **Buffer type NPD (non-posted header):** Buffer space for Data fields of I/O write and configuration write requester transaction packets.
- **Buffer type CPLH (completer header):** Buffer space for Header fields of completer transaction packets.
- **Buffer type CPLD (completer data):** Buffer space for Data fields of completer transaction packets.

Table 7.2 Initialization with Minimum FCC values for Transaction packets

Transaction Packet Type	Receiving Port Buffer Type Used Header/Data	FCCs for Header Field (1) Minimum for Initialization (2)(3)	FCCs for Data Field Minimum for Initialization (2)
Memory Write requester or Message requester	PH/PD	01h	Maximum_Payload_Size in bytes/16 (4) (Rounded up)
Memory Read requester, I/O Read requester or Configuration read Requester	NPH	01h	000h (5)
I/O Write Requester or Configuration Write Requester	NPH/NPD	01h	001h
Completer for a memory read requester (data) Receiving port is switch, Root Complex (7), or bridge (8)	CPLH/CPLD	01h	Maximum_Payload_Size in bytes/16 or the largest read request, whichever is smaller (Rounded up)

Table 7.2 Initialization with Minimum FCC values for Transaction packets *(continued)*

Transaction Packet Type	Receiving Port Buffer Type Used Header/Data	FCCs for Header Field (1) Minimum for Initialization (2)(3)	FCCs for Data Field Minimum for Initialization (2)
Completer for a memory read requester (data) Receiving port is Root Complex (6) or endpoint	CPLH/CPLD	00h (3)	000h (3)
Completer for I/O or configuration read requester (data) Receiving port is switch or PCI Express/PCI-X Bridge in PCI-X mode (5)	CPLH/CPLD	01h	001h
Completer for I/O or configuration read requester (data) Receiving port is Root Complex, endpoint, or PCI Express/PCI Bridge (5)	CPLH/CPLD	00h (3)	000h (3)
Completer for I/O or configuration write requester (data) Receiving port is switch, Root Complex (7), or bridge (8)	CPLH	01h	000h (5)
Completer for I/O or configuration write requester (data) Receiving port is Root Complex (6) or endpoint	CPLH	00h (3)	000h (5)

Table Notes:

1. The FCC value does not include buffer space required for the Digest field.
2. Initialization of VC0 after reset is done by hardware and these values are defined as default.
3. The 000h defines a value of infinite buffer space at the receiver port for use by the transmitting port. The assumption is that the Root Complex or endpoint that sourced the memory read requester transaction packet will have reserved sufficient buffer space to accept the data of the associated completer transaction packet.
4. For message requester transactions with no data the value = 000h.
5. There is no specific value for these buffers for the associated TLPs because there is no actual data to be buffered. However, for the calculation in the transfer equations in Chapter 11, the value of 000h is defined as the initial value, which indicates infinite available buffer space.
6. The specific downstream port of the Root Complex is not connected internally to another downstream port to support the optional peer-to-peer transactions through the Root Complex.
7. The specific downstream port of the Root Complex is connected internally to another downstream port to support the optional peer-to-peer transactions through the Root Complex. Optionally, the Root Complex can define CPLH and CPLD to be 000h if it can be designed such that deadlocks are avoided.
8. These are the basic bridge requirements. The *PCI Express to PCI/PCI-X Bridge Specification revision 1.0,* released February 18, 2003, defines these additional specific requirements for the initial FCC value.

The Data Link Layer uses Flow Control DLLPs to communicate the available buffer space at the receiving port. The Flow Control DLLPs that are used for initialization information are InitFC1 and InitFC2. The information is the FCC values contained in the HdrFC and DataFC fields of the Flow Control DLLPs for Header and Data fields of the TLPs, respectively.

The maximum amount of available buffer space that LLTPs' receiving port can indicate at any time is 128d for the Header field and 2048d for the Data field. The exception is the previously noted buffers with an effective infinite available buffer space (00h and 000h). This requirement is due to the protocol of the transfer equations discussed in Chapter 11. The LLTPs' transmitting ports (receivers of the DLLP) can optionally check for this error and optionally report it as **Flow Control Protocol Error (FCPE)**. Obviously, this should never be an issue for the initial buffer values of the InitFC1 and InitFC2 DLLPs, but may become an issue for the Update FC DLLP. See Chapter 9 for more information about **FCPE**.

Flow Control Initialization Protocol

The link activity states related to Flow Control Initialization are summarized in Figure 7.2. Its primary focus is the Flow Control Initialization of the minimum VC number, which by definition is VC0. Once VC0 has completed Flow Control Initialization in the DL_Init link activity state, the Flow Control Initialization continues for the VC numbers other then VC0 in the DL_Active link activity state.

An example of the Flow Control Initialization protocol is illustrated in Figures 7.3 through 7.8. The example focuses on the upstream port being initialized by the downstream port and vice versa for VC0. VC0 is hardwired to be enabled and Flow Control Initialization must first be applied to VC0 (the PCI Express specification requires that VC0 is always operational). The Flow Control Initialization is applied to the other VC numbers (VCx in this example) after it has been successfully applied to VC0 and if the associated VC number has been enabled by software. The software must enable a specific VC number via Bit [31] = 1b of the VC Resource Control Register in the configuration register bank of a PCI Express device. The software must enable and disable the specific VC number at each end of the link as a pair. That is, for a specific VC number, it cannot be enabled on one end of the link and disabled on the other end, except for the time it takes for the software to program one and then the other. Once Flow Control Initialization is applied to VC0 in both directions, the link activity state transitions from DL_Init to DL_Active and the Flow Control Initialization of the other enabled VCx can be done in any order. The Flow Control Initialization of VC0 occurs by default. Configuration transactions within LLTPs during the DL_Active link activity state use VC0 to enable VC numbers at different ports. If VC0 is not able to complete the Flow Control Initialization in both directions the link activity state transitions from DL_Init to DL_Inactive.

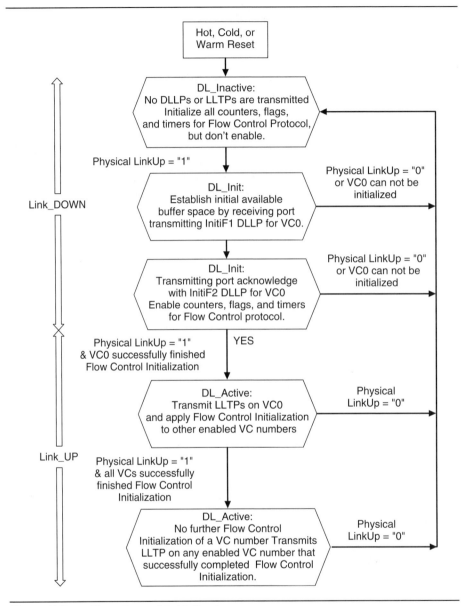

Figure 7.2 Flow Control Initialization

> The FI1 flag discussed in this section actually represents three sub-flags, one for each buffer of the set of buffers for a specific VC number. It takes the successful reception of three InitFC1 DLLPs to initialize the set of three buffers for each VC number. When all three sub-flags are set to "1" (the associated initialized buffer sets the sub-flag) for a specific VC number, then the FI1 flag for that VC number is also set to a "1" value. The FI2 flag discussed in this section is a single flag; three InitFC1 DLLPs need not be received to set the FI2 flag to a "1" value.
>
> Any violation of the Flow Control Initialization protocol may be reported as a **Data link Layer Error (DLLPE)**. See Chapter 9 for more information about **DLLPE**.

Flow Control Initialization Sub-state Protocol

Figures 7.3 through 7.8 illustrate an example of the Flow Control Initialization of VC0. Part of the example is the application of two sub-states: FC_Init1 and FC_Init2. The two sub-states provide a handshake protocol that can be applied to determine whether InitiFC1 DLLPs or InitiFC2 DLLPs are being transmitted. The actual implementation may vary provided the resulting transmission of InitiFC1 and InitiFC2 DLLPs look the same.

Sub-state FC_Init1

- After Hot, Cold, or Warm Reset the FL1 and FL2 flags are set to "0" values, one set in the upstream port and one set in the downstream port.

- Three InitFC1 DLLPs are transmitted by a port in the sub-state FC_Init1 (one in the upstream port and one independently in downstream port) uninterrupted in the following exact sequence for VC0 (the port transmits no other DLLPs):
    - InitiFC1 DLLP for PH and PD buffers.
    - InitiFC1 DLLP for NPH and NPD buffers.
    - InitiFC1 DLLP for CPLH and CPLD buffers.

- For VC0 this sequence is retransmitted by the port at the maximum possible rate.
- InitiFC1 or InitiFC2 DLLPs for VC0 received by a port in sub-state FC_Initi1 record the values in the HdrFC and DataFC fields.
- Once a port has successfully received a sequence of InitFC1 or InitFC2 DLLPs for VC0, it sets the FI1 flag to "1" and transitions to the sub-state FC_Init2. The transition to the sub-state FC_Init2 by one port does not mean that the port on the other end of the link has also transitioned to sub-state FC_Initi2.

Sub-state FC_Initi2

- When the FL1 flag is set = "1" and FL2 flag is set = "0" in a port, it is defined as being in the sub-state FC_Initi2.
- Three InitFC2 DLLPs are transmitted by a port in the sub-state FC_Initi2 (one in upstream port and one independently in downstream port) uninterrupted in the following exact sequence for VC0 (the port transmits no other DLLPs):
  - InitiFC2 DLLP for PH and PD buffers.
  - InitiFC2 DLLP for NPH and NPD buffers.
  - InitiFC2 DLLP for CPLH and CPLD buffers.
- For VC0 this sequence is retransmitted by the port at the maximum possible rate.
- InitiFC1 or InitiFC2 DLLPs for VC0 received by a port in sub-state FC_Initi2 ignore the values in the HdrFC and DataFC fields.
- Once a port in the sub-state FC_Initi2 receives one InitFC2 DLLPs, LLTP, or UpdateFC DLLP for VC0; it sets the FI2 flag to "1" and transitions out of sub-state FC_Init2. The transition from the sub-state FC_Init2 indicates that the port has completed Flow Control Initialization.

Example of Flow Control Initialization Sub-state Protocol for VC0

As illustrated by Figures 7.3 and 7.4, the downstream port begins transmitting the InitF1 DLLP sequences first, followed independently by the upstream port. As illustrated by Figure 7.5, the upstream port is first to receive the entire InitF1 DLLP sequences and successfully record the available buffer space in the downstream port for VC0. Consequently, the upstream port sets FL1 to "1", transitions to sub-state FC_Init2, begins transmitting InitF2 DLLP sequences, and Gate circuitry no longer blocks the transfer TLPs from the transmit buffer to the retry buffer for VC0. The downstream port remains in sub-state FC_Init1.

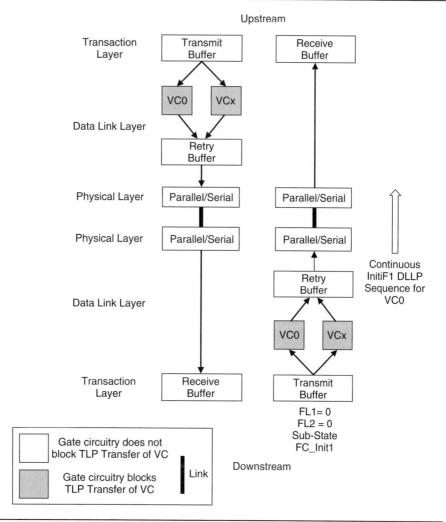

Figure 7.3 Initialization of VC0, Part 1

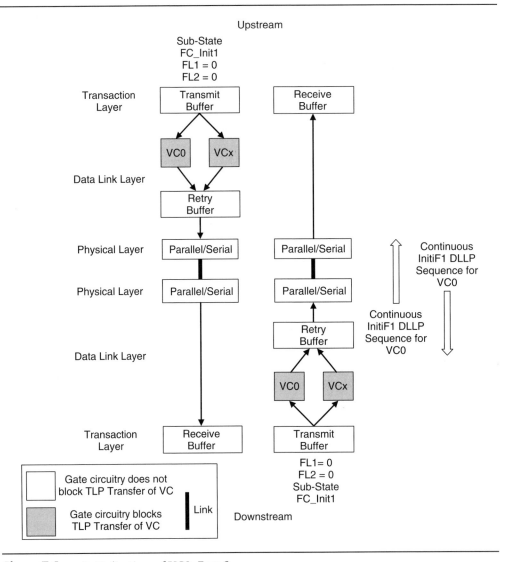

Figure 7.4 Initialization of VC0, Part 2

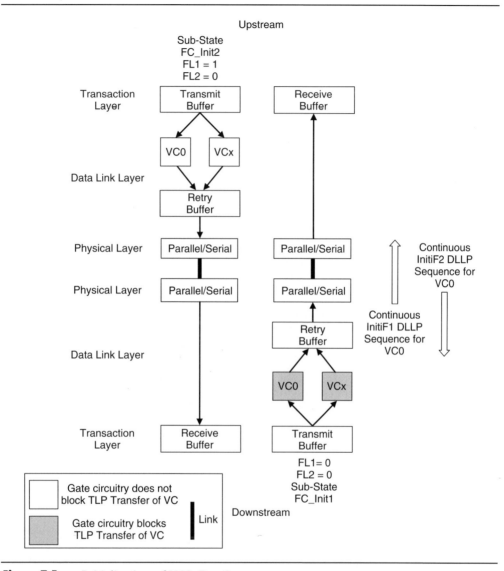

Figure 7.5 Initialization of VC0, Part 3

As illustrated in Figure 7.6, the downstream port is second to receive the entire InitF1 DLLP or InitF2 DLLP sequences and successfully record the available buffer space in the upstream port for VC0. Consequently, the downstream port sets FL1 to "1", transitions to sub-state FC_Init2, begins transmitting InitF2 DLLP sequences, and Gate circuitry no longer blocks the transfer of TLPs from the transmit to retry buffer for VC0.

It does not matter whether the downstream port's transition to sub-state FC_Init2 was due to the InitF1 DLLP sequences in process when the upstream port began transmitting InitF2 DLLPs, or due to the InitF2 DLLP sequences (in whole or in part).

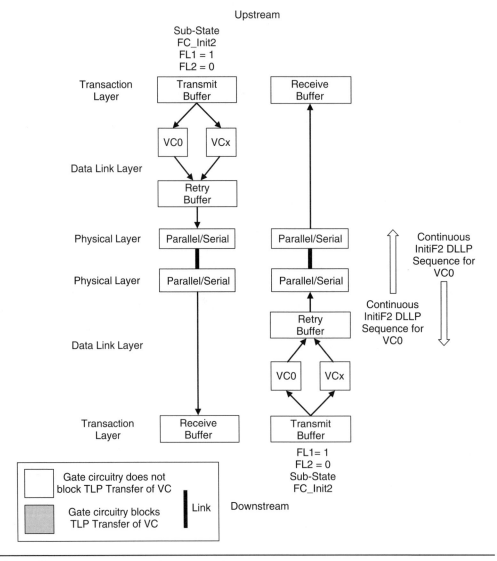

Figure 7.6 Initialization of VC0, Part 4

Chapter 7: Data Link Layer and Packets ■ 337

As illustrated in Figure 7.7, the downstream port is the first to receive at least one InitF2 DLLP once it has transitioned to sub-stateFC_Init2. Consequently, the downstream port is the first to exit sub-stateFC_Init2 and cease transmitting the InitiFC2 DLLP sequences. Alternatively, as illustrated in Figure 7.8, the upstream port is the first to receive at least one InitF2 DLLP once it has transitioned to sub-stateFC_Init2. Consequently, the upstream port is the first to exit sub-stateFC_Init2 and cease transmitting the InitiFC2 DLLP sequences.

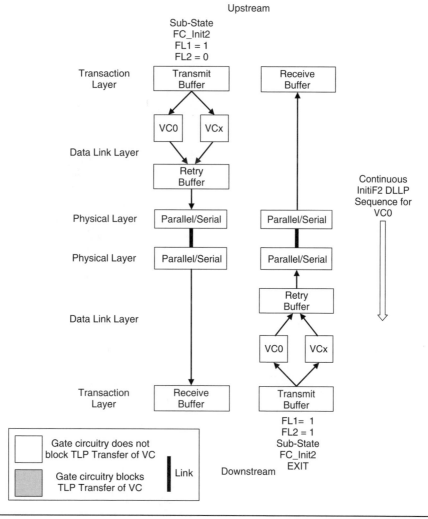

Figure 7.7 Initialization of VC0, Part 5

338 ■ The Complete PCI Express Reference

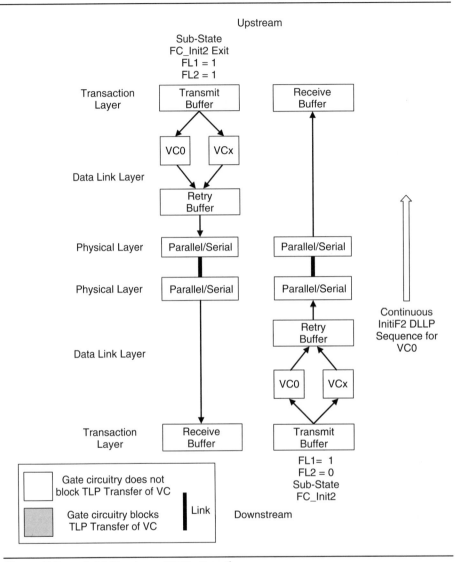

Figure 7.8   Initialization of VC0, Part 6

In either situation illustrated in Figures 7.7 and 7.8, one port has ceased transmitting the InitiF2 DLLP sequences and the other port has not exited sub-state FC_Init2. Typically, sufficient InitiF2 DLLP sequences are in transit such that the other port receives the one InitF2 DLLP to permit it to exit from sub-state FC_Init2. However, the previously discussed protocol for sub-state FC_Init2 permits a port to exit from this sub-state with the receipt of at least one InitF2 DLLP, LLTP, or UpdateFC DLLP. Consequently, for VC0 both upstream and downstream ports eventually exit sub-state FC_Init2. For a specific VC number, in this example VC0, when both ports have successfully exited for sub-state FC_Init2 the associated VC number has successfully completed Flow Control Initialization.

Flow Control Initialization Applied to VCx

The Flow Control Initialization protocol discussed in the previous section for VC0 also applies to each VC number (VCx) as it becomes enabled. If multiple VC numbers are enabled, the receiving portion of the LLTPs (containing TLPs) of each port on each end of a link sequentially applies the Flow Control Initialization protocol in a round robin sequence. For example, it possible for the downstream port on a link to transmit an InitFC1 DLLP sequence for VC1 when the downstream port of the same link transmits an InitFC1 DLLP sequence for another VC2. If each port transmits the next InitFC1 DLLP sequence for the next VC number by using a round robin, eventually both ports transmit an InitFC1 DLLP sequence for all enabled VC numbers. Similarly, the round robin approach assures that both ports transmit an InitFC2 DLLP sequence for all enabled VC numbers. As each VC number successfully completes the Flow Control Initialization, it is taken out of the round robin sequence. Applying Flow Control Initialization to VC0 differs from applying Flow Control Initialization to non-VC0 (VCx) in two major ways.

The first difference is that VC0 is the first VC number that Flow Control Initialization is applied to and it must be successful. If VC0 does not successfully complete Flow Control Initialization, the link transitions to the DL_Inactive link activity state. When Flow Control Initialization is applied to VCx and it is not successful, NO LLTPs can be transmitted on VCx, but the link remains in the DL_Active link activity state, and LLTPs are transmitted on the other VC numbers that successfully completed Flow Control Initialization. It is possible for one port of the two ports on the link to successfully transition from sub-state InitF1 to sub-state InitF2 and thus begin transmitting LLTPs. The other port may not successfully transition from sub-state InitF1 to sub-state InitF2. Eventually an error will occur because LLTPs requiring responses will not receive them.

The second difference is that during the execution of Flow Control Initialization on VC0, the transmission of InitFC1 and InitFC2 DLLP are at the maximum rate. During the execution of a Flow Control Initialization on VCx, the transmission of InitFC1 and InitFC2 are governed by the resend timer. The resend timer times out and allows an InitFC1 or InitFC2 DLLP to be transmitted as follows. Upon entry into sub-state FC_Init1 or sub-state FC_Init1 for VCx a resend timer is started and runs continuously throughout the sub-state. For enabled VC numbers other than VC0, the retransmission of a sequence occurs whenever transmission of LLTPs' other DLLPs (that is, DLLPs not having to do with InitiFC1 and InitiFC2 DLLPs) are not pending. The one exception for a sequence transmission to take priority over pending LLTPs of other DLLPs not related to InitiFC1 and InitiFC2 DLLPs) is when transmission of the sequence is needed to maintain minimum frequency. Minimum frequency is defined as a pending sequence transmission at 17 to 34 microseconds from the start of the last sequence transmission. Note: According to the *PCI Express Specification revision 1.0* it is not clear whether the minimum frequency is between InitiFC1 or InitiFC2 DLLP sequences of a specific VC number *or* between the InitiFC1 or InitiFC2 DLLP sequences of any VC number undergoing Flow Control Initialization.

Flow Control Protocol Update of Available Buffer Space

As discussed in the previous section, during Flow Control Initialization the InitFC1 and InitFC2 DLLPs are transmitted to establish the initial available buffer space at the LLTPs' receiving ports of a specific VC number. After Flow Control Initialization of a specific VC number, the Flow Control protocol includes the transmission of UpdateFC DLLPs to update the available buffer space information as LLTPs are transmitted and received. The LLTPs' receiving ports transmit the most recent FCC values contained in UpdateFC DLLPs to the transmitting ports following the protocol discussed in Chapter 11.

Consider the buffer pairs defined as PH and PD buffers, NPH and NPD buffers, or CPLH and CPLD buffers. If during Flow Control Initialization an infinite FCC value was transmitted (00h or 000h in the InitiFC1 and InitiFC2 DLLPs) for both buffers of a specific buffer pair, subsequent transmissions of UpdateFCC DLLPs are not required. If a subsequent transmission of an UpdateFCC DLLP does optionally occur, it must contain an infinite FCC value (00h and 000h) for both buffers of the specific buffer pair, and the UpdateFCC DLLP is discarded These buffer pairs are PH and PD buffers, NPH and NPD buffers, or CPLH and CPLD buffers. If during Flow Control Initialization an infinite FCC value was transmitted (00h or 000h in the InitiFC1 and InitiFC2 DLLPs) for only one buffer of a specific buffer pair (PH and PD buffers or NPH and NPD buffers or CPLH and CPLD buffers), subsequent transmissions of UpdateFCC DLLPs for both buffers of the specific buffer are required. In the UpdateFCC DLLP, the FCC value of the buffer of the buffer pair that was initially infinite via InitFC1 and InitFC2 DLLPs must have an infinite value (00h or 000h) in the UpdateFC DLLP and that value is ignored. The FCC value of the other buffer of the buffer pair is used. The transmitting port of the LLTP can optionally check to determine whether these protocols have been followed; if not, a **Flow Control Protocol Error (FCPE)** is optionally reported. See Chapter 9 for more information on this error.

The maximum amount of available buffer space that LLTPs' receiving port can indicate at any time is 128d for the Header field and 2048d for the Data field. The exceptions are the previously noted buffers with an effective infinite available buffer space (00h and 000h). This requirement is due to the protocol of the transfer equations discussed in Chapter 11. The LLTPs' transmitting ports (receivers of the DLLP) can optionally check for this error and optionally report it as **Flow Control Protocol Error (FCPE)**. Obviously this should never be an issue for the initial buffer values of the InitFC1 and InitFC2 DLLPs, but may become an issue for the Update FC DLLP. See Chapter 9 for more information about **FCPE**.

DLLP Formats

As previously discussed, the two types of link packets are the link packet for link management called the Data Link Layer Packet (DLLP), and the link packet containing a TLP called the Link Layer Transaction Packet (LLTP). The ability to distinguish between DLLPs and LLTPs on the link is integral to the FRAMING information of the Physical Packets that contain either the DLLPs or LLTPs. The FRAMING K symbols are attached by the Physical Layer of a port based on the direction of the Data Link Layer. The Physical Layer in the port on the other end of the link can only distinguish LLTPs versus DLLPs by the FRAMING K symbols. Once the Physical Layer has made the distinction, the relevant information is provided to the Data Link Layer. See Chapter 8 for more information about the FRAMING K symbols and the Physical Layer.

As previously discussed, DLLPs are part of the link management activities for the Flow Control, Power Management, and Vendor-Specific activities. Each of these activities has a corresponding set of DLLPS as follows:

- **Flow Control:** InitFC1, InitFC2, and UpdateFC DLLPs.
- **Acknowledgement:** Ack and Nak DLLPs.
- **Power Management:** PM_Active_State_Request_L1, PM_Enter_L1, PM_Enter_L23, and PM_Request_Ack DLLPs.
- **Vendor-Specific:** Vendor Specific DLLPs.

Each DLLP contains a set of packet words (PAWs). The exact set of PAWs and the fields within the PAWs that are fixed versus programmable varies dependent on the DLLP transmitted. The values of the fixed fields and the range of programmable fields within the PAWs are discussed in the following section. All other fixed and programmable field values are RESERVED.

Reserved

A RESERVED field value is logical "0" and is marked with an "R." The use of a RESERVED for individual bits or collectively the entire field for any purpose is prohibited. Unless otherwise specified, the PCI Express device creating the field must implement "0" in the RESERVED field. The PCI Express device that is processing the RESERVED field must ignore it except for cyclic redundancy check calculations.

The RESERVED field is included in the CRC calculations with no preconceived notion by the receiving PCI Express device that the RESERVED field has a "0" value. This permits the same hardware to be used with future revisions of the PCI Express specification when some of the RESERVED bits or fields may be redefined. See Chapter 9 for more information.

Flow Control DLLPs

The purpose of Flow Control (FC) DLLPs is to provide buffer information between the two ports on a link. The Gate circuitry at the boundary of the Transaction and Data Link Layers will not transfer TLPs from the Transaction Layer unless the receiving port has sufficient buffer space as discussed in Chapter 11. The InitiFC1 and InitiFC2 DLLPs are only used for Flow Control Initialization discussed earlier in this chapter. The UpdateFC DLLP is only used for available buffer space updates following the Flow Control protocol. See Chapter 12 for more information.

The FC DLLPs are transmitted across the link without acknowledgement. Without an acknowledgement, the primary protection of the integrity of the DLLPs is the CRC that is required for each DLLP. See Chapter 9 for more information. The FC DLLPs are transmitted independent of the enabled VC number(s). A protocol prioritizes transmission between DLLPs, and between DLLPs and LLTPs.

Flow Control DLLPs consist of fields within the PAWs, as illustrated in Figure 7.9.

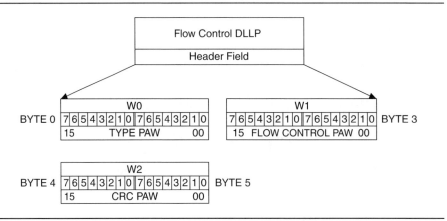

Figure 7.9   Flow Control DLLP

InitFC1 DLLP Header Field

7	6	5	4	3	2	1	0	7	6	5	4	3	2	1	0
7		LINK TYPE					0	R	R	7		HdrFC			2
HdrFC 1 0		R	R	11				DataFC							0
15							CRC								00

15	TYPE PAW	00
15	FLOW CONTROL PAW	00
15	CRC PAW	00

TYPE PAW

LINK TYPE [7::0]: Select link management

- **LINK TYPE [7::6]** = 01b defines an InitFC1 DLLP.
- **LINK TYPE [5::4]** defines which type of TLP the DLLP is associated with:
    - LINK TYPE [5::4] = 00b for memory write or message requester transactions (PH,PD).
    - LINK TYPE [5::4] = 01b for memory read, I/O read and write, and configuration read and write requester transactions (NPH, NPD).
    - LINK TYPE [5::4] = 10b for completer transactions (CPLH, CPLD).
- **LINK TYPE [3::0]** = 0v2v1v0b. Select the respective virtual channel undergoing Flow Control Initialization.
- All other values are RESERVED unless defined for other DLLPs.

HdrFC [7::2]: Used in conjunction with HdrFC [1::0].

FLOW CONTROL PAW

HdrFC [1::0]: Used in conjunction with HdrFC [7::2] TYPE PAW form HdrFC[7:0]. HdrFC[7:0] contains the FCC value for the buffer of the Header field of the associated type of TLP.

DataFC [11::0]: Contains the FCC value for the buffer of the Data field of the associated type of TLP. If no associated Data field, the value is RESERVED = 000h.

CRC PAW

CRC [15:00]: Provides the cyclic redundancy check information for the TYPE and FLOW CONTRL PAW. (See Chapter 9.)

InitFC2 DLLP Header Field

7	6	5	4	3	2	1	0	7	6	5	4	3	2	1	0
7		LINK TYPE					0	R	R	7		HdrFC			2
HdrFC 1 0		R	R	11				DataFC							0
15							CRC								00
15							TYPE PAW								00
15							FLOW CONTROL PAW								00
15							CRC PAW								00

TYPE PAW

LINK TYPE [7::0]: Select link management

- LINK TYPE [7::6] = 11b defines an InitFC2 DLLP.

- LINK TYPE [5::4] defines which type of TLP the DLLP is associated with:
    - LINK TYPE [5::4] = 00b for memory write or message requester transactions (PH,PD).
    - LINK TYPE [5::4] = 01b for memory read, I/O read and write, and configuration read and write requester transactions (NPH, NPD).
    - LINK TYPE [5::4] = 10b for completer transactions (CPLH, CPLD).

- LINK TYPE [3::0] = 0v2v1v0b. Select the respective virtual channel undergoing Flow Control Initialization.

- All other values are RESERVED unless defined for other DLLPs.

HdrFC [7::2]: Used in conjunction with HdrFC [1::0].

FLOW CONTROL PAW

HdRFC [1::0]: Used in conjunction with HdrFC [7::2] TYPE PAW form HdrFC[7:0]. HdrFC[7:0] contains the FCC value for the buffer of the Header field of associated type of TLP.

DataFC [11::0]: Contains the FCC value for the buffer of the Data field of the associated type of TLP. If no associated Data field, the value is RESERVED = 000h.

CRC PAW

CRC [15:00]: Provides the cyclic redundancy check information for the TYPE and FLOW CONTRL PAW. (See Chapter 9.)

UpdateFC DLLP Header Field

7	6	5	4	3	2	1	0	7	6	5	4	3	2	1	0
7			LINK TYPE				0	R	R	7		HdrFC			2
HdrFC 1 0		R	R	11						DataFC					0
15								CRC							00

15	TYPE PAW	00
15	FLOW CONTROL PAW	00
15	CRC PAW	00

TYPE PAW

LINK TYPE [7::0]: Select link management

- LINK TYPE [7::6] = 10b defines an UpdateFC DLLP.
- LINK TYPE [5::4] = defines which TLP the DLLP is associated with:
  - LINK TYPE [5::4] = 00b for memory write or message requester transactions (PH,PD).
  - LINK TYPE [5::4] = 01b for memory read, I/O read and write, and configuration read and write requester transactions (NPH, NPD).
  - LINK TYPE [5::4] = 10b for completer transactions (CPLH, CPLD).

- LINK TYPE [3::0] = 0v2v1v0b. Select the respective virtual channel is having its available buffer space updated.

- All other values are RESERVED unless defined for other DLLPs.

HdrFC [7::2]: Used in conjunction with HdrFC [1::0].

FLOW CONTROL PAW

HdrFC [1::0]: Used in conjunction with HdrFC [7::2] TYPE PAW to form HdrFC[7:0]. HdrFC[7:0] contains the FCC value for the buffer of the Header field of the associated type of TLP.

DataFC [11::0]: Contains the FCC value for the buffer of the Data field of the associated type of TLP. If no associated Data field, the value is RESERVED = 000h.

CRC PAW

CRC [15:00]: Provides the cyclic redundancy check information for the TYPE and FLOW CONTRL PAWs. See Chapter 9 for more information.

Acknowledgement DLLPs

The purpose of the Acknowledgement DLLPs Ack and Nak is to provide the LLTPs' transmitting ports information from the receiving ports on whether receipt of transmitted LLTPs was successful or unsuccessful. The Ack and Nak DLLPs only apply to LLTPs and do not provide information related to other DLLPs.

An Ack DLLP acknowledges successful receipt of a single LLTP or a collection of LLTPs. Successful is defined as receipt of a correct SEQ# and correct CRC in the LLTP. See Chapter 11 for more information on the Flow Control protocol. Successful is also defined for receipt of correct FRAMING K symbol in the Physical Layer. See Chapter 8 for more information.

A Nak DLLP acknowledges unsuccessful receipt of a LLTP. Unsuccessful is defined as receipt an incorrect SEQ# and/or an incorrect LCRC. See Chapters 9 and 11 for more information. Unsuccessful is also defined for receipt of incorrect FRAMING K symbols in the Physical Layer.

When the LLTP received contains a nullified TLP there are two important considerations. First, if the LLTP is received with no LCRC error, the packet is discarded and no Ack DLLP is returned related to the LLTP. Second, if the LLTP is received with an LCRC error, it must be treated as a not successfully received LLTP and a Nak DLLP must be scheduled for transmission to the transmitting port of the LLTP. See Chapter 9 for more information about processing LLTPs received.

The Ack or Nak DLLPs are transmitted across the link without acknowledgement. Without an acknowledgement, the primary protection of the integrity of the DLLPs is the LCRC that is required for each DLLP. See Chapter 9 for more information. The Ack and Nak DLLPs are transmitted independent of the enabled VC number(s). A protocol prioritizes transmission protocol between DLLPs, and between DLLPs and LLTPs. See Chapter 12 for more information.

Ack and Nak DLLPs consist of fields within the PAWs, as illustrated in Figure 7.10.

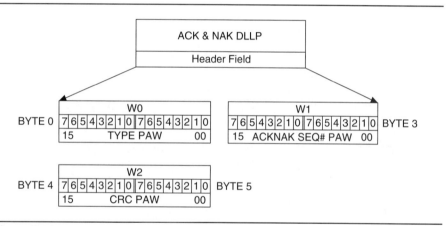

Figure 7.10   Ack DLLP and Nak DLLP

Ack DLLP and Nak DLLP Header Field

7	6	5	4	3	2	1	0	7	6	5	4	3	2	1	0
7	LINK TYPE						0	R	R	R	R	R	R	R	R
R	R	R	R	11				AckNakSEQ#							0
15								CRC							00

15	TYPE PAW	00
15	ACKNAK SEQ PAW	00
15	CRC PAW	00

TYPE PAW

LINK TYPE [7::0]: Select link management.

- LINK TYPE [7::0] = 0000 0000b defines ACK DLLP and 0001 0000b defines NAK DLLP.
- All other values are RESERVED unless defined for other DLLPs.

ACKNAK SEQ PAW

ACKNAK SEQ [15::0]: AckNakSEQ# [11::00] contains the SEQ# for the last successfully received link packet. See Chapter 11 for more information.

CRC PAW

CRC [15:00]: Provides the cyclic redundancy check information for the TYPE and ACKNAK SEQ PAWs. See Chapter 9 for more information.

Power Management (PM) DLLPs

Power Management (PM) DLLPs are part of the PCI Express Power Management protocol at the link level and support the transition from one link state to another. (For more detailed information on PCI Express Power Management, see Chapter 15.) The different link states define different power levels relative to the link. The PM DLLPs are defined as follows:

- **PM_Enter_L1**: A downstream PCI Express device sends this DLLP to request an upstream PCI Express device to transition to the L1 link state. This request is required by the PCI-Power Management protocol.
- **PM_Enter_L23**: A downstream PCI Express device sends this DLLP to request an upstream PCI Express device to transition to the L2/3 ready link state. This request is required by the PCI-Power Management protocol.

- **PM_Active_State_Request_L1:** A downstream PCI Express device sends this DLLP to request an upstream PCI Express device to transition to the L1 link state. This request is required by the Active-State Power Management protocol.

- **PM_Request_Ack:** Allows an upstream PCI Express device to acknowledge to the downstream PCI Express device the receipt of a PM_Enter_L1, PM_Enter_L23, or PM_Active-State_Request_L1 DLLP transmitted by the downstream PCI Express device. This acknowledgement is required by both the request is required by the PCI-Power Management and Active-State Power Management protocols.

The ports transmitting PM_Request_Ack DLLP acknowledge the PM_Enter_L1 DLLP, PM_Enter_L23 DLLP, or PM_Active_State_Request_L1 DLLPs. The PM_Request_Ack DLLP is not acknowledged. Without an overall acknowledgement, the primary protection of the integrity of the PM_Request_Ack DLLP is the CRC that is required for each DLLP. See Chapter 9 for more information. The PM DLLPs are transmitted independent of the enabled VC number(s). A protocol prioritizes transmission between DLLPs, and between DLLPs and LLTPs. See Chapter 12 for more information on this protocol.

PM DLLPs consist of fields in Figure 7.11.

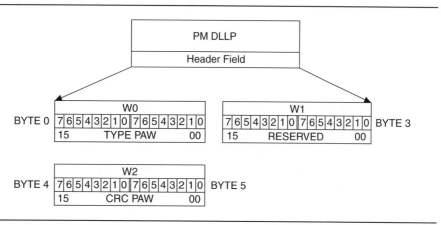

Figure 7.11 PM DLLP

PM DLLPs Header Field

7	6	5	4	3	2	1	0	7	6	5	4	3	2	1	0
7			Link Type				0	R	R	R	R	R	R	R	R
R	R	R	R	R	R	R	R	R	R	R	R	R	R	R	R
15							CRC PAW								00

15	TYPE PAW	00
15	RESERVED PAW	00
15	CRC PAW	00

TYPE PAW

LINK TYPE [7::0]: Select link management.

- Link [7::0] = 0010 0000b defines PM_Enter_L1.
- Link [7::0] = 0000 0001b defines PM_Enter_L23.
- Link [7::0] = 0010 0011b defines PM_Active_State_Request_L1.
- Link [7::0] = 0010 0100b defines PM_Request_Ack.
- All other values are RESERVED unless defined for other DLLPs.

RESERVED PAW

RESERVED [15::0]: RESERVED = 0000 0000 0000 0000b

CRC PAW

CRC [15:00]: Provides the cyclic redundancy check information for the TYPE and RESERVED PAWS. (See Chapter 9.)

Vendor-Specific DLLPs

> The Vendor-Specific DLLPs are defined as one of the link management packets. This is a loose association. The Vendor-Specific DLLP is by definition defined by the vendor. Its implementation may have nothing to do with link management, but is included in the set of DLLPs for two reasons — One: The format is the same as the other link management link packets. Two: It is only defined for each link and is not defined for flowing beyond the ports connected to a specific link.

The Vendor-defined DLLP is transmitted across the link without acknowledgement. Without an acknowledgement, the primary protection of the integrity of the Vendor-Specific DLLP is the CRC that is required for each DLLP. The Vendor-Specific DLLP is transmitted independent of the enabled VC number(s). A protocol prioritizes transmission between DLLPs, and between DLLPs and LLTPs. See Chapter 12 for more information on this protocol.

Vendor-specific link packets consist of fields in Figure 7.12.

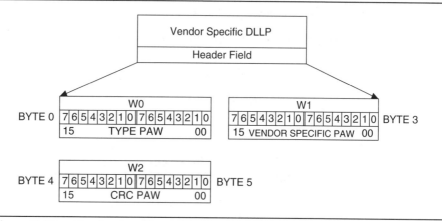

Figure 7.12 Vendor-Specific DLLP

Vendor Specific DLLP Header Field

7	6	5	4	3	2	1	0	7	6	5	4	3	2	1	0
7			Link Type				0	7			Vendor-Specific				0
15							VENDOR-SPECIFIC								00
15							CRC								00

15	TYPE PAW	00
15	VENDOR-SPECIFIC PAW	00
15	CRC PAW	00

TYPE PAW

LINK TYPE [7::0]: Select link management.

- Link [7::0] = 0011 0000b defines Vendor-Specific DLLP.
- All other values are RESERVED unless defined for other DLLPs.

Vendor-Specific [7::0]: RESERVED and specific to vendor.

VENDOR-SPECIFIC PAW

Vendor-Specific [15::0]: RESERVED and specific to vendor.

CRC PAW

CRC [15:00]: Provides the cyclic redundancy check information for the TYPE and Vendor-Specific PAWS. (See Chapter 9.)

LLTP Format

As previously discussed, the two types of link packets are the link packet for link management called the Data Link Layer Packet (DLLP) and the link packet containing a TLP called the Link Layer Transaction Packet (LLTP). The ability to distinguish between DLLPs and LLTPs on the link is integral to the FRAMING information of the physical packets. The FRAMING K symbols are attached by the Physical Layer of a port based on the direction of the Data Link Layer. The Physical Layer in the port on the other end of the link can only distinguish LLTPs versus DLLPs by the FRAMING K symbols. Once the Physical Layer has made the distinction, the relevant information is provided to the Data Link Layer. See Chapter 8 for more information about the FRAMING K symbols and the Physical Layer.

Each LLTP contains three packet words (PAWs): SEQ#, LCRC (upper order), and LCRC (lower order). The fields within the PAWs are all programmable. The values and the range of programmable fields within the PAWs are discussed in the following section.

Reserved

A RESERVED field value is logical "0" and is marked with an "R." The use of a RESERVED for individual bits or collectively the entire field for any purpose is prohibited. Unless otherwise specified, the PCI Express device creating the field must implement "0" in the RESERVED field. The PCI Express device that is processing the RESERVED field must ignore it except for cyclic redundancy check calculations. For a LLTP the only RESERVED fields are within the TLP it contains.

The RESERVED field is included in the LCRC calculations with no preconceived notion by the receiving PCI Express device that the RESERVED field has a "0" value. This permits the same hardware to be used with future revisions of the PCI Express specification when some of the RESERVED bits or fields may be redefined.

LLTPs

LLTPs packets consist of PAWs attached to *one* TLP, as illustrated in Figure.7.13. Unlike transaction requester packets, no completer transaction packets are associated with the LLTPs directly. The Requester/Completer protocol only applies to the TLPs contained within the LLTPs, but not the LLTPs directly. However, as discussed earlier in this chapter, the Ack and

Nak DLLPs provide information to the LLTPs transmitting ports about whether receipt of the LLTPs at the receiving ports was successful or unsuccessful.

The DLLPs are transmitted independent of the enabled VC number(s). However, a protocol prioritizes transmission between DLLPs and other DLLPs, and between DLLPs and LLTPs.

The LLTPs received at a port are checked by the Data Link Layer for the proper sequence (SEQ#) and for any errors in the LLTPs (LCRC). The Data Link Layer responds with an Ack or Nak DLLPs to acknowledge successful or unsuccessful receipt of the LLTP, respectively. If the port transmitting the LLTPs does not receive Ack DLLPs in a timely fashion, the Flow Control protocol includes timers and retry transmissions. See Chapters 9 and 11 for more information on this procedure.

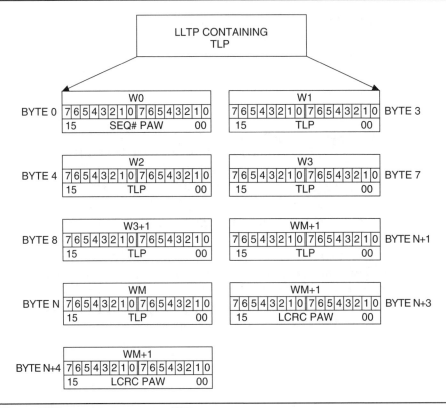

Figure 7.13 LLTP Containing TLP

Chapter 7: Data Link Layer and Packets ■ 357

LLTP Format with TLP

7	6	5	4	3	2	1	0	7	6	5	4	3	2	1	0
15							SEQ#								00
15							TLP								00
15							TLP								00

v
v

Multiple DWORDS associated with one TLP

v
v

15	TLP	00
31	LCRC	16
15	LCRC	00
15	SEQ# PAW	00
15	TLP	00

v
v

15	TLP	00
31	LCRC PAW	16
15	LCRC PAW	00

SEQ# PAW

SEQ# [11::00]: Sequence number for the associated transaction packet. See Chapter 11 for more discussion about the SEQ#.

7	6	5	4	3	2	1	0	7	6	5	4	3	2	1	0
R	R	R	R	\multicolumn{3}{c}{11}		\multicolumn{6}{c}{SEQ#}		\multicolumn{2}{c}{00}

TLP

The TLP is the Transaction Layer Packet that is defined by the Transaction Layer. Only one TLP is contained in each LLTP. The TLP is further defined in Chapter 6.

LCRC PAW

LCRC [31::00]: Contains the LCRC value that provides the required cyclic redundancy checking the other SEQ# and transaction packet included in the DLLP. This is different that the LCRC associated with DLLPs. (See Chapter 9.)

Chapter 8

Physical Layer and Packets

This is the third of three chapters that focus on the specific elements of each of the three layers introduced in Chapter 2: Transaction Layer, Data Link Layer, and Physical Layer. As shown in Figure 8.1, the Transaction Layer interfaces with the PCI Express device core and provides the basis for the layer's Transaction Layer Packets (TLPs). The Data Link Layer converts the TLPs and any link management information into Link Layer Transaction Packets (LLTPs) and Data Link Layer Packets (DLLPs), respectively. The Physical Layer transmits onto the link LLTPs and DLLPs within Physical Packets and extracts them from the Physical Packets received from the link. Each type of packet is supported by an associated PCI Express layer. This chapter discusses the Physical Layer and the associated Physical Packets. Chapters 6 and 7 discuss the Transaction Layer and Data Link Layer with associated packets, respectively.

Several activities support the Physical Layer's primary purpose of providing the interface between the parallel orientation of the Transaction and Data Link Layers and the serial orientation of the link. The other activities related to the Physical Layer, which have to do with the link states, are discussed in Chapter 14.

The first part of this chapter examines the interaction between the parallel orientation of the Data Link Layer Packets and the series of bytes of the Physical Packets, covering two key concepts.

The first key concept is that the Physical Layer must encapsulate the LLTPs and DLLPs in Physical Packets. The second key concept is that the Physical Layer must include a reference clock in the byte and bit streams that comprise the Physical Packets from the transmitting port to be used at the receiving port. As discussed in Chapter 1, a clock or a strobe signal line as the effective reference clock running in parallel with information and data signal lines introduces settling time issues. The settling time limits the bit transfer rates. To maximize the bit transfer rate, the settling time issue is eliminated by integrating the reference clock with the information and data bytes being transmitted onto each differential driven pair of signal lines of each lane. The Physical Layer at the transmitting port must include the reference clock in the sbyte stream of Physical Packets. At the receiving port *phase lock loops* (PLLs) extract the reference clock and use it to determine the valid sampling points of the incoming sbyte stream bits. The actual transfer of the reference clock with the information and data across a link is via a sbyte stream.

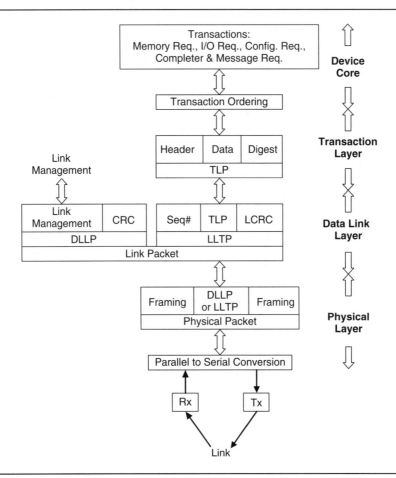

Figure 8.1 PCI Express Link Layers

> This book uses the term *sbytes* to reference the 10-bit bytes used for special or data symbols of the Physical Packets that result from the integrating of the reference clock with 8-bit bytes of the FRAMING, LLTPs, and DLLPs. See the following discussion.

The second part of this chapter examines the parsing of the symbol streams of the Physical Packets to the multiple lanes that can be configured in a link. As discussed in Chapters 1 and 2, each link consists of 1 to 32 lanes. Each lane consists of two signal line pairs, one pair for each direction. Each signal line pair supports a symbol stream for the Physical Packets. In order to maximize the bandwidth of the link the Physical Layer parses single symbol stream Physical Packet across multiple lanes of the link for each symbol period. At the receiving port, the Physical Layer must deparse the symbol stream on the multiple lanes into one Physical Packet symbol stream. Several symbols must be used as lane fillers if the lane does not contain a Physical Packet.

The third part of this chapter discusses Ordered Sets. The primary purpose of Ordered Sets is to transmit specific byte patterns across the link to cause transitions in the Link Training and Status State Machines (LTSSM) and the link states of the associated link. Also, the Ordered Sets (Training Sequence and Fast Training OSs) are used for Link Training and configuration.

The fourth part of this chapter discusses the serial orientation of the bits transmitted. Special transmitting and receiving considerations address electromagnetic emission issues and provide easy circuit board routing.

The last part of this chapter reviews Link Training. The lanes of each link must be configured before the transfer of Physical Packets across the link. The discussion of the configuration protocol includes the application of some Ordered Sets, K symbols, and other items discussed in the chapter.

Parallel LLTPs and DLLPs versus Serial Physical Packets

For DLLPs and LLTPs to be transmitted across a link, their inherently parallel structure must be converted to a serial protocol and vice versa. The serial protocol permits the transfer of DLLPs and LLTPs across each lane of a link via a pair of differentially driven signal lines. The Physical Layer at the transmitting ports assembles the Physical Packets from the DLLPs and LLTPs, and disassembles the Physical Packets at the receiving ports. Each Physical Packet contains one LLTP or one DLLP; but not two LLTPs, two DLLPs, or both. Physical Packets only exist on the links; they are not ported through switches, and are not posted at any PCI Express device.

As illustrated in Figure 8.2, the preparation of the Physical Packet is done in three steps. First, the LLTPs' and DLLPs' SEQ# and Header fields are defined in WORDs, the LLTPs' Data field is defined in DWORDs, and the LCRC and CRC are defined in WORDs. The DWORDs and WORDs comprise a series of bytes. Second, FRAMING bytes are added to identify the Physical Packet as containing either an LLTP or a DLLP, and to provide the boundaries of the Physical Packet for the eventual transmission in a serial symbol stream. Third, the Physical Packet is then encoded using the 8/10b encoding protocol to integrate a reference clock in the serial symbols and bit streams representing the series of bytes, after being parsed to multiple lanes.

As illustrated in Figure 8.3 the receipt of the Physical Packet byte stream from the deparser after the symbol stream on each lane is first decoded with the 8/10b encoding protocol (also known as 10/8b decoding). The resulting Physical Packet byte stream is processed in two steps: First, the FRAMING bytes distinguish the beginning and end of Physical Packet. The FRAMING bytes also identify the Physical Packet as containing an LLTP or DLLP. Second, the series of bytes is converted from a serial to a parallel orientation for transfer to the Data Link Layer.

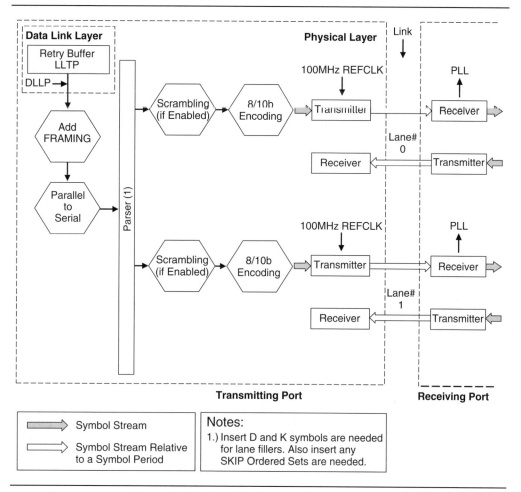

Figure 8.2 Transmitting Physical Packets

364 ■ The Complete PCI Express Reference

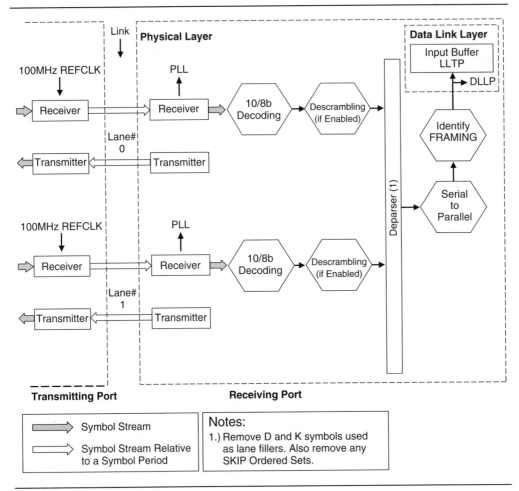

Figure 8.3 Receiving Physical Packets

> The focus of the following two sections is on these two steps at the transmitting port. Though not specifically discussed, the two steps defined for the receiving port are simply the reverse of the protocol applied to the transmitting port.

Step 1: Conversion of LLTP and DLLP Portion of Physical Packet

As previously discussed Physical Packets contain either Link Layer Transaction Packets (LLTPs) or Data Link Layer Packets (DLLPs). It is only the Physical Packets that are transmitted across the link with the symbols of the lane fillers and Ordered Sets.

LLTP Conversion

To begin the discussion, assume that a memory requester transaction is being executed. Figure 8.4 illustrates the Header field of the TLP for a 32 address bit memory write requester transaction mapping one-to-one to a series of bytes aligned in WORDs. Figure 8.5 illustrates the series of bytes expanded with the inclusion of the TLP's Data and Digest fields to the Header field. Figure 8.5 represents the mapping of this TLP into a series of bytes; other TLPs will have similar representations. Some of the other TLPs will have a different number of bytes in the Header Field. The Data field may not be included in all TLPs. Also, the Digest field is optional for any of the TLPs No bytes are associated with the Data and Digest fields if either of these fields is not implemented. See Chapter 6 for more information.

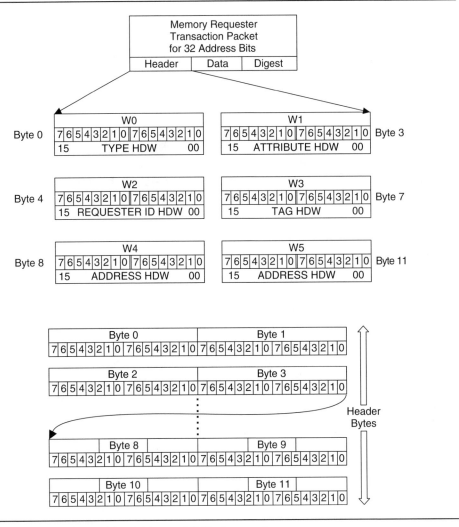

Figure 8.4 Header Field WORDs to BYTEs Conversion

Chapter 8: Physical Layer and Packets ■ **367**

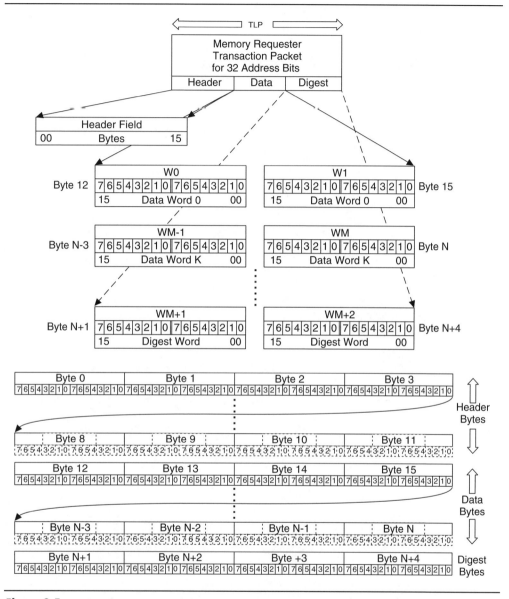

Figure 8.5 Header, Data, and Digest Field WORDs to BYTEs Conversion

The TLP created by the Transaction Layer is converted into an LLTP by the Data Link Layer. This conversion is simply the addition of the SEQ# and the LCRC. As illustrated in Figure 8.6, this adds a total of six bytes. See Chapter 7 for more information.

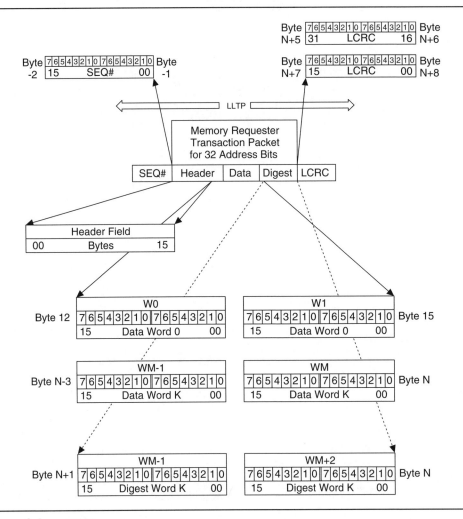

Figure 8.6 LLTP

The adding of bytes associated with the SEQ# and LCRC to the series of bytes associated with the TLP results in the LLTP as a series of bytes as illustrated in Figure 8.7.

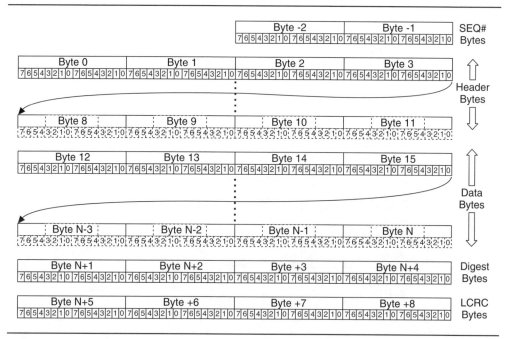

Figure 8.7 LLTP as a Series of Bytes

DLLP Conversion

Physical Packets can also contain DLLPs. Figure 8.8 illustrates the Header field of the DLLP for Flow Control mapping one-to-one to a series of bytes aligned in WORDs. The Flow Control is one of the link management entities. All of the link management entities consist of four bytes as shown. As discussed in Chapter 7, the addition of the CRC to the Header field completes the DLLP. Figure 8.8 also illustrates the DLLP as a series of bytes.

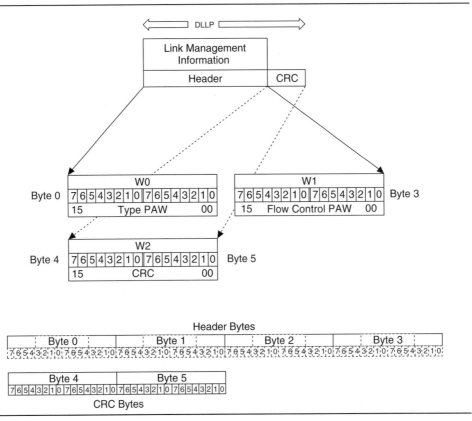

Figure 8.8 DLLP WORDs to BYTEs Conversion

Step 2: Addition of FRAMING to Complete the Physical Packet

The previous discussions have focused on the conversion of LLTPs and DLLP to a series of bytes. To complete the Physical Packets in terms of a series of bytes requires the addition of FRAMING. FRAMING consists of one of four possible bytes that are 8/10b encoded into K symbols discussed later in this chapter. The K symbols STP and END are FRAMING for LLTP. K symbols SDP and END are FRAMING for DLLP. The EBD K symbol is also used at the end of a nullified LLTP. The resulting series of bytes prior to 8/10b encoding for a Physical Packet containing an LLTP or DLLP is shown in Figure 8.9.

Chapter 8: Physical Layer and Packets ■ 371

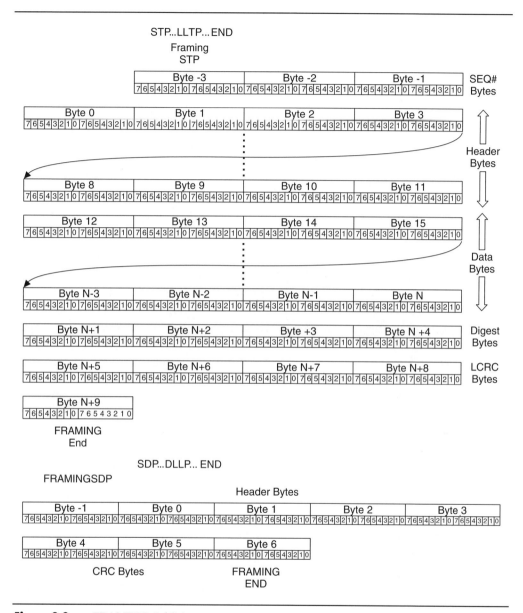

Figure 8.9 FRAMING Addition

As discussed at the beginning of this chapter, the attachment of FRAMING bytes by the transmitting port to an LLTP or a DLLP to create a Physical Packet has two purposes. First, FRAMING distinguishes the beginning and end of a Physical Packet. This is important given that a

Physical Packet is a series of bytes to be transmitted over a serial link. Second, FRAMING distinguishes between a Physical Packet containing an LLTP versus a DLLP. At the receiving port, FRAMING is used for the same reason it is used for the transmitting port.

The bytes of the DLLPs and LLTPs resemble the FRAMING bytes. However, the 8/10b encoding discussed in the next section encodes the DLLPs' and LLTPs' bytes into D symbols and the FRAMING bytes in K symbols. To provide the distinction between bytes to be encoded into D symbols versus K symbols a control bit is assigned to each byte. The control bit is implementation-specific and will not be discussed further in this book.

Step 3: Encoding and Decoding Protocols and Reference Clock

The Physical Packet's series of bytes and associated bits are not simply serialized into a stream of bytes and bits and directly transmitted on the link. If the bits were directly transmitted onto the link, there would be the need for a reference clock or strobe. That is, a reference must be delivered at the receiver on adjacent signal lines for when the individual bits that comprise the bytes are valid. As discussed at the beginning of this chapter, this would limit bit transfer rates across the link. In order to eliminate the problems associated with such a reference, the reference clock needs to be integrated into the bit stream that comprises the byte stream for the Physical Packet to be transmitted. At the receiver, the integrated reference clock must be extracted from the Physical Packet sbyte stream in order to permit a phase lock loop (PLL) to "lock" onto the integrated reference clock. At the receiver the PLL provides a reference clock to the Physical Packet byte stream to extract the DLLP or LLTP byte stream.

The integration of the reference clock into the Physical Packet's 8-bit byte stream creates a Physical Packets consisting of 10-bit byte streams providing ever-changing bit inversions. The ever-changing bit inversions provide transition points for the PLL. The PCI Express specification has adopted the ANSI x 3.230-1994, clause 11 (also known as IEEE 802.3z, 36.2.4) encoding protocol to integrate and extract a reference clock in the Physical Packet byte stream. At the transmitter of the Physical Packet, the byte stream is sourced as 10-bit sbytes encoded from the 8-bit bytes. At the receiver the Physical Packet byte stream of 10-bit sbytes are decoded to the 8-bit bytes.

> This book uses the term *sbytes* to reference the 10-bit bytes used for special or data symbols of the Physical Packets that result from the encoding of the 8-bit bytes per ANSI x 3.230-1994, clause 11 encoding protocol (herein called the *encoding protocol*). The ANSI x 3.230-1994, clause 11 decoding protocol is herein called simply the *decoding protocol.*

As illustrated in Figures 8.2 and 8.3, the encoding and decoding protocol can be viewed in two parts. One part is the encoding and decoding of the bits of the byte stream. The other part is the introduction of a reference clock at the transmitting port and the extraction of it by the phase lock loop (PLL) at the receiving port. Figures 8.2 and 8.3 also identify scrambling, which is discussed later in this chapter.

Encoding and Decoding of the Physical Byte Stream

As illustrated in Figure 8.9, Physical Packets consists of a byte stream. The byte stream must be encoded into a stream of sbytes to integrate a reference clock. Each 8-bit byte of the Physical Packet byte stream is encoded to the 10-bit sbyte called a symbol. The Physical Packet symbol stream is a result of the 8/10b encoding protocol and flows into the parser as illustrated in Figure 8.2. The Physical Packet symbol stream ensures that sufficient transitions are in the bits of the sbytes to allow the receiving port to extract the reference clock. See Chapter 21 for more information.

As illustrated by Figure 8.3, the reception of the Physical Packet symbol stream requires the extraction of a reference clock. The encoding protocol provides a reference clock that a PLL (phase lock loop) in the receiver can synchronize to (lock). This permits the receiver to determine the valid portion of each bit period of the sbytes' waveform. The resulting Physical Packet symbol stream parsed across two lanes is de-parsed into a single Physical Packet symbol stream. The Physical Packet symbol stream is subsequently decoded into a Physical Packet byte stream with an 8-bit byte.

Encoding Protocol The encoding protocol divides each 8-bit byte of the link packets into 3 bits ([7::5]) and 5 bits [4::0]. The 3 bits of each byte are encoded to 4 bits [9::6] of the sbyte (symbol byte) and the 5 bits of each byte are encoded to 6 bits [5::0] bits of the sbyte, as illustrated in Figure 8.10. The encoding process has four key components:

- **D symbols:** The encoding of the bytes to sbytes (symbols) for all bytes associated with the LLTPs and DLLPs is designated by the "D" prefix.

- **K symbols:** The encoding to create sbytes for special symbols is designated by the "K" prefix. The FRAMING of the Physical Packets consists of K symbols. Special symbols are discussed in next section.

- **Encoding and Decoding:** The encoding of bytes to sbytes and vice versa is based on the table provided in Appendix B of the PCI Express specification.

- **Alpha equivalent:** To assist in the explanation of sbytes' serial transmission of the D and K symbol streams, the alpha equivalent to the bit number within each sbyte is included in subsequent figures.

Figure 8.10 is simply the byte stream illustrated in Figure 8.9 for an LLTP Physical Packet with two additions. First, the 8-bit bytes have been mapped to 10-bit sbytes. Second, the sbytes are referenced to an alpha equivalent, to assist in the discussion of the Physical Packet symbol stream's bit transmission. Also shown in Figure 8.10 is the reorientation for the alpha equivalent to align with the order of the sbyte transmission order in Figures 8.11 and 8.13.

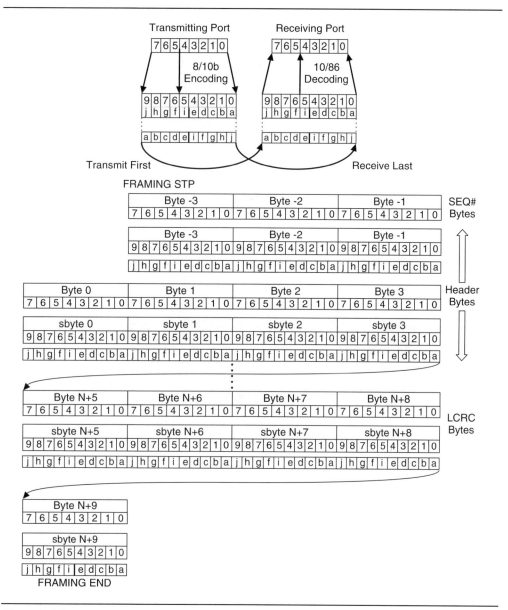

Figure 8.10 8-bit to 10-bit Encoding of LLTP Physical Packet

Illustrated in Figure 8.11 is a simplification of Figure 8.10, with the focus on just the sbytes used in the Physical Packet symbol stream containing an LLTP (SEQ# + TLP + LCRC).

376 ■ The Complete PCI Express Reference

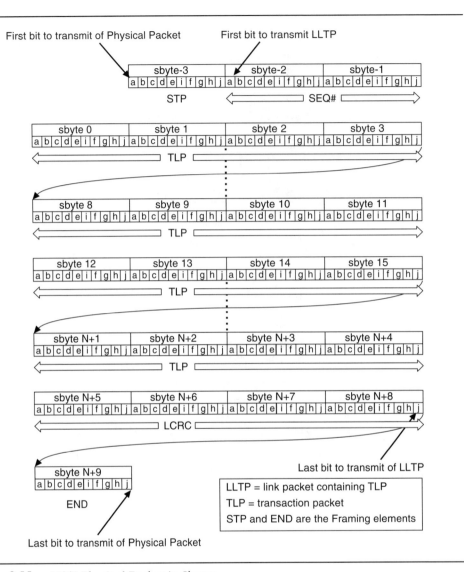

Figure 8.11 LLTP Physical Packet in Sbytes

Similar to Figure 8.10, Figure 8.12 is simply the series of bytes illustrated in Figure 8.9 for a DLLP Physical Packet with two additions. First, the 8-bit bytes have been mapped to 10-bit sbytes. Second, the sbytes are referenced to an alpha equivalent to assist in the discussion of the Physical Packet symbol stream's transmission. Illustrated in Figure 8.13 is a simplification of Figure 8.12, with the focus on just the sbytes of a Physical Packet symbol stream containing a DLLP (link management + CRC).

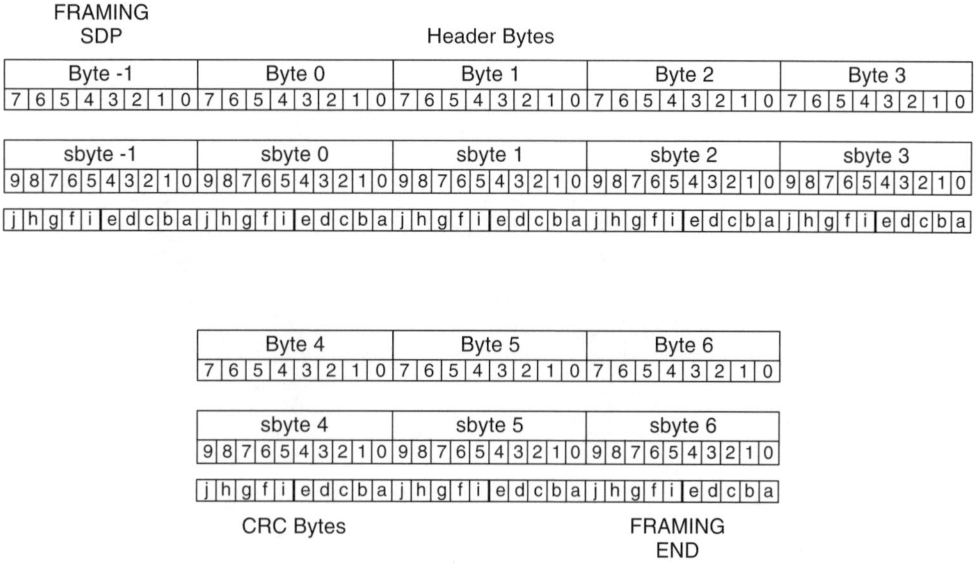

Figure 8.12 8-bit to 10-bit Encoding of DDLP Physical Packet

The decoding protocol at the receiving port is just the reverse of the encoding protocol at the transmitting port.

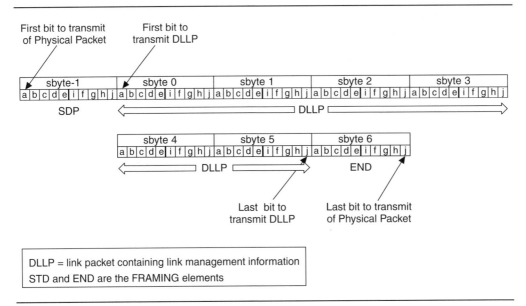

Figure 8.13 DLLP Physical Packet in Sbytes

REFCLK and PLL

Each sbyte of the Physical Packet symbol stream defines a symbol period, which is inversely proportional to the symbol rate. Within each symbol period are 10 bit periods. The bit period is inversely proportional to the date bit rate. The PCI Express specification defines this rate as the data rate and it is specified in the Data Rate Identifier in the TS1 and TS2 Ordered Sets (discussed later in this chapter). Presently, the PCI Express specification revision 1.0a only defines a data bit rate of 2.5 gigabits per second. This translates into a symbol rate of 250 megasymbols per second with a symbol period of 4.0 nanoseconds per lane.

As discussed in Chapter 21, the differentially driven waveform representing the symbol stream has transition points for each bit of the symbols (sbytes). These transition points are defined by a bit period. As illustrated in Figure 8.14, the parsed Physical Packet symbol stream is transmitted over each lane with a symbol period of 4.0 nanoseconds. The transmitters are driven by a REFCLK common to each transmitter of each lane. As will be discussed in the next section, Order Sets and other D and K symbols are also transmitted with the same symbol period.

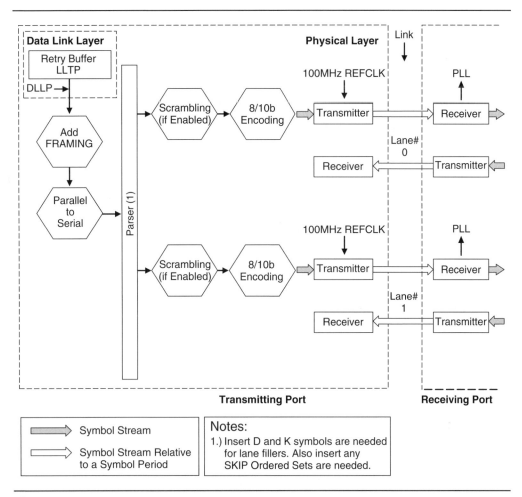

Figure 8.14 Transmitting Physical Packets

As illustrated in Figure 8.15, the parsed Physical Packet symbol stream is received on each lane. The receivers of each lane synchronize a phase lock loop (PLL) to the incoming symbol stream on each lane to provide a reference clock. This provides the receivers of each lane with a reference clock to determine the valid portion of each bit period of the symbol (sbyte). The resulting symbol streams from the two lanes are de-parsed into a single symbol stream after being decoded into a stream of 8-bit bytes. As discussed in the next section, Order Sets and other D and K symbols are also received with the same symbol period.

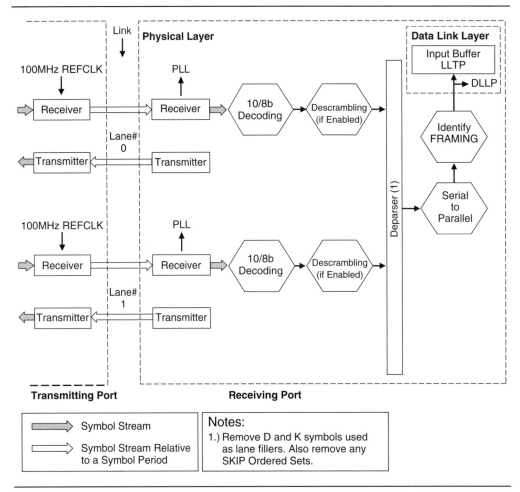

Figure 8.15 Receiving Physical Packets

The bit periods and symbol periods of the sbytes among the lanes of each specific link are aligned at the transmitter. Later in this chapter are additional discussions relative to the transmission and reception of the bit streams associated with the symbol streams. Also see Chapter 21 for more information about the transmitted and received bit waveforms.

> This book uses the terms *data bit rate* and *Data Bit Rate Identifier* instead of the terms used in the PCI Express specification, *data rate* and *Data Rate Identifier,* respectively. This is to emphasize that the rate is related to bits and not bytes.

Physical Packet and Other Symbol Transmission

Transmission order of the LLTP and DLLP occurs in the Transaction and Data Link Layer as discussed in other chapters. The Physical Layer simply takes the LLTPs and DLLPs from the Data Link Layer, attaches the appropriate FRAMING at the beginning and the end, applies 8/10b encoding, and then transmits. The FRAMING consists of certain K symbols (STP, SDP, END, and EDB) discussed later in this chapter. The FRAMING simply defines the beginning and end of Physical Packet and whether the Physical Packet contains an LLTP or DLLP.

The discussion to this point has focused on the transmission of Physical Packets constructed of D symbols representing DLLPs and LLTPs. As stated above, the Physical Packets also include certain K symbols at the beginning and end (FRAMING). Other K symbols are transmitted between the Physical Packets. Consequently, the symbol stream observed on the link consists of Physical Packets of FRAMING K symbols and D symbols of LLTP and DLLP. The observed symbol streams also include other symbols between the Physical Packets. The other symbols of the symbol stream are PAD K symbols, Ordered Sets, and IDLE D symbols.

Later in the chapter configured lanes and configure links are discussed. The symbol stream that transfers across a link discussed here can only do so on configured lanes of a configured link. Consequently, it is implied in the following discussions that lanes and links referenced are configured.

Receiver Error

The successful transmission of Physical Packets, Ordered Sets, PAD K symbols, and IDLE D symbols across a link must satisfy many requirements. For example, K symbols used for FRAMING and lane filling must be correct, the binary pattern for the K and D symbols must be defined and supported, the LLTP and DLLP must be parsed across lanes, and so on. The receiving port can optionally check for *any* combination of violations of the terms of these requirements and other requirements discussed later on. If checked and an error is found, the symbol or Physical Packet is discarded and a **Receiver** error is reported by a message error requester transaction packet. See Chapter 9 for more information on errors.

Transmission of Physical Packets

The transmission of the sbyte bits of the symbol stream associated with Physical Packets is strictly low order to higher order in two levels. As illustrated in Figure 8.11, a Physical Packet containing an LLTP begins and ends with the sbytes for K symbols used for FRAMING: STP and END. First, the sbyte encoded for STP is transmitted, then the sbyte encoded from byte −2, sbyte, then the sbyte encoded from byte −1, and so on. Second, the bits within each sbyte are transmitted in low to high order. As illustrated in Figure 8.11, bit a is the lowest order bit of sbyte 0 and is transmitted first, then bit b, then bit c, and so on. A similar statement is made for the Physical Packet containing a DLLP illustrated in Figure 8.13. First, the sbyte encoded for STD is transmitted first, then the sbyte encoded from byte −1, sbyte, then the sbyte encoded from byte 0, and so on. Second, the bits within each sbyte are transmitted in low to high order. As illustrated in Figure 8.13, bit a is the lowest order bit of sbyte 0 and is transmitted first, then bit b, then bit c, and so on.

Transmission of Ordered Sets, PAD K Symbols, and IDLE D Symbols

The transmission of the sbyte bits of the symbol stream associated with Ordered Sets, PAD K symbols, and IDLE D symbols are also transmitted using the same strict low order to higher order in two levels discussed earlier for Physical Packets.

Lane Parsing

As discussed earlier in this book, each link can have 1, 2, 4, 8, 12, 16, or 32 lanes. The LANE# values are 0 to 31. The determination by the link of just how many lanes are configured, and thus how many can support Physical Packet transfer, is discussed at the end of this chapter. If the link is x1, only LANE# 0 exists. If the link is x2, LANE# 0 and 1 exists. If the link is x4, LANE# 0, 1, 2, and 3 exists, and so on. The symbol stream is parsed in parallel across the all the lanes for transmission.

As illustrated in Figure 8.11, in its *simplest* form the transmission of the first s-byte (The STP K symbol) begins on LANE# 0, and the balance is parsed across the other lanes as follows:

- **One Lane:** With only one lane (x1), obviously all sbytes are transmitted on LANE# 0.

- **Two Lanes:** With two lanes (x2): sbyte-3, sbyte-1, sbyte1, sbyte3, and so on are on LANE# 0. sbyte-2, sbyte0, sbyte2, sbyte4 and so on are on LANE# 1.

- **Four Lanes:** With four lanes (x4): sbyte-3, sbyte1, sbyte5, sbyte9, and so on are on LANE# 0. sbyte-2, sbyte2, sbyte6, sbyte10 and so on are on LANE# 1. sbyte-1, sbyte3, sbyte7, sbyte11, and so on are on LANE# 2. Sbyte0, sbyte4, sbyte8, sbyte12, and so on are on LANE# 3.

- **Eight Lanes:** With eight lanes (x8): sbyte-3, sbyte5, sbyte13, sbyte21, and so on are on LANE# 0. Sbyte-2, sbyte6, sbyte14, sbyte22 and so on are on LANE# 1. Sbyte-1, sbyte7, sbyte15, sbyte23,and so on are on LANE# 2. Sbyte0, sbyte8, sbyte16, sbyte24, and so on are on LANE# 3. Sbyte1, sbyte9, sbyte17, sbyte25, and so on are on LANE# 4. Sbyte2, sbyte10, sbyte18, sbyte26, and so on are on LANE# 5. Sbyte3, sbyte11, sbyte19, sbyte27, and so on are on LANE# 6. Sbyte4, sbyte12, sbyte20, sbyte28, and so on are on LANE# 7.

- **Twelve Lanes and so on:** The pattern is the same for additional numbers of lanes.

Parsing Example with Four Lanes

Figure 8.16 illustrates the parsing of the Physical Packet symbol stream onto a link with four lanes. As previously discussed, each lane is a differentially driven pair of signal lines. The 10 bits within each sbyte that defines a symbol are transmitted in a serial fashion beginning with bit "a" of each sbyte. All the bits of a specific sbyte are transmitted on the same lane.

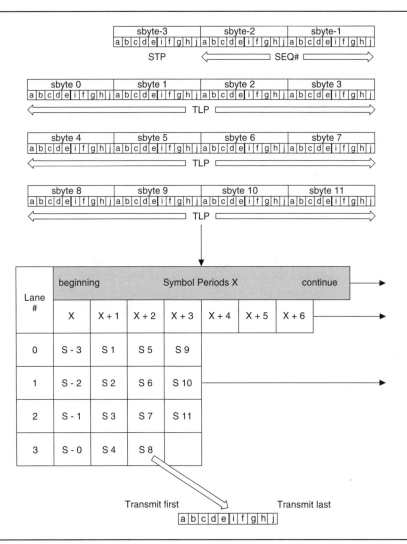

Figure 8.16   Parsing onto Four Lanes

The deparsing at the receiving port is just the reverse of the above parsing at the transmitting port.

K and D Symbols' Application and Lane Parsing

The above discussion illustrates the *simplest* parsing and transmission of a Physical Packet symbol stream transmission. It focused on the transmission of sbytes that represent the K symbols of the FRAMING and the D symbols of the LLTP or DLLP. As previously discussed other entities are also transmitted on the lanes: PAD K symbols, Ordered Sets, and IDLE D symbols.

K Symbols

The implementation of special symbols (K symbols) can be defined in three categories: Physical Packet FRAMING, PAD and Ordered Sets. As previously illustrated, a Physical Packet containing an LLTP or DLLP includes the sbytes representing K symbols and D symbols. The K symbols STP and END for FRAMING at the beginning and end of the stream of D symbols of the LLTP, respectively. If the Physical Packet contains a DLLP, it includes the K symbols SDP and END for FRAMING at the beginning and end of the stream of D symbols that represent the DLLP, respectively. As summarized in Table 8.1, STP, SDP, and END K symbols are only used for FRAMING. In addition, for the nullified LLTP, the END K symbol used for FRAMING of the LLTPs is replaced by the EDB K symbol.

Except for the PAD K symbol, the other K symbols are used to create Ordered Sets. The Ordered Sets are used to support the transitions in the Link Training and Status State Machines (LTSSM) and the link states of the associated link. (See Chapter 14 for more information on link states.) The PAD K symbol is used as lane filler and as part of the link width protocol in the Configuration link states; both uses are discussed later in the chapter.

Special Symbols (K symbols)

As discussed earlier in this chapter, Appendix B of the PCI Express specification provides unique encoding of sbytes for K symbols. A summary of the K symbols is listed in Table 8.1.

Table 8.1   K Symbols

Symbol Name	Code Name	Common Name	Description
COM	K28.5	Comma	Used to identify the beginning of a K symbol sequence for Ordered Sets. Also used for compliance pattern.
STP (FRAMING)	K27.7	Start LLTP	Identifies the beginning of a Physical Packet containing an LLTP.
SDP (FRAMING)	K28.2	Start DLLP	Identifies the beginning of a Physical Packet containing a DLLP.
END (FRAMING)	K29.7	End	Identifies the end of a Physical Packet containing an LLTP or DLLP.
EDB (FRAMING)	K30.7	End Bad	Identifies the end of a Physical Packet containing a nullified LLTP.
PAD	K23.7	PAD K symbol	Used for symbol time filler in x8 and greater links. Also used in the Link Width negotiations.
SKP	K28.0	Skip	Used as part of the SKIP OS to compensate for frequency bit rate difference between transmitter and receiver on a link.
FTS	K28.1	Fast	Used as part of the FTS OS to transition from L0s to L0 link states.
IDL (1)	K28.3	Idle	Used as part of the Electrical Idle OS used prior to entering electrical idle.
	K28.4 K28.6 K28.7	Unused	Reserved.

Table Notes:

1. This is *not* the IDLE encoded into D symbols.

IDLE D symbol

In addition to the transmission of the Physical Packet symbol streams, Ordered Sets, and PAD K symbols, the transmitting port must also include IDLE D symbols. The IDLE D symbol provides the phase lock loops in the receiver a stream of sbytes bits in the IDLE D symbol to retain synchronization lock of the integrated reference clock when no other K or D symbols are being transmitted. IDLE is idle data character 000 00000b that is encoded using 8/10b encoding to 1101 000110b (high to low), which is transmitted as 011000 1011b (abc ... ghi low order first).

As discussed below, the transmission of IDLE D symbols is simultaneous across all lanes of a link for each symbol period. The Logical Idle period is defined as one or more symbol periods when only IDLE D symbols are transmitted. The IDLE D symbols must be transmitted when no Physical Packets or Ordered Sets are being transmitted and the link is not in electrical idle. (Note, this IDLE is a D symbol and is not the IDL K symbol that is part of the Electrical IDLE Ordered Set.)

Application of K and D Symbols Relative to Physical Packets and Actual Lane Parsing

As previously discussed, in a Physical Packet, FRAMING K symbols are attached to the stream of D symbols representing the 8/10b encoding of the DLLP and LLTP. The link may consist of several lanes that operate in parallel to increase the overall bandwidth of the link. The *simplest* parsing example discussed earlier in this chapter with four lanes aligned the beginning of the Physical Packet symbol stream to LANE# 0. The *actual* parsing protocol provides greater latitude to the lane alignment of Physical Packet symbol stream with the transmission of IDLE D symbols and PAD K symbols.

Parsing Protocol with IDLE D Symbol

Whenever a Physical Packet, Ordered Set, or PAD K symbol is not being transmitted, an IDLE D symbol is transmitted to retain the reference clock lock at the receiving port. Table 8.2, Case A and Case B illustrate the transmission of an LLTP with FRAMING. The D symbols associated with the LLTP and DLLP are parsed across these links. In a similar consideration, the FRAMING K symbols are also parsed across the lanes. In Table 8.2, Cases A and B the K and D symbols of the Physical Packet neatly aligned to the four lanes (in these examples sbytes S 5 to S N+1 and the associated symbol periods are implied but not shown). In Case A, the next Physical Packet or Ordered Sets are not ready for transmission with the completion of the present Physical Packet. The transmitters must use the IDLE D symbols as filler ("I" in the Tables). In Case B, the next Physical Packet (which could have been an Ordered Set) is ready for transmission with the completion of the present Physical Packet and no IDLE D symbols are used (symbol period X+2).

Table 8.2   Transmission Parsing CASE A

LANE #	beginning		Symbol Periods		continue	
	0	1	2	X	X+1	X+2
0	I D symbol	S−3 STP K symbol	S 1 LLTP D symbol	S N+2 LLTP D symbol	S N+6 LCRC D symbol	I D symbol
1	I D symbol	S−2 SEQ# D symbol	S 2 LLTP D symbol	S N+3 LLTP D symbol	S N+7 LCRC D symbol	I D symbol
2	I D symbol	S−1 SEQ# D symbol	S 3 LLTP D symbol	S N+4 LLTP D symbol	S N+8 LCRC D symbol	I D symbol
3	I D symbol	S 0 LLTP D symbol	S 4 LLTP D symbol	S N+5 LCRC D symbol	S N+9 END K symbol	I D symbol

Table 8.2   Transmission Parsing CASE B

LANE #	beginning		Symbol Periods		continue	
	0	1	2	X	X+1	X+2
0	I D symbol	S−3 STP K symbol	S 1 LLTP D symbol	S N+2 LLTP D symbol	S N+6 LCRC D symbol	S−3 STP K symbol
1	I D symbol	S−2 SEQ# D symbol	S 2 LLTP D symbol	S N+3 LLTP D symbol	S N+7 LCRC D symbol	S−2 SEQ# D symbol
2	I D symbol	S−1 SEQ# D symbol	S 3 LLTP D symbol	S N+4 LLTP D symbol	S N+8 LCRC D symbol	S−1 SEQ# D symbol
3	I D symbol	S 0 LLTP D symbol	S 4 LLTP D symbol	S N+5 LCRC D symbol	S N+9 END K symbol	S 0 LLTP D symbol

In the two examples in Table 8.2, Cases A and B, the Physical Packets within the lanes were "neatly" aligned to LANE# 0 and LANE# 3. The use of IDLE D symbols as filler allows Physical Packets and Ordered Sets to be transmitted in a manner that is not back-to-back. Whenever the IDLE D symbol is transmitted as part of a Logical Idle period, the subsequent transmission of Physical Packets and Ordered Sets must begin on LANE# 0.

Parsing Protocol with FRAMING

In actual implementations, the Physical Packets will not always "neatly" align into the lanes. The PCI Express specification defines additional protocols where a Physical Packet or Ordered Set can begin. The discussion here first focuses on Physical Packets. The alignment of Ordered Sets is discussed later in this chapter. The protocol of the FRAMING (STP, SDP, END, and EBD) K symbols, and thus the beginning and end of Physical Packets relative to lane position illustrated in above table and following tables are as follows:

- **FRAMING begins at LANE# 0:** The STP or SDP K symbols are placed at the beginning of the transmission of a Physical Packet containing an LLTP or a DLLP, respectively. In a x1 link implementation, the STP or SDP K symbols must begin the transmission on LANE# 0 by default. For links x2 or wider implementations the STP or SDP K symbols must begin the transmission on LANE# 0 when preceded by a Logical Idle period ("I" in the tables).

- **FRAMING begins at LANE# 0, LANE# 4, and so on:** If the Physical Packets are not preceded by a Logical Idle period, the link is x2 or x4, and the end of one Physical Packet naturally aligns with the beginning of the next; the beginning of the next Physical Packet is on LANE# 0. In this situation, no IDLE D symbol lane filler is required, as illustrated in Table 8.3 for a x4 link. Link x8 or wider has additional considerations when a Physical Packets do not "neatly" align with the lanes to begin on LANE# 0.

These additional considerations avoid the use of IDLE D symbols as lane filler and thus allow Physical Packets to waive the requirement to begin on LANE# 0. These considerations only apply to links x8 and wider as follows:

- **Not Preceded by Logical Idle period**: As illustrated in Table 8.4 for a link x8, if the next Physical Packet is not preceded by IDLE D symbol, the end of one Physical Packet naturally aligns with the beginning of the next; the beginning of the next can optionally be on LANE# 0 (symbol period X + 1) or LANE# 4 (symbol period X). If the link is x12, the optional beginning points of back-to-back Physical Packets are LANE# 0, LANE# 4, or LANE# 8, as shown in Table 8.5, and so forth.

■ **FRAMING ends**: The END K symbol is placed at the end of the transmission of a Physical Packet containing an LLTP or DLLP. An EBD K symbol is placed at the end of the transmission of a Physical Packet containing a nullified LLTP.

■ The alignments to the beginning of the Physical Packets to LANE#0, LANE#4, and so forth, reflect that Physical Packets containing DLLPs are eight bytes in size and Physical Packets containing LLTPs are multiples of four bytes in size.

Table 8.3 Transmission Parsing

LANE #	beginning	Symbol Periods			continue	
	0	1	2	X	X+1	X+2
0	I D symbol	S –3 STP K symbol	S 1 LLTP D symbol	S N+4 LLTP D symbol	S N+8 LCRC D symbol	S –3 STP K symbol
1	I D symbol	S –2 SEQ# D symbol	S 2 LLTP D symbol	S N+5 LLTP D symbol	S N+9 LCRC D symbol	S –2 SEQ# D symbol
2	I D symbol	S –1 SEQ# D symbol	S 3 LLTP D symbol	S N+6 LLTP D symbol	S N+10 LCRC D symbol	S –1 SEQ# D symbol
3	I D symbol	S 0 LLTP D symbol	S 4 LLTP D symbol	S N+7 LCRC D symbol	S N+11 END K symbol	S 0 LLTP D symbol

Table 8.4 Transmission Parsing

LANE #	beginning		Symbol Periods	continue		
	0	1	X	X+1	X+2	X+3
0	I D symbol	S − 3 STP K symbol	S N LCRC D symbol	S 3 DLLP D symbol	S −3 STP K symbol	S 5 LLTP D symbol
1	I D symbol	S − 2 SEQ# D symbol	S N+1 LCRC D symbol	S 4 CRC D symbol	S − 2 SEQ# D symbol	S 6 LLTP D symbol
2	I D symbol	S − 1 SEQ# D symbol	S N+2 LCRC D symbol	S 5 CRC D symbol	S − 1 SEQ# D symbol	S 7 LLTP D symbol
3	I D symbol	S 0 LLTP D symbol	S N+3 END K symbol	S 6 END K symbol	S 0 LLTP D symbol	S 8 LLTP D symbol
4	I D symbol	S 1 LLTP D symbol	S -1 SDP K symbol	PAD K symbol	S 1 LLTP D symbol	S 9 LLTP D symbol
5	I D symbol	S 2 LLTP D symbol	S 0 DLLP D symbol	PAD K symbol	S 2 LLTP D symbol	S 10 LLTP D symbol
6	I D symbol	S 3 LLTP D symbol	S 1 DLLP D symbol	PAD K symbol	S 3 LLTP D symbol	S 11 LLTP D symbol
7	I D symbol	S 4 LLTP D symbol	S 2 DLLP D symbol	PAD K symbol	S 4 LLTP D symbol	S 12 LLTP D symbol

Additional Parsing Protocol with FRAMING The beginning of a Physical Packet is identified by a STP or SDP K symbol. Certain requirements must be met when placing STP and SDP K symbols simultaneously within a specific symbol period. The protocol for the placement of these K symbols to indicate the beginning of a Physical Packet is as follows:

- **STP K Symbol:** Only one STP K symbol can be transmitted on any lane during a specific symbol period.

- **SDP K Symbol:** Only one SDP K symbol can be transmitted on any lane during a specific symbol period.

- **STP and SDP Symbol:** Only one STP K symbol and one SDP K symbol can be transmitted in the same symbol period on different lanes. These are illustrated for a x12 wide link in Table 8.5 for a

DLLP Physical Packet. If the next Physical Packet in symbol period X + 1 LANE# 8 contains a DLLP, a SDP K symbol would have to be used, which is not permitted (LANE# 0 of symbol period X + 1 contains SDP K symbol). It would violate the protocol that only one SDP K symbol can be transmitted per symbol period. In this situation, the transmitters must use the PAD K symbols on LANE# 8 to LANE# 11 in symbol period X + 1 (see the following discussion of the PAD K symbol protocol). Alternatively, if an LLTP Physical Packet is ready to transmit it could begin on LANE# 8 symbol period X + 1 (as illustrated in symbol period X + 2).

Table 8.5 Transmission Parsing

LANE #	beginning	Symbol Periods		continue		
	0	1	X	X+1	X+2	X+3
0	I D symbol	S – 3 STP K symbol	S N LLTP D symbol	S -1 SDP K symbol	S -1 SDP K symbol	S 1 LLTP D symbol
1	I D symbol	S – 2 SEQ# D symbol	S N+1 LLTP D symbol	S 0 DLLP D symbol	S 0 DLLP D symbol	S 2 LLTP D symbol
2	I D symbol	S – 1 SEQ# D symbol	S N+2 LLTP D symbol	S 1 DLLP D symbol	S 1 DLLP D symbol	S 3 LLTP D symbol
3	I D symbol	S 0 LLTP D symbol	S N+3 LLTP D symbol	S 2 DLLP D symbol	S 2 DLLP D symbol	S 4 LLTP D symbol
4	I D symbol	S 1 LLTP D symbol	S N+4 LLTP D symbol	S 3 DLLP D symbol	S 3 DLLP D symbol	S 5 LLTP D symbol

Table 8.5 Transmission Parsing *(continued)*

LANE #	beginning		Symbol Periods	continue		
	0	1	X	X+1	X+2	X+3
5	I D symbol	S 2 LLTP D symbol	S N+5 LLTP D symbol	S 4 CRC D symbol	S 4 CRC D symbol	S 6 LLTP D symbol
6	I D symbol	S 3 LLTP D symbol	S N+6 LLTP D symbol	S 5 CRC D symbol	S 5 CRC D symbol	S 7 LLTP D symbol
7	I D symbol	S 4 LLTP D symbol	S N+7 LCRC D symbol	S 6 END K symbol	S 6 END K symbol	S 8 LLTP D symbol
8	I D symbol	S 5 LLTP D symbol	S N+8 LCRC D symbol	PAD K symbol	S – 3 STP D symbol	S 9 LLTP D symbol
9	I D symbol	S 6 LLTP D symbol	S N+9 LCRC D symbol	PAD K symbol	S – 2 SEQ# D symbol	S 10 LLTP D symbol
10	I D symbol	S 7 LLTP D symbol	S N+10 LCRC D symbol	PAD K symbol	S – 1 SEQ# D symbol	S 12 LLTP D symbol
11	I D symbol	S 8 LLTP D symbol	S N+11 END K symbol	PAD K symbol	S 0 LLTP D symbol	S 13 LLTP D symbol

Parsing Protocol with PAD K Symbol As previously discussed, the FRAMING K symbols for the beginning of Physical Packets must begin on LANE# 0 with the possibility of beginning or LANE# 0, LANE# 4, and so on for links of x8 or wider.

For links x8 or wider, the parsing protocol also requires the use of the PAD K symbol. If the link implementation is x1, x2, or x4, the PAD K symbol is not used. The PAD K symbol must be used if the END or EDB FRAMING K symbols are not transmitted on LANE# N−1 of a xN link and the STP or STD K symbols are not immediately transmitted after END or EDB K symbols. In this situation, a PAD K symbol must be used for filler from the END or EDB K symbols to LANE# N−1 of that symbol period. As illustrated in Table 8.4, this may occur if the END or EDB K symbols are on LANE# 3 and the next Physical Packet contains an LLTP that is not ready for back-to-back transmission (that is, STP or SDP K symbols not ready to transmit on LANE# 4). The transmitters must complete lanes (LANE# 4 to LANE# 7) in symbol period X + 1 with PAD K symbols as filler. Another use for the PAD K symbol filler (as discussed in Table 8.5) is when the next Physical Packet is ready but it is another Physical Packet containing a DLLP, which violates the restriction of only one SDP per symbol period (X + 1 in Figure 8.5). Note: In symbol period X + 2, the Physical Packet containing an LLTP can begin on LANE# 8 (STP K symbol) because the only other FRAMING K symbol is SDP on LANE# 0.

Once the PAD K symbols are used as filler, the next Physical Packet or Ordered Sets must begin (if ready) in the next symbol period (X + 2 in Table 8.4). If the next Physical Packet or Ordered Set is not ready, the transmitters must use the IDLE D symbols are filler as illustrated in symbol period X + 2 in Table 8.6. Note: The PAD K symbol is used to fill lanes in a specific symbol period because the IDLE D symbols of a Logical Idle period can only be transmitted on *all* lanes simultaneously. Such a requirement does not apply to PAD K symbols.

Table 8.6 Transmission Parsing

LANE #	beginning		Symbol Periods	continue		
	0	1	X	X+1	X+2	X+3
0	I D symbol	S – 3 STP K symbol	S N LCRC D symbol	S 3 DLLP D symbol	I D symbol	S –3 STP K symbol
1	I D symbol	S – 2 SEQ# D symbol	S N+1 LCRC D symbol	S 4 CRC D symbol	I D symbol	S – 2 SEQ# D symbol
2	I D symbol	S – 1 SEQ# D symbol	S N+2 LCRC D symbol	S 5 CRC D symbol	I D symbol	S – 1 SEQ# D symbol
3	I D symbol	S 0 LLTP D symbol	S N+3 END K symbol	S 6 END K symbol	I D symbol	S 0 LLTP D symbol
4	I D symbol	S 1 LLTP D symbol	S -1 SDP K symbol	PAD K symbol	I D symbol	S 1 LLTP D symbol
5	I D symbol	S 2 LLTP D symbol	S 0 DLLP D symbol	PAD K symbol	I D symbol	S 2 LLTP D symbol
6	I D symbol	S 3 LLTP D symbol	S 1 DLLP D symbol	PAD K symbol	I D symbol	S 3 LLTP D symbol
7	I D symbol	S 4 LLTP D symbol	S 2 DLLP D symbol	PAD K symbol	I D symbol	S 4 LLTP D symbol

Receiver Error

The successful transmission of Physical Packets, Ordered Sets, PAD K symbols, and IDLE D symbols across a link must satisfy many requirements. For example, K symbols used for FRAMING and lane filling must be correct, the binary pattern for the K and D symbols must be defined and supported, the LLTP and DLLP must be parsed across lanes, and so on. The receiving port can optionally check for *any* combination of violations of the terms of these requirements and other requirements discussed later on. If checked and an error is found, the symbol or Physical Packet is discarded and a **Receiver** error is reported by a message error requester transaction packet. See Chapter 9 for more information on errors.

Application of Ordered Sets

The discussion to this point has mentioned Ordered Sets as one of the entities that is transmitted onto the lanes of the link. As previously stated the Ordered Sets consist of K symbols that can be parsed onto the link. Before discussing the inclusion of the Order Sets into the lane parsing, this book introduces the Ordered Sets and discusses each Ordered Set in detail.

Ordered Sets (OSs) are streams of K and in some cases D symbols used to implement transitions between states of the Link Training and Status State Machines (LTSSM) and the link states of the associated link. The link states provide different operational levels for the link. One of these operations is establishing the link width (configuring lanes and links) discussed at the end of this chapter. The other link states are discussed in detail in Chapter 14.

The one exception to the above statements is the Skip Ordered Set. Its purpose is to compensate for the differences between reference clocks at the two ends of the link.

The sbytes associated with the K symbols for the OSs are transmitted in the order listed (read left to right, transmitted first to last) The Ordered Sets are as follows (the "OS" suffix means "Ordered Set"):

- Skip OS = COM SKIP SKIP SKIP

- Fast Training Sequence OS (also known as FTS OS) = COM FTS FTS FTS

- Electrical Idle OS = COM IDL IDL IDL

The other OSs are the Training Sequence 1 and 2 OSs (also known as TS1 and TS2 OSs), which contain both K and D symbols, and are discussed later in this chapter.

Skip OS

A single Skip OS consists of four K symbols: COM SKP SKP SKP. The Skip OS is transmitted on a regular basis to address the following consideration.

The frequency of the data bit rate and timing tolerance among all symbols transmitted on parallel configured lanes is 0 parts per million. However, the frequency of the data bit rate for the transmitting end of the link has a nonzero tolerance relative to the frequency bit rate at the receiving end of the link. The worst case frequency difference between the clocks on the link is 600 parts per million; consequently, a potential for a one clock period error can exist every 1666 clock periods. Each clock period represents a transmission period for one bit of the sbyte. The total clock period to transmit one sbyte is the symbol period. To compensate for this frequency difference, a Skip OS is transmitted and received using the following protocol:

- On all configured lanes, a single Skip OS is transmitted between every 1180 to 1538 symbol periods when possible. The COM K symbols begin simultaneously on *ALL* configured lanes. When the link is in the L0s, L1, or L2 link state, the counting of the symbol period does not occur and the counter itself is held in reset during these link states.

- Whenever a transmission of the Skip OS is scheduled, there may be a transmission of a Physical Packet or OSs underway. In these cases, the Skip OSs are accumulated and transmitted at the next Physical Packet or OS boundaries. No Skip OSs are transmitted during a Physical Packet or other OSs. The one exception is *no* Skip OS can be sent in the stream of "N" FTS OSs, but after the stream of "N" FTS OSs are transmitted, *one* Skip OS must be transmitted.
  - Under the above consideration, the receiver must be able to receive and process Skip OS with up to 5664 symbol periods from beginning of one COM K symbol associated with a Skip OS to the beginning of the next COM K symbol associated with the next Skip OS.

- The Skip OS is defined as four K symbols: COM SKP SKP SKP. The SKIP OS is defined as received when the receiver senses COM K symbol and one to five consecutive SKP K symbols.

- Skip OS are transmitted during the transmission of the IDLE D symbols during the Logical Idle period to correct for timing. However, all configured lanes must have either *ALL* IDLE D symbols or *ALL* the same K symbols of a Skip OS for each symbol period.

- As discussed in more detail in Chapter 14, some of the transitions from one link state to another require receipt of a sequence multiple "consecutive" OSs. The inclusion of a Skip OS in the sequence is not considered an interruption of the sequence.

- No Skip OS can be transmitted in a Compliance Pattern during Polling.Compliance link sub-state.

Fast Training Sequence OS

The Fast Training Sequence (FTS) protocol is required to achieve bit and symbol lock in the transition from L0s to L0 link states. The transition from L0s to L0 link states occurs in the Link Training and Status State Machines (LTSSM) and the link states of the associated link. Receivers of a port detect an exit of electrical idle by the transmitters on the other end of the link with the receipt of a FTS OS. Each FTS OS consists of four K symbols: COM, FTS, FTS, and FTS. The receivers receive "N" number of FTS OSs from these transmitters followed by a single Skip OS defined as part of FTS protocol. No Skip OS can be sent in the stream of "N" FTS OSs, but after the stream of "N" FTS OSs are transmitted, one Skip OS must be transmitted. The receivers will attempt to establish bit and symbol lock based upon receipt of the "N" FTS OSs. See Chapter 14 for more information.

The number "N" is defined in the exchange of TS1 and TS2 OSs in a TS1/TS2 training sequence after link initialization. The number of FTS OSs that can be defined for the FTS protocol is 1 to 255. Four sbytes per OS and 10 bits per sbyte render between 16 to 10,200 bits. Consequently the total time to achieve a bit and symbol lock is 16 nanoseconds to 4.08 microseconds (Based on Unit Interval (UI) = 3999.88 to 400.12 picoseconds; the typical time is 400 picoseconds per bit). In an extension of this protocol, the Extend Synch bit can be set to "1" in the configuration register blocks, which results in the receiver transmitting 4096 FTS OSs

(N= 4096) with a total time of 65,536 microseconds. The purpose of this extension to the protocol is to allow external link monitoring equipment time to synchronize (test and debug).

> For value "N" to be established, two or more TS1 or TS2 OS must be received with the same value "N". Also, this same value "N" must be the same on all configured lanes. If both of these conditions are not met, the resulting behavior is undefined.

Electrical Idle OS

As previously stated, Electrical Idle OS = COM IDL IDL IDL. The Electrical IDLE OS is considered successfully received when two of the three IDLE K symbols are received. The transmitters must transition to electrical idle after they transmit the last K symbol of an Electrical Idle OS. The term *electrical idle* simply refers to the electrical state in which the transmitters and receivers are the same voltage and are held in high or low impedance.

> The transition from one link state to another (or sub-states) requires the transmitters of one of the ports to transmit a pattern that is detected by the receivers of the other port on the link.

Electrical Idle OS and Electrical Idle Protocol

The electrical idle is simply a steady state condition with the primary purpose of providing power savings during specific link states. Prior to the transmitters transitioning to electrical idle they are required to transmit an Electrical Idle OS. The Electrical Idle OS indicates the pending electrical idle to the other port on the link with exceptions noted later on. Once the Electrical Idle OS is transmitted, the transmitters must transition to electrical idle within 20 Unit Intervals after the last K symbol of the Electrical Idle OS is transmitted. Once the transmitters transition to electrical idle, they must remain in it for a minimum of 50 Unit Intervals (Unit Interval = 3999.88 to 400.12 picoseconds).

The protocol of Electrical Idle OS and electrical idle is as follows:

- **Transmit Electrical Idle OS:** The transmitter transmits an Electrical Idle OS, the transmitter subsequently enters electrical idle.
  - The transition from the last bit of the Electrical Idle OS to entering electrical idles is a maximum of 20 Unit Intervals.
  - The transmitter must remain in electrical idle for a minimum of 50 Unit Intervals.
  - A transmitter that enters electrical idle is in either high or low impedance (as discussed later).

- **Transmitter in electrical idle**: The transmitter in electrical idle must periodically poll for a receiver at the other end of the link (Receiver Detection). The transmitter changes the common voltage level of the + and − signal lines (defines electrical idle) to a different common level. A receiver is not at the other end of the link if the rate of voltage change is correct for transmitter impedance, capacitance of the interconnect, and a series capacitor. This value for voltage changes if a receiver that is attached to the link will be different relative to the receivers load impendence of 40 to 60 ohms on each signal line relative to ground. Once no receiver is detected the transmitter ceases this periodic polling. See Chapter 14 for more information.

- **Receipt of an Electrical Idle OS:** After receipt of an Electrical Idle OS by the receivers, the receivers enter an electrical idle exit detection. Electrical idle exit detection is entered 50 Unit Intervals after the first detection of the transmitters on the other end of the link entering electrical idle. The receivers define an exit from electrical idle when they detect a voltage difference between the two differentially driven signal lines greater than 65 millivolts.

- **UI:** The times discussed above reference in terms of UI. The UI is typically 400 picoseconds with a minimum and maximum of value of 399.88 to 400.12 picoseconds, respectively.

Electrical Idle OS and Electrical Idle Protocol Exceptions

As defined in most of the link state discussions, Electrical Idle OS must be transmitted or received prior to electrical idle. Electrical idle *can be* entered without Electrical Idle OS; this is an exception where one end of the link may have been disconnected or disabled. The entrance into electrical idle in this situation can be detected by the receivers still attached and enabled on the other end of the link as follows:

- During L0, Recovery, Configuration, or Loopback link states, the detection of electrical idle by the receivers causes the link to immediately enter the Detect link state. Upon detection of electrical idle without an Electrical Idle OS, the receivers at the end of the link that is still connected and enabled enters electrical idle.

- The receivers must detect electrical idle without an Electrical Idle OS within 10 milliseconds.

Impendence Levels of Electrical Idle

The use of high or low impedance at the transmitter and receiver is determined by the following protocol:

- Unless otherwise stated as "required," the transmitter can be in either the high or low impedance states during electrical idle

- The transmitter is required to be in the high impedance whenever a hot plug / removal or a power up event may occur.

- The transmitter is required to be in the low impedance whenever differential data is to be transmitted.

- The receiver must be in low impedance all the time except when it does not have power and by definition must be high impedance.

Training Sequence 1 and 2 OSs (also Known as TS1 and TS2 OSs)

The Skip, FTS, and Electrical Idle OSs are each simply a stream of four symbols. The TS1 and TS2 OSs are more complex with a stream of 16 symbols. For the TS1 or TS2 OS, a K symbol is first transmitted with the remaining being D symbols dependent on the information to be conveyed. These TS1 and TS2 OS are used within link states to convey the specific information listed in Table 8.7. The TS1 and TS2 OS are also used to establish the link width as part of link configuration as discussed at the end of this chapter.

Table 8.7 TS1 and TS2 Ordered Sets

Order in Stream & Symbol Number (First to last transmitted)	Symbol Name	Allowed Values to be Encoded per Appendix B of PCI Express Specification Rev. 1.0 (1)	Description
0 (First)	COM	K28.5	COMMA to begin stream
1	Data	D0.0 to D31.7 Values of 0 to 255 Or PAD (K23.7)	Link number
2	Data	D0.0 to D31.0 Values of 0 to 31 Or PAD (K23.7)	Lane number within port
3	Data	D0.0 to D31.7 Values of 0 to 255	N_FTS: This provides the number "N" of FTS Oss required by the receiver to establish a bit and symbol lock.
4	Data	D2.0 (2)	Data Bit Rate Identifier [2::0] Bit 1 = 1 (defines 2.5 Gb/s) Bit 0, 2-7 = Reserved ("0")

Table 8.7 TS1 and TS2 Ordered Sets *(continued)*

Order in Stream & Symbol Number (First to last transmitted)	Symbol Name	Allowed Values to be Encoded per Appendix B of PCI Express Specification Rev. 1.0 (1)	Description
5	Data	D0.0, D1.0, D2.0, D4.0, or D8.0	Training Control Bits [7::0]: Hot Reset Bit [0]: De-assert Reset = "0", Assert Reset = "1" Disable Link Bit [1]: Enable Link = "0", Disable Link = "1" Loopback Bit [2]: No Lookback = "0", Enable Loopback = "1" Disable Scrambling Bit [3]: Bit 3: Enable Scrambling = "0", Disable Scrambling = "1" Bits [7::4] = Reserved ("0")
6 to 15 (Last)	Data	D10.2 (TS1) or D5.2 (TS2)	Simply identifies the stream as either TS1 or TS2

Table Notes:

1. Of the 16 symbols in the stream, only the first one is a K symbol. All of the others are D symbols with two exceptions: Symbol 1 and 2 (link # and lane #) are defined be either D symbol or PAD (K23.7) K symbol.

2. The PCI Express specification revision 1.0 defines only one Data Bit Rate Identifier of 2.5 gigabits per second is defined. Future revisions of the specification may define other data bit rates.

Inclusion of Ordered Sets into Lane Parsing

When an OS is transmitted, the associated stream of the K and D symbols is not parsed across the lanes as done for Physical Packet symbol streams. The OS is transmitted as multiple individual streams simultaneously on all configured lanes (if x2 or wider). The same symbol of the stream is transmitted in the same symbol time across all the lanes. Consequently, if a Physical Packet of a xN link does not end (END or EBD) on LANE# N-1, the PAD K symbol is used, and IDLE D symbol lane filler is used as needed to allow the OS stream to begin on LANE# 0 simultaneously with beginning on LANE# 1, LANE#2, LANE# 3, and so on. As illustrated in Tables 8.8 and 8.9, the OSs begins at LANE# 0 and the first K symbol of the OS is transmitted on all parallel lanes in the same symbol period.

Table 8.8 Transmission Order of Ordered Sets

LANE #	beginning	Symbol Periods			continue		
	0	1	2	X	X+1	X+2
0	I D symbol	S−3 STP K symbol	S 1 LLTP D symbol	S N+2 LLTP D symbol	S N+6 LCRC D symbol	OS 1 K symbol
1	I D symbol	S−2 SEQ# D symbol	S 2 LLTP D symbol	S N+3 LLTP D symbol	S N+7 LCRC D symbol	OS 1 K symbol
2	I D symbol	S−1 SEQ# D symbol	S 3 LLTP D symbol	S N+4 LLTP D symbol	S N+8 LCRC D symbol	OS 0 K symbol
3	I D symbol	S 0 LLTP D symbol	S 4 LLTP D symbol	S N+5 LCRC D symbol	S N+9 END K symbol	OS 0 K symbol

Table 8.9 Transmission Order of Ordered Sets

LANE #	beginning		Symbol Periods		continue	
	0	1	X	X+1	X+2	X+3
0	I D symbol	S – 3 STP K symbol	S N LCRC D symbol	S 3 DLLP D symbol	OS 0 K symbol	OS 1 K or D symbol
1	I D symbol	S – 2 SEQ# D symbol	S N+1 LCRC D symbol	S 4 CRC D symbol	OS 0 K symbol	OS 1 K or D symbol
2	I D symbol	S – 1 SEQ# D symbol	S N+2 LCRC D symbol	S 5 CRC D symbol	OS 0 K symbol	OS 1 K or D symbol
3	I D symbol	S 0 LLTP D symbol	S N+3 END K symbol	S 6 END K symbol	OS 0 K symbol	OS 1 K or D symbol
4	I D symbol	S 1 LLTP D symbol	S -1 SDP K symbol	PAD K symbol	OS 0 K symbol	OS 1 K or D symbol
5	I D symbol	S 2 LLTP D symbol	S 0 DLLP D symbol	PAD K symbol	OS 0 K symbol	OS 1 K or D symbol
6	I D symbol	S 3 LLTP D symbol	S 1 DLLP D symbol	PAD K symbol	OS 0 K symbol	OS 1 K or D symbol
7	I D symbol	S 4 LLTP D symbol	S 2 DLLP D symbol	PAD K symbol	OS 01 K symbol	OS 1 K or D symbol

Receiver Error

The successful transmission of Physical Packets, Ordered Sets, PAD K symbols, and IDLE D symbols across a link must satisfy many requirements. For example, K symbols used for FRAMING and lane filling must be correct, the binary pattern for the K and D symbols must be defined and supported, the LLTP and DLLP must be parsed across lanes, and so on. The receiving port can optionally check for *any* combination of violations of the terms of these requirements and other requirements discussed later on. If checked and an error is found, the symbol or Physical Packet is discarded and a **Receiver** error is reported by a message error requester transaction packet. See Chapter 9 for more information on errors.

Special Transmitting and Receiving Considerations

The transmission of a K and D symbols stream across a link requires a couple of additional modifications at the transmitter and the receiver.

Scrambling

When a binary bit stream has a high frequency of bit transition points, measurable electromagnetic interference (EMI) may be generated. If certain binary patterns are generated in a repetitive fashion, it is possible for certain frequencies to be reinforced. The PCI Express specification has defined a scrambling protocol to address this issue. The bit stream that comprises the sbytes of the symbol stream is transmitted on each lane and is passed through a scrambling mechanism. The scrambling mechanism prevents the occurrence of repetitive binary patterns. The scrambling mechanism defined by the PCI Express specification is a Linear Feedback Shift Register (LFSR) for each transmitting lane. When multiple lanes are supported, the maximum output skew of the LFSRs across all the lanes must be less than 20 nanoseconds.

On the transmitting side of a link, the scrambling is applied to the bytes prior to the 8/10b encoding into the sbytes (D or K symbols), illustrated in Figure 8.14. On the receiving side of the link, the de-scrambling is applied after the bit stream has undergone 10/8b decoding, illustrated in Figure 8.15.

Figure 8.17 is a graphical representation of the scrambling and de-scrambling mechanism with a polynomial of $G(x) = X^{16} + X^{15} + X^{13} + X^{4} + 1$. The initialized seed value is FFFFh for the D flip-flops. The initialization of the LFSR in the transmitter occurs after the transmission COM K symbol. The initialization of the LFSR in the receivers occurs with the reception and entrance of the COM K symbol at the LFSR.

Appendix C of the PCI Express specification provides the full details for the LFSR.

The scrambling mechanism and protocol are not applied to the following:

- K symbols in the FRAMING, PAD or Ordered Sets
- D symbols used in Training Sequence Ordered Sets and Compliance Pattern

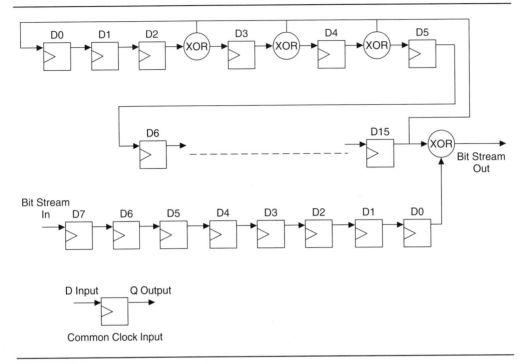

Figure 8.17 Scrambling Mechanism

The scrambling mechanism and protocol are applied to the following:

- IDLE (from the logical idle data character)
- D symbols used for DLLP and LLTP

The scrambling mechanism by default is enabled by the Detect link state. The TS1 and TS2 OSs do provide the Disable Scramble Bit to permit disabling the scrambling mechanism at the end of the Configuration link state. Scrambling is never implemented by the Loopback link state. Basically, the disabling of scrambling is used to support test and debugging. See Chapter 14 for more information.

Lane Polarity Inversion

One additional consideration with the transmission of the TS1 or TS2 OS is the information included in the TS1 and TS2 identifiers. It is possible to design the lanes within the links with "easy routing." Easy routing enables the differential pair of signal wires to interconnect with the different polarities. The transmitters within a port make no judgment as to the routing, and thus the polarity, of the lanes. It is only the receivers within the port that determine whether a lane polarity inversion is required to compensate for the received symbol stream. For example, as the signal lines are routed, the + and − signal lines may exchange the physical traces used as they are fed between printed circuit board layers. The receivers use the value the TS1 and TS2 identifiers to determine whether the routing requires a specific lane to undergo lane polarity inversion at the receivers. If at the receiver TS1 identifier is symbol D10.2 and TS2 identifier is symbol D5.2, the associated lane does not need lane polarity inversion at the receivers. If at the receiver TS1 identifier is symbol D21.5 and TS2 identifier is symbol D26.5, the associated lane requires lane polarity reversal at the receivers. No further action or consideration relative to lane polarity is done except for the lane polarity inversion on all incoming signals at the receiving port. The lane polarity inversion is determined and applied independently on each lane.

Disparity

Appendix B of the PCI Express specification defines positive and negative disparity bit patterns for the K and D symbols. The transmitting port selects either positive or negative disparity with the first symbol upon exiting of electrical idle until it enters the next electrical idle. The transmitting port can optionally change the disparity upon exiting of "next" electrical idle.

Upon sensing an exit from electrical idle, the receiving port sets its disparity, defined as *running disparity*, to the first OS it uses to establish symbol and bit lock. If the symbol and bit lock is subsequently lost, the receiving port sets its disparity, which may be different, to the first OS it uses to establish symbol and bit lock. Once the receiving port establishes the running disparity, the symbols received must be of the correct running disparity and valid symbols per Appendix B of the PCI Express specification. If the received symbol is not a valid symbol or violates the running disparity, the receiver must report a **Receiver** error. See Chapter 9 for more information.

Link Training and Link Configuration

Before the normal link operation of transferring Physical Packets between two PCI Express devices can begin, the Link Training and Status State Machines (LTSSM) within each port must execute Link Training. As detailed in Chapter 14, part of the Link Training and Status State Machine (LTSSM) in the Physical Layer executes Link Training and the other parts execute non-Link Training link states. The execution of Link Training determines that PCI Express devices exist on both ends of the link, establishes the data bit rate of the link, and establishes the lanes in the link that are common to the PCI Express devices on the link.

These determinations by Link Training are accomplished by sequencing through the Detect, Polling, and Configuration link states. The portion of Link Training accomplished by each of these link states is as follows:

- **Detect link states:** Establish the existence of a PCI Express device on each end of link. That is, detect the lanes within a link common to both PCI Express devices on the link.

- **Polling link state:** Establish the bit and symbol lock, lane polarity inversion, and highest common data bit rate on the detected but yet-to-be- configured lanes that exist between the two PCI Express devices.

- **Configuration link state:** Some or all of the detected lanes that successfully complete the Polling link states are processed into configured lanes. Those lanes that can be detected but cannot successfully complete the Polling link state cannot be processed onto a configured lane. Configured lanes are collected into configured links in the link sub-states of the Configuration link state. These link sub-states consequently establish the link width and lane ordering with support of lane reversal. Finally, these link sub-states also implement lane-to-lane de-skew. The N_FTS value is also established.

Chapter 14 discusses the details of these link states in terms of next state tables. The Configuration link state is actually an iterative process of several sub-states. The iterative process is executed in the Link Training and Status State Machine (LTSSM) associated with the Physical Layer and is implementation-specific. This book uses the following example to examine this iterative process. This iterative process includes the application of the Training Sequence (TS) Order Sets and the PAD K symbols.

In order to retain focus on this iterative process, this discussion will assume that the Detect and Polling link states have established a set of detected un-configured lanes common to both PCI Express devices on the link. For a detailed discussion on the Detect and Polling link states, please see Chapter 14.

The PCI Express specification states that a PCI Express device can optionally check for any violation of the Link Training protocol and report them as report **Training Errors** accordingly. Elements of Link Training are also executed in the Link Retraining during the Recovery link sub-states. The ability to report a Training Error requires the transmission of a message baseline requester transaction (message transaction in Table 9.3).

Link Configuration Protocol Iterative Process Discussion

The link between two PCI Express devices contains multiple lanes. Each lane consists of two differentially driven signal line pairs, one pair for each direction. A minimal link supports one lane and additional lanes may be added to provide additional link bandwidth. Each lane provides additional hardware for signal lines in the link to operate in parallel with other signal lines in other lanes in the link. From 1 to 32 possible lanes can be defined in a specific link, but the PCI Express specification only defines seven link widths: x1, x2, x4, x8, x12, x16, and x32. The lane notation is as follows: x1 means one lane, x2 means 2 lanes, x4 means 4 lanes and so on.

It is possible that a downstream port of a PCI Express device may be attached to the upstream port of a PCI Express device with an incompatible number of lanes. The number of lanes that can possibly be implemented by one PCI Express device on a link may not be the same as the number of lanes that can possibly be implemented by the other PCI Express device on the same link. The two PCI Express devices may both be on the platform or one is on an add-in card connected through a slot. Part of the Link Training has detected which lanes are common on the link between two PCI Express devices. Another part of Link Training has established the highest common data bit rate. What remains is to configure the link of configured lanes. The Physical Layer of each PCI Express device establishes the subset of detected but yet to be configured (un-configured) lanes that can be configured. The mechanisms to establish the configured lanes, and thus each configured link, are the TS1 and TS2

OSs. The establishment of the number of configured lanes also establishes what is defined as the link width.

Topology of Links and PCI Express Devices

The next section focuses on a single link configured between the downstream port of the upstream PCI Express device on the upstream side of the link and an upstream port of the PCI Express device on the downstream side of the link. This single link is configured by the Link Training and Status State Machine (LTSSM) of the Physical Layer in each of the PCI Express devices on a link. The PCI Express specification does permit a single downstream port of a PCI Express device to connect to two or more upstream ports of PCI Express devices. That is, one downstream port is connected to two or more upstream ports with a separate configured link defined for each upstream port. If the downstream port supports multiple configured links, a separate LTSSM must be provided for each link.

The topology of links has three additional considerations:

- Unlike a downstream port, only one link can be configured per each upstream port of each PCI Express device.

- Connectors of slots and add-in cards predefine the possible lanes that can be configured. The possible lanes only permit a single link to be configured between the platform and each add-in card.

- An advanced peer-to-peer link may have similar multiple link considerations, but the present PCI Express specification does not provide sufficient information.

Without any implementation-specific information, a single downstream port of a PCI Express device assumes a single link with a specific LINK# is being considered for the collection of detected un-configured lanes. During the iterative process of the Configuration link sub-states the upstream port of the other PCI Express device on the link responds by acknowledging the specific LINK#. If only one PCI Express device is connected on each end of the link, the iterative process establishes a single configured link with LANE#s beginning with 0 and sequentially increasing. The resulting configured link between the two PCI Express devices has a single specific LINK# and a sequence of LANE#s that represent a subset of the detected un-configured lanes that have been configured. The subset may be all of or part of the detected un-configured lanes

If a downstream port of a PCI Express device is connected to two upstream ports of two PCI Express devices, the result of the iterative process is different. First, the single link assumed by the downstream port is actually two links, one for each upstream port. The downstream port first determines that possibly two links exist with reception of only a subset of LANE#s it is transmitting. For a specific LINK# the downstream port transmits LANE#s beginning with 0. An upstream port can only transmit LANE#s beginning with 0 and sequentially increasing on the lanes it is receiving a LINK#. One upstream port receives LANE#s beginning 0 and responds by transmitting LANE#s beginning with 0 with the single specific LINK#. The other upstream port receives LANE#s not beginning with 0, and thus can only respond with LINK# and LANE# equal to the PAD K symbol. The receipt by the downstream port of the LINK# and LANE# equaling the PAD K symbol on certain detected un-configured lanes results in these lanes no longer being part of the present iterative process of the LSTTM. Independently, an additional LSTTM of the downstream port can apply the iterative process to the lanes the PCI Express device on the upstream side of a link is receiving with LANE# equaling PAD K symbol. The additional LSTTM assigns a different specific LINK# while the other LSTTM is completing the present iterative process. These concepts can be expanded for multiple upstream ports connected to a single downstream port. The following discussion focuses on the iterative process for *only* one link between the upstream and downstream ports.

It may be the case that only one link is possible for a downstream port that represents a subset of the detected un-configured lanes. The existence of detected un-configured lanes after configured lanes for a single specific configured link have been established does not always mean a second link exists.

Lanes That Can Be Configured

The downstream port of the upstream PCI Express device is the "leader" in the establishment of configured lanes, and is defined as the "upstream device." It provides the link numbers (LINK#) and lane numbers (LANE#) to be considered by the upstream port of the downstream PCI Express device, which is defined as the "downstream device." The lanes of a link that are determined to be common to both devices and operational by sequencing through the iterative process of the Configuration link substates are defined as *configured lanes* of a *configured link*.

As previously discussed, the number of possible lanes is x1, x2, x4, x8, x12, x16, and x32. The lanes are numbered LANE# 0, LANE# 1, LANE# 2, and so on, up to LANE# 31. The LANE# ordering must be consecutive and must begin with LANE# 0. For example, if the link width is configured as x4, LANE#0, 1, 2, and 3 are defined as configured lanes. The protocol is discussed relative to a specific link with a single LINK#. An upstream device like the Root Complex or a switch could have several downstream ports, each with a different LINK# or set of LINK#s. In the following discussion, assume the upstream device is designed to implement link widths of x1, x4, x12, and x32. A downstream device like switch, endpoint, or bridge can each have only one upstream port with a specific LINK#. In the following discussion, assume the downstream device is designed to implement link widths of x1, x4, x8, and x16. For this discussion, assume the upstream device is configuring a single specific link with LINK# = 4.

Detected Un-Configured Lanes Per this discussion, assume only LANE#s 0 to 15 (x16) have successfully completed bit and symbol lock and lane polarity inversion via Detect and Polling link states. These lanes are defined as *detected un-configured*. The x16 support of the downstream device has limited the possible lanes to 16. If the upstream device had not supported x32, the greatest number of lanes would be 12 per x12 of the upstream device. The collection of detected un-configured lanes only defines that signal lines exist between an upstream device and one or more downstream devices.

Configured Lanes In this discussion the LANE#s 0 to 15 have been detected by the upstream and downstream devices. However, the upstream device is designed to implement x1, x4, x12, and x32 and downstream device is designed to implement x1, x4, x8, and x16. Consequently, the two PCI Express devices on this link only have LANE#s = 0, 1, 2, and 3 in common (x4). Notice in this example x1 is by default supported by both PCI Express devices; but the iterative process determines that greatest number of lanes that can be configured to a specific LINK# defines a link width of x4. Also note that even though the upstream device supports x12, it is not inclusive of the x8 supported by the downstream device. Similarly the upstream device's support of x32 is not inclusive of downstream device's support of x8 and x16.

Link Configuration Protocol Discussion

The LTSSM in the upstream and downstream devices begin the iterative process on *all* detected un-configured lanes. In this example, LANE# 0 to 15 are part of the beginning iterative process. LANE#s 16 to 31 are in electrical idle relative to the upstream device and relative to the downstream device, these lanes do not exist. Per the Polling link state both PCI Express devices have determined that LANE#s 16 to 31 did not succeed in bit and symbol lock and are disabled. Per the Disabled link state the signal lines are in electrical idle.

The basic protocol is that the upstream device takes the lead providing the LINK# and LANE#. The downstream device echoes these LINK# and LANE#. An advanced peer-to-peer link has one unique consideration relative to this protocol. In an advanced peer-to-peer link the downstream ports of two switches are interconnected. Consequently, two upstream devices are trying to take the "lead." The resolution of this situation is discussed later in this chapter.

This discussion refers to the different link *sub-states* of the configuration link state discussed in Chapter 14. The purpose of these references is to aid the reader in referencing this discussion with the Configuration link state next state tables in Chapter 14. Also, the next state tables in Chapter 14 provide more detailed information about the requirements of each link sub-state and the associated transitions. The reader does not need knowledge of the Configuration link sub-states to understand the link configuration protocol discussed below. However, Chapter 14 does provide the detailed information about the Configuration link sub-states.

The discussion uses the terms *USD* and *DSD* to refer to upstream device and downstream device, respectively. The USD is on the upstream side of the link and includes the downstream port. The DSD is on the downstream side of the link and includes the upstream port.

USD Iterative Process

- **Configuration.linkwidth.start link sub-state:** The USD begins continuous transmission of TS1 OSs on each lane it believes could be configured into a specific link. The TS1 OSs contain LINK# equal to 4 and LANE# equal to PAD K symbol and are transmitted on LANE # 0 to 11. USD knows that only x16 or less can possibly be part of a configured link by the Detect and Polling link states, but it only supports x12 and does not support x32. Consequently, the USD knows that the possible lanes that can be configured are LANE# 0 to 11. The USD transmits TS1 OSs containing LINK# and LANE# equal to PAD K symbol on LANE# 12 to 15.

- **Configuration.linkwidth.accept link sub-state:** The USD transitions to this link sub-state upon receipt of TS1 OSs containing LINK# equal to 4 and LANE# equal to PAD K symbol on ANY detected un-configured lane. As part of the iterative process the DSD is transmitting TS1 OSs with the valid LINK# equal to 4 and the LANE# equal to PAD K symbol on LANE# 0 to 7. Even though the DSD had defined x16 as the number of possible lanes that could be configured into a link, the receipt of a valid LINK# (not PAD K symbol) on LANE# 0 to 11 indicates that x16 is not supported by the USD. In addition, by definition the next largest number of lanes supported by the DSD is x8. Consequently, the DSD only transmits a valid LINK# equal to 4 and LANE# equal to PAD K symbol on LANE# 0 to 7. The DSD continues transmission of TS1 OSs with LINK# and LANE# = PAD K symbol on LANE# 8 to 15. In this link sub-state the USD begins continuous transmissions on each lane it believes could be configured into a specific link with TS1 OSs. TS1 OSs contain LINK# equal to 4 and LANE# equal to 0 to 3 on LANE # 0 to 3. Even though the DSD has indicated it could configure LANE# 0 to 7, the USD can only configure a link of width x4 or less. Also per this link sub-state the USD transmits TS1 OSs containing LINK# and LANE# equal PAD K symbol on LANE # 4 to 15.

- **Configuration.lanenum.wait link sub-state:** The USD transitions to this link sub-state if it was successful in determining that a link can be configured and began transmission of the TS1 OSs discussed in the **configuration.linkwidth.accept link sub-state** or **configuration.lanenum.accept link sub-state**. The USD is waiting to determine whether the DSD can accept the LANE# proposed for LANE# 0 to 3 being transmitted in the TS1 OSs.

- **Configuration.lanenum.accept link sub-state:** The USD transitions to this link sub-state if the LANE# in the TS1 OSs it is receiving is different than when it entered the **configuration.lanenum.wait link sub-state**. The DSD has been transmitting TS1 OSs with LANE# equal to PAD K symbol and changes to LANE# equal to 0 to 3 on LANE 0 to 3. The transition from PAD K symbol to LANE# 0 to 3 by the DSD is what causes the transition to this link sub-state by the USD.

- **Configuration.complete link sub-state:** The USD transitions to this link sub-state upon receiving TS1 OSs containing LINK# equal to 4 and LANE# equal to 0 to 3 on exactly the same four lanes in the same LANE# order that it is transmitting TS1 OSs. In this link sub-state the USD begins continuous transmissions of TS2 OSs. The TS2 OSs contain LINK# equal to 4 and LANE# equal to 0 to 3 on LANE # 0 to 3. On the other lanes the TS2 OSs contain LINK# and LANE# equal to PAD K symbol. The transition to this link sub-state reflects the DSD transmitting TS1 OSs containing LINK# equal to 4 and LANE# equal to 0 to 3 in response to receiving the same TS1 OSs on the same lanes.

- **Configuration.idle link sub-state:** The USD transitions to this link sub-state upon receiving TS2 OSs containing LINK# equal to 4 and LANE# equal to 0 to 3 on exactly the same four lanes in the same LANE# order that it is transmitting TS2 OSs. In this link sub-state the USD begins continuous transmissions of IDLE D symbols. The IDLE D symbols are transmitted on LANE# 0 to 3, which are defined as configured lanes of a configured link. On the other lanes the USD transitions to electrical idle. On the other lanes the USD may transmit the Electrical Idle OSs before transitioning to electrical idle.

DSD Iterative Process

- **Configuration.linkwidth.start link sub-state:** The DSD begins continuous transmission of TS1 OSs on each lane it believes could be configured into a specific link. The TS1 OSs contain LINK# and LANE# equal to PAD K symbol and are transmitted on LANE# 0 to 15. The DSD knows that only x16 or less can possibly be part of a configured link by the Detect and Polling link states, and this is one of the link widths it can support.

- **Configuration.linkwidth.accept link sub-state:** The DSD transitions to this link sub-state upon receipt of TS1 OSs containing LINK# equal to 4 and LANE# equal to PAD K symbol on ANY detected un-configured lane. As part of the iterative process, the USD is transmitting TS1 OSs with the valid LINK# equal to 4 and the LANE# equal to PAD K symbol on LANE# 0 to 11. The USD defines x12 as the number of possible lanes that it could configure even though the Detect and Poll link states indicate that x16 could be configured. The USD also transmits TS1 OSs with LINK# and LANE# equals PAD K symbol on LANE# 12 to 15. In this link sub-state, the DSD begins continuous transmission on each lane it believes could be configured into a specific link reflective of which lanes it received TS1 OSs with a LINK# not equal to PAD K symbol. Even though the USD has indicated it could configure LANE# 0 to 11, the DSD can only configure a link of width x8 or less. Consequently, the DSD transmits TS1 OSs containing LINK# equal to 4 and LANE# equal PAD K symbol LANE # 0 to 7. Also, per this link sub-state the DSD continues transmitting TS1 OSs containing LINK# and LANE# equal PAD K symbol on LANE# 8 to 15.

- **Configuration.lanenum.wait link sub-state:** The DSD transitions to this link sub-state if it receives TS1 OSs with the LINK# it is transmitting and with at least one LANE# 0 in the **configuration.linkwidth.accept link sub-state** or **configuration.lanenum.accept link sub-state**. With the transtion to this link sub-state, DSD begins transmitting TS1 OSs containing LINK# equal to 4 and LANE# equal 0 to 3 on LANE# 0 to 3. Even though the DSD had been transmitting TS1 OSs containing LINK# equal to 4 and LANE# equal PAD K symbol on LANE # 0 to 7, the USD responds with a TS1 OSs containing LINK# equal to 4 and LANE# equal to 0 to 3 on LANE# 0 to 3. The USD is transmitting TS1 OS contain LINK # and PAD# equal to PAD K symbol on LANE# 5 to 15.

- **Configuration.complete link sub-state:** The DSD transitions to this link sub-state upon receiving TS2 OSs containing LINK# equal to 4 and LANE# equal to 0 to 3 on exactly the same four lanes in the same LANE# order that it is transmitting TS1 OSs. In this link sub-state the DSD begins continuous transmissions of TS2 OSs. The TS2 OSs contain LINK# equal to 4 and LANE# equal to 0 to 3 on LANE # 0 to 3. On the other lanes the the TS2 OSs contain LINK# and LANE# equal to PAD K symbol. The transition to this link sub-state reflects the USD transmitting TS2 OSs containing LINK# equal to 4 and LANE# equal to 0 to 3 in response to receiving the same TS1 OSs on the same lanes.

- **Configuration.idle link sub-state:** The DSD transitions to this link sub-state upon receiving TS2 OSs containing LINK# equal to 4 and LANE# equal to 0 to 3 on exactly the same four lanes in the same LANE# order that it is transmitting TS2 OSs. In this link sub-state the DSD begins continuous transmissions of IDLE D symbols. The IDLE D symbols are transmitted on LANE# 0 to 3, which are defined as configured lanes of a configured link. On the other lanes, the DSD transitions to electrical idle. Optionally on the other lanes, the DSD may transmit the Electrical Idle OSs before transitioning to electrical idle.

Downstream Port Connected to Downstream Port

In the previous discussion, the downstream device was in full agreement with the LINK# proposed by the upstream device. Per the link configuration protocol, the upstream device on the link takes the lead in proposing LANE# and LINK#. It is possible that an advanced peer-to-peer link (also known as a cross link) is implemented, thus two downstream ports are interconnected, as illustrated in Figure 8.18. The two downstream ports both assume they are the upstream device per the link configuration protocol and try to lead. When both upstream devices are in the **configuration.linkwidth.start sub-state**, both expect to receive TS1 OSs containing the LINK# = PAD K symbol on all possible common lanes. Both upstream devices are transmitting TS1 OSs with proposed LINK# not equal to the PAD K symbol. Consequently, both upstream devices recognize the cross link and are required to redefine themselves as downstream devices and re-enter the **configuration.linkwidth.start substate**. Each of these PCI Express devices has selected a random timeout of 0 to 1 milliseconds when the cross link is recognized. The first device that times outs redefines itself as an upstream device and proceeds per the **configuration.linkwidth.start sub-state** to begin transmitting TS1 OSs containing proposed LINK# not equal to the PAD K symbol. During the period when both PCI Express devices have defined themselves as a downstream device, both are transmitting TS1 OSs containing LINK# equal to PAD K symbol per the **configuration.linkwidth.start sub-state**. Consequently, when one PCI Express device begins receiving TS1 OSs containing proposed LINK# not equal to the PAD K symbol, it must affirm itself as the downstream device and ignore its timeout. Once one PCI Express device redefines itself as the upstream device and the other device affirms itself as the downstream device, the link configuration protocol can proceed as described in the previous discussion.

> **Important Note**
>
> The inclusion of the advanced peer-to-peer link is to reflect future enhancements to the PCI Express specification. The PCI Express specification was first released with the advanced peer-to-peer link. The errata available at the time this book was printed removed most references to this entity, but retained it relative to the link configuration protocol. However, the errata had not received final approval by the engineering review board of the PCISIG at the print time of this book. Please check www.pcisig.com and www.intel.com for more information.
> See Chapter 18 for more information.

LANE# Disagreement and Lane Reversal

In the above discussion, the downstream device was in full agreement with the LANE#s proposed by the upstream device. The LANE# 0, 1, 2, and 3 were proposed by the upstream device transmitting TS1 OSs and were immediately accepted by the downstream device transmitting TS1 OSs with LANE# 0, 1, 2, and 3.

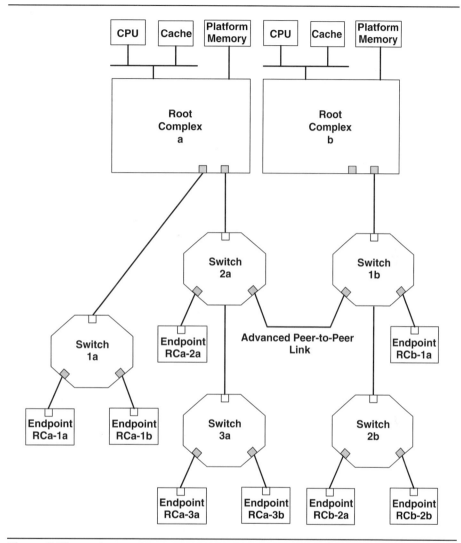

Figure 8.18 Advanced Peer-to-Peer Link

It is possible that the mechanical implementation of the downstream device does not align the LANE# with the same orientation as the upstream device. The simplest resolution is for the downstream device to accept the upstream device's proposed LANE# ordering. That is, the downstream device reverses (internally) its LANE# order and begins transmitting TS1 OSs with the same LANE# order it is receiving. This is defined as Lane Reversal.

The ability for a downstream device to support Lane Reversal internally is *optional.* Consequently, if Lane Reversal by the downstream device is not possible, the downstream device begins transmitting in the **config.linkwidth.accept** link sub-state TS1 OSs with the LANE# ordering it can accommodate, even though it is not the lane order it is receiving. If the upstream device supports Lane Reversal, it begins transmitting in the **config.complete** link sub-state TS2 OSs with a reversed LANE# ordering after receiving TS1 OSs with a lane order different than it had had been transmitting.

The ability for an upstream device to support Lane Reversal internally is optional as well. Consequently, if it does not support Lane Reversal and receives at least two consecutive TS1 OSs with the same reversed LANE# ordering, the upstream device transitions to the Detect link state. Also, it is required that the LANE# ordering is consecutive and must begin with LANE# 0. Consequently, when lane reversal is implemented, it is only a simple reversal in LANE# ordering.

Chapter 9

Errors

The PCI Express platform has three types of errors: correctable, nonfatal, and fatal. The PCI Express platform responds to each of these error types either by hardware or by software. The response by hardware means that the error is detected by hardware and is handled by hardware without any intervention by the Error software. The response by software means that there error was detected by hardware, reported to Error software, and it is the Error software that handles the error. The term *Error software* applies to any platform software that can respond to an error condition.

- **Correctable Errors:** As the name implies a correctable error can be processed without any loss of information. They are responded to by hardware and not by the Error software. Optionally, logging of correctable errors may be of value to Error software, but this is not defined by the PCI Express specification.

- **Nonfatal Errors:** Nonfatal errors are *not* correctable. They are responded to by the Error software and not by the hardware. Nonfatal errors are isolated to a specific link, do not render the specific link inoperative, and do not disturb activity on the specific link.

- **Fatal Errors:** Fatal errors are *not* correctable. They are responded to by the Error software and not by the hardware. Fatal errors may or may not be isolated to a specific link and the connected PCI Express devices. Fatal errors may require a Hot Reset of the link.

Relationship to PCI

Some of the terms discussed in the following section are defined and discussed later in this chapter. This discussion summarizes the relationship of the PCI Express error protocol to PCI as a background to following discussions specific to only PCI Express.

The basic software model for PCI Express is that of a PCI platform. Indeed, configuration register blocks contain both PCI-compatible registers and registers specifically compatible only with PCI Express. In order to support PCI software, the PCI configuration registers that report error status must reflect the associated (similar) PCI Express error status. That is, error detection causes bits to be set in error status registers compatible with PCI Express. Similar bits in PCI-compatible error status registers must also be set to indicate the detected error. However, the PCI-compatible Error software clearing the bits within the PCI-compatible registers does not result in the clearing of the bits in registers compatible with PCI Express. Similarly, the software that is compatible with PCI Express and is clearing bits in registers compatible with PCI Express does not result in the clearing of the bits in PCI-compatible registers.

Virtual PCI/PCI Bridges exist within the downstream ports of the Root Complex and switches, and the upstream ports of the bridges. These virtual PCI/PCI Bridges contain PCI-compatible configuration registers, some of which are the PCI Status Registers. The association between PCI's Status Registers, and PCI Express's Uncorrectable and Correctable Error Status Registers is as follows:

- When a bit is set to 1b (due to a detected error) in the PCI Express Uncorrectable or Correctable Error Status Register, a bit defining the same error in the PCI Status Register is also set to a corresponding "1" value. This protocol provides error information to the PCI-compatible Error software.

- The *clearing* (setting to 0b) of a bit in the PCI Status Register by Error software does not clear (set to "0") the associated bit in the PCI Express Uncorrectable or Correctable Error Status Register.

- The clearing (setting to 0b) of a bit in the PCI Express Uncorrectable or Correctable Error Status Register by Error software does not clear the associated bit in the PCI Status Register.

Additional considerations:

- The bits in the PCI Command Register associated with error reporting and the PCI Status Register do not have any affect on any of the PCI Express registers associated with error detecting, reporting, or logging.

- A Transaction Layer Packet (TLP) that is poisoned is ported through the switch unchanged from upstream side to the downstream side. The TLP is identified as poisoned by the EP bit in Header field if set to 1b. If a poisoned TLP is ported through a switch the following bits must be set to 1b in the PCI Status Registers of the virtual PCI/PCI Bridges:
  - The Detected Parity bit in the Status Register on the receiving side.
  - The Master Data Parity Error bit in the Secondary Status Register on the transmitting side. This applies if Parity Error Response bit is set to 1b in the Bridge Control Register.

- A TLP that is poisoned is ported through the switch unchanged from downstream side to the upstream side, but the following bits must be set to 1b in the PCI Status Registers of the virtual PCI/PCI Bridges:
  - The Detected Parity bit in the Secondary Status register on the receiving side.
  - The Master Data Parity Error bit in the Status Register (If Parity Error Response bit is set to 1b in the Bridge Control Register) on the transmitting side.

Error Protocol Specific to PCI Express

The remainder of this chapter is divided into four parts. The first part discusses the possible errors at the Transaction Layer and PCI Express device core. The errors detected are reported via a mixture of completion status in the completer transaction packet (where applicable), configuration registers, and message error requester transaction packets. The major sections related to this part are:

- **Errors Related Transaction Layer ... Basic**
- **Errors Related to Transaction Layer ... Completion Timeout Error**

- Errors Related to Transaction Layer ... Receiver Overflow Error
- Errors Related to Transaction Layer ... Flow Control Protocol Error (FCPE)
- Transaction Layer ... TLP Fields Checked for Errors

The second part discusses the possible errors at the Data Link Layer and Physical Layer. The errors detected in these layers are reported via a mixture of configuration registers and message error requester transactions. The major sections related to this part are:

- Errors Related to Data Link Layer
- Errors Related to Physical Layer

The third part discusses how the integrity of the Transaction Layer Packets (TLPs), Data Link Layer Packets (DLLPs), and Link Layer Transaction Packet (LLTPs). These packets are covered by the ECRC, CRC, and LCRC, respectively. ECRC, CRC, and LCRC are based on cyclic redundancy checking. The major section related to this part is:

- CRC, LCRC, and ECRC Calculations, and Digest Field

The fourth part discusses how all PCI Express errors are reported and logged within the PCI Express platform. The major section related to this part is:

- Error Reporting and Logging

Error Reporting

In the following discussions, it is noted when an error is reported via a message error requester transaction packet. As discussed in Chapter 6, the possible errors that be can reported are: ERR_FATAL, ERR_NONFATAL, and ERR_COR. These message error transaction packets are transmitted to the downstream port of the Root Complex by the requester source, requester destination, completer source, and completer destination. When the PCI Express device transmitting the message error requester transaction packet is the Root Complex's downstream port, it is a virtual message transmitted internally to itself. Two other considerations must be taken into account for any PCI Express device to transmit a message error requester transaction packet to report an error. One consideration is that the PCI Express device must be configured by the Error software to implement message error requester transactions to report errors. The other consideration is that if Advance Error Reporting protocol is implemented, the Header field of the TLP associated with the error is logged. See the "Error Reporting and Logging" section later in this chapter.

Basic Transaction Layer Errors

The possible errors at the Transaction Layer protocol are either directly related to the Transaction Layer Packets (TLPs) or indirectly at the PCI Express device core. All of the possible errors are summarized in Table 9.1. In some cases, there are multiple sources for a specific error. This chapter discusses all of the sources of these errors with the errors printed in **<u>bold underline</u>** in reference to the table. The **<u>bold underline</u>** typographical convention applies to errors described in all chapters in the book.

> Unless otherwise stated as "optional," the errors discussed in this section *must* be checked by the associated PCI Express device.

An important way that Transaction Layer errors distinguish themselves is that they interact with the sources or destinations of the TLPs, not specifically with the intervening switches or bridges porting the TLPs, with a couple of exceptions noted below. As previously discussed, this book uses the following terms:

- *Requester source* and *requester destination:* The two participants in the transmitting and receiving of the requester transaction packets.

- *Completer source* and *completer destination:* The two participants in the transmitting and receiving of the completer transaction packets.

- Requester and completer transactions that port through switches or bridges do not define these switches or bridges as participants. The switches and bridges (exceptions) are participants for accesses to their configuration register blocks and for some messages requester transactions.

Table 9.1 Transaction Layer Protocol Errors

Error Name	Message Error Requester Transactions (1)(2) [Severity]	Participant Reporting the Error	Logged? If Advance Error Reporting Header Log Registers Implemented
Malformed	ERR_FATAL or ERR_NONFATAL (4) [Uncorrectable]	Requester or completer destination via message transaction	Log Header field of TLP (3)
ECRC	ERR_NONFATAL or ERR_FATAL (5) [Uncorrectable]	Requester or completer destination via message transaction	Log Header field of TLP (3)
Poisoned TLP (Error Forwarding)	ERR_NONFATAL or ERR_FATAL (5) [Uncorrectable]	Requester or completer destination via message transaction	Log Header field of TLP (3)
Completer Abort (6)	ERR_NONFATAL or ERR_FATAL (5) [Uncorrectable]	Requester destination via message transaction	Log Header field of TLP (3)
Unsupported Requester (6)	ERR_NONFATAL or ERR_FATAL (5) [Uncorrectable]	Requester destination via message transaction	Log Header field of TLP (3)
Unexpected Completer	ERR_NONFATAL or ERR_FATAL (5) [Uncorrectable]	Completer destination via message transaction	Log Header field of TLP (3)
Completion Timeout	ERR_NONFATAL or ERR_FATAL (5) [Uncorrectable]	Requester source via message transaction	Not logged
Receiver Overflow	ERR_FATAL or ERR_NONFATAL (4) [Uncorrectable]	Requester or completer destination via message transaction	Not logged
FCPE (Flow Control Protocol Error)	ERR_NONFATAL or ERR_FATAL (5) [Uncorrectable]	Requester or completer destination via message transaction	Not logged

Table Notes:

1. The type of error in the first column is reported by the appropriate message error requester transactions in the second column.
2. In addition to the message error requester transaction, the appropriate bit in the Uncorrectable Status Register is set to "1" independent of the value of the associated bit in the Uncorrectable Mask Register (if Advance Error Reporting protocol is implemented).
3. If appropriate bit in the Uncorrectable Mask Register is *not* set to "1" and an error is detected (according to the associated bit in the Uncorrectable Status Register), the error is not logged in the Header Log Registers, the First Error Pointer is not updated, and the error is not reported column 1.
4. Reported as ERR_FATAL unless the Uncorrectable Error Severity Register is implemented and the appropriate bit is set to a "0" value. If the Uncorrectable Error Severity Register is implemented and the appropriate bit is set to a "0" value, the error is reported as ERR_NONFATAL. The default value for Minimum Error Reporting protocol (Advance Error Reporting protocol not implemented) is ERR_FATAL.
5. Reported as ERR_NONFATAL unless the Uncorrectable Error Severity Register is implemented and the appropriate bit is set to a "1" value. If the Uncorrectable Error Severity Register is implemented and the appropriate bit is set to a "1" value, the error is reported as ERR_FATAL. The default value for Minimum Error Reporting protocol (Advance Error Reporting protocol not implemented) is ERR_NONFATAL.
6. The requester destination reports this error to the Error Software via the message error requester transaction and configuration registers (as qualified by configuration registers) if no completer transaction is associated with the requester transaction packet. If a completer transaction is associated with the requester transaction packet, the PCI Express completer source (PCI Express requester destination) transmits a completer transaction with appropriate STATUS (UR or CA). The completer destination (which is the original PCI Express requester source) can retransmit the requester transaction or report the error to the Error Software via the message error requester transaction and configuration registers (as qualified by configuration registers).

Errors Related to TLPs and Checked at Requester and Completer Destinations

Figure 9.1 shows an example of how the Transaction Layer can be implemented to check for errors related to this layer. The discussion of this implementation is divided into four parts. The focus of each part is as follows:

- **Malformed and Type Supported:** The preliminary processing that applies to both requester and completer transaction packets.
- **Further Processing of Requester Transaction Packets**: The core processing of requester transaction packets.

- **Further Processing of Completer Transaction Packets:** The core processing of completer transaction packets.

- **Continued Processing of Unsupported and Completer Abort Errors for Requester Transaction Packets:** The additional processing of requester transaction packets associated with unsupported and completer abort errors.

Malformed, Type Defined, and Type Supported

If no errors are found in the Physical or Data Link Layers at the requester or completer destination, illustrated in Figure 9.1, there may be errors in the TLP itself. As summarized in Table 9.1, the Transaction Layer checks for errors related to TLPs at several different levels. As illustrated in Figure 9.1, the requester or completer destination must first determine whether the TLP is properly formed. A TLP is properly formed if the bit values in the packet's Header field are correct for a specific type of transaction. The details for the incorrect bits in the Header field are listed in the section "TLP Fields Checked for Errors" later in this chapter.

If any PCI Express device receives a TLP with a TC number mapped to a non-enabled VC number for the port, the packet is discarded, is defined as malformed, and a **malformed** error is reported via a message error requester transaction packet.

Once the TLP is determined as properly formed, the requester or completer destination must determine whether the type of transaction is defined as a TLP. This determination is for whether the type is defined and not whether the requester or completer destination supports it. For non-message requester transactions and completer transactions, the type of requester transaction packet is defined in the Type field (Bits [4::0]) in the TYPE HDW. For message requester transactions the type of packet is defined in the Type field (Bits [4::0]) in the TYPE HDW. A TLP's Type field is defined if the bit values in the packet's Header field are correct for a specific type of transaction. The details for the incorrect bits in the Header fields are listed in the section "TLP Fields Checked for Errors" later in this chapter. As outlined in Figure 9.1, when a TLP is not properly formed or if the type defined is not correct, the TLP is defined as malformed and a **malformed** error is reported via message error requester transaction packet.

In the ATTRIBUTE HDW of a TLP's Header field, the error forwarding bit (Error Poison Bit 14 of ATTRIBUTE HDW) may only = 1b for certain types of TLPs. If the type of TLP does not support error forwarding and EP = 1b, it is defined as malformed and thus a **malformed** error has occurred. If the type of TLP does support error forwarding and EP = 1b, there may be an issue of an **unsupported requester** error, discussed later in this chapter. See also the section "Error Forwarding (also Known as Data Poisoning)" for more information.

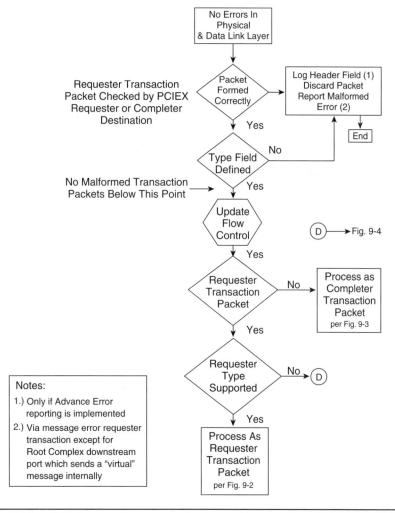

Figure 9.1 PCI Express TLP Error Check at Requester or Completer Destination

> In Figure 9.1 and the following figures and discussions, a TLP that is discarded has its Header field logged if the appropriate Header Log Registers have been implemented according to the Advance Error reporting protocol. If Advance Error reporting is not implemented, the TLP that is discarded does not have its Header field logged.

As illustrated by Figure 9.1, once the requester or completer transaction packet is determined to be properly formed and the type is defined, the Transaction Layer updates the Flow Control information and determines whether the type of requester transaction is supported. See Chapters 11 and 12 for more information about the Flow Control protocol.

In the example in Figure 9.1, it is assumed that all completer transactions processed through the Update Flow Control point are supported by the completer destination. Consequently, for completer transactions the processing begins. For requester transactions the associated TLPs' Type field is either supported or unsupported depending on the bit values in the TLPs' Header field for a specific type of transaction. The details for the unsupported bits in the Header fields are listed in the section "TLP Fields Checked for Errors." When a requester transaction packet is not supported by the requester destination, the requester transaction packet is defined as unsupported and thus an **unsupported requester** error is reported.

Further Processing of Requester Transaction Packets

As illustrated in Figure 9.2, prior to a requester transaction being processed, two additional things must be considered. One is the programming model for a non-message requester transaction and the other the determination of whether the Message Code is supported by the requester destination.

For non-message requester transactions, the requester destination can also optionally check for programming model violations, as illustrated in Figure 9.2. If the requester destination determines that a programming model violation has occurred, the requester transaction packet can either be processed (the manner in which it is processed is implementation-specific) as if no violation has occurred, or processed with violation. If processed as if no violation has occurred, the next consideration is whether it can be processed. If processed with a violation a **completer abort** error is executed.

Message requester transactions also have a Message Code field (Bits [7::0]) in the TAG HDW that is checked by the requester destination to determine whether message codes are supported. The details for the unsupported bits in the Header fields are listed in the section "TLP Fields Checked for Errors." As illustrated in Figure 9.2, when a message requester transaction packet is not supported by the requester destination the message requester transaction packet is defined as unsupported and thus an **unsupported requester** error is reported.

Continuing with the flow of the example illustrated in Figure 9.2, the next determination is whether the requester transaction packet can be processed or not. That is, there may be an internal problem (requester destination not fully functional) and the requester destination must execute a **completer abort** error.

If the requester transaction packet is supported, if no programming violations have occurred (excluding message requester transactions), and if the requester transaction packet can be processed by the requester destination (resource is fully functional), then the requester transaction packet is defined as successfully received. For successfully received memory read, I/O read and write, and configuration read or write requester transaction packets, the associated completer transaction packets contain STATUS = SC. For successfully received configuration read or write requester transaction packets for which the PCI Express device is not ready to indicate completion (temporarily not able due to reset), the associated completer transaction packets contain STATUS = SCR, but configuration read or write requester transaction packets are still defined as successfully received.

If no "Other Errors" are reported (see the following paragraph), the requester destination has defined the TLP as successfully received. However, the requester destination must check two other possible error conditions and report them to the Error software, illustrated in Figure 9.2, as follows:

- **Poison TLP Error:** In the ATTRIBUTE HDW of a TLP's Header field, the error forwarding bit (Error Poison Bit 14 of ATTRIBUTE HDW) may only = 1b for certain types of TLPs. If the type of TLP does not support error forwarding and EP = 1b, it is defined as malformed and thus a **malformed** error has already occurred, as illustrated in Figure 9.1. If the type of TLP supports EP but is writing to a control space, the TLP is defined as unsupported and thus an **unsupported requester** error is reported. Otherwise, if EP = 1b the TLP is discarded and a **Poisoned TLP** error is reported via message error requester transaction packet. See the section "Error Forwarding (also known as Data Poisoning)" later in this chapter for more information.

- **Transaction Digest**: If the TD bit = 1b in the Header field (Transaction Digest Bit 14 of ATTRIBUTE HDW) indicates the Digest field is attached and must be checked. The Digest field contains the ECRC for the TLP. If an **ECRC** error is detected, the TLP is discarded and an ECRC is reported via a message error requester transaction packet. See the "CRC, LCRC, ECRC, and Digest Field" section later in this chapter.

Other Errors It is possible for the requester destination to be fully functional, for the requester transaction packet to be received without error, and to be supported (by the bits in the Header field), and yet still have an error. Additional factors must be considered when assessing a response of the requester transaction, as illustrated in Figure 9.2. For example, if the requester destination is a bridge, it is possible that a Master Abort (Status = UR) or a Target Abort (Status = CA) on the PCI or PCI-X bus segment may occur. A Master or Target Abort requires the completer source (requester destination) to transmit a completer transaction packet with the appropriate Abort status. If a completer transition is not associated with the requester transaction, the requester destination must report the error to the Error software via the message error requester transaction and configuration registers (as qualified by configuration registers). In the example later in this chapter in Figure 9.4, the requester destination selects whether the Other Errors are defined as **unsupported requester** or **completer abort** errors (implementation-specific).

As is covered in more detail in Chapter 15, a function in a PCI Express device in the D1 state cannot initiate any TLPs except for message PM_PME requester transaction and the configuration completer transaction. The only transactions that can be received by the function are configuration requester transactions. All other transactions received are defined as **unsupported requester** error.

A function in a PCI Express device in the D3hot state can participate in the message PME_Turn_Off/PME_TO_Ack requester transaction protocol. A function cannot initiate or receive any other TLPs. If a transaction other than message PME_Turn_Off requester transaction is received, it is defined as an **unsupported requester** error.

Chapter 9: Errors **433**

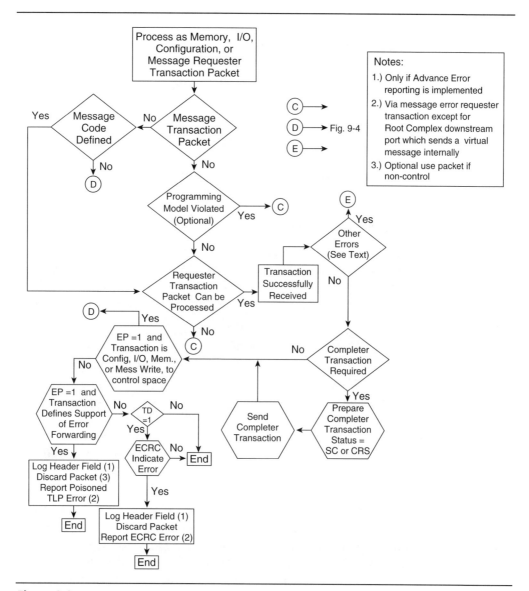

Figure 9.2 Processing Requester Transaction Packet

Further Processing of Completer Transaction Packets

As illustrated in Figure 9.3, after initial processing, the completer transaction packets are further processed differently than requester transaction packets. Prior to defining a completer transaction as successfully received, three unique error conditions may exist for completer transaction packets and must therefore be searched for:

- **Completer Transaction Expected:** The completer destination must determine whether the completer transaction packet is expected. There must be an associated requester transaction in the completer destination (requester source); if not the completer destination discards the TLP and reports an **unexpected completer** error via a message error requester transaction packet.

- **Status Field is Reserved:** If Status field value is RESERVED the completer destination processes the completer transaction packet as if the Status [2::0] value = 001b (Unsupported Request). The TLP is further processed following the Re-transmit protocol.

- **Status Field is Unsupported Requester or Completer Abort:** If the Status field value equals unsupported requester or completer abort, the completer destination that is also the requester source further processes the TLP following the Re-transmit protocol.

If the above three unique error detection considerations do not apply to the completer transaction packet, the TLP is defined as successfully received. However, the completer destination must check two other possible error conditions and report them to the Error software as follows:

- **Error Poison:** In the ATTRIBUTE HDW of a TLP's Header field, the error forwarding bit (Error Poison Bit 14 of ATTRIBUTE HDW) may only = 1b for certain types of TLPs. If the type of TLP does not support error forwarding and EP = 1b, it is defined as malformed and thus a **malformed** error has already occurred. Otherwise, if EP = 1b the TLP is discarded and a **Poisoned TLP** error is reported via message error requester transaction packet. Also, see the "Errors Related to Transaction Layer: Error Forwarding (also Known as Data Poisoning)" section later in this chapter.

- **Transaction Digest:** If the TD bit = 1b in the Header field (Transaction Digest Bit 14 of ATTRIBUTE HDW), this indicates that the Digest field is attached and must be checked. The Digest field contains the ECRC for the TLP. If an **ECRC** error is detected, the TLP is discarded and an ECRC is reported via a message error

Chapter 9: Errors ■ 435

TLP is discarded and an ECRC is reported via a message error requester transaction packet. See the "CRC, LCRC, ECRC, and Digest Field" section later in this chapter.

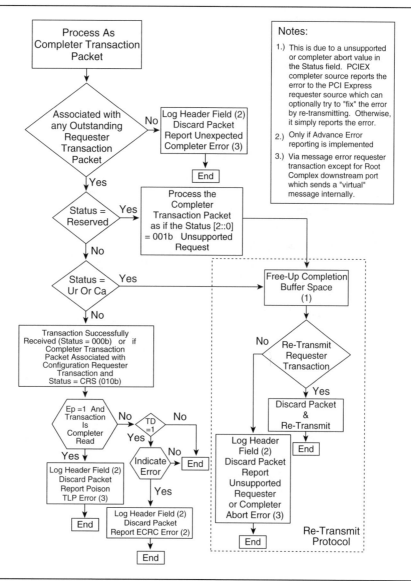

Figure 9.3   Processing Completer Transaction Packet

Re-transmit Protocol of Completer Transaction Packets As illustrated in Figure 9.3 and discussed previously, a completer transaction packet can be received with the Status field indicating an **unsupported requester** or **completer abort** error. In this situation, the completer destination is also the requester source of the associated requester transaction packet. It has the option to re-transmit the requester transaction packet after completion buffer space is freed up. If the requester source does not want to re-transmit the requester transaction packet, the completer transaction packet is discarded, and the **unsupported requester** or **completer abort** error is reported via the message error requester transaction. Otherwise the completer transaction packet is discarded and the associated requester transaction packet is re-transmitted.

Continued Processing of Unsupported and Completer Abort Errors for Requester Transaction Packets As discussed earlier, the requester destination may determine that an **unsupported requester** error or a **completer abort** error has occurred in a requester transaction packet. As illustrated in Figure 9.4, if no completer transaction packet is associated with the requester transaction packet; the requester transaction packet is discarded and the error is reported by the requester destination via message error requester transition packets. If the requester transaction packet is associated with a completer transaction packet, the requester destination transmits a completer transaction packet with the Status field = UR or CA to provide the requester source an opportunity to re-transmit (possibly after taking other action). If the requester source determines that it cannot successfully re-transmit the associated requester transaction packet, it reports the error via message error requester transaction packet.

Chapter 9: Errors ■ 437

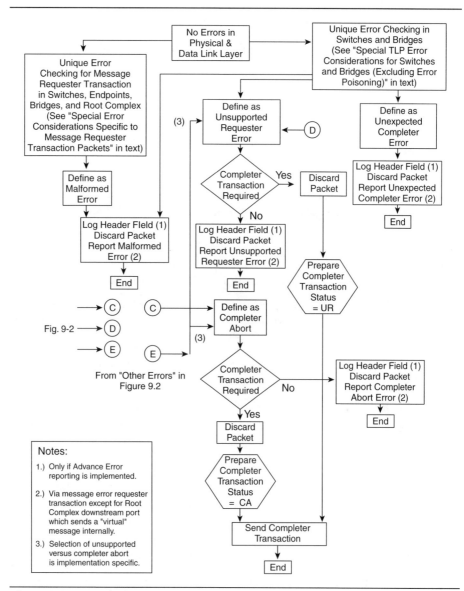

Figure 9.4 Processing of Unsupported Requester and Completer Abort Errors

Special Error Considerations Specific to Message Requester Transaction Packets

The following section examines special error considerations for message requester transaction packets.

Switches

Message Transaction Related A switch (not defined as a requester or completer destination that is not an access to its configuration register block) does not determine whether a TLP is for a non-message requester transaction it is porting between ports is properly formed; it simply ports the TLPs from one port to another with one exception. However, switches must determine whether the TLP is for a message baseline or vendor-defined requester packet. If it is message baseline or vendor-defined requester packet, the switch can optionally check for the following errors in the r [2::0] (not n or c field defined for message advanced switching requester transaction) in the TYPE HDW of the Header field.

- The upstream port of a switch receives a message baseline or vendor-defined requester transaction with r [2::0] = 000b or 101b.

- A downstream port of a switch receives a message baseline requester or vendor-defined requester transaction with r [2::0] = 011b.

- The upstream port of a switch receives a message interrupt or interrupt acknowledge requester transaction with r [2::0] = 100b.

If one of these errors is found, the switch defines the message requester transaction packet as malformed, discards the TLP, and a **malformed** error is reported via a message requester transaction packet, as illustrated in Figure 9.4. If not checked, the TLP cannot be properly routed and is thus simply discarded.

Endpoints and Bridges

Endpoints and bridges must determine whether the TLP is a message baseline or vendor-defined requester packet. If it is message baseline or vendor-defined requester packet, the endpoint or bridge can optionally check for the following errors in the r [2::0] (not n or c field defined for message advanced switching requester transaction) in the TYPE HDW of the Header field.

- The upstream port of an endpoint or a bridge receives a message baseline or vendor-defined requester transaction with r[2::0] = 000b or 101b.

- The upstream port of an endpoint or a bridge receives a message interrupt or interrupt acknowledge requester transaction with r[2::0] = 100b.

If one of these errors is found, the endpoint or bridge defines the message requester transaction packet as malformed, discards the TLP, and a **malformed** error is reported via a message requester transaction packet, as illustrated in Figure 9.4. If not checked, the TLP cannot be properly routed and is thus simply discarded.

Root Complex

The Root Complex's downstream ports do not determine whether a TLP is properly formed, they simply port the TLPs between the Root Complex proper and the downstream PCI Express devices with one exception. The Root Complex's downstream ports *must* determine whether the TLP is a message baseline requester packet, and if so must check for the following error in the r [2::0] (not n or c field)].

- A downstream port of the Root Complex receives a message baseline or vendor-defined requester transaction with r [2::0] = 011b.

If this error is found, the Root Complex's downstream port defines the message requester transaction packet as malformed, discards the TLP, and a **malformed** error is reported via a virtual message requester transaction packet to itself, as illustrated in Figure 9.4.

Special TLP Error Considerations for Switches and Bridges (Excluding Error Poisoning)

The following section examines special TLP error considerations for switches and bridges.

Switches

As discussed earlier, the error checking by the Transaction Layer is done by the PCI Express device that is the destination of the TLP. In the case where the TLPs' destination is the configuration register block in a switch, the error checking activities at the Transaction Layer level within the switch are the same as non-switch PCI Express devices. In that the switch does port TLPs between ports, some other error conditions must be considered.

TD and ECRC The value of the TD bit in the Header field (Transaction Digest Bit 14 in the ATTRIBUTE HDW) and the ECRC value in the Digest field are used for data integrity of the TLP. These values are not monitored (with the exception noted in the following paragraph) or modified by switches when porting TLPs from one port to another port (that is, when the switch is not the requester or completer destination). If the switch is the requester or completer destination, the TD bit and ECRC value are processed as they are for any other PCI Express device that is the requester or completer destination.

The PCI Express specification does permit switches that are porting TLPs from one port to another port (not as the requester or completer destination) to *optionally* monitor the TD bit and check the ECRC value. The switch reports the **ECRC** error via a message error transaction packet, if found. However, the switch ports the TLP to the transmitting port without modifying the TD bit or the ECRC value. See the "CRC, LCRC, ECRC, and Digest Field" later in this chapter.

Unsupported and Unexpected A LLTP is received at a switch's port and without any Data Link Layer error the TLP component is ported to one of the switch's transmitting ports. As discussed in this book, each port of the switch is associated with a virtual PCI/PCI Bridge. For all requester transaction packets (except message requester transaction packets) the address space in the ADDRESS HDW must be in the address range of one of the switches' transmitting ports as defined by the PCI/PCI Bridge protocol. For configuration requester transaction packets the BUS# in the CONFIGURATION HDW must route to one of the switches' transmitting ports. For completer transaction packets the BUS# in the REQUESTER ID HDW must route to one of the switches' transmitting ports. Also, all TLPs have a mapping based on traffic class (TC) numbers in the TYPE HDW in the Header field to the virtual channel (VC) numbers assigned to each transmitting port. If any of these addresses or routings is not correct, the switch's receiving port must define the TLP as **unsupported requester** or **unexpected completer** error as illustrated in Figure 9.4.

TC and VC Related The TLPs received and ported through the switch must have TC numbers assigned to the enabled VC numbers of the correct transmitting ports. Otherwise, the packet is discarded, the packet is defined as malformed, and a **malformed** error is reported via a message error transaction packet as illustrated in Figure 9.4.

SERR# Enable In order for the message error transaction packets to be ported from a switch's downstream port to its upstream port, the SERR# Enable bit must be set to "1" in the virtual PCI/PCI Bridges' Command and Bridge Control Registers (secondary side to primary side interface). If the appropriate SERR# Enable bits are not set to a "1" value, the message error transaction is simply discarded by the virtual PCI/PCI Bridge. Consequently, the message transaction packet does not arrive at the Root Complex's downstream ports and the error is not reported.

Bridges

As discussed earlier, the error checking by Transaction Layer is done by the PCI Express device that is the destination of the TLP. In the case where the TLPs' destination is the configuration register block in a virtual PCI/PCI Bridge with the bridge, the error checking activities at the Transaction Layer level within the virtual PCI/PCI Bridge is the same as for non-switch PCI Express devices.

The bridge may receive TLPs that cannot be ported to the downstream PCI or PCI-X bus segment due to address space issues. That is, for address ranges within the memory, I/O, and configuration address spaces, the downstream bus segments are not defined for the accessed address range. The receipt of such a TLP results in the bridges' upstream ports defining it as unsupported and the TLP as **unsupported requester** or **unexpected completer** error as illustrated in Figure 9.4.

Errors that occur on the PCI or PCI-X bus segments are mapped to one of the message error requester transaction packets to be transmitted by the bridges' upstream ports. The only requirement is that the SERR# Enable bit is set to "1" in the Command and Control Registers in the bridges' downstream virtual PCI/PCI Bridges.

Errors Related to Transaction Layer: Error Forwarding (also Known as Data Poisoning)

This section describes errors related to the Transaction Layer.

PCI Express Devices (That Are Requester Source, Requester Destination, Completer Source, or Completer Destination)

Error forwarding provides PCI Express devices the means to transmit a valid TLP that is error-free according to the standard PCI Express protocol, but with the indication that an error has been identified beyond the PCI Express TLP protocol. Error forwarding can be used optionally for many reasons. Here are some examples:

- **Corrupted:** Data to be read or written has been corrupted outside of the PCI Express transaction transfer.

- **Device Fails:** A PCI Express device is operating properly at the port transmitting and receiving level, but the portion of the resource that is responsible for data has failures.

- **Buffer Overflow:** A PCI Express device buffer is being read, but there have been overflows behind the buffer and some data inside the resource has been discarded so that the data sequence is broken.

Error forwarding can be implemented by write requester transaction packets with data (memory, I/O, and configuration) and with completer transaction with data (completer transaction for memory read, I/O read, configuration read). Error forwarding can also be implemented with message baseline, message vendor-defined, and message advanced switching requester transactions; providing the TLP includes data.

As discussed in Chapter 6, PCI Express provides the PCI Express requester or completer source the opportunity to forward an error (indicate that the data is poisoned) by the use of the EP bit (Error Poison) in the ATTRIBUTE HDW in the Header field. EP is defined for requester write transaction and completer read transaction packets. If EP = 0, no errors are forwarded. If EP = 1 and is defined for the TLP, a **poisoned TLP** error is reported. The value of the ECRC in the Digest field is irrelevant. The use of the EP bit by the source of the TLP is optional, but if used (EP = 1) the packet's destination must report the error as **poisoned TLP**, **unsupported requester**, or **malformed**. The reporting of any of these errors is via a message requester transaction packet.

As illustrated in Figures 9.2 and 9.3, when error forwarding is detected, the response is dependent on the TLP as follows:

- For all TLPs that define Error forwarding, when EP = 1b the associated TLP is discarded (except as noted in the following bullets) and a **poisoned TLP** error is reported.

- The above bullet does not apply when EP = 1b and the TLP is accessing a control register or control structure. The TLP must contain a configuration write, memory write, I/O write, or message baseline, message vendor-defined, and message advanced switching (when Data field is included). In this situation, the TLP is discarded and an **unsupported requester** error is reported.

- In addition to the first bullet when EP = 1b for memory write requester transaction packets to a non-control memory address space, the data in the packet may be used optionally by the PCI Express requester destination.

- In addition to the first bullet when EP = 1b for message requester transaction packets to a non-control PCI Express device, the data in the packet may be used optionally by the PCI Express requester destination.

Error forwarding implementation is not defined for read requester transaction packets (memory read, I/O read, configuration read), message baseline requester transaction packets without Data field, message vendor-defined requester transaction packets without Data field, message advanced switching requester transaction packets without Data field, and completer transaction packets without Data field. If one of these TLPs is received with EP = 1b, the TLP is discarded and a **malformed** error is reported. This is determined very early in the processing of the TLP illustrated in Figure 9.1.

> Error forwarding is never used as a means to indicate other errors defined by the PCI Express Physical, Data Link, or Transaction Layers. Also, error forwarding is never used as a means to indicate errors related to CRC, ECRC, and LCRC.
>
> The value of the EP is never monitored by the Physical or Data Link Layers.

Within Switches (Not Requester Source, Requester Destination, Completer Source, or Completer Destination)

The value of the EP bit (in the ATTRIBUTE HDW of the Header field of the TLP) is used for Error forwarding (also known as data poisoning) between the requester source and destination or completer source and destination. It is not monitored by switches when porting TLPs from one port to another port (that is, when the switch is not the requester or completer destination). However, if the switch is able to determine that it has introduced some error to the TLP it is porting, it must set the EP = 1b for those TLPs that define Error forwarding. If the switch is the requester or completer destination, the EP bit is processed as it is for any other PCI Express device that is the requester or completer destination.

Within Bridges (Not Requester Source, Requester Destination, Completer Source, or Completer Destination)

The PCI Express definition of Error forwarding is not defined for PCI or PCI-X bus segments. The receipt of a TLP with Error forwarding (EP = 1b) from the link must be terminated at the bridge. The receiving side of the virtual PCI/PCI Bridge must set the Detected Parity Error bit in the Status Register to a "1" value. Also, if the Parity Error Response bit in the virtual PCI/PCI Bridge's Control register is set to a "1" value, the Master Data Parity bit in the Secondary Status Register is set to 1b.

Error forwarding is not defined in the PCI or PCI-X bus segments; however parity errors are supported. To indicate EP Error forwarding (EP = 1b) in the TLP from a parity error, the virtual PCI/PCI Bridges must have the following bits set to 1b: the Detected Parity Error bit in the Secondary Status Register and the Master Data Parity Error bit in the Status Register (only if the Parity Error Response bit is also set to 1b).

Errors Related to Transaction Layer: Completion Timeout Error

When PCI Express devices source a requester transaction packet that is associated with completer transaction packet, the PCI Express devices assume a completer transaction packet will be returned. The PCI Express devices that are defined as possible sources of requester transaction packets that must implement the Completion Timeout protocol are the Root Complex, bridges, and endpoints. The PCI Express specification requires the implementation of a Completion Timeout timer as described in the following section to support the Completion Timeout protocol.

The Completion Timeout timer is enabled when a requester transaction packet is transmitted onto the link (as a TLP within the LLTP and Physical Packet). The associated completer transaction packet must be received before the Completion Timeout time expires. The PCI Express specification requires the expiration to be 50 milliseconds or less. The PCI Express specification requires the expiration to not be less than 50 microseconds, with a recommendation of 10 milliseconds if possible. If the completer transaction packet is not successfully returned prior to the Completion Timeout timer expiring, a **Completion Timeout** error is reported via a message error requestor transaction packet.

Following the Requester/Completer protocol, there may be several requester transaction packets transmitted prior to the reception of the associated (outstanding) completer transaction packets. Consequently, each of the outstanding requester transaction packets has a Completion Timeout timer. The one exception to the Completion Timeout protocol is configuration requester transaction packets during the initialization.

One other consideration is that a single memory read requester transaction packet may have a set of completer transaction packets returned. The Completion Timeout protocol requires that the entire set of completer transaction packets successfully return all of the data prior to the expiration of the Completion Timeout timer. If all of the data is not successfully returned prior to the Completion Timeout timer expiring, a **Completion Timeout** error is reported via a message error requestor transaction packet. The requester source can either retain or discard the data that was received prior to the expiration of the Completion Timeout timer.

Errors Related to Transaction Layer: Receiver Overflow Error

The ports of the PCI Express devices that are receiving TLPs provide the available buffer space information to the ports of the PCI Express devices transmitting. This information is provided in the form of Flow Control Credits (FCC). Chapter 11 discusses how the available buffer space information is exchanged between the two PCI Express devices on a specific link. The possibility of a receiver overflow can occur at *any* receiving port of a PCI Express device.

In addition to the values maintained as detailed in Chapter 11, the TLPs' receiving ports can optionally check for receiving buffer overflow. The basic requirement of the Flow Control protocol requires maintaining FCC_Allocated and FCC_Processed Counters. The TLPs' receiving port may also maintain an FCC_Received Counter. Chapter 11 discusses the comparison of these three counters in determining whether a receiver buffer overflow has occurred. If the receiver buffer overflows, the following must occur:

- Discard the TLP.
- Do not change contents of any of the aforementioned counters.
- De-Allocate any resources allocated to the TLP.
- Optionally report a **Receiver Overflow** error via a message error requester transaction packet.

The PCI Express specification does not define any specific reaction by the TLPs' receiving port when a **Receiver Overflow** error has occurred. It recommends that the receiving port continue as if the error did not occur.

Errors Related to Transaction Layer: Flow Control Protocol Error (FCPE)

The following is a summary of the sources of FCPEs, please see the referenced chapters for more detailed information.

During Flow Control Initialization of a link, the minimal FCCs for each type of buffer are listed in Table 7.2. The transmitting port of the LLTP can optionally check to determine whether the initial buffer space available is equal to or greater than the minimum; if not, a **(Flow Control Protocol Error) FCPE** is optionally reported. See Chapter 7 for more information.

During Flow Control Initialization of a link, some of the buffers in the LLTP receiving port may indicate an infinite amount of available buffer space to the LLTP transmitting port. A protocol exists for updating the available buffer space when any buffer of a buffer set is defined as infinite. If this protocol is not followed, the LLTP transmitting port can optionally report a **(Flow Control Protocol Error) FCPE**. See Chapter 7 for more information.

Transaction Layer: TLP Fields Checked for Errors

This chapter's discussion to this point has focused on error checking of TLPs. Some of the TLP errors are related to the contents of the Header Field as follows: malformed, poisoned TLP, completer abort, unsupported requester, and unexpected completer.

The following discussions define the specific contents of specific HDWs within the Header field that cause these errors. The following discussions do not include the ADDRESS HDW of the different TLPs. The value of the associated address bits is not checked for other than for porting through a switch. Refer to the section "Special TLP Error Considerations for Switches and Bridges (excluding Error Poisoning)" earlier in this chapter for other ADDRESS# error considerations.

> Unless otherwise stated as optional, the errors listed below in the TLPs *must* be checked by the PCI Express device that is the requester or completer destination. As discussed earlier in this chapter under certain conditions, a switch that is not the source or destination of the TLP can check the value of the TC, TD, and ECRC. These switch-specific considerations are not discussed here, nor are any error issues relating to addressing or routing. See the earlier discussions in this chapter for specifics.
>
> The requester or receiver destination ignores the value of fields or bits defined as RESERVED. However, the value if these bits and fields are included in the calculation for CRC, LCRC, and ECRC.

MEMORY and I/O REQUESTER TRANSACTIONs contain TYPE, ATTRIBUTE, REQUESTER, and TAG HDWs:

7	6	5	4	3	2	1	0	7	6	5	4	3	2	1	0
R	FMT 1	0	4	TYPE		1	L	R	2	TC	0	R	R	R	R
T D	E P	ATTRI 1	0	R	R	9				LENGTH					0
7		BUS NUMBER					0	4		DEVICE #		0	2	FUNC#	0
7		TAG					0	3	LAST BE		0	3	FIRST BE		0

15	TYPE HDW	00
15	ATTRIBUTE HDW	00
15	REQUESTER ID HDW	00
15	TAG HDW	00

TYPE HDW

FMT [1:0]: Not checked for errors.

TYPE [4::0]

- If not defined as a valid non-message requester transaction type, the TLP is **malformed**.

- If the not supported by the requester destination, the TLP defined as **unsupported requester**.

- Type [0] = 1b means LOCK function and 0b means no LOCK function. This is only defined for memory transactions If TYPE [0] = 1b and the PCI Express device does not support the LOCK function or if it is an I/O requester transaction packet, the TLP is defined as **unsupported requester**. If an I/O requester transaction packet is supported by requester destination and TYPE [0] = 1b, the TLP is defined as **malformed**.

TC [2::0]

- Memory read requester and the associated completer transactions used for the Lock protocol must use TC [2::0] = 000b. If the memory read requester transaction is for the LOCK function and TC [2::0] does not = 000b, the TLP is defined as **malformed**.

- If an I/O requester transaction packet and TC [2::0] not = 000b, the TLP is defined as **malformed**.

Refer to the section "Special TLP Error Considerations for Switches and Bridges (excluding Error Poisoning)" earlier in this chapter for other TC error considerations.

ATTRIBUTE HDW

TD

- TD = 0b signifies that no Digest DWORD is included in packet. The requester destination must check to ensure that no Digest field is included in packet. If one is included, the TLP is defined as **malformed**.

- TD = 1b signifies that Digest DWORD is attached. The requester destination must check to ensure Digest field is included in packet. If one is not included, the TLP is defined as **malformed**.

Refer to the section "Special TLP Error Considerations for Switches and Bridges (excluding Error Poisoning)" earlier in this chapter for other TD error considerations.

EP = Error and Poisoning

- If EP = 0b, then no errors are forwarded, or error information from the source is not implemented for the TLP.

- The EP is defined for only for requester write transaction and completer read transaction packets.

- If EP = 1b and is not defined for the TLP, the TLP is defined as **malformed**.

Refer to the section "Errors Related to Transaction Layer: Error Forwarding (also Known as Data Poisoning)" earlier in this chapter for other EP error considerations.

ATTRI [1:0]: The Attributes field defines snoop and ordering information.

- The PCI Express specification does not define whether these bits are checked for errors. However, these bits are similar to TC, which are defined as requiring error checking. Consequently, this book assumes the value must be checked and if an error is found the TLP is defined as **malformed**. (Note: there are special requirements for isochronous transfers. See Chapter 12 for more information.)
    - ATTRI [1:0] must = 00b if the memory requester transaction packet is used for a message-signaled interrupt.
    - The ATTRI [1::0] must = 00b for I/O requester transaction packets.

LENGTH [9::0]: LENGTH in DWORDs to be read or written.

- For memory write requester transactions packets, the LENGTH [9::0] must agree with the size of the Data field. This agreement is checked by the requester destinations; if not in agreement the TLP is defined as **malformed**.

- For memory write requester transaction packets, the LENGTH [9::0] must specify a Data field size that is less than or equal to (must be in agreement with) the Max_Payload_Size defined in the Device Control Register of the requester destination. This agreement is checked by the requester destinations; if not in agreement the TLP is defined as **malformed**.
    - For memory read requester transaction packets, the LENGTH [9::0] is not compared to the Max_Payload_Size defined in the Device Control Register of the requester destination.

- For memory requester transaction packets, the combination of the ADDDRESS field and LENGTH [9::0] should define data transfers that do not cross the 4 kilobyte address boundary. The requester destination can optionally check for this boundary crossing. If the boundary is crossed, the TLP is defined as **malformed**.

- The I/O requester read transaction packets are only defined to have a LENGTH [9::0] = 00 0000 0001b; if not this value, then the TLPs are defined as **malformed**.

- The I/O requester write transaction packets are only defined for a Data field of one DWORD and the LENGTH [9::0] = 00 0000 0001b; if not these values, then the TLPs are defined as **malformed**.
- For memory read requester transaction packets the LENGTH [9::0] is not related to the Data field size; consequently, agreement between Data field and the LENGTH field is not checked for errors.

REQUESTER ID HDW

BUS# [7::0]: BUS NUMBER. Not checked for errors.

DEVICE# [4::0]: DEVICE NUMBER. Not checked for errors.

FUNC# [2:0]: FUNCTION NUMBER. Not checked for errors.

TAG HDW

TAG# [7::0]

- Undefined behavior if the TAG#s assigned by one requester source for TLP that requires a completion transaction are not unique.
- The TAG#s assigned by one requester source for a TLP that does not require a completion transaction can be of any value.

LAST BE [3::0] and FIRST BE [3::0] — Optionally checked for compliance based on definitions in Chapter 5. If not in compliance, the TLP is defined as **malformed**.

LOWER ADDRESS [6::0] — Optionally checked for compliance to definitions in Chapter 5. If not in compliance, the TLP is defined as **malformed**.

CONFIGURATION REQUESTER TRANSACTIONs contain TYPE, ATTRIBUTE, REQUESTER, and TAG HDWs:

7	6	5	4	3	2	1	0	7	6	5	4	3	2	1	0
R	FMT 1 0		4 TYPE 1				T	R	2 TC 0			R	R	R	R
T D	E P	ATTRI 1 0		R	R	9		LENGTH							0
7	BUS NUMBER						0	4	DEVICE #			0	2 FUNC# 0		
7	TAG						0	3	LAST BE		0	3 0	FIRST BE		
7	BUS NUMBER						0	4	DEVICE #			0	2 FUNC# 0		
R	R	R	R	EXT. REGISTER				REGISTER ADDRESS						R	R
				3 ADDRESS			0	5					0		

15	TYPE HDW	00
15	ATTRIBUTE HDW	00
15	REQUESTER ID HDW	00
15	TAG HDW	00
15	CONFIGURATION ID HDW	00
15	REGISTER HDW	00

TYPE HDW

FMT [1:0]: Not checked for errors.

TYPE [4::0]

- If not defined as a valid non-message requester transaction type, the TLP is **malformed**.

- If the not supported by the requester destination, the TLP defined as **unsupported requester**.

TC [2::0]

- If configuration requester transaction packet and TC [2::0] not = 000b, the TLP is defined as **malformed**.

Refer to the section "Special TLP Error Considerations for Switches and Bridges (excluding Error Poisoning)" earlier in this chapter for other TC error considerations.

ATTRIBUTE HDW

TD

- TD = 0b signifies that no Digest DWORD is included in the packet. The requester destination must check to ensure that no Digest field is included in packet. If one is included, the TLP is **malformed**.

- TD = 1b signifies that a Digest DWORD is attached. The requester or destination must check to ensure that a Digest field is included in the packet. If one is not included, the TLP is **malformed**.

Refer to the section "Special TLP Error Considerations for Switches and Bridges (excluding Error Poisoning)" earlier in this chapter for other TD error considerations.

EP = Error and Poisoning

- If EP = 0b, no errors are forwarded, or error information from the source is not implemented for the TLP.

- The EP is defined for only requester write transaction and completer read transaction packets.

- If EP = 1b and is not defined for the TLP, the TLP is defined as **malformed**.

Refer to the section "Errors Related to Transaction Layer: Error Forwarding (also Known as Data Poisoning)" earlier in this chapter for other EP error considerations.

ATTRI [1:0]: The Attributes field defines snoop and ordering information.

- The PCI Express specification does not define whether these bits are checked for errors. However, these bits are similar to TC, which are defined as requiring error checking. Consequently, this book assumes the value must be checked and if an error is found the TLP is defined as **malformed**. (Note: there are special requirements for isochronous transfers. See Chapter 12 for more information.)
  - The ATTRI [1::0] must = 00b for configuration requester transaction packets.

LENGTH [9::0]: LENGTH in DWORDs to be read or written.

- The configuration requester read transaction packets are only defined to have a LENGTH [9::0] = 00 0000 0001b; if not this value, then the TLPs are defined as **malformed**.
- The configuration requester write transaction packets are only defined for a Data field of one DWORD and the LENGTH [9::0] = 00 0000 0001b; if not these values, then the TLPs are defined as **malformed**.

REQUESTER ID HDW

BUS# [7::0]: BUS NUMBER. Not checked for errors.

DEVICE# [4::0]: DEVICE NUMBER. Not checked for errors.

FUNC# [2:0]: FUNCTION NUMBER. Not checked for errors.

TAG HDW

TAG# [7::0]

- Undefined behavior if the TAG#s assigned by one requester source for TLP that requires a completion transaction are not unique.
- The TAG#s assigned by one requester source for a TLP that does not require a completion transaction can be of any value.

LAST BE [3::0] and FIRST BE [3::0]: Optionally checked for compliance based on definitions in Chapter 5. If not in compliance, the TLP is defined as **malformed**.

LOWER ADDRESS [6::0]: Optionally checked for compliance to definitions in Chapter 5. If not in compliance, the TLP is defined as **malformed**.

CONFIGURATION ID HDW

BUS# [7::0]: BUS NUMBER. Not checked for errors.

DEVICE# [4::0]: DEVICE NUMBER. Not checked for errors.

FUNC# [2:0]: FUNCTION NUMBER. Not checked for errors.

Refer to the section "Special TLP Error Considerations for Switches and Bridges (excluding Error Poisoning)" earlier in this chapter for other BUS# error considerations.

REGISTER HDW

REGISTER ADDRESS [5::0]: Combined with the lowers order Bits [1:0] of the EXTENDED REGISTER ADDRESS. Not checked for errors.

EXTENDED REGISTER ADDRESS [3::0]: Used with the REGISTER ADDRESS [5::0]. Not checked for errors.

COMPLETER TRANSACTIONs contain TYPE, ATTRIBUTE, COMPLETER, STATUS, REQUESTER, and TAG HDWs:

7	6	5	4	3	2	1	0	7	6	5	4	3	2	1	0
R	FMT 1 0		4 TYPE 1				L	R	2 TC 0			R	R	R	R
TD	EP	ATTRI 1 0		R	R	9		LENGTH							0
7	BUS NUMBER						0	4	DEVICE #			0	2 FUNC# 0		
STATUS 2 0			BCM	11				BYTE COUNT					00		
7	BUS NUMBER						0	4	DEVICE #			0	2 FUNC# 0		
7	TAG #						0	R	6 LOWER ADDRESS						0

15	TYPE HDW	00
15	ATTRIBUTE HDW	00
15	COMPLETER ID HDW	00
15	STATUS HDW	00
15	REQUESTER ID HDW	00
15	TAG HDW	00

TYPE HDW

FMT [1:0]: Not checked for errors.

TYPE [4::0]

- If not defined as a valid non-message requester transaction type, the TLP is **malformed**.

- Type [0] = 1b means LOCK function and 0b means no LOCK function. This is only defined for memory transactions If TYPE [0] = 1 and the completer destination (requester source) is not executing the associated requester transaction with no LOCK function, the TLP is defined as **malformed**.

TC [2::0]

- Memory read requester and the associated completer transactions used for the Lock protocol must use TC [2::0] = 000b. If the completer transaction is for the LOCK function and TC [2::0] does not = 000b, the TLP is defined as **malformed**.

- If completer transaction is associated with an I/O or configuration requester transaction packet and TC [2::0] not = 000b, the TLP is defined as **malformed**.

Refer to the section "Special TLP Error Considerations for Switches and Bridges (excluding Error Poisoning)" earlier in this chapter for other TC error considerations.

ATTRIBUTE HDW

TD

- TD = 0b signifies that no Digest DWORD is included in packet. The completer destination must check to ensure that no Digest field is included in packet. If one is included, the TLP is defined as **malformed**.

- TD = 1b signifies that a Digest DWORD is attached. The requester or completer destination must check to ensure that a Digest field is included in packet. If one is not included, the TLP is defined as **malformed**.

Refer to the section "Special TLP Error Considerations for Switches and Bridges (excluding Error Poisoning)" earlier in this chapter for other TD error considerations.

EP = Error and Poisoning

- If EP = 0, no errors are forwarded, or it is not implemented for the TLP.

- The EP is defined for requester write transaction and completer read transaction packets.

- If EP = 1 and is not defined for the TLP, the TLP is defined as **malformed**.

Refer to the section "Errors Related to Transaction Layer: Error Forwarding (also Known as Data Poisoning)" earlier in this chapter for other EP error considerations.

ATTRI [1:0]: The Attributes field defines snoop and ordering information.

- The PCI Express specification does not define whether these bits are checked for errors. However, these bits are similar to TC, which are defined as requiring error checking. Consequently, this book assumes the value must be checked and if an error is found the TLP is defined as **malformed**. (Note: there are special requirements for isochronous transfers. See Chapter 12 for more information.)

    – ATTRI [1:0] = 00b to 11b for completer transaction. The ATTRI [1::0] for completer transaction packets must be as same as the associated requester transaction packets. If the completer transaction was expected (an associated requester transaction is identified via the Transaction ID), the ATTRI [1:0] must equal the ATTRI [1:0] of associated requester transaction.

LENGTH [9::0]: LENGTH in DWORDs if Data field is included in packet.

- For completer transaction packets, the LENGTH [9::0] must agree with the size of the Data field. This agreement is to be checked by the completer destinations. If not in agreement, the TLP is defined as **malformed**.

- For completer requester transaction packets associated with memory read requester transaction packets, the combination of the Address field of the requester transaction and the LENGTH [9::0] should define data transfers align with the RCB (Read Completion Boundary). The completer destination can optionally check for this boundary crossing. If the boundary is crossed, the TLP is defined as **malformed**.

- For completer transaction packets associated with memory read requester transaction packets, the LENGTH [9::0] must specify a Data field size that is less than or equal to (that is in agreement with) the Max_Payload_ Size defined in the Device Control Register of the PCI Express completer source (requester destination). This is checked by the completer destinations. If not in agreement, the TLP is defined as **malformed**.

- Completer transaction packets with data associated with I/O read or configuration read are only defined for a Data field of one DWORD and the LENGTH [9::0] = 00 0000 0001b. If they do not have these values then the TLPs are defined as **malformed**.

- For completer transaction packets with no Data field, the LENGTH [9::0] is defined as RESERVED; consequently, it is not checked for errors.

COMPLETER ID HDW

BUS# [7::0]: BUS NUMBER. Not checked for errors.

DEVICE# [4::0]: DEVICE NUMBER. Not checked for errors.

FUNC# [2:0]: FUNCTION NUMBER. Not checked for errors.

STATUS HDW

STATUS [2::0]: Completer status defines the result of the associated requester transaction packets as follows:

- 000b = Successful Completion (SC). No error.
- 001b = Unsupported Request (UR). An **unsupported requester** is defined for the associated requester transaction packet.
- 010b = Configuration Request Retry Status (CRS). If the completer transaction packet with this status is *not* associated with an outstanding configuration requester transaction packet, the TLP is defined as **malformed**.
- 100b = completer abort (CA). A **completer abort** is defined for the associated requester transaction packet.
- All other values are RESERVED. If the completer transaction packet has this RESERVED status, the completer transaction packet is processed by the PCI Express completer destination as if a STATUS [2::0] = 001b were returned. That is, an **unsupported requester** is defined for the associated requester transaction packet.

BCM: Byte count modified. This bit is not set to 1b by the completer source and is ignored by the completer destination (requester source). It is not checked for errors.

BYTE COUNT [11::00]

- Each I/O requester or configuration requester transaction packet must be responded to with a single completer transaction packet. For a reader requester, transaction the associated completer transaction packet contains one DWORD in the Data field. For a write requester transaction, the associated completer transaction packet contains no Data field. The BYTE COUNT [11::00] for the completer transaction packets for I/O requester and configuration requester transactions must equal 0000 0000 0100b. It is not checked for errors.
- Each memory read requester transaction can be responded to in a single completer transaction packet or multiple completer transaction packets. It is not checked for errors.

REQUESTER ID HDW

BUS# [7::0]: BUS NUMBER. Not checked for errors.

DEVICE# [4::0]: DEVICE NUMBER. Not checked for errors.

FUNC# [2::0]: FUNCTION NUMBER. Not checked for errors.

Refer to the section "Special TLP Error Considerations for Switches and Bridges (excluding Error Poisoning)" earlier in this chapter for other BUS# error considerations.

TAG HDW

TAG# [7::0]

- The behavior is undefined if the TAG#s assigned by one requester source for TLP that requires a completion transaction are not unique.
- The TAG# assigned by requester source for the TLP that does not require a completion transaction can be of any value.
- If the TAG# of a completer transaction packet does not match the TAG# of an outstanding requester transaction, the completer transaction is defined as **unexpected completer**.

LOWER ADDRESS [6::0]: Optionally checked for compliance with definitions in Chapter 5. If not in compliance, the TLP is defined as **malformed**.

MESSAGE BASELINE REQUESTER TRANSACTIONs contain TYPE, ATTRIBUTE, REQUESTER, and TAG HDWs

7	6	5	4	3	2	1	0	7	6	5	4	3	2	1	0
R	FMT 1 0		4	TYPE			0	R	2	TC	0	R	R	R	R
TD	EP	ATTRI 1 0		R	R	9		LENGTH							0
7	BUS NUMBER						0	4	DEVICE #			0	2	FUNC#	0
7	TAG						0	7	MESSAGE CODE						0

15	TYPE HDW	00
15	ATTRIBUTE HDW	00
15	REQUESTER ID HDW	00
15	TAG HDW	00

TYPE HDW

FMT [1:0]: Not checked for errors.

TYPE [4::0]

- If not defined as a valid message baseline requester transaction type, the TLP is defined as **malformed**.
- TYPE [4::0] = 10r2r1r0b ... r [2::0] used for Implied Routing.
  - The "r" values are used for Implied Routing information of message baseline requester transactions. TYPE is not checked for errors, except for the unique error checking in the switch (see the section "Special Error Considerations for Switches and Bridges").
- If not supported by the requester destination, the TLP is defined as **unsupported requester**.

Refer to the section "Special Error Considerations for Message Requester Transaction Packets" earlier in this chapter for other r [2::0] error considerations.

TC [2::0]

- For message baseline requester transaction packets, if the TC [2::0] does not equal 000b the TLP is defined as **malformed**.

Refer to the section "Special TLP Error Considerations for Switches and Bridges (excluding Error Poisoning)" earlier in this chapter for other TC error considerations.

ATTRIBUTE HDW

TD

- TD = 0b signifies that no Digest DWORD is included in packet. The requester destination must check to ensure that no Digest field is included in the packet. If one is included, the TLP is defined as **malformed**.

- TD = 1b signifies that a Digest DWORD is attached. The requester destination must check to ensure a Digest field is included in the packet. If one is not included, the TLP is defined as **malformed**.

Refer to the section "Special TLP Error Considerations for Switches and Bridges (excluding Error Poisoning)" earlier in this chapter for other TD error considerations.

EP = Error and Poisoning

- If EP = 0, no errors are forwarded or Error and Poisoning is not implemented for the TLP.

- The EP is defined for requester write transaction and completer read transaction packets.

Refer to the section "Errors Related to Transaction Layer: Error Forwarding (also Known as Data Poisoning)" earlier in this chapter for other EP error considerations.

ATTRI [1:0]: The Attributes field defines snoop and ordering information.

- The PCI Express specification does not define whether these bits are checked for errors. However, these bits are similar to TC, which are defined as requiring error checking. Consequently, this book assumes the value must be checked and if an error is found the TLP is defined as **malformed**. (Note: there are special requirements for isochronous transfers. See Chapter 12 for more information.)
    - ATTRI[1:0] must = 00b for message baseline requester transaction packets.

LENGTH [9::0]: LENGTH in DWORDs to be written.

- For message baseline requester transactions packets that do not define a Data field the LENGTH [9::0] is defined as RESERVED; consequently, it is not checked for errors.

- For message baseline requester transactions packets that defines a Data field it is one DWORD and the LENGTH [9::0] = 00 0000 0001b, if these are not the values, then the TLPs are defined as **malformed**.

REQUESTER ID HDW

BUS# [7::0]: BUS NUMBER. Not checked for errors.

DEVICE# [4::0]: DEVICE NUMBER. Not checked for errors.

FUNC# [2:0]: FUNCTION NUMBER. Not checked for errors.

TAG HDW

TAG# [7::0]

- The TAG#s assigned by one requester source for a TLP that does not require a completion transaction can be of any value.

Message Code [7::0]

- If a requester destination receives a message code it does not support or is not defined, the requester destination defines the TLP as **unsupported requester**.

MESSAGE VENDOR-DEFINED REQUESTER TRANSACTIONs contain TYPE, ATTRIBUTE, REQUESTER, and TAG HDWs

7	6	5	4	3	2	1	0	7	6	5	4	3	2	1	0
R	FMT 1 0		1	0	2 r FIELD 0			R	2 TC 0			R	R	R	R
T D	E P	ATTRI 1 0		R	R	9		LENGTH							0
7	BUS NUMBER						0	4	DEVICE #				0	2 FUNC# 0	
7	TAG						0	7	MESSAGE CODE 0						
7	BUS NUMBER						0	4	DEVICE #				0	2 FUNC# 0	
15	VENDOR-DEFINED ID HDW														00
15	VENDOR ID HDW														00

15	TYPE HDW	00
15	ATTRIBUTE HDW	00
15	REQUESTER ID HDW	00
15	TAG HDW	00
15	VENDOR-DEFINED ID HDW	00
15	VENDOR ID HDW	00

TYPE HDW

FMT [1:0]: Not checked for errors.

TYPE [4::0]

- ■ If not defined as a valid message baseline requester transaction type, the TLP is defined as **malformed**.

- TYPE [4::0] = 10r2r1r0b ... r [2::0] used for Implied Routing.
  - The "r" values are used for Implied Routing information of message baseline requester transactions. They are not checked for errors, except for the unique error checking in the switch (see the section "Special Error Considerations for Switches and Bridges").
- If the not supported by the requester destination, the TLP defined as **unsupported requester**.

TC [2::0] = Traffic Class

- For message vendor-defined requester transaction packets, all values are defined and the value is not checked for errors.

ATTRIBUTE HDW

TD

- TD = 0 signifies that no Digest DWORD is included in packet. The requester destination must check to ensure that no Digest field is included in packet. If one is included, the TLP is defined as **malformed**.
- TD = 1 signifies that a Digest DWORD is attached. The requester destination must check to ensure that a Digest field is included in packet. If one is not included, the TLP is defined as **malformed**.

Refer to the section "Special TLP Error Considerations for to Switches and Bridges (excluding Error Poisoning)" earlier in this chapter for other TD error considerations.

EP = Error and Poisoning

- If EP = 0, no errors are forwarded or Error and Poisoning is not implemented for the TLP.
- The EP is defined for requester write transaction and completer read transaction packets.

Refer to the section "Errors Related to Transaction Layer: Error Forwarding (also Known as Data Poisoning)" earlier in this chapter for other EP error considerations.

ATTRI [1:0]: The Attributes field defines snoop and ordering information.

- ATTRI[1:0] can be 00 to 11b; it is not checked for errors.

LENGTH [9::0]: LENGTH in DWORDs to be written.

- For message vendor-defined requester transactions packets that do not define a Data field, the LENGTH [9::0] is defined as RESERVED; consequently, it is not checked for errors.

- For message vendor-defined requester transaction packets that define a Data field, the LENGTH [9::0] must agree with the size of the Data field. This agreement is checked by the requester destinations. If not in agreement the TLP is defined as **malformed**.

REQUESTER ID HDW

BUS# [7::0]: BUS NUMBER. Not checked for errors.

DEVICE# [4::0]: DEVICE NUMBER. Not checked for errors.

FUNC# [2:0]: FUNCTION NUMBER. Not checked for errors.

TAG HDW

TAG# [7::0]

- The behavior is undefined if the TAG#s assigned by one requester source for TLP that requires a completion transaction are not unique.

- The TAG#s assigned by one requester source for TLP transaction that does not require a completion transaction can be of any value.

- If the TAG# of a completer transaction packet does not match the TAG# of an outstanding requester transaction, the completer transaction is defined as **unexpected completer**.

Message Code [7::0]

- If a requester destination receives a message code it does not support or that is not defined, the requester destination defines the TLP as **unsupported requester** with one exception:
  - If the message requester transaction is a Vendor-Defined Type 1 that is not supported by the PCI Express requester destination, the PCI Express requester destination discards the message requester transaction and does not report any error.

VENDOR-DEFINED ID HDW

BUS# [7::0]: BUS NUMBER. Not checked for errors.

DEVICE# [4::0]: DEVICE NUMBER. Not checked for errors.

FUNC# [2:0]: FUNCTION NUMBER. Not checked for errors.

Refer to the section "Special TLP Error Considerations for Switches and Bridges (excluding Error Poisoning)" earlier in this chapter for other BUS# error considerations.

VENDOR ID HDW

This is a unique number for each vendor that is provided by the PCI-SIG. It is not checked for errors.

RESERVED FOR VENDOR HDW

Not checked for errors.

MESSAGE ADVANCED SWITCHING REQUESTER TRANSACTIONs contain TYPE, ATTRIBUTE, REQUESTER, and TAG HDWs

7	6	5	4	3	2	1	0	7	6	5	4	3	2	1	0
R	FMT 1 0			1	0	2 n FIELD 0 2 c FIELD 0		R	2	TC	0	R	R	R	R
TD	EP	ATTRI 1 0		R	R	9		LENGTH							0
15					RESERVED HDW FOR MsgAS & MsgASD										00

15	TYPE HDW	00
15	ATTRIBUTE HDW	00
15	RESERVED HDW FOR MsgAS & MsgASD	00

TYPE HDW

FMT [1:0]: Not checked for errors.

TYPE [4::0]

- TLP is defined as **malformed**.
- TYPE [4::0] = 11n2n1n0b or 11c2c1c0b; "n" is used with no data, "c" is used with data.
- Not checked for errors.
- If not supported by the requester destination, the TLP defined as **unsupported requester**.

TC [2::0] = Traffic Class

- For message advanced switching requester transaction packets all values are defined and the value is not checked for errors.

ATTRIBUTE HDW

TD

- TD = 0 signifies that no Digest DWORD is included in packet. The requester destination must check to ensure that no Digest field is included in packet. If one is included, the TLP is defined as **malformed**.
- TD = 1 signifies that a Digest DWORD is attached. The requester or completer destination must check to ensure that a Digest field is included in packet. If one is not included, the TLP is defined as **malformed**.

Refer to the section "Special TLP Error Considerations for Switches and Bridges (excluding Error Poisoning)" earlier in this chapter for other TD error considerations.

EP = Error and Poisoning

- If EP = 0, no errors are forwarded or Error and Poisoning is not implemented for the TLP.
- The EP is defined for requester write transaction and completer read transaction packets.

Refer to the section "Errors Related to Transaction Layer: Error Forwarding (also Known as Data Poisoning)" earlier in this chapter for other EP error considerations.

ATTRI [1:0]: The Attributes field defines snoop and ordering information.

- ATTRI [1:0] can be 00 to 11b; it is not checked for errors.

LENGTH [9::0]: LENGTH in DWORDs to be written.

- For message advanced switching requester transactions packets that do not define a Data field, the LENGTH [9::0] is defined as RESERVED; consequently, it is not checked for errors.

- For message advanced switching requester transaction packets that define a Data field, the LENGTH [9::0] must agree with the size of the Data field. This agreement is checked by the requester destinations; if not in agreement the TLP is defined as **malformed**.

RESERVED HDW FOR MsgAS & MsgASD

Not checked for errors.

Data Link Layer Errors

The possible errors at the Data Link Layer protocol are either directly related to the link packets or indirectly related at a higher level to Gate circuitry for Flow Control protocol. As previously discussed there are two types packets associated with the Data Link Layer: Data Link Layer Packets (DLLPs) used for link management and Link Layer Transaction Packets (LLTPs) that contain TLPs.

All of the possible errors are summarized in Table 9.2. In some cases there a multiple sources for a specific error. This chapter discusses all of the sources of these errors with a **bold underline** to signify an error in reference to the table. The **bold underline** applies to all chapters in the book.

> Unless otherwise stated (optional), the errors discussed below in this section *must* be checked by the associated PCI Express device.

Table 9.2 Errors Related to Data Link Layer Protocol

Error Name	Message Error Requester Transactions (1) [Severity]	Participant That Reports the Error (1)	Logged? If Advance Error Reporting Header Log Registers implemented
BAD LLTP (link packet containing TLP)	ERR_COR [Correctable]	Requester or completer destination via message transaction (3)	Not logged
BAD DLLP (link management packet)	ERR_COR [Correctable]	Requester or completer destination via message transaction (3)	Not logged
Retry_Timer Timeout	ERR_COR [Correctable]	Requester or completer source via message transaction (3)	Not logged
Retry_Num# Rollover	ERR_COR [Correctable]	Requester or completer source via message transaction (3)	Not logged
DLLPE (Data Link Layer Protocol Error)	ERR_NONFATAL or ERR_FATAL (4) [Uncorrectable]	Requester or completer destination via message transaction (2)	Not logged

Table Notes:

1. The type of error in the first column is reported by the appropriate message error requester transactions in the second column.
2. The appropriate bit in the Uncorrectable Error Status Register is set to "1" independent of the value of the associated bit in the Uncorrectable Mask Register. If the appropriate bit in the Uncorrectable Error Mask Register is *not* set to "1" and an error is detected (based on the associated bit in the Uncorrectable Status Register), the First Error Pointer is not updated and the error not reported.
3. The appropriate bit in the Correctable Error Status Register is set to "1" independent of the value of the associated bit in the Correctable Mask Register. If the appropriate bit in the Correctable Error Mask Register is *not* set to "1" and an error is detected (based on the associated bit in the Correctable Error Status Register), the error is not reported.
4. Reported as ERR_FATAL unless the Uncorrectable Error Severity Register is implemented and the appropriate bit is set to a "0" value. If the Uncorrectable Error Severity Register is implemented and the appropriate bit is set to a "0" value, the error is reported as ERR_NONFATAL. The default value for Minimum Error Reporting protocol (Advance Error Reporting protocol not implemented) is ERR_FATAL.

Special Error Considerations for Switches

The error checking at the Data Link Layer level is by the ports connected to each link. The DLLPs and LLTPs do not port through the switches. LLTPs (containing TLPs) are received at the ports of switches and TLPs are extracted. If DLLPs are received, the link management information in the DLLPs is only used by the receiving ports of the switches. It is the TLPs that are ported through the switches to the transmitting ports. At the transmitting ports of the switches, completely new link packets are created (either DLLP or LLTP). Consequently, the error detection and reporting protocols for Root Complex's downstream ports, endpoints, and bridges also apply to switches as discussed in the following section.

Errors Related to CRC, LCRC, and Nullified TLP Packets

At the link packets' receiving port, even if no errors are found in the Physical Packets by the Physical Layer, there may still be errors in the DLLPs or LLTPs themselves. As illustrated in Figure 9.5, the Physical Layer first determines whether the Physical Packet is successfully received, which is discussed in detail later this chapter. If the Physical Packet is successfully received it is processed on several levels. If there are no Physical Layer errors, the DLLP or LLTP is extracted from the Physical Packet. The FRAMING K symbols at the beginning of the symbol stream of the Physical Packet identify the DLLP versus LLTP.

BAD DLLP

For a DLLP, the remaining error check is determining whether a cyclic redundancy error (CRC) is found in the CRC included in the DLLP. If no CRC error is found, the DLLP information is used by the Data Link Layer. If a CRC error is found, the DLLP is discarded and a **BAD DLLP** error is reported via a message error requester transaction packet.

The PCI Express device that receives a BAD DLLP may or may not receive another DLLP that provides superceding information for the discarded DLLP. No protocol provides for the re-transmission of the discarded DLLP. The exception to this concept is the transmission and receipt of Ack and Nak DLLPs. As will be discussed in Chapter 12 in more detail, these two DLLPs are used to provide the LLTPs' transmitting port information on whether the LLTP was successfully received. The LLTPs' transmitting port discards the Ack or Nak DLLP it received if a CRC error occurs and a **BAD DLLP** error is reported via a message error requester transaction packet. In this situation the LLTPs' transmitting port may also re-transmit the LLTP. If a LLTP is re-transmitted, another Ack or Nak DLLP is transmitted by the LLTPs' receiving port.

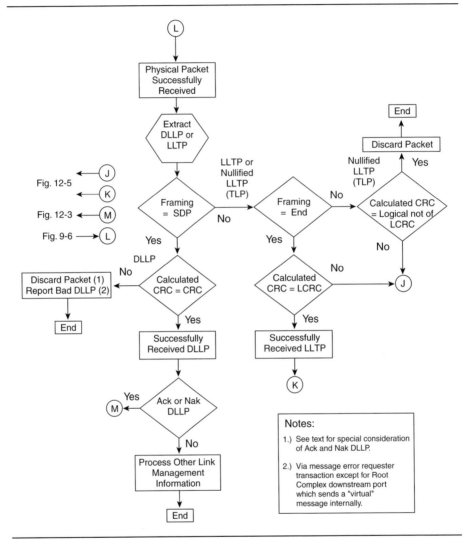

Figure 9.5   PCI Express Data Link Layer

BAD LLTP

For a LLTP or nullified LLTP (LLTP containing nullified TLP), the remaining error check is to determine whether a cyclic redundancy error (CRC) is found in the LCRC. If no CRC is found, the LLTP is further processed by the Transaction Layer unless it contains a nullified TLP. A nullified TLP without a CRC error in the associated LLTP is simply discarded, as illustrated in Figure 9.5. However, if the LLTP containing the nullified TLP is received with an LCRC error it must be treated as a not successfully received LLTP and a Nak DLLP must be scheduled for transmission to the transmitting port of the LLTP. See Chapter 12 for more information on scheduling of a Nak DLLP.

If a CRC error is found in the LLTP, the LLTP is discarded and a **BAD LLTP** error is reported via a message error requester transaction packet.

A BAD LLTP error will also be reported if the SEQ# in the LLTP is out of sequence. See Chapter 12 for more information.

Errors Related to Retry_Timer Timeout and Retry_Num# Counter Rollover

The LLTPs' transmitting port will retry the transmission until successful acknowledgement of receipt, or until a Retry timer expires. The transmitting port of the LLTP reports the Retry timer expiration as **Retry Timer Timeout** error in response to the Ack and NaK DLLPs received. If the LLTPs' transmitting port has too many Retry_Timer Timeouts, the port reports a **Retry_Num# Rollover** error. See Chapter 12 for more information.

Data Link Layer Protocol Errors (DLLPEs)

The following is a summary of the sources of DLLPEs. Please see the chapters referenced in the following paragraphs for more detailed information.

Any violations of the Flow Control Initialization protocol are defined as Data Link Layer Protocol Errors (DLLPEs). The checking and reporting of **DLLPE**s is optional. See Chapter 7 for more information.

When a LLTP is transmitted it contains SEQ#, TLP, and LCRC. In response, an Ack or Nak DLLP will be received. The value of the AckNakSEQ# in these DLLPs must equal an unacknowledged LLTP in the retry buffer or the most recently acknowledged LLTP, otherwise the following occurs (See Chapter 12 for more information):

- The Ack or Nak DLLP is discarded
- A **DLLPE** must be error is reported.

An exception to this error condition is when a Ack or Nak DLLP is received and there are *no* unacknowledged LLTPs in the retry buffer. Under this condition no error occurs provided the AckNakSEQ# = ACKD_Seq# Register.

Physical Layer Errors

The possible errors at the Physical Layer protocol are divided into two types: Receiver and Training. These two types of errors are summarized in Table 9.3. In some cases, there are multiple sources for a specific error. This chapter discusses all of the sources of these errors with a **bold underline** to signify the error in reference to the table. The **bold underline** applies to all chapters in the book.

> Unless otherwise stated (optional), the errors discussed in this section *must* be checked by the associated PCI Express device.

Table 9.3 Errors Related to Physical Layer Protocol

Error Name	Message Error Requester Transactions (1) [Severity]	Participant That Reports the Error (1)	Logged? If Advance Error Reporting Header Log Registers implemented
Receiver Error	ERR_COR [Correctable]	Requester or completer destination via message transaction (5)	Not logged
Training Error	ERR_FATAL or ERR_NONFATAL (2) [Uncorrectable]	PCI Express device on the link via message transaction (3) (4)	Not logged

Table Notes:

1. The type of error in this column links to the appropriate message error requester transaction to be possibly transmitted and other associated actions as defined in Tables 9.4 and 9.5.
2. Reported as ERR_FATAL unless the Uncorrectable Error Severity Register is implemented and the appropriate bit is set to a "0" value. If the Uncorrectable Error Severity Register is implemented and the appropriate bit is set to a "0" value, the error is reported as ERR_NONFATAL. The default value for Minimum Error Reporting protocol (Advance Error Reporting protocol not implemented) is ERR_FATAL.
3. This error is detected when Link Training has failed; consequently, only the PCI Express device on the upstream side of the link can transmit a message to the Root Complex.

4. The appropriate bit in the Uncorrectable Error Status Register is set to "1" independent of the value of the associated bit in the Uncorrectable Mask Register. If the appropriate bit in the Uncorrectable Error Mask Register is *not* set to "1" and an error is detected (based on the associated bit in the Uncorrectable Error Status Register), the First Error Pointer is not updated and the error is not reported.
5. The appropriate bit in the Correctable Error Status Register is set to "1" independent of the value of the associated bit in the Correctable Error Mask Register. If the appropriate bit in the Correctable Error Mask Register is *not* set to "1" and an error is detected (based on the associated bit in the Correctable Status Register), the error is not reported.

> In the following discussions, the expression *to report an error* is used. For any PCI Express device to report an error requires consideration of other qualifications. See the section "Error Reporting and Logging" for these qualifications.

Special Error Considerations for Switches

The ports connected to each link perform the error checking at the Physical Layer level. The Physical Packets do not port through the switches. Physical Packets are received at switches' ports and TLPs are extracted. It is the TLPs that are ported through the switches to the transmitting ports. At the switches' transmitting ports, completely new Physical Packets are created. Consequently, the error detection and reporting protocols for the Root Complex's downstream ports, endpoints' ports, and bridges' ports also apply to switches' ports.

Receiver Errors

The transmission of Physical Packets, TS Ordered Sets, K symbols, and D symbols across a link has many requirements. The **Receiver** errors that are checked vary depending on the link states and the associated states of the Link Training and Status State Machine (LTSSM). For Detect, Polling, Recovery, L0s, L1, L2, and Loopback link states there are no **Receiver** errors. For Configuration, Disabled, and Hot Reset link states the **Receiver** errors that can occur and must be checked have 8/10b encoding and decoding.

When the link is in the L0 link state, as illustrated in Figure 9.6, the Physical Packet's receiving port must determine whether there are any <u>Receiver</u> errors on two different levels: Basic Reception and Physical Packet Reception.

Receiver Errors in L0 Link State

Basic Reception As discussed in Chapter 8, a transmitted Physical Packet contains a LLTP or DLLP encoded into Data Symbols via 8/10b encoding (D symbols). The series of D symbols are framed by special symbols via 8/10b encoding (K symbols). These K symbols are called FRAMING. In addition to FRAMING, other K and D symbols are transmitted as lane fillers (PAD K symbol and IDLE D symbol) and to encoded Ordered Sets. All of these symbols must be as listed in Appendix B of the PCI Express specification in order for the receiving port to use the symbols. In addition to actual K and D symbols transmitted, there are possibilities of 8/10b encode or decode errors, loss of a symbol during transmission across the link, and loss of lane to lane de-skew. Also as discussed in Chapter 8, there are specific requirements for parsing the K and D symbols onto the lanes of a link. Parsing errors can be optionally checked.

Obviously, if the any of the above situations occurs or the requirements are not met, the receiving port may not be able to use the stream of K or D symbols received.

Note: If a symbol is received with incorrect running disparity, the symbol is defined as having a *disparity error*. Also any symbol received must be a valid symbol as defined in Appendix B of the PCI Express specification. If a symbol is received that is not valid or has a disparity error, a **Receiver** error must be reported. See Chapter 8 for more information.

As illustrated in Figure 9.6, the receiving port can optionally check for any combination of violations of these requirements. If the receiving port is checking for any violations and an error is found, the symbols are discarded and a **Receiver** error is reported via message error requester transaction packet.

Physical Packet Reception The previous discussion did not make distinctions whether the K and D symbols were associated with a Physical Packet, lane filling (PAD K symbol and IDLE D symbol), or Ordered Sets. If the K or D symbols are received with a **Receiver** error that cannot be discerned, there are still two error conditions that must be checked prior to transferring the Physical Packet to the Data Link Layer, illustrated in Figure 9.6. First, the FRAMING K symbols at the beginning and end of the Physical Packet must match. Second, the symbols between the FRAMING K symbols must all be correct D symbols. If an error is found, the Physical Packets are discarded and a **Receiver** error is reported via message error requester transaction packet.

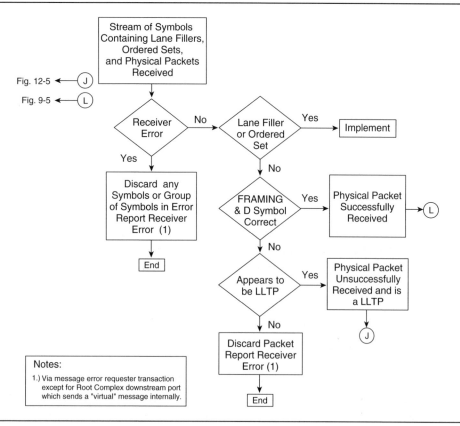

Figure 9.6 PCI Express Physical Layer

Training Errors

Chapter 8 defines the protocol for Link Training. Link Training occurs by sequencing through the Detect, Polling, and Configuration link states. The portion of Link Training accomplished by each of these link states is:

- **Detect link state:** Establish the existence of a PCI Express device on each end of link. That is, detect the lanes within a link common to both PCI Express devices on the link.

- **Polling link state:** Establish the bit and symbol lock, lane polarity inversion, and highest common data bit rate on the detected but as-yet-to-be-configured lanes that exist between the two PCI Express devices.

- **Configuration link state:** Some or all of the detected lanes that successfully complete the Polling link states are processed into configured lanes. Those lanes that can be detected but cannot successfully complete the Polling link state cannot be processed onto a configured lane. Configured lanes are collected into configured links in the link sub-states of the Configuration link state. These link sub-states consequently establish the link width and lane ordering with support of lane reversal. Finally, these link sub-states also implement lane-to-lane de-skew, and the N_FTS value is established.

The PCI Express specification states that a PCI Express device may optionally check for any violation of the Link Training protocol and report them as report **Training Errors** accordingly. Elements of Link Training are also executed in the Link Retraining during the Recovery link sub-states The ability to report a Training Error requires the transmission of a message baseline requester transaction (message transaction in Table 9.3).

CRC, LCRC, ECRC, and Digest Field

In order to protect the control, address (address bits and routing) and data flowing throughout a PCI Express platform, cyclic redundancy checking (CRC) is applied at two levels. One level is at the Data Link Layer and the other is at the Transaction Layer.

CRC and LCRC

For the Data Link Layer there are two sizes: For Data Link Layer Packets (DLLPs), the CRC is one WORD (16 bits) and is called *CRC*. For Link Layer Transaction Packets (LLTP), the CRC is one DWORD (32 bits) called *LCRC*. For a DLLP, the CRC is calculated over the combination of all (PAWs in a DLLP including RESERVED), but not including the CRC itself. The result of the CRC calculation is stored in the CRC PAW. For a LLTP the CRC is calculated over the combination of SEQ PAW and the TLP for the specific LLTP, but not including the LCRC itself. The result of the CRC calculation is stored in two LCRC PAWs. The calculation and checking of the CRC and LCRC are *required* activities of the Data Link Layer at each port connected to a link. Note: For a TLP contained in an LLTP, LCRC calculated and checked includes as all fields of the TLP plus the SEQ PAW.

ECRC

Independent of the calculation and checking of the CRC and LCRC, for Transaction Layer Packets (TLPs) a CRC is *optionally* calculated and checked. This CRC is one DWORD in size (32 bits), stored in the Digest field of the TLP, and the CRC is called *ECRC*. The ECRC is calculated and checked over Header field (combination of HDWs) and Data field (if applicable).

The calculation and checking of the ECRC is an optional activity of the Transaction Layer at the source and destination of the TLP, respectively. If the ECRC (Digest field) is included by the TLP source, the TD bit in the ATTRIBUTE HDW of the Header field must be set to 1b.

If a PCI Express device calculates the ECRC and attaches the Digest field for one TLP it sources, it must calculate the ECRC and attach the Digest field for all TLPs it sources. The destination of the TLP with a Digest field can optionally check the ECRC in the Digest with the following qualifications:

- The destination of the TLP cannot check the ECRC unless the destination implements the Advance Error Reporting protocol.

- If the destination is able to check the ECRC for one TLP, it must be apply and execute checking for ECEC for all TLPs received with TD Bit = 1b.

- If EP= 1b in the ATTRIBUTE HDW, the TLP is poisoned (Error Forwarding) and ECRC is irrelevant. As discussed in this chapter, a TLP with Error Forwarding does not contain reliable information; consequently, the ECRC is not checked.

The ECRC contained in the Digest field provides end-to-end coverage TLP integrity. End-to-end coverage recognizes that the Physical and Data Link Layers have successfully transferred the associated packets between the PCI Express devices. As discussed in this book, the requester source and requester destination or the completer source and the completer destination are the actual PCI Express devices that use the contents of the TLPs. As discussed earlier in this chapter, the PCI Express specification permits switches that are porting TLPs from one port to another port (not as the requester or completer destination) to optionally check the ECRC value in the Digest field. The switch reports the error if found and ports the TLP to the transmitting port without modifying the ECRC value in the Digest field. The ECRC checking and reporting qualifications listed for the destination of the TLPs also apply to switches.

Protocols for Calculating and Checking CRC, LCRC and ECRC

The integrity of the defend packets is protected by cyclic redundancy checking. For Link Layer Transition Packets and Data Link Layer Packets CRC is required. For the Transaction Layer Packets, protecting their integrity by CRC is optional.

Calculating CRC with 16 bits

The calculation and checking of the CRC for DLLPs is over all of the bits of the PAWs. The calculation and checking includes all of the fields and bits defined as RESERVED. The protocol for CRC is as follows:

- The initial value at the beginning of the calculation and checking is FFFFh in the flip-flops of the CRC circuitry, illustrated in Figure 9.7.

- The first bit to be input into the CRC circuitry is the lowest order of the lowest order byte. The next bit is the next higher order bit of the lowest order byte. Once all the bits of the lowest order byte are input, the lowest order bit of the next higher byte is input, and so forth.

- The polynomial used has a coefficient expressed as 100Bh.

- The final output of the calculation (for initial calculation or checking) is complemented and mapped to the LCRC as listed in Table 9.4.

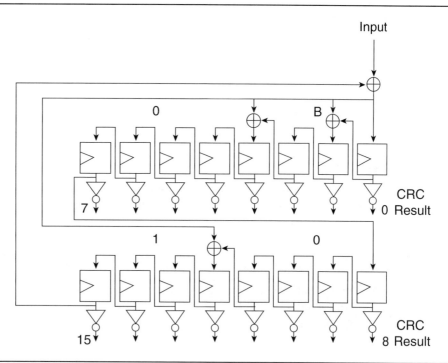

Figure 9.7 CRC Circuitry

Table 9.4 LCRC Mapping

CRC Output	LCRC Bit Position	CRC Output	LCRC Bit Position
0	7	8	15
1	6	9	14
2	5	10	13
3	4	11	12
4	3	12	11
5	2	13	10
6	1	14	9
7	0	15	8

Calculating LCRC with 32 bits

The calculation and checking of the LCRC is over all of the bits of the SEQ PAW and TLP within a LLTP. The calculation and checking includes all of the fields and bits defined as RESERVED. The protocol for LCRC is as follows:

- The initial value at the beginning of the calculation and checking is FFFF FFFFh in the flip-flops of the CRC circuitry, as illustrated in Figure 9.8.

- The first bit to be input into the CRC circuitry is the lowest order of the lowest order byte. The next bit is the next higher order bit of the lowest order byte. Once all the bits of the lowest order byte are input, the lowest order bit of the next higher byte is input, and so forth.

- The polynomial used has a coefficient expressed as 04C1 1DB7h.

- The final output of the calculation (for initial calculation or checking) is complemented and mapped to the LCRC as listed in Table 9.5.

Table 9.5 LCRC and ECRC Mapping

CRC Output	LCRC or ECRC Bit Position	CRC Output	LCRC or ECRC Bit Position
0	7	16	23
1	6	17	22
2	5	18	21
3	4	19	20
4	3	20	19
5	2	21	18
6	1	22	17
7	0	23	16
8	15	24	31
9	14	25	30
10	13	26	29
11	12	27	28
12	11	28	27
13	10	29	26
14	9	30	25
15	8	31	24

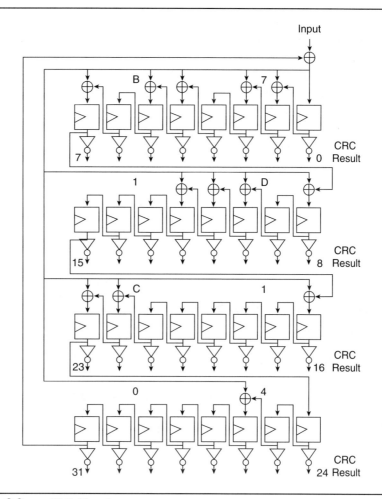

Figure 9.8 CRC Circuitry

Calculating ECRC with 32 bits

The calculation and checking of the ECRC is over all of the bits of the Header field and Data field (if included) of the TLP. The calculation and checking includes all of the fields and bits defined as RESERVED. The protocol for ECRC is the same as discussed for the LCRC with 32 bits with the following differences:

- There are two bits in the Header field that will have a value of "0" or "1" that *must* be "assumed" to be "1" for the purposes of ECRC calculation and checking. These two bits are Bit 0 of the TYPE field in the TYPE HDW and EP bit of the ATTRIBUTE HDW in the Header field.

- The final output of the calculation (for initial calculation or checking) is complemented and mapped to the ECRC contained in the Digest field as listed in Table 9.5.

Error Reporting and Logging

As discussed previously in this chapter, errors can be detected in the Physical, Data Link, and Transaction Layers' packets. Once a PCI Express device has detected an error in addition to the actions discussed earlier and in other chapters, there are certain requirements to report it to the HOST bus segment CPU (or an equivalent system resource).

The basic concept is that an error is detected by a PCI Express device (or internal function), and error information is stored in appropriate registers within the configuration address block of the PCI Express device. PCI Express defines two levels of error reporting information: Minimal and Advance. The PCI Express device notifies the downstream port of the Root Complex of the error via message requester transaction packets. If the Root Complex's downstream port detects the error, it sends itself a virtual message requester transaction packet. Once the downstream port has knowledge of the error, it causes either a system error or an interrupt to the HOST bus segment to alert Error software.

The later part of this chapter focuses on three key entities related to error reporting: message error requester transactions, Minimal Error Reporting protocol, and Advance Error Reporting protocol.

General Error Considerations

Before discussing in detail message error requester transactions, Minimal Error Reporting protocol and Advance Error Reporting protocol, there are a couple of general error considerations for the Error software executed by a CPU on the HOST bus segment.

Reporting by System Error versus Interrupt

A distinction between Minimal Error Reporting versus Advance Error Reporting is system error versus using interrupts for error reporting, respectively. Minimal Error Reporting protocol either does not report the error to Error software or reports the error via system error. Traditionally, system error as implemented by PCI and PCI-X (via SERR# signal line) can cause a major reset of the platform. For PCI Express, a system error is implementation-specific based on the Root Complex and HOST bus segment. Advance Error Reporting expands the error reporting options to include interrupt of the Root Complex and HOST bus segment to report an error. Both methods for error reporting are internal to the Root Complex.

Pollution

The Error software that establishes the error detection and reporting protocols must be careful to avoid error pollution. Error pollution occurs when the same basic error associated with a particular packet (TLP, DLLP, LLTP, or Physical Packet) is detected and reported by multiple layers (Transaction, Data Link and Physical). For example, an error detected and reported by the Physical Layer for a Physical Packet must not also be reported (it may still be detected) by the Data Link or Transaction Layers for the associated LLTP and TLPs; respectively.

Message Error Requester Transaction Packets

In order for the downstream PCI Express devices to report errors to the Root Complex's downstream ports, messages error requester transactions are defined. Messages error requester transaction packets are transmitted upstream, as illustrated in Table 9.6, to a downstream port of the Root Complex in response to error detection at the function level of a downstream PCI Express device. It is also possible for one of the Root Complex's downstream ports to detect an error and to report it to itself without the external transmission of a message error requester transaction packet. Conceptually, the Root Complex's downstream port transmits a message error requester transaction to itself. The protocol of the message error requester transaction packets is discussed in Chapter 6.

Both the Minimum and Advance Error Reporting protocols implement the message error requester transactions to report the occurrence of the error (if enabled) to the Root Complex's downstream ports. As previously discussed at the beginning of this chapter, there are three categories of errors and each category has a corresponding message error requester transaction packet. The notations and the associated message error requester transactions are follows:

- **ERR_COR:** Correctable errors are reported by message ERR_COR requester transaction. Corrected by hardware and not by Error software. Optionally, logging of correctable errors may be of value to Error software, but is not defined by the PCI Express specification.

- **ERR_NONFATAL:** Nonfatal errors are reported by message ERR_NONFATAL requester transaction. Nonfatal errors are not correctable, are isolated to a specific link, do not render the specific link inoperative, and do not disturb activity on the specific link. They are responded to by the Error software and not by the hardware.

- **ERR_FATAL:** Fatal errors are reported by message ERR_FATAL requester transaction. Fatal errors are not correctable. Fatal errors may or may not be isolated to a specific link and the connected PCI Express devices, they may require a Hot Reset of the link. They are responded to by the Error software and not by the hardware.

As discussed earlier in this chapter, the source of an error will vary depending on whether it is related to the Physical, Data Link, or Transaction Layers. Once a PCI Express device detects an error, it must be mapped to the appropriate message error requester transaction, which also defines its severity (type of error) (Tables 9.1 through 9.3). The PCI Express device subsequently transmits the appropriate message error requester transaction packet. The requester source and destination of the message error requester transaction packets is summarized in Table 9.6.

Table 9.6 Error Related Messages

Message Error Requester Transactions [Severity]	Destination of Message Transaction Packet	Source of Message Transaction Packet Data Packet In ATTRI. DWORD Length [9::0]	Requester ID Field Entries
ERR_COR [Correctable]	Root Complex	Legacy EP PCI Express EP PCI Express/PCI Bridge PCI Express/PCI-X Bridge No Data Length = 00 0000 0000b	BUS# = Actual DEVICE# =Actual FUNC# =Actual or Reserved (1)
ERR_NONFATAL [Uncorrectable]	Root Complex	Legacy EP PCI Express EP PCI Express/PCI Bridge PCI Express/PCI-X Bridge No Data Length = 00 0000 0000b	BUS# = Actual DEVICE# =Actual FUNC# =Actual or Reserved (1)
ERR_FATAL [Fatal Error]	Root Complex	Legacy EP PCI Express EP PCI Express/PCI Bridge PCI Express/PCI-X Bridge No Data Length = 00 0000 0000b	BUS# = Actual DEVICE# =Actual FUNC# =Actual or Reserved (1)

Table Notes:

1. In some cases the actual FUNCTION# cannot be returned. For example, the access is to a multiple FUNCTION device and the access is to a FUNCTION# that is not enabled (does not exist). The device will return a completer transaction packet with the Status field = unsuccessful and the FUNCTION# = 000b.

Message Error Requester Transaction Requester and Functions

As summarized in Table 9.6, when an error message requester transaction is transmitted it may include the FUNCTION# within a specific PCI Express device. The following errors' sources cannot provide a specific FUNCTION#. Consequently, the error status and any error logging (if applicable) must be recorded in the configuration register blocks of all functions of a multi-function PCI Express device. The function(s) supporting Advance Error Reporting protocol will have the most error information, but the other functions whether implementing Minimum or Advance Error Reporting, must also provide the same information (or subset of information if Minimum Error Reporting). If all the function(s) only implement Minimum Error Reporting, all the functions must provide the same information. The Requester ID provided (including FUNCTION#) identifies the function with the most information. The error sources are as follows:

- Any error sourced from the Physical or Link Layers.
- The following errors sourced from the Transaction Layer.
    - ECRC error in TLP
    - Malformed TLP
    - When UR (unsupported requester transaction packet) is reported
    - Receiver Overflow
    - Flow Control Protocol Error (FCPE)

Minimum Error Reporting Protocol

An error detected at the function level of a downstream PCI Express device connected to a downstream port of the Root Complex can be reported in the configuration register blocks on two levels: Minimum and Advance. For the Minimum Error Reporting protocol, implemented in the Root Complex's downstream port are the Root Complex Control, Device Control, and Device Status Registers. On the downstream PCI Express devices, the Device Control and Device Status Registers are implemented. These are the configuration registers associated with the Minimum Error Reporting protocol, and must be implemented by all PCI Express devices.

As implied in Table 9.7, the error information provided by the Minimum Error Reporting protocol is very limited. For example, this protocol does not log the sources of errors. At most, the Minimum Error Reporting protocol reports the error to Error Software by generating a system error. Additional error information and methods for reporting to the Error Software are provided by the optional Advance Error Reporting protocol discussed in the next section. Note that each bit defined in the Device Control and Device Status register the defined per each function within a PCI Express device.

Configuration Registers for Root Complex's Downstream Ports and PCI Express Devices (Functions)

When one of the three types of errors (correctable, nonfatal, or fatal) is detected as outlined in Tables 9.1 through 9.3, the appropriate message error requester transaction packet is transmitted if the associated bit of the function's Device Control register is set = 1b. Also, the appropriate bits are set = 1b in the function's Device Status register independent whether a message error requester packet is transmitted or not. A system error is generated upon receipt of the message error transaction packet if BIT# 0, BIT# 1, or BIT #2 in Root Complex Control register is set = 1b. The method by which the system error is generated in the Root Complex to the HOST bus segment is implementation-specific. Note: If the proper bit is not set to 1b in the Device Control Register the associated message error requester transaction packet is *not* transmitted. If the proper bit is not set to 1b in the Root Control Register, the associated message error requester transaction packet received is not seen and thus no system error is generated.

Table 9.7 Summary of Minimum Error Reporting Configuration Registers (Also used as a subset of the Advance Error Reporting protocol)

Row	Type of Error	Root Complex Control Register (1) Also used for Advance Error Reporting	Device Control Register (2) Also used for Advance Error Reporting	Device Status Register (2) (3) Also used for Advance Error Reporting
1	Correctable Error	Bit 0 = 1b: System error generated when message ERR_COR requester transaction packet received by Root Complex's downstream port (5)	Bit 0 = 1b: Message ERR_COR requester transaction packet transmitted by PCI Express device when error detected	Bit 0 = 1b: Correctable error detected
2	Nonfatal Error	Bit 1 = 1b: System error generated when message ERR_NONFATAL requester transaction packet received by Root Complex's downstream port	Bit 1 = 1b: Message ERR_NONFATAL requester transaction packet transmitted by PCI Express device when error detected	Bit 1 = 1b: Nonfatal error detected

Table 9.7 Summary of Minimum Error Reporting Configuration Registers (Also used as a subset of the Advance Error Reporting protocol) *(continued)*

Row	Type of Error	Root Complex Control Register (1) **Also used for Advance Error Reporting**	Device Control Register (2) **Also used for Advance Error Reporting**	Device Status Register (2) (3) **Also used for Advance Error Reporting**
3	Fatal Error	Bit 2 = 1b: System error generated when message ERR_FATAL requester transaction packet received by Root Complex's downstream port	Bit 2 = 1b: Message ERR_FATAL requester transaction packet transmitted by PCI Express device when error detected	Bit 2 = 1b: Fatal error detected
4	Unsupported TLP received by PCI Express device	Bit 1 = 1b: System error generated when message ERR_NONFATAL requester transaction packet received by Root Complex's downstream port.	Bit 3 = 1b: An unsupported requester transaction error can be reported. (4)	Bit 3 = 1b: The PCI Express device received an unsupported requester or completer transaction packet

Table Notes:

1. Root Complex's ports that sense the associated error will generate a system error if the associated bit in Root Complex Control register is set to 1. No message error requester transaction packet is transmitted by the Root Complex port. The method by which a system error is generated to the Host CPUs is system-specific.
2. Defined for each function in each PCI Express device
3. The bit is set to 1b for an error detected independent of the value of the associated Root Complex Control or Device Control bits.
4. If Bit 3 = 1b and an unsupported requester is detected, the message ERR_NONFATAL or ERR_FATAL requester transaction is transmitted per bit 01 and bit 1. Reported as ERR_NONFATAL unless the Uncorrectable Error Severity Register is implemented and the appropriate bit is set to a "1" value. If the Uncorrectable Error Severity Register is implemented and the appropriate bit is set to a "1" value, the error is reported as ERR_FATAL. The default value for Minimum Error Reporting protocol (Advance Error Reporting protocol not implemented) is ERR_NONFATAL.
5. The reporting of a correctable error is only for purposes of error counting. By definition, the error has already been corrected by hardware at the link.

Figure 9.9 summarizes the locations of the configuration registers implemented for support of the Minimum Error Reporting protocol. These registers are required to be implemented. The locations of the configuration registers for Minimum Error Reporting protocol are from two viewpoints. The first viewpoint is that the Root Complex's downstream ports are receiving the message error transaction packets from the downstream PCI Express devices. Consequently, the Root Complex's downstream ports contain the Root Control Register. The second viewpoint is that each of the Root Complex's downstream ports can also be the error detection entity. Consequently, each function of each of the Root Complex's downstream ports contain the Device Control and Device Status Registers. As previously discussed, the Root Complex's downstream port that detects the error transmits a message error requester packet internal to itself. For the Root Control Register, it is as if a downstream PCI Express device transmitted the message error requester packet.

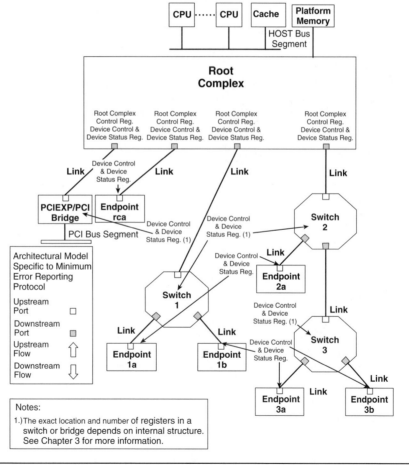

Figure 9.9   Minimum Error Reporting Protocol Registers

Advance Error Reporting Protocol

As discussed above, an error detected at the function level of a downstream PCI Express device connected to a downstream port of the Root Complex can be reported in the configuration register blocks on two levels: Minimum and Advance.

For the Minimum Error Reporting protocol, the configuration registers implemented in the Root Complex's downstream port are the Root Complex Control, Device Control, and Device Status Registers. For the Advance Error Reporting protocol these registers also are implemented. In addition, the Advance Error Reporting protocol also implements the following configuration registers in the Root Complex's downstream port (with qualifications): Root Complex Error Command, Root Complex Error Status Registers, and Error Source Identification.

For the Minimum Error Reporting protocol, the downstream PCI Express devices implement the Device Control, and Device Status Registers are implemented. For the Advance Error Reporting protocol, these registers also are implemented. In addition, the Advance Error Reporting protocol also implements the following configuration registers in the downstream PCI Express devices implement (with qualifications): Header Log, Correctable Error, Uncorrectable Error, and Advance Error Capabilities and Control Registers.

Qualifications for Configuration Register Implementations

The configuration registers implemented by the Minimum Error Reporting protocol are required, and thus always available to the Advance Error Reporting protocol. The Advance Error Reporting protocol provides more information about the detected errors; consequently, more configuration registers are implemented in the Root Complex's downstream ports and the downstream PCI Express devices. Implementation of these additional configuration registers is optional. As illustrated in Figure 9.10, not all downstream ports of the Root Complex and not all downstream PCI Express devices (internal functions) may implement the additional configuration registers required to support the Advance Error Reporting protocol and, even if implemented, may not be able to use them. This places restrictions on the use of the Advance Error Reporting protocol as follows:

- **Bridges:** The PCI Express/PCI or PCI Express/PCI-X Bridge *can only implement* the Minimum Error Reporting protocol. If the bridge is connected to a Root Complex's downstream port that only implements the Minimum Error Reporting protocol configuration registers, only the Minimum Error Reporting protocol is possible between these two participants. If the bridge was connected to a Root Complex's downstream port that implemented the optional additional registers for Advance Error Reporting protocol, these optional registers could not be used and only the Minimum Error Reporting protocol is possible between these two participants. These same concepts apply to a bridge connected to a switch.

- **Endpoint and Switch with Minimum Error Reporting protocol:** Endpoint rca and the upstream port of switch 2 only implement the Minimum Error Reporting protocol configuration registers. Both of these PCI Express devices are connected to downstream ports of the Root Complex that implements the Advance Error Reporting protocol. Consequently, only the Minimum Error Reporting protocol is possible between these two participants.

    - Endpoints 3a and 3b also only implement the Minimum Error Reporting protocol configuration registers. As with the upstream port of switch 2, only Minimum Error Reporting protocol is possible even though the associated downstream port of the Root Complex implements the Advance Error Reporting protocol configuration registers.

- **Endpoints and Root Complex Downstream Port with Advance Error Reporting protocol:** Endpoint 2a and the associated downstream port of the Root Complex both implement the Advance Error Reporting protocol configuration registers. Consequently, the Advance Error Reporting protocol is supported between these participants. Note: The support of Advance Error Reporting protocol between these participants is independent of the intervening switches' implementation of Minimum or Advance Error Reporting protocols. In this example, switch 2 only implements the Minimum Error Reporting protocol.

- **Endpoints and Switches with Advance Error Reporting protocol but Root Complex without Advance Error Reporting protocol:** Endpoint 1a, endpoint 1b, and the upstream port of switch 1 implement the Advance Error Reporting protocol configuration registers. The associated downstream port of the Root Complex implements the Minimum Error Reporting protocol configuration registers. Consequently, only the Minimum Error Reporting protocol is possible between the endpoints and the Root Complex, or between the switch 1 and the Root Complex.

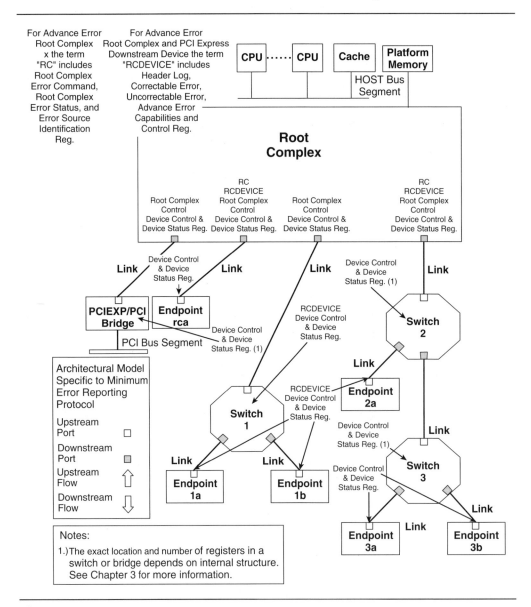

Figure 9.10 Advance Error Reporting Protocol Registers

Advance Error Reporting Registers' Definition and Use

As previously discussed, the optional support of the Advance Error Reporting protocol requires additional configuration registers to those shared with the Minimum Error Reporting protocol. The locations of the configuration registers specific to the Advance Error Reporting protocol is from two viewpoints. The first viewpoint is that the Root Complex's downstream ports are receiving entities of the message error transaction packets from the downstream PCI Express devices. Consequently, the Root Complex's downstream ports contain the Root Complex Error Command, Root Complex Error Status Registers, and Error Source Identification. The second viewpoint is that each of the Root Complex's downstream ports can also be the error detection entity. Consequently, each function of each Root Complex's downstream ports contains the Header Log, Correctable Error, Uncorrectable Error, and Advance Error Capabilities and Control Registers. These registers are also implemented by the downstream PCI Express devices that implement Advance Error Reporting protocol.

Advance Error Reporting Protocol Configuration Registers Only for Root Complex's Downstream Ports

- **Root Complex Control and Root Complex Error Command Registers:** These registers are defined in Tables 9.7 and 9.8. The Root Error Command register permits the error to be reported by interrupt instead of by system error. As previously discussed for the Minimal Error Reporting protocol, the Root Control Registers only supported the report of an error via system error.

- **Root Complex Error Status Register:** As defined in Table 9.8, the Root Complex Error Status Register provides additional information about the error for the benefit of the Error software.

- **Error Source Identification Register:** This register simply contains the Requester IDs of the first uncorrectable (per message ERR_FATAL or ERR_NONFATAL requester transaction packet) and the first correctable error (message ERR_COR requester transaction) reported in the Root Complex Error Status Register. The Error Source Identification Register is updated independent of the values in the Root Control or Root Error Registers. The Requester IDs placed in the Error Source Identification Register are from the message EER_COR, ERR_FATAL, or ERR_NONFATAL requester transaction packets. See the further discussion in the section "Logging and Multiple Error Reporting."

Table 9.8 Summary of Root Complex's Error Command and Status Configuration Block Registers for Advance Error Reporting

Row	Name	Root Complex Error Command Register Advance Error Reporting ONLY	Root Error Status Register (1) Advance Error Reporting ONLY
1	Correctable Error	Bit 0 = 1b: Interrupt generated when message ERR_COR requester transaction packet received by Root Complex's downstream port.	Bit 0 = 1b: When message ERR_COR requester transaction packet received by Root Complex's downstream port and bit not already set to 1b.
			Bit 1 = 1b: When message ERR_COR requester transaction packet received by Root Complex's downstream port AND Bit 0 already set to 1b.
2	Nonfatal Error	Bit 1 = 1b: Interrupt generated when message ERR_NONFATAL requester transaction packet received by Root Complex's downstream port.	Bit 2 = 1b: When message ERR_FATAL or ERR_NONFATAL requester transaction packet is received Root Complex's downstream port and bit is not already set to 1b.
			Bit 3 = 1b: When message ERR_FATAL or ERR_NONFATAL requester transaction packet received by Root Complex's downstream port? AND Bit 2 already set to 1b.
			Bit 5 = 1b: When one or more message ERR_NONFATAL requester transaction packets received by Root Complex's downstream port.

Table 9.8 Summary of Root Complex's Error Command and Status Configuration Block Registers for Advance Error Reporting *(continued)*

Row	Name	Root Complex Error Command Register Advance Error Reporting ONLY	Root Error Status Register (1) Advance Error Reporting ONLY
3	Fatal Error	Bit 2 = 1b: Interrupt generated when message ERR_FATAL requester transaction packet received by Root Complex's downstream port.	Bit 2 = 1b: When message ERR_FATAL or ERR_NONFATAL requester transaction packet is received Root Complex's downstream port and bit is not already set to 1b.
			Bit 3 = 1b: When message ERR_FATAL or ERR_NONFATAL requester transaction packets received by Root Complex's downstream port? AND Bit 2 already set to 1b.
			Bit 4 = 1b: When the cause of Bit 2 being set to 1 is message ERR_FATAL requester transaction packet received by Root Complex's downstream port.
			Bit 6 = 1b: When one or more message ERR_FATAL requester transaction packet received by Root Complex's downstream port.
4	Advance Error Interrupt Message Number		Bits [31:27] contains the number to use for Advance error MSI.

Table Notes:

1. Updated independent of the value of associated bits in the Root Control or Root Error Command registers.

Advance Error Reporting Protocol Configuration Registers for Root Complex's Downstream Ports and PCI Express Devices (Functions)

- **Device Control Register:** Shared with Minimum Error Reporting protocol and defined in Table 9.7.

- **Device Status Register:** Shared with Minimum Error Reporting protocol and defined in Table 9.7.

- **Header Log Registers:** These registers contain the Header field of the TLPs associated with the detected uncorrectable errors. The logging is required when the Advance Error Reporting protocol is implemented and logging is specified per Table 9.1 (independent of the bit values in the Device Control register). The uncorrectable errors are reported to the Root Complex's downstream ports by message ERR_NONFATAL or ERR_FATAL requester transaction packets if enabled by the Device Control Register. See further discussion in the section "Logging and Multiple Error Reporting."

- **Correctable Error Registers:** Collectively there are two registers: Correctable Error Status and Correctable Error Mask Registers. The correctable errors are reported to the Root Complex's downstream ports by message ERR_COR requester transaction packets if enabled by the Device Control Register. See further discussion in the section "Logging and Multiple Error Reporting." If the severity of the error is defined as correctable according to Tables 9.2 and 9.3, these registers are used as follows:
    - The appropriate bit is set to 1b in the Correctable Error Status Register if the associated error was detected. (These bits are set independent of the value of the associated bit in the Correctable Error Mask Registers.)
    - If a bit set to 0b in the Correctable Error Mask Register and the associated bit is set to "1" in the Correctable Error Status Register (indicating an error detected), the error is reported as defined in Tables 9.2 and 9.3, and the message ERR_COR requester transaction packet is transmitted if the Device Control Register permits. Otherwise the detected error is not reported.

- **Uncorrectable Error Registers:** Collectively there are three registers: Uncorrectable Error Status, Uncorrectable Error Mask, and Uncorrectable Error Severity Registers. The uncorrectable errors are reported to the Root Complex's downstream ports by message ERR_FATAL or ERR_NONFATAL requester transaction packets if enabled by the Device Control Register. See further discussion in the section "Logging and Multiple Error Reporting."

If the severity of the error is defined as uncorrectable according to Tables 9.1 through 9.3, the registers are used as follows:

- The appropriate bit is set to 1b in the Uncorrectable Error Status Register if the associated error was detected. (These bits are set independent of the value of the associated bit in the Uncorrectable Error Mask Registers.)
- If a bit set to 0b in the Uncorrectable Mask Register and the associated bit is set to 1b in the Uncorrectable Error Status Register (indicating an error detected), the error is reported as defined in Tables 9.1 through 9.3, and the message ERR_NONFATAL or ERR_FATAL requester transaction packet is transmitted, if Device Control Register permits. Otherwise the detected error is not reported. The Header Log Registers and First Pointer (in the Advance Error Capabilities and Control Register) are updated.
- If appropriate bit in the Uncorrectable Error Mask Register is set to 1b and an error is detected (based on the associated bit in the Uncorrectable Error Status Register), the error is not reported. The Header Log Registers and the First Error Pointer (in the Advance Error Capabilities and Control Register) are not updated.
- The Uncorrectable Error Severity Register permits the Error software to modify the level of reporting eleven different errors identified in Tables 9.1 through 9.3. The selection determines whether a message ERR_NONFATAL or ERR_FATAL requester transaction packet is transmitted, if Device Control Register permits.

■ **Advance Error Capabilities and Control Register:** This register has two purposes: First, it contains the five bits that identify the bit position of the first uncorrectable error reported via the Uncorrectable Error Status Register (First Error Pointer). That is, there are 32 possible bits in the Uncorrectable Error Status Register. According to the present revision of the PCI Express specification, eleven of these bits are associated with a specific uncorrectable error. See further discussion in the section "Logging and Multiple Error Reporting." Second, it controls the generation and defines the support of an error related to the ECRC.

Logging and Multiple Error Reporting

As previously discussed in this chapter and summarized in Table 9.1, some of the TLPs can be the source of an uncorrectable error. The Header field of the TLP associated with the error is logged if the Advance Error Reporting protocol is implemented. Uncorrectable errors related to the Physical and Data Link Layers are not logged. The values of the Header fields are logged in Header Log Registers. Each Header Log Register is 16 bytes in size; consequently only one Header field of an uncorrectable error can be logged in each function's configuration register block within a PCI Express device. The First Error Pointer in the Advance Error Capabilities and Control Register points to the uncorrectable error in the Uncorrectable Error Status Register associated with the contents of the Header Log Register. After reset, the First Error Pointer and the Header Log Registers provide information for the first uncorrectable error. If another uncorrectable error is detected prior to the Error software servicing the first uncorrectable error (or a reset), the value of the First Error Pointer and the Header Log Registers remain unchanged. The subsequent uncorrectable error is noted in the Root Error Status Register (Rows 2 and 3 of Table 9.8). The Header Log Register contains valid information only when the First Error Pointer is pointing to a valid error bit (bit set to 1b, defined as enabled) in the Uncorrectable Error Status Register.

The servicing of the first uncorrectable error by the Error software (or a reset) enables the First Error Pointer and the Header Log Registers to provide information about the next uncorrectable error (which becomes the new "first uncorrectable error"), but Header field information for uncorrectable errors detected between the first uncorrectable error and its serving are not logged. Consequently, Error software must quickly service uncorrectable errors in order to not miss Header field information of other uncorrectable errors.

> The First Error Pointer and the Header Log Register can only be updated to report uncorrectable errors when the associated bit in the Uncorrectable Error Mask Register is set to 0b.

Chapter 10

Transaction Ordering

Transaction Ordering is a protocol first developed in PCI platforms. It recognized two key attributes of PCI bus transactions flowing through a platform. First, bus transactions can begin but not immediately complete. Second, the bus transactions can flow through buffers situated throughout the platform. The combination of these two attributes gives rise to possible livelock and deadlock conditions.

The implementation of the Requester/Completer protocol in PCI Express platforms and the buffers situated throughout the PCI Express fabric also give rise to livelock and deadlock conditions. As will be discussed in this chapter, the PCI Express specification defines a version of the PCI and PCI-X Transaction Ordering protocol to address livelock and deadlock concerns.

Before detailing the Transaction Ordering protocol in the next section, this section will briefly discuss these key attributes.

Transactions Begin and End

This section examines the execution of simple and complex transactions.

Simplest Transaction Execution

On PCI and PCI-X platforms the simplest execution of a bus transition occurs when the PCI or PCI-X bus master and target are on the same bus segment. The simplest execution of a write transaction is for a PCI or

PCI-X bus master (that is, the source of the transaction) to send address/data to a target, the target acknowledges receipt of address/data and requests completion of transaction, then the PCI or PCI-X bus master completes the transaction. The simplest execution of a read transaction is for a PCI or PCI-X bus master to send address to a target, the target acknowledges by returning data and requests completion of transaction, then the PCI or PCI-X bus master completes the transaction. In both executions, it is assumed that the bus segment is only executing the one transaction between the two participants (bus master and target) at any given time. Consequently, any data or completion information is specific between the participants for the specific transaction. It is also assumed that the participants retain full use of the bus segment (that is, the bus segment is dedicated).

Extending this discussion to PCI Express platforms, assume the Root Complex and endpoint are on the same link. The simplest execution of transactions between the two participants (Root Complex and endpoint) follows the same protocol as discussed for PCI or PCI-X bus segment. That is, only one transaction between the two participants at any given time and the participants retain full use of the link (that is, the link is dedicated).

More Complex Transaction Execution

In actual implementations of PCI and PCI-X platforms, the participants are usually on different bus segments. Similarly, in actual implementations of the PCI Express platform, the two participants are usually separated by multiple switches and links.

Focusing just on a PCI Express platform, if the two participants are not on the same link, the simplest execution of a transaction is not possible unless all intervening switches and links are dedicated to the transaction's execution. Complete dedication is not possible, in order to provide platform-wide bandwidth. Consequently, in order to maximize platform-wide bandwidth, multiple transactions executed by one or more PCI Express devices must share the intervening switches and links. As each transaction begins execution, the intervening switches and links must be available for other transactions to begin execution and for completion acknowledgements to be returned.

As discussed in Chapters 1 and 2, the PCI Express specification has defined the Requester/Completer protocol to allow multiple transactions to begin execution (requester transactions) and complete later, with completion acknowledgements (completer transaction) arriving out of order. PCI Express's implementation of Requester/Completer protocol gives rise to livelock and deadlock situations. Consider the following example: The Root Complex is accessing several endpoints via one downstream port, and the transmitting and receiving buffers of the downstream port are FIFO (first-in and first-out). The transmission FIFO of the downstream port may not have yet transmitted requester transaction because the FIFO is waiting for completion (receipt) of an earlier requester transaction. The completer transactions of the earlier requester transactions may be in the receiver FIFO, but not received in the same order as their associated requester transactions were transmitted. The Root Complex has not yet accepted other completer transactions received ahead of a specific completer transaction in the FIFO. The non-acceptance may be based on the Root Complex waiting for completer transactions to be processed in the same order the requester transactions were transmitted.

If the Root Complex's transmission of a requester transaction is dependent on the completion of another requester transaction, livelocks and deadlocks may occur. In a livelock situation, the transmission of a requester transaction or completion of a transaction (receipt of completer transaction) is delayed temporarily for completion of a dependent earlier transaction. In a deadlock situation, the transmission of a requester transaction or completion of a transaction (receipt of completer transaction) is delayed permanently because the completion of a dependent earlier never occurs.

Buffers in the PCI Express Fabric

Chapters 1 and 2 discussed the Requester/Completer protocol. As specified by the Requester/Completer protocol, all transactions begin as requester transaction packets. The requester transaction packets move between PCI Express devices as Transaction Layer Packets (TLPs) contained within Link Layer Transaction Packets (LLTPs) and Physical Packets. At each switch between the participants (requester transaction source and destination), the requester transaction packets can be posted and merged with other requester transaction packets moving in the same direction. Once TLPs containing the requester transaction packets arrive at the destination (Root Complex, bridge, or endpoint), the destination resource responds with completer transactions. Completer transaction packets also move between PCI Express devices as TLPs contained within LLTPs and Physical Packets. At each switch between the participants (completer transaction source and destination), the completer transaction packets can be posted and merged with other completer transaction packets moving in the same direction. Within each switch, there are buffers at the transmitting ports.

If the flow of TLPs through the buffers in the same direction are based on FIFO protocol (first-in first-out order), TLPs associated with one set of participants may be delayed in flowing through the PCI Express fabric due to TLPs associated with another other set of participants. If for some reason the TLPs cannot successfully flow to their destinations, the TLPs moving to other destinations will be stopped in the switches' buffers.

Buffers versus FIFOs

There are two types of buffers discussed in this chapter. One type is defined simply as *buffer* and the other as *FIFO*. The term *buffer* refers to a set of registers in ports that must implement combinations of Transaction Ordering and Port Arbitration, and VC Arbitration protocols.

The term *FIFO* refers to a unique version of a buffer. It consists of a set of registers implemented with strong ordering. That is, the first TLP from the link to be stored in the FIFO is the first TLP removed. This book has adopted the use of a FIFO in the different ports throughout a PCI Express fabric to simplify the discussion of the Transaction Ordering protocol in this chapter and the Flow Control protocol in Chapters 11 and 12. The FIFOs are only defined at the receiving portion of the ports. The use of FIFOs at the receiving portion of the ports ensures the results of Transaction Ordering, Port Arbitration, and VC Arbitration protocols implemented by the associated transmitting ports on the other end of the links are maintained. In the actual implementation, the FIFOs can be designed without true strong ordering. For example, if the FIFO is in the upstream port of a switch, the stored TLPs may be ported to the transmitting portion of different downstream ports of the switch. Consequently, the need to maintain true strong ordering in the FIFO between TLPs porting to different downstream ports is not needed or is indeed practical.

The exact design implementations specific to buffers and FIFOs provided by different vendors are beyond the scope of this book. Here the focus is simply on the concepts of buffers and FIFOs to illustrate Transaction Ordering, Port Arbitration, and VC Arbitration protocols.

Solution to Livelock and Deadlock Situations

In order to avoid livelocks and deadlocks within the transmit and receive buffers of all PCI Express devices, a PCI Express Transaction Ordering protocol is implemented by all the buffers of PCI Express devices.

In the development of PCI, there were extensive behind-the-scenes discussions about bus transaction ordering. To improve platform performance, elaborate protocols were considered, but because of difficulty in understanding and applying them across the platform, they were dropped. PCI finally settled on a relatively simple bus transaction ordering protocol that could be understood and implemented. PCI-X essentially adopted this protocol with minor modifications. As discussed in the previous section, the execution of transactions within a PCI Express platform faces the same issues as bus transaction in PCI and PCI-X platforms; consequently, slightly modified versions of the PCI and PCI-X bus transaction ordering protocol were adopted by PCI Express for its Transaction Ordering protocol. There were two key concepts to the Transaction Ordering protocol selection and modification: First, keep it simple and uniform throughout the platform. Second, make sure the protocol avoids most livelocks and prevents all deadlock situations. For more detailed information and historical background, see the book and specification references in Chapter 1.

PCI Express Transaction Ordering Protocol

Here is a brief description of the PCI Express Transaction Ordering protocol.

Inclusion of Endpoints

As discussed earlier, this book assumes that the buffers of all PCI Express devices implement the PCI Express Transaction Ordering protocol. For the same considerations that applied to PCI designs, this book assumes that this requirement also applies to the buffers in the endpoints' interfaces to the links to ensure compatibility with other PCI Express devices (that is, port interface circuitry between the link and internal ASIC circuitry of any PCI Express device).

Attribute

PCI Express's Transaction Ordering protocol is based on that of PCI and PCI-X; consequently there are two types of transaction ordering: "PCI-like" strong ordering and PCI-X-like" relaxed ordering. The type of transaction ordering to be applied to the packet is selected in the Attribute (ATTRI [1:0] field in the ATTRIBUTE HDW of Header field of packet). ATTRI [1] = 1b defines the "PCI-X-like" relaxed ordering for the packet. ATTRI [1] = 0b defines the "PCI-like" strong ordering for the packet. Strong ordering is the default.

Assignments of relaxed versus strong ordering in the TLPs are as follows:

- **Memory requester transaction packets:** ATTRI [1] = 0b or 1b.
  - ATTRI [1] must = 0b if the memory requester transaction packet is used for a message signaled interrupt.
- **I/O and configuration requester transaction packets:** ATTRI [1] must = 0b.
- **Completer transaction packets:** ATTRI [1] = 0b or 1b.
  - The completer transaction ATTRI [1] must equal the ATTRI [1] of the associated requester transaction packet.
- **Message baseline requester transaction packets:** ATTRI [1] must = 0b.
- **Message vendor-defined and advanced switching transaction packets:** ATTRI [1] = 0b or 1b.

Traffic Class

The transaction ordering is only among the TLPs assigned the same Traffic Class number (TC [2::0] field in the TYPE HDW of Header field of packet). There is no transaction ordering considerations between TLPs assigned different TC numbers. This part of the PCI Express Transaction Ordering protocol gives rise to the following two considerations:

- A PCI Express requester source may assign either relaxed or strong ordering to each requester transaction of a specific TC number.
- The TC number of a completer transaction must be the same as the associated requester transactions.

Transaction Ordering Protocol Implementation

The PCI Express Transaction Ordering protocol (or simply *transaction ordering*) is implemented in several locations within the PCI Express fabric. Transaction ordering is defined as part of the interface between the PCI Express device core (Root Complex, endpoint, or bridge) and Transaction Layer. At the PCI Express device core that will source the requester or completer transaction packet, the associated transactions are transferred to the transmit buffer and processed from the transfer buffer in accordance with transaction ordering. As illustrated in Figures 10.1 (an example of downstream flow) and 10.2 (an example of upstream flow), the transmit buffer is associated with the Transaction Layer. The Gate circuitry must transfer the TLPs from the transmit buffer to the retry buffer with transaction ordering considerations. Once in the retry buffer, the TLPs (contained within LLTPs) are transmitted by strict ordering onto the link. See Chapters 7 and 11 for more information.

At the receiving port of the TLPs' destination (Root Complex, endpoint, or bridge), the TLPs are extracted from the Physical Packets and LLTPs, and placed into the receive buffer. The PCI Express device core retrieves the transactions contained in the TLP in the receive buffer with transaction ordering considerations.

Transaction ordering is also a consideration in the transmit buffers in switches. Figures 10.1 and 10.2 focus on the PCI Express devices of the Root Complex, endpoints, and bridges with the PCI Express device core. The PCI Express device core is implementation-specific. In switches there are entities in the switch that can collectively be viewed as equivalent to the PCI Express device core. These entities in the switches are the receiving portion of the ports and the internal Address Mapping & TC to VC Mapping module between the receiving and transmitting ports as discussed in Chapter 11. As with the Root Complex, endpoints and bridges, the transmit buffer in a switch is associated with the Transaction Layer. The transmit buffer is filled with TLPs ported from other ports of the switch via the Address Mapping and TC to VC Mapping module. The Gate circuitry must transfer the TLPs from the transmit buffer to the retry buffer with transaction ordering considerations. Once in the retry buffer, the TLPs (contained within LLTPs) are transmitted by strict ordering onto the link. See Chapters 7 and 11 for more information.

At the receiving port of a switch, the TLPs are extracted from the Physical Packets and LLTPs, and placed into the receive buffer. For the receive buffer of a switch it is essentially a FIFO. The switch must port the TLPs to the appropriate transmitting port buffer via the Address Mapping and TC to VC Mapping module.

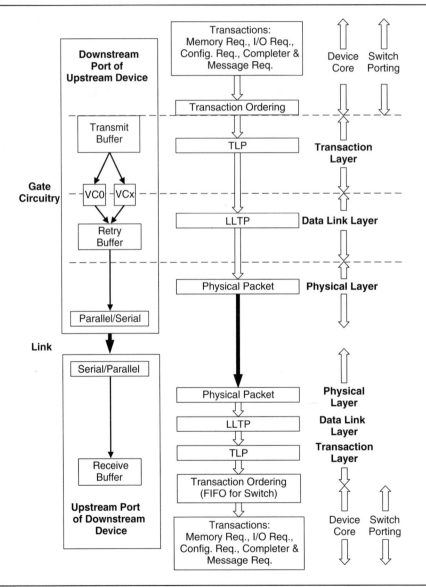

Figure 10.1 Transactions and Packet Downstream Flows in a Link

Chapter 10: Transaction Ordering ■ 509

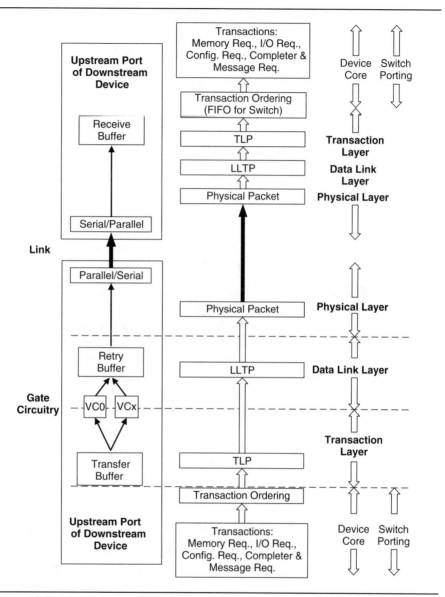

Figure 10.2 Transactions and Packet Upstream Flows in a Link

The switches port TLPs from one port to another. Switches port the TLPs without altering ATTRI [1] or TC [2::0]. As illustrated in Figure 10.3, the TLPs entering the receiving port of a switch may be stored in a FIFO. When sufficient space is available, the TLPs port to the transmitting port buffer of the switch. At the transmitting port buffer of the switch, TLPs of the same TC number are merged and transaction ordering is applied to the transmission order from this buffer.

In Figure 10.3, the buffers of the upstream ports and downstream ports are marked as unshaded and shaded, respectively. Adjacent to the receiving ports (arrows pointing outward) and the receiving ports (arrows pointing inward) are TO, FIFO, VA, and PA references. The references are defined as follows:

- **TO**: TLPS retrieved for transmission onto the link or for use internal to the PCI Express device from the buffer must adhere to transaction ordering (TO).

- **FIFO**: TLPs are stored and retrieved with strong ordering; effectively the TLPs that are first in are the first out. As previously stated, this strong ordering is for discussion simplification. The actual implementation may not have strong ordering among all the TLPs of a FIFO. For example, the TLPs stored in the FIFO in the upstream port of a switch may preserve strong ordering among all TLPs destined for the same downstream port. However, strong ordering may not be preserved between TLPs destined for different downstream ports. Indeed, the FIFO of a switch's upstream port may be divided into independent FIFOs with one FIFO assigned for each group of TLPs destined to a specific downstream port. It is also possible to implement these FIFOs as buffers that adhere to the Transaction Ordering protocol.

- **VA**: TLPs are transmitted onto the link with regard to Virtual Channel Arbitration.

- **PA**: TLPs are transmitted onto the link with regard to Port Arbitration.

At each of the buffers at the transmitting portion of the ports in Figure 10.3 marked with TO, transaction ordering must be considered for transmission onto the link. In the Root Complex and endpoints TO must be considered for internal buffers and the buffers in the receiving ports, respectively. Additionally, the buffers in receiving ports in Figure 10.3 are

Chapter 10: Transaction Ordering ■ 511

marked as FIFO to reflect the strong ordering defined to simplify the discussion in this book.

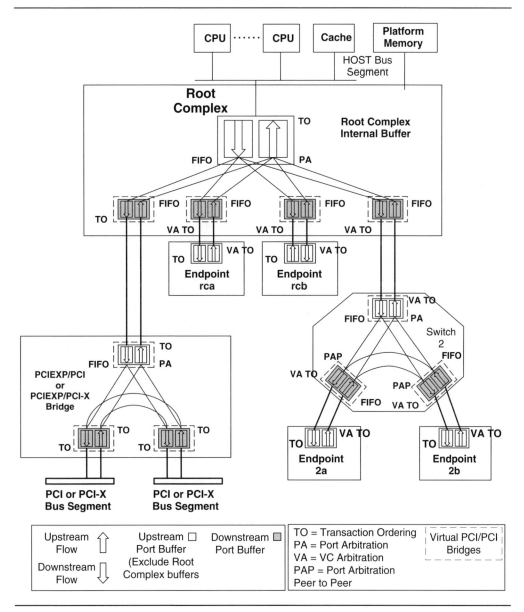

Figure 10.3 TLPs' Buffers throughout PCI Express Fabric

PCI Express/PCI and PCI Express/PCI-X Bridges must never alter the transactions ordering bit ATTRI [1] of transactions being ported through the bridges. PCI Express/PCI and PCI Express/PCI-X Bridges must treat the transactions ordering for their own porting buffers protocol as if the transactions ordering bit ATTRI [1] = 0b. That is, they must always execute by strong ordering.

The transaction ordering is only defined for TLPs assigned the same TC number. See Chapters 11 and 12 for other considerations relative to the exact transmission order onto a link due to other elements of Virtual Channel Arbitration and Port Arbitration related to the Flow Control protocol. These are identified by the VAs and PAs in Figure 10.3, but will not be discussed in this chapter.

Specifics

Part of the consideration is the type of transaction ordering assigned to each of the TLPs. As previously discussed, the ATTRI [1] bit in the Header field of the TLPs define "PCI-like" strong ordering or "PCI-X-like" relaxed ordering. The following discussion and Tables are provided to explain the transaction ordering implementation for buffers containing a mixture of TLPs with strong ordering and relaxed ordering.

The focus of the following discussions is requester transactions between upstream (U) PCI Express requester sources and downstream (D) PCI Express requester destinations. This includes completer transactions between upstream (U) PCI Express completer sources and downstream (D) PCI Express completer destinations.

For the following discussion of transaction ordering, the example is of TLPs moving from upstream PCI Express devices to downstream PCI Express devices. The transaction ordering also applies to transaction packets moving from downstream PCI Express devices to upstream PCI Express devices, except as noted in the tables. Note: Peer-to-peer transaction packets have upstream considerations until the switch inflection point when the transaction packets have downstream considerations.

The tables determine which buffered TLPs may pass, cannot pass, or must pass other buffered TLPs when sufficient link bandwidth is available. Tables 10.1 and 10.2 are used as follows:

- The Reference Transaction columns list the possible different transaction packet types that have been buffered.

- Subsequent Transaction rows list the possible different transaction packet types that must be placed in the buffer relative to the existing Reference transactions.

- The transaction ordering of the Subsequent Transaction determines its ordering relative to those transaction packets in the buffer associated with the Reference Transaction column. Once a transaction packet is placed in the buffer it is defined by the Reference Transaction column.

- Table 10.1 is used when the subsequent transaction packet is labeled as strong ordered (ATTRI [1] = 0b). Table 10.2 is used when the subsequent transaction is labeled as relaxed ordering (ATTRI [1] = 1b).

- The *May Pass* entries indicate a totally optional interpretation of the transaction ordering by the buffer.

Table 10.1 Strong Ordering in Subsequent Transaction

Subsequent Transaction vvv	Reference Transaction				
	Requester (Posted) Memory Write or Message (U) > (D)	Requester Memory Read (U) > (D)	Requester I/O or Config. Write (U) > (D)	Completer for Read (U) > (D)	Completer for Write (U) > (D)
Requester Transaction: (Posted) Memory Write or Message (U) > (D)	Cannot pass	Must pass	Must pass	Must pass (1) (2)	Must pass (1) (2)
Requester Transaction: Memory Read (U) > (D)	Cannot pass	May pass	May pass	May pass	May pass
Requester Transaction: I/O or Config. Write (U) > (D)	Cannot pass	May pass	May pass	May pass	May pass
Completer Transaction for Read (U) > (D)	Cannot pass	Must pass	Must pass	May pass (3)	May pass
Completer Transaction for Write (U) > (D)	May pass	Must pass	Must pass	May pass	May pass

Table Notes:

1. For the memory write and message transactions flowing downstream (PCI Express link to PCI or PCI-X bus segment) through a PCI Express/PCI or PCI Express/PCI-X Bridge, the transaction ordering must be set as strong (ATTRI[1] = 0b).

2. For the memory write and message transactions sourced from endpoints, switches or the Root Complex flowing upstream or downstream the transaction ordering can optionally be set as relaxed ordering (ATTRI[1] = 1b).

3. This is a special consideration: Completer transactions associated with a "specific" requester read transaction defined by the Transaction ID = Requester ID plus TAG#; cannot pass other completer transactions associated to the same "specific" requester read transaction. Completer transactions associated with a "specific" requester read transaction may optionally pass other completer transactions associated with the other requester read transactions.

Table 10.2 Relaxed Ordering in Subsequent Transaction

Subsequent Transaction Activity vvv	Reference Transaction				
	Requester (Posted) Memory Write or Message (U) > (D)	Requester Memory Read (U) > (D)	Requester I/O or Config. Write (U) > (D)	Completer Transaction for Read (U) > (D)	Completer Transaction for Write (U) > (D)
Requester Transaction: (Posted) Memory Write or Message (U) > (D)	May pass	Must pass	Must pass	May Pass (1) (2)	May Pass (1) (2)
Requester Transaction: Memory Read (U) > (D)	Cannot pass	May pass	May pass	May pass	May pass
Requester Transaction: I/O or Config. Write (U) > (D)	Cannot pass	May pass	May pass	May pass	May pass
Completer Transaction for Read (U) > (D)	May pass	Must pass	Must pass	May pass (3)	May pass
Completer Transaction for Write (U) > (D)	May pass	Must pass	Must pass	May pass	May pass

Table Notes:

1. For the memory write and message transactions flowing downstream (PCI Express link to PCI or PCI-X bus segment) through a PCI Express/PCI or PCI Express/PCI-X Bridge, the transaction ordering must be set as strong (ATTRI[1] = 0b).

2. For the memory write and message transactions sourced from endpoints, switches, or the Root Complex flowing upstream or downstream the transaction ordering can optionally be set as relaxed ordering (ATTRI[1] = 1b).

3. This is a special consideration: Completer transactions associated to a "SPECIFIC" requester read transaction (defined by the Transaction ID = Requester ID plus TAG#; see Chapter 4 for more information) cannot pass other completer transactions associated to the same "SPECIFIC" requester read transaction. Completer transactions associated to a "SPECIFIC" requester read transaction may optionally pass other completer transactions associated with the other requester read transactions. See the following section "Special Considerations" for more information.

Special Considerations

There are special considerations that arise under certain conditions that affect how the data is read or written.

Completer Transactions that May Pass

As discussed in Note 3 of Table 10.2, a set completer transaction packets associated with a "SPECIFIC" requester read transaction will be received by the source of the requester read transaction packet in the exact order transmitted by the source of the completer transaction packets. If a "SPECIFIC" requester read transaction is accessing 1024 bytes and the set of completer transaction packets consists of four packets with each containing a block of 128 bytes, the order of the blocks received will be in strict increasing addressing order. If the "SPECIFIC" requester read transaction is done as two requester read transactions, the set of two completer transaction packets associated with one requester transaction will be received before, after, or between the set of two completer transaction packets associated other requester transaction. That is, not all of the 128 byte blocks received will be in strict increasing addressing order.

Granularity of Reading

The PCI Express specification does not place any requirement on read granularity, but it should be noted. Consider the following example of reading platform memory by a PCI Express device. The PCI Express device executes a single read transaction accessing to a very large block of platform memory. As portions of the block are read from the platform memory, the CPU on the HOST bus segment may be writing to other portions of this large block of platform memory. Consequently, some of the data returned to the PCI Express device may be old data and some will be data just written, but all data of the large block will not be of the same relative time frame.

Granularity of Writing

The PCI Express specification does not place any requirement on write granularity, but it should be noted. The consideration is the same as in the above read example except the PCI Express device is writing and the CPU on the HOST bus segment is reading Platform Memory.

When one PCI Express device writing to another PCI Express device with a single memory write requester transaction with multiple DWORDs and Relaxed Ordering bit = 0b, the subsequent update to the associated buffers must consider the following. If a bridge is in the path of the write requester transactions, the write data may be combined per the PCI or PCI-X specifications. Consequently, the updates to the address range of the buffer must be in increasing address order. The granularity of these updates is not defined by the PCI Express specifications.

Cachelines

The cacheline boundaries and associated sizes defined by the CPUs on the HOST bus segment relative to the Platform Memory and other PCI Express resources are not defined or used as a boundary references by other PCI Express devices for transactions.

Chapter 11

Flow Control Protocol: Part 1

Chapters 11 and 12 describe the Flow Control protocol of PCI Express.

Introduction to the Flow Control Protocol, Part 1

As previously discussed in this book, it may not be possible to connect all endpoints and bridges directly to the single Root Complex. In order to provide sufficient endpoints and bridges for all applications, switches are added in a hierarchical fashion via links, as shown in Figure 11.1. Each link can be constructed of 1 to 32 lanes. Each lane contains a set of two differentially driven pairs of signal lines. One pair drives an 8/10-bit encoded bit stream as symbols in one direction, the other pair drives the 8/10-bit encoded bit stream as symbols in the other direction. Each link in the PCI Express fabric may implement a different number of lanes, and thus the available link bandwidth varies from link to link.

As discussed in Chapter 2, Transaction Layer Packets (TLPs) are contained within Link Layer Transaction Packets (LLTPs), which in turn are contained in Physical Packets. Only the LLTPs and Physical Packets traverse the individual links. The TLPs are the only entity that traverses the transaction's source and transaction's destination. As discussed in Chapter 10, the TLPs are sourced from different PCI Express devices and are merged at the buffers of transmission ports within switches. Buffers are contained within the downstream ports of the Root Complex, and the upstream ports of the endpoints and bridges. The TLPs within these transmission buffers must vie for link bandwidth.

The Flow Control protocol addresses the issues of different bandwidths of different links and the impact of TLPs within buffers of transmission ports vying for link bandwidth throughout the PCI Express fabric.

As discussed in Chapter 8, it is possible to configure more than one link between one downstream port and the upstream ports of two or more PCI Express devices. In order to simplify the discussion, this chapter assumes only a single link of multiple lanes between any two PCI Express devices in a PCI Express fabric.

The number of lanes per each link is a physical implementation of the platform. Part of the Flow Control protocol that will be discussed in this chapter is the implementation of Virtual Channels (VCs) on each link. There are one or more VCs that are virtually defined to each link, each with a unique number (VC0, VC1, ...VCN).

Flow Control Protocol Elements Overview

The transactions defined by the Requester/Completer protocol are executed between PCI Express devices contained within packets. As shown in Figures 11.2 and 11.3, the requester or completer transactions are first processed at the sources through the PCI Express layers before conversion to a serial bit for transmission onto the links. At the destinations, the received serial bit stream is processed through the PCI Express layers to recover the transaction.

Assume that the Root Complex is connected to an endpoint as illustrated by Figure 11.2. In the *simplest implementation* at the requester transaction's source, only transaction ordering is considered at the interface between the PCI Express device core and the PCI Express Layers. At the requester transaction's destination the PCI Express device core only considers transaction ordering as transactions are retrieved from the receive buffer. The same consideration applies in the reverse direction for the completer transaction.

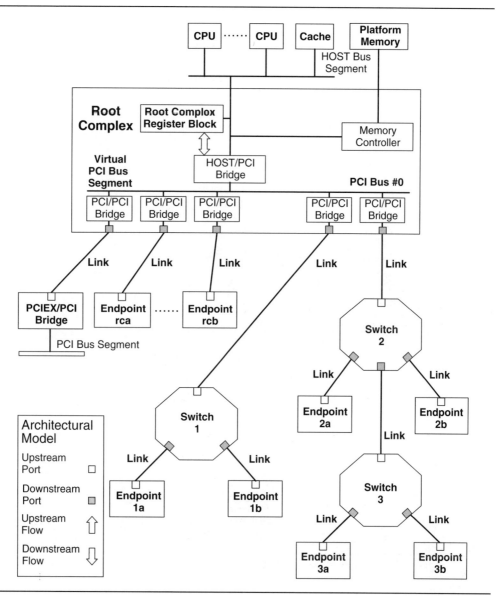

Figure 11.1 PCI Express PC Platform Model with Virtual Channels

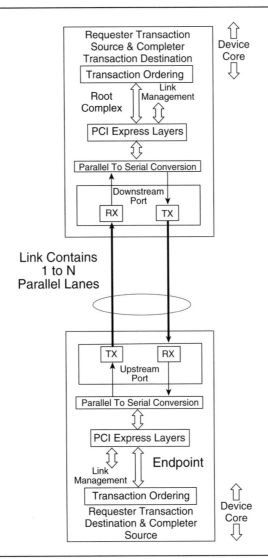

Figure 11.2 TLPs Flow Downstream to One Endpoint

Assume that the Root Complex and two endpoints are separated by a switch as illustrated in Figure 11.3. In the *simplest implementation* at the requester transaction's source, only transaction ordering is considered between the PCI Express device core and the PCI Express Layers. The associated TLPs are ported through a switch to one of two endpoints. At the requester transaction's destination, the PCI Express device core only considers transaction ordering as transactions are retrieved from the receiver buffer. The considerations that apply in the reverse direction for the completer transactions sourced from the two endpoints are more complex. Within each endpoint the *simplest implementation* at these completer transaction sources, only transaction ordering is considered between the PCI Express device core and the PCI Express Layers. At the switch, TLPs are received by the downstream ports and stored in FIFOs to preserve the transaction ordering implemented by the endpoints, as shown in Figure 11.4. The two FIFOs port TLPs to the upstream port buffer. A Flow Control protocol issue arises as the TLPs are transmitted upstream to the Root Complex. As discussed in Chapter 10, the transaction ordering between TLPs of the same TC numbers in the upstream port's buffer must be considered by the transmitting port.

In these examples, the simplest implementation considered only the transaction ordering, including only transaction ordering at the upstream port of the switch. As discussed in Chapter 10, transaction ordering is only between TLPs with the same Traffic Class (TC) number. But what if the buffer in the upstream port of the switch contains many TLPs with different TC numbers? What if the upstream link to the Root Complex has limited bandwidth for the TLPs in the buffer in the upstream port waiting to be transmitted? One solution is to make the switch's upstream port buffer a FIFO for transmission considerations other than transaction ordering. However, doing this would prevent TLPs associated with certain transactions to be ported and transmitted through the switch more quickly. If preference cannot be given to certain transactions it would be impossible to implement Quality of Service considerations in the PCI Express platform. For example, isochronous transfers could not be implemented unless the associated TLPs were given priority in flowing across the PCI Express fabric.

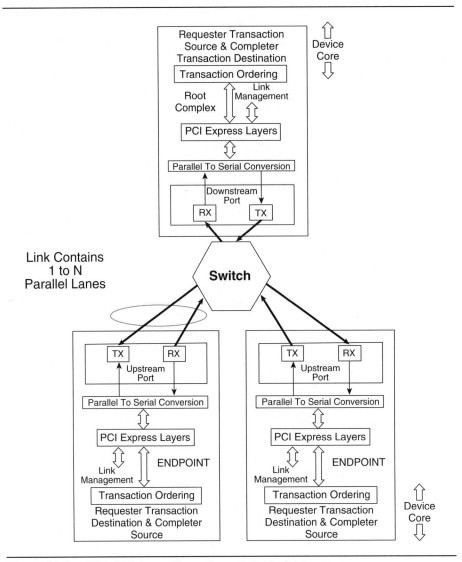

Figure 11.3 TLPs Upstream Flows from Two Endpoints

Another consideration is the buffer space available at the receiving ports. As illustrated in Figure 11.2, the requester and completer sources considered only transaction ordering without taking into account the buffer space available at the receiving port. There was no consideration of the buffer space available at the receiving port of the Root Complex. Obviously, a specific TLP cannot be transmitted across the link if it is larger than the available buffer space at the Root Complex. The available buffer space at the receiving ports of switches, endpoints, and bridges have similar considerations.

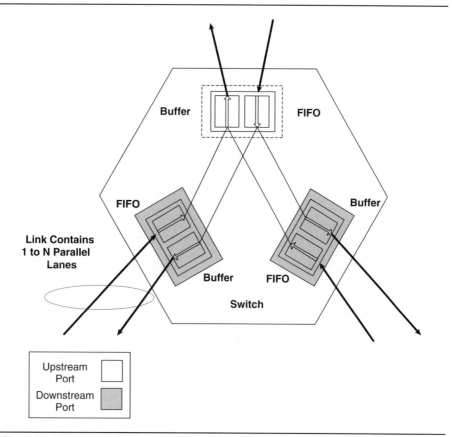

Figure 11.4   TLPs within a Switch

If the transaction ordering permits several different transactions to be processed through the PCI Express layers, selecting one that would fit in the buffer space at the receiving port would be important. As illustrated in Figures 11.3 and 11.4, the switch only considers the transaction ordering between the buffered TLPs.

Consequently, when examining TLPs flowing through a PCI Express fabric, other factors beyond transaction ordering must be considered. These other factors must be addressed by the Flow Control protocol.

Elements of the Flow Control Protocol

The Flow Control protocol elements are Traffic Classes, Virtual Channels, Buffer Management, LLTP tracking, Port Arbitration, Virtual Channel Arbitration, and Application for Quality of Service. This chapter focuses on the foundation of these protocol elements of the Flow Control protocol. That is, before TLPs can flow throughout a PCI Express fabric each one must be assigned a Traffic Class number. Also, each link must be assigned a set of Virtual Channels to provide virtual parallel paths across the link. The switches the TLPs will port through must have TLPs assigned to specific Virtual Channels (via Traffic Classes) on the connected links for subsequent transmission from the switch. Finally, prior to a PCI Express device's port transmitting a Physical Packet containing an LLTP that in turn contains a TLP across a link, there must be sufficient buffer space at the receiving port of the PCI Express device connected on the other end of the link.

Once the foundation elements of the Flow Control protocol are established, the next step is to transmit TLPs across the link as LLTPs contained within the Physical Packets. To ensure the correct receipt of the LLTPs at the other end of the link, the Flow Control protocol includes the LLTP tracking. Next the Flow Control protocol must consider the flow of TLPs throughout the PCI Express fabric beyond a specific link. For each transmitting port of a PCI Express device there may be buffered TLPs with bandwidth needs greater than the instantaneous bandwidth of the connected link. Consequently, in addition to transaction ordering there must be an arbitration protocol. The Flow Control protocol elements that provide arbitration are Port Arbitration and Virtual Channel Arbitration. Associated with the link bandwidth and arbitration for link bandwidth is their application to achieve a Quality of Service. These elements will be discussed in Chapter 12.

This chapter is divided into three parts: The first part introduces the concepts of TLPs flowing through a PCI Express fabric and the associated elements of Virtual Channels and Traffic Classes to assist in this flow. The second part considers that buffers that exist on both ends of a link. A TLP contained in one buffer of the transmitting port on the link cannot be transmitted to a buffer in the receiving port on the other end of the link unless buffer space is available. Consequently there has to be buffer management. The third part discusses other considerations relative to which TLP in a transmit buffer is selected to be the next TLP transmitted across the link. The TLP is contained within a LLTP which in turn is transmitted across the link in a Physical Packet.

Referring to Figure 10.3 reproduced here as Figure 11.5, part of the Flow Control protocol is characteristics of the TLP buffers and FIFOs throughout the PCI Express fabric. As discussed in Chapter 10, the TO of each buffer defines adherence to transaction ordering (TO). The VA of each buffer defines adherence to Virtual Channel Arbitration. The PA of each buffer defines adherence to Port Arbitration. The PA concept is also applied to peer-to-peer transaction (PAP). The FIFO defines adherence to strong ordering.

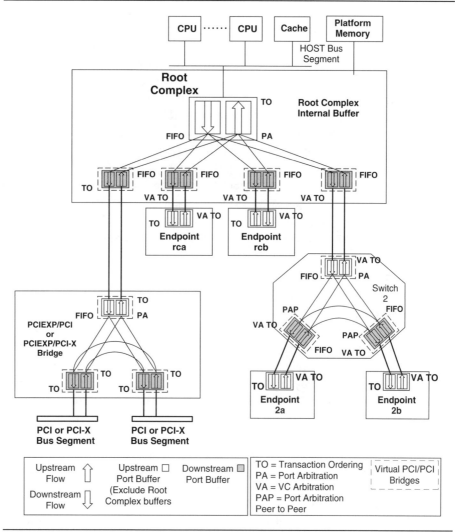

Figure 11.5   TLPs' Buffers throughout the PCI Express Fabric

Flow Control Protocol: Virtual Channels and Traffic Classes

Integral to the discussion of the Flow Control protocol are Virtual Channels and Traffic Classes. These concepts are unique to the PCI Express specification and are not defined in PCI and PCI-X platform specifications.

Flow of Transaction Packets and Definition of *Packet* in This Chapter

The discussion in this chapter focuses on the flow of Transaction Layer Packets (TLPs). As previously discussed, the Physical Packets contain Link Layer Transaction Packets (LLTPs), which in turn contain the TLPs. The Physical Packets and LLTPs only traverse ports on a link. The Physical Packets and LLTPs do not port through a switch or a bridge. It is only the TLPs that traverse a link, port through a switch, and pass through the port of the endpoint, bridge, or Root Complex. Consequently, the term *packet* in this chapter only refers to the TLP(s) unless otherwise noted. See Chapter 7 and 8 for more information about LLTPs and Physical Packets, respectively.

The possible flows of packets within a PCI Express fabric are upstream, downstream, and peer-to-peer through switches. Additional considerations are packets flowing between endpoints and the Root Complex (no intervening switches), and packets flowing between bridges and the Root Complex (no intervening switches).

Packets Flowing Upstream

Consider the upstream flow of packets from the upstream ports of endpoints to the Root Complex's downstream ports via switches, as illustrated in Figure 11.1. A stream of packets flow from endpoint 1a upstream to switch 1 and another stream of packets flow from endpoint 1b upstream to switch 1. The packets flow on different links. The number of lanes within each of these two links may be the same or different. The flow rates of the packet streams of each endpoint may be the same or different; this is particularly true with the inclusion of data in the packet. Consequently, the flow rates of the two packet streams into switch 1's downstream ports are most likely different. The two streams of packets into switch 1's downstream ports merge at switch 1's upstream port. Flow Control protocol is needed to determine which packet has priority to be transmitted by switch 1's upstream port.

The packets' upstream flow is further complicated by the hierarchy of switches. The upstream flow rates of the packets from endpoints 2a and 2b and their relative priority to be transmitted by switch 2's upstream port is one level of flow control complexity. However, the upstream flow of the packets from switch 3's upstream port into switch 2's downstream port may make flow control in switch 2 more complex. The upstream packets from switch 3 are from sourced endpoints 3a and 3b. Prior to transmission of any packets from switch 2 upstream to the Root Complex, the relative priority of packets merged from four endpoints (2a, 2b, 3a, and 3b) must be understood. Part of the Flow Control protocol must determine which packets have priority to be transmitted by switch 2's upstream port.

Packets porting from a switch's downstream ports to the upstream port cannot be snooped by other downstream ports of the switch. That is, the upstream flowing packet is not copied or observed by other downstream ports of a switch.

Packets Flowing Downstream

Consider the downstream flow of packets from the Root Complex to an endpoint via switches in Figure 11.1. Packets flowing to endpoints 1a and 1b are received from the Root Complex's downstream port to switch 1's upstream port. Switch 1 ports the packets to its downstream ports connected to endpoints 1a and 1b. The Flow Control protocol considerations are minimal.

Packets porting from a switch's upstream to a specific downstream port cannot be snooped by other downstream ports of the switch. That is, the downstream flowing packet is not copied or observed by other downstream ports of the switch that are not the downstream port selected by addressing, Implied Routing, or ID Routing with one exemption: one type of Implied Routing broadcasts a message transactions packet porting downstream from a switch's upstream port to all of the switch's downstream ports. See Chapter 4 for more information on Implied Routing.

Packets Flowing Peer-to Peer

An additional packet downstream flow consideration is peer-to-peer packets flowing between two downstream ports of a switch, illustrated in Figure 11.1. For switch 1, peer-to-peer packet flow is between endpoints 1a and 1b. For switch 2, peer-to-peer packet flow is between endpoints 2a and 2b. In the case of switch 2, a third link is connected to switch 3 and not directly to an endpoint. Switch 3 represents endpoint 3a and 3b, and can cause peer-to-peer packet flow to endpoints 2a or 2b.

Peer-to-peer packet flows are possible between endpoints 2a and 3a, endpoints 2a and 3b, endpoints 2b and 3a, and endpoints 2b and 3b.

The packets flowing downward from switch 1's upstream port, and packets flowing peer-to-peer merge at switch 1's downstream ports. Part of the Flow Control protocol must determine which packets have priority to be transmitted by switch 1's downstream ports.

As packets port through a switch from one downstream port to another, these packets cannot be snooped by a third downstream port or the upstream port. For example, peer-to-peer packets flowing between endpoint 2a and 2b cannot be snooped to the switch 3's downstream port representing endpoints 3a and 3b and its upstream port.

Packet Flow Summary

Although the discussion of packet flow, packet merging, and packet transmission priorities up to this point has focused on switches and endpoints, the same packet flow behavior applies to the Root Complex and bridges. For packet flows upstream, downstream, and peer-to-peer:

- For an endpoint or bridge connected directly to a Root Complex's downstream port
  - Packets flowing to the Root Complex are defined as flowing upstream.
  - Packets flowing from the Root Complex are defined as flowing downstream.
- For an endpoint or bridge connected directly to a switch's downstream port and not a Root Complex's downstream port
  - Packets flowing to the switch are defined as flowing upstream.
  - Packets flowing from the switch are defined as flowing downstream.
- The peer-to-peer packet flow outside of the switch can be defined as part of a packet flow upstream and a packet flow downstream. For example, a packet from endpoint 3a flows upstream to switch 3, it ports between switch's 3 downstream ports, and it flows downstream to endpoint 3b. The packet flow upstream and downstream can also be collectively defined as a peer-to-peer packet flowing between endpoints 3a and 3b.

Virtual Channels

The PCI Express specification has introduced a new concept called the *Virtual Channel* (VC). The ports of the Root Complex, switches, bridges, and endpoints map all the VCs into the connected links. Each link's hardware, in the form of lanes, provides the link's bandwidth. VCs are a concept and are not directly implemented in hardware. The purpose of VCs is to prioritize the packets that are flowing through the port and link. Part of this prioritization is dependent on Traffic Classes. The present discussion focuses on VCs. Traffic Classes will be discussed later.

Consider each link as a set of VCs, as shown in Figure 11.1. VCN represents the VC highest number defined for the link. The number of VCs defined in this revision of the PCI Express specification is one to five, as follows: VC0, VC1, VC2, VC4, and VC8. VC0 must be supported by all links. For simplicity, the PCI Express specification requires matched pairs of VCs in each link. A downstream VC0 is complemented by an upstream VC0, a downstream VC1 is complemented by an upstream VC1, and so forth. As previously mentioned, the physical size of the link hardware is reflected in the number of lanes, the number of VCs associated with a link is concept and independent of the number of lanes in a link. VCs are created at each port by the VC ID (in the configuration register block) with the mapping of one or more Traffic Class numbers to each VC number by configuration software.

VCs are used to establish the priority of packets flowing out of switches' ports (upstream or downstream), Root Complex's downstream ports, or endpoints' upstream ports (defined as transmitting). In switches, the packets are ported to the transmitting ports from the switches' other ports (receiving). The switches' receiving ports are the downstream ports receiving packets from endpoints, bridges, and other switches. Also, switches' receiving ports include their upstream port receiving packets from the Root Complex and other switches. For the Root Complex, the packets are transmitted and received by its downstream ports. For the endpoints and bridges, the packets are transmitted and received by their upstream ports.

For the upstream port of a bridge, only VC0 can be defined; consequently, no VCs can be used for prioritizing the packets transmitted.

VC Protocol for Ports and Links

The basic protocol that defines VCs at the ports and connected links throughout the PCI Express fabric can be summarized as follows:

- Each VC number is a virtual interconnect between an endpoint or bridge and a switch, an endpoint or bridge and the Root Complex (no intervening switches), between two switches (no intervening switches), or the Root Complex and a switch (no intervening switches).

- Within a specific port of the Root Complex, switches, bridges, or endpoints each VC number is independent of other VCs with different numbers except for VC Arbitration. For example, VC0 is independent of VC2 and VCN.

- The minimum defined VC number for a port is VC0. If another VC number is defined for a port it can be VC1, VC2, or VCN; but it does not have to be contiguous. Presently the possible VC numbers are defined as VC0, VC1, VC2, VC4, and VC8. The upstream port of a bridge can only implement VC0 and no other VC numbers.

- The VC numbers supported at the downstream port of a link map one-to-one with the upstream port connected to the same link. That is, no VCs are created, destroyed, or merged within a link.

VC Protocol within Root Complex

The basic protocol that defines VCs within the Root Complex can be summarized as follows:

- VCs on the Root Complex's downstream ports with the same or different VC numbers are not connected internally.

- No peer-to-peer packet flow is typically defined between downstream ports of the normal Root Complex as there would be for a switch. As previously discussed, to support packet peer-to-peer flow, a virtual switch must be built into the Root Complex. If the Root Complex optionally implements peer-to-peer transition flow the operation for the internal virtual switch is the same as for the external switches. To simplify the discussion, this book assumes that the Root Complex does not support peer-to-peer transactions.

VC Protocol within Switches

The basic protocol that defines VCs within switches can be summarized as follows:

- Each VC number on a switch's upstream port operates independent of the VCs of the same number on the downstream ports. For example, packets flowing through the link connected on the switch's downstream port on VC1 may or may not be mapped to on the switch's upstream port VC1. As illustrated in Figure 11.6, VC numbers are *not* unique virtual interconnections through a switch.

- Within a switch, the packets are mapped to a specific VC number on each upstream port and each downstream port is dependent on the mapping of Traffic Class (TC) numbers to VC numbers for each port. Collectively, this mapping is done in the Address Mapping and TC to VC Mapping module of the switch. This causes packets associated with a specific TC to be transmitted by the port on a specific VC. The purpose of the Traffic Class numbers will be discussed in the next section.

> The figures in this chapter identify the packets by their Traffic Class numbers (TC numbers). In Figure 11.6, a generic range of [M::N] is listed for each VC number. This represents a range of numbers and does *not* represent any specific TC number on one VC number of one port associated with a specific TC number on a VC number of another port.

534 ■ The Complete PCI Express Reference

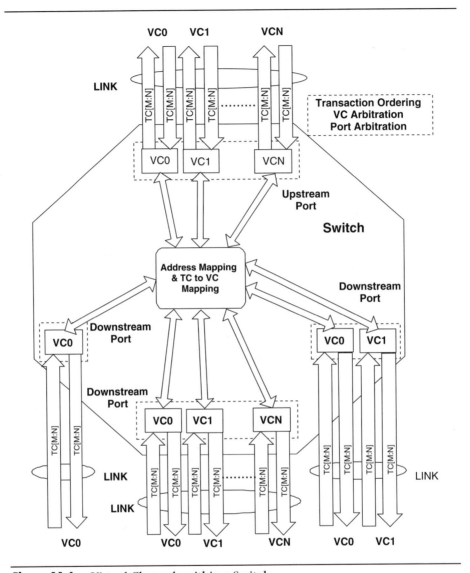

Figure 11.6 Virtual Channels within a Switch

Assignment of Traffic Class Numbers to VC Numbers in Switches

The Traffic Class (TC) numbers assigned to a specific VC number on a link are not always assigned to the same VC number on another link. The assignment of TC numbers to VC numbers is part of the Flow Control protocol. The following examples focus on switches. As illustrated in Figures 11.6, 11.7, 11.8, and 11.9, the assignment of TC numbers to VC numbers is done in the Address Mapping and TC to VC Mapping module. Address Mapping within the module is a separate activity and is not part of TC to VC Mapping. The mapping of TC numbers to VC numbers is only associated with the transmitting ports of the switches. An example of this assignment (or mapping) in a switch follows.

Symmetrical Mapping Figure 11.7 illustrates an example of the symmetrical approach to TC to VC Mapping within a switch. The same set of TC numbers is assigned to the same VC number on each port. The packets assigned a specific TC number exit the switch on the same VC number on which they entered the switch. This is defined as *symmetrical mapping*. Symmetrical mapping results in VC symmetry (all VC numbers on the downstream ports exist on the upstream port) and TC symmetry (assignment of specific TC numbers to specific VC numbers is retained at all ports). VC symmetry does not always define TC symmetry. For example, if TC3 on the upstream port is mapped to VC4 instead of VC1 as in Figure 11.7, the packets' upstream flow is received on the switch's downstream port on VC1 and is transmitted on the switch's upstream on VC4. The changing of VCs in the switch used by packets assigned to a specific TC number is defined as *asymmetrical mapping*.

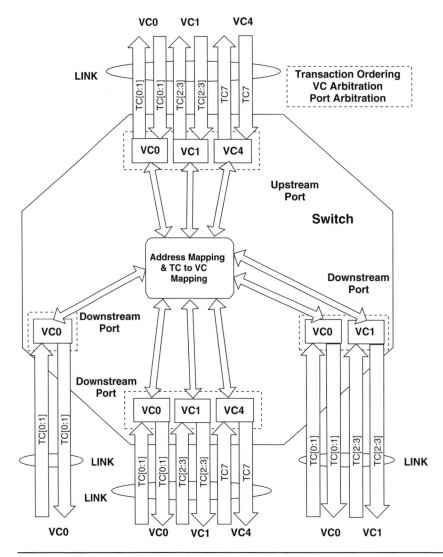

Figure 11.7 Symmetrical Mapping of Virtual Channels and Traffic Classes

Asymmetrical Mapping Figure 11.8 illustrates an example of an asymmetrical mapping of TCs to VCs within a switch. TC numbers may or may not be mapped to the same VC number on each port. In this example, TC [6:7] are mapped to VC6 on the upstream port and VC4 on one of the downstream ports. Also, TC4 is mapped to VC4 on the upstream port and VC1 on the downstream port. In a packet flow upstream, the packets associated with TC6 or TC7 are received by the switch on VC4 and are transmitted on VC6. Also, in an upstream flow, packets associated with TC4 are received by the switch on VC1 and are transmitted by the switch on VC4. The reverse is true for the packet flow in the downstream direction.

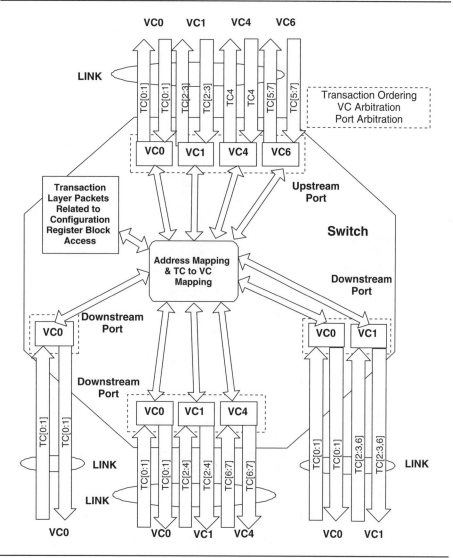

Figure 11.8 Asymmetrical Mapping of Virtual Channels and Traffic Classes

Figure 11.9 shows an example of another asymmetrical mapping of TCs to VCs within a switch. TC numbers may or may not be mapped to the same VC number on each port. In this example, TC [6:7] are mapped to VC1 on the upstream port. On the downstream ports, TC [6:7] are mapped to VC4. Packets flowing upstream associated with TC6 or TC7 are received by the switch on VC4 and are transmitted by the switch on VC1. The reverse is true for packets flowing downstream.

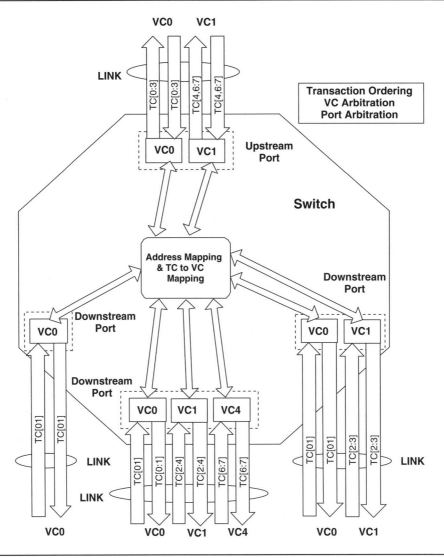

Figure 11.9 Asymmetrical Mapping of Virtual Channels and Traffic Classes

Peer-to-Peer Mapping This TC to VC Mapping for packets flowing upstream and downstream also applies to peer-to-peer packet flows. The assignment of TC numbers to VC numbers for peer-to-peer flows is done in the Address Mapping and TC to VC Mapping module independent of the upstream and downstream flows. Address Mapping within the module is a separate activity and is not part of TC to VC Mapping. The mapping of TC numbers to VC numbers is only associated with the transmitting ports of the switches.

Assignment of Traffic Class Numbers to VC Numbers in Non-Switch PCI Express Devices

As shown in Figure 11.1, the multiple VCs assigned to ports on switches also exist on the ports of the Root Complex, endpoints, and bridges. Obviously, only the switch contains the Address Mapping and TC to VC Mapping module. However, the ports of the Root Complex and the endpoints do map packets with specific TC numbers to a specific VC number. In the case of the bridge, only VC0 can be defined for the connected link; consequently, no TC to VC mapping occurs except that the default of all TC numbers maps to VC0. There is further discussion of TC number mapping in non-switch PCI Express devices later in this chapter.

Additional Considerations

The discussion to this point has provided the basics of the concepts. There are some additional considerations in order to understand these concepts fully and correctly.

Packet Sources

The discussion so far has focused on packets flowing through switches sourced from the Root Complex, bridges, and endpoints. Switches may also source packets (internally) and not simply port packets from other PCI Express devices. The packets sourced by switches are those containing message baseline requester transaction packets and completer transaction packets for a configuration register block access. The sourced packets of switches are assigned TC numbers and merged with other packets received by the switch for transmission from the switch. Also, other message baseline requester transaction packets are received on the switches' downstream ports and combined into representative message baseline requester transaction packets to be transmitted by the switches' upstream port.

Address Mapping

As illustrated in Figures 11.6 through 11.9, for switches, the TC to VC mapping is centered within the switch to map packets from any receiving port to any transmitting port. In addition to TC to VC mapping, switches must also map packets from receiving ports based on their addresses (Address Mapping). To support both of these functions, an Addressing Mapping and TC to VC Mapping module is defined for each switch.

Packets contain address information to identify the destinations of the packets. In addition to the TC to VC mapping for each port, the module also contains information about each port for the address information that is provided in each packet.

For memory and I/O requester transaction packets, the address information is the actual address within an address space. For configuration and completer transaction packets, and some types of message requester transaction packets, the address information is based on ID Routing. Consequently the module must know the address space and BUS# ranges of each link connected to the switch.

The module must also be aware of address information based on Implied Routing information in the r field of the Type HDW in the Header field. Implied Routing is used by some types of message requester transaction packets. Also, the module must also be aware of address information based on RID (Route Identifier). Message advanced switching requester transaction packets use RID.

The ports of the Root Complex, bridges, and endpoints support TC to VC mapping but do not support address mapping.

Priority and VC Arbitration

The packets to be transmitted are assigned to a specific VC number on the link. The link bandwidth is limited; consequently, a priority between VC numbers assigned to a link must be established. The arbitration between VC numbers for priority transmission on a link is called VC Arbitration.

As shown in Figures 11.6 through 11.9, for switches, VC Arbitration occurs in the transmission portion of the switches' ports. For other PCI Express devices sourcing packets, VC Arbitration also occurs in the transmission portion of their ports.

The VC Arbitration protocol is discussed in more detail in Chapter 12.

Traffic Classes

As discussed in the first part of this chapter, packets (TLPs) come from multiple sources. As the packets flow through the PCI Express fabric, they are merged with other packets in buffers. All of the packets within a buffer ready for transmission onto a link must vie for link bandwidth. The method to determine transmission priority (that is, use of link bandwidth) between VC numbers is VC Arbitration. Packets are assigned by TC to VC mapping to a specific VC number based on the TC numbers assigned to those packets. In addition, the TC numbers assigned by the packets' sources determine transaction ordering of the packets at any port.

The TC number assigned by the packet's source therefore provides a priority indicator for the packet that can be used throughout the PCI Express fabric. Also, software can program the VC numbers assigned to a link and to program the mapping of TC numbers to specific VC numbers. In this way, the software can fine-tune the performance of the entire PCI Express fabric.

Traffic Classes to Virtual Channel Mapping Protocol

PCI Express devices that source packets (and not simply port them) are the Root Complex, bridges, endpoints, and switches. In that the primary function of switches is to port packets from link to link, this discussion focuses on the Root Complex, endpoints, and bridges as packet sources. It is implied that switches can also internally source packets for completer transactions' access to their configuration register blocks or for message baseline requester transactions as previously discussed.

The Root Complex, endpoints, and bridges assign each packet (TLP) a Traffic Class (TC) with a three-bit field contained in the Header field, as shown in Figure 11.10. The TC numbers are therefore available in all types of packets.

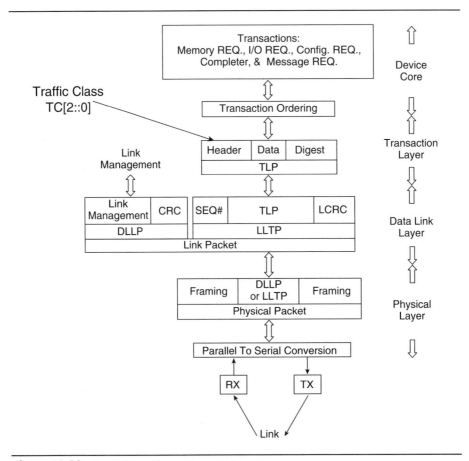

Figure 11.10 Overview of PCI Express Layers

The three bits define eight TC numbers ([TC0::TC7]). The minimum required TC number for the packet is TC0 and it is hardwired by all ports to VC0. The use of all other TC numbers and VC numbers is optional, and their implementation is defined by the PCI Express *Virtual Channel Capability Structure* (PVCCS) in the configuration register block.

Each port of the Root Complex, switches, and endpoints independently supports or does not support the PCI Express Virtual Channel Capability Structure. If supported, the configuration software must ensure that the VC numbers at each port on a link are symmetrical, and also must ensure that the TC numbers assigned are compatible between the two ports on the link. If a port does not support the PCI Express Virtual Channel Capability Structure, only TC0 mapped to VC0 is defined. All the possibilities of the VC numbers assigned for a port are discussed in detail in the following section.

The protocols for the Root Complex, endpoints and switches also apply to bridges with one limitation: bridges can only support VC0. VC1 to VC7 are not supported for bridges.

Root Complex and Endpoint without Intervening Switches The simplest implementation of the TC to VC Mapping protocol is for direct connection of endpoints to the Root Complex. Switches add additional considerations to the TC to VC Mapping protocol. One important consideration is whether the PCI Express Virtual Channel Capability Structure (PVCCS) is supported by each port on a link.

Note: In the following discussion, the term *port* is used. Obviously, the port is the focus of the link, but it is the PCI Express device core behind the port that actually makes the VC and TC assignments.

VC Selection: The selection of the VC numbers by the TC to VC Mapping protocol for the ports on the link interconnecting a Root Complex and an endpoint is as follows:

- **No Support:** If neither port supports PVCCS, only VC0 can be defined for the link and the ports.

- **Mixed Support:** If one port supports the PVCCS and one port does not, only VC0 can be defined on the link. The port that supports the PVCCS must map all TC numbers to VC0.

- **Both Support:** If both ports do support PVCCS, multiple VCs can be defined for the link. The configuration software must symmetrically match the VC numbers and TC numbers in the ports on the link.

TC Selection: The selection of TC numbers by the TC to VC Mapping protocol for the ports on the link interconnecting a Root Complex and an endpoint is as follows:

- **Minimum TC Number:** The TC0 must be mapped to VC0 and is the minimum TC number that must be supported by the ports.

- **No Support:** If neither port supports PVCCS, only VC0 can be defined for the link and the ports.
  - Such ports must assign TC0 to all requester transaction packets.
  - Such ports must accept all packets on VC0 containing a requester transaction independent of the TC number and the ports must preserve the TC number received. Such ports must assign the preserved TC number to the associated completer transaction packets and map the associated packets to VC0.
- **Mixed Support:** If one port supports the PVCCS and one port does not: VC0 must be supported in both the upstream flow and downstream flow (symmetrically). TC0 is mapped only to VC0 and is symmetrical. The other TC numbers are not assigned to requester transaction packets but are symmetrically assigned to completer transaction packets. For example, the endpoint does not support the PVCCS, but the Root Complex does with the mapping of multiple TC numbers to VC0, as illustrated in Figure 11.11. The endpoint can only assign TC0 to the requester transaction's packet and the Root Complex responds with a completer transaction packet assigned TC0. The Root Complex can assign TC0 or TC7. For example, the Root Complex transmits with TC7 to the requester transaction's packet and the endpoint responds with a completer transaction packet assigned TC7. In this example, the Root Complex must map TC0 and TC7 to VC0 because the endpoint only supports VC0. The endpoint must preserve the number TC7 and use it in the completer transaction's packet even though it does not support PVCCS. Not supporting PVCCS requires the endpoint to use only TC0 in requester transaction packets it sources.

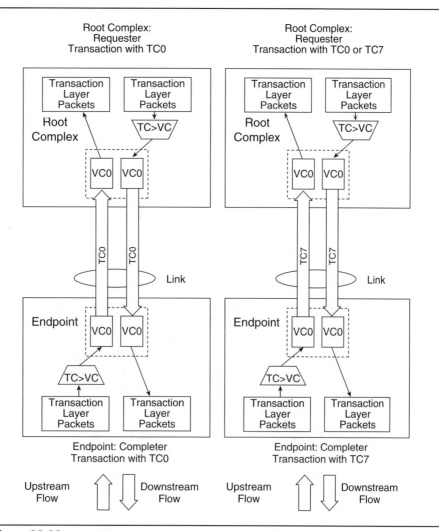

Figure 11.11 Single Virtual Channel: One Port Does Not Support PCI Express Virtual Channel Capability Structure

- **Both Support:** If both ports support the PVCCS, the ports can implement more VCs than VC0. If such ports were to map all packets with TC numbers other than TC0 to VC0, they would need to support the PVCCS. Ports that support the PVCCS must define VC0 and TC0, and TC0 must always be mapped to VC0.
  - Such ports can assign TC numbers other than TC0 to requester transaction packets.

- Such ports can only receive packets assigned TC numbers mapped to VC numbers defined for the receiving port. Such resources must filter out (not respond to) the associated requester transaction packets, discard the packets, and define them as **malformed**. See TC discussions in Chapters 6 and 9 for more information.
- The support of the PVCCS by both ports on a link places restrictions on mapping of TC numbers to VC numbers of the transmitting portion of the ports. The mapping restrictions are:
  - The specific VC numbers supported in the packet flow upstream must equal the specific VCs in the packet flow downstream (it must be symmetrical). TC numbers mapping to specific VC numbers are also symmetrical. For example, in Figure 11.12 PVCCS is supported by both ports with a single VC number defined. VC0 is the default VC number that must be supported, and in this example TC [0:3] is mapped to it and must be symmetrically supported. TC [4:7] are not supported in either packet flow direction. In the multiple VCs example in Figure 11.13, TC [0::4,7] are supported in both packet flow directions. TC [5,6] are not supported in either direction packet flow. Also note that the mapping of TC [0:1] toVC0, TC [2:4] to VC1, and TC7 to VC4 is symmetrical.
  - A specific TC number can only be assigned to a specific VC and not to multiple VC numbers.
  - If TCb is greater than TCa, both can be mapped to same VC or TCb can be mapped to a VC with a higher number. TCb cannot be mapped to a VC of a lower number than the VC to which TCa is mapped. As illustrated in Figures 11.14 and 11.15, TC4 can be mapped to VC1 or VC4, but cannot be mapped VC0, as illustrated by Figure 11.16.

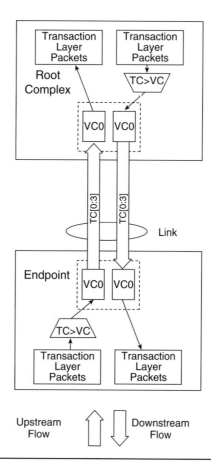

Figure 11.12 Single Virtual Channel: Both Ports Support the PCI Express Virtual Channel Capability Structure

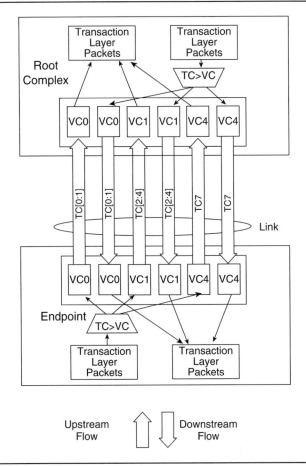

Figure 11.13 Multiple Virtual Channels: Both Ports Support the PCI Express Virtual Channel Capability Structure

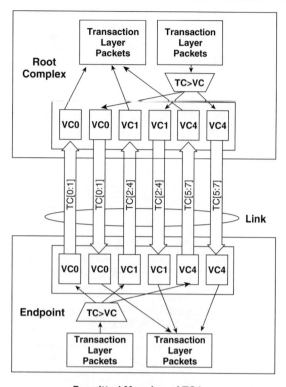

Permitted Mapping of TC4

Figure 11.14 Multiple Virtual Channels: Both Ports Support the PCI Express Virtual Channel Capability Structure, Permitting TC4 Mapping

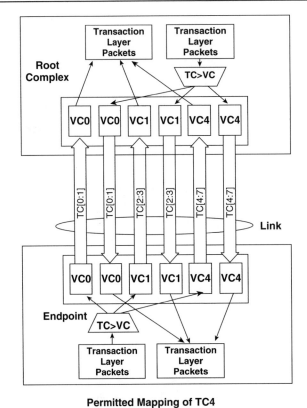

Figure 11.15 Multiple Virtual Channels: Both Ports Support the PCI Express Virtual Channel Capability Structure, Permitting TC4 Mapping

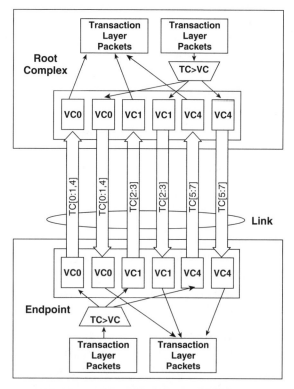

Not Permitted Mapping of TC4

Figure 11.16 Multiple Virtual Channels: Both Ports Support the PCI Express Virtual Channel Capability Structure, Not Permitting TC4 Mapping

Root Complex to Switch versus Switch to Endpoint versus Switch to Switch The previous section described a single link between the Root Complex and endpoint. In most implementations, the link is between a Root Complex and switch, a switch and endpoint, or a switch and a switch.

VC Selection: The selection of the VC numbers by the TC to VC Mapping protocol for the ports on links under each of these interconnects is the same as discussed in the previous section for the Root Complex and endpoint.

TC Selection: The selection of the TC numbers by the TC to VC Mapping protocol for the ports on links under each of these interconnects is the same as discussed in the previous section for the Root Complex and endpoint. However, additional TC to VC Mapping protocol considerations for the switch port end of the link are:

- **No Support:** A switch port that does not support the PVCCS can only support VC0.
    - A switch port that is a receiving port must accept all packets received independent of the TC number, preserve the TC number, and forward the unchanged packet to Address Mapping and TC to VC Mapping Module in the switch.
    - A switch port that is a transmitting port must transmit all packets received from Address Mapping and TC to VC Mapping Module with all TC numbers mapped to VC0. The Address Mapping and TC to VC Mapping Module simply force TC numbers received by the other receiving ports of the switch to be mapped to VC0 on the transmitting port of the switch.
- **Support:** A switch port that supports the PVCCS can implement more VCs than VC0. If the switch port were to map all packets with TC numbers other than TC0 to VC0, it would need to support the PVCCS. Ports that support the PVCCS must define VC0 and TC0, and TC0 must always be mapped to VC0.
    - A receiving port can only receive packets that have been assigned TC numbers mapped to the VC numbers enabled for the port. The port must filter out (not respond to) the associated requester transaction packets, discard the packet, and define it as **malformed**. See TC discussions in Chapters 6 and 9 for more information.
    - A transmitting port can only transmit packets that have been assigned TC numbers mapped to the VC numbers defined for the port. The Address Mapping and TC to VC Mapping Module ensures that transmission port never receives a packet with an improper TC number to VC number mapping.

Other TC Mis-Mapping Considerations A TC number must be assigned to a specific VC number. If any PCI Express device receives a packet (TLP) with a TC number mapped to a VC number that is not enabled for the port, the packet is discarded and the packet is defined as **malformed**. In the case of switches, the packets received and ported through the switch must have TC numbers assigned to the enabled VC numbers of the correct transmitting ports. Otherwise, the packet is discarded and the packet is defined as **malformed**.

Flow Control Protocol: Buffer Management

As shown in Figure 11.17, buffers are located throughout the PCI Express platform. The buffers for upstream packet flow are separate from the buffers for downstream packet flow. Prior to any port transmitting a Physical Packet containing a LLTP and thus a TLP (either upstream or downstream) onto a link, the available amount of buffer space at the receiving port must be known. In a transmitting port a TLP in the Transaction Layer is not transferred to the Data Link Layer for subsequent transmission onto the link if sufficient buffer space is not available at the receiving port. The transmitting port must have an acceptable amount of available buffer space at the receiving port.

Consider two things: First, what is the acceptable amount and how does it relate to available buffer space? Second, how is available buffer space information communicated between the ports on a specific link? The acceptable amount is called the Flow Control Credit (FCC). The communication method is information contained in the Flow Control Data Link Layer Packets (DLLPs).

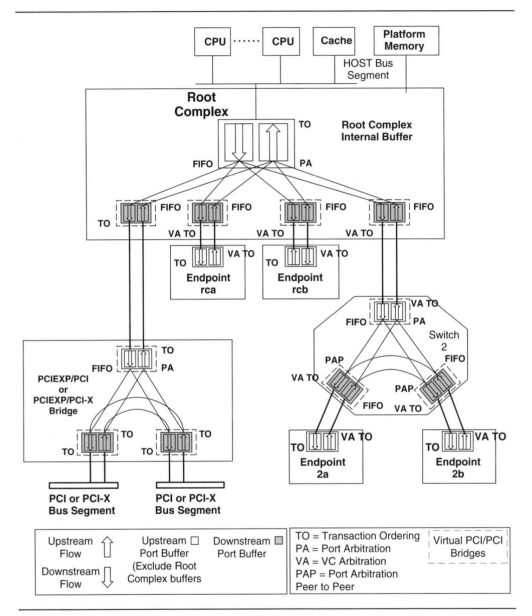

Figure 11.17 TLPs' Buffers throughout PCI Express Fabric

A Note on Terms

TLP versus *LLTP* and *FCC:* As discussed throughout this book, the actual packets transmitted across the link are Physical Packets containing link packets. The link packets consist of Data Link Layer Packets (DLLPs) used for link management and Link Layer Transaction Packets (LLTPs) containing Transaction Layer Packets (TLPs). The focus of this book in other chapters has been the transmission of the LLTPs across the links given the TLPs they contain. The TLPs, however, are what are actually stored within the buffers and FIFOs associated with the ports, not the LLTPs. The Flow Control Credits (FCCs) are associated with the TLPs as well. Consequently, the focus here relative to the buffers and FIFOs within PCI Express devices is on TLPs.

Receiving ports versus *transmitting ports:* In the following discussion, the terms *receiving port* and *transmitting port* are used in reference to the TLPs. Receiving port refers to the portion of the port that receives the TLPs. Transmitting port refers to the portion of the port that transmits TLPs. In that the LLTPs contain the TLPs, the concept of receiving and transmitting ports also extends to the LLTPs.

Buffers versus *FIFOs:* Figure 11.17 shows two types of buffers. One is defined simply as *buffer* and the other as *FIFO*. The term *buffer* refers to a set of registers in ports that must implement combinations of Transaction Ordering, Port Arbitration, and VC Arbitration protocols. The term *FIFO* refers to a unique version of a buffer. It consists of a set of registers implemented with strong ordering. That is, the first TLP from the link to be stored in the FIFO is the first TLP removed. This book has adopted the use of a FIFO in the different ports throughout a PCI Express fabric to simplify the discussion of Flow Control. As noted in the Figure 11.17, the FIFOs are only defined at the receiving portion of the ports. The use of FIFOs at the receiving portion of the ports ensures that the results of Transaction Ordering, Port Arbitration, and VC Arbitration protocols implemented by the associated transmitting ports on the other end of the links are maintained. In the actual implementation, the FIFOs can be designed without true strong ordering. For example, if the FIFO is in the upstream port of a switch; the stored TLPs may be ported to the transmitting portion of different downstream ports of the switch. Consequently, the need to maintain true strong ordering in the FIFO between TLPs porting to different downstream ports is not needed or is indeed practical.

The exact design implementations of buffers and FIFOs provided by different vendors are beyond the scope of this book.

Please see the further discussion about FIFOs in Chapter 10.

FCC and Set of Buffers

The determination to transfer a TLP in the PCI Express Transaction Layer to the Data Link Layer in the transmitting port is done by the Gate circuitry in the transmitting port. The Gate circuitry determines the buffer space needed at the receiving port to transmit a specific transaction packet by its FCC. The FCC assigned to each type of TLP is summarized in Table 11.1. Each FCC requires 16 bytes of buffer.

FCCs are defined for both the Header and Data fields of TLPs. The buffer at the receiving port is conceptually a set of six buffers for each VC number. Each of these six buffers is assigned a group of specific types of TLPs:

- **Buffer type PH (posted header)**: Buffer space for Header fields of memory write and message requester transaction packets.

- **Buffer type PD (posted data)**: Buffer space for Data fields of memory write and message requester transaction packets.

- **Buffer type NPH (non-posted header)**: Buffer space for Header fields of memory read, I/O read and write, and configuration read and write requester transaction packets.

- **Buffer type NPD (non-posted header)**: Buffer space for Data fields of I/O write and configuration write requester transaction packets.

- **Buffer type CPLH (completer header)**: Buffer space for Header fields of completer transaction packets.

- **Buffer type CPLD (completer data)**: Buffer space for Data fields of completer transaction packets.

> Though not part of the FCC Header field value, it is implied that each transaction with or without data also has a Digest field. Each Digest field of each TLP requires 4 bytes of buffer.

Set of Buffers and Virtual Channel

When examining the buffering at each receiving port, the Virtual Channel (VC) numbers assigned to the ports on a specific link must also be considered. Each VC number is assigned set of buffers in the receiving port connected to each link. The buffer space information exchanged on the link is specific for each VC number.

So in summary, each VC number assigned to a receiving port contains a set of buffers. For example, if four different VC numbers are assigned to the receiving port, four independent sets of buffers are established. Each set of buffers consists of six buffers, with each one of these six buffers assigned the Header or Data field of a specific type of TLP.

In Figure 11.17, the upstream and downstream buffers collectively represent the set of buffers for all VC numbers assigned to the ports and associated links.

FCC

When the link is first initialized, each VC number that is enabled for the ports on an associated link (in the configuration register block) must provide information on the size of the set of buffers (specific to each enabled VC number) at the receiving port to the transmitting port. Due to the bi-directional structure of the link, each link has two receiving ports and two transmitting ports.

As discussed in Chapter 7, the minimum initial available buffer space at the receiving port is communicated by Flow Control Initialization for each enabled VC number. The available buffer space is measured by FCC values. During Flow Control Initialization the initial available buffer space is transmitted in the InitFC1 and InitFC2 DLLPs. Once the initial available buffers spaces are communicated from the receiving ports to the transmitting ports for each enabled VC number, the TLP transmitting ports must use this information to determine whether sufficient buffer space exists at the receiving port to accept a specific TLP. The FCC value of each specific type of TLP transaction to be transmitted across the link is summarized in Table 11.1. It should be noted that each FCC unit is 16 bytes. Also, the buffer space for the Digest field is implied but not specified in the table.

Table 11.1: FCCs for TLPs

Transaction Packet Type	Receiving Port Buffer Type Used Header/Data	FCCs for Header Field per TLP (1)(2)	FCCs for Data Field per TLP
Memory Write requester or Message requester	PH/PD	01h	LENGTH field in Header / 4 (3) (Rounded up)
Memory Read requester, I/O Read requester or Configuration read Requester	NPH (5)	01h	000h (5)
I/O Write Requester or Configuration Write Requester	NPH/NPD	01h	001h (4)
Completer for a memory read requester (data)	CPLH/CPLD	01h	LENGTH field in Header / 4 (Rounded up)
Completer for I/O or configuration read requester (data)	CPLH/CPLD	01h	001h (4)
Completer for I/O or configuration write requester (data)	CPLH (5)	01h	000h (5)

Table Notes:

1. The FCC value does not include buffer space required for the Digest field.
2. The 000h defines a value of infinite buffer space at the receiver port from the transmitter's viewpoint.
3. For message requester transaction with no data the value = 000h.
4. The FCC value reserves a buffer space of 16 bytes but only 4 bytes are used.
5. This FCC value is used in all calculations discussed in the following section for determination of available buffer space and in the UpdateFC DLLP. Effectively it defines the associated available buffer space as infinite because no data is actually buffered. This applies to the NPD and CPLD.

The transmitting and receiving ports must keep track of the available buffer space for each specific VC number at the receiving port. The Data Link Layer uses Flow Control DLLPs to communicate the available buffer space at the receiving port. As discussed in Chapter 7, the Flow Control DLLPs that are used for initialization information are the InitFC1 and InitFC2 DLLP. The update information of available buffer space at the receiving port is conveyed to the transmitting port by UpdateFC DLLPs. The initialization or update information consists of the FCC values contained in the HdrFC and DataFC fields of the Flow Control DLLPs for Header and Data fields of the TLPs, respectively.

Other Considerations Associated with FCCs

- The memory (except Lock related) transaction packets, message vendor-defined transaction packets, and message advanced switching requester transaction packets can be assigned any TC number and they can therefore have different VC numbers. Lock memory, I/O, configuration, and message baseline requester transaction packets can only be assigned TC0 and TC0 can only be mapped to VC0. Consequently the available buffer space assigned to VC0 at the receiving port is critical.

- FCCs and the associated buffers are only used to buffer the TLPs with LLTPs. DLLPs are not buffered.

- If an InitiFC1, InitiFC2, or UpdateFC DLLP is received with a VC number that is not enabled, the DLLPs are discarded without any error.

- VC0 is always defined as enabled. The other VC numbers are enabled or disabled by a software setting of "1" or "0" in the configuration register block.

Flow Control Protocol Execution

Once the Flow Control Initialization is completed for a specific VC number, the transmitting port on the link knows the initial available buffer space for the set of buffers at the receiving port. If the available buffer space is large enough, TLPs contained within LLTPs and Physical Packets can then be transmitted. The discussion now focuses on the TLPs and LLTPs.

As discussed earlier in this chapter, the transmitting portion of the ports is aligned with virtual channels (VCs). As shown in the example in Figure 11.18, the TLPs flow from the Transaction Layer to the Physical Layer via the Data Link Layer. The Gate circuitry selects TLPs from the transmit buffer to transfer to the retry buffer specific to VC0 (always enabled) and any other enabled VC numbers.

How the TLPs placed in the transmit buffer arrive depends on the PCI Express device. If the PCI Express device is a Root Complex, bridge, or endpoint; the PCI Express device core places TLPs into the transmit buffer with an execution order. If the PCI Express device is a switch, the TLPs in the transmit buffer are those ported from other ports of the switch via the Address Mapping and TC to VC Mapping module of the switch.

Transmitting ports of TLPs determine whether sufficient available buffer space exists at the receiving port prior to Gate circuitry, transferring the TLP to the retry buffer, which results in subsequent transmission. The following section examines how this occurs. It is assumed for the purposes of this discussion that the associated VC numbers have been enabled and the associated available buffer space information has already been initialized. See Chapter 7 for more information on the Flow Control Initialization.

Determining Available Buffer Space

The Gate circuitry is the logic between the Transaction Layer and Data Link Layer within the transmitting port, as shown in Figure 11.18. It must determine whether buffer space is available at the receiving port prior to transferring, or "gating," the TLP from the Transaction Layer to the Data Link Layer. The Data Link Layer part of the transmitting port contains a retry buffer that temporarily holds the TLPs that have been modified into LLTPs with the addition of SEQ# and LCRC.

As discussed above and in Chapter 7, after Flow Control Initialization the transmitting port knows the FCC value of available buffer space in the receiving port for each VC number enabled for each link. As each VC number completes Flow Control Initialization, the Gate circuitry can begin transferring TLPs to the Data Link Layer if buffer space is available in the receiving port for that specific VC number.

The quantities and associated equations that follow have been selected to explain how the Gate circuitry determines when a TLP can be placed into the retry buffer (as a LLTP) for subsequent transmission. These quantities and equations may differ for the exact Gate circuitry implementation.

The quantities and equations are only for explanatory purposes. The only requirement is that values in the InitiFC1, InitiFC2, and UpdateFC DLLPs between the implementations of the transmitting port and the receiving port are compatible with the PCI Express specification.

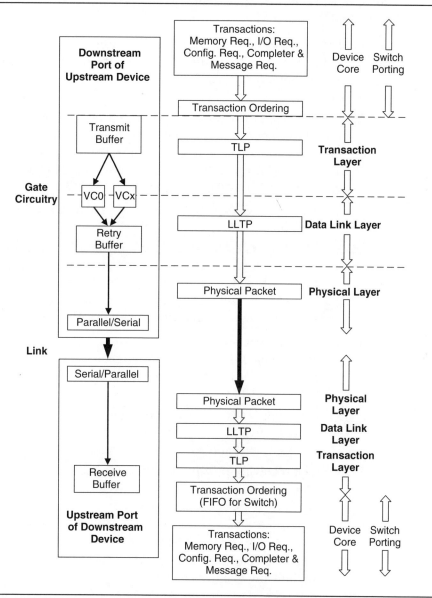

Figure 11.18 PCI Express Layers, VC Channels, and Gate Circuitry

Also, the quantities and equations are determined for a specific VC number for each of the six buffer types. That is, the quantities and the Transfer Equations are maintained and applied individually to PH, NPH, CPLH, PD, NPD, and CPLD buffers, respectively. As previously stated, there must be sufficient available buffer space for both Header fields and Data fields for each TLP.

At the Transmitting Port of the TLP To determine the available buffer space at the receiving port, the Gate circuitry in the transmitting port tracks the following quantities for each of the buffer types for each of the enabled VC numbers.

- **FCC_Consumed Counters:** These counters contain the current total FCC value of TLPs that have been transferred to the retry buffer in the Data Link Layer since Flow Control Initialization. They are 8-bit counters for PH, NPH, and CPLH; they are 12-bit counters for PD, NPD, and CPLD.
  - The initial value is "0" after Flow Control Initialization is completed.
  - The FCC_Consumed counter increases by the value of the FCC_Pending value with the actual transfer of the TLP to the retry buffer. It rolls over once maximum count is reached. The FCC_Pending value is the FCC value for every TLP transferred to the Data Link Layer from the Transaction Layer.

- **FCC_Limit Registers:** These registers contain the FCC values of the available buffer space in the receiving port buffer. They are 8-bit registers for PH, NPH, and CPLH; they are 12-bit registers for PD, NPD, and CPLD.
  - The initial FCC values of these registers are established during Flow Control Initialization with receipt of InitFC1 and InitFC2 DLLPs. See Chapter 7 for more information.
  - The receiving port periodically sends an UpdateFC DLLP containing the FCC value of the current available buffer space at the receiving port (maintained at the receiving port as FCC_Allocated). The FCC Limit values in these registers are always set to the values provided in the UpdateFC DLLP.
  - The FCC values are contained in the HdrFC and DataFC fields of the Flow Control DLLPs.

- **FCC_Pending Values:** These are the FCC values of the TLP under consideration for transfer to the retry buffer of the Data Link Layer.
  - The FCC values assigned to each type of TLP Header field and Data fields are summarized in Table 11.1. The value is an 8-bit value for PH, NPH, and CPLH; it is a 12-bit value for PD, NPD, and CPLD.
  - As discussed below, the Transfer Equations use the FCC_Pending value to determine whether sufficient buffer space is available to transfer the TLP to the retry buffer. If sufficient space is not available, the FCC_Pending value is discarded, the TLP is not transferred, and the value of the FCC_Consumed remains unchanged. If space is available, the FCC_Consumed counter is increased by the FCC_Pending value and the TLP is transferred to the retry buffer.

For the Gate circuitry in the transmitting port to transfer a TLP to a retry buffer of the Data Link Layer, the following transfer equations must be true: If the TLP contains a Data field, both of the following conditions for each buffer type must be met for the transfer to the retry buffer in the Data Link Layer to occur. If a TLP is a nullified TLP as identified by FRAMING of Physical Packet of STP and EDB K symbols, its associated FCC value is not considered in any counters or equations.

- **For TLP types with PH, NPH, and CPLH:** The transfer equation that must be true is [FCC_Limit Register: (FCC_Consumed Counter + FCC_Pending Value)] $<= 2^{[Fieldsize]} / 2$ ... Fieldsize of 8 bits.
  - If the FCC value of the FCC_Limit Register is infinite (00h) per Flow Control Initialization, this calculation is not done and the condition is by definition always met. That is, the comparison is always true.
  - The binary equivalents of FCC_Consumed Counter and FCC_Pending Value are added together. This total is complemented and "1" is added to achieve one's complement. Any overflow is discarded.
  - The one's complement of (FCC_Consumed Counter + FCC_Pending Value) is equivalent to − (FCC_Consumed Counter + FCC_Pending Value) in the transfer equation.

- The transfer equation is completed by adding the binary equivalent of the FCC_Limit Register to the one's complement of (FCC_Consumed Counter + FCC_Pending Value). Any overflow is discarded.
- The transfer equation is considered true if the resulting addition is = < 128d. If the transfer equation is true, the TLP is transferred to the retry buffer. Note: This is unsigned arithmetic such that the comparison binary number to 128d is always assumed to be a number equal to or greater than 0000 0000b.

- **For TLP types with PD, NPD, and CPLD:** The transfer equation that must be true is [FCC_Limit Register: (FCC_Consumed Counter + FCC_Pending Value)] <= $2^{[Fieldsize]} / 2$ ... Fieldsize of 12 bits.
  - If the FCC value of the FCC_Limit is infinite (000h) based on Flow Control Initialization of InitFC1 and InitFC2 DLLPs, this calculation is not done and the condition is by definition always met. That is, the comparison is always true.
  - The balance of the transfer equation is the same as above except the comparison of the resulting addition is = < 2048 instead of 128.

Example of Quantities and Transfer Equations Here is an example of how the quantities are tracked at the transmitting port. Assume that based on the Flow Control Initialization, the FCC_Limit Register = 4d for the PH buffer and FCC_Limit Register = 128d for the PD buffer. The FCC_Consumed Counters for both PH and PD buffers = "0" after Flow Control Initialization.

- **Transfer of First TLP Testing for PH:** The Gate circuitry wants to transfer a TLP to the retry buffer with PH = 1d and PD = 16d.
  - FCC_Consumed Counter + FCC_Pending Value = 00000000b + 0000001b = 00000001b.
  - The one's complement of (FCC_Consumed Counter + FCC_Pending Value) = 11111110b + 1b = 11111111b.
  - Adding the FCC_Limit Register's binary equivalent to one's complement of (FCC_Consumed Counter + FCC_Pending Value) = 0000 0100b + 11111111b = 00000011b.
  - The transfer equation is true because 00000011b = < 128d.

- **Transfer of First TLP Testing for PD:** The Gate circuitry wants to transfer a TLP to the retry buffer with PH = 1d and PD = 16d.
    - FCC_Consumed Counter + FCC_Pending Value = 000000000000b + 000000010000b = 000000010000b.
    - The one's complement of (FCC_Consumed Counter + FCC_Pending Value) = 111111101111b + 1b = 111111111110b.
    - Adding the FCC_Limit Register's binary equivalent to one's complement of (FCC_Consumed Counter + FCC_Pending Value) = 0000 1000 0000b + 111111111110b = 000001111110b.
    - The transfer equation is true because 000001111110b = < 2048d.

  The transfer equations for both PH and PD are true, so the first TLP is transferred to the retry buffer, the FCC_Consumed Counters for PH and PD are increased by the FCC_Pending Values resulting in FCC_Consumed Counter = 1d for PH and FCC_Consumed Counter = 16d for PD. The FCC_Limit Register remains = 4d for PH buffer and FCC_Limit Register remains = 128d for PD buffer.

- **Transfer of Second TLP:** The Gate circuitry wants to transfer a second TLP with PH = 1d and PD = 32d. Beginning with the FCC_Consumed Counters and FCC_Limit Registers from the previous transfer, the same protocol is applied to the transfer equations. Both transfer equations are true and the second TLP is transferred to the retry buffer. The FCC_Consumed Counters for PH and PD are increased by the FCC_Pending Values resulting in FCC_Consumed Counter = 2d for PH and FCC_Consumed Counter = 48d for PD. The FCC_Limit Register remains = 4d for PH buffer and FCC_Limit Register remains = 128d for PD buffer.

The transfer equations with the one's complement arithmetic and comparison to <= $2^{[Fieldsize]}$ / 2 is applied to each TLP under consideration for transfer to the retry buffer. The Gate circuitry continues the transfers of TLPs until either the transfer equation for the PH or PD buffer is not true. If the transfer equation is not true for either PH buffer or PD buffer, the Gate circuitry can consider other TLPs related to PH and PD based on the qualifications of transmission priority requirements discussed later in this chapter.

Unique Activities The application of the transfer equations with one's complement arithmetic, comparison to <= $2^{[Fieldsize]}$ / 2, and infinite numbers defined as 00h and 000h result in the transmitting port correctly tracking the available buffer space at the receiving port because of three unique activities. The first activity is the incrementing by the FCC_Consumed Counters for each TLP transferred to the retry buffer. Eventually they will roll over, but can they still be used as part of the transfer equations due to the comparison conditions of the transfer equations. In the second activity, the FCC_Limit Registers provide relative values of the available buffer space at the receiving port. The FCC_Limit Registers begin with the actual value of available buffer space at the receiving port. TLPs transferred to the retry buffer are subsequently transmitted to the receiving port, so the values of the FCC_Limit Registers become out of date. However, the FCC values of the TLPs transferred to the retry buffer are incorporated in the FCC_Consumed Counters. The transfer equations' use of both FCC_Limit Registers and FCC_Consumed Counters compensates for the FCC_Limit Registers being out of date. In the third activity, the receiving port eventually must provide updated information on the available buffer space. The UpdateFC DLLP is transmitted from the receiving port to the transmitting port, and the associated FCC values are set as the FCC values in the FCC_Limit Registers. The FCC values provided by the UpdateFC DLLP permit the FCC_Consumed Counters to simply be incremented with the FCC values of the TLPs transferred to the retry buffer and they roll over as needed. Consequently, the FCC values provided by the UpdateFC DLLPs from the receiving port must be of a specific format to ensure compatibility between the transmitting and receiving ports. This specific format of the FCC value is discussed in the next section.

At the Receiving Port of the TLP Complementary to the operation of the Gate circuitry in the transmitting port is the reception of TLPs within LLTPs and Physical packets. As the receiving port accepts TLPs, an acknowledgement is returned to the transmitting port and thus the TLP is removed from the retry buffer. The purpose of the retry buffer and the form of acknowledgement are discussed in detail in Chapter 12.

The receiving port must track the following quantity for each of the buffer types for each of the enabled VC numbers.

- **FCC_Allocated Counters:** These counters contain the current total FCC value of TLPs that have been allocated since Flow Control Initialization. They are 8-bit counters for PH, NPH, and CPLH, and 12-bit counters for PD, NPD, and CPLD.
    - The initial FCC values are transmitted during Flow Control Initialization within InitFC1 and InitFC2 DLLPs. Thus, immediately after the Flow Control Initialization of each VC number, the contents of the FCC_Limit Registers in the transmitting port equal the contents of the FCC_Allocated Counters in the receiving port. See Chapter 7 for more information.
    - As TLPs are received, processed, and sent to the Transaction Layer the FCC_Allocated Counters (update) = [FCC_Allocated Counters (present) + FCC_Processed Counters]. The FCC_Allocated Counters are updated just prior to receiving port transmitting an UpdateFC DLLP.
    - The receiving port transmits an UpdateFC DLLP to the transmitting port containing the FCC values in the FCC _Allocated Counters based on the buffer type of a specific enabled VC number. The FCC values in the UpdateFC DLLP are used to set FCC values in the FCC_Limit Registers. Thus, the contents of the FCC_Limit Registers in the transmitting port equal the contents of the FCC _Allocated Counters in the receiving port based on buffer type from a specific enabled VC number. The transmission of the UpdateFC DLLP is according to a specific protocol discussed in the following section. The result of the UpdateFC DLLP protocol is that the contents of the transmitting port FCC_Limit Registers are effectively equal to or less than the receiving port FCC_Allocated Counters. Thus, the FCC_Limit Registers can be used in the transfer equations to determine whether a TLP can be transferred to the retry buffer. That is, the receiving port has sufficient available space.

- **FCC_Processed Counters:** These counters contain the current FCC values of TLPs that have been received, acknowledged as received, and transferred to the Transaction Layer since the last update of the FCC_Allocated Counters. As will be discussed further in the Chapter 12, the receipt of a group of TLPs is indicated to the TLPs' transmitting port with a single acknowledgement. They are 8-bit counters for PH, NPH, and CPLH and 12-bit counters for PD, NPD, and CPLD.
  - The initial value is "0" after Flow Control Initialization is completed.
  - The counter is increased by the FCC value for every TLP that has been received, acknowledged as received, and transferred to the Transaction Layer since the last update of the FCC_Allocated Counters.
  - When the FCC_Allocated Counters are updated, the FCC_Processed Counters are returned to a "0" value.
  - If a TLP received is a nullified TLP (that is, the FRAMING of the Physical Packet is STP and EDB), its associated FCC value is not considered in any counters or equations.

Continuing Example of Quantities and Transfer Equations Here is a continuation of the example discussed for the transmitting port. Assume that based on the Flow Control Initialization the FCC_Allocated Counter = 4d for PH buffer and FCC_Allocated Counter = 128d for PD buffer. The FCC_Processed Counters for both PH and PD buffers = "0" after Flow Control Initialization.

- **Receipt of the First TLP from the Transmitting Port:** The receiving port has received a TLP with PH = 1d and PD = 16d.
  - Once the first TLP been received, acknowledged as received, and transferred to the Transaction Layer, the FCC_Processed Counter for PH = 1d and the FCC_Processed Counter for PD = 16d.
  - The FCC_Allocated Counter for PH = 4d and the FCC_Allocated Counter for PD = 128d.
  - Prior to the receiving port transmitting an UpdateFC DLLP, FCC_Allocated Counters are updated with the FCC_Processed Counters for the type of buffer for a specific VC number. After the update the FCC_Allocated Counter for PH = 5d, the FCC_Allocated Counter for PD = 144d, the

FCC_Processed Counter for PH = 0d, and the FCC_Processed Counter for PD = 0d. The FCC values used in the Update DLLP are those of the updated FCC_Allocated Counters. The FCC_Processed Counters are initialized to "0" with the transmission of the UpdateFC DLLP.

– In this example, the UpdateFC DLLP was transmitted prior to the receipt of the second TLP.

- **Receipt of the Second TLP from the Transmitting Port:** The receiving port has received a TLP with PH = 1d and PD = 32d. Beginning with the FCC_Allocated Counters and FCC_Processed Counters from the previous example, the same protocol is applied.

    – Once the second TLP been received, acknowledged as received, and transferred to the Transaction Layer, the FCC_Processed Counter for PH = 1d and the FCC_Processed Counter for PD = 32d.

    – The FCC_Allocated Counter for PH = 5d and the FCC_Allocated Counter for PD = 144d.

Rollover of FCC_Allocated Counters In this example, the updates of the FCC_Allocated Counters did not cause rollovers. If the update of the FCC_Allocated Counter caused it to rollover, the rollover would occur with no side effects. The FCC values used in the UpdateFC DLLP are those of the rolled-over FCC_Allocated Counter. This is one of the unique activities discussed with the transmitting port. This rollover is allowed because the rollover of the FCC_Consumer Counters and the nature of the transfer equations are tailored to this activity.

Nullified TLP (Cut-Through) It is possible to send a nullified TLP across the link without affecting the transmitting and receiving tracking mechanisms. The protocol to support the transmission of a nullified TLP is called *cut-through*. A nullified TLP and its associated LLTP are simply ignored by the receiving port. The nullified TLP is created as a normal TLP except the LCRC has a value that is the inverted value of what it would normally have been and the END K symbol in the Physical Packet FRAMING is EDB K symbol instead. Using the cut-through protocol, the transmission of a nullified TLP does not increment the FCC_Consumed Counter and its reception at the receiving port does not increment the FCC_Allocated or FCC_Processed Counters.

UpdateFC DLLP Protocol

As discussed earlier, the receiving port must transmit UpdateFC DLLPs to provide information on available buffer space to the transmitting port. If the UpdateFC DLLPs are not transmitted often enough, the transmitting port assumes that no further buffer space is available at the receiving port and stops transfers of TLPs to the retry buffer. This assumption may be wrong and precious link bandwidth may be lost. Consequently, the Update DLLPs must be sent with a frequency that minimizes the impact of a wrong assumption.

Another consideration of the UpdateFC DLLP protocol is that of infinite available buffer space. This may be due to a very large relative buffer at the receiving port. That is, the buffer for PH may be very large compared to the buffer for the PD. Consequently, the PD buffer will be filled long before the PH buffer and the PH buffer can be viewed as infinite.

Frequency of UpdateFC DLLP Under certain conditions the receiving port must transmit to the transmitting port an Update FC DLLP. The frequency of an UpdateFC DLLP transmission is based on two criteria. One of the criteria is simply a set of boundary conditions for the FCC_Allocated Counters in the receiving port, and the other is a timed interval. The transmission of an UpdateFC DLLP is defined for each buffer type from each enabled VC number. Please note, if both types of buffer of a pair of buffers (PH and PD, or NPH and NPD, or CPLH and CPLD) are defined as infinite, an UpdateFC DLLP need not be transmitted for that buffer pair. If one or both buffers of the buffer pair is not infinite the UpdateFC DLLP must be transmitted by the receiving port.

The receiving port transmits an UpdateFC DLLP with the most recent FCC values in the (updated) FCC_Allocated Counters to the transmitting port with the following requirements:

- For each buffer type PH, NPH, CPLH, and NPD:
    - Whenever the (present) FCC_Allocated Counter = FCC_Received Counter.

- For each buffer type PD and CPLD:
  - Whenever a FCC_Allocated Counter (present) < Max_Payload_Size.
- When the link state is L0 or L0s, the UpdateFC DLLP must be transmitted at least every 30 to 45 microseconds.

In addition to the above requirements the receiving port must transmit an UpdateFC DLLP with the most recent FCC values in the (update) FCC_Allocated Counters to the transmitting port at timed intervals. The timed intervals are defined relative to a maximum period between transmissions.

- **Update Transmission Latency value**: The maximum period between UpdateFC DLLPs being transmitted is defined by the Update Transmission Latency value in units of symbol time. The maximum period is measured from the last bit of the previous UpdateFC DLLP to the first bit of the present UpdateFC measured at the receiving port.
- **Symbol time units:** Each symbol time unit equals the bit period × 10. Presently this is defined for only the one Data Rate Identifier of 2.5 gigabits per second per TS1 and TS2 OSs. The maximum Data Rate Identifier possible and the maximum Data Rate Identifier agreed to by the ports on a specific link are defined in the Link Capabilities and Link Status Registers, respectively. The value in the Link Status Register is used to calculate the symbol time. For the value of 2.5 gigabits per second the bit period is 399.88 to 400.12 picoseconds. See Chapter 8 for more information about bits and symbols.

The Update Transmission Latency value is dependent on the link width and the maximum payload size. The Update Transmission Latency values are summarized in Table 11.2 in symbol time values. The "Max Payload Size" in Table 11.2 is defined on the basis of the TLPs' transmitting port. The value of the "Max Payload Size" is defined on the basis of the Device Control Register as follows:

- Bits [7::5] of Device Control Register define the maximum payload a PCI Express device can support in any TLP except for a memory read requester transaction (related to the PD buffer).
- Bits [14::12] of Device Control Register define the maximum payload a PCI Express device can request in a memory read requester transaction (related to the CPLD buffer).

Table 11.2 UpdateFC Transmission Latency Value

	Link Width							
		x1	x2	x4	x8	x12	x16	x32
	128 bytes	237	128	73	67	58	48	33
Max Payload Size	256 bytes	416	217	118	107	90	72	45
	512 bytes	559	289	154	86	109	86	52
	1024 bytes	1071	545	282	150	194	150	84
	2048 bytes	2095	1057	538	278	365	278	148
	4096 bytes	4143	2081	1050	543	706	534	276

Infinite Buffers Any receiving port that defines either the Header or Data buffer of a buffer pair (PH and PD, or NPH and NPD, or CPLH and CPLD) as infinite (00h or 000h) during the Flow Control Initialization (InitFC1 and InitFC2 DLLPs) must transmit UpdateFC DLLPs to the TLPs transmitting port with the following qualifications.

- **Both Buffers Infinite:** If both the Header and Data field buffers are defined as infinite, the receiving port has the option not to transmit UpdateFC DLLPs to the transmitting port. If an UpdateFC DLLP is transmitted, it must define the available buffer space as infinite.

- **Both Buffers Not Infinite:** If either or both the Header or Data field buffers are defined as not infinite, the receiving port is required to transmit UpdateFC DLLPs to the transmitting port for both buffers. The UpdateFC DLLP for the infinite buffer (if any) must define the available buffer space as infinite. The UpdateFC DLLP for the *finite* buffer (if any) must define the available buffer space based on the value of the FCC_Allocated Counter.

- **Error:** If during initialization an infinite FCC value was transmitted (00h or 000h in the InitiFC1 and InitiFC2 DLLPs) for a specific buffer, any subsequent transmission of an UpdateFCC DLLP must contain an infinite FCC value. If the FCC value in the UpdateFCC DLLP is not infinite, the TLP transmitter (receiver of UpdateFCC DLLP) must discard it. Optionally, the receiver of UpdateFCC DLLP can report it as **FCPE (Flow Control Protocol Error)**. See Chapter 9 for more information about FCPE.

> A discussed in Chapter 6, the TC number = TC0 must be assigned to all transactions that cannot be posted. TC0 must be mapped to VC0. Thus, all non-posted transactions must be mapped to VC0. No non-posted transactions can be mapped to any VC other than VC0. The buffers of VCs other than VC0 associated with non-posted transactions can be effectively defined as infinite.

Optional Checking of TLPs' Receiving Port

The discussion to this point has focused on the TLPs' transmitting port tracking the available buffer space at the TLPs' receiving port. The receiving port also has the option to track the TLPs received in the LLTPs and Physical Packets to check for input buffer overrun. If the receiving port does this, the FCC_Received Counters for each of the buffer types for each of the enabled VC numbers are checked.

- **FCC_Received Counters:** These counters contain the current total FCC value of TLPs that have been received (acknowledgement and transfer to Transaction Layer has not occurred) since Flow Control Initialization. They are 8-bit counters for PH, NPH, and CPLH and 12-bit counters for PD, NPD, and CPLD.
    - The initial value is "0" after Flow Control Initialization is completed.
    - It is increased by the FCC value for every TLP received (acknowledgement and transfer to Transaction Layer has not occurred).
    - The counter rolls over once maximum count is reached.
- **FCC_Allocated Counters:** These are the same counters discussed earlier.
- **FCC_Processed Counters:** These are the same counters discussed earlier.

The receiving port can determine whether the transmitting port has transmitted TLPs that are causing overrun conditions by applying the error equations. If either error equation is true for a buffer type for a specific enabled VC numbers, a **Receiver Overflow** error has occurred. The TLP that caused the overflow error is discarded and this is not transferred to the Transaction Layer. The PCI Express specification does not define how the TLPs' receiving port should operate once an overflow error has occurred. It is assumed that the receiving port will continue as normal. If a TLP received is a nullified TLP (the FRAMING of the Physical Packet is STP and EDB), its associated FCC value is not considered in any counters or equations. See Chapter 9 for more information on errors.

The one's complement arithmetic protocol applied to (FCC_Consumed Counter + FCC_Pending Value) in the TLPs' transmitting port transfer equations is also applied to the FCC_Received Counter in the TLPs' receiving port error equations. The error equations are:

- **For TLP types with PH, NPH, and CPLH:** The error equation that must be true is [(FCC_Allocated Counter (present) + FCC_Processed Counter) - FCC_Received Counter] $>= 2^{[Fieldsize]} / 2$ ... Fieldsize of 8 bits.

- **For TLP types with PD, NPD, and CPLD:** The error equation that must be true is [(FCC_Allocated Counter (present) + FCC_Processed Counter) - FCC_Received Counter] $<= 2^{[Fieldsize]} / 2$ ... Fieldsize of 12 bits.

Other Considerations for Transferring TLPs from the Transaction Layer to the Data Link Layer

As discussed earlier, Gate circuitry must determine available buffer space at the receiving port prior to "gating" the TLP from the Transaction Layer to the retry buffer of the Data Link Layer. Once in the retry buffer of the Data Link Layer, the LLTP that contains the TLP is transmitted in a Physical Packet using strict ordering.

Gate circuitries implemented at the different ports have other considerations besides available buffer space at the receiving port. In the simplest implementation, the Transaction Layer would select the next TLP to be gated and the Gate circuitry would determine whether available buffer space is at the receiving port before gating it the Data Link Layer, connecting SEQ# and LCRC, and placing the result into the retry buffer. In reality, the Transaction Layer presents several TLPs to the Gate circuitry in the transmit buffer. The Gate circuitry must determine whether the transmit buffer contains a TLP of a size that is compatible with the available buffer space at the receiving port as qualified by transaction ordering, which is between TLPs of the same TC number. See Chapter 10 for more information about transaction ordering. The transmission priority selected by the Transaction Layer and just what the Gate circuitry must consider are summarized here and illustrated in Figure 11.17. For the Root Complex, endpoints, and switches an additional consideration is the transmission priority between the TLPs of different VC numbers based on the VC Arbitration protocol. Switch transmission priority and the Root Complex internal buffer must also consider Port Arbitration in addition to transaction ordering and VC Arbitration. Finally, bridges must also consider Port Arbitration.

The transmission priority selected by the Transaction Layer and just what the Gate circuitry must consider are summarized here and illustrated in Figure 11.17:

- Each downstream port of the Root Complex or switch must consider VC Arbitration and transaction ordering. For a Root Complex, the execution order is also a consideration of the PCI Express device core. For a switch, the Port Arbitration protocol is also a consideration of the switch porting.

- Each upstream port of an endpoint must consider VC Arbitration, transaction ordering, and execution order.

- Each upstream port of a switch must consider VC Arbitration, transaction ordering, and Port Arbitration protocol.

- Each upstream port of a bridge must consider transaction ordering and execution order (that is, what arrives from the downstream bus segments).

Transmission Order Details

Figure 11.19 illustrates an example of the first part of transmission priority considerations for the Root Complex's downstream ports, endpoint's upstream port, and bridges' upstream port. The TLPs assigned to VCx are used for this example. The PCI Express device core provides a selection of TLPs based on execution order that is implementation-specific. The TLPs are first parsed to the correct VC (VCx, in this example) based on the TC number assignments. Note: this discussion does not assume Port Arbitration is applied to bridges; see further discussion about the application of Port Arbitration in Chapter 12.

576 ■ The Complete PCI Express Reference

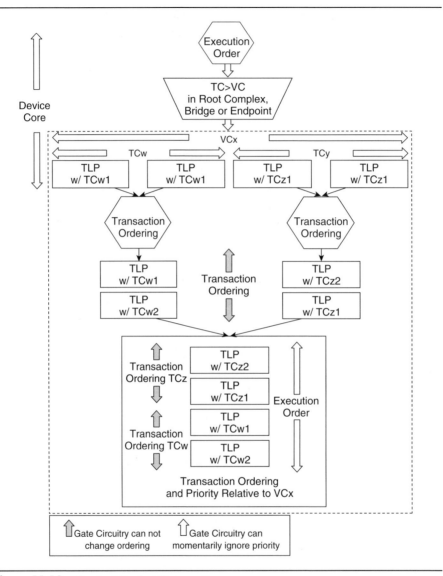

Figure 11.19 Physical Packet Preparation

Figure 11.20 shows an example of the first part of transmission priority considerations for a switch. The TLPs in the FIFOs of the receiving ports of the switches are ported by the Address Mapping and TC to VC Mapping Module to the correct transmitting ports based on the address and routing information in the TLPs. At the transmitting port, the TLPs are first parsed to the correct VC (VCx in this example) based on the TC number assignment and Port Arbitration considerations. Port Arbitration is based on which receiving port of the switch was the source of the TLP. See Chapter 12 for more information on Port Arbitration protocol.

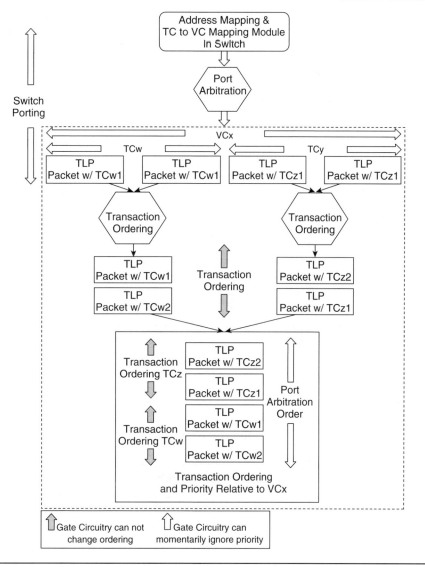

Figure 11.20 Physical Packet Preparation

Once TLPs are parsed by VCx, the result of execution order and Port Arbitration are integral to the sets of TLPs assigned to each VCx. As illustrated in Figures 11.19 and 11.20, TLPs with TCw are ordered (1 and 2 in this example) independently of the ordering of TLPs with TCz (1 and 2 in this example). As discussed in Chapter 10, transaction ordering is only applied to the TLPs of the same TC number. The resulting transmission priority of TLPs shown in Figure 11.19 is based on the transaction ordering within a specific TC number within the range of execution order for a specific VC number. The resulting transmission priority of TLPs in Figure 11.20 is based on transaction ordering within a specific TC number within the range of Port Arbitration for a specific VC number.

Figure 11.21 shows the second part of the Flow Control protocol to prepare the TLPs to be transmitted. Added in this example is a block of TLPs for VCy, which was formed by the same protocol discussed for VCx. VC Arbitration selects the transmission priority between the TLPs mapped to one VC number versus the other. In this example, assume a simple round robin for VC Arbitration. (Chapter 12 has more information about VC Arbitration.) A TLP is selected by the Gate circuitry from VCx, VCy, VCx, VCy, VCx, and so on. As discussed previously, the receiving port buffers are aligned to VC numbers. If the receiving port buffer for one VC number is temporarily full, the Gate circuitry selects TLPs from other VC numbers. In this example if the receiving buffer for VCy is full, Gate circuitry selects all transactions from VCx until the receiving buffer for VCy has space. Two additional considerations having to do with Gate circuitry selection of the TLP to transmit specific to a VC number are:

- For a specific TC number of a specific VC number, if the receiving buffer space is too small for the TLP with the highest priority for transmission, a smaller TLP with lower priority can be selected provided the transaction ordering is no violated.

- For a specific VC number, if the receiving buffer space is too small for all the TLPs of a specific TC number for transmission, a smaller TLP from a TC number with lower priority can be selected.

Once a TLP is selected by the Gate circuitry, it is transferred from the Transaction Layer to the Data Link Layer and the order of transmission becomes fixed, as shown in Figure 11.21. In the Data Link Layer, the SEQ# and LCRC are added and the resulting LLTP is placed in the retry buffer. The contents of the retry buffer are transmitted by the Physical Layer with strict ordering.

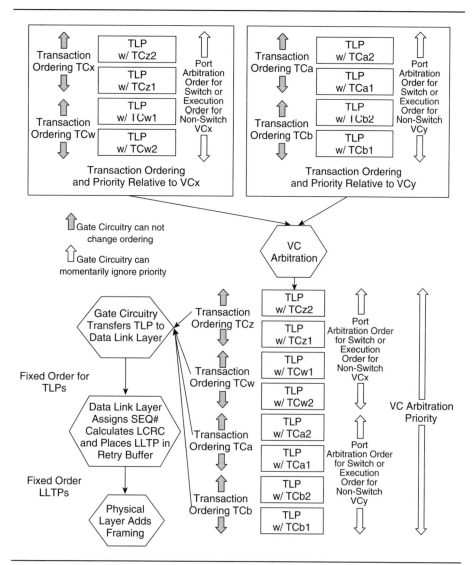

Figure 11.21 Physical Packet Preparation

Chapter 12

Flow Control Protocol: Part 2

This chapter, in conjunction with Chapter 11, discusses elements of the Flow Control protocol. Chapter 11 focuses on the foundation elements of the Flow Control protocol. This chapter examines the application of the foundation elements via LLTP tracking, Port Arbitration, Virtual Channel Arbitration, and Application for Quality of Service. The key concept is that the Flow Control protocol provides software a means to fine-tune the PCI Express fabric, to optimize the flow of transaction packets.

Introduction to the Flow Control Protocol: Part 2

This chapter is divided into three parts: The first part of this chapter discusses the LLTP tracking, which is the next consideration once the initial available buffer space has been established via Flow Control Initialization protocol and the Gate circuitry discussed in Chapter 11 has determined that currently buffer space is available at the LLTPs' receiving port. This part focuses on tracking the TLPs contained within the LLTPs and Physical Packets across the link. The second part of this chapter discusses the consideration that each transmitting port has buffers containing TLPs that have been assigned different Traffic Class (TC) numbers and are mapped to different Virtual Channel (VC) numbers. The transmission priorities of these TLPs, governed by the Port and VC Arbitrations, are examined in this part. The third part of this chapter is the application of the Flow Control protocol to establish a Quality of Service (QoS). One of the most important uses of the QoS is isochronous transfers, which will be discussed in detail.

Receiving Ports versus Transmitting Ports

As introduced in Chapter 11, in the following discussion the terms *receiving port* and *transmitting port* are used in reference to the TLPs. *Receiving port* refers to the portion of the port that receives the TLPs. *Transmitting port* refers to the portion of the port that transmits TLPs. In that the LLTPs contain the TLPs, the concept of receiving and transmitting ports also extends to the LLTPs.

VC0 versus VCx

As discussed in this book, virtual channels (VCs) are defined for each link but are indeed virtual. The VCs are part of the determination of transmission priority for the LLTPs. Once the transmission priority has been established, the LLTP is transmitted and the receiving port's only consideration is the maintenance of a strong sequence order identified by the SEQ#.

Once Flow Control Initialization has been successfully completed by for VC0, LLTPs can be transmitted according to the Flow Control protocol discussed in this chapter. The Flow Control protocol incrementally applies to the different VC numbers as these VCs are enabled and successfully complete Flow Control Initialization.

LLTP Tracking

As discussed in the previous chapter, the Gate circuitry in a transmitting port transfers TLPs to the Data Link Layer based on available buffer space at the receiving port. The Data Link Layer, in conjunction with the Physical Layer, is responsible for transmitting the Physical Packet containing the TLP within a Link Layer Transaction Packet (LLTP). For each TLP transferred to the Data Link Layer, the Data Link Layer adds a sequence number (SEQ#) and calculates a LCRC over the SEQ# and the TLP but not including the LCRC. This creates the LLTP that is stored in a retry buffer within the Data Link Layer. The SEQ# ensures that any LLTPs "lost" during transmission are identified and the LCRC ensures the integrity of the LLTP. Once the LLTP is placed into the retry buffer, the Physical Layer attempts to transmit the LLTP.

The other packets associated with the Data Link Layer are Data Link Layer Packets (DLLPs) used for link management. DLLPs do not include a SEQ# like an LLTP and are transmitted immediately without regard to available buffer space. The format of the LLTPs and DLLPs is discussed in detail in Chapter 7.

> This book assumes that the TLP transferred from the Transaction Layer is transformed to an LLTP (by addition of SEQ# and LCRC by the Data Link Layer), and it is the LLTP that is stored in the retry buffer. It is possible to only store the TLP in the retry buffer, and reattach the SEQ#, and recalculate the LCRC for a retry transmission. However, the Data Link Layer circuitry would still have to store the SEQ# assigned to that particular TLP so it is the same SEQ# for the retry transmission. Consequently, for ease of discussion and the probable implementation, the entire LLTP is assumed to be stored in the retry buffer.

LLTP Tracking at Transmitting and Receiving Ports

The LLTP tracking protocol includes support by counters and timers in the LLTPs' transmitting and receiving ports. The LLTP tracking protocol also includes Ack and Nak DLLPs. Before explaining the actual mechanism for LLTP tracking protocol, each of these entities will be defined.

The only other interaction between the transmitting and receiving ports of the LLTPs are Ack and Nak DLLPs; consequently, the exact mechanism to track LLTPs may vary per actual implementation of the Ack and Nak DLLPs within each port. The key concept is that the information and reasons for the transmission of the Ack and Nak DLLPs are compatible with the PCI Express specifications.

Finally, the entire focus is the retry buffer in the Data Link Layer of the transmitting port. As discussed in Chapter 11, the Gate circuitry determines if a TLP can be transferred to the retry buffer. The chapter discusses how TLPs contained in the retry buffer as LLTPs are transmitted and how their receipt at the receiving port is tracked. This tracking of the LLTPs provides acknowledgment of which LLTPs in the retry buffer and thus TLPs have been successfully transmitted.

Transmitting Port Counters, Registers, and Timers

The Data Link Layer is responsible for tracking the transmission of the LLTP retrieved from the retry buffer and for this purpose implements the following counters and timer in the transmitting port.

- **Next_Transmit_Seq# Counter:** Contains the next SEQ# to be attached to the next TLP transferred from the Transaction Layer by the Gate circuitry.
  - 12 bits in size counter ... (Field size).
  - Initialized to 000000000000b during L_Inactive link activity state.
  - Enabled with the transition from L_Init to L_Active link activity state. The transition represents the successful Flow Control Initialization for VC0.
- **ACKD_Seq# Register:** Contains the SEQ# of the most recently successfully received LLTP. The Ack or Nak DLLP always returns the SEQ# equal to the value of the ACKD_Seq# Register.
  - 12 bits in size register ... (Field size).
  - Initialized to 111111111111b during L_Inactive link activity state.
  - Enabled with the transition from L_Init to L_Active link activity state. The transition represents the successful Flow Control Initialization for VC0.
- **Retry_Num# Counter:** Counts the number of times the retry buffer has been retransmitted within the Retry protocol.
  - 2 bits in size counter ... (Field size).
  - Initialized to 00b during L_Inactive link state.
  - Enabled with the transition from L_Init to L_Active link activity state. The transition represents the successful Flow Control Initialization for VC0.
  - The PCI Express specification calls this counter REPLAY_NUM. This book adopts Retry_Num to affirm the associated with the retry buffer.

- **Retry_Timer:** Contains the time since the most recent Ack or Nak DLLP was received by the LLTPs' transmitting port.
    - 12 bits in size timer ... (Field size).
    - Initializes to 000000000000b during L_Inactive.
    - Disabled with the transition from L_Init to L_Active link activity state. The transition represents the successful Flow Control Initialization for VC0.
    - The input clock frequency for the timer is defined the subsequent "Other Considerations" section.
    - The PCI Express specification calls this counter REPLAY_TIMER. This book adopts Retry_Timer to affirm the associated with the retry buffer.

Receiving Port Counters, Timers, and Flags

In conjunction with the aforementioned counters and timer in the transmitting ports, each receiving port implements the following counter, timer, and flag:

- **Next_Rcv_Seq# Counter:** Contains the value that the SEQ# that should be included in the next LLTP received, if the sequence is correct and not a duplicate LLTP.
    - 12 bits in size counter ... (Field size)
    - Initialized to 000000000000b during L_Inactive link activity state.
    - Enabled with the transition from L_Init to L_Active link activity state. The transition represents the successful Flow Control Initialization for VC0.
    - The Next_Rcv_Seq# Counter is incremented by 1b whenever an LLTP is successfully received, is in the correct sequence, and is not a duplicate LLTP. It naturally rolls over.

- **AckNak_Latency Timer:** Counters time since the most recent ACK or NAK DLLP link management packet was scheduled to be transmitted.
  - 12 bits in size timer ... (Field size)
  - Initializes to 000000000000b during L_Inactive link activity state.
  - Enabled with the transition from L_Init to L_Active link activity state. The transition represents the successful Flow Control Initialization for VC0.
  - The input clock frequency for the timer is defined the subsequent "Other Considerations" section.
- **Nak_Sch Flag:** This is a flag set to 1b to ensure only one Nak DLLP is scheduled for transmission at any given time.
  - 1 bit in size flag.
  - Initializes to 0b during L_Inactive link activity state.
  - Enabled with the transition from L_Init to L_Active link activity state. The transition represents the successful Flow Control Initialization for VC0.

Ack and Nak DLLPs

The Ack and Nak DLLPs are unique to affirming the receipt of LLTPs but are not used for affirming receipt of DLLPs.

Ack DLLP An Ack DLLP acknowledges successful receipt of a single LLTP *or* a collection of LLTPs. Successful is defined as receipt of a correct SEQ# and correct LCRC in the LLTP. Successful is also defined for receipt of correct FRAMING in the Physical Layer. An Ack DLLP contains the SEQ# of a specific LLTP to acknowledge a successful receipt at the LLTP's receiving port. It is possible for the receiving port to have successfully received multiple LLTPs without transmitting an Ack DLLP. Eventually, when an Ack DLLP is transmitted to the LLTPs' transmitting port, the LLTP in the retry buffer with SEQ# equal to the AckNakSEQ# plus all of the older LLTPs in the retry buffer are defined as successfully received, and thus all are defined as acknowledged. The AckNakSEQ# is contained in the Ack DLLP equals the Next_Rcv_Seq# -1 Counter in LLTP's receiving port.

Nak DLLP A Nak DLLP acknowledges unsuccessful receipt of a single LLTP. Unsuccessful is defined as receipt an incorrect SEQ# and/or an incorrect LCRC. Unsuccessful is also defined for receipt of incorrect FRAMING in the Physical Layer. A Nak DLLP contains the SEQ# of a specific LLTP to acknowledge the unsuccessful receipt of the most recently received LLTP. It is possible for the LLTP's receiving port to have successfully received multiple LLTPs without transmitting an Ack DLLP. Consequently, the LLTPs' transmitting port that receives the Nak DLLP assumes that the LLTP in the retry buffer with SEQ# equal to the AckNakSEQ#, plus all of the other older LLTPs in the retry buffer were defined as successfully received and thus defined as acknowledged. The AckNakSEQ# that is contained in the Nak DLLP equals the Next_Rcv_Seq# -1 Counter in LLTP's receiving port.

> The value of the AckNakSEQ# in the Ack and Nak DLLPs must equal an unacknowledged LLTP in the retry buffer or the most recently acknowledged LLTP, otherwise the following occurs:
>
> - The Ack or Nak DLLP is discarded.
> - A **DLLPE** must be reported. See Chapter 9 for more information on errors.
>
> Exception: It is not an error if an Ack or Nak DLLP is received and no unacknowledged LLTPs are in the retry buffer, provided the AckNakSEQ# = ACKD_Seq# Register.

Retry Buffer and Acknowledgement

Until an Ack or Nak DLLP is received by the transmitting port, the LLTPs in the retry buffer that have been transmitted are defined as unacknowledged. "Transmitted" here means that the LLTP is contained in a Physical Packet and transmitted by the Physical Layer. Once the Ack or Nak DLLP is received, the retry buffer is purged of all unacknowledged LLTPs that were successfully received at the LLTPs' receiving port. The AckNakSEQ# value in the Ack or Nak DLLPs indicates the SEQ# of the most recently successfully received LLTP. The unacknowledged LLTPs with this SEQ# number and older SEQ#s are purged from the retry buffer. It is possible for two types of LLTPs to remain in the retry buffer after the most recent purging. One type is the LLTP that has not yet been transmitted and thus has not yet been defined as unacknowledged, but does in fact exist in the retry buffer. The other type is the LLTP that has been transmitted and is unacknowledged, and the associated SEQ# is newer than the AckNakSEQ# value in the last received Ack or Nak DLLP.

Protocols for Transmitting LLTPs, and Ack and Nak DLLPs

The protocol for how the counters, timers, flags, LLTPs, Ack DLLPs, and Nak DLLPs interact is illustrated in the flow charts and the associated text. The discussion is divided into two protocols: LLTPs Transmitting protocol and Ack and Nak DLLPs Transmitting protocol. The LLTPs Transmitting protocol includes how to respond to the Ack and Nak DLLPs received from the LLTPs' receiving port. The Ack and Nak DLLPs Transmitting protocol includes how to respond to the LLTPs received from the LLTPs' transmitting port.

It is possible to implement the flow charts differently than shown, but key elements must be retained it to ensure compatibility between two ports on the same link provided by two different vendors.

The LLTP Transmitting protocol, Ack, and Nak DLLPs Transmitting protocol are discussed in several parts.

LLTP Transmitting Protocol: Transfer of TLPs to Retry Buffer As previously discussed, the Gate circuitry transfers the TLP to the retry buffer in the Data Link Layer if sufficient available buffer space is available at the TLPs' receiving port. This determination is done for each buffer type for each enabled VC number. Additional criteria beyond the buffer type and VC number exist for the availability of buffer space at the LLTPs' receiving port. As illustrated in Figure 12.1, the Gate circuitry continuously transfers TLPs into the retry buffer with attachments of the SEQ# and LCRC to create LLTPs. In addition to the availability of buffer space at the TLPs' receiving port (not shown in the figure) are two other criteria: one is that transfers do not occur during the execution of the Retry Protocol. The other criterion relates to the retry buffer space. The transfers do not occur whenever the following equation is true, which is a measure of how full the retry buffer is:

- [Next_Transmit_Seq# Counter − ACKD_Seq# Register] $\geq 2^{[Fieldsize]}$ / 2 ... Fieldsize of 12 bits
    - The binary equivalents of ACKD_Seq# Register complemented and "1" is added to achieve one's complement. Any overflow is discarded.

- The one's complement of "ACKD_Seq# Register" is equivalent to " - ACKD_Seq# Register" in the equation.
- The equation is completed by adding binary equivalent of the Next_Transmit_Seq# Counter to the one's complement of "ACKD_Seq# Register". Any overflow is discarded.
- The equation is considered true if the resulting addition is = > 2048d. Note: This is unsigned arithmetic such that the comparison of the binary number to 2048d is always assumed to be a number = > 000000000000b.

LLTP Transmitting Protocol: Transmit LLTPs and Retry_Timer ... not part of Retry Protocol Once an LLTP is placed into the retry buffer, the Data Link Layer transmits it as soon as possible via Physical Packets and the Physical Layer. Once an LLTP is transmitted within a Physical Packet by the Physical layer, it becomes defined as an "unacknowledged" LLTP in the retry buffer. Until the Retry protocol is executed, each LLTP in the retry buffer is transmitted once.

As previously discussed, the acknowledgement of the LLTP's receipt by the LLTPs' receiving port purges the LLTP from the retry buffer. The intent of the LLTP Transmitting protocol is to receive Ack and Nak DLLPs to purge unacknowledged LLTPs in the retry buffer. Until a specific transmitted LLTP is purged from the retry buffer, it is defined as unacknowledged. The Ack and Nak DLLPs indicate which LLTPs have been successfully received by the LLTPs' receiving port and defined as acknowledged. This part of the LLTP Transmitting protocol is discussed in the next section. As illustrated Figure 12.2, the LLTP Transmitting protocol also implements a Retry_Timer to periodically check that the contents of the retry buffer. If any LLTPs are contained in the retry buffer at the Retry_Timer time out, the LLTP Transmitting protocol implements the Retry protocol to retransmit the LLTPs. The protocol of the Retry_Timer is as follows:

- Retry_Timer is disabled and initialized whenever the retry buffer contains no LLTPs.

- Retry_Timer is enabled and starts counting or remains enabled and continues counting with the transmission by the Physical Layer of the last symbol of a Physical Packet containing an LLTP. The transmission may be for the first time an LLTP is transmitted or for an LLTP that has previously been transmitted. If a group of LLTPs are being transmitted back-to-back, the last symbol is defined by the first LLTP of the group.

590 ■ The Complete PCI Express Reference

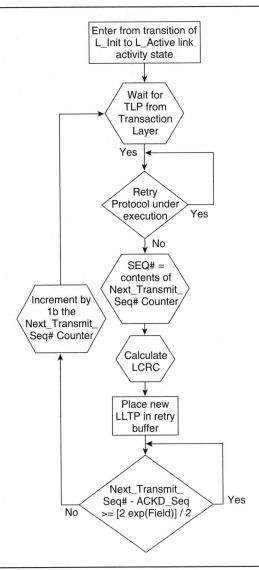

Figure 12.1 Transfer of TLPs

- When an Ack DLLP is received, it purges one or more of the unacknowledged LLTPs in the retry buffer to acknowledge successful receipt at the LLTPs' receiving port. The Retry_Timer is disabled and initialized to "0" if no unacknowledged LLTP(s) remain in retry buffer. Otherwise, for receipt of an Ack DLLP, the Retry_Timer continues counting if no LLTP was purged or is initialized to "0" and continues counting if an LLTP was purged. Ack DLLP processing is illustrated in Figure 12.3.

- When a Nak DLLP is received, it purges one or more of the unacknowledged LLTPs in the retry buffer, and the Retry_Timer is disabled and initialized to "0." Nak Processing is illustrated in Figure 12.3 and the Retry Protocol is illustrated in Figure 12.4.

LLTP Transmitting Protocol: Receipt of Ack and Nak DLLPs ... not part of Retry Protocol As discussed in the previous section, the intent of the LLTP Transmitting protocol is to receive Ack and Nak DLLPs and to purge unacknowledged LLTPs in the retry buffer. The Ack and Nak DLLPs indicate which unacknowledged LLTPs in the retry buffer can be purged to reflect acknowledgement of successful reception by the LLTPs' receiving port. As illustrated in the Figure 12.3, these activities can be viewed in four parts: DLLP Processing, Purging, Ack Processing, and Nak Processing.

- **DLLP Processing**: The Physical Packet containing the Ack or Nak DLLP must be received without error. If an error is found, the content of the Ack or Nak DLLP is not used to purge unacknowledged LLTPs in the retry buffer. See Chapter 9 for more information on errors.

- **Purging**: When an Ack or Nak DLLPs is received without error it contains an AckNakSEQ# value to be tested. Test A: If no unacknowledged LLTP exists in the retry buffer *but* AckNakSEQ# value = ACKD_Seq#, the Register simply discards the DLLP without error. Otherwise proceed to Test B. Test B: If AckNakSEQ# value does not equal a SEQ# of an unacknowledged LLTP in the retry buffer *and* is not equal to the ACKD_Seq# Register, the DLLP is discarded with a **Data Link Layer Protocol Error (DLLPE)**. Otherwise, the DLLP is defined as "acceptable." As illustrated in Figure 12.3, an acceptable DLLP is used to purge unacknowledged LLTPs in the retry buffer with SEQ# = or older than AckNakSEQ# value. Subsequently, the ACKD_Seq# Register is set = AckNakSEQ# value and determination is made whether Retry_Num Counter is initialized to 00b.

- **Ack Processing:** As illustrated in Figure 12.3 and previously discussed, the balance of the Ack DLLP processing is determining how the Retry _Timer continues.

- **Nak Processing:** As illustrated in Figure 12.3, if a Nak DLLP was received, it must be determined whether Retry protocol is executed.

LLTP Transmitting Protocol ... Retry Protocol The Retry protocol is entered as illustrated in Figures 12.2 and 12.3. The transfer of TLPs to the retry buffer is blocked during the execution of the Retry protocol (Figure 12.1). The Retry protocol is illustrated by Figure 12.4.

The purpose of the Retry protocol is to retransmit all unacknowledged LLTPs in the retry buffer. The transmission is in the original order. As previously discussed, an LLTP in a retry buffer that has *not* yet been transmitted within a Physical Packet of the Physical Layer is *not* defined as unacknowledged. Consequently, the retransmission may not be of all the LLTPs in the retry buffer. The Retry_Timer is initialized to "0," is enabled, and starts counting with the retransmission by the Physical Layer of the last symbol of a Physical Packet containing an LLTP. If a group of LLTPs are being retransmitted back-to-back, the last symbol is defined by the first LLTP of the group.

Part of the Retry protocol is to track the Retry_Num Counter. If the Retry protocol is entered multiple times without intervening initialization of the Retry_Num Counter, the LLTPs' transmitting port must consider that the link is not fully operational. Consequently, when the Retry_Num Counter rolls over an error is reported and retraining of the link is requested.

An additional consideration when executing the Retry protocol is receipt of Ack or Nak DLLP. If an "acceptable" Ack or Nak DLLP is received during the Retry protocol execution it can be processed in two ways. One way is for the Retry protocol to "ignore" these DLLPs until the Retry protocol is complete. If the DLLP(s) is ignored, it must be preserved and subsequently processed upon completion of the Retry protocol. The other way is for the Retry protocol to process the Ack or Nak DLLP immediately, and simply to remove any unacknowledged LLTPs purged by the DLLP from the list of LLTPs to be retransmitted. It is required that once an LLTP is transferred to the Physical Layer its retransmission within a Physical Packet must be completed.

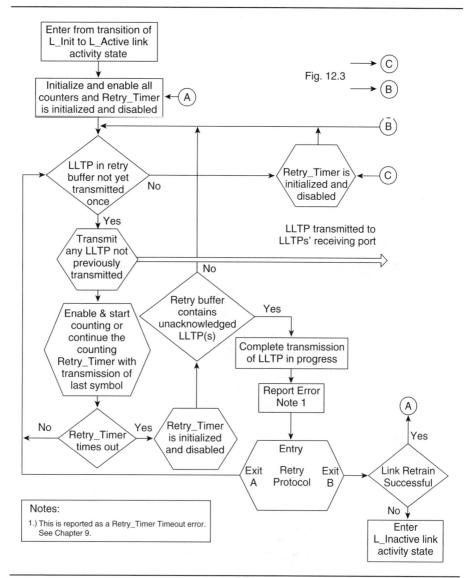

Figure 12.2 Transmit LLTPs and Retry_Timer

594 ■ The Complete PCI Express Reference

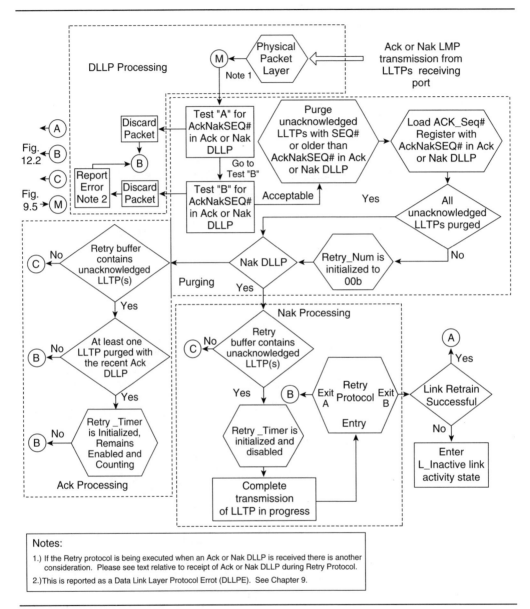

Figure 12.3 Receipt of Ack and Nak DLLPs

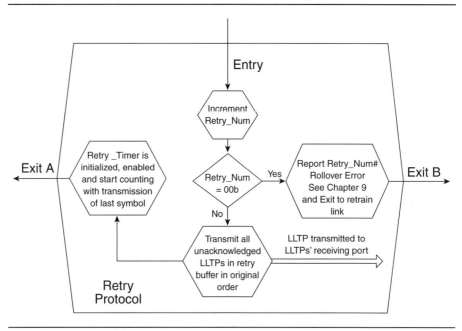

Figure 12.4 Retry Protocol

DLLP Transmitting Protocol: AckNak Latency Timer and the Nak_Sch Flag Protocol As illustrated in Figures 12.5 and 12.6, under the typical operation of successfully receiving LLTPs in the correct sequence, an Ack DLLP is periodically scheduled for transmission to the LLTPs' transmitting port. It contains the SEQ# of the last successfully received LLTP. The maximum period between transmissions of Ack DLLPs under typical operation is defined by the AckNak_ Latency Timer.

The Ack DLLP is scheduled for earlier transmission than defined by the AckNak_ Latency Timer if a duplicate LLTP is received. In addition to the transmission of Ack DLLP, a Nak DLLP is scheduled for transmission when an LLTP is unsuccessfully received or an out-of-sequence LLTP is received. If a Nak DLLP is scheduled, another Nak DLLP is not scheduled for transmission via the use of the Nak_Sch Flag.

The term *scheduled* reflects the fact that the limited link bandwidth does not permit the Ack or Nak DLLP to be transmitted immediately. Later in this chapter is a summary of the transmission priority between LLTPs and DLLPs.

DLLP Transmitting Protocol: Receipt of LLTP As illustrated in Figure 12.5, the receipt of Physical Packets must first be checked for errors, and the CRC of the LLTP itself is checked. Due to the complexity, the checking for these errors and CRC are discussed in Chapter 9. Relative to Figure 12.5, the LLTP is provided to the Data Link Layer as unsuccessfully or successfully received. If successfully received by the protocol discussed in Chapter 9, the LLTP is processed to determine whether the SEQ# is correct, duplicate, or out of sequence. This is discussed in the next section.

As illustrated in Figure 12.5, if an LLTP is unsuccessfully received per the protocol of Chapter 9, the TLP is discarded, any allocated buffer space is freed up, and a **BAD LLTP** error is reported. Subsequently, the Nak_Sch Flag is tested; if a Nak DLLP is not currently scheduled for transmission, one is scheduled.

DLLP Transmitting Protocol: Processing LLTP As illustrated in Figure 12.6, once an LLTP is successfully received the SEQ# number must be checked. If the SEQ# in the LLTP = Next_Rcv_Seq# Counter, the sequence of the LLTP is correct and the TLP is transferred to the Transaction Layer. The Next_Rcv_Seq# Counter is incremented and any scheduled transmission of a Nak DLLP is canceled. If this is the first TLP to transfer to the Transaction Layer since Flow Control Initialization, the ACKNak Latency Timer is initialized, enabled, and started. If the SEQ# in the LLTP does not equal Next_Rcv_Seq# Counter, the LLTP is either a duplicate or out of sequence. The associated TLP is discarded and any allocated buffer space is freed up.

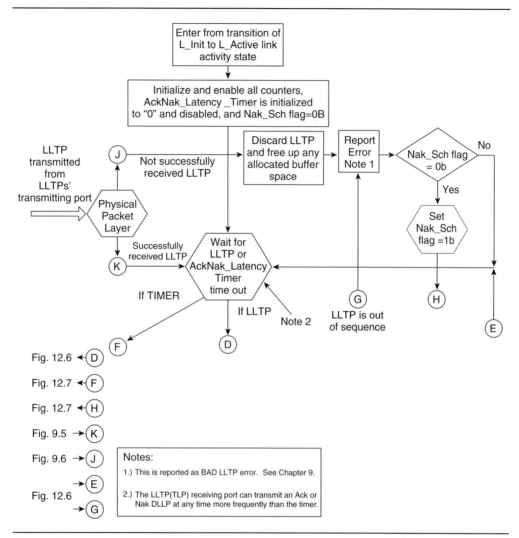

Figure 12.5 LLTP Receiving

- [Next_Rcv_Seq# Counter - SEQ#] <=2 [Field size] / 2 ... Field size of 12 bits.
    - The binary equivalent of SEQ# is complemented and "1" is added to achieve one's complement. Any overflow is discarded.
    - The one's complement of "SEQ#" is equivalent to "- SEQ#" in this equation.
    - This equation is completed by adding the binary equivalent of the Next_Rcv_Seq# Counter to the one's complement of "SEQ#." Any overflow is discarded.
    - The equation indicates a duplicate LLTP if the result < = 2048d. Otherwise, the LLTP is defined as out of sequence. Note: This is unsigned arithmetic such that the comparison binary number to 2048d is always assumed to be a number = > 000000000000b.

If the LLTP is a duplicate, an ACK DLLP is immediately scheduled for transmission. If the LLTP is out of sequence, an error is reported, and a Nak DLLP is scheduled for transmission if one is not already scheduled (see "G" of Figure 12.5).

DLLP Transmitting Protocol: Scheduling LLTP Transmission Once it is determined that an Ack or Nak DLLPs needs to be transmitted, it must be scheduled. As illustrated in Figure 12.7, several things must be considered when an Ack or Nak DLLPs is scheduled. First, once an Ack or Nak DLLP is scheduled the AckNak_Latency Timer is initialized, it is enabled, and it starts counting in parallel to the port waiting to transmit these DLLPs. The port must respect the transmission priority for LLTPs and DLLPs. Second, if an LLTP is successfully received and it is in the right sequence; any scheduled Nak DLLP is canceled (set Nak_Sch Flag = 0b in Figure 12.6). Third, when an Ack or Nak DLLP is scheduled for transmission, the current Next_Rcv_Seq# Counter value is attached to the scheduled DLLP. It is possible for the Next_Rcv_Seq# Counter value to change while the DLLP is waiting for transmission, but the DLLP must use the Next_Rcv_Seq# Counter value when the DLLP's transmission is scheduled.

Chapter 12: Flow Control Protocol: Part 2 ■ 599

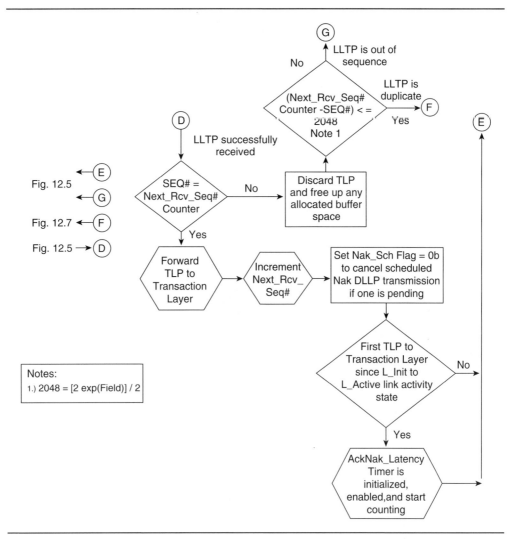

Figure 12.6 Processing of LLTP

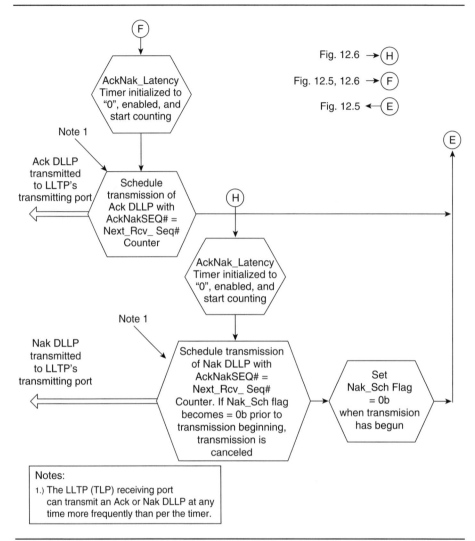

Figure 12.7 Scheduling of DLLPs and Transmission

Other Considerations

In addition to the protocols illustrated in Figure 12.1 through 12.7, several other things must be considered in transferring LLTPs between the two ports on a link: Nullified TLP, Retry_Timer and AckNak_Latency Timer, and retry buffer size.

Nullified TLP (Cut-through) As discussed in Chapter 11, it is possible to send a nullified TLP across the link without affecting the transmitting and receiving tracking mechanisms. The protocol to support the transmission of a nullified TLP is called *cut-through*. A nullified TLP and its associated LLTP are simply ignored by the receiving port. The nullified TLP is created as a normal TLP, except the LCRC has a value that is the inverted value of what it would normally have been and the END K symbol in the Physical Packet FRAMING is the EDB K symbol instead. Per the Cut-through protocol, the transmission of a nullified TLP does not increment the FCC_Consumed Counter and its reception at the receiving port does not increment the FCC_Allocated or FCC_Processed Counters. Relative to LLTP tracking, it does not increment the Next_Transmit_Seq# Counter at the LLTPs' transmitting port. As illustrated in Figure 12.5, it has no effect on the counters, timers, or flags of the LLTPs' receiving port other than and a **CRC Error** in the Physical Packet. See more information in Chapter 9 on errors.

Retry_Timer and AckNak_Latency_Timer In the above discussions two timers invoke certain actions when the timeout occurs. The two timers are Retry_Timer and AckNak_Latency_Timer and their respective timeouts are determined as follows:

- **Retry_Timer Timeout:** For a specific PCI Express device relative to the Retry_Timer, timeout is from the last bit of the previous LLTP to the first bit of the present LLTP measured at the transmitter port of the LLTP:
  - Retry_Timer timeout (in units of symbol time) = [(Ack Transmission Latency Value) x 3] + Rx_L0s_ Adjustment.
  - Rx_L0s_ Adjustment is the amount of time (in units of symbol time) for receiver circuits to transition from L0s link state to L0 link state. The transition time is defined by the L0s Exit Latency bits of the Link Capabilities Register in the configuration register block of the PCI Express device. If L0s is not enabled, the transition time is "0."
  - Symbol time equals to the bit rate x 10 (see bullet below).

- **AckNak_Latency_Timer Timeout:** For a specific PCI Express device relative to the AckNak_Latency_Timer timeout is from the last bit of the previous LLTP to the first bit of the present LLTP measured at the receiver port of the LLTP:
  - AckNak_Latency_Timer timeout (in units of symbol time) = [Ack Transmission Latency Value] + Tx_L0s_ Adjustment.
  - Tx_L0s_ Adjustment is the amount of time (in units of symbol time) for transmitter circuits to transition from L0s link state to L0 link state. The transition time is defined by the L0s Exit Latency bits of the Link Capabilities Register in the configuration register block of the PCI Express device. If L0s is not enabled, the transition time is "0."
  - Symbol time equals the bit period x 10. Presently, this is defined for only the one Data Rate Identifier of 2.5 gigabits per second per TS1 and TS2 OSs. The maximum Data Rate Identifier possible and the maximum Data Rate Identifier agreed to by the ports on a specific link are defined in the Link Capabilities and Link Status Registers, respectively. The value in the Link Status Register is used to calculate the symbol time. For the value of 2.5 gigabits per second the bit period is 399.88 to 400.12 picoseconds. See Chapter 8 for more information.

The Ack Transmission Latency Value is dependent on the link width and the maximum payload size. The Ack Transmission Latency values are summarized Table 12.1 in symbol time values. The "Max Payload Size" in Table 12.1 is defined by the LLTP transmitter for AckNak_Latency_Timer timeout or by the LLTP receiver for Retry_Timer timeout. The value of the "Max Payload Size" for the transmitter and receiver are defined by the Device Control Register as follows:

- Bits [7::5] of the Device Control Register define the maximum payload a PCI Express device can support in any LLTP except for memory read requester transactions. This is related to the PD buffer.
- Bits [14::12] of the Device Control Register define the maximum payload a PCI Express device can request in a memory read requester transaction. This is related to the CPLD buffer.

The tolerance for the values in Table 12.1 are defined for −0% / +100%

Table 12.1 Ack Transmission Latency Value

		Link Width						
		x1	x2	x4	x8	x12	x16	x32
Max Payload Size	128 bytes	237	128	73	67	58	48	33
	256 bytes	416	217	118	107	90	72	45
	512 bytes	559	289	154	86	109	86	52
	1024 bytes	1071	545	282	150	194	150	84
	2048 bytes	2095	1057	538	278	365	278	148
	4096 bytes	4143	2081	1050	543	706	534	276

Retry Buffer Size As previously discussed, the Gate circuitry in the transmitting port must place the LLTP into the retry buffer, and relies on the Data Link Layer to subsequently transmit the LLTP from the retry buffer. The rate of filling the retry buffer should be less than the rate of emptying the retry buffer. One important value to keep in mind when determining the retry buffer size is related to the AckNak_Latency_Timer timeout. For the protocol discussed in this chapter, the Ack-Nak_Latency_Timer timeout determines when an Ack or Nak DLLP is transmitted to acknowledge successful receipt at the LLTPs' receiving port of a portion of the LLTPs in the retry buffer of the LLTPs' transmitting port. Consequently, the number of maximum-sized LLTPs that can be transmitted referenced to the AckNak_Latency_Timer timeout is a value. Table 12.2 provides the summary of the Retry Buffer Size Factor values. The Retry Buffer Size Factor values represent the number of LLTPs that can be transmitted (of the Max Payload Size stated in the table) between Ack or Nak DLLPs sent by the AckNak_Latency_Timer timeout.

The Retry Buffer Size Factor provides a retry buffer "basic" size. The size must be adjusted to take the following into account:

- Any processing times for LLTPs passing through the Data Link Layer and Physical Layer from the retry buffer to the link must be considered.

- The delays associated with a link's transition time from the L0s to the L0 link states must be considered. That is, the transmitting port transitions from L0s to L0 link state and begins transmitting LLTPs. The receiving port must transition from L0s to L0 link state during the time the LLTPs are being received. Eventually, the receiving port can send an Ack or Nak DLLP to acknowledge receipt of the LLTPs. However, there may be a large number of LLTPs in the retry buffer (due to a long transition time) representing those that were transmitted. Prior to the receipt of an Ack or Nak DLLP by the transmitting port, the retry buffer must be large enough so as not to have filled up.

- If a transmitting port connected to the link does not transmit LLTPs as large as stated by the Max Payload Size versus Link Width in the table, it may reduce the size of its retry buffer below the minimum size required.

Table 12.2 Retry Buffer Size Factor

		Link Width					
		x1	x2	x4	x8	x12	x16	x32
Max Payload Size	128 bytes	1.4	1.4	1.4	2.5	3.0	3.0	3.0
	256 bytes	1.4	1.4	1.4	2.5	3.0	3.0	3.0
	512 bytes	1.0	1.0	1.0	1.0	2.0	2.0	2.0
	1024 bytes	1.0	1.0	1.0	1.0	2.0	2.0	2.0
	2048 bytes	1.0	1.0	1.0	1.0	2.0	2.0	2.0
	4096 bytes	1.0	1.0	1.0	1.0	2.0	2.0	2.0

PCI Express Arbitration

As discussed in Chapter 2, PCI Express devices are all interconnected point-to-point with a link. Each link can support simultaneous bi-directional operation; consequently, neither arbitration for ownership of the link nor central arbitration circuitry is required as they would be with PCI and PCI-X bus segments. However, the architecture of PCI Express's fabric dictates additional arbitration considerations for link bandwidth. Consider Transaction Layer Packets (TLPs) flowing downstream split off to different downstream PCI Express devices at each switch. Increasingly fewer TLPs at each downstream port vie for link bandwidth on links farther and farther downstream. Conversely, as TLPs flow upstream from multiple PCI Express devices, they are merged in each switch and must vie for upstream link bandwidth. More TLPs vie for link bandwidth on links farther and farther upstream (Note: This is a simple introductory observation. Downstream link bandwidth issues are due to peer-to-peer transactions, which will be discussed later).

Figure 12.8 summarizes the buffers in a PCI Express fabric required for TLPs. As discussed in Chapter 3, the ports consist of functions within PCI Express devices. Each function contains a virtual PCI/PCI Bridge with associated buffers for TLPs. The Root Complex has a general purpose buffer for TLPs associated with virtual HOST/PCI Bridge.

Two primary elements of PCI Express arbitration are *Port Arbitration* and *Virtual Channel Arbitration*. These are identified with the applicable buffers using the labels *PA* and *VA* in Figure 12.8. Two associated elements are *Port Arbitration: Peer-to-Peer* and *Transaction Ordering*. These are identified with the applicable buffers by the labels *PAP* and *TO* in Figure 12.8.

606 ■ The Complete PCI Express Reference

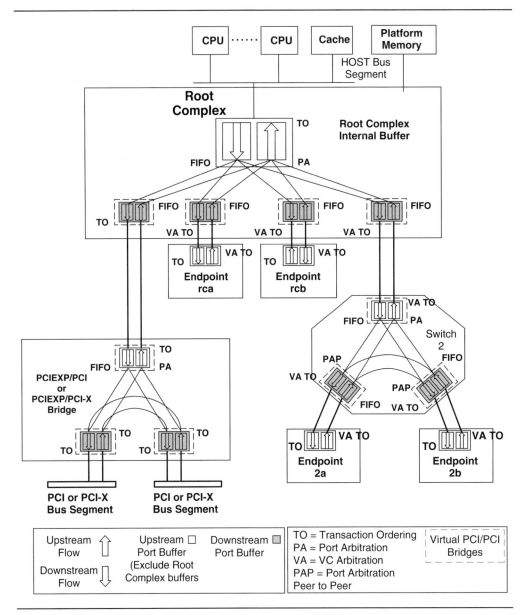

Figure 12.8 TLPs' Buffers throughout the PCI Express Fabric

Other Arbitration Considerations

Before discussing the PCI Express arbitration protocol specific to TLPs, first consider that Physical Packets traverse the links and contain two types of link packets. As discussed in Chapter 8, one type of link packet is the LLTP, which contains one TLP, and the other type is the DLLP, which contains link management packets. When the associated LLTPs are transmitted onto the link, there may be DLLPs also vying for link bandwidth. Consequently, each Physical Layer of each transmitting port must consider arbitration between Physical Packets containing LLTPs and DLLPs. These Physical Packets arrive at the Physical Layer to be transmitted in no particular order except for the LLTPs, which are internally ordered at the Data Link Layer. The priority (strong order) of the Physical Packets to be transmitted is as follows:

- Complete current transmission of any LLTP or DLLP (Highest priority)
- Nak DLLPs
- Ack DLLPs
- UpdateFC DLLPs
- LLTPs due to retry transmission
- LLTPs not previously transmitted
- FC DLLPs that are not UpdateFC DDLPs
- All other link management link packets (Lowest priority)

> **Note:** Ack or Nak link packets can be transmitted by the layer more frequently than by the AckNak_Timer timeout; these link packets have to be part of the lowest priority group).

Transaction Layer Packet (TLP) Arbitration

As discussed on Chapter 11, one consideration of the Flow Control protocol was related to the TLPs' available buffer space at the receiving port based on Flow Control Credits. Another consideration of the Flow Control protocol is determining the next TLP to transmit onto a link, to port through a switch, port through a bridge, internal porting in a Root Complex, or internal porting in an endpoint. The most complex implementation of the Flow Control is related to TLPs sourced from different PCI Express devices merging at switches' buffers for transmission onto upstream links and eventually into the Root Complex. This chapter focuses on TLP transmission and priority schemes relative to Port Arbitration and VC arbitration within a Root Complex, switch, bridges, and endpoints as illustrated in Figure 12.8. Transaction Ordering is an allied issue referenced in this chapter but discussed in detail in Chapter 10.

The two types of buffers are defined as *buffer* and *FIFO*. The term *buffer* here refers to a set of registers in ports that must implement combinations of Transaction Ordering and Port Arbitration, and VC Arbitration protocols. The term *FIFO* refers to a set of registers implemented with strong ordering. This book has adopted the use of FIFO in the different ports and internal buffers throughout a PCI Express fabric to simplify the discussion of Flow Control protocol For example, the use of FIFOs at the receiving portion of the ports ensures that the results of Transaction Ordering, Port Arbitration, and VC Arbitration protocols implemented by the associated transmitting ports on the other end of the links are maintained.

The Flow Control protocol discussion begins with arbitration and transaction ordering within switches, which is the most complex. The Flow Control protocol discussion concludes with arbitration and transaction ordering in other PCI Express devices.

Switches with Upstream TLP Flow

The protocol for TC to VC Mapping is discussed extensively in Chapter 11. As indicated by that discussion, the Root Complex, switches, bridges, and endpoints the TC to VC Mapping assign specific TLPs to specific Virtual Channel (VC) numbers based on the Traffic Class (TC) numbers assigned to the TLPs. As the TLPs flow through the PCI Express fabric, the TC numbers provide arbitration information, particularly in switches where TLPs from multiple sources with other TC numbers are merged to specific VC numbers.

In switches, the complexity of multiple TLPs received on multiple ports makes TC to VC Mapping very complex. In addition, switches must consider address, ID Routing, and Implied Routing. As previously discussed these functions are combined in the Address Mapping and TC to VC Mapping Module in the switch.

Consider a switch that implements three downstream ports with a range of VCs such as the one illustrated in Figure 12.9. Each VC number is assigned a set of upstream flowing TLPs based on their TC numbers.

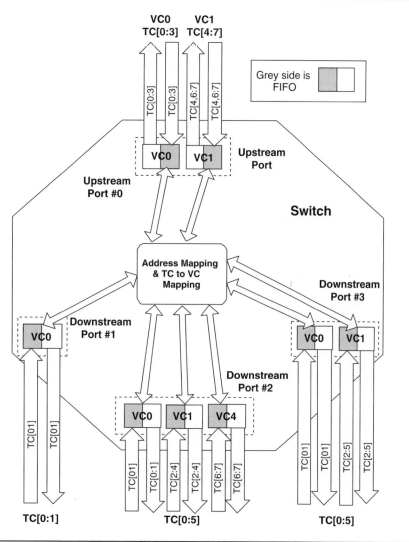

Figure 12.9 Generic Switch, VC, and TC

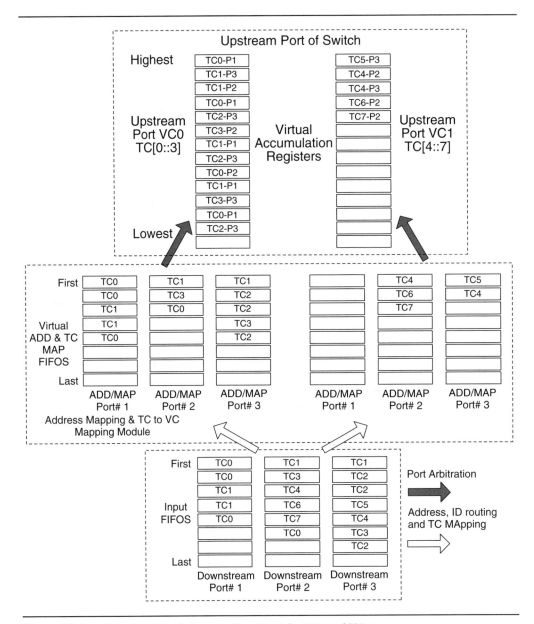

Figure 12.10 Internal Switch Transaction Switch, VC, and TC

Upon arrival, the TLPs for the different VCs are placed in FIFOs to retain the transaction ordering and transmission priority determined by the downstream PCI Express devices. As previously discussed in this book, the VCs only exist as a virtual concept relative to transmitting ports. Thus, for this discussion, the buffer for each receiving port is viewed as a single FIFO containing the TLPs from all VCs. As illustrated in Figure 12.10, TLPs (identified by their TC numbers) must be parsed to the correct upstream port VC by the Address Mapping and TC to VC Mapping Module. To illustrate this parsing, a Virtual Add and TC Map FIFO is shown in Figure 12.10; it is only for discussion purposes. As illustrated by Figure 12.10, the Address Mapping & TC to VC Mapping Module has determined that the TLPs received on the downstream ports are to be ported to the upstream port per address or routing information in the TLPs. The alternate destination per address and routing information is another downstream port (peer-to-peer transactions). Once the destination of the TLPs has been established to the upstream port, the Address Mapping and TC to VC Mapping Module must map the TLPs to the appropriate virtual FIFOs related to one of the VC numbers on the upstream port. The mapping is accomplished by the TC number of the TLP matching the range of TC numbers mapped to each VC number. In this example, TLPs assigned TC [0::3] are mapped to VC0 and TLPs assigned TC [4,6,7] are mapped to VC1.

TLPs are now ready to arbitrate for priority porting through the switch (Port Arbitration) and arbitrate for link bandwidth at the upstream transmitting port (VC Arbitration and Transaction Ordering).

Port Arbitration

As previously discussed, the process of determining the priority of eventual transmission of the merging TLPs among the TLPs assigned to a specific VC number at the transmitting port requires arbitration. This arbitration is defined as Port Arbitration and it partly determines priority for link bandwidth among the TLPs of a specific VC number. In this example, for each upstream port VC the TLPs identified by their TC numbers "exist" in the virtual Add and TC Map FIFOs, sorted by the down receiving ports (Port# [1::3]). The Port Arbitration defines the priority for a TLP by its receiving port. In this example, assume a simple round robin arbitration that provides the following priority: Port# 1, Port# 3, Port #2, Port# 1, Port# 3, and so on. The resulting priority order of the TLPs is shown in the virtual Accumulation Registers for each upstream port VC number. The virtual Accumulation Registers shown in Figure 12.10 are only for discussion purposes. The suffix "Px" in each entry is only to indicate which receiving port is its source. Port# 0 is reserved for a switches' upstream port or the Internal Buffer of the Root Complex. The downstream port numbering begins with Port# 1. At this point, Port Arbitration has been completed, but the TLPs are not ready for transmission to the upstream link. VC Arbitration and Transaction Ordering each have priority issues, both discussed in the next section.

Port Arbitration Protocol

Port Arbitration is only defined by the PCI Express specification for the transmitting portion of switches' ports. It is not defined for the Root Complex's downstream ports or endpoints' upstream ports. It is defined by this book for the Root Complex's internal buffer and transmitting portion of bridges' upstream port to complete the discussion (see the following discussion).

The balance of the Port Arbitration protocol is as follows:

- Port Arbitration within a specific VC number of a specific port is independent of Port Arbitration within other VC numbers of the same port.
- Port Arbitration within a specific port is independent of Port Arbitration of other ports.
- Port Arbitration is independent of VC Arbitration.
- The priority schemes supported for Port Arbitration are (discussed in detail later):
    - Hardware Fixed (also known as Round Robin)
        - Any Round Robin like
    - Programmable Weighted (also known as Programmable Weighted Round Robin)
        - 32 phases
        - 64 phases
        - 128 phases
        - 256 phases
    - Timed Based Weighted (also known as Time Based Weighted Round Robin)
        - 128 phases

VC Arbitration

As illustrated in Fig 12.10, the Port Arbitration has provided a priority between the TLPs from downstream receiving ports. It is required because the switch is merging TLPs from different downstream PCI Express devices to a specific VC. The VC Arbitration determines the priority of the transmission between VCs on the upstream port. In the example illustrated in Figure 12.11, assume two virtual Accumulation Registers with the result of Port Arbitration. Remember, the virtual Accumulation Registers are only for discussion purposes. The "P" suffix is removed from the entries of the virtual Accumulation Registers to emphasize that the source of the TLP is not a consideration, only the TC number assigned to each TLP is a consideration. The small letter suffixes are added to aid following the discussion.

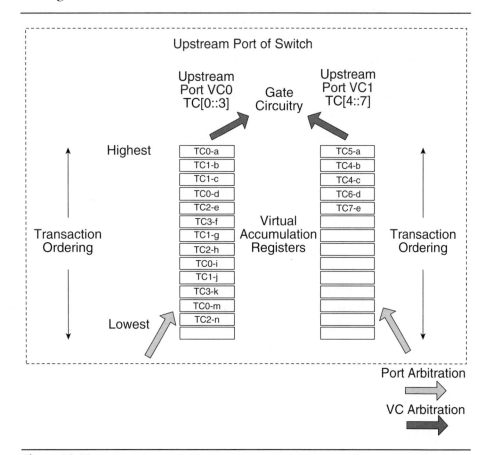

Figure 12.11 Internal Switch Transaction Switch, VC, and TC

In this example, assume a simple round robin arbitration that provides the following priority: VC0, VC1, VC0, VC1, and so on. If one *disregards* transaction ordering, the TLPs are transmitted in the following order first to last using their TC numbers as identification: TC0-a, TC5-a, TC1-b, TC4-b, TC1-c, TC4-c, TC0-d, TC6-d, TC2-e, TC7-e, TC3-f, TC1-g, TC2-h, and so on.

If transaction ordering is *included*, which it must be per the PCI Express specification, all TLPs of a specific TC number are considered when any one is transmitted. The PCI Express specification does not provide any further discussion on this topic. One interpretation is to select the next highest priority TLP per a specific TC, but allow another TLP of the same TC number to be transmitted instead per Transaction Ordering protocol.

VC Arbitration Protocol

VC Arbitration only occurs at the transmission portion of the ports. It is only defined for the Root Complex's downstream ports, all ports of switches, and upstream ports of endpoints. It is not defined for the Root Complex's Internal Buffer or the upstream ports of bridges. In the case of bridges, only one VC number can be enabled and that is the default VC0.

The balance of the VC Arbitration protocol is as follows:

- VC Arbitration is between VC numbers of the same port.

- VC Arbitration within a specific port is independent of VC Arbitration of other ports.

- VC Arbitration is independent of Port Arbitration.

- The VC Arbitration priority schemes are divided into two groups:
    - VC numbers VCn+1 to VC7 are defined as the VC upper group.
    - VC numbers VCn to VC0 are defined as the VC lower group. The VC lower group can use any of the priority schemes listed below.
    - VC upper group as a group has a high priority than the VC lower group as a group.
    - The number "n" is programmable via the configuration address space and is the value of the Low Priority Extended VC Count field ([2::0] = "n" in the Port VC Capability Register 1.
    - If n = 1 to 6, both VC upper and VC lower groups are defined.

- If n = 0, only the VC upper group is defined and the strict priority scheme of the VC upper group applies to all VCs of the port.
- If n = 7, only the VC lower group is defined and the priority scheme of the VC lower group applies to all VCs of the port.

■ The priority scheme of the VC upper group is only strict priority with VC7 being the highest.

■ The priority schemes defined for the VC lower group are (discussed in detail later):
- Hardware Fixed (also known as Round Robin)
  ■ Any Round Robin like
- Programmable Weighted (also known as Programmable Weighted Round Robin)
  ■ 32 phases
  ■ 64 phases
  ■ 128 phases

> For n = 0 to 6, VC7 is defined with the highest priority. At each port in the PCI Express fabric, the transmitting of TLPs mapped to VC7 occurs before TLPs mapped to VC0 to VC6. Consequently, VC7 is reserved for Physical Packets for isochronous transfers.

> VC upper group as a group has a higher priority than the VC lower group as a group, and the VC upper group only uses strict priority. Consequently, all Physical Packets associated with low latency transfers should be mapped to VCs of the VC upper group.

> The VC lower group as a group is lower priority per strict priority than the VC upper group. The TLPs associated with the VC lower group cannot be transmitted until no TLPs in the VC upper group are ready for transmission. TLPs may not be ready for transmission because no TLPs are ready, or because insufficient buffer space is available at the receiving port.

Switches with Downstream Flowing and Peer-to-Peer TLP

The previous discussion focused on Port Arbitration and VC Arbitration within a switch for an upstream TLP flow. Two other TLP flows must be considered: downstream flowing and peer-to-peer transactions (also known as downstream flowing TLP and peer-to-peer TLP).

A switch's downstream port must transmit TLPs flowing to the connected downstream link. The sources of these transaction packets are the switch's other downstream ports (peer-to-peer TLPs) and its upstream port (downstream TLPs). The transaction ordering and VC Arbitration protocols as discussed for upstream flowing TLPs also apply to those TLPs transmitting on the downstream port. The Port Arbitration defined for upstream flowing TLPs is defined for these peer-to-peer TLPs and downstream flowing TLPS from the upstream port in the form of Peer-to-Peer Port Arbitration (PAP). The PAP protocol is the same as the Port Arbitration protocol discussed above, except one of the downstream ports is replaced by the upstream port as the source for the TLPs. The PAP is merging downstream flowing TLPs from the switch's upstream port with the peer-to-peer TLPs packets from downstream ports.

Upon arrival, the TLPs for the different VCs are placed in FIFOs to retain the transaction ordering and transmission priority determined by the downstream and upstream PCI Express devices. As previously discussed in this book, the VCs only exist as a virtual concept relative to transmitting ports. Thus for this discussion, the buffer for each receiving port is viewed as a single FIFO containing the TLPs from all VCs. Similar to the discussion for downstream flowing TLPs, TLPs (identified by their TC numbers) must be parsed to the correct downstream port VC by the Address Mapping & TC to VC Mapping Module. To illustrate this parsing, a Virtual Add and TC Map FIFO is assumed as before; it is only for discussion purposes. The Address Mapping & TC to VC Mapping Module has determined that the TLPs received on the downstream ports and the upstream port are to be ported to a specific downstream port per address or routing information in the TLPs. Once the destination of the TLPs has been established to a specific downstream port, the Address Mapping and TC to VC Mapping Module must map the TLPs to the appropriate virtual FIFOs related to one of the VC numbers on the downstream port. The mapping is accomplished by the TC number of the TLP matching the range of TC numbers mapped to each VC number.

The balance of the protocol to transmit TLPs from a switches' downstream port is the same as that of the upstream port with one exception.

If there are no peer-to-peer TLPs, the only source of TLPs is the upstream port and no Port Arbitration is defined.

Non-Switch Port Arbitration, VC Arbitration, and Buffer Considerations

The previous discussion focused on Port Arbitration and VC Arbitration within a switch as an example. As illustrated in Figure 12.12 and briefly discussed earlier, the application Port Arbitration (PA), VC Arbitration (VA), and Transaction Ordering (TO) occurs at *other non-switch* buffers within a PCI Express fabric and is summarized as follows:

- For TLPs flowing upstream:
  - PA can apply to the flow of TLPs from the Root Complex downstream ports into the Root Complex's internal buffer. This was not specifically defined by the PCI Express specification and is highly implementation-specific. It has been included in this book to complete the discussion. The Root Complex may merge TLPs from the downstream ports by other methods including a hardwired version of Port Arbitration.
  - PA does not apply to endpoints because there are no downstream ports.
  - PA can also apply to the flow of TLPs from bridges' downstream ports. This was not specifically defined by the PCI Express specification and is highly implementation-specific. It has been included in this book to complete the discussion. The bridge may merge TLPs from the downstream ports by other methods including a hardwired version of Port Arbitration.
  - VA does not apply to the flow of TLPs from the Root Complex's downstream ports into the Root Complex's internal buffer.
  - VA applies to endpoints because multiple VC numbers can be defined for the upstream links.

- VA does not apply to the flow of TLPs from bridges. Only VC0 can be defined for any upstream link of a bridge, so VA is not necessary.
- TO does not apply to the flow of TLPs from the Root Complex's downstream ports into the Root Complex's internal buffer with the implementation a FIFO at the downstream ports. However, the flow from the Root Complex's internal buffer must consider BT livelock and deadlock conditions.
- TO applies to endpoints to avoid livelock and deadlock conditions.
- TO applies to bridges to avoid livelock and deadlock conditions.

- Relative to TLPs flowing downstream:
    - PA does not apply to the flow of TLPs from the Root Complex downstream ports, since the TLPs come from only one source: the internal buffer.
    - PA does not apply to the flow of TLPs into endpoints or bridges; the upstream link buffer is the only source for the TLPs.
    - VA applies to the flow of TLPs from the Root Complex downstream ports when multiple VC numbers are defined on the downstream link.
    - TO applies to the flow of TLPs from the Root Complex downstream ports to avoid livelock and deadlock conditions.
    - TO applies to the flow of TLPs into endpoints to avoid livelock and deadlock conditions.
    - TO applies to the flow of TLPs at the downstream port of a bridge to avoid livelock and deadlock conditions.

Chapter 12: Flow Control Protocol: Part 2 ■ 619

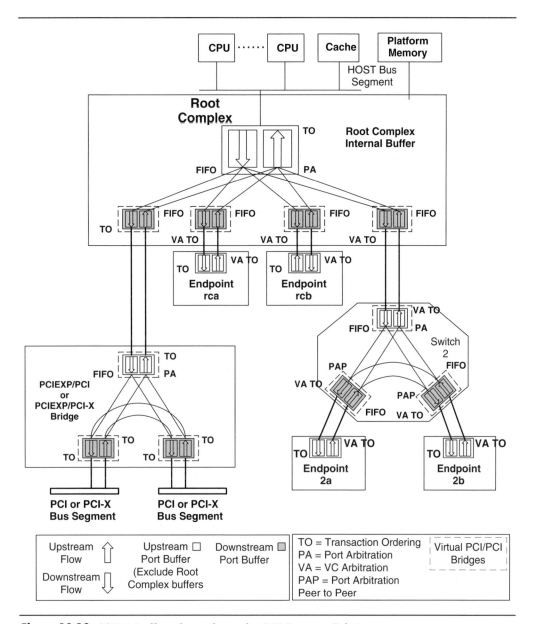

Figure 12.12 TLPs' Buffers throughout the PCI Express Fabric

Priority Schemes

The priority schemes applied to Port Arbitration and VC Arbitration are defined in the configuration register block of the associated port.

Within each virtual channel, the priority scheme for the Port Arbitration is defined and implemented. The priority schemes defined for Port Arbitration of the VC is located in Bits [7::0] of the VC Resource Arbitration Register. If the Programmable Weighted (also known as Programmable Weighted Round Robin) priority scheme is implemented, Bits [31::24] of the VC Resource Capability Register indicates the offset location in the configuration register block of the Port Arbitration Table. The size defined for each entry in the Port Arbitration Table is located in Bits [11::10] of the Port VC Capability Register 1. If the Time Based Weighted (also known as Time Based Round Robin) priority scheme is implemented, Bits [22:16] of the VC Resource Capability Register indicate the maximum number of time slots minus one.

The boundary between the aforementioned VC lower group versus upper group for VC Arbitration is defined in the Port VC Capability Register 1. The priority schemes defined for VC Arbitration of the aforementioned VC lower group is located in Bits [7::0] of the Port VC Capability Register 2. The priority scheme implemented for VC Arbitration of the aforementioned VC lower group is located in Bits [7::0] of the Port VC Capability Register 2. If the Programmable Weighted (also known as Programmable Weighted Round Robin) priority scheme is implemented, Bits [31::24] of the Port Capability Register 2 indicate the offset location in the configuration register block of the VC Arbitration Table. The next section examines implementation considerations.

Strict Priority for VC Arbitration Only

This priority scheme only applies to the VC upper group of VC Arbitration. VC7 is assigned the highest priority (reserved for isochronous transfers), VC6 has the second highest (VC Arbitration), and so on. The strict priority protocol requires all TLPs assigned to a higher priority VC number to be transmitted before any TLPs of a lower priority VC number. If sufficient link bandwidth is not provided, TLPs assigned to a VC number with lower priority may not be transmitted in a consistent fashion (starvation).

> The strict priority starvation considerations also indirectly apply to the VC lower group because VC upper group as a group has a higher priority (per strict priority) than the VC lower group as a group.

Hardware Fixed (also Known as Round Robin) for Port and VC Arbitration

This priority scheme treats all assigned downstream port numbers equally with other downstream port numbers (Port Arbitration), and all assigned VC lower group numbers equally with other VC lower group numbers (VC Arbitration). For example, the VC lower group consists of VC0 through VC5. For this discussion, assume the VC upper group consists of VC6 and VC7, but are inactive. VC5 is offered link bandwidth and one TLP is transmitted if available. VC4 is then offered link bandwidth to transmit one TLP, whether a TLP mapped to VC5 is available or not. This protocol continues until all VCs (VC0 through VC5) have been offered to transmit one TLP, whether TLPs mapped to higher priority VCs were available or not. The protocol repeats with VC5 being offered link bandwidth to transmit one TLP. Other protocols similar to a round robin are possible that may give limited priority among a subset of VCs.

Weighted Round Robin

The Weighted Round Robin priority scheme is a more complex arbitration than a simple round robin; consequently, the discussion is in several parts.

Arbitration Table The Weighted Round Robin (WRR) priority scheme requires the use of an Arbitration Table. When applicable, an Arbitration Table is contained in the configuration requester block of the port and consists of an array of arbitration port entries (APE). As will be discussed below, each APE can be 2, 4, or 8 bits in size. The beginning address is from the offset in the aforementioned entries in the configuration register block. The lower order APE is contained in the lowest order byte of the lowest order DWORD. The highest order APE is contained in the highest order byte of the highest order DWORD. The APE ordering is sequential between the lowest and highest order APE.

The number of phases defined for each application defines the number of APE entries in the Arbitration Table. For example, 128 phases defines an Arbitration Table of 128 APEs.

The entities for the Arbitration Table for VC Arbitration are VC numbers enabled for the port.

Programmable Weighted (also Known as Programmable Weighted Round Robin) for Port and VC Arbitration The number of phases that can be defined for the Arbitration Table by the port's configuration address space is:

- 32 phases
- 64 phases
- 128 phases
- 256 phases (only for Port Arbitration)

For VC Arbitration, the entities in the arbitration port entry (APE) of an Arbitration Table are VC numbers enabled for the port. For the round robin aspect of a WRR priority scheme, the VC number referenced in each APE is given the opportunity to transmit its associated TLP. However, with Programmable Weighted Round Robin (PWRR), the software can program any VC number into any APE. When an APE is referenced, a TLP is immediately transferred. If the TLP is not ready for immediate transmission, or upon completion of the transmission, the VC number in the next sequentially higher order APE is referenced. Once the highest order APE is referenced, the transmission mechanism returns to the lowest order APE and continues. Any time when no TLP is ready to transmit by the referenced entity, the transmission mechanism proceeds to the next highest order APE, once the phase period has elapsed. The software can program into each APE any VC number enabled for the port. Other considerations are:

- Each APE size is fixed with 4 bits [4::0]. Bits [3::0] define the VC number. Bit [4] is RESERVED and equals 0b.
- The value of each APE is programmable with the default value (prior to software initialization) of each APE set to VC0.
- If the VC number in the APE is not enabled it is ignored and treated as if no TLP were ready for transmission.

For Port Arbitration the entities in the arbitration port entry (APE) of an Arbitration Table are downstream port numbers of a switch or the Root Complex. For the round robin aspect of a WRR priority scheme, the port number referenced in each APE is given the opportunity to port its associated TLP from a receiving buffer to an accumulation buffer. However, with Programmable Weighted Round Robin (PWRR), the software can program any port number into any APE. When an APE is referenced a

TLP is immediately transferred. When the TLP is not ready for immediate transfer, or when the transfer is complete, the port number in the next sequentially higher order APE is referenced. Once the highest order APE is referenced, the transfer mechanism returns to the lowest order APE and continues. When no TLP is ready to transfer by the referenced entity, the transfer mechanism proceeds to the next highest order APE once the phase period has elapsed. The software can program into each APE any port number defined for a switch or the Root Complex. Other considerations are:

- The APE size of a specific Port Arbitration Table is programmable to 2, 4, or 8 bits. The size is selected dependent on the number of downstream ports.

- The value of each APE is programmable with the default value (prior to software initialization) to provide equal round robin to all downstream ports.

- If the port number in the APE is not enabled it is ignored and treated as if no TLP were ready for transmission.

Timed Based Weighted (also Known as Time Based Weighted Round Robin) for Port Arbitration Only This priority scheme only applies to Port Arbitration with an Arbitration Table size of 128 phases. An Arbitration Table associated with Time-Based Weighted Round Robin (TBWRR) priority scheme defines the APE in the same fashion as WRR. The one difference is that the time interval to progress from APE to APE is fixed. Even if TLPs are ready to transfer, or the port selected is not enabled, the progression to the next is still fixed. The implementation of the TBWRR priority scheme has two additional parameters to consider:

- In the Port VC Capability Register 1 bits [9:8] defines the Reference Clock. Per PCI Express revision 1.0 the only value all is 00b, which provides a Reference Clock of 100 nanoseconds. This is the time interval to progress from APE to APE.

- In the VC Resource Capability Register bits [22:16] defines the Maximum Time Slots. The values 000 0000b to 111 1111b represent 1 to 127 time slots.

The application of the TBWRR with these parameters is best explained by application to isochronous transfers. Please see the discussion in the next section.

Quality of Service

Consider a PCI Express device such as a video endpoint executing a memory write transaction that needs to begin data transfer by a specific time (latency) and maintain a specific transfer rate (effective bandwidth) for the inrush of data from outside platform. The endpoint must have a buffer large enough to accommodate the latency to begin the data transfer. Similarly, once data transfer begins, the endpoint relies on the buffering of the data to accommodate the differences in platform bandwidth versus the inrush of data. In both considerations, if the buffer is too small, data will be lost. Another bandwidth consideration is the importance of timely delivery. The buffer at the endpoint sourcing the data may prevent the loss of data for the effective bandwidth, but inconsistent bandwidth may provide poor transfer of the data. For example, if this inrush of data from outside the platform is a video stream, inconsistent bandwidth will make the image look jerky or make the viewer experience blank screens.

One of the important improvements of PCI Express platforms over PCI and PCI-X platforms is that the PCI Express specification supports Quality of Service (QoS) quantitatively. Earlier this chapter detailed Flow Control protocol elements of VC and Port Arbitrations. Chapter 11 discussed other Flow Control protocol elements for traffic class assignment and TC to VC mapping. Collectively, all of these Flow Control protocol elements provide software the opportunity to fine-tune the path of specific Transaction Layer Packets (TLPs) from specific PCI Express devices. This permits a PCI Express platform to support a QoS unattainable with PCI and PCI-X platforms. It also permits PCI Express platforms to support true isochronous transfers. As the following section illustrates, isochronous transfers are a good example of the extreme use of the QoS.

Historically, PCI and PCI-X platforms considered latency (time to begin the transfer), effective bandwidth, and bandwidth consistency. Applying these considerations to PCI Express and isochronous transfers provides the following expanded definition of these parameters and new parameters.

Latency

Latency is the time to begin the transfer. The link is point-to-point bi-directional and there is no arbitration for link ownership. However, as discussed in Chapter 11, the link's point-to-point interconnect relies on sufficient buffer space at the receiving port to ensure TLPs can be transmitted as planned. If the buffer space at the receiving port is not available, or the updated flow control credit information is not provided in a timely fashion, "backpressure" can occur at the transmitting port. Thus, latency to begin the transmission of TLPs is in the form of link backpressure on a PCI Express fabric. (See a further discussion about backpressure below.)

Effective Bandwidth

This parameter identifies the actual bandwidth available for data transfer over a link. Effective bandwidth defines the portion of the total link bandwidth available for data transfer once the non-data overhead is removed. The definition of effective bandwidth assumes that the vast majority of the link bandwidth is needed for TLPs that contain data.

Bandwidth Consistency

Typically link bandwidth and effective link bandwidth represent performance over a time period. However, within the time period consideration must be given to whether the data is transferring mainly at the beginning, middle, or end of the time period versus evenly across the time period. Obviously, the value of the effective link bandwidth assigned to a data transfer is reduced if the transfer is not evenly distributed. As will be discussed in the next section, the Port Arbitration protocol and general PCI Express definition of isochronous transfers ensures even distribution of TLP transmission.

Backpressure

As discussed throughout this book, TLPs are trying to make forward progress across links and switches between their sources and destinations. TLPs flowing upstream must merge at buffers in switches' upstream ports and the Root Complex's Internal Buffer. TLPs flowing downstream must merge at buffers in switches' downstream ports with peer-to-peer TLPs. As discussed in this chapter, the Port Arbitration protocol controls the selection of downstream port buffers for TLPs to be ported to the upstream or downstream port buffers of switches or the Internal Buffer of the Root Complex). If the buffers the TLPs are porting to are very large or are being emptied quickly, the TLPs make forward progress using the Port Arbitration protocol. If the buffers the TLPs are porting to are too small or are being emptied slowly, the TLPs forward progress is impaired independent of the Port Arbitration protocol. This latter condition results in internal backpressure in the Root Complex, and switches, which in turn prevents downstream port buffers (Root Complex and switches) and upstream port buffers (switches) from receiving additional TLPs from the link once the receiving port buffers are full. This in turn causes backpressure on the link, which prevents the transmitting port buffers on the other end of the link from emptying as quickly, which causes backpressure in other switches or the Rot Complex, and so forth. An additional cause of link backpressure is when the flow control credit information discussed in Chapter 11 is not updated at the transmitting port frequently enough. A final additional cause of link backpressure occurs when the receiving port has space available and it is properly reported to the transmitting port, but is the wrong size for the next transaction that must be transmitted. Thus backpressure can occur internally to the Root Complex and switches, which manifests itself externally as link backpressure along with the other causes mentioned. It is also possible for bridges and endpoints to cause link backpressure.

Total Latency

The total elapsed time for a transaction to begin execution and for it to be entirely complete is defined as of *total latency*. For example, a memory read requester transaction begins the transaction execution (requester source), but the completion of the transaction execution is not until *all* associated completer transactions have returned from completer source to the requester source. In another example, a memory write requester transaction begins the transaction execution but the completion of the transaction execution is not until *all* associated data has been written at the requester destination. Isochronous transfers have further expanded the total latency definition by defining it as the sum of three parts: PCI Express fabric requester latency, Completer latency, and PCI Express fabric completer latency. The upper bounds of a successful isochronous transfer must be defined and implemented relative to the total latency.

- **PCI Express Fabric Requester Latency:** PCI Express fabric requester latency is measured from the time the requester transaction leaves the Transaction Layer of the requester source to the time it completely arrives at the Transaction Layer of the requester destination (completer source). For memory write requester transactions, it includes the transfer of all of the data. The requester source must consider the delay for wakeup, the effective bandwidth of all links, the possible impact of backpressure, and possible retries of the retry buffer.

- **Completer Latency:** Completer latency is defined in two ways. For a read requester transaction, it is the sum of the time measured from the time it is received at the requester destination to the time the last data of the associated completer transaction is sourced from the Transaction Layer of the completer source (requester destination). For a memory write transaction, it is the sum of the time measured from when last data arrives at the transaction layer to the time for all of the received data to be "seen" (that is, written) by all PCI Express devices that need to "see" it.

- **PCI Express Fabric Completer Latency:** PCI Express fabric completer latency is only defined for completer transaction associated with memory read requester transactions. It is measured from the time the last completer transaction (if a set of completer transactions is returned other just the one) leaves the Transaction Layer of the completer source to the time it arrives at the Transaction Layer of the completer destination (requester source). It must consider the effective bandwidth of all links, the possible impact of backpressure, and possible retries of the retry buffer.

Isochronous Transfers

One of the unique applications of QoS is isochronous transfers. Isochronous transfers are illustrated by the video stream example discussed earlier. That is, the goal is to provide the latency (to start the data transfer and for the duration of the transfer) to enable video that displays smoothly and with no blank screen. All of the QoS parameters discussed earlier are considerations for isochronous transfers.

> Isochronous transfers make full use of PCI Express fabric, link bandwidth, and the associated Flow Control protocol (Traffic Classes, Virtual Channels, and so on). The PCI Express specification does not outline the implementation of non-isochronous transfers except for certain transactions being assigned specific TC numbers. Otherwise, flow of TLP not associated with isochronous transfers is implementation-specific.

The Isochronous Transfer protocol is as follows:

- Between Root Complex and endpoint or between endpoints as peer-to-peer transactions:
    - Isochronous transfers are only defined for memory address space.
    - The memory read or write requester transactions for isochronous transfers can only be sourced by downstream endpoints.
    - The completer transactions for isochronous transfers can only be sourced by the Root Complex or an endpoint as part of peer-to-peer transactions.
- Transactions associated with isochronous transfers must be assigned TC numbers that map to VC numbers that permit latency, bandwidth, and bandwidth consistency to be achieved at all times.
    - TLPs executed for different isochronous transfers can be assigned the same or different TC numbers that permit the TLPs to be mapped to different or the same VC number.
    - TLPs not associated with isochronous transfers can be assigned TC numbers that are shared by TLPs associated with isochronous transfers.
    - Transactions not associated with isochronous transfers can be mapped to VC numbers that are shared by transactions associated with isochronous transfers.
- Transaction Ordering cannot distinguish between memory transaction associated with isochronous transfers and those memory transactions not associated with isochronous transfers. Consequently, this is a consideration when TLPs for isochronous transfers are assigned the same TC number as TLPs that are not part of isochronous transfers.

- Isochronous transfers are not supported across advanced peer-to-peer links.

> Bridges only implement VC0; consequently, other VC numbers are not available. Given that a specific VC number can only support isochronous transfers (isochronous and non-isochronous transfers cannot be mapped to the same VC number), no isochronous transfers are defined for bridges.

The intermixing of TLPs for isochronous and non-isochronous transfers makes quantifying the QoS difficult. It is recommended that the PCI Express path fine-tuned to support isochronous transfers does not support non-isochronous transfers. The best implementation is to assign the TLPs for a specific isochronous transfer a specific non-shared TC number mapped to a specific non-shared VC number. The second best implementation is to assign the TLPs for a specific isochronous transfer to a specific TC number shared only by other isochronous transfers and mapped to a specific VC shared only by other isochronous transfers. Either implementation or other variations are possible, provided the software has determined that the QoS for all isochronous transfers can be maintained.

Quantitative Parameters

Independent of how the software fine-tunes the path between two participants, the two participants must meet quantitative parameters. For this discussion, assume an endpoint is the requester source and the Root Complex is the completer source. Also, for this discussion, no distinction is made for a memory read transaction versus a memory write transaction. Due to the symmetric nature of the link (which defines the same VC and TC in both directions) and the bandwidth of a memory read requester/completer transaction combination, this is essentially the same as a memory write requester transaction.

To establish the concepts of the basic quantitative isochronous information, first assume a simple Root Complex to endpoint interconnected (two participants) over one direct link (no intervening switches). The discussion will later expand into the considerations for intervening switches.

Quantitative Parameters for Isochronous Transfers One of the basic concepts of isochronous transfers is that over an isochronous time period (T seconds) a minimum amount of data (Dbytes) must be transferred between the two participants. The participants must always transfer if possible the maximum amount of data bytes defined for a TLP. This value is the Max_Pay_Size (Y), and is defined in the configuration register blocks of the participants. Obviously not all transactions transfer the maximum possible bytes of data, but the Max_Payload_Size value provides the worst-case calculation.

The number of TLPs (N) required to transfer all of the data in a specific time period is:

N = Dbytes /Y

The maximum bandwidth (BWmax) required to be allocated to the isochronous transfer to support the transfer on the link is:

BWmax = (N × Y)/T

The value is defined as BWmax for isochronous transfer. The value BWmax also provides a measure for the minimum effective link bandwidth if no other transactions are sharing the link. The effective link bandwidth has to increase above this value to accommodate any sharing of the link.

A TLP containing a Data field equal to Y must be transmitted with maximum period of *tp* (virtual time slot period):

tp = T/N

Assuming one transaction for the isochronous transfer per virtual time slot Nlink = N.

Consequently, within a time period T the number of Nlink (virtual time slots) that must be allocated to the isochronous transfer is:

Nlink = T/*tp*

If only one TLP for the isochronous transfer is allowed per virtual time slot, then Nlink = N.

Substituting for *tp* yields the following:

Nlink = (BWmax x T)/Y

BWmax = (Nlink x Y)/T

Example 1: Assume that a link is set up to support TLPs for an isochronous transfer with Max_Payload_Size = 256 bytes (Y). Further assume that 32,768 bytes must be transferred in 12.8 microseconds (T). Also assume that tp = 0.1 microseconds (the virtual time slot period) for the isochronous transfer:

- Nlink = 128

- BWmax = 2560 megabytes per second

Application of Quantitative Parameters for Isochronous Transfers
As previously discussed, the purpose of the Port Arbitration protocol is to provide a priority scheme to the downstream ports of the Root Complex or a switch. In this simple example of an endpoint directly connected to the Root Complex, the Port Arbitration protocol provides a priority scheme to select a specific downstream port to transfer a TLP flowing upstream into the Root Complex's Internal Buffer, as illustrated in Figure 12.12.

Only isochronous transfers can be defined for a specific VC number. Isochronous transfers and non-isochronous transfers cannot be mixed on the same VC number. For a VC number assigned to isochronous transfers, the Port Arbitration priory scheme must be Time-Based Weighted Round Robin (TBWRR). Per PCI Express specification revision 1.0, the TBWRR priority scheme can only implement an Arbitration Table of 128 phases (that is, 128 APE). Each APE is programmable and selects one of the downstream ports of the Root Complex or a switch. Each APE is referenced by the interval defined by the Reference Clock. Per PCI Express specification 1.0, only 0.1 microsecond is defined for the Reference clock. Thus the T (isochronous time period) is fixed at 12.8 microseconds. T reflects the time to reference all of the APEs in the Arbitration Table. For Example 1, BWmax = 2560 megabytes per second and as shown in Table 12.3, a link x12 or larger will be sufficient. If link x12 is selected, the other calculation of Example 1 indicates that Nlink = 128. Consequently, every APE in the Arbitration Table must select the one downstream port with this isochronous transfer.

Table 12.3 PCI Express Signal Lines and Link Bandwidth

PCI Express Bit Stream Rate versus Bandwidth Default 2.5Gb/sec per Differential Pair (Each Direction)							
Lanes per Link	x1	x2	x4	x8	x12	x16	x32
Signal Lines each direction	2	4	8	16	24	32	64
Raw Bit Stream per Second	2.5 Gb/s	5.0 Gb/s	10.0 Gb/s	20.0 Gb/s	30.0 Gb/s	40.0 Gb/s	80.0 Gb/s
Bandwidth Bytes per Second	250 MB/s	500 MB/s	1000 MB/s	2000 MB/s	3000 MB/s	4000 MB/s	8000 MB/s

In Example 1, the one downstream port of the Root Complex or the switch must be selected at all times, and it must have a link with x12 lanes. The port may have multiple VC numbers, which may have isochronous and non-isochronous transfers, and each VC number can have one to eight TC numbers assigned to it. All other transactions mapped to the VC number with an isochronous transfer via TC number must be for isochronous transfers. Thus, the transfer of 32,768 bytes from the one downstream port can include several independent isochronous transfers. Note two issues in this example. First, only the one downstream port of the Root Complex or the switch can be selected. Second, the inclusion of only one isochronous transfer on one port requires the Port Arbitration protocol with the TBWRR priority scheme for all the ports of the Root Complex or the switch. Thus, the non-isochronous transfers are also subject to this priority scheme.

Example 2: Consider another example. Assume that a link is set up to support TLPs for an isochronous transfer with Max_Payload_Size = 512 bytes (Y). Further assume that 32,768 bytes must be transferred in 12.8 microseconds (T). Also assume that tp = 0.2 microseconds (the virtual time slot period) for the isochronous transfer:

- Nlink = 64
- BWmax = 2560 megabytes per second

For Example 2, the BWmax and the total number of bytes to transfer have remained the same as Example 1, however Nlink = 64. Consequently, only every other APE in the Arbitration Table must select the one downstream port with this isochronous transfer. The other APEs in the Arbitration Table can select the other downstream ports of the Root Complex or a switch.

In these two examples, the term BWmax was compared to the link bandwidth of Table 12.3, which implies that there is no overhead to the bandwidth for the Data field. The Data field is transmitted in the TLPs with Header and Digest fields as overhead. Consequently, the *effective* bandwidth available for data is less than the bandwidth summarized in Table 12.3. Table 12.4 provides the effective bandwidths that take into consideration the overhead. With the exception of Y = 128 bytes, the effective bandwidth is at least 90 percent of link bandwidth, and greater than the BWmax referred to in the two examples.

Table 12.4 PCI Express Effective Bandwidth

	PCI Express Effective Bandwidth for a 2.5Gb/sec (x1) with Raw 250 MB/s versus 80 Gb/sec (x32) with Raw 8000MB/s Differential Pair (Each Direction)					
Data Bytes Payload per Packet	128	256	512	1024	2048	4096
Efficiency 32-Bit Address	84.21%	91.42%	95.52%	98.84%	98.84%	99.42%
64-Bit Address	83.12%	90.78%	95.17%	97.52%	98.75%	99.37%
Effective Data Bandwidth per Second for 32-Bit Address	210.5 MB/s versus 6736.8 MB/s	228.6 MB/s versus 7315.2 MB/s	237.5 MB/s versus 7600 MB/s	244.2 MB/s versus 7817.6 MB/s	247.1 MB/s versus 7907.2 MB/s	248.6 MB/s versus 7955.2 MB/s
Effective Data Bandwidth per Second for 64-Bit Address	207.5 MB/s versus 6640 MB/s	227.0 MB/s versus 7264 MB/s	237.9 MB/s versus 7613.6 MB/s	243.8 MB/s versus 7801.6 MB/s	246.9 MB/s versus 7900 MB/s	247.5 MB/s versus 7920 MB/s

Considerations beyond Quantitative Parameters for Isochronous Transfers

The previous example assumed that the endpoint was directly connected to the Root Complex. Isochronous transfers can also occur between endpoints or between an endpoint and the Root Complex with intervening links and switches. The effective link bandwidth considerations apply to all intervening links and application of the TBWRR Port Arbitration protocol to downstream ports of the Root Complex (if a participant) and all intervening switches. The requirements and considerations for implementation of isochronous transfers are as follows:

Requester Sources (Endpoints Only)

- The Length field in the memory requester TLP must be equal to or less than the Max_PayLoad_Size.

- If the requester destination is the Root Complex and the Root Complex does not support cache coherency for isochronous transfers, the requester source must set ATTRI[0] = 1b in the Header field of the memory requester TLP. If this bit is not set to "1," the memory requester TLP is "rejected." The Root Complex can only support snooping (cache coherency) of isochronous transfers if it can meet completer latency component of the total latency requirements. The term "rejected" is used by PCI Express specification revision 1.0, but does specify whether the memory requester TLP is discarded and/or whether an error is reported. This book assumes that the response is the same as for a **malformed** TLP.

Completer Sources (Root Complex or Endpoint)

- The Port Arbitration priority scheme for the Root Complex's Internal Buffer implemented must be the TBWRR.

- Total latency must be met at all times.

- Under normal operating conditions, the completer must not use the Flow Control protocol to induce backpressure to the memory write requester transaction source (requester source) to force uniform TLP transmission.

Intervening Switches

- The Port Arbitration priority scheme for the upstream port of a switch implemented must be the TBWRR.

- Total latency must be met at all times.

- The VC Arbitration protocol priority scheme selected in conjunction with the software must provide the highest priority to the VC numbers associated with the isochronous transfers.

- Under normal operating conditions, the switch must not use the Flow Control protocol to induce backpressure to the memory write requester transaction source (requester source) to force uniform TLP transmission. Also, under normal operating conditions the switch must not use the Flow Control protocol to induce backpressure to completer transaction source for memory read to force uniform TLP transmission.

Software Considerations

- All the Attributes bits in the Header fields of the requester TLPs with the same destination must be set identically.

- Any of the VC Arbitration priority schemes for the upstream ports of the endpoint and the downstream ports of the Root Complex and switches can be implemented. The only restriction is that the selected VC Arbitration priority scheme meets all of the isochronous transfer performance requirements.

- The Port Arbitration priority scheme for the upstream port of a switch and the Root Complex's Internal Buffer must be the TBWRR.

- A specific VC number can only support isochronous transfers. Isochronous and non-isochronous transfers cannot be mapped via the assigned TC numbers to the same VC number.

- The memory request TLPs for isochronous transfers can be assigned to one or more TC numbers.

- The entire link bandwidth cannot be allocated exclusively to isochronous transfers.

- The Max_PayLoad_Size defined for the participants must be the same and limited to a value that meets the isochronous transfers' total latency requirements.

- The Max_Read_Request_Size must be equal to or less than the Max_PayLoad_Size.

- The VC Arbitration protocol priority scheme selected in the switch must provide the highest priority to the VC numbers associated with the isochronous transfers.

Other Considerations

- The isochronous transfers and non-isochronous transfers must be assigned to different VC numbers; consequently, there is no Transaction Ordering between isochronous transfers and non-isochronous transfers.

- The two examples assumed a simple interconnect between an endpoint and Root Complex. The same basic concepts discussed relative to links connected to the downstream of the Root Complex also apply to links connected to switches' downstream ports, with only two differences. First, the Port Arbitration protocol applies to the Root Complex's Internal Buffer in the examples. In switches, the Port Arbitration protocol applies to the upstream port with Root Complex or endpoint as completer source, or a downstream port with endpoint as completer source. Second, the VC Arbitration protocol is not a consideration in the Root Complex. In switches, the VC Arbitration protocol applies to the upstream port with Root Complex or endpoint as completer source, or a downstream port with endpoint as completer source.

- Consider the transmission events of TLPs by requester or completer sources and the reception events of the same packets by the requester or completer destinations for isochronous transfers. The PCI Express devices assume an even distribution of these events across the isochronous time period (T). The requester and completer sources must transmit the transactions evenly across the isochronous time period T with the Data field sizes defined earlier. TLPs port through switches or are ported to the Root Complex's Internal Buffer; consequently their flow is governed by the TBWRR Port Arbitration priority scheme.

Chapter 13

PME Overview, Wake Events, and Reset

The chapter discusses several diverse but related topics. The relationship of these topics is based on conventions established by this book for purposes of explaining these topics.

Introduction

This book has divided Power Management Events (PMEs) into three categories: Power Management protocol, Hot Plug protocol, and Slot Power protocol.

- **Power Management protocol:** The purpose of the Power Management (PM) protocol is to provide PME software a means to control the power level of the PCI Express devices and the associated links. The element of the PME software related to the PM protocol is Power Management (PM) software. One set of tasks of the PM protocol is to transition from sleep or powered-off to powered-on the PCI Express devices and associated links. Another set of tasks of the PM protocol and software is to lower the power consumed by the PCI Express devices and associated links, Another set of tasks of the PM protocol and software is to place into sleep, or prepare to be powered-off, the PCI Express devices and associated links.

- **Hot Plug protocol:** The purpose of the Hot Plug (HP) protocol is to permit PCI Express add-in cards to be inserted or removed from the platform's slot without powering off the entire platform. The element of the PME software related to the HP protocol is Hot Plug (HP) software. The HP protocol and software are not discussed in this chapter, but are discussed in detail in Chapter 16.

- **Slot Power protocol:** The purpose of the Slot Power (SP) protocol is to provide Power Management Event (PME) software a means to control power allocation to add-in cards. The element of the PME software related to the SP protocol is slot power (SP) software. The SP protocol and software are not discussed in this chapter, but are discussed in detail in Chapter 17.

This chapter details the power cycling topics of sleep, powered-off, powered-on, and the associated transitions. Associated with these topics are the Wake Events and Resets, which are also discussed in detail in this chapter. To tie-in these topics with Chapters 15, this chapter includes an overview of the PMEs related to PM protocol.

This chapter is divided into the following topics:

- Power Management Protocol Overview
- Power Cycling
- Wake Events: Wakeup Protocol and Wakeup Event
- Resets

Other Background Information

Before continuing to the main topics of this chapter, there are several terms that need to be defined and explained.

Sleep versus Powered Off

In this book, the term *powered off* means that the PCI Express platform or add-in card in a Hot Plug compatible slot is totally without main power or Vaux, and can not implement the Wakeup protocol. This book defines this as the L3 link state for Link Training and Status State machine (LTSSM) and link. The term *sleep* means that the PCI Express platform and any add-in cards operate only with the retention of Vaux when the main power is off. This book defines this as the L2 link state for LTSSM and link. For any PCI Express device that can execute the Wakeup protocols, the PCI Express platform and add-in cards must be in sleep.

Link Training and Flow Control Initialization

Prior to any normal PCI Express link operation within the L0 link state, the PCI Express devices on the link must complete Link Training and Flow Control Initialization. Link Training is executed by the Link Training and Status State machine (LTSSM) by sequencing through the Detect, Polling, and Configuration link states. The Detect link state detects the lanes within a link common to both PCI Express devices. The Polling link state establishes the bit and symbol lock, lane polarity inversion, and highest common data bit rate. The configuration link state processes a subset of the detected lanes that successfully complete the Polling link states into configured lanes. Configured lanes have established LINK# and LANE#, implemented lane-to-lane de-skew, and have established the N_FTS value. See Chapter 14 for more information of Link Training.

Once the LTSSM and link have transitioned to L0 link state, only DLLPs and IDLE D symbols can be transmitted until the available buffer space at the receiving port of LLTPs has been established. The process of establishing the available buffer space is called Flow Control Initialization. See Chapters 8 and 12 for more information.

Link Retraining, and N_FTS OSs

Link Retraining will occur when bit and symbol lock may be lost or higher layer activities as defined in Chapter 14 have requested it. Link Retraining is implemented by the transmission and reception of TS1 and TS2 OSs within the Recovery link state. The Link Retraining reestablishes the bit and symbol lock, reestablishes the lane-to-lane skew, and sets a new N_FTS value for use of future L0s. The Link Retraining is accomplished by successfully sequencing through the Recovery.RcvrLock and Recovery.RcvrCfg link sub-state.

Not part of the definitions of Link Training or Link Retraining is the transmitting and receiving FTS OSs on the configured link for transitioning from L0s to L0 link states. On the configured link established by the Configuration link state, the transmission of a set of N_FTS reestablishes the bit and symbol lock. "N" represents the number of FTS OSs defined in the TS2 OSs as part of Link Training and Link Retraining. Also, the Link Training bit in the Link Status Register of the configuration register block is set to 1b when TS1 and TS2 OSs are being transmitted by the downstream port of the Root Complex or switches. As defined in this book, the Link Training bit only reflects the portion of the Link Training and Link Retraining when TS1 and TS2 OSs are transmitted.

Link_UP, Link_DOWN, Physical LinkUp = "1", and Physical LinkUp = "0"

For PCI Express device cores of the two PCI Express devices on a link to communicate with Transaction Link Packets (TLPs) via their respective Transaction Layers, their respective Data Link and Physical Layers must be in specific logic states. The logic state of the Data Link Layer is maintained by Data Link Control and Management State Machine (DLCMSM) within each PCI Express device, and is defined by the link activity states. The logic state of the Physical Layer is maintained by the Link Training and Status State Machine (LTSSM) within each PCI Express device, and is defined by the link states. Also, for the Data Link Layer to transmit Data Link Layer Packets (DLLPs) for purposes of link management, the LTSSM of the Physical Layer must be in a specific logic state. See Chapter 7 for more information on the Data Link Layer.

Link_UP and Link_DOWN Link_UP is a status indication from the Data Link Layer to the Transaction Layer that TLPs contained within LLTPs *can* be transmitted across the link. Link_DOWN is an indication from the Data Link Layer to the Transaction Layer that TLPs contained within LLTPs *cannot* be transmitted across the link. The Link_UP or Link_DOWN status is a result Data Link Control and Management State Machine (DLCMSM) and the link activity states. Once the Physical Layer reports to the Data Link Layer a Physical LinkUp = "1" the Data Link Layer applies link activity states to execute Flow Control Initialization on VC0 to transition the link from Link_DOWN to Link_UP. See Chapter 7 for more information about link activity states.

The status of the DLCMSM is Link_DOWN when the Physical LinkUP = "0" with two boundary conditions. When the PCI Express device is exiting from the Configuration link state to the L0 link state the Physical LinkUp = "1," but the status of the DLCMSM must remain as Link_DOWN until the Flow Control Initialization is successfully completed. The other boundary is when the transition is from the L0 to Recovery to L0 link states. Even though the Physical LinkUP = "0" in the Recovery link state, the status of the DLCMSM remains as Link_Up during this sequence.

As discussed earlier, the transition from Link_DOWN to Link_UP permits the transmission of TLPS within LLTPs to begin. In addition, the transition from Link_DOWN to Link_UP results in the Data Link Layer executing the following:

- The downstream port, the Root Complex, or a switch transmits a message Set_Slot_Power_Limit requester transaction packet downstream. The data associated with this packet provides slot power limit value and slot power limit scale.
    - Note: The transmission also occurs when the Slot Capabilities Register is written to in the downstream port, the Root Complex, or a switch. See Chapters 5 and 17 for more information.

As the link states transition from one to another, a link is defined as Link_UP or Link_DOWN. As discussed earlier, the transition from Link_UP to Link_DOWN prevents the continued transmission of LLTPs. In addition, the transition from Link_UP to Link_DOWN results in the Data Link Layer executing the following:

For a switch

- When the link is connected to the upstream port of a switch, the resources specific to PCI Express within the upstream port (registers, timers, counters, and port state machines) are placed in the default and initial conditions. The exceptions are the registers in the configuration register block that are defined as "sticky," which are not modified or initialized. The sticky registers are labeled ROS, RWS, and RW1CS. All LLTPs and the LLTPs contained within that are being processed (including buffered) are discarded. Note: It is assumed that DLLPs are processed immediately and are not buffered.

- When the link is connected to the downstream port of the switch, the resources specific to PCI Express within the downstream ports (registers, timers, counters, and port state machines) are placed in the default and initial conditions. The exceptions are the registers in the configuration register block that are defined as "sticky," which are not modified or initialized. The sticky registers are labeled ROS, RWS, and RW1CS. All LLTPs and the LLTPs contained within (including buffered) are processed as follows:
    - Requester transaction packets that have been transmitted but cannot be completed are discarded. Determination of whether requester transaction packets can be completed is

implementation-specific for requester transaction packets with associated completer transaction packets. Those requester transaction packets not associated with completer transaction packets are transmitted.
- Completer transaction packets that are already "formed" and ready for transmission are discarded.
- Completer transaction packets that are being "formed" such that the Status can be set to unsupported, transmitted, and associated requester transaction packets are discarded.
- It is assumed that DLLPs are processed immediately and are not buffered.

For an endpoint

- When the link is connected to the upstream port of an endpoint, the resources specific to PCI Express within the upstream port (registers, timers, counters, and port state machines) are placed in the default and initial conditions. The exceptions are the registers in the configuration register block that are defined as "sticky," which are not modified or initialized. The sticky registers are labeled ROS, RWS, and RW1CS. All LLTPs and the LLTPs contained within that are being processed (including buffered) are discarded. Note: It is assumed that DLLPs are processed immediately and are not buffered.

For a bridge

- When the link is connected to the upstream port of a bridge, the resources specific to PCI Express within the upstream port (registers, timers, counters, and port state machines) are placed in the default and initial conditions. The exceptions are the registers in the configuration register block that are defined as "sticky," which are not modified or initialized. The sticky registers are labeled ROS, RWS, and RW1CS. All LLTPs and the LLTPs contained within being processed (including buffered) are discarded. Note: It is assumed that DLLPs are processed immediately and are not buffered.

For a Root Complex

- When the link is connected to a downstream port of the Root Complex, the PCI Express specific resources within the downstream ports (registers, timers, counters, and port state machines) are placed in the default and initial conditions. The exceptions are the registers in the configuration register block that are defined as "sticky," which are not modified or initialized. The sticky registers are labeled ROS, RWS, and RW1CS. All LLTPs and the LLTPs contained within (including buffered) are processed as follows:
    - Requester transaction packets that have been transmitted but cannot be completed are discarded. Determination of whether requester transaction packets can be completed is implementation-specific for requester transaction packets with associated completer transaction packets. Those requester transaction packets not associated with completer transaction packets are transmitted.
    - Completer transaction packets that are already "formed" and ready for transmission are discarded.
    - Completer transaction packets that are being "formed" such that the Status can be set to unsupported, transmitted, and associated requester transaction packets are discarded.
    - It is assumed that DLLPs are processed immediately and are not buffered.

Physical LinkUP = "1" and Physical LinkUP = "0" The LTSSM must report to the Data Link Layer when the Physical Layer is ready to transmit LLTPs and DLLPs via Physical Packets. The LLTPs are part of the Data Link Layer that contains the TLPs of the Transaction Layer. The DLLPs are part of the Data Transaction Layer used for link management.

The ability of the Physical Layer to transmit LLTPs and DLLPs is established by Link Training and the associated link states of the LTSSM as indicated. When the Physical LINKUP = "0," the Physical Layer indicates to the Data Link Layer that *no* Physical Packets containing DLLPs or LLTPs can be transmitted on the link. When the Physical LINKUP = "1," the Physical Layer indicates to the Data Link Layer that Physical Packets containing DLLPs or LLTPs can be transmitted on the link. The transition from Physical LINKUP = "0" to Physical LINKUP = "1" by the LTSSM of the Physical Layer occurs with the successful completion of Link Training or Link Retraining.

> The logic level of the Link_UP, Link_DOWN, and Physical LinkUP indicators have no affect on the transmission of ordered sets and symbols by the Physical Layer.

"Sticky Bits"

When main power is powered-off (Cold Reset) and Vaux is maintained as part of the sleep definition, specific bits in the configuration register blocks are also maintained. These bits are call "sticky bits" and are designated in the PCI Express specification as ROS, RWS, and RW1CS. The Vaux must be supplied to maintain the value of these sticky bits. As will be discussed later, the values of the sticky bits are also retained when main power is not powered-off and a Warm Reset or Hot Reset is executed.

Power Management Protocol Overview

The Power Management (PM) protocol and the associated PM software overview is divided into three parts: Transition of PCI Express devices and associated links from the sleep or powered-off link states, lowering of the power consumed by PCI Express devices and associated links, and placement of PCI Express devices and associated links into sleep or powered-off link states. Chapter 15 has more detailed information about the transitions between the L0, L0s, L1, L2/3 Ready, L2, and L3 link states.

Transition from Sleep (L2) or Powered Off (L3) to Powered On (L0)

Summarized in Figure 13.1 are the link level activities related to the transition from a sleep (L2) or powered-off (L3) condition to powered-on (L0 link state). For a PCI Express platform that has been powered-off, the PCI Express devices and associated links have only one path to transition from the L3 link state to the L0 link state. This one path is the entire PCI Express platform to be as powered-on with the application of main power. As the PCI Express devices on each link complete Link Training by the LTSSM in the Physical Layer and Flow Control Initialization by the DLCMSM in the Data Link Layer, each PCI Express device and link transitions to the L0 link state.

If the entire PCI Express platform is in sleep, two possible paths may transition the PCI Express devices and links from sleep (L2) to powered-on (L0). One path is the path discussed for an entire PCI Express platform that had been powered-off. If all or a part of the PCI Express platform is in sleep (L2), the application of main power transitions all or part of the PCI Express platform to powered-on (L0). The other path is executed if any function of any PCI Express device has been programmed to execute the Wakeup protocol. The Wakeup protocol can be executed if the PCI Express platform had been powered on, the appropriate registers in the configuration register block programmed, and subsequently the main power is powered off but Vaux is retained. The specifics of the Wakeup protocol and power cycling are discussed later.

The application of the Wakeup protocol is defined on a hierarchy domain basis. As discussed in Chapter 2, each downstream port of a Root Complex connects to an independent hierarchy domain. With each hierarchy domain it is possible for to implement a PCI Express platform that places individually or collectively the hierarchy domains into sleep (L2). The execution of the Wakeup Protocol as discussed below is by a hierarchy domain. If the Root Complex applies main power to the hierarchy domain executing the Wakeup Protocol, the Root Complex may or may not apply main power to other hierarchy domains.

As illustrated in Figure 13.1, when a PCI Express device executes the Wakeup protocol, the path to transition from L2 link state to the "normal operation" powered-on L0 link state is as follows:

- The execution of the Wakeup protocol results in the main power being applied, the execution of Link Training, and execution of Flow Control Initialization by hierarchy domain. As discussed in the "Wake Events: Wakeup Protocol and Wakeup Event" section later in the chapter, the Wakeup Protocol includes the WAKE# signal line, or the Beacon. The WAKE# signal line, or the Beacon, causes the power controller(s) of the hierarchy domain to apply main power. A hierarchy domain may have more than one power controller. Also, a single power controller may be defined for multiple hierarchy domains.

- The execution of the Wakeup Protocol enables the PCI Express devices and links within a hierarchy domain to transition from the L2 link state to the L0 link state with the application of main power. The application of main power initiates Link Training and Flow Control Initialization by the associated PCI Express devices and links. Other PCI Express devices and links in the hierarchy domain in the L1 and L0s link state transition to the L0 link state as needed. With the application of main power, Link Training, and Flow Control Initialization, the downstream PCI Express device that caused the execution of the Wakeup Protocol can transmit a message PM_PME requester transaction packet.

- The last part of the Wakeup Protocol is the transmission of the message PM_PME requester transaction packet to the Root Complex. The purpose of the message PM_PME requester transaction packet is to request service by the PM software. The PCI Express device that requested the execution of the Wakeup Protocol continues to transmit the message PM_PME requester transaction packet using the PME Service Timeout protocol. See Chapter 15 for more information.

As illustrated in Figure 13.1, it is possible that when the entire PCI Express platform is powered on, a Warm Reset will force all PCI Express devices and links to follow the main power applied as discussed earlier.

The HP Protocol discussed in Chapter 16 will also implement the Wakeup Protocol in order to transmit a message PM_PME requester transaction packet for a wakeup event. A wakeup event occurs when one of the four Hot Plug (HP) events related to hardware occurs: attention button pressed, power fault detected, MSI sensor changed, and add-in card present state changes.

Lowered Power

Once the PCI Express devices and associated links are in the normal L0 link state, it is possible to transition the PCI Express devices and associated links to links states consuming less power. As illustrated in Figure 13.1, several possible transactions depend on the execution of the Active-State Power Management and PCI-PM Software protocols. The PCI-PM Software protocol also includes support The PCI power management standards are defined in the *PCI Bus Power Management Interface Specification Rev. 1.1* (PCI-PM*)* and *Advance Configuration and Power Interface Specification Rev. 2.0*.

Chapter 13: PME Overview, Wake Events, and Reset **647**

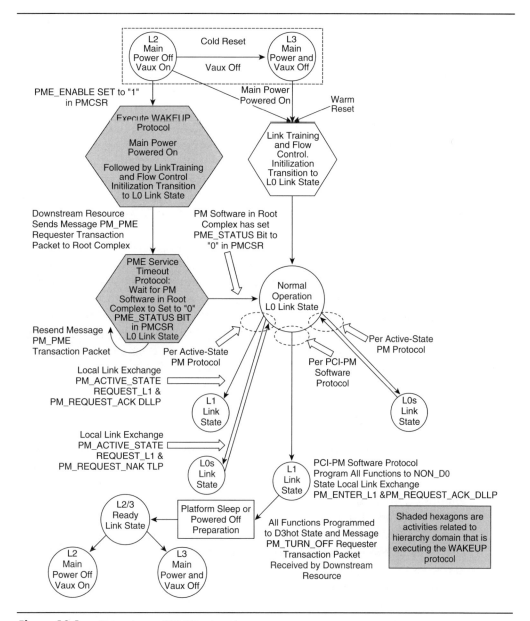

Figure 13.1 Overview of PM Protocol

The Active-State Power Management protocol only supports transitions between the L0 and L0s link states, and between L0 and L1 link states. The transition between the L0 and L0s link states is specific to the link, and is executed by the Physical Layer of either PCI Express device on the link. The transition from the L0 and L0s link states on a specific link is also possible with the exchange of the PM_Active_State_Request_L1 DLLP and the PM_Active_State_Nak LTTP. This transition is an alternative to transitioning to the L1 link state requested by the PCI Express device on the downstream side of the link that began transmitting the PM_Active_State_Request_L1 DLLPs. The transition from the L0 link state to a L1 link state is specific to the link and executed with the exchange of the PM_Active_State_Request_L1 and PM_Request_Ack DLLPs.

The execution of the PCI-PM Software protocol requires the PM software to program the functions in the PCI Express device on the downstream side of a link to specific "D" states. As illustrated in Figure 13.1, if all the functions in the PCI Express device on the downstream side of the link are programmed to non-D0 states, the transition is to the L1 link state. The transition occurs with the PCI Express device on the downstream side of the link transmitting PM_Enter_L1 DLLPs and receiving PM_Request_Ack DLLPs from the other PCI Express on the link. These transitions are discussed in detail in Chapters 14 and 15.

Placement into Sleep or Powered off Link State

The execution of the PCI-PM Software protocol requires the PM software to program the functions in the PCI Express device on the downstream side of the link to specific "D" states. As illustrated in Figure 13.1, if all the functions in the PCI Express device are in the non-D0 state, the transition is to the L1 link state, as discussed earlier. If all the functions in the PCI Express device are programmed to the D3hot states, it is possible for the PCI Express devices on the link to transition from the L1 link state to the L2/3 Ready link state. The protocol to transition from the L1 link state begins with the reception by the PCI Express device on the downstream side of a link of a message PM_Turn_Off requester transaction packet. As illustrated in Figure 13.1, the reception of message PM_Turn_Off requester transaction packet transitions the PCI Express devices on the link from the L1 to L2/3 Ready link state via Platform Sleep or powered-off preparation. Note: the message PM_Turn_Off requester transaction packet can be sent only to a PCI Express device on the downstream side of the link with all the functions in the D3hot state, otherwise the packet is discarded but no error is reported.

The Platform Sleep or Powered-off Preparation is illustrated in Figure 13.2, and consists of several activities. Subsequent to the PCI Express devices on the link transitioning from L1 link state the Root Complex or switch sends a message PM_Turn_Off requester transaction packet to the PCI Express device on the downstream side of the link. In response to the message PM_Turn_Off requester transaction packet, the PCI Express device transmits a message PM_TO_Ack requester transaction packet upstream to the source of the message PM_Turn_Off requester transaction packet. In order to transmit these message baseline transaction packets, the intervening PCI Express devices and links must transition to the L0 link state via the Recovery link state. The exchange of the message PM_Turn_Off and message PM_TO_Ack requester transaction packet must be done using the Power Fence Protocol discussed in Chapter 15.

Once the message PM_TO_Ack requester transaction packet is transmitted, the PCI Express device on the downstream side of the link begins transmitting PM_Enter_L23 DLLPs. In response to the PM_Enter_L23 DLLP, the PCI Express device on the other side of the link begins transmitting PM_Request_Ack DLLPs. The PCI Express devices on the link transition to the L2/3 ready link state. Once the PCI Express devices and link transition to the L2/3 Ready link state, the subsequent transition to the L2 link state or L3 link state is dependent on the main power and Vaux. These transitions are discussed in detail in Chapters 14 and 15.

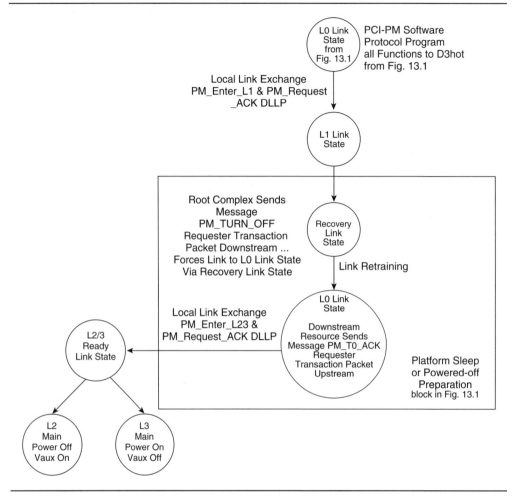

Figure 13.2 Platform Sleep or Powered-off Preparation

Power Cycling

There are two sources of power for a PCI Express device: main power and Vaux. The PCI Express specification defines specific reactions by PCI Express devices to the application and removal of main power, and with the retention of Vaux when main power is removed. Typically, when main power is "removed," it is termed "powered-off," but this book will use both terms interchangeably. Also, when main power is "applied," it is termed "powered-on," but this book will use both terms interchangeably.

Main Power and Vaux

Each PCI Express device receives 3.3 volts and 12 volts, collectively defined as main power. Optionally, each PCI Express device can receive 3.3 volts independent of the main power defined as Vaux. The intent of Vaux is to maintain the operation and conditions of a specific functions, or collection of functions, within a PCI Express device when main power is off. Per the discussion in Chapter 16, in the configuration register block of each function is a Power Management Capabilities Register (PMCR) that contains AUX Current bits that define the amount of current to draw from Vaux. (For more information, see Chapter 16.) These bits are used in conjunction with the PME_En bit in the Power Management Control/Status Register (PMCSR). These registers are part PCI power management standards defined in the *PCI Bus Power Management Interface Specification Rev. 1.1* (PCI-PM). The PCI Express specification defines an additional bit that permits the auxiliary current to be drawn from Vaux by a function not implemented by the PCI-PM protocol. This additional bit is the Auxiliary Power PM Enable bit in the Device Control Register. When the Auxiliary Power PM Enable bit is set to "1" in *any* function, the entire PCI Express device can draw the current specified in the AUX Current bits of the PMCR from Vaux independent of the PME_En bit in the PMCR.

The discussion in this and other chapters in this book implies that Vaux supplied PCI Express device is *also* supplied to each function. As discussed earlier, it is possible for Vaux to be supplied to the PCI Express device provided, but not supplied to every function in the PCI Express device. In order to simplify the discussion, the concept of Vaux supplied or maintained is relative to all functions within a PCI Express device. Also to simplify the discussion, this book assumes that Vaux is the *only* auxiliary power supplied when main power is off.

Powered On to Powered Off or Sleep

Per the discussion in Chapter 16, the PME software can place the links into the L2/3 Ready link state and the associated PCI Express devices are prepared to lose power. (See Chapter 16 for more information.) When the PME software powers off the main power of a platform or portion of a platform, the link and PCI Express devices transition to the L2 or L3 link states and sleep or powered-off, respectively. If Vaux is maintained, L2 link state and sleep applies. If Vaux is not maintained, L3 link state and powered-off applies. Associated with these link states is the electrical protocol defined in the *PCI Express Card Electromechanical Specification*. Summarized in Figure 13.3 are some of the basics of this electrical protocol. The summary is provided in this book to assist in understanding the PME and resets. Please see the *PCI Express Card Electromechanical Specification* for more detailed information.

For Figure 13.3, the electrical protocol is as follows:

- When main power is being powered off, the PERST# signal line transitions to "0" (active) prior to the voltage level of the main power being allowed to become unstable and drop below minimum voltage levels. The minimum time is T3 = 100 milliseconds.

- The REFCLK must be at the proper voltage and frequency for a minimum of T4 = 0.0 microseconds after the main power becomes unstable, and is allowed to drop below minimum voltage levels.

- The JTAG signal lines, and thus component testing, are active for a maximum of T5 = 0.0 microseconds after the PERST# signal line transitions to active.

- The SMBus is active for a maximum of T1 = 0.0 microseconds after Vaux becomes unstable, and is allowed to drop below minimum voltage levels when main power is off. If Vaux is off, SMBus is active for a maximum of T2 = 0.0 microseconds after the main power becomes unstable, and is allowed to drop below minimum voltage levels.

- Link activity (that is, PME software completes preparation for powered-off or sleep) and transitioning of the links to the L2/3 Ready link state prior to the power control turning off main power must be completed a minimum of T6 = 100 nanoseconds prior to the PERST# signal line transitions to active. The PERST# signal line is viewed at the worst case indicator.

- Note: If surprise main power powered-off occurs, the PERST# signal line transitions to active for a maximum of 500 nanoseconds after the main power becomes unstable, and is allowed to drop below minimum voltage levels. The minimum period for PERST# signal line is 100 microseconds.

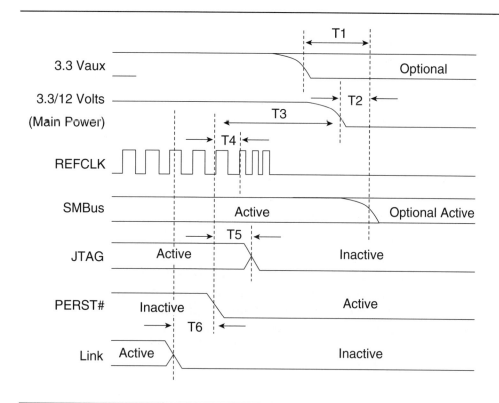

Figure 13.3 Powered Off or Sleep

Powered Off to Powered On

As discussed in Chapter 16, if main power is off and Vaux is also off, the associated links and PCI Express devices are defined as in the L3 link state and powered-off, respectively. See Chapter 16 for more information. The functions within PCI Express devices transition from the D3cold state to the D0 un-initialized state, the PCI Express devices transition to powered-on, and the link states transition to the L0 link state (via the Detect link states). Associated with the protocols discussed in more detail in Chapters 14 and 16 is the electrical protocol defined in the *PCI Express Card Electromechanical Specification*. Summarized in Figure 13.5 are some of the basics of this electrical protocol. The summary is provided in this book to assist in understanding the PME and reset. Please see the *PCI Express Card Electromechanical Specification* for more detailed information.

For Figure 13.4 the electrical protocol is as follows:

- When main power is being powered on the PERST# signal line transitions to inactive after the main power becomes stable and above minimum voltage levels. The minimum time is T3 = 100 milliseconds.

- The REFCLK must be at the proper voltage and frequency for a minimum of T4 = 100 microseconds prior to the PERST# signal line transitioning to inactive.

- The JTAG signal lines and thus component testing become active for a maximum of T5 = 0.0 microseconds after the PERST# signal line transitioning to inactive.

- The SMBus is active for a minimum of T1 = 0.0 microseconds after Vaux becomes stable and is above minimum voltage levels when main power is off. If Vaux is off, SMBus is active for a minimum of T2 = 0.0 microseconds after the main power becomes stable and is above minimum voltage levels. Note: The SMBus specification may have additional timing restrictions relative to main power or Vaux supplied versus an active state. See the *System Management Bus Specification, Revision 2.0* for more information.

Sleep to Powered On

If main power is powered off and Vaux is retained, the associated links and PCI Express devices are defined as L2 link state and sleep, respectively. The power-on protocol is the same as discussed earlier for powered-off to powered-on with the following differences.

- The Vaux has remained on.
- The sticky bits are valid.
- The Wakeup protocol can be executed.

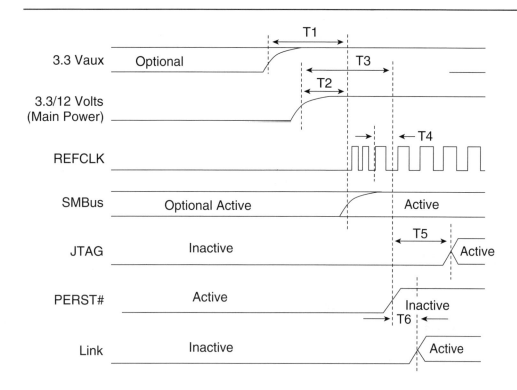

Figure 13.4 Powered On

Wake Events: Wakeup Protocol and Wakeup Event

The mechanisms for wake events are defined in two parts: Wakeup Protocol and wakeup event. The first part of the Wakeup Protocol results in the main power being applied, the execution of Link Training, and the execution of Flow Control Initialization by the hierarchy domain. The WAKE# signal line, or the Beacon, causes the power controller(s) of the hierarchy domain to apply main power, the application of main power causes the LTSSM in the PCI Express devices and links to transition from the L2 link state to the Detect link state. Once in the Detect link state, the LTSSM in the PCI Express devices and the links execute Link Training. Successful completion of Link Training transitions the LTSSM in the PCI Express devices and links to the L0 link state, and Flow Control Initialization can be executed. The successful completion of Flow Control Initialization allows the execution of the other element of the Wakeup Protocol. The other element is the transmission of the message PM_PME requester transaction.

Before discussing in detail the Wakeup Protocol, the other wake event and significance of the PCI PME# signal line needs to be examined.

The Other Wake Event

The other wake event is defined by the Hot Plug protocol and called the wakeup event. The four Hot Plug (HP) events related to hardware are attention button pressed, power fault detected, MSI sensor changed, and add-in card present state changes. In response to these HP hardware events, a wakeup event occurs. The wakeup event is simply the sourcing of a message PM_PME requester transaction packet by the switch. In the case of a Root Complex downstream port, it is the "virtual" internal sourcing of this message to itself. See Chapter 16 for more information.

During the actual Hot Plug protocol the WAKE# signal line must be "1" (inactive).

PCI Legacy PME# Signal Line

As will be discussed in Chapter 16, there are elements in the functions of the PCI Express devices defined by the PCI legacy power management (herein called *PCI power management standards*). The compatible portion of the Power Management protocol is the implementation of "D" states within the functions of PCI Express devices. Another element of the PCI legacy power management is the PME# signal line. The purpose of the PME# signal line is to indicate a PME has occurred. However, the PCI Express specification did not adopt this particular element from the PCI legacy power management. Any PCI or PCI-X devices (bridge or legacy endpoint) that implement the PME# signal line must rely on the generation of an interrupt or the porting of the PME# signal to a Root Complex designed to support it. For PCI and PCI-X devices downstream of a bridge, the bridge can provide an interrupt or convert the PME# signal line to the message PM_ PME requester transaction packets. The PCI Express specification does not provide any additional guidance other than what is detailed in the Chapters 15, 16, and 17.

Wakeup Protocol

When main power is off and Vaux is on, PCI Express devices transition to sleep and the associated links transition to the L2 link state. The LTSSM of the PCI Express devices and associated link are able to transition from the L2 link state to the Detect link state occurs by executing the first part of the Wakeup protocol.

The first part of the Wakeup protocol includes the assertion of the WAKE# signal line or the sourcing of the Beacon by a PCI Express device to cause main power to be applied to all or part of the hierarchy domain. The WAKE# signal line is the traditional physical wire used to request servicing from the power controller, and is defined in Chapter 21. The Beacon is the non-typical approach to request servicing from the power controller. The Beacon is a unique waveform transmitted onto the link. See the PCI Express specification for more information about the Beacon

The second part of the Wakeup protocol is the PCI Express device that caused the first part of the Wakeup protocol also sources a message PM_PME requester transaction.

The support of the Wakeup protocol for PCI Express devices on PCI Express platforms and add-in cards is optional, but if implemented, the requirements for the WAKE# signal line or Beacon are as follows:

- PCI Express devices that are not mounted on an add-in card must support the Beacon and may optionally support the WAKE# signal line.

- A Root Complex on a PCI Express platform that does not have any add-in card slots must support the Beacon and may optionally support the WAKE# signal line.

- For a PCI Express platform that implements add-in card slots and for which all PCI Express platform devices implement the Wakeup protocol with WAKE# signal line, the Root Complex is only required to implement the WAKE# signal line and is not required to the support of the Beacon.

- Add-in cards and PCI Express devices on the add-in cards that implement the Wakeup protocol must implement the WAKE# signal line and may optionally implement the Beacon. PCI Express platforms that implement the Wakeup protocol and slots must implement the WAKE# signal and may optionally implement the Beacon. Consequently, the add-in card must execute the Wakeup protocol on both the WAKE# signal line and Beacon, or just the WAKE# signal line.

- A switch that supports WAKE# signal line on its downstream side must translate to a Beacon if needed to be compatible upstream. A switch that supports a Beacon signal line on its downstream side must translate to the WAKE# signal line if needed to be compatible upstream.

- The power controllers for main power of the entire PCI Express platform or each hierarchy domain are typically controlled by the Root Complex. Consequently, the WAKE# signal line(s) are inputs to the Root Complex and the final destination of the Beacon is also the Root Complex. Some power controllers for main power of a portion of the hierarchy domain can be controlled by a switch. For these power controllers, the associated WAKE# signal lines are inputs to the switch and the final destination of the Beacon moving from a downstream PCI Express device on the hierarchy domain.

The Wakeup protocol is only defined for the transitions from the L2 link state to the Detect link state. While in the L2 link state, the retention of the sticky bits and the ability to transition to the Detect link state via WAKE# signal line or Beacon can only be supported by power from Vaux. The support of the Vaux on a platform *does not* mean that the Wakeup protocol is supported. For the Wakeup protocol to be executed, the Vaux must be maintained since the main power was powered off and the functions within PCI Express devices that assert the WAKE# signal line or source the Beacon must be programmed to do so. For a function to execute the Wakeup protocol, the PME software must have set the PME_EN bit of the Power Management Control/Status Register to 1b.

The WAKE# signal line is *not* defined as equivalent to the PME# signal line defined by the PCI specification.

WAKE# Signal Line

A shown in Figure 13.5, the WAKE# signal line is an open collector "OR" that connects as an input to power controllers connected to the Root Complex and switches. These power controllers supply main power to downstream PCI Express devices. By definition, the power controller is upstream of the PCI Express devices whose power is being controlled. Some items of note in Figure 13.5 are as follows:

- The WAKE# signal line can be fed directly to the power controller or can feed a switch that must convert it to a Beacon. Per Note 1 of Figure 13.5, the WAKE# signal line is connected to endpoints 1a and 1b. Alternatively, if the PCI Express platform is so designed, the switch can also drive a WAKE# signal line on behalf of the WAKE# signal line shared by endpoints 1a and 1b.

- Per Note 2 of Figure 13.5, a switch that controls the power controller is the input for the WAKE# signal line shared by downstream devices on the hierarchy domain.

- PCI Express platforms that implement the Wakeup protocol and slots must implement the WAKE# signal on *all* slots. The implementation of the WAKE# signal line can be a common line shared by *all* slots and interconnected with the power controller, or the implementation can be individual WAKE# signal lines between some of the slots and power controller. The exception is slots that support the Hot Plug (HP) protocol: the WAKE# signal line for each HP slot must be connected individually to the power controller (Note 3 of Figure 13.5). During the insertion and removal of the add-in card, the associated WAKE# signal must be driven inactive.

- The WAKE# signal line can never be an input to bridges or endpoints.

- The PCI Express device that drives the WAKE# signal line = "0" (asserted) must continue to assert it until the main power is powered on for the associated hierarchy domain and has returned to a stable level meeting minimum voltage levels based on PERST# signal line being set to a "1" value.

Beacon

Another element of the Wakeup protocol is the Beacon. The WAKE# signal line is defined as a sideband signal. That is, platform events and conditions are reported between two PCI Express devices without using the signal lines in the link used for Physical Packets. Instead of using the WAKE# signal line, a Beacon can be used. The Beacon uses the signal lines in the link used for Physical Packets; consequently, it is defined as an in-band signal. As shown in Figure 13.3, the Beacon only flows upstream to the Root Complex and switches that the associated power controllers are connected to. The Beacon does not flow upstream any farther than the PCI Express device that is connected to the power controller. By definition, the power controller is upstream of the PCI Express devices whose power is being controlled.

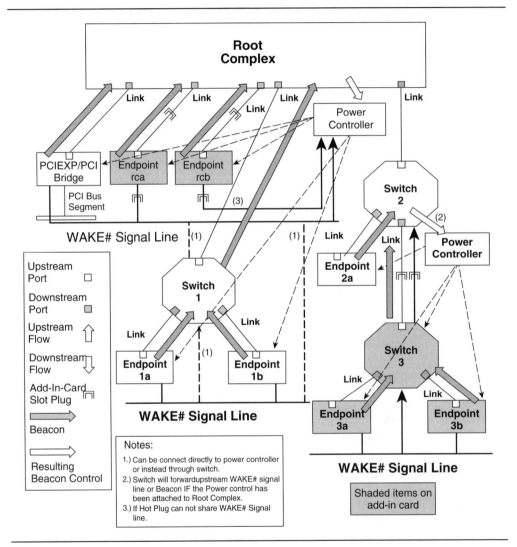

Figure 13.5 WAKE# Signal Line and Beacon

The PCI Express device that drives the Beacon must continue to drive it until the main power is powered on and has returned to a stable level meeting minimum voltage levels (effectively PERST# signal line = "1") for the associated hierarchy domain.

The transmission of the Beacon is only from the upstream ports of endpoints, switches, and bridges. The transmission of the Beacon is from a downstream device to an upstream device on LANE# 0. The Beacon can also be transmitted on other configured lanes of a multi-lane link provided it is also transmitted on LANE# 0. The existence of configured lanes is left over from a previous configuration link state, and thus requires that Vaux be applied to the PCI Express devices when main power is off to retain the configured lane information. Obviously, because Beacon is used as part of the Wakeup protocol, the Vaux is the source of power for the transmitters to drive the Beacon.

The profile of the Beacon is defined as any periodic electrical pattern that has a pulse width between 2 nanoseconds and 16 microseconds in width. The PCI Express specification and the associated PCI Express electrical specification provide the full electrical and timing requirements for a Beacon. These particulars are sufficiently covered by these specifications, and are not discussed here.

Message PM_PME Message Requester Transaction

The last part of the Wakeup Protocol is the transmission of the message PM_PME requester transaction packet to the Root Complex. The purpose of the message PM_PME requester transaction packet is to request service by the PM software via the PME software. The PCI Express device that requested the execution of the Wakeup Protocol by assertion of the WAKE# signal line or Beacon continues to transmit the message PM_PME requester transaction packet using the PME Service Timeout protocol. The message PM_PME requester transaction packet is also source for a wakeup event as part of the Hot Plug protocol. For more information, including discussion of the PME Service Timeout protocol, see Chapter 15.

The message PM_PME requester transaction is also executed as part of the Power Management protocol and the Hot Plug protocol, discussed in Chapters 15 and 16, respectively.

Wakeup Protocol and the D3 State

The D3 state is required to be supported by all functions in PCI Express devices. When functions in PCI Express devices are programmed to the D3hot state, it is with the main power on. The PCI Express devices in the D3hot state are connected to links transitioning from the L0 link state to L2/3 Ready link state. When main power is powered off, the functions within the PCI Express devices transition to the D3cold state.

In the D3cold state, the Vaux power may or may not be retained during the period that main power is powered off. If Vaux power is not retained, the functions of a PCI Express device in the D3 cold state have not retained any PME information. If Vaux power is retained, the PCI Express device has preserved the PME-related information. The preservation of the PME-related information permits the execution of the Wakeup protocol, if a function of the PCI Express device is so programmed in the D3hot state.

Resets

Three different resets are defined by the PCI Express reset specification for a PCI Express platform and add-on cards: Cold, Warm, and Hot. The Cold and Warm Resets are defined as the Fundamental Resets, and they affect PCI Express devices and links. The Hot Reset is not defined as a Fundamental Reset and only affects links.

The Cold Reset is the type of Fundamental Reset that occurs after main power has cycled off then on, with Vaux optionally remaining on or being turned off. If the Vaux remains on, the transition from the Cold Reset is additionally associated with the WAKE# signal line and Beacon. The Cold Reset is associated with the functions within PCI Express devices transitioning from D3cold to D0 un-initialization.

The Warm Reset is the type of Fundamental Reset that occurs without the main power being cycled off then on. The PCI Express specification does not define exactly how the Warm Reset is implemented. Consequently, this book focuses on the Fundamental Reset from the viewpoint of a Cold Reset.

The Hot Reset is only defined for a specific link, is not defined as part of the Fundamental Reset protocol, and is not related to the main power cycling except that main power is maintained.

The exiting from a Fundamental Reset or Hot Reset is to the Detect link state. The Detect link state is the beginning point for Link Training. As discussed in Chapters 8 and 14, Link Training begins with the detection of the lanes between to PCI Express devices and completes with the configuration of lanes into configured links. The completion of Link Training occurs with the transition to the L0s link state. Within the L0 link state, the Flow Control Initialization protocol is executed in as discussed in Chapter 7. The Link Training and Flow Control Initialization protocol includes the PCI Express ports' registers being initialized to the default values and the initialization of state machines including LTSSM

(Link Training Status State Machine) and DLCMSM (Data Link Control and Management State Machine). In the transition from Fundamental Reset, the values of the sticky bits are retained if Vaux has been retained. In the transition from Hot Reset, the values of the sticky bits are retained because main power has been on. Under these circumstances, the sticky bits are used to initialize specific registers within configuration register block to non-default levels. If Vaux has not been retained during Fundamental Reset, the values of the sticky bits are not retained and these values are not used in place of default values.

PERST# Signal Line Considerations

Before discussing the particulars of Fundamental and Hot Resets, further discussion of the PERST# signal line is required. Typically the PERST# signal line is used as a reference point for Cold Reset as an indicator for the power level of main power. The PCI Express specification also permits the PERST# signal line to be used as a reference point for a Warm Reset, and not just as an indicator for the power level of main power. The use of the PERST# signal line beyond an indicator for main power is implementation-specific. Both Cold Reset and Warm Reset are defined as a Fundamental Reset; consequently, a Fundamental Reset occurs when the PERST# signal line transitions to "0" (active)" to "1" (inactive).

The PCI Express specification does not require the PERST# signal line to be *the one and only* reset input to a PCI Express device. If a PCI Express device is supplied with the PERST# signal line, it must be used based on the definition of a Fundamental Reset. If the PERST# signal line is not provided to a PCI Express device, a Fundamental Reset must be executed by an implementation-specific mechanism. This implementation-specific mechanism must include an indication of main power cycling. The PERST# signal line is a required on the connector between the chassis and add-in card.

> To simplify discussions, this book assumes that the only input to a PCI Express device for a Fundamental Reset is the PERST# signal line, and that it is linked to the power level of main power.

Fundamental Reset Protocol

The Fundamental Reset is the collective name for the Cold and Warm Resets. The following attributes of the Fundamental Reset protocol apply to both Cold and Warm Resets.

Fundamental Reset Active

The execution of a Fundamental Reset (active) begins with the PERST# signal line = "0" (active). When the Fundamental Reset is active, the ports' transmitters and receivers are in the following states:

- Receivers termination is Z $_{RX\text{-}HIGH\text{-}IMP\text{-}DC}$ (Table 4-6 of PCI Express specification).

- Transmitters termination is Z$_{TX\text{-}DC}$ (Table 4-6 of PCI Express specification).

- Transmitters hold at a constant DC common voltage but do not have to V$_{TX\text{-}CM\text{-}DC\text{-}ACTIVE\text{-}IDLE\text{-}DELTA}$ (Table 4-5 of PCI Express specification).

During the period of Fundamental Reset Active the Physical LinkUp = "0" and the link is defined as Link_DOWN.

Transitioning from Fundamental Reset Active to Inactive

The exiting from the Fundamental Reset (transition to inactive) occurs with the transition of the PERST# signal line = "1" (inactive). Upon the transition from Fundamental Reset Active to Inactive, the LTSSMs and DLCMSMs in the ports are initialized. The LTSSMs must transition to the Detect link state to begin Link Training within 80 milliseconds. The transition within 80 milliseconds is in conjunction with software waiting a minimum of 100 milliseconds from Fundamental Reset Active to Inactive for any accesses to configuration address space. All PCI Express devices on the platform must adhere to the 100 milliseconds requirement and be ready to accept a configuration access. In addition, the Root Complex and software must wait 1.0 to 1.5 seconds after the transition from a Fundamental Reset before the defining a PCI Express device as broken due to non-receipt of successful completer transaction packets (STATUS = 000b) in response to configuration requester transaction packets. In addition, if the access is to the configuration address space of PCI or PCI-X devices downstream of a bridge, the Trhfa time requirement defined by the PCI specification specific must be adhered to.

The Detect link state is the beginning point for the sequence of link states of the Link Training and Status State Machines (LTSSMs) in the Physical Layer and links for Link Training. The Link Training includes the Detect, Polling, and Configuration link states (See Chapters 8 and 14.) The successful completion of the Link Training results in the transition of Physical LinkUP = "0" to Physical LinkUp = "1" and the LTSSMs and links transition to the L0 link state. The Physical LinkUp indicator is used by the Physical Layer to inform the Data Link Layer that the link is ready to transmit Physical Packets that contain the Link Layer Transaction Packets (LLTPs) or Data Link layer packets (DLLPs) from the Data Link Layer. With the transition to Physical LinkUp = "1," the Data Link Control and Management State Machines (DLCMSMs) in the Data Link Layer begin execution of Flow Control Initialization protocol as defined by the link activity states. The Flow Control Initialization protocol establishes the available buffer space at each end of the links for transmission of Transaction Layer Packets (TLPs). If the Flow Control Initialization of VC0 is successful and Physical LinkUp = "1," then the link is defined as Link_UP. See Chapter 7 and 8 for more information.

The PCI Express specification does not completely define the overall absolute PERST# signal line protocol across a specific PCI Express fabric and across multiple PCI Express fabrics. Obviously, the PERST# signal line levels and the associated timing for Fundamental Reset transitions from active and inactive within a PCI Express fabric must ensure all resources are properly reset. This includes ensuring reference clocks are stable prior to PERST# signal line being inactive (T$_{PERST-CLK}$).

When multiple PCI Express fabrics are interconnected, the PCI Express specification defines several parameters that must be established:

- TPERST-INACTPOWER: PERST# signal line is inactive a minimum time after main power is at or above valid power levels. The PCI Express specification has defined this as 100 milliseconds minimum.

- TPERST-INACT: PERST# signal line is inactive a minimum time.

- TPERST-ACT: PERST# signal line is active a minimum time. The PCI Express specification has defined this as 100 microseconds minimum.

- TPERST-FAIL: PERST# signal line must become active within a maximum time after main power is below valid power levels if powered-off was not planned. The PCI Express specification has defined this as 500 nanoseconds maximum.

Hot Reset Protocol

The Hot Reset is specific to the PCI Express device and links and is not associated with the hardware elements of the PCI Express platform as is the case in Fundamental Reset.

Hot Reset Active

The Hot Reset link state places the configured lanes in a specific link into a reset condition. The Hot Reset link state protocol can be explained by defining two types of ports on a link: *transmitting* and *receiving*. One port on the link must be directed to the Hot Reset link state from the Recovery link state by a higher layer or, in the case of a switch, if the upstream port reports Link_DOWN or begins receiving TS1 OSs with Training Control Bits [2::0] = 001b on *all* configured lanes. See Chapter 14 for more information. This type of port is called the *transmitting port*. The transmitting port must always be the port on the upstream side of a link. That is, only the downstream ports of the Root Complex and switches can become transmitting ports for a Hot Reset. In addition, in the case of switches, their downstream ports become transmitting ports when the upstream ports transitioned to Hot Reset link state as discussed in the following section.

The other type of port is the *receiving port*. It is directed to the Hot Reset link state by receipt of the two consecutive TS1 OSs with Training Control Bits [2::0] = 001b on *any* configured lanes. The receiving port must always be the port on the downstream side of a link. That is, only the upstream ports of the endpoints, switches, and bridges can become receiving ports for a Hot Reset link state. The link states associated with the Hot Reset protocol are discussed in more detail in Chapter 14.

During the period of Hot Reset link state the Physical LinkUp = "0" and the link is defined as Link_DOWN. When a Hot Reset is executed on a link, it impacts the PCI Express devices connected to the link in various ways:

Receiving Port on a Switch: When the receiving port is also the upstream port of a switch, all of the switch's downstream ports become the transmitting port of those links and transition to Hot Reset link state as soon as possible. The resources specific to PCI Express within the upstream port (registers, timers, counters, DLCMSM, and LTSSMs) are placed in the default and initial conditions. The exception is the sticky bits in the configuration register block, which are not modified or initialized by Hot Reset. All Transaction Layer Packets (TLPs) being buffered in the receiving port to be transmitted upstream are discarded, including TLPs in the retry buffer.

Transmitting Port on a Switch: When the transmitting port is a downstream port of a switch, the Hot Reset link state propagates farther downstream. The downstream port of the switch becomes a transmitting port by a higher layer or due to the upstream port being the receiving port as per the previous discussion. The PCI Express specific resources within the downstream ports (registers, timers, counters, DLCMSM, and LTSSMs) are placed in the default and initial conditions. The exception is the sticky bits in the configuration register block that are not modified or initialized by Hot Reset. All Transaction Layer Packets (TLPs) being buffered in the transmitting port to be transmitted downstream are processed as follows:

- Requester transaction packets that have been transmitted but cannot be completed are discarded. Determination of whether requester transaction packets can be completed is implementation-specific for requester transaction packets with associated completer transaction packets. Those requester transaction packets not associated with completer transaction packets are transmitted.

- Completer transaction packets that are already "formed" and ready for transmission are discarded.

- Completer transaction packets that are being "formed" such that the Status can be set to unsupported, transmitted, and associated requester transaction packets are discarded.

Receiving Port on an Endpoint: When the receiving port is also the upstream port of an endpoint, the resources within the endpoint not related to the port are unchanged. The resources specific to PCI Express within the upstream port (registers, timers, counters, DLCMSM, and LTSSMs) are placed in the default and initial conditions. The exception is the sticky bits in the configuration register block, which are not modified or initialized by Hot Reset. All Transaction Layer Packets (TLPs) being buffered in the receiving port to be transmitted upstream are discarded, including TLPs in the retry buffer.

Receiving Port on a Bridge: When the receiving port is also the upstream port of a bridge, the non-PCI Express devices within the bridge and downstream of the bridge are unchanged. The resources specific to PCI Express within the upstream port (registers, timers, counters, DLCMSM, and LTSSMs) are placed in the default and initial conditions. The exception is the sticky bits in the configuration register block, which are not modified or initialized by Hot Reset. All Transaction Layer Packets (TLPs) being buffered in the receiving port to be transmitted upstream are discarded, including TLPs in the retry buffer.

Transmitting Port on the Root Complex: When the transmitting port is a downstream port of the Root Complex, the Hot Reset propagates farther downstream. The downstream port of the Root Complex becomes a transmitting port by a higher layer. The resources specific to PCI Express within the downstream ports (registers, timers, counters, DLCMSM, and LTSSMs) are placed in the default and initial conditions. The exception is the sticky bits in the configuration register block, which are not modified or initialized by Hot Reset. All Transaction Layer Packets (TLPs) being buffered in the transmitting port to be transmitted downstream are processed as follows:

- Requester transaction packets that have been transmitted but cannot be completed are discarded. Determination of whether requester transaction packets can be completed is implementation-specific for requester transaction packets with associated completer transaction packets. Those requester transaction packets not associated with completer transaction packets are transmitted.

- Completer transaction packets that are already "formed" and ready for transmission are discarded.

- Completer transaction packets that are being "formed" such that the Status can be set to unsupported, transmitted, and associated requester transaction packets are discarded.

Transitioning from Hot Reset Active to Inactive

The exiting from the Hot Reset (transition to inactive) differs for the transmitting port and the receiving port. The transmitting ports remains in the Hot Reset link state until directed to the Detect link state by direction of a high layer. The receiving ports remain in the Hot Reset link state until reception of TS1 OSs ceases. See Chapter 14 for more information.

Upon the transition from Hot Reset Active to Inactive, the LTSSMs and DLCMSMs in the ports are initialized as discussed earlier. The LTSSMs must transition to the Detect link state to begin Link Training immediately. The Detect link state is the beginning point for the sequence of link states of the LTSSM and links for Link Training. The Link Training includes the Detect, Polling, and Configuration link states. (See Chapters 8 and 14 for more information.) The successful completion of the Link Training results in the transition of Physical LinkUP = "0" to Physical LinkUp = "1" and the LTSSMs and links transition to the L0 link state. The Physical LinkUp indictor is used by the Physical Layer to inform the Data Link Layer that the link is ready to transmit Physical Packets that contain the Link Layer Transaction Packets (LLTPs) or Data Link layer packets (DLLPs) from the Data Link Layer. With the transition to Physical LinkUp = "1," the Data Link Control and Management State Machines (DLCMSMs) in the Data Link Layer begin execution of Flow Control Initialization protocol as defined by the link activity states. The Flow Control Initialization protocol establishes the available buffer space at each end of the links for transmission of Transaction Layer Packets (TLPs). If the Flow Control Initialization of VC0 is successful and Physical LinkUp = "1," then the link is defined as Link_UP. See Chapter 7 and 8 for more information.

Chapter 14

Link States

The operation of the Physical Link Layer within each PCI Express device is defined by different logic states of the Link Training and Status State Machine (LTSSM) and the associated link. These logic states of the LTSSM are defined as "link states" of the Physical Layer of the PCI Express device, the LTSSM, and the associated link.

Introduction to Link States

As discussed in Chapter 13, the LTSSMs are initialized to a known state by the transition from Fundamental (Cold and Hot) or Hot Resets. Some of these link states have sub-states. The sub-states only interact with each other; some are entry points for the overall link state, and others are exit points from the link states.

A transition to a link state is actually to its entry point sub-state, if one is defined. The entry points into a specific link state are as follows:

- **L0 link state:** entry is via L0 link state
- **L0s link state:** entry is via Tx_L0s.Entry sub-state and Rx_L0s.Entry link sub-state
- **L1 link state:** entry is via L1.Entry link sub-state
- **L2 link state:** entry is via L2.Idle link sub-state
- **L3 link state:** entry is via L3 link state

- **L2/3 Ready link state:** entry is via L2/3 Ready.Entry link sub-state
- **Detect link state:** entry is via Detect.Quiet link sub-state
- **Polling link state:** entry is via Polling.Active link sub-state
- **Configuration link state:** entry is via Config.Linkwidth.Start or Config.Linkwidth.Accept link sub-state
- **Hot Reset link state:** entry is via Hot Reset link state
- **Disabled link state**: entry via Disabled link state
- **Recovery link state:** entry is via Recovery.RcvrLock link sub-state
- **Loopback link state:** entry is via Loopback.Entry link sub-state

The transition between link states or link sub-states is for various conditions summarized in the states tables in this chapter. Figure 14.1 illustrates the transition between link states of the LTSSM of the Physical Layer of PCI Express devices and associated links.

This chapter has divided the discussion of the link state into five categories of link states as follows:

- Normal Operation: L0 link state
- Power Management: L0s, L1, L2, L3, and L2/3 Ready link states.
- Link Training: Detect, Polling, and Configuration link states.
- Other: Recovery, Hot Reset, and Disabled link state
- Testing and Compliance
  - **Loopback link state:** Used for PCI Express device testing, debug, and fault detection.
  - **Polling.Compliance sub-state:** Part of the Polling link state used for compliance testing.

Other Considerations

The primary focus of this chapter is the definition and the conditions for transitions between link states. These link states reflect logic states of the LTSSM in the Physical Link Layer. Another set of logic states are the link activity states associated with the Data Link Control and Management State Machine (DLCMSM) of the Data Link Layer for each PCI Express device.

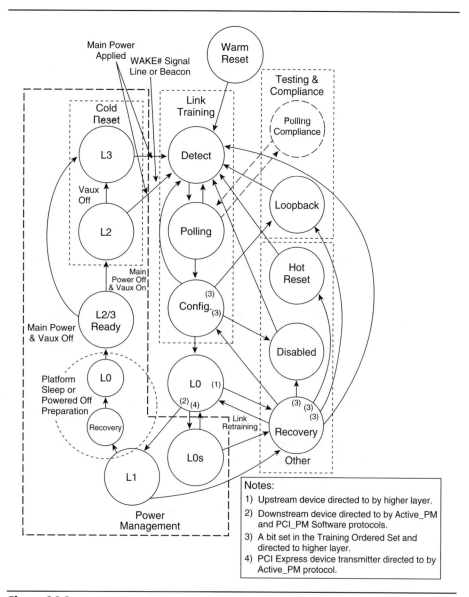

Figure 14.1 Overall Link States

Link States versus Link Activity States

Link states are not to be confused with the *link activity states* discussed in other chapters. The link activity states are: DL_Inactive, DL_Init, and DL_Active and are specific to the Flow Control Initialization protocol and the DLCMSM of the Data Link Layer for each PCI Express device. The purpose for the Flow Control Initialization protocol is to establish available buffer spaces in the LLTPs' receiving ports on the link. The Flow Control Initialization protocol is implemented by the DLCMSM sequencing through the link activity states once the LTSSM is in the L0 link state. See Chapter 7 for more information.

Physical LinkUp = "0" or "1"

The tables in this chapter list the value of Physical LinkUp, which is the status of the LTSSM. When Physical LinkUp = "1" the Physical Layer indicates to the Data Link Layer that the link has been configured and Physical Packets *can* be transmitted across the link. When Physical LinkUp = "0," the Physical Layer indicates to the Data Link Layer that Physical Packets *cannot* be transmitted across the link. The value of the Physical LinkUp does not affect the transmission of K and D symbols used for ordered sets and lane fillers.

USD and Downstream Port versus DSD and Upstream Port

In the following discussions the terms *USD, downstream port, DSD,* and *upstream port* are used. The term USD (upstream device) refers to the PCI Express device on the upstream side of the link. The port of this PCI Express device is defined as the downstream port. The term DSD (downstream device) refers to the PCI Express device on the downstream side of the link. The port of this PCI Express device is defined as the upstream port.

Advanced Peer-to-Peer Link

In the case of an advanced peer-to-peer link (also known as cross-link), the downstream port of a switch may be configured as an upstream port. Consequently, one downstream port of one switch on the cross-link operates as a downstream port of a USD and the other downstream port of the other switch on the cross-link must operate as if it were an upstream port of a DSD. In this chapter, the terms *upstream ports* and *DSDs* apply to the downstream port of the other switch on the cross-link, which must operate as if it were an upstream port of a DSD. See Chapter 8 for more information.

The operation of the USDs and DSDs connected via advanced peer-to-peer links is the same as USDs and DSDs connected via non-advanced peer-to-peer links with the following exceptions:

- Operation for the Configuration is implementation-specific.

- Operation for Hot Reset link state for an advanced peer-to-peer link is implementation-specific.

Link State Transitions and Higher Layer Direction

The transitioning through the link states of the Link Training and Status State Machines (LTSSMs) of the Physical Link Layer of the PCI Express devices, and associated links, begins with a Fundamental Reset. The Fundamental Reset results in transition to the Detect link state. As discussed in Chapter 13, the Fundamental Reset is related to the hardware reset signal line. Several other conditions cause link state transitions in addition to Fundamental Reset. Some of these transitions are caused by the defined interaction between link states and link sub-states. Other transitions are caused by the removal of main power or Vaux, or the execution of Wake Events. Finally, some of the transitions in the link states of the LTSSM and associated link are due to higher layer direction. The higher layer direction is defined as activities that occur at the PCI Express device core, Transaction Layer, or the Data Link Layer. The activities defined as higher layer direction are as follows:

- **Retrain Bit:** The transition from the L0 link state to Recovery link state occurs when the Retrain bit is set to 1b in the Link Control Register in the configuration register block of the downstream port. By definition only downstream ports of the Root Complex and switches implement this bit. The Retrain bit is programmable by software.

- **Disable Bit:** The transition from the Configuration or Recovery link state to the Disable link state occurs when the Disable bit is set to 1b in the in the Link Control Register in the configuration register block of the downstream port. By definition, only downstream ports of the Root Complex and switches implement this bit. The Disable bit is programmable by software.

- **Loopback link state:** The transition from Configuration or Recovery link state to the Loopback link state is due to one PCI Express device becoming the Loopback Master. The determination by a PCI express device becoming a Loopback Master is implementation-specific and not defined by the PCI Express specification.

- **Configuration link state:** The transition from the Recovery link state to the Configuration link state is implementation-specific and not defined by the PCI Express specification. By definition, only downstream ports of the Root Complex and switches make the request to transition to this link state.

- **Reset link state:** The transition from the Recovery link state to the Hot Reset link state is implementation-specific and not defined for all possible conditions by the PCI Express specification. Conditions are defined for the downstream ports' LTSSM and link for transition to the Hot Reset link state by direction of a higher layer programming as follows:
    - For the virtual PCI/PCI Bridge associated with the downstream port of the Root Complex or a switch, its Secondary Bus Reset bit is set to "1" within the Bridge Control Register.
    - For a switch that is implemented with a virtual PCI/PCI Bridge on its upstream port (that is, upstream of all the downstream virtual PCI/PCI Bridges within the switch), its Secondary Bus Reset bit is set to 1b within the Bridge Control Register.

- **PCI-PM Software protocol:** In the L0 link state the transition to the L1 and L2/3 Ready link states occurs by Power Management software programming the "D" states of the functions with PCI Express devices. See Chapter 15 for more information.

Link transitions have two other considerations in addition to the transitions due to higher layers. The first consideration is that transition between the L0 and L0s link states, and the transition between L0 link state and L1 link state, are due to the Active-PM protocol. By convention, the Active-PM protocol is not defined as direction by a higher layer. See Chapter 15 for more information. The second consideration is specific to switches. The downstream ports of a switch transition to the Reset link states when the switch's upstream port's Data Link Layer reports Link_DOWN or the upstream port begins receiving TS1 OSs with Training Control Bits [2::0] = 001b on *ALL* configured lanes. See Chapter 13 for related considerations to Hot Reset protocol in a switch.

Normal Operation Link State

The following discussions for L0, L1, L0s, L2/3 Ready L2, and L3 link state focus on the transitions between these different link states of the LTSSM and the link state. A more detailed discussion about the transitions between the link states of the Power Management category and with link states outside of the Power Management category are in Chapter 15.

L0 Link State

The purpose of this link state is for the normal transfer of Physical Packets containing DLLPs and LLTPs between the two ports connected to the link. As illustrated in Figure 14.1, the L0 link state is entered from the Configuration, L0s, and Recovery link states. When the transition is from the Configuration link states, the transmission of DLLPs is immediate. The Flow Control Initialization protocol upon transition to the L0 link state must be successfully completed before LLTPs can be transmitted. The Flow Control Initialization protocol establishes available buffer space at the LLTPs' receiving ports. Part of the Flow Control Initialization is the execution of the link activity states by the Data Link Control and Management State machine (DLCMSM). See Chapter 12 for more information. When the transition is from the Recovery or L0s link states, the transmission of DLLPs and LLTPs is immediate; the values for available buffer space at the LLTPs' receiving ports are retained from previous L0 link state.

Transition to Recovery Link State

L0 link state can only transition to link states that are of the Power Management category or the Recovery link state within the Other category. Transitions between the L0 link state and other link states in the Power Management category are governed by the PCI-PM Software and Active-State PM protocols detailed in Chapter 15. Even though Training OSs are transmitted to transition to the Recovery link state, the LTSSM cannot use the Training Control Bits [2::0] in the Training OS to transition to the Hot Reset, Loopback, or Disabled link states directly from the L0 link state; the LTSSM of the PCI Express devices and the associated link must always transition through the Recovery link state. The value of the Training Control Bits [3::0] for the transition from the L0 to Recovery link state is undefined by the PCI Express specification. This book assumes that the value of each bit reflects the value to be used for transition from the Recovery link state.

Transition to L0s Link State

The transition occurs according to direction of the Active-State PM protocol as outlined in Chapter 15.

Transition to L1 Link State

The transition to the L1 link state occurs according to the Active-State PM protocol or PCI-PM Software protocol detailed in Chapter 15. The transition from the L0 link state to the L1 link state occurs in three parts as detailed in Table 14.1:

- **Part 1:** The DSD's transmitters are directed by a higher level to transmit and Electrical Idle OS on *ALL* configured lanes and then enter "electrical idle."

- **Part 2:** The USD's receivers detect an Electrical Idle OS on *ANY* configured lanes from the transmitters at the other end of the link. In response, the USD's transmitters send an Electrical Idle OS on *ALL* configured lanes and then enter "electrical idle." The LTSSM and associated transmitters and receivers of the USD transition to the L1 link state.

- **Part 3:** When the DSD's receives an Electrical Idle OS on *ANY* configured lanes, the DSD's LTSSM, transmitters, and receivers transition to the L1 link state.

Transition to L2/3 Ready Link State

The transition to the L2/3 Ready link state is governed by PCI-PM Software as outlined in Chapter 15. The transition from the L0 link state to the L2/3 Ready link state occurs in three parts (details in Table 14.5):

- **Part 1:** The DSD's transmitters are directed by a higher level (see Chapter 15) to transmit and Electrical Idle OS on *ALL* configured lanes and then enter "electrical idle."

- **Part 2:** The USD's receivers detect an Electrical Idle OS on *ANY* configured lanes from the transmitters at the other end of the link. In response, the USD's transmitters transmit an Electrical Idle OS on *ALL* configured lanes and then enter "electrical idle." The LTSSM and associated transmitters and receivers of the USD transition to the L2/3 Ready link state.

- **Part 3:** When the DSD's receivers an Electrical Idle OS on *ANY* configured lanes, the DSD's LTSSM, transmitters, and receivers transition to the L2/3 Ready link state.

Table 14.1 L0 Link State

Present State Attributes:	
DLLP and LLTP are transferred across the link. Link_UP is indicated from Data Link Layer and Transaction Layer once Flow Control Initialization has been successfully completed upon the transition from the Configuration link state. Until the Flow Control Initialization has been successfully completed, Link_DOWN is indicated. See Chapter 12 for more information about Flow Control Initialization. Physical LinkUp = "1" is indicated from Physical Layer to Data Link Layer.	
Cause of Transition to Next State	**Next State Transitioned To**
Transition if the activities in one of the following items occurs at the USD, otherwise remain in L0 link state. • Directed to by higher layer (Retrain bit), the downstream port's transmitters transitioning to "electrical idle" on *ALL* configured lanes without transmitting Electrical Idle OS on *ANY* configured lanes prior to transitioning to Recovery. RcvrLock link sub-state. • A TS1 or TS2 OS is received on *ANY* configured lane. This results in the downstream port's transmitters transitioning to "electrical idle" on *ALL* configured lanes without transmitting Electrical Idle OS on *ANY* configured lanes prior to transitioning to Recovery. RcvrLock link sub-state. • The downstream port's receivers detect "electrical idle" with *ALL* configured lane and no Electrical Idle OS received on *ANY* configured lane. This results in the downstream port's transmitters transition to "electrical idle" on *ALL* configured lanes without transmitting Electrical Idle OS on *ANY* configured lanes prior to transitioning to Recovery. RcvrLock link sub-state. (1). Transition if the activities in one of the following items occurs at the DSD; otherwise remain in L0 link state. • A TS1 or TS2 OS is received on *ANY* configured lane. This results in the upstream port's transmitters transitioning to "electrical idle" on *ALL* configured lanes without transmitting of Electrical Idle OS on *ANY* configured lanes prior to transitioning to Recovery. RcvrLock link sub-state. • The upstream port's receivers detect "electrical idle" with *ALL* configured lane and no prior reception of Electrical Idle OS on *ANY* configured lane. This results in the upstream port's transmitters transitioning to "electrical idle" on *ALL* configured lanes without transmitting Electrical Idle OS on *ANY* configured lanes prior to transitioning to Recovery. RcvrLock link sub-state. (1).	Recovery. RcvrLock
The transition occurs as per direction of the Active-State PM protocol as outlined in Chapter 15.	L0s

Table 14.1   L0 Link State *(continued)*

Cause of Transition to Next State	Next State Transitioned To
Transition if the following occurs at the USD; otherwise remain in L0 link state. • The downstream port's receivers detect reception of Electrical Idle OS on *ANY* configured lanes. This results in the downstream port's transmitters transitioning to "electrical idle" on *ALL* configured lanes after transmitting Electrical Idle OS on *ALL* configured lanes prior to transitioning to L1 link state. Transition if one of the following occurs at the DSD; otherwise remain in L0 link state. • The transition occurs as per direction of the Active-State PM protocol or higher layer (PCI-PM Software protocol), and the following exchange occurs: the upstream port's transmitters transition to "electrical idle" on *ALL* configured lanes after transmitting Electrical Idle OS on *ALL* configured lanes. Prior to the LSTTM and link transitioning to the L1 link state the upstream port's receivers detect reception of Electrical Idle OS on *ANY* configured lanes.	L1
Transition if the following occurs at the USD; otherwise remain in L0 link state. • The downstream port's receivers detect "electrical idle" on *ALL* configured lanes after reception of Electrical Idle OS on *ALL* configured lanes. This results in the downstream port's transmitters transitioning to "electrical idle" on *ALL* configured lanes after transmitting Electrical Idle OS on *ALL* configured lanes prior to transitioning to L2/3 Ready link state. Transition if the following occurs at the DSD; otherwise remain in L0 link state. • The transition occurs as per direction of higher layer (PCI-PM Software protocol). The upstream port's transmitters transitioning to "electrical idle" on *ALL* configured lanes after transmitting Electrical Idle OS on *ALL* configured lanes. Prior to the LSTTM and link transitioning to the L2/3 Ready link state the upstream port's receivers detect "electrical idle" with *ALL* configured lanes after reception of Electrical Idle OS on *ALL* configured lanes.	L2/3 Ready

Table Notes:

1. The PCI Express device may complete any LLTP or DLLP in progress.

Power Management Category of Link States

The previous discussion of the L0 link state and the following discussions for L1, L0s, L2/3 Ready L2, and L3 link state focus on the transitions between these different link states of the LTSSM and the link state. A more detailed discussion about the transitions between the link states of the Power Management category, and with link states outside of the Power Management category, are in Chapter 15.

L0s Link State

The basic purpose for this link state is to provide a standby power state for the link and the associated PCI Express devices. PCI Express defines three low power link states (assuming L3 link state is simply "off" and not a low power state): L0s, L1, and L2. The L0s link state provides shorter transition latency than other low power link states to the fully active L0 link state. The L1 and L2 link states require transitioning to the Recovery link state and Detect link state, respectively. L0s link state transitions to the L0 link state without using the Detect link state and typically without the Recovery link state.

Another important consideration is that a transmitter on one end of a link transitions from the L0 link state to the L0s link state when the LTSSM has determined that no Physical Packets containing DLLPs or LLTPs are ready for immediate transmission. The receiver on the other end of the link notes the L0s link states but does not force the LTSSM or its transmitter of the same port to transition to the L0s link state. That is, the transmitter and receiver of the same port are in the L0s link state independently.

As summarized in Figure 14.2 and Tables 14.2 through 14.4, the LTSSM and the associated transmitters enter and remain in the L0s link state until a Physical Packet is ready to transmit. When a Physical Packet is ready to transmit, the transmitters execute the Fast Training Sequence (FTS) protocol and transition to the L0 link state. Notice that the transmitters do not transition through the Recovery link state to transition to the L0 link state.

Also as summarized in Figure 14.2 and Tables 14.5 through 14.7, at the other end of the link the LTSSM and the associated receivers note the transition to the L0s link state, transition to that L0s link state, and wait for the transmitters at the other end of the link transition to from the L0s link state per the FTS protocol. Depending on the conditions summarized in the following table, the LTSSM and the associated receivers may or may not transition through the Recovery link state to transition to the L0 link state.

The FTS protocol is required to achieve bit and symbol lock in the transition from L0s to L0 link states. Receivers of a port detect an exit of "electrical idle" by the transmitters on the other end of the link by the receipt of "N" number of FTS OSs from these transmitters followed by a single Skip OS (defined as FTS protocol). NO Skip OS can be sent in the stream of N_FTS OSs. The receivers that establish the bit and symbol lock and lane-to-lane de-skew with the receipt of the N_FTS OSs transition to the L0 link state. The transmitters of the same port also transition to the L0 link state. The LTSSM is in the L0 link state with the receivers and transmitters the L0 link state. If the receivers do not establish bit lock, symbol lock, and lane-to-lane de-skew, the receivers transition to the Recovery link state. The transmitters in the L0s link state may have independently transitioned to the L0 link state. Consequently, the transmitters must transition from the L0s or L0 link state (they may optionally complete any Physical Packets being transmitted) to Recovery link state if the receivers in the same port transitions to the Recovery link state. The LTSSM is in the Recovery link state with the receivers and transmitters the Recovery link state. See Chapter 8 for more information.

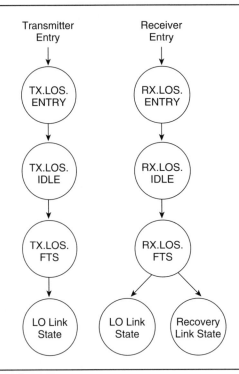

Figure 14.2 L0s Link State

Transmitter and Associated LTSSM

Table 14.2 Tx_L0s.Entry Link Sub-state: Transmitter

Present State Attributes:	
Upon entry into the link sub-state, the transmitters transmit one Electrical Idle OS and then transition to "electrical idle" within 20 UI on *ALL* configured lanes. The "electrical idle" must meet the $V_{TX-CM-DC-ACTIVE-IDLE-DELTA}$ requirement.	
This transition is from the L0 link state and by direction of the Data Link Layer (No DLPP or LLTP and thus no Physical Packets to transmit).	
Physical LinkUp = "1" indicated from Physical Layer to Data Link Layer (UI = 399.88 to 400.12 psec.).	
Cause of Transition to Next State	**Next State Transitioned To**
Transition occurs after entry into this link sub-state plus a minimum of 50 UI of "electrical idle" had elapsed.	Tx_L0s.Idle

Table 14.3   Tx_L0s.Idle Link Sub-state: Transmitter

Present State Attributes: Upon entry into the link sub-state, the transmitters are in "electrical idle." The "electrical "idle" must meet the $V_{TX-CM-DC-ACTIVE-IDLE-DELTA}$ requirement	
Physical LinkUp = "1" is indicated from Physical Layer to Data Link Layer.	
Cause of Transition to Next State	**Next State Transitioned To**
Transaction occurs when the Data Link Layer wants to transmit a DLLP or LLTP and thus a Physical packet is ready for transmission.	Tx_L0s.FTS

Table 14.4   Tx_L0s.FTS Link Sub-state: Transmitter

Present State Attributes: Upon entry into the link sub-state, the transmitters transmit "N" number of FTS OSs *ALL* configured lanes.	
"N" is the N_FTS value specified in the TS2 OS in the Config.Complete link sub-state *but* if the Extended Synch bit = 1b in the Link Control Register "N" = 4096. *No* Skip OSs are inserted before the "N" number of FTS OSs.	
Physical LinkUp = "1" is indicated from Physical Layer to Data Link Layer.	
Cause of Transition to Next State	**Next State Transitioned To**
Transition occurs after the transmission of the "N" number of FTS OSs *and* then transmission of one Skip OS on *ALL* configured lanes. It is permitted to send one full FTS OS before the set of "N."	L0

Relative to Receiver and Associated LTSSM

Table 14.5   Rx_L0s.Entry Link Sub-state: Receiver

Present State Attributes:	
This transition is from the L0 link state due to the receipt of an Electrical Idle OS on *ALL* configured lanes.	
Physical LinkUp = "1" is indicated from Physical Layer to Data Link Layer (UI = 399.88 to 400.12 psec.).	
Cause of Transition to Next State	**Next State Transitioned To**
Transition occurs after entry into this link sub-state plus a minimum of 50 UI of detecting "electrical idle" on *ALL* configured lanes. The "electrical "idle" must meet the $V_{TX-CM-DC-ACTIVE-IDLE-DELTA}$ requirement.	Rx_L0s.Idle

Table 14.6 Rx_L0s.Idle Link Sub-state: Receiver

Present State Attributes:	
Upon entry into the link sub-state, the receivers are monitoring for an exit from "electrical idle" on *ANY* configured lane.	
Physical LinkUp = "1" is indicated from Physical Layer to Data Link Layer.	
Cause of Transition to Next State	**Next State Transitioned To**
Transition occurs if on *ANY* configured lanes the receivers detects an exit from "electrical idle."	Rx_L0s.FTS

Table 14.7 Rx_L0s.FTS Link Sub-state: Receiver

Present State Attributes:	
The transmitting port on the other end of the link is transmitting a set of "N" FTS OSs as part of the FTS protocol. At the completion of the set a Skip OS is transmitted. The receiving port that receives this set is looking for the Skip OS on *ALL* configured lanes.	
Physical LinkUp = "1" is indicated from Physical Layer to Data Link Layer	
(UI = 399.88 to 400.12 psec.).	
Cause of Transition to Next State	**Next State Transitioned To**
Transition occurs if the transaction to the L0 link state did not occur within the timeout of entry into this link sub-state plus a N_FTS timeout (1). If the receivers transition to the Recovery. RcvrLock link sub-state, the associated transmitters must also transition to the Recovery. RcvrLock link sub-state.	Recovery. RcvrLock
Transition occurs if a Skip OS is received on *ALL* configured lanes. Lane-to-lane de-skew must be completed before leaving this link sub-state if transitioning to the L0 link state.	L0

Table Notes:

1. The N_FTS timeout is equal to or longer than 40(N_FTS +3) UI but less than 80 (N_FTS +3) UI. When the Extend Sync bit is set to 1b the N_FTS timeout is equal to or longer than 40(2048) UI but less than 40(4096) UI (2048 or 4096 FTS OS.).

L1 Link State

The basic purpose for this link state is to provide a low power state for the link and the associated PCI Express device. PCI Express defines three low power link states (assuming L3 link state is simply "off" and not a low power state): L0s, L1, and L2. The L1 link state provides longer transition latency than the L0s link state to the L0 link state, but shorter transition latency than the L2 link state to the L0 link state. The power level of the L1 link state is lower than that of the L0s link state and higher than the L2 link state. The L1 requires transitioning to the Recovery link state and cannot transition to the L0 link state directly.

As summarized in Figure 14.3 and Tables 14.8 and 14.9, the LTSSMs and the associated transmitter and receivers remain in the L1 link state until the receiver detects and exit from "electrical idle" or the LTSSM is directed by a higher level as discussed in Chapter 15.

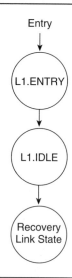

Figure 14.3   L1 Link State

Table 14.8   L1.Entry Link Sub-state

Present State Attributes:	
Upon entry into this link sub-state from the L0 link state, the transmitters must be in "electrical idle" within 20 UI after transmitting Electrical Idle OS on *ALL* configured lanes.	
The "electrical "idle" must meet the $V_{TX-CM-DC-ACTIVE-IDLE-DELTA}$ requirement	
Physical LinkUp = "0" is indicated from Physical Layer to Data Link Layer	
(UI = 399.88 to 400.12 psec.).	
Cause of Transition to Next State	**Next State Transitioned To**
Transition occurs after entry into this link sub-state plus a minimum of 50 UI.	L1.Idle

Table 14.9 L1.Entry Link Sub-state

Present State Attributes:	
Upon entry into the link sub-state, the receivers are monitoring for an exit from "electrical idle" on *ANY* configured lane. The transmitters are in "electrical idle" on *ALL* configured lanes.	
Physical LinkUp = "0" is indicated from Physical Layer to Data Link Layer.	
Cause of Transition to Next State	**Next State Transitioned To**
Transition occurs for the LTSSM and both transmitters and receivers on the same port when the receivers detect exit from "electrical idle" on *ANY* configured lane *or* are directed by a higher level as discussed in Chapter 15.	Recovery. RcvrLock

L2/3 Ready Link State

This link state provides a PCI Express device the time to prepare the PCI Express device for loss of main power and Vaux or just main power with Vaux retained. As summarized in Figure 14.4 and Tables 14.10 and 14.11, the LTSSMs and the associated transmitter and receivers remain in the L2/3 Ready link state until a change in the main power and/or Vaux levels. Also, as outlined in Table 14.11 it is possible for the LTSSMs and the associated transmitters and receivers in the L2/3 Ready to transition to the Recovery link state if main power is retained.

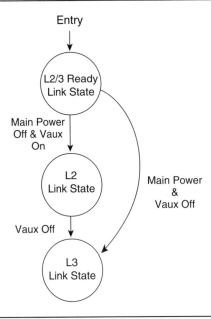

Figure 14.4 L2/3 Ready Link State

Table 14.10 L2/3 Ready.Entry Link Sub-state

Present State Attributes:	
Upon entry into this link sub-state from the L0 link state, the transmitters must be in "electrical idle" within 20 UI after transmitting Electrical Idle OS on *ALL* configured lanes.	
The "electrical "idle" does not meet the $V_{TX-CM-DC-ACTIVE-IDLE-DELTA}$ requirement	
Receivers must be in the low impedance.	
Physical LinkUp = "0" is indicated from Physical Layer to Data Link Layer	
(UI = 399.88 to 400.12 psec.).	
Cause of Transition to Next State	**Next State Transitioned To**
Transition occurs after entry into this link sub-state plus a minimum of 50 UI.	L2/3 Ready .Idle

Table 14.11 L2/3 Ready.Idle Link Sub-state

Present State Attributes:	
Main power is on and Vaux is on (if Vaux provided).	
Physical LinkUp = "0" is indicated from Physical Layer to Data Link Layer.	
Cause of Transition to Next State	**Next State Transitioned To**
Main power is powered-off but Vaux remains on. See Chapter 15 for more information.	L2.Idle
Main power is powered-off and Vaux powered-off. See Chapter 15 for more information.	L3

L2 Link State

The basic purpose for this link state is to provide a sleep state for the link and the associated PCI Express devices. PCI Express defines three low power link states (assuming L3 link state is simply "off" and not a low power state): L0s, L1, and L2. The L2 link state provides a longer transition latency than the L0s or L1 link states to the L0 link state. The power level of the L2 link state is lowest of link states. The L2 link state requires that the Vaux remain on after the main power is powered off.

A transition from the L2 link state can occur in one of three situations. See Chapter 13 for more detailed information.

- **Vaux Removed:** One situation is when Vaux is removed; the PCI Express device and associated link transition to the L3 link state.

- **Main Power Applied and WAKE# Signal line:** Another situation is when main power is applied; the PCI Express device and associated link transition to the Detect link state. In some cases the application of main power is due to the assertion of the WAKE# signal line.

- **Beacon:** The third situation is when a Beacon is received on a downstream port of a USD or an exit from "electrical idle" is detected on an upstream port of a DSD.

Only the third situation is summarized in Tables 14.12 through 14.14. The transitions due to Vaux being removed, main power being applied, and WAKE# signal line cause immediate transition to the Detect link state as illustrated in Figure 14.5. See Chapter 13 for more information.

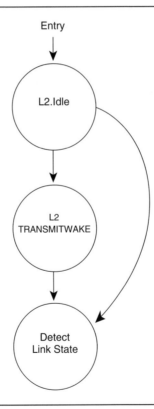

Figure 14.5 L2 Link State

Table 14.12 USD for L2.Idle Link Sub-state

Present State Attribute:	
Upon entry into the link sub-state the receivers in the downstream port of the Root Complex or switch are enabled and in low impedance. The transmitters are in "electrical idle" on *ALL* configured lanes. The "electrical idle" *does not* have to meet the $V_{TX-CM-DC-ACTIVE-IDLE-DELTA}$ requirement Physical LinkUp = "0" is indicated from Physical Layer to Data Link Layer.	
Cause of Transition to Next State	**Next State Transitioned To**
Transition occurs if a Beacon is detected on at least LANE# 0 of ANY downstream port and when the main power has been applied to the Root Complex or switch. The transition also occurs if Root Complex or switch is directed to transition by other means. For example, WAKE# signal line into the Root Complex or switch to enable main power. For a switch the upstream port transitions to the L2.TransmitWake link sub-state when one of the above two occurs.	Detect.Quiet

Table 14.13 DSD for L2.Idle Link Sub-state

Present State Attribute:	
Upon entry into the link sub-state, the receivers in the upstream port of a switch, endpoint, or bridge are enabled and in low impedance. The transmitters are in "electrical idle" on *ALL* configured lanes. The "electrical idle" *does not* have to meet the $V_{TX-CM-DC-ACTIVE-IDLE-DELTA}$ requirement Physical LinkUp = "0" is indicated from Physical Layer to Data Link Layer.	
Cause of Transition to Next State	**Next State Transitioned To**
Transition occurs if exit from "electrical idle" is detected on *ANY* configured lane *and* the main power has been applied. Note: By convention, the main power is applied prior to or with exit. In the case of a switch *ALL* downstream ports also transition to the Detect.Quiet when the upstream port of the switch transitions.	Detect.Quiet

Table 14.14 DSD for L2.TransmitWake Link Sub-state

Present State Attribute *(Defined only for the purpose of transmitting a Beacon upstream)*: Upon entry into the link sub-state, the receivers in the upstream port of a switch, endpoint, or bridge are enabled and in low impedance. The transmitters are in "electrical idle" on *ALL* configured lanes that are not transmitting a Beacon. The transmitter on LANE# 0 (at a minimum) transmits the Beacon until transition from this link sub-state. Physical LinkUp = "0" is indicated from Physical Layer to Data Link Layer.	
Cause of Transition to Next State	**Next State Transitioned To**
Transition occurs if exit from "electrical idle" is detected on ANY receivers in the upstream port on *ANY* configured lane. Note: By convention the main power is applied prior to or with exit from the "electrical idle."	Detect.Quiet

L3 Link State

The basic purpose for this link state is to represent the powered-off condition (no main power or Vaux). It is summarized in Table 14.15. See Chapters 13 and 15 for more information.

Table 14.15 L3 Link State

Present State Attributes: Main power and Vaux are off.	
Cause of Transition to Next State	**Next State Transitioned To**
Main power is applied and perhaps Vaux is applied.	Detect.Quiet

Link Training Link States

Before normal link operation of the L0 link state can begin, or under certain conditions resume, the state machines of the two ports on the link must establish several things. The ports of the PCI Express devices must determine whether another port exists on the other end of the link, the data bit rate of the link, and the lanes in the link that are common to both ports. The overall process to establish these things is called Link Training. The Link Training is accomplished by sequencing through the Detect, Polling, and Configuration link states. The state machine in the ports that sequences through these link states is the Link Training and Status State Machine (LTSSM) and resides in the Physical Layer. Part of the LTSSM specifically addresses Link Training, the other part addresses the non-training link states.

The Link Training sequence is divided among the three link states as follows:

- **Detect link state**: Establishes the existence of a port on each end of link. That is, the Detect link state "detects" the lanes within a link common to both ports on the link.

- **Polling link state:** Establishes the bit and symbol lock, lane polarity inversion, and highest common data bit rate on the lanes that exist between the two ports that are detected but have yet to be configured.

- **Configuration link state:** The detected lanes that successfully complete the Polling link states are processed into configured lanes. Configured lanes are collected into configured link(s) in the link sub-states of the Configuration link state. These link sub-states consequently establish the link width and lane ordering with support of lane reversal. Finally, these link sub-states also implement lane-to-lane de-skew and establish the N_FTS value.

Detect Link State

The purpose of this link state is for a port's transmitters to detect the presence of receivers on the other end of the link on lanes common to both. The transition to the Detect link state is from other different link states, some of which are related to Warm and Cold Resets. See Chapter 13 for more information.

The transmitters are in "electrical idle" upon entry into the Detect link state. The assumption for this link state is that there may have been no previously configured link between the two ports. Consequently, transmission of any bits and associated symbols in this link state by the LTSSM in the Physical Layer implements the minimum data bit rate (2.5 gigabits per second). Other possible data bit rates will be implemented as defined in the Polling link state.

Receiver Detection Sequence

The purpose of the Receiver Detection sequence is for transmitters of one port to detect on *ANY* lane receivers in the other port of the link. Once a group of lanes are detected, they can be configured into a link and become configured lanes.

The Receiver Detection sequence must be executed by all transmitters of a port not assigned to a configured lane as follows:

- The transmitters begin in "electrical idle" by changing common mode voltage of D+ and D• to a different voltage level. The transmitters of each lane must detect whether the load's impedance is Z_{RX-DC} or lower.

- The transmitters detect the presence of receivers if the voltage change rate reflects transmitter impedance, the series capacitor, interconnect capacitance, and receiver termination.

- The transmitters detect the absence of receivers if the voltage change rate reflects only transmitter impedance, series capacitor, and interconnect capacitance.

The Receiver Detection sequence is not executed by a transmitters if "electrical idle" is exited or the detection of the associated lane is aborted.

Detect Link State Protocol

Summarized in Figure 14.6 and Tables 14.16 and Table 14.17 are the transitions of the LTSSMs, and the associated transmitters and receivers on each end of the link.

Table 14.16 Detect.Quiet Link Sub-state

Present State Attributes:	
Upon entry into this link sub-state, the transmitters of un-configured lanes are in "electrical idle."	
The data bit rate is set to 1b.	
The "electrical idle" *does not* have to meet the $V_{TX-CM-DC-ACTIVE-IDLE-DELTA}$ requirement	
Physical LinkUp = "0" is indicated from Physical Layer to Data Link Layer.	
Cause of Transition to Next State	**Next State Transitioned To**
Transition occurs 12 msec. after entering Detect.Quiet or if receivers detect an exit from "electrical idle" on *ANY* lane.	Detect.Active

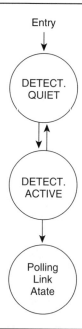

Figure 14.6 Detect Link State

Table 14.17 Detect.Active Link Sub-state

Present State Attributes:	
Upon entry into this link sub-state, the transmitters on *ALL* un-configured lanes execute the Receiver Detection sequence.	
Physical LinkUp = "0" is indicated from Physical Layer to Data Link Layer.	
Cause of Transition to Next State	**Next State Transitioned To**
Transition occurs if a receiver is detected on *ALL* un-configured lanes.	Polling.Quiet
Transition occurs if a receiver is not detected on *ANY* un-configured lanes.	Detect.Quiet sub-state
If receivers are detected on a *set* of (one or more) un-configured lanes but not *ALL* un-configured lanes, wait for 12 msec. and then execute the following: • Transmitters execute the Receiver Detection sequence on *ALL* un-configured lanes of the *set*. If the entire *set* of un-configured lanes detect a receiver transition to Polling.Quiet. • Otherwise transition to Detect.Quiet.	Polling.Quiet (1) or Detect.Quiet

Table Notes:

1. The transition to the Polling.Quiet in this situation may still have un-configured lanes that *did not* have any receivers detected. These lanes must be associated with a new LTSSM if this optional feature is supported *or* the lanes that cannot be associated with the new LTSSM must transition to "electrical idle" and re-associated with the next LTSSM immediately after the LTSSM in progress transitions back to Detect link state. Under this later situation, an Electrical Idle OS is *not* transmitted prior to the transition to "electrical idle."

Polling Link State

The purpose of this link state is to establish bit and symbol lock on the all detected un-configured lanes defined by the Detect link state. It is also used to establish lane polarity inversion and to establish data bit rate for these detected un-configured lanes.

One of the polling link state sub-states is also used for compliance testing with test equipment. The Polling.Compliance link sub-state is discussed in detail at the end of this chapter.

Data Bit Rate and Data Bit Rate Identifier

All lanes of a specific link are required to transmit the K and D symbols at the same transmission rate. These symbols are composed of 10 bit sbytes. The first revision of the PCI Express specification only defines one data bit rate for transmission of the bits within the sbytes. In the future, the Polling link state will establish the highest common data bit rate that can be supported bi-directionally on all configured lanes of a specific configured link. All lanes of a specific configured link must have the same data bit rate. The data bit rate information is transmitted in the TS1 and TS2 OSs in the Data Bit Rate Identifier Bits [7::0].

In the execution of the Polling link state protocol TS1 and TS2 OSs will be transmitted at different data bit rates. When the Polling link state is entered from non-Polling link states, the data bit rate of transmissions is at the minimum possible rate (2.5 gigabits per second). This minimum data bit rate for transmissions is maintained until adjusted by the Polling.Speed link sub-state based on the information provided in the Data Bit Rate Identifier Bits [7::0] in the TS1 and TS2 OSs. Consequently, as the link sub-states within the Polling link state are executed multiple times the data bit rate of transmission changes. When the Polling link state is exited, the data bit rate for transmission is established until reentry into the Polling link state.

The only transmission data bit rate permitted by the first revision of the PCI Express specification is Data Bit Rate Identifier Bit [1] = 1b, which defines a 2.5 gigabit/second data bit rate. All of the other Data Bit Rate Identifier Bits = "0" (Reserved).

Polling Link State Protocol

Summarized in Figure 14.7 and Tables 14.18 through 14.21 are the transitions of the LTSSMs and the associated transmitters and receivers on each end of the link.

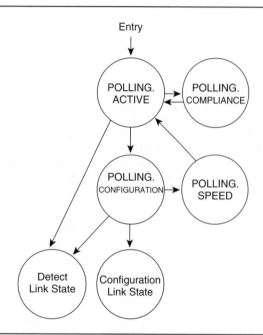

Figure 14.7 Polling Link State

Table 14.18 Polling.Active Link Sub-state

Present State Attributes:	
Upon entry into this link sub-state, the transmitters begin transmitting TS1 OSs with the LANE# and LINK# = PAD K symbol on *ALL* detected un-configured lanes. Physical LinkUp = "0" is indicated from Physical Layer to Data Link Layer.	
Cause of Transition to Next State	**Next State Transitioned To**
Within the first 24 msec. (entry timeout) upon entering this link sub-state: The transition occurs if receivers on *ALL* detected un-configured lanes successfully received eight consecutive TS1 or TS2 OS with the LANE# and LINK# = PAD K symbols (1). The transmitters in the same port as the receivers must have transmitted at least 1024 TS1 OS on *ALL* detected un-configured lanes after receiving the first TS1 or TS2 OS. The transmissions of course began with entry into this link sub-state.	Polling.Config.
After a 24 msec. upon entering this link sub-state: The transition occurs if receivers on *ANY* detected un-configured lanes successfully received eight consecutive TS1 or TS2 OSs with the LANE# and LINK# = PAD K symbols were received during entry timeout (1). The transmitters in the same port as the receivers must have transmitted at least 1024 TS1 OS on *ALL* detected un-configured lanes after receiving the first TS1 or TS2 OS during entry timeout *and ALL* detected un-configured lanes have been detected be the receivers to have exited from "electrical idle" at least once since entering this link sub-state.	Polling.Config.
After 24 msec. upon entering this link sub-state: The transition occurs if receiver on at least *one* detected un-configured lane has never detected an exit from "electrical idle" during entry timeout.	Polling.Compliance
After 24 msec. upon entering this link sub-state: The transition occurs if receivers on *ANY* detected un-configured lanes have not successfully received eight consecutive TS1 or TS2 OS (1) with the LANE# and LINK# = PAD K symbols received during entry timeout. As part of the transition the data bit rate defined by Data Bit Rate Identifier Bits [7::0] is reduced unless already at the minimum (1).	Detect.Quiet

Table Notes:

1. The transition also occurs if the complement value of the TS1 or TS2 OS is received because lane polarity inversion does not occur until Polling.Config. link sub-state.

Table 14.19 Polling.Compliance Link Sub-state

Present State Attributes:

This link sub-state is used only for compliance testing and is not part of normal operation. See discussion at the end of this chapter. It permits compliance testing to check voltage and timing specific to Table 4-5 and 4-6 of the PCI Express specification.

Physical LinkUp = "0" is indicated from Physical Layer to Data Link Layer.

Cause of Transition to Next State	Next State Transitioned To
Transition occurs if an exit from "electrical idle" is detected by receivers on *ALL* detected un-configured lanes.	Polling.Active

Table 14.20 Polling.Configuration Link Sub-state

Present State Attributes:

Upon entry into this link sub-state, the receivers must invert polarity (lane polarity inversion) if needed. The transmitters begin transmitting TS2 OSs with the LANE# and LINK# = PAD K symbol on *ALL* detected un-configured lanes.

Physical LinkUp = "0" is indicated from Physical Layer to Data Link Layer.

Cause of Transition to Next State	Next State Transitioned To
Within the first 48 msec. (entry timeout) upon entering this link sub-state: The transition occurs if receivers on *ANY* detected un-configured lanes successfully received eight consecutive TS1 or TS2 OS with the LANE# and LINK# = PAD K symbols *and* none of these same lanes are transmitting or receiving a Data Rate Identifier [7::0] value greater than the data bits rate used for transmission of the TS2 OSs. The transmitters in the same port as the receivers must have transmitted at least 16 TS2 OS on *ALL* detected un-configured lanes after receiving the first TS1 or TS2 OS. The transmissions of course began with entry into this link sub-state.	Config. RcvrCgf
Within the first 48 msec. (entry timeout) upon entering this link sub-state: The transition occurs if receivers on *ALL* detected un-configured lanes successfully received eight consecutive TS1 or TS2 OS with the LANE# and LINK# = PAD K symbols *and* at least one of these same lanes are transmitting or receiving a Data Rate Identifier [7::0] value greater than the data bits rate used for transmission of the TS2 OSs. The transmitters in the same port as the receivers must have transmitted at least 16 TS2 OS on *ALL* detected un-configured lanes after receiving the first TS1 or TS2 OS. The transmissions of course began with entry into this link sub-state.	Polling.Speed
Transition occurs if none of the other transitions occur within 48 msec. of entering this link sub-state.	Detect.Quit

Table 14.21 Polling.Speed Link Sub-state

Present State Attributes:
In this link sub-state, the data bit rate for transmissions is changed to highest common rate supported by both sides of the link as defined in the Data Rate Identifier in the TS OS. Upon entry into this link sub-state, the transmitter sends an Electrical Idle OS and then enters "electrical idle" for a minimum of TTx-Idle-Min and no longer that 2 msec. on *ALL* detected un-configured lanes.
The "electrical idle" *does not* have to meet the $V_{TX-CM-DC-ACTIVE-IDLE-DELTA}$ requirement
Physical LinkUp = "0" is indicated from Physical Layer to Data Link Layer.

Cause of Transition to Next State	Next State Transitioned To
Transition 2 msec. after entering this link sub-state *and* after the data bit rate for subsequent transmissions has been adjusted to the highest common value per the Data Rate Identifier [7::0] in the TS OSs.	Polling.Active

Configuration Link State

Prior to entering the Configuration link state, the link has achieved bit and symbol lock, lane reversal polarity has been completed, and the highest common data bit rate has been established on *ALL* detected un-configured lanes. In the typical configuration is one link between two PCI Express devices with only one link connected to each port. This single link is configured by the Link Training and Status State Machine (LTSSM) of the Physical Layer in each of the PCI Express devices. The PCI Express specification does permit a single port on a PCI Express device on the upstream side of the link to connect to two or more downstream PCI Express devices. That is, one downstream port is connected to two or more upstream ports with only one upstream port and configured link defined for each PCI Express device on the downstream side for the link. If the downstream port supports multiple configured links, a separate LTSSM must be provided for each link. See Chapter 8 for more information.

Without any implementation-specific information, a single downstream port of a PCI Express device on the upstream side of the link assumes a single link with a specific LINK# is being considered for the collection of detected un-configured lanes. During the iterative process of the Configuration link sub-states, the upstream port of the PCI Express device on downstream side of the link responds by acknowledging the specific LINK#. If only one PCI Express device is connected on each end of the link, the iterative process establishes LANE#s beginning with 0 and sequentially increasing. The resulting configured link between the two

PCI Express devices has a single specific LINK# and a sequence of LANE#s that represent a subset of the detected un-configured lanes that have been configured. The subset may be all of or part of the detected un-configured lanes.

The collection of configured lanes with the same LINK# becomes part of a specific configured link. Chapter 8 discusses the overall protocol to configure a link. As discussed in Chapter 8, the downstream port of a Root Complex or switch takes the lead in processing the detected lanes into configured lanes within a configuration. The downstream port of the upstream PCI Express device on a link is referred to here as the *upstream device* (USD). It provides the link numbers and lane numbers to be considered by the upstream port of the downstream PCI Express device. The upstream port of the downstream PCI Express device is referred to here as the *downstream device* (DSD). The lanes of a link that are determined to be common to both devices and operational are defined as *configured lanes*.

Summarized in Figure 14.8 and Tables 14.22 through 14.30 are the transitions of the LTSSMs and the associated transmitters and receivers on each end of the link. The entries in these tables show the configuration of a specific link and the associated lanes. There may be other detected and un-configured lanes that can be configured into another link with a different LINK#. Consequently, in the following tables the term *all* references the lanes being processed into a specific configured link and not *ALL* of the lanes detected between the two ports.

702 ■ The Complete PCI Express Reference

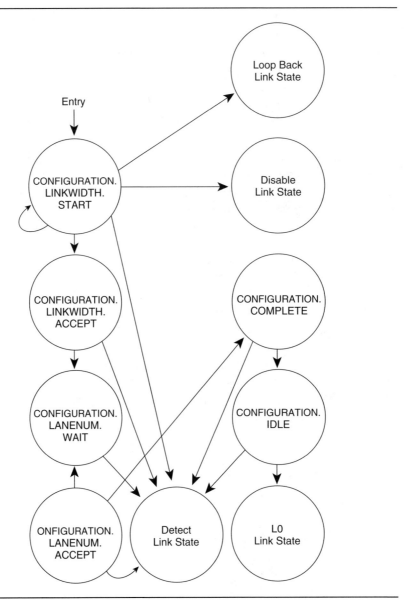

Figure 14.8 Configuration Link State

Table 14.22 USD and Config.Linkwidth.Start Link Sub-state

Present State Attributes:	
Upon entry into this sub-state, the downstream port of the Root Complex or switch begins continuous transmission of TS1 OSs on *ALL* detected un-configured lanes and begins checking for reception of TS1 OSs on *ALL* detected un-configured lanes. The TS1 OSs transmitted contain a specific LINK#, LANE # = PAD K symbol, and Training Control Bits [7::0] = "0"for those lanes it believes can be configured into a specific link. The TS1 OSs transmitted on the remaining lanes have LINK# and LANE# = PAD K symbol(2).	
or	
TS1 will be continuously transmitted on *ALL* detected un-configured lanes with LINK# and LANE# = PAD K symbol and Training Control Bit [2] or [1] = 1b as directed by a higher layer than the Physical Layer(1)(2).	
Transmission of the TS1 OSs ceases with the transition from this link sub-state.	
Physical LinkUp = "0" is indicated from Physical Layer to Data Link Layer.	
Cause of Transition to Next State	**Next State Transitioned To**
The transaction occurs if the downstream port receives one or more TS1 OSs on *ANY* of the detected un-configured lanes with LINK# and LANE# equal PAD K symbol followed by the receipt on the *same* detected un-configured lanes of two consecutive TS1 OSs with the LINK# equal the number transmitted (not PAD K symbol) and the LANE# = PAD K symbol (4).	Config. Linkwidth. Accept
This transition occurs 24 msec. after entering this link sub-state if the other transitions did not occur.	Detect.Quit
Directed to by a higher layer (Disable bit), the downstream port transitions after beginning transmission of TS1 OSs on *ALL* detected un-configured lanes with Training Control Bit [1] = 1b has begun (2).	Disabled (not advance peer-to-peer link)
Directed by higher layer (implementation-specific), the downstream port transitions after beginning transmission of TS1 OSs with specific LINK# and LANE# equal to PAD K symbol, and Training Control Bits [2::0] = 100b on *ANY* detected lanes. TS1 OS transmission continues into the Loopback link state. This establishes the USD as the Loopback Master (2) (3).	Loopback.Entry
It is also possible for an USD to be the Loopback Slave. Transition occurs after two consecutive TS1 OSs are received on *ANY* detected lanes with Training Control Bits [2::0] = 100b. Response is undefined if two or more Training Control Bits [2::0] = 1b. Transmission of IDLE D symbols or TS1 OS cease at transition (2)(3).	

Table Notes:

1. The downstream port of the Root Complex or switch is the only entity that transmits the TS1 or TS2 with Disabled bit set = 1b. The receipt of TS1 OSs by the downstream port of a switch means it is connected to an advance peer-to-peer link (between the downstream ports of two switches).
2. The PCI Express specification is inconsistent: in the Disabled, Loopback, and Hot Reset link state sections, only the TS1 OSs is defined as part of the protocol. The USD in the Recovery and Configuration link state are defined as transmitting TS1 And TS2 OSs, yet the DSD is defined as only monitoring TS1 OSs. Consequently, for consistency only TS1 OSs are stated as transmitted.
3. For a transition from the Configuration link state this may be any combination and number of detected lanes. The combination is implementation-specific and selected by the Loopback Master.
4. The receipt of TS1 OS with LINK# and LANE# both not equal to PAD K symbol by the downstream port of a switch means that the port is connected to an advance peer-to-peer link (between the downstream ports of two switches). The port assumes the role of a downstream device until a random timeout has occurred. See "Downstream Port Connected to Downstream Port" in Chapter 8 for more information.

Table 14.23 DSD and Config.Linkwidth.Start Link Sub-state

Present State Attributes:	
Upon entry into this sub-state, the upstream port of a switch, endpoint, or bridge begins continuous transmission of TS1 OSs on *ALL* detected un-configured lanes and begins checking for reception of TS1 OSs on *ALL* detected un-configured lanes. The TS1 OSs transmitted contain LINK# and LANE# = PAD K symbol and Training Control Bits [7::0] = "0." *or* TS1 or TS2 OSs is transmitted with LINK# and LANE# = PAD K symbol and Training Control Bit [2] = 1b as directed by a higher layer than the Physical Layer(1)(2). Transmission of the TS1 OSs ceases with the transition from this link sub-state. Physical LinkUp = "0" is indicated from Physical Layer to Data Link Layer.	
Cause of Transition to Next State	**Next State Transitioned To**
The transition occurs if the upstream port receives two or more TS1 OSs on *ANY* of the detected un-configured lanes with a LINK# = not PAD K Symbol and LANE # = PAD K symbol.	Config.Linkwidth. Accept
This transition occurs 24 msec. after entering this link sub-state if the other transitions did not occur.	Detect.Quit
Transition occurs if two consecutive TS1 OSs are received on *ANY* detected un-configured lanes with Training Control Bit [1] = 1b. If more than one of the Training Control Bits is set to "1" in a specific TS1 and TS2, the response to the OS is not defined.	Disabled

Table 14.23 DSD and Config.Linkwidth.Start Link Sub-state *(continued)*

Cause of Transition to Next State	Next State Transitioned To
Transition occurs after two consecutive TS1 OSs are received on *ANY* detected lanes with Training Control Bits [2::0] = 100b. Response is undefined if two or more Training Control Bits [2::0] = 1b. Transmission of IDLE D symbols cease at transition. The DSD becomes the Loopback Slave (2).	Loopback.Entry
It is also possible for a DSD to be a Loopback Master. Directed by higher layer (implementation-specific) the downstream port transitions after beginning transmission of TS1 OSs with specific LINK# and LANE# equal to PAD K symbol, and Training Control Bits [2::0] = 100b on *ANY* detected lanes. TS1 OSs continues into Loop (2).	

Table Notes:

1. The downstream port of the Root Complex or switch is the only entity that transmits the TS1 or TS2 with Disabled bit set = 1b.
2. For a transition from the Configuration link state this may be any combination and number of detected lanes. The combination is implementation-specific and selected by the Loopback Master.

Table 14.24 USD and Config.Linkwidth.Accept Link Sub-state

Present State Attributes:
Upon entry into this sub-state, the downstream port of the Root Complex or switch begins continuous transmission of TS1 OSs on the detected un-configured lanes that received two consecutive TS1 OSs with LINK# equal to the specific LINK# transmitted in the previous sub-state *if a link can be configured. If a link can be configured, these TS1 OSs also contain the specific LINK# and LANE# that equal specific numbers representing the corresponding x1 lanes, x2 lanes, and so on, that are supported.* On the remaining detected un-configured lanes is the continuous transmission TS1 OSs with LINK# and LANE# = PAD K symbol.
Transmission of the TS1 OSs begun for transition to Config.Lanenum.Wait link sub-state continues to the next link sub-state. Otherwise, it ceases transmission of the TS1 OSs with transition from this link sub-state.
Physical LinkUp = "0" is indicated from Physical Layer to Data Link Layer.

Cause of Transition to Next State	Next State Transitioned To
Transition occurs if the downstream port was able to begin transmitting TS1 OSs because it determined it could configure a link with a unique LINK# (1).	Config.Lanenum.Wait

Table 14.24 USD and Config.Linkwidth.Accept Link Sub-state *(continued)*

Cause of Transition to Next State	Next State Transitioned To
This transition occurs 2 msec. after entering this link sub-state if the other transition did not occur. The transition occurs immediately if a link cannot be configured or if the downstream port receives two consecutive TS1 OSs with LINK# and LANE# = PAD K symbol on *ALL* detected un-configured lanes.	Detect.Quit

Table Notes:

1. The LANE#s must be sequential beginning with 0 and must only be transmitted on the detected un-configured lanes that are also receiving LINK# and equaling the specific non PAD K symbol LINK# being transmitted.

Table 14.25 DSD and Config.Linkwidth.Accept Link Sub-state

Present State Attributes:
Upon entry into this sub-state, the upstream port of a switch, endpoint, or bridge begins continuous transmission of TS1 OSs with LINK# equal to non-PAD K symbols it is receiving and LANE# equal to PAD K symbol on those detected un-configured lanes it supports and that are receiving the LINK# equal to non-PAD symbol. It also begins continuous transmission of TS1 OSs with LINK# and LANE# equaling PAD K symbols on those detected un-configured lanes it does not support.
Transmission of the TS1 OSs begun for transition to Config.Lanenum.Wait link sub-state continues to the next link sub-state. Otherwise, it ceases transmission of the TS1 OSs with transition from this link sub-state.
Physical LinkUp = "0" is indicated from Physical Layer to Data Link Layer.

Cause of Transition to Next State	Next State Transitioned To
Transition occurs if on the lanes the upstream port it is transmitting the LINK# (not PAD K symbol) and LANE# equal PAD K symbol receives two consecutive TS1 OSs with LINK# equaling the one it is transmitting and at least one LANE# = "0." Prior to transitioning from this link sub-state, the upstream port begins transmitting TS1 OSs on these lanes with LANE# in the same order or in reverse order that it supports (1). On the remaining lanes TS1 OSs continue transmitting LINK# and LANE# equal PAD K symbol.	Config. Lanenum.Wait
Transition occurs if the other transition does not occur within 2 msec. after entering this link sub-state. The transition occurs immediately if a link cannot be configured, or if the upstream port receives two consecutive TS1 OSs with LINK# and LANE# = PAD K symbol on *ALL* detected un-configured lanes.	Detect.Quit

Table Notes:

1. The LANE#s received from the upstream device may not be in the order assumed by the downstream port. The downstream port can optionally be designed to respond with a LANE# order that matches the order received from the USD. If this lane reversal feature is not supported by the downstream device, the LANE# order transmitted by the downstream device may or may not match the order received. In any situation, the LANE#s must be sequential beginning with 0.

Table 14.26 USD and Config.Lanenum.Wait Link Sub-state

Present State Attributes:	
For the downstream ports of the Root Complex or a switch, this is an interactive link state used in conjunction with the Config.Lanenum.Accept link sub-state.	
The TS1 OSs that began transmission during in the Config.Linkwidth.Accept link sub-state or Config.Lanenum.Accept link sub-state continue transmission in this link sub-state and into the next link sub-state except for transition to Detect link sub-state, which ceases TS1 OSs transmission.	
Physical LinkUp = "0" is indicated from Physical Layer to Data Link Layer.	
Cause of Transition to Next State	**Next State Transitioned To**
Transition occurs if two consecutive TS1 OSs are received on *ANY* detected un-configured lanes that have a LANE# different than the LANE# when the upstream port first entered from the Config.Lanenum.Wait link sub-state (1) *and* not *ALL* detected un-configured lanes are receiving TS1 OSs with both LINK and LANE numbers = PAD K symbol.	Config. Lanenum. Accept
Transition occurs if the other transition does not occur within 2 msec. after entering this link sub-state or if the downstream port receives two consecutive TS1 OSs with LINK# and LANE# = PAD K symbol on *ALL* detected un-configured lanes.	Detect.Quit

Table Notes:

1. This change in the LANE# represents the downstream device that was transmitting LANE# = PAD K symbol begins transmitting some LANE# not PAD K symbol.

Table 14.27 DSD and Config.Lanenum.Wait Link Sub-state

Present State Attributes:	
For the upstream ports of an endpoint, switch, or bridge, this is an interactive link state used in conjunction with the Config.Lanenum.Accept link sub-state.	
The TS1 OSs that began transmission in the Config.Linkwidth.Accept link sub-state or Config.Lanenum.Accept link sub-state continue transmission in this link sub-state and into the next link sub-state, except for transition to Detect link sub-state which ceases TS1 OSs transmission.	
Physical LinkUp = "0" is indicated from Physical Layer to Data Link Layer.	
Cause of Transition to Next State	**Next State Transitioned To**
Transition occurs if two consecutive TS1 OSs are received on *ANY* detected un-configured lanes that have a LANE# different than the LANE# received when the upstream port first entered from the Config.Lanenum.Wait link sub-state *and* not *ALL* detected un-configured lanes are receiving TS1 OSs with LINK# and LANE# = PAD K symbol *or* if *ANY* of the detected un-configured lanes receive two consecutive TS2 OSs.	Config. Lanenum. Accept
Transition occurs if the other transition does not occur within 2 msec. after entering this link sub-state or if the upstream port receives two consecutive TS1 OSs with LINK# and LANE# = PAD K symbol on *ALL* detected un-configured lanes.	Detect.Quit

Table 14.28 USD and Config.Lanenum.Accept Link Sub-state

Present State Attributes:
For the downstream ports of the Root Complex or a switch, this is an interactive link state used in conjunction with the Config.Lanenum.Wait link sub-state.
The TS1 OSs transmitted during in the Config.Lanenum.Wait link sub-state continues transmission in this link sub-state and ceases with the transition to the Config.Complete link sub-state.
The TS1 OSs transmitted during in the Config.Lanenum.Wait link sub-state continues transmission in this link sub-state until the upstream port begins transmitting new TS1 OSs to transition to the Config. Lanenum.Wait link sub-states and continues transmission into this next link sub-state.
Otherwise, it ceases transmission of the TS1 OSs with transition from this link sub-state to the Detect.Quiet link sub-state.
Physical LinkUp = "0" is indicated from Physical Layer to Data Link Layer.

Cause of Transition to Next State	Next State Transitioned To
Transition occurs if conditions for transition to Config.Complete does not occur *and* the following two conditions occur: 1. Two consecutive TS1 OSs are received on *ANY* detected un-configured lanes that have a specific LINK# and at least one LANE# equals 0. 2. The downstream port can begin transmission of TS1 OSs with the specific LINK# received on a subset of the LANE# (non PAD K symbol) received in condition 1 with matching LANE# on the detected un-configured lanes it supports (may be a match or a subset). The upstream port begins transmitting TS 1 OSs with LINK# and LANE# equals PAD K symbol on the remaining detected un-configured lanes. On the remaining detected un-configured lanes the upstream port transmits TS1 OSs with both LINK and LANE numbers = PAD K symbol.	Config. Lanenum. Wait
Transition occurs if two consecutive TS1 OSs are received with the LINK# and LANE# (not PAD K symbol) that exactly matches on *ALL* detected un-configured lanes the TS1 OSs that the upstream device is transmitting with a specific LINK# and LANE# (not PAD K symbol). A reverse LANE# order is considered a match if the upstream port *optionally* supports lane reversal.	Config. Complete
Transition occurs if the no link can be configured *or* if the downstream port receives two consecutive TS1 OSs with LINK# and LANE# = PAD K symbol on *ALL* detected un-configured lanes.	Detect.Quit

Table Notes:

1. A subset means that the largest LANE# number of the new transmission is less than the LANE# received. If the lane order received is reversed for the present transmission, the new transmission will include a lane reversal if supported by the downstream port.

Table 14.29 DSD and Config.Lanenum.Accept Link Sub-state

Present State Attributes:
For the upstream ports of an endpoint, switch, or bridge, this is an interactive link state used in conjunction with the Config.Lanenum.Wait link sub-state.
The TS1 OSs transmitted during in the Config.Lanenum.Wait link sub-state continues transmission in this link sub-state and ceases with the transition to the Config.Complete link sub-state.
The TS1 OSs transmitted during in the Config.Lanenum.Wait link sub-state continues transmission in this link sub-state until the upstream port begins transmitting new TS1 OSs to transition to the Config.Lanenum.Wait link sub-states and continues transmission into this next link sub-state.
Otherwise, it ceases transmission of the TS1 OSs with transition from this link sub-state to the Detect.Quiet link sub-state.
Physical LinkUp = "0" is indicated from Physical Layer to Data Link Layer.

Cause of Transition to Next State	Next State Transitioned To
Transition occurs if conditions for transition to Config.Complete do not occur *and* the following two conditions occur: 1. Two consecutive TS1 OSs are received on *ANY* detected un-configured lanes that have a specific LINK# and at least one LANE# equals 0. 2. The upstream port can begin transmission of TS1 OSs with the specific LINK# received on a subset of the LANE# (non PAD K symbol) received in condition 1 with matching LANE# on the detected un-configured lanes it supports (may be a match or a subset). The upstream port begins transmitting TS 1 OSs with LINK# and LANE# equals PAD K symbol on the remaining detected un-configured lanes. On the remaining detected un-configured lanes, the upstream port transmits TS1 OSs with both LINK and LANE numbers = PAD K symbol.	Config. Lanenum. Wait
Transition occurs if two consecutive TS2 OSs are received with the LINK# and LANE# (not PAD K symbol) that exactly matches the TS1 OSs that the downstream device is transmitting with a specific LINK# and LANE# (not PAD K symbol) on *ALL* detected un-configured lanes.	Config. Complete
Transition occurs if no link can be configured *or* if the downstream port receives two consecutive TS1 OSs with LINK# and LANE# = PAD K symbol on *ALL* detected un-configured lanes.	Detect.Quit

Table 14.30 USD and Config.Complete Link Sub-state

Present State Attributes:
Upon entry into this sub-state, the downstream ports of the Root Complex or a switch begin transmitting TS2 OSs with "same" LINK# and LANE# (non PAD K symbol) on as the TS1 OSs it received in the Config.Lanenum.Accept link sub-state. The *remaining* detected un-configured lanes begin transmitting TS2 OSs with LINK# and LANE# equal PAD K symbol. The term "same" includes lane reversal when lane reversal is possible.
Before leaving this link sub-state, the TS2 OSs transmitted must contain the correct value for the N_FTS value for use in the L0s link state.
Before leaving this link sub-state, the lane-to-lane skew must be completed.
The scrambling mechanism in upstream device is disabled if two consecutive TS2 OSs with Training Control Bit [4] = 1b are received. The downstream port transmitting TS2 OSs with Training Control Bit [4] = 1b must have also disabled its scrambling mechanism.
Transmission of the TS2 OSs ceases with the transition from this link sub-state.
Physical LinkUp = "0" is indicated from Physical Layer to Data Link Layer.

Cause of Transition to Next State	Next State Transitioned To
Transition occurs if eight consecutive TS2 OSs are received on *ALL* detected un-configured lanes with LINK# and LANE# (not PAD K symbol) *matching* those of the TS2 OSs being transmitted on the *same* lanes and after receiving one TS2 OSs with *matching* LINK# and LANE# on these *same* lanes and the subsequent transmitting of sixteen TS2 OSs with *matching* LINK# and LANE# on these *same* lanes.	Config.Idle
With the transition the group of detected un-configured lanes with *matching* LINK# and LANE# (not PAD K symbol) become defined as *configured lanes* of a *configured link*(1).	
Transition occurs if the other transition does not occur within 2 msec. after entering this link sub-state.	Detect.Quit

Table Notes:

1. Note: Once the conditions for the transition to Config.Idle link sub-state are met the transmitters on the downstream port of the *remaining* lanes transition to control of a new LTSSM if implemented. These *remaining* lanes are defined as detected *un-configured lanes,* or if the *remaining* lanes cannot be assigned to a new LTSSM, these lanes transitions to "electrical idle" without the requirement that an Electrical Idle OS is first transmitted (becomes optional). The common mode being driven by the transmitters does not have to meet the $V_{TX-CM-DC-ACTIVE-IDLE-DELTA}$ requirement. Receivers' terminations must meet the $Z_{RX-HIGH-IMP-DC}$ requirement. These *remaining* lanes transition to the Detect link state and are defined as *un-detected lanes.*

Table 14.31 DSD and Config.Complete link sub-state

Present State Attributes:	
Upon entry into this sub-state, the upstream port of a switch, endpoint, or bridge begins transmitting TS2 OSs on the same detected un-configured lanes with same LINK# and LANE# (non PAD K symbol) as the TS2 OSs it received in the Config.Lanenum.Accept link sub-state. The remaining detected un-configured lanes begin transmitting TS2 OSs with LINK# and LANE# equal PAD K symbol.	
Before leaving this link sub- state, the TS2 OSs transmitted must contain the correct value for the N_FTS value for use in the L0s link state.	
Before leaving this link sub-state, the lane-to-lane skew must be completed.	
The scrambling mechanism in upstream device is disabled if two consecutive TS2 OSs with Training Control Bit [4] = 1b are received. The downstream port transmitting TS2 OSs with Training Control Bit [4] = 1b must have also disabled its scrambling mechanism.	
Transmission of the TS2 OSs ceases with the transition from this link sub-state.	
Physical LinkUp = "0" is indicated from Physical Layer to Data Link Layer.	
Cause of Transition to Next State	**Next State Transitioned To**
Transition occurs if eight consecutive TS2 OSs are received on *ALL* detected un-configured lanes with LINK# and LANE# (not PAD K symbol) *matching* those of the TS2 OSs being transmitted on the *same* lanes *and* after receiving one TS2 OSs with matching LINK# and LANE# on these *same* lanes *and* the subsequent transmitting of sixteen TS2 OSs with *matching* LINK# and LANE# on these *same* lanes. With the transition the group of detected un-configured lanes with *matching* LINK# and LANE# (not PAD K symbol) become defined as *configured lanes* of a *configured link*(1).	Config.Idle
Transition occurs if the other transition does not occur within 2 msec. after entering this link sub-state	Detect.Quit

Table Notes:

1. Note: Once the conditions for the transition to Config.Idle link sub-state are met, the transmitters on the downstream port of the *remaining* lanes transition to control of a new LTSSM, if implemented. These *remaining* lanes are defined as detected *un-configured lanes, or* if the *remaining* lanes cannot be assigned to a new LTSSM, these lanes transitions to "electrical idle" without the requirement that an Electrical Idle OS is first transmitted (becomes optional). The common mode being driven by the transmitters does not have to meet the $V_{TX\text{-}CM\text{-}DC\text{-}ACTIVE\text{-}IDLE\text{-}DELTA}$ requirement. Receivers' terminations must meet the $Z_{RX\text{-}HIGH\text{-}IMP\text{-}DC}$ requirement. These *remaining* lanes transition to the Detect link state and are defined as *un-detected lanes.*

Table 14.32 USD and Config.Idle Link Sub-state

Present State Attributes:

Upon entry into this sub-state, the downstream port of the Root Complex or a switch begin transmitting the IDLE D symbol on *ALL* the configured lanes of the configure link. Also, the Physical LinkUp finally transitions to "1." *The remaining detected un-configured lanes state transition to "electrical idle." Optionally, the Electrical Idle OSs can be transmitted prior to the transition to 'electrical idles." These detect un-configured can participate in the configuration of another link by the LTSSM.*

The transmission of the IDLE D symbols in this link sub-state continues into the next link sub-state if transition is to L0 link state.

The transmission of the IDLE D symbols in this link sub-state ceases and the associated lanes transition to "electrical idle" if transition to the Detect.Quiet link state.

Physical LinkUp = "1" indicated from Physical Layer to Data Link Layer

Cause of Transition to Next State	Next State Transitioned To
Transition occurs if eight consecutive IDLE D symbols are received on *ALL* configured lanes AND after receiving one IDLE D symbol on *ALL* configured lanes AND the subsequent transmitting of sixteen IDLE D symbols on *ALL* configured lanes of the configure link.	L0
Transition occurs if the other transition does not occur within 2 msec. after entering this link sub-state.	Detect.Quit

Table 14.33 DSD and Config.Idle Link Sub-state

Present State Attributes:

Upon entry into this sub-state the upstream port of a switch, endpoint, or bridge begin transmitting the IDLE D symbol on *ALL* the configured lanes of the configure link. Also, the Physical LinkUp finally transitions to "1." The remaining detected un-configured lanes state transition to "electrical idle. Optionally the Electrical OSs can be transmitted prior to the transition to 'electrical idles" These detect un-configured can participate in the configuration of another link by the LTSSM.

The transmission of the IDLE D symbols in this link sub-state continues into the next link sub-state if transition is to L0 link state.

The transmission of the IDLE D symbols in this link sub-state ceases and the associated lanes transition to "electrical idle" if transition to the Detect.Quiet link state.

Physical LinkUp = "1" indicated from Physical Layer to Data Link Layer

Cause of Transition to Next State	Next State Transitioned To
Transition occurs if eight consecutive IDLE D symbols are received on *ALL* configured lanes AND after receiving one IDLE D symbol on *ALL* configured lanes AND the subsequent transmitting of sixteen IDLE D symbols on *ALL* configured lanes of the configure link.	L0
Transition occurs if the other transition does not occur within 2 msec. after entering this link sub-state.	Detect.Quit

Other Link States

The following are other link states that simply cannot be placed in any specific category.

Recovery Link State

The Recovery link state has four uses: first, as a means to transition to the Hot Reset, Loopback, Disable, Detect, and Configuration link states from the L0, L1, and L0s link states; second, as a means to transition to the L0 link state from the L1 link state and under certain conditions from the L0s link state; third, as a means to reestablish bit and symbol lock; and fourth, to complete lane-to-lane de-skew.

Part of the Recovery link state is the execution of the Link Retraining with the successful sequencing through the Recovery.RcvrLock and Recovery.RcvrCfg link sub-states. Link Retraining reestablishes the bit and symbol, reestablishes the lane-to-lane skew, and sets a new N_FTS value for use of future L0.

As illustrated in Figure 14.9, the link sub-states of the Recovery link state may transition to either Configuration or Detect link sub-state if the activities in the sub-states of the Recovery link state are not successful.

As summarized in Figure 14.9 and Tables 14.34 through 14.37, the LTSSMs and the associated transmitter and receivers of the link interact within three distinct sub-links as follows:

- **Recovery.Rcvrlock:** This link sub-state tries to re-establish the bit and symbol lock between the two PCI Express devices on the link.

- **Recovery.RcvrCfg:** This link sub-state reflects that bit and symbol lock was reestablished, and the value of the N_FTS can optionally be changed for future use in the L0s link state. Lane-to-lane de-skew is also reestablished.

- **Recovery.Idle:** This link sub-state reflects that bit and symbol lock was reestablished, the value of the N_FTS was changed (if desired), and the lane-to-lane de-skew is complete. Consequently the link is ready to return to normal operation in the L0 link state or transition to the Hot Reset, Configuration, Loopback, or Disabled link states per direction of a higher layer.

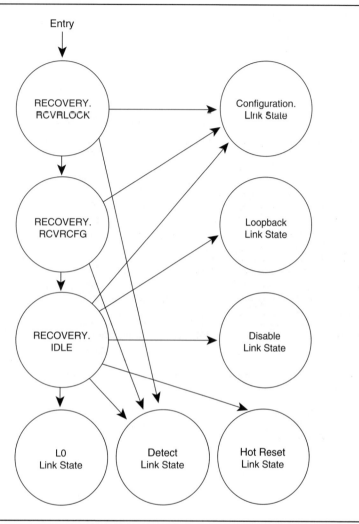

Figure 14.9 Recovery Link State

Table 14.34 Recovery.RcvrLock Link Sub-state

Present State Attributes:	
Upon entry into this link sub-state, the port's transmitters begin continuous of TS1 OSs on *ALL* configured lanes with LINK# and LANE# equal to the values from the last configuration link state and Training Control Bits [2::0] = 000b. All of the other values in the TS1 are the same as per the last configuration link state.	
The transmission of the TS1 OSs ceases with the transition from this link sub-states. (1).	
Physical LinkUp = "0" is indicated from Physical Layer to Data Link Layer.	
Cause of Transition to Next State	**Next State Transitioned To**
Transition occurs if eight consecutive TS1 or TS2 OSs are received on *ALL* configured lanes, and the link and lane values match those transmitted by the transmitters of the same port within 24 msec. after entry into this link sub-state.	Recovery.RcvrCfg
After a 24 msec. timeout from the entry into this link sub-state, the transition occurs if during the timeout at least one TS1 or TS2 OS is received on *ALL* configured LINK# and LANE# values that match those transmitted by the transmitters of the same port.	Config.Linkwidth.Start
Transition to Detect.Quiet after a 24 msec. timeout from the entry into this link sub-state, if transaction to the Recovery.RcvrCfg or Config.Linkwidth.Accept has not occurred.	Detect.Quiet

Table Notes:

1. If the Extended Synch Bit = 1b, a total of 1024 TS1 OS must be transmitted before the transmitters transition to the Recovery.RcvrCfg link sub-link state to allow link monitoring tools (logic analyzers) time to achieve bit and symbol lock.

Table 14.35 Recovery.RcvrCfg Link Sub-state

Present State Attributes:		
Upon entry into this link sub-state, TS2 OSs are continuously transmitted on *ALL* configured lanes with LINK# and LANE# equal to the values from the last configuration link state and Training Control Bits [2::0] = 000b.		
The transmission of the TS2 OSs ceases with the transition from this link sub-states. (1) (2)		
Physical LinkUp = "0" is indicated from Physical Layer to Data Link Layer.		
Cause of Transition to Next State		**Next State Transitioned To**
Transition occurs if eight consecutive TS2 OSs are received on *ALL* configured lanes with the LINK# and LANE# .that match those transmitted by the transmitters of the same port *and* sixteen TS2 OSs are transmitted on *ALL* configured lanes after the receipt of one TS2 OS by the receivers of the same port (1) (2).		Recovery.Idle
Transition occurs if eight consecutive TS1 OSs are received on *ALL* configured lanes with LINK# and LANE# that do not match the LINK# and LANE# in the TS2 OSs. *and* sixteen TS2 OSs are transmitted on *ALL* configured lanes after the receipt of one TS1 OS by the receivers of the same port (1).		Config. Linkwidth. Accept
Transition to Detect.Quiet occurs after a 48 msec. timeout from the entry into this link sub-state, if transaction to the Recovery.Idle or Config.Linkwidth.Accept has not occurred.		Detect.Quit

Table Notes:

1. If the N_FTS value is changed, this new value must be used for future L0s link states.
2. Before transitioning from the Recovery.RcvrCfg link sub-state to the Recover.Idle link sub-state the lane-to-lane de-skew must be completed.

Table 14.36 USD and Recovery.Idle Link Sub-state

Present State Attributes for Transition to Non-L0 Link State:

Upon entry into this link sub-state, the transmitters of the downstream port begin continuous transmission TS1 OSs on *ALL* configured lanes as defined below *and* simultaneously transition to the next link state. The transition also affects the receivers of the same port and the associated LTSSM (1).

Present State Attributes for Transition to L0 Link State: Upon entry into this link sub-state, the transmitters of the downstream port begin continuous transmission of IDLE D symbols on *ALL* configured lanes as defined below, assuming the transition will be to the L0 link state. The transition also affects the receivers of the same port and the associated LTSSM (1).

Physical LinkUp = "0" is indicated from Physical Layer to Data Link Layer.

Cause of Transition to Next State	Next State Transitioned To
Transition occurs if none of the other transitions occur within 2 msec. after entering this link sub-state. IDLE D symbols or TS1 OS transmission cease at transition.	Detect.Quit
Transition occurs after eight consecutive IDLE D symbols are received on *ALL* configured lanes *and* sixteen consecutive IDLE D symbols are transmitted on *ALL* configured lanes after the first IDLE symbol is received. Transmission of the IDLE D symbols continue into the L0 link state.	L0
Directed by higher layer (implementation-specific), the downstream port transitions after beginning transmission of TS1 OSs with LANE# = PAD and LINK# = correct value, and Training Control Bits [2::0] = 000b on *ALL* configured lanes. Optionally the LINK# can be retained or set equal to PAD symbol. TS1 OS transmission ceases at transition.	Config.Linkwidth. Start
Directed to by higher layer (Disable bit), the downstream port transitions after beginning transmission of TS1 OSs with LANE# and LINK# = any values, and Training Control Bits [2::0] = 010b on *ALL* configured lanes. TS1 OS transmission ceases at transition.	Disabled
Directed by higher layer (implementation-specific), the downstream port transitions after beginning transmission of TS1 OSs with LANE# and LINK# = correct values, and Training Control Bits [2::0] = 100b on *ANY* configured lanes. TS1 OS transmission continues transmitting until a single identical TS1 OS is received from the PCI Express device that is established as the Loopback Slave. This establishes the USD as the Loopback Master (2). It is also possible for an USD to be the Loopback Slave. Transition occurs after two consecutive TS1 OSs are received on *ANY* configured lanes with Training Control Bits [2::0] = 100b. Response is undefined if two or more Training Control Bits [2::0] = 1b. Transmission of IDLE D symbols or TS1 OS cease at transition (2).	Loopback.Entry

Continued

Table 14.36 USD and Recovery.Idle Link Sub-state *(continued)*

Cause of Transition to Next State	Next State Transitioned To
Directed by higher layer (implementation-specific), the downstream port transitions after beginning transmission of TS1 OSs with LANE# and LINK# = correct values, and Training Control Bits [2::0] = 001b on *ALL* configured lanes. TS1 OS continues transmission into the Hot Reset link state.	Hot Reset

Table Notes:

1. The PCI Express specification is inconsistent; in the Disabled, Loopback, and Hot Reset link state sections, only the TS1 OSs are defined as part of the protocol. The USD in the Recovery and Configuration link state are defined as transmitting TS1 And TS2 OSs, yet the DSD are defined as only monitoring TS1 OSs. Consequently, for consistency only TS1 OSs are stated as transmitted.
2. For a transition from the Recovery link state this may be any combination and number of configured lanes. The combination is implementation-specific and selected by the Loopback Master.

Table 14.37 DSD and Recovery.Idle Link Sub-state

Present State Attributes for Transition to L0 Link State:	
Upon entry into this link sub-state, the transmitters of the upstream port begin continuous transmission of IDLE D symbols on *ALL* configured lanes as defined below assuming the transition will be to the L0 link state. The transition also affects the receivers of the same port and the associated LTSSM.	
Present State Attributes for Transition to Non-L0 Link State: Upon entry into this link sub-state, the transmitters of the upstream port begin continuous transmission TS1 OSs if the PCI Express device is selected as Loopback Master. Selection is implementation-specific.	
Physical LinkUp = "0" is indicated from Physical Layer to Data Link Layer.	
Cause of Transition to Next State	**Next State Transitioned To**
Transition occurs if none of the other transitions occur within 2 msec. after entering this link sub-state. Transmission of IDLE D symbols or TS1 OS cease at transition.	Detect.Quit
Transition occurs after eight consecutive IDLE D symbols are received on *ALL* configured lanes *and* sixteen consecutive IDLE D symbols are transmitted on *ALL* configured lanes after the first IDLE symbol is received. Transmission of the IDLE D symbols continues into the L0 link state.	L0

Continued

Table 14.37 DSD and Recovery.Idle Link Sub-state *(continued)*

Cause of Transition to Next State	Next State Transitioned To
Transition occurs after two consecutive TS1 OSs are received on *ANY* configured lanes with the LANE # = PAD symbol and Training Control Bits [2::0] = 000b. Optionally, the LINK# can be retained or set equal to PAD symbol. Response is undefined if two or more Training Control Bits [2::0] = 1b. Transmission of IDLE D symbols or TS1 OS cease at transition (1).	Config.Linkwidth. Start
Transition occurs after two consecutive TS1 OSs are received on *ANY* configured lanes with Training Control Bits [2::0] = 010b. Response is undefined if two or more Training Control Bits [2::0] = 1b. Transmission of IDLE D symbols or TS1 OS cease at transition (1).	Disabled
Transition occurs after two consecutive TS1 OSs are received on *ANY* configured lanes with Training Control Bits [2::0] = 100b. Response is undefined if two or more Training Control Bits [2::0] = 1b. Transmission of IDLE D symbols cease at transition. The DSD becomes the Loopback Slave (2).	Loopback.Entry
It is also possible for a DSD to be a Loopback Master. Directed by a higher layer (implementation-specific), the downstream port transitions after beginning transmission of TS1 OSs with LANE# and LINK# = correct values, and Training Control Bits [2::0] = 100b on *ANY* configured lanes. TS1 OSs continues into Loop (2).	
Transition occurs after two consecutive TS1 OSs are received on *ANY* configured lanes with Training Control Bits [2::0] = 100b. Response is undefined if two or more Training Control Bits [2::0] = 1b. Transmission of IDLE D symbols or TS1 OS cease at transition (1).	Hot Reset

Table Notes:

1. It is assumed that if the DSD was selected as Loopback Master, it will begin transmission of TS1 OSs. However, if the USD is requesting a transition to the Disabled, Config.Linkwidth.Start, or Hot Reset, the DSD will cease trying to become Loopback Master.
2. For a transition from the Recovery link state this may be any combination and number of configured lanes. The combination is implementation-specific and selected by the Loopback Master.

Disabled Link State

The transition to the Disabled link state is either from the Recovery link state or the Configuration link state. The Disabled link state can only be requested by downstream ports of the Root Complex or switches. The upstream ports of switches, endpoints, and bridges must monitor for the transition to the Disable link state. One application for the Disabled link state is to permit the PCI Express devices associated with a slot to implement the Hot Plug protocol.

The USD device requests the Disabled link state by transmitting TS1 OSs with the Training Control Bits [2:0] =010b as directed by higher layer. The transition of the LTSSMs and the associated transmitters and receivers transition on each end of the link is summarized in Tables 14.38 and 14.39.

Table 14.38 USD and Disabled Link State

Present State Attributes:	
Entered when a downstream port of the Root Complex or switch begins transmitting in previous link state on "*ALL* lanes" TS1 OSs with the Training Control Bits [2::0] = 010b. After sixteen of these TS1 OSs are transmitted, the transmitters transition to "electrical idle" after transmitting one Electrical OS on "*ALL* lanes."	
Response is undefined if two or more Training Control Bits [2::0] = 1b.	
The "electrical idle" *does not* have to meet the $V_{TX-CM-DC-ACTIVE-IDLE-DELTA}$ requirement.	
Physical LinkUp transitions from "1" to "0" relative to a port when an Electrical Idle OS is both transmitted and received on "*ALL* lanes" by the port. If Electrical Idle OS is not received on "*ALL* lanes," Physical LinkUp transitions from "1" to "0" within 2 msec. after entry into this link state as part of transition to the Detect link state. Indication is from Physical Layer to Data Link Layer.	
"*All* lanes" mean *ALL* detected un-configured lanes when transition is from the Configuration link state.	
"*All* lanes" mean *ALL* configured lanes when transition is from the Recovery link state.	
Physical LinkUp = "0"	
Cause of Transition to Next State	**Next State Transitioned To**
If the downstream port has received "*ALL* lanes" a single Electrical Idle OS (even while transmitting TS1 and Electrical Idle OSs) within 2 msec after entry into this link state, the downstream port remains in the Disabled link state link. The transition occurs when the downstream ports' receivers detect an exit from "electrical idle" on *ANY* lanes or when the downstream port is directed by a higher layer.	Detect.Quiet
The transition occurs if the downstream port has not received an Electrical Idle OS on "*ALL* lanes" within 2 msec. after entry into this link state.	

Table 14.39 DSD and Disabled Link State

Present State Attributes:
Entered by the upstream port of a switch, endpoint, or bridge after two consecutive TS1 OSs are received with Training Control Bits [2::0] = 010b on "*ANY* lanes."
Response is undefined if two or more Training Control Bits [2::0] = 1b.
The "electrical idle" *does not* have to meet the $V_{TX-CM-DC-ACTIVE-IDLE-DELTA}$ requirement.
Physical LinkUp transitions from "1" to "0" relative to a port when an Electrical Idle OS is both transmitted and received on "*ALL* lanes" by the port. If Electrical Idle OS is not received on "*ALL* lanes," Physical LinkUp transitions from "1" to "0" within 2 msec. after entry into this link state as part of transition to the Detect link state. Indication is from Physical Layer to Data Link Layer.
"*ALL* lanes" mean *ALL* detected un-configured lanes when transition is from the Configuration link state.
"*ALL* lanes" mean *ALL* configured lanes when transition is from the Recovery link state.
"*Any* lanes" mean *ANY* detected un-configured lanes when transition is from the Configuration link state.
"*ANY* lanes" mean *ANY* configured lanes when transition is from the Recovery link state.
Physical LinkUp = "0"

Cause of Transition to Next State	**Next State Transitioned To**
If the upstream port has received "*ALL* lanes" a single Electrical Idle OS (even while transmitting TS1 and Electrical Idle OSs) within 2 msec after entry into this link state, the upstream port remains in the Disabled link state link. The transition occurs when the upstream ports' receivers detect an exit from "electrical idle" on *ANY* lanes or when the downstream port is directed by a higher layer.	Detect.Quiet
The transition occurs if the upstream port has not received an Electrical Idle OS on "*ALL* lanes"within 2 msec. after entry into this link state.	

Hot Reset

The Hot Reset link state places the link into a reset condition. The transition to the Hot Reset link state is only from the Recovery link state by a higher layer. The Hot Reset protocol discussed in Chapter 13 presents other considerations as well. The Hot Reset link state can only be requested by downstream ports of the Root Complex or switches. The upstream ports of switches, endpoints, and bridges must monitor for the transition to the Hot Reset link state.

The transition of the LTSSMs and the associated transmitters and receivers transition on each end of the link is summarized in Tables 14.40 and 14.41.

Table 14.40 USD and Hot Reset Link State

Present State Attributes:	
Entered from the Recovery link state when a downstream port of the Root Complex or switch begins transmitting on *ALL* configured lanes with the Training Control Bits [2::0] = 001b and the LINK# and LANE# equal the configured value. The transmission of the TS1 OSs continues until transition to the Detect link state. Response is undefined if two or more Training Control Bits [2::0] = 1b. Physical LinkUp = "0"	
Cause of Transition to Next State	**Next State Transitioned To**
If the downstream port has received on *ANY* configured lanes two consecutive TS1 OS equal to those transmitted within 2 msec after entry into this link state, the downstream port remains in the Hot Reset link state. The transition occurs when the downstream port is directed by a higher layer.	Detect. Quiet
The transition occurs if the downstream port has not received on *ANY* configured lane two consecutive TS1 OS equal to those transmitted within 2 msec. after entering into this link state.	

Table 14.41 Hot Reset Link State

Present State Attributes:	
Entered from the Recovery link state by the upstream port after two consecutive TS1 OSs are received with Training Control Bits [2::0] = 001b with LINK# and LANE# equal the configured value on *ANY* configure link. The upstream port subsequently begins transmitting TS1 OSs with Training Control Bits [2::0] = 001b on *ALL* configured lanes with the Training Control Bits [2::0] = 001b and the LINK# and LANE# equal the configured value. The transmission of the TS1 OSs continues until transition to the Detect link state. Response is undefined if two or more Training Control Bits [2::0] = 1b. Physical LinkUp = "0"	
Cause of Transition to Next State	**Next State Transitioned To**
Two msec. after entering this link sub-state, the transition occurs if the upstream port is not continuing to receive two consecutive TS1 OSs on *ANY* configured lanes with Training Control Bits [2::0] = 001b with LINK# and LANE# equal the configured value, effectively waiting for downstream port to transition to the Detect link state and cease transmission of TS1 OSs.	Detect.Quiet

Testing and Compliance

Two link states are used purely for testing a specific component against another and for checking for compliance with electrical and timing specifications.

Loopback Link State

The purpose of the Loopback link state is to provide a mechanism for a PCI Express device to be tested and for any faults to be detected. The transition to the Loopback link state is either from the Recovery link state or the Configuration link state. Per the Loopback protocol, a Loopback Master and a Loopback Slave must be established. The Loopback Master can be established on the downstream ports of the Root Complex or switches. Thus the upstream ports of switches, endpoints, and bridges must monitor for the transition to the Loopback link state and be established as the Loopback Slave. The Loopback Master can also be established on upstream ports of switches, endpoints, and bridges. Thus the downstream ports of the Root Complex or switches monitor for the transition to the Loopback link state and be established as the Loopback Slave.

The method by which a PCI Express device is directed to transition to the Loopback link state, and by which it establishes itself as the Loopback Master, is implementation-specific. The direction to the Loopback link state for PCI Express device that becomes the Loopback Slave occurs with the receipt of TS1 OS with the Training Control Bits [2:0] =100b.

When the transition is from the Configuration link state, the Loopback link state can be applied by a detected lane or combination of detected lanes. The TS1 OSs discussed in this section contain LANE# and LINK# equal to PAD K symbol. When the transition is from the Recovery link state, the Loopback link state can be applied by a configured lane or by configured link. These TS1 OSs contain LANE# and LINK# equal to the correct configured value. In either case, the Loopback protocol is implemented only on the lanes that the Loopback Master transmitted TS1 OSs. For a transition from the Configuration link state, this may be any combination and number of detected lanes. For a transition from the Recovery link state, this may be any combination and number of configured lanes. The combination of lanes selected to execute the Loopback protocol is implementation-specific and selected by the Loopback Master. In the following tables, the term "*ANY* lanes" refers to ANY of this combination of lanes. The term "*ALL* lanes" means ALL of the lanes of this combination.

As summarized in Figure 14.10 and Tables 14.42 through 14.44, the LTSSMs and the associated transmitters and receivers transition to the Loopback link state and interact as either a Loopback Master or a Loopback Slave within three distinct sub-links as follows:

- **Loopback.Entry:** This link sub-state establishes the Loopback Master and the Loopback Slave.

- **Loopback.Active:** This link sub-state executes the Loopback protocol discussed in the next section.

- **Loopback.Exit:** This link sub-state provides the transition from the Loopback link state to the Detect Link state.

Loopback Protocol

The Loopback protocol is executed within the Loopback.Active link sub-state.

The Loopback Master must transmit a stream of symbols that follows the 8/10b encoding protocol. See Chapter 8 for more information about the 8/10b encoding protocol. The Loopback Slave echoes the stream of symbols as 10 bits per symbol exactly as received from the Loopback

Master without applying any checking to determine whether the symbol is valid or not. The inclusion of the Skip OSs is as follows:

- The Loopback Master must implement Skip OSs as per normal operation but on a lane by lane basis. The inclusion of a Skip OSs can be done on each lane independent of another lane and thus does not have to be simultaneous between the lanes.

- The Loopback Slave echoes (retransmits) the same number of Skip OSs received with addition or deletion of the Skip OSs re-transmitted to accommodate timing tolerance corrections. If a Skip OS is added, it must be of the same disparity as received and connected to the re-transmitted Skip OS(s) following the COM K symbol.

- Data scrambling is not relevant because what is transmitted is echoed (retransmitted) back.

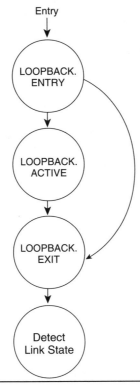

Figure 14.10 Loopback Link State

Table 14.42 Loopback.Entry Link Sub-state

Present State Attributes:	
Entered when a PCI Express device establishes itself as the Loopback Master and begins transmitting TS1 OSs with the Training Control Bits [2::0] = 100b and continues transmitting until a single identical TS1 OS is received from the PCI Express device that is established as the Loopback Slave.	
It is the designers' responsibility to make sure that only one PCI Express device is defined as the Loopback Master.	
Physical LinkUp = "0" is indicated from Physical Layer to Data Link Layer.	
Cause of Transition to Next State	**Next State Transitioned To**
Transition occurs for the Loopback Slave once it has received a TS1 OS with Training Control Bits [2::0] = 100b and responded with a transmission of an identical TS1 OS. Transition occurs for the Loopback Master once it has received a TS1 OS with Training Control Bits [2::0] = 100b (1) (2).	Loopback. Active
Transaction occurs after an implementation-specific timeout less than 100 msec. relative to the entry into this link sub-state by the Loopback Master and *no* TS1 OS is received. Consequently, no Loopback Slave has been established.	Loopback.Exit

Table Notes:

1. The transition to the Loopback.Active link state may cause a boundary condition where the Loopback slave discards a Skip OSs and does not echo it back to the Loopback Master. Consequently, the Loopback Master does not see a Skip OSs on a lane for twice the normal scheduled interval between Skip OSs.
2. The Loopback Slave can transition symbol boundary and thus may truncate any OS being received.

Table 14.43 Loopback.Active Link Sub-state

Present State Attributes:

The Loopback Slave echoes the pattern of symbols received from the Loopback Master back to the Loopback Master per the Loopback protocol.

Physical LinkUp = "0" is indicated from Physical Layer to Data Link Layer.

Cause of Transition to Next State	Next State Transitioned To
Transition occurs for the Loopback Slave when a single Electrical Idle OS is received or it detects "electrical idle." The Loopback Slave must be able to detect "electrical idle" within 1 msec. of detecting an Electrical Idle OSs on *ANY* lane. In the time between receiving an Electrical Idles OSs and detecting "electrical idle" the Loopback Slave may receive undefined 10 bit symbols that it will echo back to Loopback. Master. The TTX-IDLE-SET-TO-IDLE does not apply in this situation. Transition of the Loopback Master is directed by a higher layer.	Loopback.Exit

Table 14.44 Loopback.Exit Link Sub-state

Present State Attributes:

Upon entry into this link sub-state, the Loopback Master transmits a single Electrical Idle OS and transitions to "electrical idle" for 2 msec. on *ALL* lanes. Entry into this link sub-state for the Loopback Slave occurs with receipt of a single Electrical Idle OS or when it detects an "electrical idle."

The Loopback Master must transition to "electrical idle" after transmitting Electrical Idle OS on *ALL* lanes within TTX-IDLE-SET-TO-IDLE.

Any symbols received by Loopback Master after transmission of the Electrical Idle OSs must be ignored.

Physical LinkUp = "0" is indicated from Physical Layer to Data Link Layer.

Cause of Transition to Next State	Next State Transitioned To
Transition occurs for the Loopback Slave once it has echoed ALL symbols received including the single Electrical Idle OS (if received) prior to detecting "electrical idle" *and* remains in "electrical idle" for 2 msec. Transition occurs for the Loopback Master once it has been in "electrical idle" for 2 msec.	Detect.Quiet

Polling.Compliance Link Sub-state

The Polling.Compliance link sub-state is used to test the transmitters of a PCI Express device for compliance with the voltage and timing requirements of the PCI Express specification. See discussion earlier in the chapter about requirements to transition to this link sub-state and exit from this link sub-state.

Compliance Pattern

The compliance pattern is transmitted in as K and D symbols using the 8/10b encoding under worst-case conditions. Part of the worst-case consideration is the maximum interference between adjacent lanes. Consequently, the pattern that is transmitted includes the following:

- The compliance sequence (CS) is four symbols (first to last) ... COM+ D21.5• COM• D10.2
    - COM+ provides binary pattern "0011111010"
    - D21.5• provide binary pattern "1010101010"
    - COM: provides binary pattern "1100000101"
    - D10.2 provides binary pattern "0101010101"
- The transmitters of a port continue with transmitting the compliance pattern until an exit from "electrical idle" is detected by the receivers on *ALL* the detected lanes of the same port.
- See the PCI Express specification for full information about this compliance pattern and its repetition across multiple detected lanes of a port.

Chapter 15

Power Management

This is the first of the three chapters related to power consumption in a PCI Express platform. Chapter 16 discusses the Hot Plug protocol of a slot and add-in card. Chapter 17 discusses the Slot Power protocol. This chapter focuses on the Power Management (PM) protocol applied to PCI Express devices on the platform or add-in card, and the associated link.

As discussed in Chapter 13, this book has grouped the topics of Chapters 15, 16 and 17 under the concept of Power Management Events (PMEs). PMEs consist of the Power Management (PM), Hot Plug (HP), and Slot Power (SP) protocols. PMEs are associated with PME software, which consists of Power Management (PM) software for the PM protocol, Hot Plug (HP) software for HP protocol, and Slot Power (SP) software for the SP protocol.

Introduction to Power Management

The purpose of the Power Management (PM) protocol and the associated PM software can be viewed as three sets of tasks: Transition of PCI Express devices and associated links *from* the sleep or powered-off to powered-on, *lowering* the power consumed by PCI Express devices and associated links, and *placing* of PCI Express devices and associated links into sleep or powered-off. Each of these tasks is supported by PM protocol independently, or in conjunction with Wakeup protocol or HP protocol as follows. The PM and SP protocols and software are part of the PME but operate independent of each other.

- **From the sleep or powered-off to powered-on:** The PM protocol defines the support of the message PM_PME requester transaction used with the Wake Events consisting of the Wakeup protocol and wakeup events of the HP protocol. As discussed in Chapter 13, the Wakeup protocol uses the WAKE# signal line or Beacon to transition a hierarchy domain or an entire PCI Express platform from the sleep to powered-on. The PM software works in conjunction with the Wakeup protocol, which includes powering up the hierarchy domain. As discussed in Chapters 13 and 16, when an HP protocol wakeup event occurs, a message PM_PME requester transaction packet is transmitted. The PM and HP software working in conjunction service the wakeup event, which includes powering up a link that has been powered off for the Hot Plug protocol.

- **Lowering the power consumed by PCI Express devices and associated links:** As introduced in Chapter 13, the PM protocol and software implements the Active-State Power Management protocol for transactions from L0 link state to the L0s or L1 link state. The PM protocol and software also implements the PCI-Power Management Software protocol for transactions from L0 link state to the L1 link state.

- **Placement of PCI Express devices and associated links into sleep or powered-off link states:** As introduced in Chapter 13, PM protocol and software implements the PCI-Power Management Software protocol for placing the PCI Express platform or a hierarchy domain into sleep or powered-off. In conjunction with HP protocol and software, the PM protocol and software also supports placing a specific link into powered-off for the purpose of add-in card removal.

Support Requirements

PCI Express platform requirements for supporting PM protocol vary depending on the PCI Express device. All PCI Express devices except for the Root Complex must support the PCI-Power Management (PCI-PM) Software protocol for the D0 through D3 states. All PCI Express devices including the Root Complex must support the message baseline requester transactions (LLTPs) and link information DLLPs used by the PCI-PM Software protocol. All PCI Express devices including the Root Complex must support the Active-State Power Management (PM) protocol.

Powered-on, Powered-off, and Sleep

The term *powered-on* means that the PCI Express devices on a PCI Express platform, individual hierarchy domains, or a portion of a hierarchy domain are provided main power with or without Vaux. The term *powered-off* means the PCI Express devices on a PCI Express platform, individual hierarchy domains, or a portion of hierarchy domain are without main power or Vaux. Consequently, these PCI Express devices cannot implement the Wakeup protocol. The term *sleep* means that the PCI Express devices on a PCI Express platform, individual hierarchy domains, or a portion of a hierarchy domain are without main power but are provided Vaux. See Chapter 13 for more information.

In addition to the typical term of *main power off,* the term *main power removed* is used to refer to the activity of turning off main power. In addition to the typical term of *main power on,* the term *main power applied* is used to refer to the activity of turning on main power.

Vaux Considerations The retention of Vaux when main power is off is defined as *sleep*. Even though Vaux may be on, each specific PCI Express device may not be drawing power from Vaux. Per the PCI-PM Software protocol, the Power Management Capabilities Register (PMCR) contains AUX Current bits that define the amount of current to draw from Vaux. These bits are used in conjunction with the PME_En bit in the Power Management Control/Status Register (PMCSR).

The PCI Express specification defines an additional bit that permits the auxiliary current to be drawn from Vaux by the PCI Express device not implemented per the PCI-PM Software protocol. This additional bit is the Auxiliary Power PM Enable bit in the Device Control Register. When the Auxiliary Power PM Enable bit is set to 1b in *any* function, the entire PCI Express device can draw the current specified in the AUX Current bits of the PMCR from Vaux independent of the PME_En bit in the PMCR.

> When this book defines *sleep* or Vaux is *retained* or *on* it means that both the platform is supplying Vaux and the PCI Express devices are programmed to use Vaux. If a specific PCI Express device is programmed not to draw current from Vaux, that PCI Express device can never be in sleep (L2).

USD and DSD

In this chapter, the term "downstream device" (DSD) means PCI Express device on the downstream side of a link (i.e. upstream ports of an endpoint, switch, or bridge). Also, the term "upstream device" (USD) means PCI Express device on the upstream side of a link (i.e. downstream port of the Root Complex or a switch).

PCI Express Reference Clock (REFCLK) and PLL

Finally, the tables in this chapter use the terms *PCI Express Reference Clock* and *PLL*. PCI Express Reference Clock is the 100-megahertz clock fed to the transmitters of each PCI Express device port to establish the bit and symbol periods for the 8/10b encoding. The PCI Express Reference Clock is also known as REFCLK. The value of 100 megahertz is based on the current revision of the PCI Express specification. PLL means the phase lock loop of the receivers of each PCI express device port are trying to maintain bit and symbol lock when ON.

L0 and Power Management Link States

The PM protocol defines five possible Power Management (PM) link states: L0s (standby), L1 (lower power standby), L2/3 ready (power removal pending), L2 (sleep), and L3 (powered-off). The L0 (normal operation) link state is the only link state the LTSSM in the PCI Express device and associated links can transition to the L0s, L1, and L2/3 Ready link states. As discussed in Chapter 13, the transition to the L2 and L3 link states from the L0 link state is via the L2/3 Ready link state. The LTSSM is defined as the Link Training State and Status Machine in the PCI Express device's Physical Layer.

Tables 15.1 through 15.6 summarize these six link states. Within each table is description of the major attributes of the link state and the connected PCI Express devices. Chapter 14 has more information on the requirements for transitions between these link states.

L0 Link State

The L0 link state is the normal operational state for the link and the connected PCI Express devices. For the PCI-PM Software protocol the "D" state of the associated functions are defined as D0, which supports normal operation.

Table 15.1 Normal Operation Link State

\multicolumn{2}{c}{L0 link state: Defined as "Normal Operation"}
\multicolumn{2}{c}{PCI Express Reference Clock and Main Power are ON}
\multicolumn{2}{c}{PLL ON}
\multicolumn{2}{c}{Vaux either ON or OFF}

"D" states of all functions within PCI Express devices on link (1)	USD = D0 when DSD = D0
Attributes of this "L0" link state	Required for Active-State PM protocol.
	Required for PCI-PM Software protocol.
	All DLLPs and LLTPs are transmitted on the link.
	Transaction, Data Link, and Physical Layers enabled.
	All USD and DSDs' timers and counters remain active.
Transitions to	L0s link state: Per Active-State PM protocol as outlined in Fig. 15.4.
	L1 link state: Per PCI-PM Software protocol and Active-State PM protocol and as outlined in Figures 15.1 and 15.3, respectively.
	L2/3 Ready link state: Per PCI-PM Software protocol as outlined in Figure 15.2.

Table Notes:

1. Upstream device must be in a "D" state that is equal to or more active that the "D" state of any of the functions of any DSD directly or indirectly through other switches or bridges.

L0s Link State

The L0s link state is the standby state. It permits power reduction by stopping the transmission of DLLPs and LLTPs, and by placing the associated transmitters in an "electrical idle" state. The purpose of the L0s is to provide some power reduction with a quick transition to the L0 link state. The PCI Express devices whose transmitters have transitioned to the L0s link state may also reduce power usage, but this is implementation-specific and not part of the PCI Express specification.

As discussed in Chapter 14, the quick transition requires the reestablishment of bit and symbol lock via the transmission of FTS OSs. If quick reestablishment of bit and symbol lock is not possible, the link transitions to the Recovery link state.

As discussed in the Chapter 14, the transmitters of one port on the link transition the L0s link state independent of the transmitters on other end of the link remaining in the L0 link state.

Table 15.2 Link Power Management State

L0s link state: Defined as "Standby State"	
PCI Express Reference Clock and Main Power are ON	
PLL ON	
Vaux either ON or OFF	
"D" states of all functions within PCI Express devices on link (1)	USD = D0 when DSD = D0
Attributes of this "L" link state	Required for Active-State PM protocol.
	Not applicable and thus not required for PCI-PM Software protocol.
	No DLLPs or LLTPs are transmitted on the link.
	Transaction and Data Link Layers enabled. Physical Layer disabled.
	Provides minimum latency to the L0 link state, typically 100 symbol times.
	USD and DSDs' timers and counters remain active.
Transitions to	L0 link state or Recovery link state and eventually L0 link state: See Active-State PM protocol discussion. Chapter 14 has more information about the Recovery link state.

Table Notes:

1. USD device must be in a "D" state that is equal to or more active that the "D" state of any of the functions of any DSD directly or indirectly through other switches or bridges.

L1 Link State

The L1 link state is the lower power standby link state. It permits power reductions greater than the L0s link state by stopping the transfer of DLLPs and LLTPs and placing the associated transmitters in an "electrical idle" state. It permits power reductions greater than the L0s link state by additionally permitting the DSD's functions' "D" states to be the reduced power levels of D1 or D2 or D3hot.

As discussed in Chapter 14, the quick transition to reestablish of bit and symbol lock for normal operation (L0 link state) via the transmission of FTS OSs is not defined; the link transitions from the L1 link state to the Recovery link state. Consequently, the lower power level provided by the L1 link state relative to the L0s link state requires a greater latency for the link and the connected PCI Express devices to attain normal operation (L0 link state).

Table 15.3 Link Power Management State

L1 link state: Defined as "Lower Power Standby State"	
PCI Express Reference Clock and Main Power are ON	
PLL either ON or OFF	
Vaux either ON or OFF	
"D" states of all functions within PCI Express devices on link (1)	USD = D0 when DSD = D0 (2)
	USD = D0 or D1 when DSD = D1
	USD = D0 or D1 or D2 when DSD = D2
	USD = D0 or D1 or D2 or D3hot when DSD = D3hot (See discussion in next section)
	D1 and D2 implementation is optional
Attributes of this "L" state	Optional for Active-State PM protocol.
	Required for PCI-PM Software protocol within a mobile platform, optional for other platforms.
	No DLLPs or LLTPs are transmitted on the link.
	Transaction, Data Link, and Physical Layers disabled.
	USD and DSDs' timers and counters are suspended.
	Provides higher latency to the L0 link state. Typically a few microseconds, but the maximum is 64 microseconds.
Transitions to	Recovery link state and eventually L0 link state. See Chapter 14 for more information about Recovery link state. Additional procedure discussed in the "Active-State Power Management Protocol" section.

Table Notes:

1. USD device must be in a "D" state that is equal to or more active that the "D" state of any of the functions of any DSD directly or indirectly through other switches or bridges.
2. This combination is only possible via the Active-State PM protocol. The PCI-PM Software protocol for the PCI-compatible power management standards does not permit the DSD function to be in the D0 state for transition to L1 link state due to this protocol.

L2/3 Ready Link State

The L2/3 Ready link state is the power removal pending state used to provide a PCI Express device the time to prepare for loss of all power (main power and Vaux) or power reduced to only Vaux.

As discussed in Chapter 14, the L2/3 Ready link state can only transition to the L2 link state (main power removed but Vaux retained) or to L3 link state (main power and Vaux removed).

Table 15.4 Link Power Management State

L2/3 Ready link state: Defined as "Power Removal Pending State"	
PCI Express Reference Clock and Main Power are ON	
PLL either ON or OFF	
Vaux either ON or OFF	
"D" states of all functions within PCI Express devices on link (1)	USD = D0 or D1 or D2 or D3hot when DSD = D3hot (See discussion in next section)
	D1 and D2 implementation is optional
Attributes of this "L" state	Not applicable to Active-State PM protocol.
	Required for PCI-PM Software protocol.
	No DLLPs or LLTPs are transmitted on the link.
	Transaction, Data Link, and Physical Layers disabled.
	USD and DSDs' timers and counters are suspended.
	PCI Express devices on the link are preparing for main power removal at the platform level.
Transitions to	L2 link state when main power is removed but Vaux is maintained.
	L3 link state when main power and Vaux are both removed.
	If main power is retained, the USD device on the link can request transition to the Recovery link state and eventually L0 link state. See Chapter 14 has more information about the Recovery link state. Additional procedure discussed in the "PCI-PM Software protocol" section.

Table Notes:

1. USD device must be in a "D" state that is equal to or more active that the "D" state of any of the functions of any DSD (immediate or indirectly through other switches or bridges).

L2 Link State

The L2 link state is the lower power sleep state. It permits power reductions greater than the L0s or L1 link states by stopping the transfer of DLLPs and LLTPs, and placing the associated transmitters in an "electrical idle" state. It permits power reductions greater than the L0s and L1 link state by additionally permitting the USD's functions' "D" states to be the reduced power levels of D1 or D2 or D3hot. Also, the DSD's functions' "D" states must be D3hot prior to main power being powered off and must be in the D3cold when only Vaux is provided.

As discussed in Chapter 14, the quick transition to reestablish of bit and symbol lock for normal operation (L0 link state) via the transmission of FTS OSs, or transition through the Recovery link state, is not defined; the link transitions from the L2 link state to the Detect link state. Consequently, the lower power sleep state level provided by the L2 link state relative to the L1 and L0s link states requires a greater latency for the link and the connected PCI Express devices to attain normal operation (L0 link state).

Table 15.5 Link Power Management State

L2 link state: Defined as "Lower Power Sleep State" PCI Express Reference Clock and Main Power are OFF PLL OFF Vaux ON	
"D" states of all functions within PCI Express devices on link (1)	USD = D0 or D1 or D2 or D3hot (main power on) or D3cold (main power off) when DSD = D3hot or D3cold (see discussion in next section).
	D1 and D2 implementation is optional.
Attributes of this "L" link state	Not applicable to Active-State PM protocol.
	Required for PCI-PM Software protocol if Vaux implemented on platform.
	No DLLPs or LLTPs are transmitted on the link.
	Transaction, Data Link, and Physical Layers disabled.
	USD and DSDs' timers and counters are suspended.
Transitions to	L3 link state when Vaux is removed (powered off).
	Detect link state when main power is applied. Chapter 14 has more information about Detect link state.

Table Notes:

1. USD device must be in a "D" state that is equal to or more active that the "D" state of any of the functions of any DSD (immediate or indirectly through other switches or bridges).

L3 Link State

The L3 link state is the powered-off state. As the name implies, it provides the ultimate power reduction.

As discussed in Chapter 14, the quick transition to reestablish of bit and symbol lock for normal operation (L0 link state) via the transmission of FTS OSs or transition through the Recovery link state is not defined; the link transitions from the L3 link state to the Detect link state. Consequently, the powered-off level provided by the L3 link state relative to the L2, L1, and L0s link states requires a greater latency for the link and the connected PCI Express devices to attain normal operation (L0 link state).

Table 15.6 Link Power Management State

L3 link state: Defined as "Off State"	
PCI Express Reference Clock and Main Power are OFF	
PLL OFF	
Vaux OFF	
"D" states of all functions within PCI Express devices on link (1)	Not applicable to PCI-PM Software protocol or Active-State PM protocol in that "off" is "off" for both.
Attributes of this "L" link state	Not applicable to PCI-PM Software protocol or Active-State PM protocol in that "off" is "off" for both.
Transitions to	Detect link state when main power is applied. Chapter 14 has more information about Detect link state.

Table Notes:

1. USD device must be in a "D" state that is equal to or more active that the "D" state of any of the functions of any DSD (immediate or indirectly through other switches or bridges).

PCI-Power Management Software Protocol

The PM protocol defined for a PCI Express platform has a portion compatible to PCI legacy power management, referred to here as *PCI power management standards*. This compatible portion of the PM protocol is called the *PCI-Power Management Software protocol*, referred to here as the *PCI-PM Software protocol*. The PCI-PM Software protocol has specific activities that result in a change in power level of the link and connected PCI Express devices. These specific activities are related to the programming the PCI Express devices' functions' "D" states specified by the PCI power management standards. The "D" states are not defined for functions within the virtual PCI devices in the Root Complex; PCI-PM Software protocol permits links and associated LTSSM of the PCI Express devices to transition between the L0, L1, L2, and L3 link states. PCI-PM Software protocol implements message baseline requester transactions and link management information. The message baseline requester transactions are contained in Transaction Layer Packets (TLPs) of the Transaction Link Layer, which in turn are contained in Link Layer Transaction Packets (LLTPs) of the Data Link Layer. The link management information is contained in Data Link Layer Packets (DLLPs) of the Data Link Layer.

This discussion focuses on the PM protocol in support of the "D" states. For more information on the "D" states of functions, refer to the PCI power management standards. The PCI power management standards are defined in the *PCI Bus Power Management Interface Specification Rev. 1.1* and *Advance Configuration and Power Interface Specification Rev. 2.0*. For more information on the PCI power management standards, see the book and specification references in Chapter 1.

Before discussing the transition between link states and the power levels of the connected PCI Express devices specified by PCI-PM Software protocol, this chapter describes the three major items associated with conditions defined by the PCI-PM Software protocol: "D" States, TLPs, and DLLPs.

"D" States for PCI-PM Software Protocol

When a function in a PCI Express device is programmed to a specific "D" state via configuration register block, certain attributes and restrictions apply to the PCI Express device as follows.

The PCI-PM Software protocol implements PCI power management standards' compatible "D" states applicable to each function with in a PCI Express device. The PM software includes software routines specified by the PCI-PM Software protocol that program the "D" states of the functions within a PCI Express device. The programming of the "D" states results in the transition of the associated link from L0 link state to L1 or L2/3 Ready link states as will be discussed in the next section.

D0 State

The D0 state is required to be supported by all PCI Express devices except the Root Complex. The D0 state has two levels. Immediately after main power is applied beyond Vaux to a PCI Express device, the functions' "D" state are defined as in D0 un-initialized. With the completion of enumeration (BUS# assignment) and configuration of the function, it enters the D0 active state or simply D0 state. Also, part of a function's transition to the D0 state requires the programming of the function's Memory Space Enable, I/O Space Enable, or Bus Master Enable bits with the configuration register block set to 1b. A function in the D0 state can initiate and respond to any transaction.

D1 State

The D1 state is optionally supported by a PCI Express device but is never supported by the Root Complex. PM software retains function information needed prior to the transition into the D1 state. PM software must not program the function into the D1 state if any outstanding transactions have yet to be completed. That is, requester transaction packets have been transmitted but an associated completer transition packet has not been received. The PM software must wait for receipt of the outstanding completer transaction packet.

Once a function is in the D1 State, it cannot initiate any transaction except for message PM_PME requester transaction and the configuration completer transactions. The only requester transaction packets that can be received by the function are configuration requester transactions. All other transaction packets received are defined as an **unsupported requester** error. See Chapter 9 for more information.

D2 State

D2 state is optionally supported by a PCI Express device but is never supported by the Root Complex. PM software retains function information needed prior to the transition into the D2 state. PM software must not program the function into the D1 state if any outstanding transactions have yet to be completed. That is, requester transaction packets have been transmitted but an associated completer transition packet has not been received. The PM software must wait for receipt of the completer transaction packet.

Once a function is in the D2 state, it cannot initiate any transaction packet except for message PM_PME requester transaction and the configuration completer transactions. The only requester transaction packets that can be received by the function are configuration requester transactions. All other transaction packets received are defined as an **unsupported requester** error. See Chapter 9 for more information.

D3 State

The D3 state is required to be supported by all PCI Express devices. The D3 state consists of two types: D3cold and D3hot. A function is programmed into the D3hot state by the PM software while main power is on. The transition D3hot state to D3cold state occurs when main power is removed but Vaux is retained to the PCI Express device.

D3hot State Prior to the transition to this state the PM software retains function information that will be lost if main power is removed to transition to D3cold. PM software must not program the function in the D3hot state if any outstanding transactions have yet to be completed. That is, requester transaction packets have been transmitted but an associated completer transition packet has not been received. The PM software must wait for receipt of the completer transaction packet. PM software must not program the function into the D3hot state if any outstanding transactions have yet to be completed. That is, requester transaction packets have been transmitted but an associated completer transition packet has not been received. The PM software must wait for receipt of the completer transaction packet.

Once a function is in the D3hot state, it cannot initiate or receive any TLPs except as follows. When in the D3hot state, the function must participate in the Power Fence protocol, which includes the handshake of message PME_Turn_Off and PME_TO_Ack requester transaction packets discussed in the following section. The function can initiate message PM_PME requester transaction and the configuration completer transactions. Other than message PME_Turn_Off requester transaction or configuration requester transactions, all other transaction packets received are defined as an **unsupported requester** error. See Chapter 9 for more information.

Functions programmed to the D3hot permit PM software to program their PMCR (Power Management Status/Control Register) in its configuration register block to reprogram the function to D0 un-initialized state. This reprogramming does not require a hardware reset internally to the PCI Express device.

The transition from the D3hot state to D0 State requires 10 milliseconds in which time a function cannot initiate or receive any transaction packets. The receipt of a transaction packet during this time period results in an **undefined** error.

> Once the Power Fence protocol is initiated with the reception of the message PME_Turn_Off requester packet, the PCI Express device that received it is guaranteed that a Fundamental Reset will occur with or without removing main power.

D3cold State When main power is removed from a PCI Express device but Vaux is retained, the functions transition from the D3hot state to the D3cold state.

The transition from the D3cold state requires the function's software to re-initialize the function. When main power is applied, the function transitions to D0 un-initialized. In the D3cold state, the Vaux power may or may not be supplied. If Vaux power is not retained, the function in the D3cold state cannot implement the Wakeup protocol. If Vaux power is provided, the function can implement the Wakeup protocol and the Power Management Status/Control Register information is retained.

TLPs for PCI-PM Software Protocol

The PCI-PM Software protocol implements three message baseline requester transaction packets (TLPs) specific to the PM protocol. These TLPs are: message PM_PME, message PME_Turn_Off, and message PME_TO_Ack requester transaction packets. The purpose of these three TLPs is to provide additional mechanisms for PM software and PCI Express devices to request transitions in the link states for power management.

Table 15.7 is a summary of the source and destination of the message baseline transactions associated with PCI-PM Software protocol.

Table 15.7 Message Transactions for PME

Name	Destination of message transaction packet	Source of message transaction packet
PM_PME (Also used by the Hot Plug protocol)	Root Complex	Switch Legacy EP PCI Express EP PCI Express/PCI Bridge PCI Express/PCI-X Bridge
PME_Turn_Off	Legacy EP PCI Express EP PCI Express/PCI Bridge PCI Express/PCI-X Bridge	Root Complex Switch (1)
PME_TO_Ack	Root Complex Switch (1)	Legacy EP PCI Express EP PCI Express/PCI Bridge PCI Express/PCI-X Bridge Switch

Table Notes:

1. According to the present PCI Express specification, the PME_Turn_Off requester transaction packet uses r = 011b for Implied Routing, which is only defined as a broadcast from the Root Complex and not from a downstream port of a switch. Similarly, the message PME_TO_Ack requester transaction packet uses r = 101b for Implied Routing, which only defines the destination as the Root Complex. See further discussion below in the section "Sources of Message PME_Turn_Off and PME_TO_Ack Requester Transaction Packets."

Message PM_PME Requester Transaction

The purpose of the message PM_PME requester transaction is for a function in a DSD on any link to request activation of the PM software via the Root Complex. Any function within a DSD can source a message PM_PME requester transaction packet, provided main power is on. Any switch must route any message PM_PME requester transactions it receives on any downstream port to its upstream port. The message PM_PME requester transaction is implemented for two situations as follows:

- **Sleep to Powered-on:** For a Wakeup protocol the Wake# signal line or Beacon is used to request main power to be applied to the

PCI Express platform, individual hierarchy domains, or to a portion of the hierarchy domain. With main power applied, the DSD must transmit a message PM_PME requester transaction packet to inform PM software that a transition from sleep (L2) to powered-on (L0) is occurring. See Chapter 13 for more information about the Wakeup protocol.

- **Wakeup Event:** If properly programmed, the Root Complex or switch connected to a Hot Plug slot will transmit a message PM_PME requester transaction packet for occurrence of a Hot Plug (HP) hardware event. In the case of the Root Complex, the message PM_PME requester transaction packet is internally sourced. The transmission of the message PM_PME requester transaction packet per the Hot Plug (HP) protocol is called a wakeup event. See Chapter 16 for more information about the Hot Plug protocol. PM software parses this request to the HP software.

PME Service Timeout Protocol and Discarding of Message PM_PME Requester Transaction It is important for the function within a PCI Express device to confirm that the message PM_PME requester transaction successfully arrived at the Root Complex, and that the PM software was activated. Consequently, functions within sources that can transmit a message PM_PME requester transaction must implement the PME Service Timeout protocol.

The reason for the PME Service Timeout protocol is to prevent deadlock conditions. The Root Complex can only buffer a limited number of message PM_PME requester transaction packets. If too many are transmitted upstream, some will be buffered in switches after the Root Complex buffer becomes full. The completer transaction packets associated with the PM software accessing the configuration register blocks for previously received message PM_PME requester transaction packets are flowing upstream. Both the message PM_PME requester transactions and completer transactions for configuration address space must be assigned a Traffic Class (TC) equal to 000b, and thus their packets must use the buffers of Virtual Channels 0 (VC0). Due to transaction ordering rules, the completer transaction packets will be "stuck" behind the buffered message PM_PME requester transaction packets, resulting in deadlock. In order to prevent completer transaction packets and other message PM_PME requester transaction packets from being "stuck" in switches, the Root Complex *must* discard any incoming message PM_PME requester transaction packets without reporting an error when its internal buffers are full. In order for a function in a PCI Express device to know that its message PM_PME requester transaction packets were discarded and that they must be retransmitted, the PME Service Timeout protocol has been defined. The PME Service Timeout protocol is as follows.

Functions within PCI Express devices that can source message PM_PME requester transaction packets must support the PME_Status bit within the PMCSR (Power Management Control/Status Register) in their configuration register block and a PME Service timer. The function must also implement the PMC (Power Management Capabilities) register in its configuration register block. When the message PM_PME requester transaction packet is transmitted, the PME_Status bit is set to 1b, and the PME service timer is set to 0b and started. The PME_Status bit is set to 0b by the PM software in acknowledgment of servicing the message PM_PME requester transaction packets received by the Root Complex. If the PME_Status bit has not been set to 0b within the range of 95 milliseconds to 150 milliseconds by the PME Service timer, the function must retransmit the message PM_PME requester transaction packet, set the PME Service timer to 0b, and restart the timer. This procedure continues indefinitely until PME_Status bit is set to 0b by the PM software prior to the PME Service timer timing out. If at anytime a function must request a wake event associated link, the PME_Status bit is set to 0b. See Chapter 13 for more information about wake events.

Message PM_Turn_Off and PME_TO_Ack Requester Transactions

The purpose of the message PME_Turn_Off requester and message PME_TO_Ack requester transactions (TO = Turn Off) is to permit PCI Express devices and associated link to transition to the L2/3 Ready link state. The message PME_Turn_Off requester and message PME_TO_Ack requester transaction packets are received to, and transmitted from, a PCI Express device, respectively.

A downstream port of the Root Complex or a switch transmits a message PME_Turn_Off requester transaction packet in a broadcast fashion on all DSDs downstream of the Root Complex or switch. Each DSD responds by transmitting a message PME_TO_Ack requester transaction packet to the Root Complex or switch that is the source of the message PME_Turn_Off requester transaction packet. Subsequently, the DSD on each link exchanges DLLPs with the USD on the same link. The USD, DSD, and link then transition to the L2/3 Ready link state.

If both the main power and Vaux supplied to any PCI Express device are removed, the PCI Express devices and associated link transitions from the L2/3 Ready link state to the L3 link state (powered-off). If the main power supplied to the PCI Express devices is removed but Vaux is retained, the PCI Express devices and associated link transitions from the L2/3 Ready link state to the L2 link state (sleep). In either case, once message PME_Turn_Off requester packet is received, the PCI Express device that received it is guaranteed that a Fundamental Reset will occur with or without removing main power.

Prior to the PCI Express devices and associated links transitioning to the L2/3 Ready link state, all message PM_PME transaction requesters packets flowing upstream to the Root Complex are required to be transmitted across all links and port through all switches. Transaction ordering discussed in Chapter 10 and the Power Fence protocol discussed later in this chapter ensure that this requirement is met.

The PM software and associated hardware that source the message PME_Turn_Off requester transaction packet is not required to receive a message PME_TO_Ack requester transaction packet prior to the associated power controller removing main power. The PM software and associated hardware may optionally implement a timer to request the associated power controller to remove main power transition when the timer times out. The timer timeout is equivalent to the source of the message PME_Turn_Off requester transaction packet receiving a message PME_TO_Ack requester transaction packet.

This power controller associated with the source of the message PME_Turn_Off requester transaction packet cannot remove reference clocks or main power until after at least 100 nanoseconds after all PCI Express devices and associated links transitioned to the L2/3 Ready link state. As specified by the Power Fence protocol discussed later in this chapter, the transition to the L2/3 Ready link state does not occur at the source of the message PME_Turn_Off requester transaction packet until all the downstream PCI Express devices and links have transitioned to the L2/3 Ready link state.

Sources of Message PME_Turn_Off and PME_TO_Ack Requester Transaction Packets The purpose of exchanging the message PME_Turn_Off and message PME_TO_Ack requester transactions is to request the PCI Express devices and the associated links to transition to the L 2/3 Ready link state. Once in the L2/3 Ready link state, the removal of main power and retention of Vaux transition PCI Express devices and the associated links to the L2 link state (sleep). Once in the L2/3 Ready link state, the removal of main power and the non-retention of Vaux transitions the PCI Express devices and the associated links to the L3 link state (powered-off). The collection of PCI Express devices and the associated links within the PCI Express fabric that can be placed in the L2/3 Ready link state is as follows:

- **Each Hierarchy Domain:** As discussed in Chapter 2, each downstream port of the Root Complex defines a hierarchy domain. The message PME_Turn_Off requester transaction packet is transmitted in broadcast fashion by the Root Complex's downstream port to all PCI Express devices on the downstream side of the links. With the receipt of the message PME_TO_Ack requester transaction packet and the transition to the L2/3 Ready link state of the link connected to downstream port of the Root Complex, the entire hierarchy domain is ready to transition to L2 link state (sleep) or L3 link state (powered-off). The placement of a specific hierarchy domain into the L2/3 Ready link state is independent of the other hierarchy domains.

- **Entire PCI Express Platform:** The Root Complex can transition all of the hierarchy domains to the L2/3 Ready link state by the protocol discussed for each hierarchy domain. Once all of the hierarchy domains transition to the L2/3 Ready link state, the Root Complex can also transition to an equivalent state. Subsequently, the entire platform can transition to L2 link state (sleep) or L3 link state (powered-off).

- **Portion of Hierarchy Domain:** Any downstream port of a switch can be the source of a message PME_Turn_Off requester transaction packet. It can broadcast the message PME_Turn_Off requester transaction packet downstream in a broadcast fashion for its own purpose and not on behalf of an upstream PCI Express device. Following the protocol discussed for each hierarchy domain, the portion of the hierarchy domain downstream of the port that sourced the message PME_Turn_Off requester transaction packet will transition to the L2/3 Ready link state. This portion of the hierarchy domain is ready to transition to L2 link state (sleep) or L3 link state (powered-off), with the remainder of the hierarchy domain remaining in the L0, L0s, or L1 link states.

The message PME_Turn_Off requester and message PME_TO_Ack requester transaction packets route through the PCI Express platform by Implied Routing. As discussed in Chapter 4, the Implied Routing defined for message PME_Turn_Off requester and message PME_TO_Ack requester transaction packets is r = 011b and r = 101b, respectively. According to the present PCI Express specification, the r = 011b Implied Routing is only defined as a broadcast from the Root Complex and not from a downstream port of a switch. In the section of the PCI Express specification that discusses power management, the downstream port of a switch and Root Complex are defined as the two possible original sources of a message PME_Turn_Off requester transaction packet as it is defined in this book. Similarly, according to the current PCI Express specification, the r = 101b Implied Routing only defines the destination of the Root Complex. If the original source of a message PME_TO_Ack requester transaction is a downstream port of a switch, the destination of the associated message PME_TO_Ack requester transaction packets must also be the same downstream port. In the section of the PCI Express specification that discusses power management, the downstream port of a switch and Root Complex are defined as the two possible destinations a message PME_TO_Ack requester transaction packet can have as it is defined in this book.

Power Fence Protocol In preparation of sleep (L2) or powered-off (L3) the PCI Express devices and associated links activities must transition to the L 2/3 Ready link states. In order to avoid any race conditions between a PCI Express devices (DSD) sourcing a message PM_ PME requester transaction packet simultaneous with a change in the power level associated for the PCI Express device, the PCI Express specification implements the Power Fence protocol. The Power Fence protocol is essentially the handshake mechanism of the aforementioned message PME_Turn_Off and PME_T0_Ack requester transaction packets prior to DSD requesting a transition to the L2/3 Ready link state with the exchange of DLLPs with the USD on the same link. This handshake provides a PCI Express device time to prepare for power level change. The specifics of the Power Fence protocol for the different PCI Express devices are as follows:

- **Endpoint:** Prior to a specific downstream port (PME_Turn_Off source port) of the Root Complex or a switch sourcing the message PME_Turn_Off requester transaction packet to an endpoint, all of the functions in the endpoint must be programmed to D3hot state. With the receipt of the message PME_Turn_Off requester transaction packet, the endpoint cannot transmit any message PM_PME requester transition packets. Prior to transmitting message PME_TO_Ack requester transaction packet to the PME_Turn_Off source port, the endpoint waits until all preparations are completed for the endpoint to transition to the L2 or L3 link state subsequent to the transition to the L2/3 Ready link state. Subsequent to the endpoint as the DSD on each link transmitting the message PME_TO_Ack requester transaction packet, it exchanges DLLPs with the USD on the same link to transition these PCI Express devices and associated link to the L2/3 Ready link state.

- **Switch:** Prior to a specific downstream port (PME_Turn_Off source port) of the Root Complex or a switch sourcing the message PME_Turn_Off requester transaction packet to a another switch, all functions of all PCI Express devices downstream of PME_Turn_Off source port must be programmed to D3hot state. Each switch downstream of PME_Turn_Off source port transmits the message PME_Turn_Off requester transaction packet received on its upstream port to all of its downstream ports. With the receipt of the message PME_Turn_Off requester transaction packet, the switch does not transmit any message PM_PME requester transition packets for internal functions, but must transmit upstream any message PM_PME requester transition packets received on its downstream ports. Other considerations specific to switches are:
    - Each switch downstream of the PME_Turn_Off source port must collect all message PME_TO_Ack requester transaction packets it receives on its downstream ports and transmit on its upstream port a single "collective" message PME_TO_Ack requester transaction packets.
    - Prior to transmitting a message PME_TO_Ack requester transaction packet on the upstream port, the switch waits until all message PME_TO_Ack requester transaction packets are received on all the downstream ports and all preparations are completed internal to the switch. The internal preparations are completed for a transition to the L2 or L3 link state subsequent to the transition to the L2/3 Ready link state.

- After the switch (acting as the DSD on each link) transmits the message PME_TO_Ack requester transaction packet, it exchanges DLLPs with the USD on the same link to transition these PCI Express devices and associated link to the L2/3 Ready link state. The switch acting as the DSD on a link cannot begin the exchange of DLLPs to transition to the L2/3 Ready link state on its upstream port until all of its downstream ports transitions to the L2/3 Ready link state. The switch's downstream ports are acting as the USD on a link to exchange DLLPs for transition to the L2/3 Ready link state.
- Other downstream switches also collectively represent all downstream PCI Express devices connected to their downstream ports.

■ **Bridge:** Prior to a specific downstream port (PME_Turn_Off source port) of the Root Complex or a switch sourcing the message PME_Turn_Off requester transaction packet to a bridge, all functions of all PCI Express devices, PCI devices, and PCI-X devices downstream of PME_Turn_Off source port must be programmed to D3hot state. With the receipt of the message PME_Turn_Off requester transaction packet, the bridge does not transmit any message PM_PME requester transition packets for internal functions, but must transmit upstream a message PM_PME requester transition packet for equivalent PCI and PCI-X operations. Once all of the PCI or PCI-X resource downstream of the bridge and internal functions of the bridge are properly prepared, the bridge transmits a message PME_TO_Ack requester transaction packet. The bridge sources a single message PME_TO_Ack requester transaction packet on behalf of all downstream PCI or PCI-X resources on the downstream bus segments. After the bridge (acting as the DSD on each link) transmits the message PME_TO_Ack requester transaction packet, it exchanges DLLPs with the USD on the same link to transition these PCI Express devices and associated link to the L2/3 Ready link state. The bridge acting as the DSD on a link cannot begin the exchange of DLLPs to transition to the L2/3 Ready link state on its upstream port until all of its PCI or PCI-X bus segments are in an equivalent state to L2/3 Ready link state.

DLLPs for PCI-PM Software Protocol

The PCI-PM Software protocol implements three link management packets (DLLPs) specific to power management. These DLLPs are: PM_Enter_L1, PM_Enter_L23, and PM_Request_Ack. These link management packets are exchanged by the PCI Express devices on the link with the PCI Express device that transmitted the message PME_TO_Ack requester transaction. These DLLPs are defined as follows.

PM_Enter_L1 DLLP

The PM_Enter_L1 DLLP is transmitted by the DSD to the other PCI Express device on the link a request to transition to the L1 link state.

PM_Enter_L23 DLLP

The PM_Enter_L23 Ready DLLP is transmitted by the DSD to the other PCI Express device on the link a request to transition to the L2/3 Ready link state.

PM_Request_Ack DLLP

The PM_Request_Ack DLLP is transmitted by USD to acknowledge to the other PCI Express device on the link the receipt of the PM_Enter_L1 or PM_Enter_L23 DLLP. The receipt of the PM_Request_Ack DLLP causes the PCI Express devices' LTSSM and associated link transition to the L1 or L2/3 Ready link state.

PCI-PM Software Protocol Application of "D" States, TLPs, and DLLPs

As discussed earlier, all PCI Express devices except for the Root Complex must support the minimum software for the PCI Bus Power Management Interface Specification Rev. 1.1. The support by the PM protocol is via the "D" states of the functions within PCI Express devices as defined by the PCI-PM Software protocol. All functions within each PCI Express device must implement the D0 and D3 states. The implementation of the D1 or D2 states is optional for each function in PCI Express devices. The "D" states are not defined for functions within the virtual PCI devices in the Root Complex. All PCI Express devices including the Root Complex support the TLPs and DLLPs defined for the PCI-PM Software protocol.

The programming of the "D" states relates to the last function in a DSD to be placed into a "D" state that is not D0. When the DSD is an endpoint, all of the functions within the endpoint must be placed into a

"D" state that is not D0. When the DSD is a switch all of the functions must be placed into a "D" state that is not D0, and PCI Express devices downstream of the switch also have additional considerations. The functions in PCI Express devices downstream of the switch must be programmed to "D" state as active or less active than the less active :D" state of functions with the switch. The same requirements apply to the functions within a bridge and the functions with PCI or PCI-X Express devices downstream of the bridge.

The discussion of PCI-PM Software protocol for link state transitions is in seven parts: L0 to L1, L1 to L0, L0 to L2/3 Ready, L2/3 Ready to L2, L2/3 Ready to L3, L2 to L0, and L3 to L0. The parts directly associated with the programming of the "D" states within functions of PCI Express devices are the transitions from the L1 to L0 and L0 to L2/3 Ready. The other transitions are also defined as part of the PCI-PM Software protocol by convention.

Transition L0 Link State to L1 Link State

> See Tables 15.1 and 15.3 for overall summary of L0 and L1 states not discussed here. Also see details of link state transitions in Chapter 14.

Figure 15.1 illustrates the protocol to transition from the L0 link state to the L1 link state. Integral to the protocol is the enabling and disabling of the Transaction, Data Link, and Physical Layers that are at different times in the transition. The transition from the L0 link state to the L1 link state begins with the programming of all functions in a PCI Express device on the downstream side of the link to the D1, D2, or D3hot state. The transition between link states only begins when all of the functions of a DSD are in non-D0 states. As discussed earlier, it is the responsibility of the PM software to ensure that the functions of all PCI Express devices downstream of a switch are programmed to a non-D0 state that is as active or less active that the non-D0 states in the switch. Similar concept applies to the PCI and PCI-X devices downstream of a bridge.

The request for transition from the L0 link state to the L1 link state can only be made by PCI Express device on the downstream end of the link. The possible PCI Express devices are the switches (upstream port), endpoints, or bridges. The PCI Express device on the upstream end of the link does not request a transition to the L1 link state on the link, but must acknowledge the request for transition to the L1 link state. In the L1 link state the PCI Express devices on the upstream and downstream side of the links suspend all counters and timers.

To request a transition from the L0 link state to the L1 link state, a series PM_Enter _L1 DLLPs are transmitted continuously by the DSD. To complete the transition to the L1 link state, a positive acknowledgement must be made with the continuous transmission of a series of PM_Request_Ack DLLPs by the USD. The complete transition protocol to L1 link state is illustrated by Figure 15.1. Integral to the protocol is the enabling and disabling of the Transaction, Data Link and Physical Layers that are at different times in the transition.

Transition L1 Link State to L0 Link State

The Root Complex or switch, as the USD, determines that it must transmit downstream a LLTP or a DLLP. The transition is first to the Recovery link state and then to L0 the link state. See Chapter 14 for the transitions conditions between the L1, Recovery, and L0 link states.

An endpoint, switch, or bridge, as the USD, determines that it must transmit upstream a message PM_PME requester transaction packet. The transition is first to the Recovery link state and then to L0 the link state. See Chapter 14 for the transitions conditions between the L1, Recovery, and L0 link states.

Transition L0 Link State to L2/3 Ready Link State

> See Tables 15.1 and 15.4 for overall summary of L0 and L1 states not discussed here. Also see details of link state transitions in Chapter 14.

Figure 15.2 illustrates the protocol to transition from the L0 link state to the L2/3 Ready link state. The protocol is similar to the L0 link state to L1 link state transition previously discussed with three major differences. First, all of the functions in the DSD must be programmed into the D3hot state. As discussed earlier, it is the responsibility of the PM software to ensure that the functions of all PCI Express devices downstream of a switch are programmed to D3hot. A similar concept applies to the PCI and PCI-X devices downstream of a bridge. Second, the PCI Express devices and the associated links first transition from the L0 link state to the L1 link state as discussed in the previous section. The Root Complex subsequently transmits a message PME_Turn_Off requester transition that is routed to one of the functions of a specific PCI Express device. The function representing all functions in the PCI Express device transmits a message PME_TO_Ack requester transition to the Root Complex. As discussed in Chapter 13, in order to transmit the LLTPs containing these

TLPs, all of the PCI Express devices and associated links must transition through the L0 link state through the Recovery link state. Third, the PCI Express device that sourced the message PME_TO_Ack requester transition subsequently begins to transmit a series of PM_Enter_L23 DLLPs. As illustrated in Figure 15.2, with the exchange of the PM_Enter_L23 DLLPs and the PM_Request_Ack DLLPs, the PCI Express devices and the associated link transition to the L2/3 Ready link state.

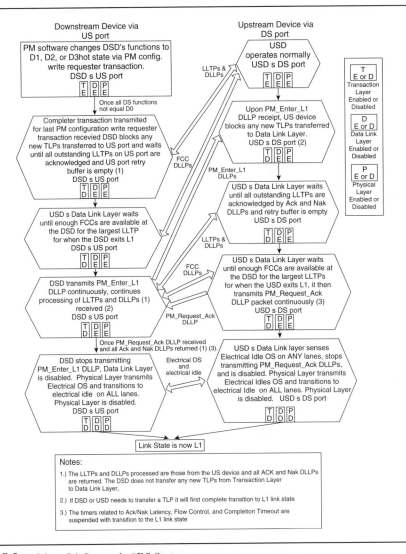

Figure 15.1 L0 to L1 State via "D" States

Chapter 15: Power Management 757

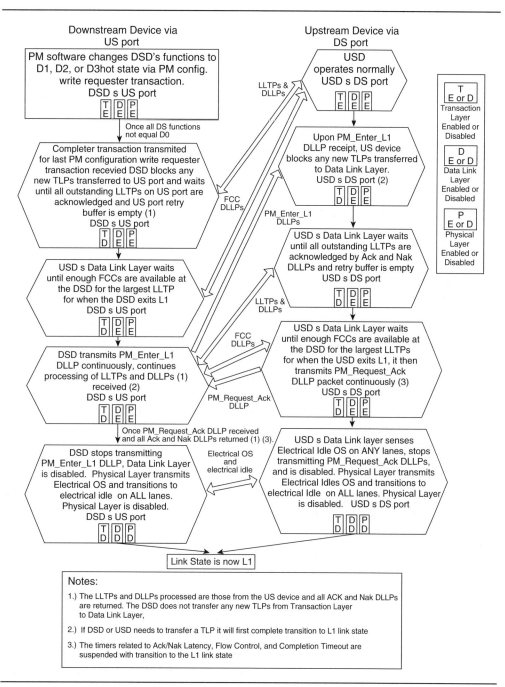

Figure 15.2 L0 to L2/3 Ready Link State

Transitions L2/3 Ready Link State to L2 Link State and L2/3 ready Link State Ready to L3 Link State

The transition from the L2/3 Ready link state to the L2 link state or L3 link state must be via the L2/3 Ready link state. Once in the L2/3 Ready link state, it is the removal of main power and the removal or not of Vaux that causes the transition as summarized in Table 15.4.

Transition L2 Link State to L0 Link State and L3 Link State to L0 Link State

The link state transition from the L2 state or L3 state to L0 state requires the application of power, initialization of the PCI Express devices, and transition to the Detect link state for Link Training. See Chapter 14 for more information on Link Training. The case of the L2 link state to L0 link state transition possibly includes the application of the Wakeup Protocol. See Chapter 13 for more information on PCI Express device initialization and the Wakeup Protocol.

Active-State Power Management Protocol

The other portion of the PM protocol is specific to PCI Express and not defined by the PCI power management standards. This PM protocol is the *Active-State Power Management protocol*, referred to here as the *Active-State PM protocol*. The Active-State PM protocol is specific to a PCI Express platform, and permits links and associated PCI Express devices to transition between the L0, L1, and L0s link states. Active-State PM protocol must be supported by all PCI Express devices, the protocol is implemented independently on each link, and it is not visible to platform resources implementing PCI-PM Software protocol.

The Active-State PM protocol implements a message baseline requester transaction and link management information. The message baseline requester transactions are contained in Transaction Layer Packets (TLPs) of the Transaction Link Layer, which in turn are contained in Link Layer Transaction Packets (LLTPs) of the Data Link Layer. The link management information is contained in Data Link Layer Packets (DLLPs) of the Data Link Layer.

Before discussing the transition between link states and the power levels of the PCI Express devices and associated links specified by the Active-State PM protocol, this chapter discusses the two major items associated with the Active-State PM protocol: TLPs and DLLPs.

TLP for Active-State PM Protocol

The one message baseline requester transaction packet implemented by Active-State PM protocol is the message PM_Active_State_Nak requester transaction packet.

The purpose of the message PM_Active_State_Nak requester transaction packet is for an USD (source in Table 15.8) to refuse to acknowledge the transition to the L1 link state as requested by the DSD (destination in Table 15.8) and instead to force the transition of the link and associated DSD to the L0s link state, if enabled. The request to transition to the L1 link state is made by the DSD transmitting a PM_Active_State_Request_L1 DLLP.

Table 15.8 Message Transaction for Active-State PM protocol

Name	Destination of message baseline requester transaction packet	Source of message baseline requester transaction packet
PM_Active_State_Nak	Legacy EP	Root Complex
	PCI Express EP	
	Switch	
	PCI Express/PCI Bridge	Switch
	PCI Express/PCI-X Bridge	

DLLPs for Active-State PM Protocol

The Active-State PM protocol implements two link management packets (DLLPs) specific to the PM protocol: PM_Active_State_Request_L1 and PM_Request_Ack. These link management packets are exchanged by the PCI Express devices on the link to transition to the L1 link state. The PM_Active_State_Request_L1 DLLP is exchanged with the message PM_Active_State_Nak requester transaction packet to transition the transmitters of one PCI Express device to the L0 s link state. These DLLPs are defined as follows.

PM_ Active-State_Request _L1 DLLP

The PM_Active_State_Request_L1 DLLP is sent by the DSD to the USD on the link a request to transition to the L1 link state.

PM_Request_Ack DLLP

The PM_Request_Ack DLLP is used by the USD to acknowledge to the DSD on the link the receipt of the PM_Active-State_Request_L1 DLLP, and to acknowledge that the transition to the L1 link state will be honored by the USD. The receipt of the PM_Request_Ack DLLP causes the DSD's LTSSM and associated link transition to the L1 link state.

Active-State PM Protocol Application of the TLP and DLLPs

As discussed earlier, Active-State PM protocol provides another method to achieve the lower power link states of L1 link states without programming all of the functions into a non-D0 state as required for the PCI-PM Software protocol. The Active-State PM protocol permits the DSD to request the link and the USD on the link to transition to the L1 link state while one or more functions in the DSD are in the D0 state. In addition, the Active-State PM protocol also permits any transmitters of a PCI Express device to transition to the L0s link states at anytime, if the associated functions in the PCI Express device are enabled for the L0s link state.

Requirements to Implement the Active-State PM Protocol

The transition between the L0 and L0s and L0 and L1 links states is controlled by three variables: The Link Control Register in the USD and DSD, the value of the "D" states of the functions in the DSD, and the pending transmissions of LLTPs and DLLPs in the DSD and USD.

Active-State Link PM Control Bits All PCI Express devices including the Root Complex must support the Active-State PM protocol and thus support the Active-State Link PM Control bits in the Link Control Registers in the configuration register blocks. Through the Link Control Register, it is possible for the PM software to enable the transmitters for only the L0s link state, enable the transmitters and receivers only for the L1 link state, enable the transmitters and receivers for both the L0 and L1 link states, or disable transmitters and receivers for both L0s or L1 link states. Independent of the values of the Active-State Link PM Control bits in the Link Control Registers, the receivers of PCI Express devices must always tolerate the L0s link state in case the transmitters on the other end of the link enter the L0s link state.

For a PCI Express device to support L0s or L1 link states as defined in the Active-State Link PM Control bits for each function, the following must be the value of these bits among all of the functions in the D0 link state. The functions in the non-D0 link state are not considered.

- If Active-State Link PM Control bits = 00b for *any* function, the PCI Express device does not execute the Active-State PM protocol.

- If Active-State Link PM Control bits = 01b for *any* function, the PCI Express device executes the Active-State PM protocol for transition to the L0s link state *only*.

- If Active-State Link PM Control bits = 11b for *all* functions, the PCI Express device does execute the Active-State PM protocol for transition to the L0s link state and L1 link states.

- If Active-State Link PM Control bits = 10b for *all* functions, the PCI Express device executes the Active-State PM protocol for transition to the L1 link state *only*.

The Active-State Link PM Control bits in the Link Control Registers in the configuration register blocks can be programmed during runtime. During runtime the Active-State Link PM Control bits of a function in the D0 can have their associated Active-State Link PM Control bits disabled. During runtime the Active-State Link PM Control bits of a function in the non-D0 can have their associated Active-State Link PM Control bits enabled for L0s link state, provided all other functions are enabled for L1 or L0s link state.

"D" States of Functions The downstream port of a switch can transition to the L0s or L1 link states independent of the "D" states of the internal functions. The downstream port of a Root Complex has no "D" states are defined for the internal functions; consequently, the transition to the L0s or L1 link states by downstream port of the Root Complex is not related to "D" states.

For the upstream port of a DSD to implement the Active-State PM protocol, "D" states of the functions within the PCI Express must be considered as follows:

- Any function within the DSD that is *not* in the D0 state is ignored for determining the Active-State PM protocol implementation, provided at least one function is in the endpoint that is in the D0 state. If no function is in the D0 state, the upstream port of the endpoint is prevented from implementing the Active-State PM protocol.

- The one or collection of functions in the D0 state must be in agreement as to which link state transition is enabled by the Active-State Link PM Control bits in the Link Control Registers in their configuration register blocks. The agreement is outlined in the "Active-State Link PM Control Bits" section.

Pending Transmissions Relative to Transition to the L0s Link State
The upstream or downstream ports on a link can consider transitions to the L0s link state if no transmissions of LLTPs or DDLPs are pending based on the following criteria.

- A port on the Root Complex or endpoint has no pending transmissions when all of the following conditions are met:
    - No DLLPs or LLTPs are ready to transmit.
    - No FCCs are available relative to the buffer at the receiving port for the TLP in the Transaction Layer.
- An upstream port on a switch or a bridge has no pending transmissions when all of the following conditions are met:
    - All of the switch's downstream ports' lanes are in the L0s state. (These are the receiving portion of the ports only — see the explanation in the following section.) This does not apply to bridges.
    - No DLLPs or LLTPs are ready to transmit.
    - No FCCs are available relative to the buffer at the receiving port for a TLP in the Transaction Link Layer.
- A downstream port on a switch has no pending transmission when all of the following conditions are met:
    - All of the switch's upstream port's lanes are in the L0s state. (These are the receiving portion of the port only — see the explanation in the following section).
    - No DLLPs or LLTPs are ready to transmit.
    - No FCCs are available relative to the buffer at the receiving port for a TLP in the Transaction Layer.

In this discussion for transition from L0 link state to L0s link state the term *pending transmission* used. The length of time for a pending transmission is implementation-specific. However, if the port of a PCI Express device is able to transition, it must do so within 7 microseconds of meeting the pending transmission conditions.

Pending Transmissions for Transition to the L1 Link State The upstream or downstream port can consider transitions to the L1 link state if no transmission of LLTPs or DDLPs are pending, based on the following criteria.

- A downstream port of a Root Complex or switch has no pending transmissions when all of the following conditions are met:
  - No LLTPs are pending to be transmitted downstream.
  - No Ack or Nak DLLPs are pending to be transmitted downstream.
- An upstream port on a endpoint has no pending transmission when all of the following conditions are met:
  - No LLTPs or DLLPs are pending to be transmitted upstream.
- The upstream port of a switch or a bridge has no pending transmission when all of the following conditions are met:
  - None of the switch's downstream ports are in the L0 or L0s link state (excludes bridge).
  - No LLTPs or DLLPs are pending to be transmitted upstream.
  - The upstream port's receivers are idle for an implementation-specific amount of time.

Transition L0 Link State to the L0s Link State

> See Tables 15.1 and 15.2 for overall summary of L0 and L0s states not discussed here. Also see details of link state transitions in Chapter 14.

For the transmitters of a port to transition from the L0 link state to L0s link state requires that at least one function be in the D0 state, that the Active-State Link PM Control bits of all the functions in the D0 state be compliant with the rules described earlier in the section "Active-State Link PM Control Bits," and that no transmissions be pending. The receivers of the downstream port are always tolerant of the L0s link state on the transmitters on the other end of link.

Transition L0s Link State to the L0 the Link State

> See Tables 15.1 and 15.2 for overall summary of L0 and L0s states not discussed here. Also see details of link state transitions in Chapter 14.

The transition from L0s link state to L0 link state occurs when the transmitters on a port on either end of the link needs to transition. The need to transition is never dependent on the availability of FCC credits, but occurs for the following reasons:

- The transmitters in an upstream port of a switch, a bridge, or an endpoint transition when a DLLP or a LLTP is to be transmitted.

- The transmitters in a downstream port of the Root Complex or a switch transition when a DLLP or a LLTP is to be transmitted.

- The transmitters in an upstream port of a switch transition when the switch detects a transition to L0 link state on any of its downstream ports (the receiving portion of the downstream ports only; see the following explanation).

- The transmitters in all downstream ports of a switch transition when the switch detects a transition to L0 link state in its upstream port (the receiving portion of the upstream port only; see the following explanation). Note: Those downstream ports in lower power state related to non-D0 states are not affected.

Receiving Portion of a Port The receiving portion (receivers) of a port that detects an Electrical Idle OS knows the transmitters of port on the other end of the link have transitioned to the L0s link state. Independent of the transmitters in the same port as the receivers, when the receivers detect an exit from "electrical idle" on *any* configured lanes it is an indication that the transmitters of the port on the other end of the link have transitioned from the L0s link state to the L0 link state. See Chapter 14 for more link state transition information.

Transition L0 Link State to the L1 Link State

> See Tables 15.1 and 15.3 for overall summary of L0 and L1 states not discussed here. Also see details of link state transitions in Chapter 14.

For the downstream port of a Root Complex or switch to transition from the L0 link state to the L1 link state requires that at least one function be in the D0 state, that the Active-State Link PM Control bits of all the functions in the D0 state be compliant with the rules described earlier in the section "Active-State Link PM Control Bits," that no transmissions be pending, and that the DSD request the transition.

For the upstream port of switch, bridge, or endpoint to transition from the L0 link state to the L1 link state, at least one function must be in the D0 state, the Active-State Link PM Control bits of all the functions in the D0 state must be compliant with the rules described earlier in the section "Active-State Link PM Control Bits," no transmissions may be pending, and the request to transition to L1 must be honored by the USD.

Assuming all of the conditions relative to "D" states, Active-State Link PM Control bits, and pending transmission are met, the actual requirements to request transaction to the L1 link state is dependent on the value of the Active-State Link PM Control bits on both ends of the link for all functions in the D0 state. Table 15.9 defines the possible exchanges between the two PCI Express devices on link. The purpose of the exchange is to determine whether the link can transition from the L0 link state to the L1 link state. The DSD transmits PM_Active_State_Request_L1 DLLPs to request transition to the L1 link state and the USD honors the request by transmitting PM_Request_Ack DLLP. If the USD cannot honor the request, it transmits a message PM_Active_State_Nak requester transaction packet.

Table 15.9 Possible TLP and DLLP Exchanges Based on the Active-State PM Protocol

Active-State Link PM Control bits in Link Control Registers 00b & 01b is *any* function in D0 state 10b & 11b is *all* functions in D0 state	PCI Express Device on Downstream Side of the Link Request (DSD) Link State Result	PCI Express Device on Upstream Side Response (USD) Link State Result
DSD = 00b L0s & L1 Disabled USD = 00b L0s & L1 Disabled	*Do not* Transmit PM_Active_State_ Request_L1 DLLP	NA
DSD = 01b L0s Enabled & L1 Disabled USD = 00b L0s & L1 Disabled	*Do not* Transmit PM_Active_State_ Request_L1 DLLP (Independently transition to L0s can occur)	NA
DSD = 10b L0s Disabled & L1 Enabled USD = 00b L0s & L1 Disabled	Transmit PM_Active_State_ Request_L1 DLLP	Transmit message PM_Active_State_ Nak requester transaction packet
DSD = 11b L0s & L1 Enabled USD = 00b L0s & L1 Disabled	Transmit PM_Active_State_ Request_L1 DLLP	Transmit message PM_Active_State_ Nak requester transaction packet
DSD = 00b L0s & L1 Disabled USD = 01b L0s Enabled & L1 Disabled	*Do not* Transmit PM_Active_State_ Request_L1 DLLP	NA (Independently transition to L0s can occur)
DSD = 01b L0s Enabled & L1 Disabled USD = 01b L0s Enabled & L1 Disabled	*Do not* Transmit PM_Active_State_ Request_L1 DLLP (Independently transition to L0s can occur)	NA (Independently transition to L0s can occur)
DSD = 10b L0s Disabled & L1 Enabled USD = 01b L0s Enabled & L1 Disabled	Transmit PM_Active_State_ Request_L1 DLLP	Transmit message PM_Active_State_ Nak requester transaction packet (Independently transition to L0s can occur)

Table 15.9 Possible TLP and DLLP Exchanges Based on the Active-State PM Protocol *(continued)*

Active-State Link PM Control bits in Link Control Registers 00b & 01b is *any* function in D0 state 10b & 11b is *all* functions in D0 state	PCI Express Device on Downstream Side of the Link Request (DSD) Link State Result	PCI Express Device on Upstream Side Response (USD) Link State Result
DSD = 11b L0s and L1 Enabled USD = 01b L0s Enabled & L1 Disabled	Transmit PM_Active_State_Request_L1 DLLP	Transmit message PM_Active_State_ Nak requester transaction packet (Independently transition to L0s can occur)
DSD = 00b L0s & L1 Disabled USD = 10b L0s Disabled & L1 Enabled	*Do not* Transmit PM_Active_State_Request_L1 DLLP	NA
DSD = 01b L0s Enabled & L1 Disabled USD = 10b L0s Disabled & L1 Enabled	*Do not* Transmit PM_Active_State_Request_L1 DLLP (Independently transition to L0s can occur)	NA
DSD = 10b L0s Disabled & L1 Enabled USD = 10b L0s Disabled & L1 Enabled	Transmit PM_Active_State_Request_L1 DLLP	Transmit PM_Request_Ack DLLP
DSD = 11b L0s and L1 Enabled USD = 10b L0s Disabled & L1 Enabled	Transmit PM_Active_State_Request_L1 DLLP	Transmit PM_Request_Ack DLLP
DSD = 00b L0s & L1 Disabled USD = 11b L0s and L1 Enabled	*do not* Transmit PM_Active_State_Request_L1 DLLP	NA
DSD = 01b L0s Enabled and L1 Disabled USD = 11b L0s & L1 Enabled	*do not* Transmit PM_Active_State_Request_L1 DLLP	NA

Table 15.9 Possible TLP and DLLP Exchanges Based on the Active-State PM Protocol *(continued)*

Active-State Link PM Control bits in Link Control Registers	PCI Express Device on Downstream Side of the Link Request (DSD)	PCI Express Device on Upstream Side Response (USD)
00b & 01b is *any* function n D0 state	Link State Result	Link State Result
10b & 11b is *all* functions in D0 state		
DSD = 10b L0s Disabled & L1 Enabled USD = 11b L0s & L1 Enabled	Transmit PM_Active_State_Request_L1 DLLP	Transmit PM_Request_Ack DLLP
DSD = 11b L0s & L1 Enabled USD = 11b L0s & L1 Enabled	Transmit PM_Active_State_Request_L1 DLLP	Transmit PM_Request_Ack DLLP

The request to transition from the L0 link state to the L1 link state can only be made by a PCI Express device on the downstream end of the link (DSD). The PCI Express device on the upstream end of the link (USD) does not request a transition to the L1 link state on the link, but must acknowledge honoring or not honoring the request for transition to the L1 link state. To request a transition from the L0 link state to the L1 link state, a series of PM_Active_State_Request_L1 DLLPs are transmitted continuously by the DSD. To complete the transition to the L1 link state, a positive acknowledgement must be made with the continuous transmission of a series of PM_Request_Ack DLLPs by the USD. The complete transition protocol to L1 link state is illustrated by Figure 15.3. Integral to the protocol is the enabling and disabling of the Transaction, Data Link, and Physical Layers at different times in the transition. Figure 15.3 assumes that both downstream and upstream devices implement the transition to the L1 link state by the Active-State PM protocol.

Chapter 15: Power Management ■ **769**

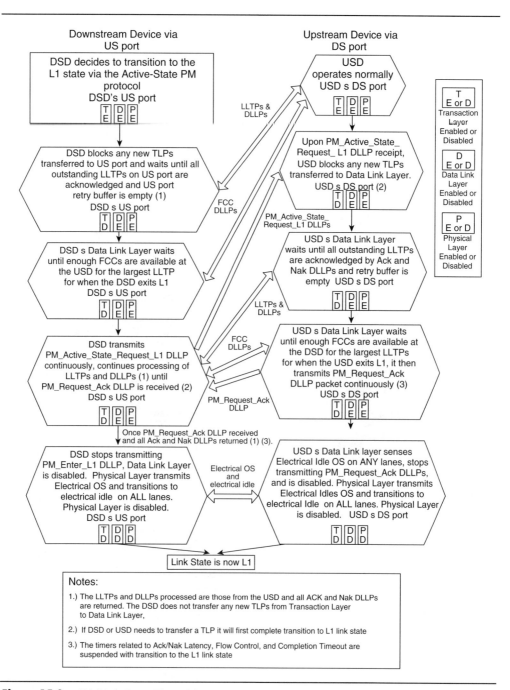

Figure 15.3 L1 Link State Transition via Active-State Power Management Protocol

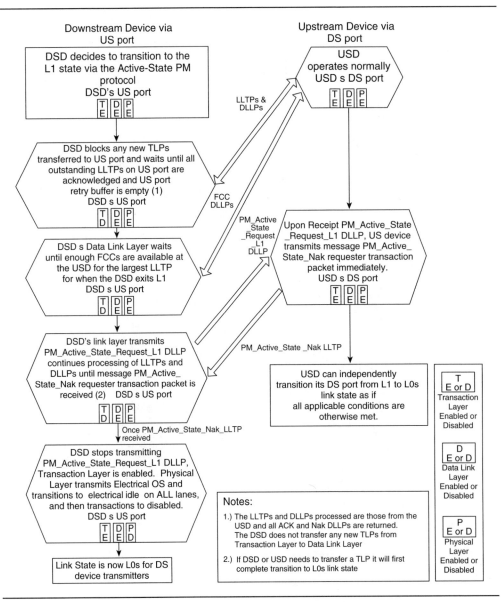

Figure 15.4 L1 Link State Transition Defaults to L0s Link State via Active-State Power Management Protocol

To request a transition to the L1 link state, a PM_Active_State_Request_L1 DLLP is transmitted by the DSD. To complete the transition to the L1 link state, a positive acknowledgement must be made by the USD. If the USD cannot provide a positive acknowledgement, it transmits a negative acknowledgement with the message PM_Active_State_Nak requester transaction packet. The complete transition protocol to the L0s link state for the DSD transmitters after failure to achieve the L1 link state is illustrated by Figure 15.4. The transmitters of the DSD transition to the L0s link state if all of the applicable conditions are met; otherwise the DSD port remains in the L0 link state. Also illustrated in Figure 15.4, the transmitters of the USD can optionally transition to the L0s link state; otherwise the USD port remains in the L0 link state.

Transition L1 Link State to the L0 Link State

> See Tables 15.1 and 15.3 for overall summary of L0 and L1 states not discussed here. Also see details of link state transitions in Chapter 14.

As illustrated in Figure 15.3, the transition to the L1 link state leaves the transmitter and receivers of the PCI Express devices on both ends of the link in the L1 link state. This is different from the L0s link state, which defines the L0s link state independently for the transmitters of the PCI Express devices on one end of the link from those on the other end of the link. The transition from the L1 link state to L0 link state can be initiated from either the DSD or USD based on the following conditions.

For the transition from the L1 link state to L0 link state by the DSD that is an endpoint, switch, or bridge:

- Transition is never dependent on the availability of FCC credits.

- The transition is requested if the DSD needs to transmit upstream a LLTP or a link manage information in a DLLP on its upstream port. The protocol to transition the DSD's transmitters through the Recovery link state to the L0 link state is discussed in Chapter 14. Once the exit from "electrical idle" is sensed by the receivers on the USD, its downstream port also transitions through the Recovery link state to the L0 link state as discussed in Chapter 14.

- If the DSD is an endpoint or bridge, it must immediately transition its upstream port to the L0 link state via the Recovery link state once it senses a transition request from the USD. The protocol to transition is discussed in Chapter 14.

- If the DSD is a switch, it must immediately transition its upstream port to the L0 link state once it senses a transition request from the USD. The switch must also immediately transition without any delay all of its downstream ports to the L0 link state that are in the L1 link state via the Recovery link state once it senses the transition to the L0 link state on its upstream port. The switch is required to transition to the L0 link state the appropriate downstream ports within 1 microsecond of the beginning transition from L1 link state on its upstream port. It is important for the switch not to wait until the transition from L1 link state to L0 link state on its upstream port is completed before beginning the transition of all of its downstream ports. Any downstream port with a function that is in the D1, D2, or D3hot must not transition to the L0 link state. Any downstream port with a function that is in the L0 or L0s link state remains in that link state. The protocol to transition is initiated by the special symbol set defined in Chapter 14.

For the transition from the L1 link state to L0 link state by the USD that is the Root Complex or a switch:

- Transition is never dependent on the availability of FCC credits.

- The transition is occurs if the USD needs to transmit downstream a LTTP or a link manage information in a DLLP on its downstream port. The protocol to transition the USD's transmitters through the Recovery link state is discussed in Chapter 14. Once the exit from "electrical idle" is sensed by the receivers on the DSD, its upstream port also transitions through the Recovery states to the L0 link state.

- If the USD is the Root Complex, it must immediately transition its downstream port to the L0 link state via the Recovery link state once it senses the transition to L0 link state from the DSD. The transition of the L1 link state to the L0 link state is not required on any of the other downstream ports of the Root Complex. The protocol to transition from the L1 link state via the Recovery link state to the L0 link state is discussed in Chapter 14.

- If the USD device is a switch, it must immediately transition its downstream port to the L0 link state via the Recovery link state once it senses the transition to L0 link state from the DSD. The switch must also immediately transition without delay its upstream port to L0 link state via the Recovery link state once it senses the transitioning from the L1 link state to the L0 link state on any of its downstream ports. The switch is required to transition its upstream port within 1 microsecond of the beginning transition from L1 link state to the L0 link on any of its downstream ports. It is important for the switch not to wait until the transition from L1 link state to L0 link state on the downstream port is completed before beginning the transition of its upstream port. The transition of the L1 link state to the L0 link state is not required on any of the other downstream ports. The protocol to transition from the L1 link state via the Recovery link state to the L0 link state as discussed in Chapter 14.

Transition Latency Values and REFCLK

L0s link state provides quick entry and exit from a low power state. L1 provides a less quick entry and exit from a lower power state. As discussed earlier, the PM software can enable the transition to the L0s link state or L1 link state. The PM software determination of which of these link states to enable for which links is partly based on the exit latency time. The two registers that the PM software must consider in determining the exit latency time are the Device Capabilities and Link Capabilities Registers. The exit latency time is only defined for the Active-State PM protocol.

Device Capabilities Register of each endpoint's or bridge's configuration register block contains the value of the transition latencies from the L0s link state to the L0 link state and the L1 link state to the L0 link state. The value is implementation-specific and dependent on the buffer size of the endpoint or bridge. These values define the longest acceptable transition latency without a negative impact on the endpoint or bridge.

The Link Capabilities Register of each endpoint's or bridge's configuration register block contains the value of the transition latencies from the L0s link state to the L0 link state and the L1 link state to the L0 link state for the associated link.

Part of the Active-State PM protocol is for the PM software to determine whether the transition from the L0 link state to L1 link state or the transition from the L0 link state to the L0s link state can be enabled. As previously discussed, the Active-State Link PM Control bits in the Link Control Registers in the configuration register blocks of the functions within each PCI Express device can be programmed to enable the transitions to the L0s or L1 link states. The PM software must determine which transitions can be enabled given the longest acceptable latency of each endpoint and the total of latencies of the links between the endpoint and the Root Complex.

The REFCLK provided to each PCI Express device at each end of a specific link is another factor when considering transition latency. The Common Clock Configuration bit of the Link Control Register in the configuration register block indicates whether the REFCLK for the PCI Express device is the same one for both ends of the link, or are different for each end. If the REFCLK to each PCI Express device on a specific link is different, the PM software must take this into consideration when calculating values of transition latency in the Link Capabilities Registers in the configuration register blocks.

Chapter 16

Hot Plug Protocol

This chapter discusses the ability for add-in cards to be inserted into and removed from a PCI Express platform. The distinction of PCI Express versus other platforms is the ability to insert and remove add-in cards while main power is maintained on the platform. This is referred to as *Hot Plug* capability.

This is the second of the three chapters related to power consumption in a PCI Express platform. Chapter 15 discusses Power Management of any PCI Express device (on the platform or add-in card) or the link. Chapter 17 discusses the Slot Power protocol. This chapter discusses the Hot Plug protocol of an add-in card slot. In order to execute this protocol, power must be removed and subsequently applied. Thus, this book has grouped the topics of these chapters under the concept of Power Management Events. Power Management Events (PMEs) consist of the Power Management (PM), Hot Plug (HP), and Slot Power (SP) protocols. PMEs are associated with PME software, which consists of Power Management (PM) software for the PM protocol, Hot Plug (HP) software for HP protocol, and Slot Power (SP) software for the SP protocol.

Introduction to the Hot Plug Protocol

The primary purpose of the Hot Plug protocol is to permit PCI Express add-in cards to be inserted or removed from a platform's slot without powering off the entire platform. The PCI Express Hot Plug protocol is a derivative of the *PCI Standard Hot-Plug Controller and Sub-System*

Specification. This book focuses on the requirements specific to PCI Express. For more information on the applicable PCI standards, see the book and specification references in Chapter 1.

If the HP protocol is implemented it can only be the HP protocol defined by the PCI Express specification—no other HP protocol is permitted on any PCI Express platform implementation. Currently, the two documented PCI Express platform implementations are non-mobile or mobile.

The definition of the term *slot* includes the connector for an add-in card and optionally any components related to the HP protocol on the platform. The implementation of the HP protocol on any slot is optional; slots with *no* support of the HP protocol are simply not Hot Plug slots. The definition of the term *chassis* is the set of mechanical elements of the platform.

The focus of this chapter is the HP protocol applied to non-mobile platforms and non-mobile add-in cards. At the print time of this book, the *PCI Express Mini Card Electromechanical Specification revision 1.0* of June 2, 2003 was just released. This specification defines the attributes of the mobile PCI Express platforms and mobile add-in cards. In general it does not provide any additional HP protocol information and by convention it must be compatible with the PCI Express HP protocol. Consequently, the chapter focuses only on non-mobile implementations of the Hot Plug protocol, but the information also applies to the mobile implementation unless otherwise noted.

Hardware Components for the HP Protocol

PCI Express implements the HP protocol via message baseline requester transactions and interrupts. Due to the nature of insertion and removal of add-in cards using an HP protocol, hardware components are also part of the protocol. The hardware components interface the user, chassis, and add-in card with the HP software.

There are seven hardware components in the HP protocol that need to be defined prior to the discussion of the associated non-hardware components of message baseline requester transaction packets, interrupts, and the HP software. These hardware components are: power indicator, MRL, MRL sensor, attention indicator, attention button, power controller (fault), and add-in card present. The hardware components are required for slot and add-in cards unless otherwise noted.

Power Indicator

As the name implies, the power indicator provides a visual indication that main power is being supplied to the slot. The power indicator (LED) operates independently of the attention indicator, is a yellow or amber light (LED) in color, and can be placed anywhere close to its associated slot on the chassis or on the add-in card. The power indicator blinks with a duty cycle of 1 to 2 hertz (45 percent to 55 percent), and is controlled by the HP software only via message baseline requester transaction packets with one exception. The power indicator can be forced to be on if the platform monitors power-on faults and a fault is detected. The operation of the power indicator is as follows:

- **Indicator off:** Main power has been removed from the slot and only Vaux power remains (if implemented). Vaux power is disconnected or connected by the MRL sensor and it occurs automatically independent of the power indicator. The implementation of the MRL sensor is optional; consequently, the use of the PRSNT# pins can disconnect or connect Vaux to the slot. Insertion or removal of the add-in card for the slot is permitted at this time.

- **Indicator blinking:** Indicates that main power is being applied or removed from the slot. Also, the blinking may be in reaction to the Attention button being pressed (see the discussion later). The insertion or removal of the add-in card for the slot is not permitted at this time.

- **Indicator on:** Indicates main power is being supplied to the slot. The insertion or removal of the add-in card for the slot is not permitted at this time.

- **Both Locations:** If the slot on the chassis and the add-in card both implement the indicator, only one is used by the HP software and the other remains off. If the add-in card is not plugged in, the chassis indicator must be used if provided.

MRL and MRL Sensor

The MRL (manually operated retention latch) is any mechanical entity that holds the add-in card in the slot when the mechanism is closed. The add-in card cannot be inserted or removed until the mechanism is opened. The opened and closed positions are monitored by an optional

MRL sensor. A single MRL can be used to provide a mechanical mechanism for two or more slots if the slots do not have the optional MSL sensors.

The MRL sensor operates in conjunction with MRL to provide information on position of the MRL. The MRL sensor is any switch, optical device, and so on, that can indicate an opened or closed MRL. When the MRL is open, the MRL sensor triggers a disconnect of the Vaux and other switched signal lines to the slot. When the MRL is closed, the MRL sensor triggers a connect of the Vaux and other switched signal lines to the slot. The MRL sensor implementation is optional. The MRL sensor is not needed if the connect and disconnect of the Vaux and switch signal lines are addressed by the PRNT# pins. If the MRL sensor is not implemented, the use of staggered pins PRNT#1 and PRSNT#2 defined by PCI can be implemented, or any other similar implementation to connect and disconnect Vaux and other switched signal lines to the slot that is required. If implemented in some or all of the slots, only one MRL sensor assigned to each slot. See the section "Add-in Card Present" later in the chapter for more information.

> It is recommended that the MSL sensor have a corresponding input to the power controller. The power controller connects and disconnects power to the slot. If a MSL sensor input is not provided, it is recommended that the MSL sensor be routed to the power fault input to the power fault detention circuit. As discussed later in this chapter, power fault detected information is relayed through a register in the configuration address space to the HP software.

If the add-in card is unexpectedly removed, the MSL sensor determines that the MRL is open, disables the port, and reports to HP software. The state of the power and attention indicators does not change automatically. After the MSL sensor reports the unexpected removal, the HP software must change the power and attention indicators appropriately.

Attention Indicator

As the name implies, the attention indicator provides a visual indication that the associated slot needs attention. The attention indicator (LED) operates independently of the power indicator, is a yellow or amber light (LED) in color, and can be placed anywhere close to its associated slot on the chassis or on the add-in card. The attention indicator blinks with a duty cycle of 1 to 2 hertz (45 percent to 55 percent), and is controlled only by the HP software via message transactions. The operation of the attention indicator is as follows:

- **Indicator off:** This is the normal operation.

- **Indicator blinking:** This is to draw attention by a software request.

- **Indicator on:** This is to indicate an operational problem specific to the add-in card. An operational error is any condition that prevents the add-in card from operating normally. For example, the add-in card or its software driver is broken, an unexpected event has occurred, or the like. An operational error is not related to the HP protocol. Examples of conditions that would not be considered operational errors are HP software that did not want to permit hot plug operation, or a platform that had insufficient power.

- **Both Locations:** If the slot on the chassis and the add-in card both implement the indicator, only one is used by the HP software and the other remains off. If the add-in card is not plugged in, the chassis indicator must be used if provided.

Attention Button

The attention button is a momentary-contact push switch that operates independently of the power or attention indicators except via HP software, can be placed anywhere close to its associated slot on the chassis or on the add-in card, and is controlled only by the HP software via message baseline requester transactions. The attention button is pressed by the operator to request insertion or removal of an add-in card for the slot. Once the HP software ready to honor the request, it causes the power indicator to blink. Once the power indicator starts blinking the operator has 5 seconds to cancel the request by pushing the attention button again. If the request is not canceled, the HP software turns off the power indicator. If request is canceled, the HP software turns on the power indicator. If the attention button is activated but the insertion or removal is not permitted (that is, the power indicator is not blinking to indicate honoring the request), it is recommended that the HP software provide error information to the application software or log an error.

Power Controller (Fault)

If an add-in card slot is implemented, a power controller can be optionally implemented. The power controller enables and disables power to the slot. Associated with power controller is a power fault detention circuit. Power fault detected information is relayed through a register in the configuration address space to the HP software as a HP hardware event. If the power fault is detected, the power indicator can be turned on independently of the HP software.

Add-in Card Present

If an add-in card slot is implemented, there must be a method by which the presence of an add-in card in the slot can be detected. The method selected by the PCI Express specification for non-mobile add-in cards is use of the PRSNT1# and PRSNT2# signal lines originally defined by the PCI local bus specification. The information of the PRSNT1# and PRSNT# pins is relayed through a register in the configuration address space to the HP software. The existence of a non-mobile add-in card as indicated by the PRSNT1# and PRSNT# signal lines is used to connect and disconnect Vaux and other signal lines as determined by the platform (for example, REFCLK, PERST#, SMBus, and so on).

As illustrated in Figure 16.1, the edge fingers on the non-mobile add-in card for the PRSNT#1 and PRSNT#2 are recessed to permit connection of the other signal lines before the presence of a non-mobile add-in card can be detected. The PRSNT1# signal line is only implemented once on the non-mobile add-in card and is the input of ground reference from the platform.

The PRSNT2# signal line is provide at multiple locations on the non-mobile connector to support the different sizes of non-mobile add-in cards. The possible non-mobile connector and add-in card sizes are x1, x4, x8, and x16. On the non-mobile add-in card, the PRSNT2# signal line represents the size. Figure 16.1 illustrates a non-mobile connector of size x4 and can accept non-mobile add-in cards of x1 or x4. The non-mobile add-in card of size x1 has its PRSNT2# edge finger (B17) connected to the single PRSNT#1 edge finger. The non-mobile add-in card of size x4 has its PRSNT2# edge finger (B31) connected to the single PRSNT#1 edge finger and does not have the PRSNT2# edge finger (B17) connected. See Chapter 21 for more information.

For the mobile add-in card defined by the PCI Express Mini Card specification, the method to detect the presence of a PCI Express Mini Card is not through the use of PRSNT1# and PRSNT#2 signal lines. These two signal lines are not defined for the non-mobile connector and PCI Express Mini Card. Please see the PCI Express Mini Card specification for further information.

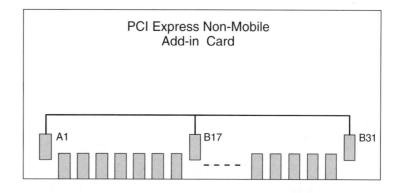

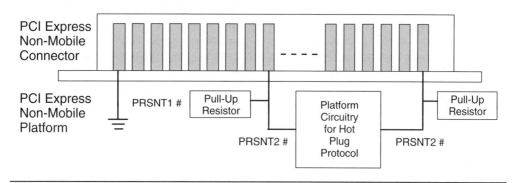

Figure 16.1   PRSNT# Signal Lines and Connector

Non-Hardware Components for the HP Protocol

The seven HP hardware components discussed in the previous section define only a part of what is needed for the HP protocol. The other parts are message baseline requester transactions and configuration address space registers.

Message Baseline Requester Transactions

Except for the message PME_PM requester transaction, all of the message baseline requester transactions are specific to the HP protocol and only transfer between the two PCI Express devices on a specific link. The message PME_PM requester transaction is also used by the Power Management protocol and flows from the add-in card slot to the Root Complex.

Message Baseline Requester Transactions Specific to HP Protocol

Table 16.1 summarizes the message baseline requester transactions related to the HP protocol. Clearly, all but one of the message baseline requester transaction packets is related to the hardware components listed in the previous section and are self-explanatory. The first seven messages listed in Table 16.1 operate the power indicator, attention indicator, and attention button. The downstream ports of the Root Complex or switches connected to slots transmit downstream indicator related message baseline requester transaction packets. The destinations of these packets are the chassis (slots) or the add-in cards (endpoints, switches, or bridges). The upstream ports of the add-in cards (endpoints, switches, or bridges) connected to slots or the chassis (slot) transmit upstream the attention button related message baseline requester transaction packets. The destinations of these packets are the downstream ports of the Root Complex or switches connected to slots.

The destinations of the message baseline requester transaction packets for the Attention and Power Indicators must support the reception even if the associated indicators are not implemented. That is, the reception when the indicator is not implemented is not an error condition. The destinations of the message baseline requester transaction packets for the Attention Button Pressed must support the reception even if the slot does not support the HP protocol. That is, the reception when the slot connected to a switch or a Root Complex that does not implement the HP protocol is not an error condition.

Table 16.1 Message Hot-Plug Requester Transactions

Name	Destination of Message Transaction Packet	Source of Message Transaction Packet
Attention_Indicator_ On	Legacy EP(7) PCI Express EP (7) Switch (2) PCI Express/PCI Bridge (7) PCI Express/PCI-X Bridge (7)	Root Complex (5) Switch (1)
Attention_Indicator_ Blink	Legacy EP (7) PCI Express EP (7) Switch (2) PCI Express/PCI Bridge (7) PCI Express/PCI-X Bridge (7)	Root Complex (5) Switch (1)
Attention_Indicator_ Off	Legacy EP(7) PCI Express EP (7) Switch (2) PCI Express/PCI Bridge (7) PCI Express/PCI-X Bridge (7)	Root Complex (5) Switch (1)
Attention_Button_ Pressed	Root Complex (6) Switch (4)	Legacy EP (8) PCI Express EP (8) Switch (3) PCI Express/PCI Bridge (8) PCI Express/PCI-X Bridge (8)
Power_Indicator_On	Legacy EP(7) PCI Express EP (7) Switch (2) PCI Express/PCI Bridge (7) PCI Express/PCI-X Bridge (7)	Root Complex (5) Switch (1)
Power_Indicator_ Blink	Legacy EP (7) PCI Express EP (7) Switch (2) PCI Express/PCI Bridge (7) PCI Express/PCI-X Bridge (7)	Root Complex (5) Switch (1)

Table 16.1 Message Hot-Plug Requester Transactions *(continued)*

Name	Destination of Message TransactionPacket	Source of Message Transaction Packet
Power_Indicator_Off	Legacy EP(7) PCI Express EP(7) Switch (2) PCI Express/PCI Bridge (7) PCI Express/PCI-X Bridge (7)	Root Complex (5) Switch (1)
PM_PME (Also used by the Power Management protocol)	Root Complex	Root Complex (9) Switch

Table Notes:

1. Defined for a platform switch with its downstream port connected to the slot and the indicator implemented on the add-in card.
2. Defined for a switch on the add-in card and the indicator connected to the switch. If the indicator is not connected, receipt of the message baseline requester transaction is ignored.
3. Defined for a switch on the add-in card and the button connected to switch.
4. Defined for switch on the platform with its downstream port connected to the slot and the button implemented on the add-in card.
5. Defined for downstream port of the Root Complex connected to the slot and the indicator implemented on the add-in card.
6. Defined for Root Complex's downstream port connected to the slot and the button implemented on the add-in card.
7. Defined for endpoint or bridge on the add-in card and the indicator connected to it. If the indicator is not connected, receipt of the message baseline requester transaction is ignored.
8. Defined for an endpoint or bridge on the add-in card and the button connected it.
9. Internally sourced to itself when slot is connected to the Root Complex directly.

Message PM_PME Requester Transaction

The message PM_PME requester transaction packet (which is also used by the Power Management protocol in Chapter 15) is unique to the previous discussion. The purpose of the message PM_PME requester transaction packet is for a downstream switch, the source of which is listed in Table 16.1, to request execution of the associated software within the PME software via the Root Complex, the destination of which is listed in Table 16.1. The sourcing of a message PM_PME requester transaction packet is in response to a HP event. If the slot is connected to a switch, any function within the downstream switch associated with the slot can be the source of the message PM_PME requester transaction packet. If the slot is connected to the Root Complex, a virtual message PM_PME requester transaction packet is sourced by the Root Complex to itself.

The message PM_PME requester transaction packet is also sourced as part of the Wakeup Protocol. See Chapter 13 for more information.

Confirmation of Receipt at Root Complex It is important for the function in the switch to confirm that the message PM_PME requester transaction packet successfully arrives at the Root Complex and that the HP software via the PME software was activated. Consequently, the function within the switch that can request support of the HP protocol implements a PME_Status bit within the PMSCR (Power Management Status/Control Register) in its configuration register block and a PME Service timer. The PME_Status bit is set to 1b and the PME service timer is set to 0b and started when the message PM_PME requester transaction packet is transmitted. The PME_Status bit will be set to 0b by the PME software. If the PME_Status bit has not been set to 0b within the range of 95 milliseconds to 150 milliseconds by the PME Service timer, the function must retransmit the message PM_PME requester transaction including setting the PME Service timer to 0b and restarting it. This procedure continues indefinitely until the PME_Status bit is set to 0b by the PME software on behalf of the HP software. The function must also implement the PMC (Power Management Capabilities) register in its configuration register block. This is defined as the PME Service Timeout protocol and it applies whenever the message PM_PME requester transaction packet is transmitted. See Chapter 15 for more detailed discussion of the PME Service Timeout protocol.

Also associated with the sourcing of the message PM_PME requester transaction packets by the HP protocol is a request to main power to the slot to be powered. The PCI Express specification requires a PCI Express device sourcing a message PM_PME requester transaction packet to also support the message requester transaction packets used for the Power Fence protocol. This requirement applies whether the PCI Express device is on the platform or on an add-in card, and is part of the HP protocol. The Power Fence protocol supports the transition from the L0 to L2/3 Ready link state. See Chapter 15 for the discussion of the Power Fence protocol.

It is possible that the PCI Express device that is the source is associated with L2 or L3 to L0 link state transitions. That is, the message PM_ PME requester transaction packet is associated with link state transition for the Power Management protocol. See Chapter 15 for more information.

Configuration Address Space Registers for Hot Plug Protocol

The message baseline requester transaction packets for the HP protocol traverse only the specific link containing the slot with the exception of the message PM_PME requester transaction packets. A set registers in the configuration register block coordinates the message baseline requester transaction packets with the HP protocol. The location of these configuration registers and their implementation depends on whether they are located on the upstream or downstream of the slot as illustrated in Figure 16.2.

Though not identified in Figure 16.2, all PCI Express devices are required to support the PCI Management Capability Structure. This structure includes the Power Management Capabilities Register and the Power Management Status/Control Register. Relative to PCI Express's HP protocol, the PME_Status bit is used to ensure the delivery of a message PM_PME requester transaction packet to the Root Complex as discussed later in this chapter. See the book and specification information in Chapter 1 for more information.

Hot Plug Related Registers

Only the downstream ports of the Root Complex and switches with add-in card slots contain the following configuration block registers relative to the HP protocol:

- PCI Express Capabilities Register: Bit [8] =1b means the downstream port has an add-in card slot (may or may not implement HP).

- Slot Capabilities Register: See Table 16.2.

- Slot Control Register: See Table 16.2.

- Slot Status Register: See Table 16.2.

The upstream ports of switches, endpoints, and bridges connected directly to the add-in card slot contain the following configuration block registers relative to HP protocol:

- Device Capabilities Register: See Table 16.2.

Chapter 16: Hot Plug Protocol ■ 787

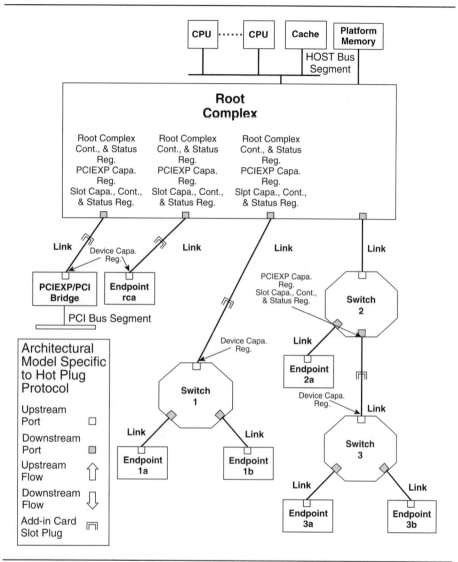

Figure 16.2 Hot Plug Registers Location

Only Root Complex ports contain the following configuration block registers relative to the HP protocol:

- Root Control Register: Bit [3] = 1b means an interrupt is generated when the message PM_PME requester transaction packet is received by the Root Status Register. The possible PME information relative to HP protocol is for the following Hot Plug protocol activities (also known as HP events):

- Attention button pressed
- Power Fault detected
- MRL sensor change
- Add-in card in slot change (PCI Express specification calls this "presence detect")

■ Root Status Register: See following discussion.

> Other configuration block registers may be associated with a specific PCI Express device but are excluded from this discussion if not specific to the HP protocol. See other chapters in this book for more information about all the registers in the configuration register block.

The configuration address space registers related to the HP protocol are summarized in Table 16.2, as is the protocol for interaction between the hardware and the HP software.

Table 16.2 Summary of Hot Plug Related Configuration Registers

Row	Name	Slot Capabilities Register	Device-Capabilities Register	Slot Control Register	Slot Status Register
1	Attention Button	Bit [0] = 1b: implement on chassis for this slot	Bit [12] = 1b:implement on add-in card (4)	Bit [0] = 1b: enables a wakeup event or interrupt when button pressed (1) (2)	Bit [0] = 1b:when button is pressed (2)
2	MRL Sensor Present (3)	Bit [2] = 1b:implement on chassis for this slot			

Table 16.2 Summary of Hot Plug Related Configuration Registers *(continued)*

Row	Name	Slot Capabilities Register	Device-Capabilities Register	Slot Control Register	Slot Status Register
3	MRL Sensor State (3)			Bit [2] = 1b: enables a wakeup event or interrupt when MRL sensor state changed	Bit [2] = 1b: when MRL sensor state changes Bit [5] = 1b MRL is open Bit [5] = 0b MRL is closed
4	Attention Indicator	Bit [3] = 1b: implement on chassis for this slot	Bit [13] = 1b:implement on add-in card (4)	Bits [7::6] Read = Status of indicator on/blink/off Write = set to on/blink/off (1) (2)	
5	Power Indicator	Bit [4] = 1b: implement on chassis for this slot	Bit [14] = 1b: implement on add-in card (4)	Bits [9::8] Read = Status of indicator on/blink/off Write = set to on/blink/off (1) (2)	
6	Power Controller Control (3)	Bit [1] = 1b: Slot has a power on/off controller		Read: Bit [10] = 1b Power is off Bit [10] = 0b Power is on Write: Bit [10] = 1b Turn Power off Bit [10] = 0bTurn Power on	
7	Hot Plug Surprise	Bit [5]= 1b:Add-in card can be removed from slot without warning			

Table 16.2 Summary of Hot Plug Related Configuration Registers *(continued)*

Row	Name	Slot Capabilities Register	Device-Capabilities Register	Slot Control Register	Slot Status Register
8	Hot Plug Capable	Bit [6] = 1b: Slot supports hot plug			
9	Add-in Card Present in slot			Bit [3] = 1b: enables a wakeup event or interrupt when add-in card present state changed (2)	Bit [3] = 1b: when add-in card present state changes (2) Bit [6] = 1b add-in card in slot (2) (6) Bit [6] = 0b slot empty (2) (6)
10	Physical slot number		Bits [31::19]: Slot number. If not associated with slot the value is 0b		
11	Power Fault detected at the slot (3)			Bit [1] = 1b: enables a wakeup event or interrupt when power fault is detected	Bit [1] = 1b: power fault is detected
12	Command Completed			Bit [4] = 1b: enables generation of hot plug interrupt when command is completed (See text) (5)	Bit [4] = 1b: When command is completed (See text) (5)
13	Hot Plug Interrupt Enable			Bit [5] = 1b enables generation of an interrupt for a HP event excluding Command Completed	

Table Notes:

1. When the activity occurs to change this bit, an associated message baseline requester transaction packet is transmitted in case the indicator exists on the add-in card.
2. Required for all slots that implement the HP protocol and are implemented on the chassis or add-in card.
3. Associated hardware is optional and only implemented on the chassis.
4. These bits are required to be implemented on all the upstream ports of switches and endpoints attached to a slot whether implementing HP protocol or not.
5. Required to be implemented independent of a slot or the HP protocol.
6. This bit is hardwired to 1b if downstream of Root Complex or switch is not connected to a slot.

Hot Plug Protocol

The PCI Express Capabilities Register has Bit [8] = 1b to indicate that a downstream port of the Root Complex or a switch has a slot. To indicate that the slot implements HP protocol, the Hot Plug Capable bit in the Slot Capabilities Register is set = 1b.

As noted in the Table 16.2, certain hardware components and their associated configuration register bits are required to be implemented on a slot that supports the HP protocol. These hardware components are: attention button, attention indicator, power indicator, and add-in card present. The locations (chassis or add-in card) of the attention button, attention indicator, and power indicator are indicated by the associated bits in the Slot Capabilities and Device Capabilities Registers. The location of the add-in card present mechanism is on the chassis.

The other hardware components MRL, MRL sensor, and power controller (fault) are only implemented on the chassis and are optional. These hardware components are also noted in Table 16.2.

One or multiple slots may be executing the HP protocol simultaneously. The HP protocol executions are independent of each other and independent of the slots not executing the HP protocol. The only interdependency is of an MRL is shared by two or more slots.

There are two levels of operation for the HP protocol. One level is when the Root Complex and switches with slots on the downstream ports and the add-in cards in the slots are operating with normal power. The normal power is relative to the D0 states of the functions in these PCI Express devices. The other level is with lower power of the D1, D2, and D3hot states of the functions in these PCI Express devices.

HP Protocol with Normal Power

The five possible HP events are: attention button pressed, power fault detected, MSI sensor changed, add-in card present state changes, and command completed. Except for the command completed HP event, the HP events are related to hardware. The occurrence of an HP hardware event sets to 1b the associated bits in the Slot Status Registers. With the exception of the attention button, the mechanisms for these hardware occurrences are on the chassis next to the slots. The attention button mechanism can be located on the chassis next to the slot or on the add-on cards. If the attention buttons are located on the add-in cards, the add-in cards transmit message Attention_Button_Pressed requester transaction packets to convey the attention button occurrence to the Bit [1] of the Slot Status Registers (Row 1 of Table 16.2). The one non-hardware HP event is command completed. Command completed is also part of the Power Management protocol. See the following further discussion.

When one of the Bits [4::0] in the Slot Control Register listed in Table 16.2 is set to 1b, an interrupt can be sourced for an HP event. However, the interrupt type cannot be distinguished. The type of interrupt (legacy or MSI) sourced is selected by other configuration registers in the downstream ports of the Root Complex and switches. See Chapter 19 for more information.

The HP interrupt is enabled when the Hot Plug Interrupt Capable Bit [5] is set to 1b in the Slot Control Registers.

When one of the Bits [3::0] in the Slot Control Register listed in Table 16.2 is set to 1b, a wakeup event is sourced for an HP event (hardware only). An HP wakeup event sourced is a message PM_PME requester transaction packet. This message is in addition to an interrupt for the associated HP event (hardware only) if an HP interrupt is enabled. The reaction by the Root Complex to the receipt of this message is dependent on other configuration registers. See Chapter 15 for more information.

Hardware Related HP Events

The four HP events related to hardware are attention button pressed, power fault detected, MSI sensor changed, and add-in card present state changes. In response to these HP hardware events, an interrupt is sourced if an HP interrupt is enabled and the associated wakeup event or interrupt bits (Bits [3::0]) in the Slot Control Register are set = 1b (Rows 1, 3, 9, 11, and 13 of Table 16.2). A wakeup event also occurs in response to these HP hardware events.

A wakeup event is simply the sourcing of a message PM_PME requester transaction packet by the switch. In the case of a Root Complex downstream port, it is the "virtual" internal sourcing of this message to itself. See Chapter 15 for more information. Note: The Wakeup protocol discussed in Chapter 15 is not a wakeup event as discussed here for the HP hardware event.

The bits in the Slot Status Registers are updated whether an interrupt or wakeup event is sourced or not. Consequently, HP software can access the Slot Status Registers at any time in a polling protocol.

In response to the interrupt or wakeup event, or from the polling of software, the HP software services the HP event. Part of the servicing clears the appropriate bits in the Slot Status Registers. In the case of the indicators, the Slot Control Registers can be read to determine present status and written to affect the operation of the indicators with the transmission of the appropriate message baseline requester transaction packets. The action of the HP software writing to the bits defined in Table 16.2 results in message baseline requester transaction packets as noted in that table. When the indicator(s) is located on the add-in card, the message baseline requester transaction packets transmitted controls the indicator(s). When the indicator(s) is located on the chassis, its control is direct and not affected by the simultaneous transmission on of the message baseline requester transaction packets.

Command Completed HP Event

The HP event not related to hardware is Command Completed. The Command Completed bit in the Slot Status Register is set to 1b by the hardware mechanisms for the attention indicator, power indicator, or power controller when the associated commands issued by the HP software are completed. The HP software sets to 0b the Command Completed bit when a command is issued relative to the attention indicator, power indicator, or power controller. The HP software issues commands to these entities via writing to the associated bits in the Slot Control Register (Rows 4, 5, 6 of Table 16.2). The HP software must wait until the Command Completed bit in the Slot Status Register is set to 1b before issuing the next command related to the attention indicator, power indicator, or power controller. One exception is if the Command Completed bit in the Slot Status Register is not set to 1b within one second of HP software setting it to 0b, the HP software may repeat the command or issue a new command for the attention indicator, power indicator, or power controller. Under this situation, the Command Completed bit in the Slot Status Register is set to 1b; it may not be related to the most recent command.

When the Command Completed bit in the Slot Status Register is set to 1b, an interrupt is sourced if the Command Completed bit in the Slot Capabilities Register is set = 1b and if the HP interrupt is enabled. Otherwise, the HP software can poll the Command Completed bit in the Slot Status Register. There is no wakeup event generated for this HP non-hardware event.

HP Protocol with Sleep or Lower Power State

Five HP events are possible: attention button pressed, power fault detected, MSI sensor changed, add-in card present in state changes, and command completed. In sleep and low power conditions, the reporting of the command completed HP event is not possible because it requires the use of an interrupt. Also, in a sleep or lower power condition, the interrupt cannot be used for the other HP events. However, the wakeup event can be used for these other HP events in the following two situations. (As discussed previously, the wakeup event is the sourcing of a message PM_PME requester transaction packet.)

One situation is when the function on the switch associated with the HP slot is in the D1, D2, or D3hot state. Of the HP events those related to hardware (HP hardware events) are attention button pressed, power fault detected, MSI sensor changed, and add-in card present state changes. None of these HP hardware events can use an interrupt but each of these events will cause a wakeup event (message PM_PME requester transaction packet).

The other situation is when the Root Complex is in sleep, or the function on the switch associated with the HP slot is in the D3cold state (Vaux retained). If any of these HP hardware events occurs, neither the interrupt or wakeup event can be used. The Root Complex or the switch will have to execute the Wakeup protocol in order for main power to be applied. Once main power is applied the HP hardware event can be reported by an interrupt and the wakeup event.

In the above two situations, the PCI Express devices on the add-in card may or may not have main power or may not may not have Vaux. In all situations, the D states of the functions in the PCI Express devices in the add-in card must be compatible with the power level of the Root Complex or the "D" states of the related function in the switch. It is only the Root Complex or the switch that is responsible for reporting the HP hardware event.

If the PCI Express device that will source the message PM_PME requester transaction packet is in sleep (L2), it will have to execute the Wakeup protocol. See Chapter 13 for more information about the Wakeup protocol.

Other Considerations for HP Protocol

The information discussed up to this point provides the basics for the HP slot. Here are some additional considerations.

Electromechanical Interlock

In addition to the previously discussed manually operated retention latch (MRL), each slot may optionally be outfitted with an electromechanical interlock (EI). The EI can operate on either the add-in card in a slot directly or via an MSL. The purpose of an EI is to prevent add-in card removal until the system software permits it. There are many reasons for the system software to prevent add-in card removal, including as part of a security protocol. Unlike the other Hot Plug hardware components, the PCI Express specification does not define a software interface for an EI. It is the responsibility of the platform designer to provide a port for software to read and write to operate the EI.

Slot Numbering

The Slot Capabilities Register also includes bits for slot numbering and Hot Plug/Removal Surprise bit.

The slot numbering is optional. Each slot for an add-in card (hot plug or not) receives a slot number (per the PCI Hot Plug 1.1 Section 1.5 and if applicable a chassis number PCI Bridge 1.1 Section 13.4). The number is reported to software via the Slot Capabilities Register. For obvious reasons, the slot numbers must be unique across the chassis.

Hot Plug Surprise Bit

The Hot Plug Surprise bit in the Slot Capabilities Register is used to indicate that the slot is designed to support the removal of an add-on card without any prior notification to the platform.

Other Registers

Other configuration registers are indirectly related to the Slot Power protocol; they are used in conjunction with other Power Management Events. See Chapter 15 for more information.

- Device Control Register: Bit [10]
- Device Status Register: Bit [4]
- PCI Express Enhanced Capability Header Register
- Data Select Register
- Data Register
- Power Budget Capability Register

ACPI and OSHP Methods

There are additional Hot Plug protocol considerations beyond the scope of the PCI Express specification. These relate to implementation of ACPI and OSHP. See section 6.7.8 of the PCI Express specification for more information.

Chapter 17

Slot Power Protocol

This is the third of the three chapters related to power consumption in a PCI Express platform. Chapter 15 focuses on Power Management of any PCI Express device on the platform or add-in card on the link. Chapter 16 focuses on the Hot Plug protocol of an add-in card. This chapter focuses on the power allocated to a slot and its associated add-in card. This book has grouped the topics of these chapters under the concept of the Power Management Event.

Introduction to Slot Power

The primary purpose of the Slot Power protocol is to provide Power Management Event (PME) software a means to control power allocation to add-in cards and associated slots.

The Slot Power (SP) protocol is supported by PME, (Power Management Event) software. In addition to supporting the SP protocol, PME software also supports Power Management and Hot Plug software for their respective protocols. As used in this chapter, the portion of the PME software relative to the SP protocol is Slot Power (SP) software.

Slot Power (SP) Protocol

The ability to program the level of current, and thus power allocated per add-in card, is unique to PCI Express. The PCI and PCI-X specifications do not have a compatible feature. In order to support the SP protocol the PCI Express specification defines a message baseline request transaction of several bits in registers in the configuration register blocks.

Power Limits

Each slot for an add-in card has a power limitation related to cooling issues. The limit is based on the size of the add-in card and the number of lanes within the connecting link in the slot plug. The maximum amount of power drawn by the add-in on the combination 3.3 volt and 12 volt power rails is listed in Table 17.1. If only one voltage (power rail) is used, the power limit applies to that one voltage.

Table 17.1 Add-in Card Cooling Power Limits

Add-in Card Size and Application	Link Size in Lanes		
	x1 (3) (Watts)	x4 or x8 (3) (Watts)	x16 (2)(3) (Watts)
Standard Height Card for non-server at powered-on Limited to half length cards (1)	10	25	25
Standard Height card for server at powered-on (1)	10	25	25
Standard Height card for server once config. write permits higher power (1)	25 For this much power add-in card is > 7 inches in length	25	75
Low profile card at powered-on Limited to half length cards	10	10	25

Table Notes:

1. The three basic add-on card sizes defined by PCI and PCI-X are retained by PCI Express: low profile, short length (variable height), and standard length. The latter two of these are collectively defined as "standard height card" in this table.
2. At the print time of this book, the connector sizes defined were x1, x4, x8, and x16. Consequently, x32 links cannot be implemented.
3. Other modules that encapsulate add-in cards over a cable have been proposed. As of the print date of this book the final specifications were not available.

The power limits defined in Table 17.1 are based on two assumptions. First, for 10 watts or lower, the cooling is done by natural means. Second, for 25 watts or greater, it is assumed that the add-in card has mechanical cooling. As listed in the table for server applications, it is possible for the SP software to program power consumption to be greater than 25 watts, with a limit of 75 watts. To accomplish this, the SP software programs specific registers in the configuration address space. These are discussed later in this chapter.

The PCI Express Mini Card is a replacement for the PCMCIA and Cardbus in mobile applications. The slot for the PCI Express Mini Card only implements an x1 link. Table 17.2 defines the power for the PCI Express Mini Card, which does not include a 12-volt power rail but does include a 1.5-volt power rail. Due to the battery operation of a mobile application, the PCI Express Mini Card includes both Peak and Normal currents, which are listed as maximum values in Table 17.2. Also, due to the battery operation, the voltage tolerance is ± 9 percent for 3.3 volts and ± 5 percent for 1.5 volts. The wattages listed in Table 17.2 reflect the associated current multiplied by the maximum voltage, with the exception of the entry for auxiliary power.

Table 17.2 Add-in Card Cooling Power Limits for PCI Express Mini Card

Add-in Card Size and Application		Link Size in Lanes x1				
		Peak		Normal		Auxiliary Power
		Current milliamps	Current milliwatts	Current milliamps	Current milliwatts	Normal Only applies to the D3 state
PCI Express Mini Card	3.3 volt	1000	3597	750	2697.8	250 milliamps @ 3.3 volts = 825 milliwatts
	1.5 volt	500	787.5	375	590.6	Not Applicable

Power Associated Message Transactions and Configuration Registers

The PCI Express specification permits Slot Power (SP) software to establish additional power limits above than those listed as the maximum in Tables 17.1 and 17.2. As shown in Figure 17.1, a slot can be connected to the downstream ports the Root Complex and any switch. These downstream ports communicate power limits to the upstream ports of the add-in cards via message Set_Slot Power_Limit requester transaction packets sourced from the Root Complex or switch. The sourcing of the message Set_Slot Power_Limit requester transaction packet and its contents are established by the SP software programming the configuration register blocks of the downstream ports associated with the add-in card slot plugs.

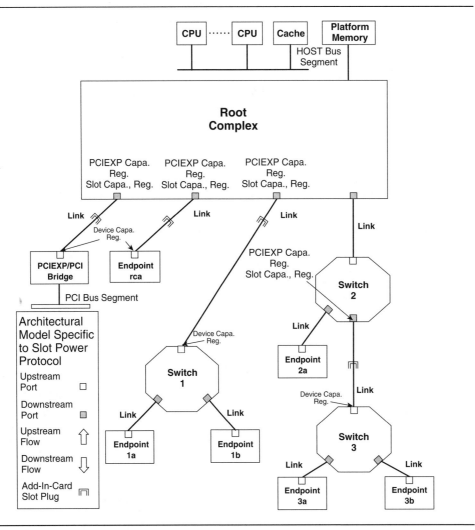

Figure 17.1 Slot Power Related Register Location

In Figure 17.1, note that a slot can be placed downstream of any downstream port. Conversely, the add-on card plug for the slot is upstream of any PCI Express device that supports an upstream port.

Also note that the until the registers related to slot power limit (outlined in Table 17.4) in the configuration register blocks for that specific slot are programmed, the power consumption of the add-in card associated with that specific slot is limited by the powered-on values in Tables 17.1 and 17.2.

Message Requester Transaction

The message Set_Slot Power_Limit requester transaction packet is summarized in Table 17.3.

Table 17.3 Message Power Requester Transactions

Name	Destination of message packet	Source of message transaction packet
Set_Slot Power_Limit	Legacy endpoint (4)	Root Complex (1)
	PCI Express endpoint (4)	Switch (2)
	Switch (3)	
	PCI Express/PCI Bridge (4)	
	PCI Express/PCI-X Bridge (4)	

Table Notes:
1. Defined for the Root Complex's downstream ports connected to the slot.
2. Defined for a switch's downstream ports connected to the slot.
3. Defined for a switch with its upstream port connected to the slot.
4. Defined for an endpoint or bridge with its upstream port connected to the slot.

See further discussion of the message Set_Slot_Limit requester transaction at the end of this section.

Configuration Address Space Registers for Slot Power

The SP software power limits are established via the registers related to slot power limit in the configuration register block of the Root Complex's and switches' downstream ports connected to the slots. The locations of these configuration registers are summarized in Figure 17.1.

> Other configuration registers may be associated with the Slot Power protocol with a specific PCI Express device but are excluded in this discussion if not specific to the Slot Power protocol. See Chapters 15 and 16 for more information about the other associated registers in the configuration register blocks.

Registers Related to Slot Power Limit Only the downstream ports of the Root Complex and switches with add-in card slots contain the following configuration registers relative to the SP protocol:

- PCI Express Capabilities Register: Bit [8] = 1 means the downstream port has an add-in card slot.
- Slot Capabilities Register: See Table 17.4.

Only the upstream port of the switches, endpoints, and bridges contain the following configuration register relative to the SP protocol:

- Device Capabilities Register: See Table 17.4.

Table 17.4 summarizes the bits within the registers related to slot power limit.

Table 17.4 Summary of Slot Power Related Configuration Registers

Row	Name	Slot Capabilities Register	Device Capabilities Register
1	Slot Power Limit Value	Bits [14::7] See Text (1)	
2	Slot Power Limit Scale	Bits [16::15] Multiplier for Slot Power Limit Value 00b = 1x 01b = 0.1x 10b = 0.01x 11b = 0.001x See Text (1)	
3	Captured Slot Power Limit Value		Bits [25::18] See Text
4	Captured Slot Power Limit Scale		Bits [27::26] Multiplier for Captured Slot Power Limit Value 00b = 1x 01b = 0.1x 10b = 0.01x 11b = 0.001x See Text

Table Notes:

1. When these bits are written to a message Set_Slot_Power_Limit requester transaction packet is transmitted from the downstream port.

Slot Power Limit Protocol Implementation

The SP software programs the Slot Capabilities Register to establish the power limits beyond the limits established per Tables 17.1 and 17.2. Writing to the Slot Power Limit Value or Slow Power Limit Scale fields (Rows 1 and 2 in Table 17.4) in the Slot Capabilities Register in the ports upstream of the slot results in the transmission of message Set_Slot Power_Limit requester transaction packets. The message Set_Slot Power_Limit requester transaction packets conveys to the upstream port of the add-in cards the additional power limits via the Captured Slot Power Limit Value or Captured Slot Power Limit Scale fields (Rows 3 and 4 in Table 17.4) in the Slot Capabilities Register. The message Set_Slot Power_Limit requester transaction packets are also transmitted whenever the associated link transitions from L_Init or L_Active link activity state (See Chapter 13).

The Slot Power Limit Value and Captured Slot Power Limit Value bits are multiplied by the Slot Power Limit Scale and Captured Slot Power Limit Scale bits, respectively. The resulting numbers represent the maximum watts available to the add-in card in the slot. The information contained within the Captured Slot Power Limit Value and Captured Slot Power Limit Scale bits (Rows 3 and 4 in Table 17.4) can be read by add-in card device driver software to adjust the power of the add-in card resources.

There are several requirements placed on the PCI Express devices on the add-in card associated with the slot:

- The add-in card must always consume less power than specified in Tables 17.1 and 17.2, until a message Set_Slot Power_Limit requester transaction packet. If the add-in card cannot meet this requirement, it must disable. The add-in card's device driver must inform the Slot Power software of this problem. The Slot Power software must address this problem if possible. Note the value for x1 link is defined as the lowest power limit for each add-in card size, and not the values associated with x4, x8, or x16 links, even if the wider links are used.

- The add-in card must always consume less power than specified by the most recently received message Set_Slot Power_Limit requester transaction packet. If the add-in card cannot meet this requirement it must disable. The add-in card's device driver must inform the Slot Power software of this problem. The Slot Power software must address this problem, if possible. If the message Set_Slot Power_Limit requester transaction packet is higher than the power than specified in Tables 17.1 and 17.2, the add-in card can operate at the higher power level, on the assumption that the proper mechanical cooling is being provided.

- If the add-in card's power is always below the lowest power limit for the add-in card size per Tables 17.1 and 17.2, the add-in card can optionally ignore and discard (no error) message Set_Slot Power_Limit requester transaction packet. In this situation, the reading of the Captured Slot Power Limit Value and Captured Slot Power Limit Scale bits have a "0" value. Note: the value for x1 link is defined as the lowest power limit for each add-in card size, and not the values associated with x4, x8, or x16 links, even if the wider links are used.

- The downstream ports of the Root Complex or a switch can never transmit a message Set_Slot Power_Limit requester transaction packet that specifies a power level less than the lowest power limit for the add-in card size per Tables 17.1 and 17.2.

Message Baseline Requester Transaction Packets: Slot Power Messages

The purpose of slot power messages is to set the power limits for add-in cards connected to a slot downstream of a Root Complex or a switch. As previously discussed, the one message slot power requester transaction packet is the Message Set_Slot_Power_Limit requester transaction packet.

The message Set_Slot_Power_Limit requester transaction packets are only transmitted downstream from the Root Complex and switches. The recipients are the upstream ports of the switches, endpoints, and bridges on the add-in cards plugged into the platform's slots. The protocol of message Set_Slot_Power_Limit requester transaction packet is discussed in Chapter 6.

The message Set_Slot_Power_Limit requester transaction packet is transmitted whenever the Slot Capabilities Register is written to in a downstream port associated with a slot, or when the link transitions from Link_DOWN to Link_UP. (See Chapter 13 for more information.) The result of the transmission is the update of the "same" bits in the Device Capabilities Register as those in the Slot Capabilities Register. The protocol one DWORD in the Data field of the packet is as follows:

- The one DWORD in the Data field is defined as follows:
    - Bits [31::10] are transmitted as "0" and are ignored by the receiving port.
    - Bits [9::8] are copies of Slot Power Limit Scale of the Slot Capabilities Register and end up in the Captured Slot Power Limit bits of the Device Capabilities Register.
    - Bits [7::0] are copies of Slot Power Limit Value of the Slot Capabilities Register and end up in the Captured Slot Power Limit Value bits of the Device Capabilities Register.

Any component that resides on the platform or on an add-in card with power less than the lowest power specified for its size can optionally hardwire the associated bits to "0" in the Device Capabilities Register. Consequently, the message Set_Slot_Power_Limit requester transaction packet data received can be ignored.

Associated Slot Power Items

The focus of this chapter has been on slot power limits. One additional power level is "off," if a power controller is implemented on the slot. This additional slot "power limit" is discussed Chapter 16 as part of the Hot Plug discussion.

Independent of the major power supplied to a slot, standby power is supplied in the form of Vaux. The Vaux permits a PCI Express device to operate at some level when main platform power is removed (the device is powered off). See Chapter 15 for more information.

Chapter 18

Advanced Switching

The advanced peer-to-peer link, and communication information based on the Advanced Switching protocol, was included in the original PCI Express specification revision 1.0. It was subsequently removed from the specification except for references in certain key software registers. The information contained in this chapter reflects some of the preliminary information that was removed from the original PCI Express specification. At the print date of this book, the publishing date for Advanced Switching protocol (also known as the "Advanced PCI Express Packet Switching Specification") had not been set. The inclusion of this chapter and other references in this book are simply placeholders to act as a reference to the eventual Advanced Switching protocol information.

Introduction to Advanced Switching

Transactions flowing through a switch can port from one downstream port to another downstream port between two endpoints, two bridges, or a combination of bridges and endpoints. This transaction flow is defined as peer-to-peer. The Root Complex does not support peer-to-peer transactions; consequently, a switch provides the inflection point. The inflection point is the switch where the transaction packets of the upstream flow become the transaction packets of the downstream flow. The two participants are endpoints or bridges that may or may not be connected to the same switch.

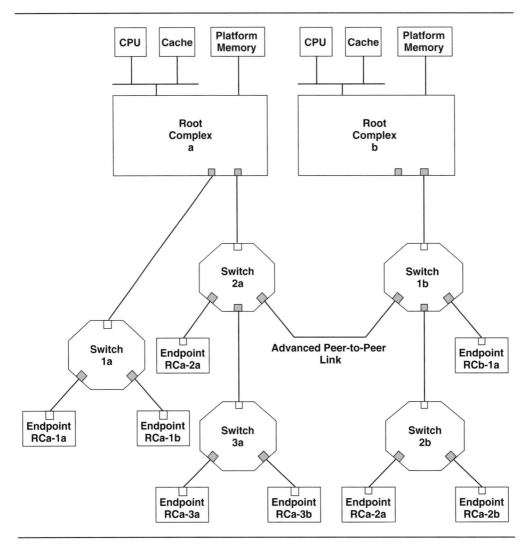

Figure 18.1 Peer-to-Peer and Advanced Peer-to-Peer

The participants of peer-to-peer transactions are shown in Figure 18.1. The peer-to-peer transaction participants for the Root Complex "a" (RCa) are endpoints RCa-1a and RCa-1b, endpoints RCa-3a and RCa-3b, endpoints RCa-2a and RCa-3b, and endpoints RCa-2a and RCa-3a. The peer-to-peer transaction participants for Root Complex "b" (RCb) are endpoints RCb-1a and RCb-2b, endpoints RCb-2a and RCb-2b, and endpoints RCb-1a and RCb-2a. These peer-to-peer transactions can only be executed within a specific PCI Express fabric. The peer-to-peer transactions access memory, I/O address spaces, and in-band communications. In-band communications are defined as message baseline and vendor-defined requester transactions. Bridges are not shown in this example, but bridges can be participants in peer-to-peer transactions.

PCI Express defines an *advanced peer-to-peer link* (also known as cross-link), which permits communication between different PCI Express fabrics. The advanced peer-to-peer link's Physical Layer is identical to the Physical Layer defined for links within the PCI Express fabric. The Physical Layer for an advanced peer-to-peer link provides FRAMING around a TLP packet, and lane width configuration of the link. The Physical Layer for an advanced peer-to-peer link also has the same electrical and timing requirements from the same protocol as the links within the PCI Express fabric. The differences between links internal to a specific PCI Express fabric and advanced peer-to-peer links are in two areas: First, the links within a specific PCI Express fabric provide communication between PCI Express devices on a specific PCI Express fabric, whereas advanced peer-to-peer links provide communication between PCI Express devices of different PCI Express fabrics. Second, PCI Express devices connected to links within a specific PCI Express fabric communicate with memory transaction, I/O transaction, and in-band communications, while PCI Express devices connected to advanced peer-to-peer links communicate with advanced peer-to-peer communications.

Advanced Peer-to-Peer Links

The implementation of the advanced peer-to-peer links is optional and the protocol is as follows:

- The advanced peer-to-peer link is between any one switch's downstream port of one PCI Express fabric to any one switch's downstream port of another PCI Express fabric. Neither of the downstream ports can be a Root Complex's downstream port.

- The advanced peer-to-peer links are implemented in the same fashion (multiple lanes, performance, electrical, and so on) as other links within a specific PCI Express fabric.

- The two participants of advanced peer-to-peer communications are endpoint(s) or bridge(s) from each PCI Express fabric. One participant is the source of the advanced peer-to-peer communications. The other participant is either a single endpoint or bridge, or multiple endpoints and bridges.
    - An endpoint or bridge of one PCI Express fabric can transmit message advanced switching requester transaction packets to one or multiple (broadcast) endpoints or bridges of the other PCI Express fabric.
    - Neither of the participants are a Root Complex.

- The message advance switching requester transaction packets between endpoints or bridges cannot traverse the Root Complex. That is, the Root Complex does not support porting of any transaction packets between its downstream ports. As illustrated in Figure 18.1 endpoints RCa-2a, RCa-3a, or RCa-3b can communicate with RCb-1a, RCb-2a, or RCb-2b. There is no communication that includes RCa-1a or RCab-1b.

- Both hierarchies linked by advanced peer-to-peer link must be initialized and configured before message advanced switching requester transaction packets can traverse the link.

- Multiple advanced peer-to-peer links can exist between multiple Root Complexes.

- Multiple advanced peer-to-peer links can exist between any two PCI Express fabrics.

Advanced Peer-to-Peer Communications

The in-band communications previously discussed for links within a specific PCI Express fabric consist of message baseline requester transactions and message vendor-defined requester transactions. The communication mechanism for advanced peer-to-peer links is defined as advanced peer-to-peer communication. Advanced peer-to-peer communication consists only of message advanced switching requester transactions.

The protocol for message advanced switching requester transaction packets is as follows:

- Message advanced switching requester transaction packets follow the same basic protocols as the message baseline requester transactions. One difference is the use of a unique addressing scheme. Unlike the addressing schemes used by the message and vendor-defined requester transactions, message advanced switching requester transactions use a 15-bit global address space defined as the Route Identifier (RID).

- Another difference between message baseline or vendor-defined requester transactions and message advance switching requester transactions is the contents. Each message advanced switching requester transaction packet contains a Header field that includes RID for addressing, TC [2::0] for traffic class, TD for error support, ATTRI[1::0] for snooping and ordering, and LENGTH [9::0] for providing length of data field if included.

- Message advanced switching requester transaction packets encapsulate other packets. These other packets can be of any format. Part of the encapsulation includes information about the encapsulation protocol. This protocol has yet to be defined.

- The other attribute not shared with all message baseline or vendor-defined requester transaction packets is the ability for message advanced switching requester transaction packets to be sourced from one point (endpoint or bridge) with a destination to one or multiple points (broadcast).

Chapter 19

Interrupts

This chapter discusses the interrupt protocol adopted from PCI Express from PCI and PCI-X specification. Elements of the PCI Express interrupt must be maintain backwards compatibility with PCI and PCI-X software. Due to the link structure of PCI Express versus the bus segment structure of PCI and PCI-X, PCI Express implements message interrupt requester transactions instead of interrupt signal lines.

Introduction to Interrrupts

Personal computers originally supported discrete interrupt signal lines (INTx#) that interconnected between PCI interrupt sources and the interrupt controller (such as an Intel® 82559 Fast Ethernet Controller). Historically, each INTx# signal line was a point-to-point signal line with one INTx# signal line per interrupt source between the interrupt controller and the interrupt source. PCI and PCI-X expanded the concept by introducing the sharing of each of the INTx# signal lines among several interrupt sources. Through open collector technology, one or more interrupt sources can pull the shared INTx# signal line low to "0" to indicate an interrupt request. The PCI and PCI-X bus protocol required the HOST bus segment CPU to execute interrupt acknowledge transactions to the interrupt controller to obtain the interrupt vector information for any pending interrupt requests. The HOST bus segment CPU would access the appropriate configuration address spaces of the interrupt controller and the interrupt source to service and acknowledge the interrupt.

The concept of the *Message Signal Interrupt* (MSI) was introduced in PCI specification 2.2, and MSI was subsequently supported by the PCI-X revisions. MSI uses memory write transactions for a PCI or PCI-X device to request interrupt service. The MSI memory write transactions can address any PCI or PCI-X device including the interrupt controller to request interrupt service.

PCI Express implements interrupt protocol on two levels: Legacy and MSI. MSI is the preferred interrupt protocol, but Legacy is supported for backward compatibility with PCI and PCI-X software and associated PCI Express devices based on PCI and PCI-X devices. The Legacy Interrupt protocol is also implemented by PCI Express to support boot devices, the associated simplicity of BIOS interrupt configuration, and interrupt servicing. Consequently, the Root Complex's internal interrupt controller must be compatible with the Legacy Interrupt protocol and MSI.

Legacy Interrupt Protocol Implementation

The Legacy Interrupt protocol (LI) can only be implemented by legacy endpoints and bridges. PCI Express endpoints and switches (except as noted in the following section) do not implement the LI protocol. The original INTx# signal lines used for interrupt requests for PCI and PCI-X have been integrated into the message baseline requester transaction protocol of PCI Express. A legacy endpoint or bridge that wants to request interrupt service transmits one of the two types of message interrupt requester transaction packets upstream towards the Root Complex. The porting of the message interrupt requester transaction packets by switches is in support of the LI protocol, and is not an implementation by the switches of the LI protocol to request interrupt service. The upstream port of each switch must maintain records of assertions and deassertions for each interrupt (A, B, C, and D) independently for each downstream port of each switch.

> The interrupt protocol for INTx# signal lines and MSI protocol are maintained downstream of the bridges, according to the PCI and PCI-X specifications. The support of the INTx# signal lines downstream of the bridge requires the interrupt controller in the Root Complex. Consequently, the bridges must translate the INTx# signal lines to message interrupt requester transaction packets flowing upstream. The message interrupt requester transaction packets flow only upstream.

Message Interrupt Requester Transaction Packets

The support of the LI protocol by the legacy endpoints and bridges is the same as for PCI and PCI-X, including the implementation of the same interrupt request-related registers in the configuration register blocks. The support of LI protocol via message interrupt requester transactions is enabled and disabled by the Command Register in the configuration register block within legacy endpoints and bridges. If Bit 10 of the Command Register is set to 1b, the PCI Express device cannot transmit message interrupt requester transactions. If Bit 3 of the Status Register in the configuration register block within a legacy endpoint and bridge is set to 1b, the PCI Express device has an interrupt pending. Interrupt pending means a message interrupt requester transaction packet was transmitted. The only unique LI protocol requirement by PCI Express is the use of message interrupt requester transaction packets to emulate the INTx# signal lines of PCI and PCI-X as follows.

The INTX# signal lines in PCI and PCI-X platforms are labeled A, B, C, and D. The LI protocol in PCI Express platforms distinguishes message interrupt requester transaction packets as representing A, B, C, and D by the encoding internal to the packet. As shown in Table 19.1, the encoding of the TAG HDW determines whether the message interrupt requester transaction packet represents interrupts A, B, C, or D. Notice also that the Requester ID field of the packet only provides the BUS# in the Header field. The DEVICE# and FUNCTION# are both RESERVED, because only DEVICE# 0 can be connected on the downstream end of the link. Also, specific interrupts for a PCI Express device's functions are internally shared, and according to the PCI and PCI-X specification, the interrupt signal line is defined by the bus segment with PCI or PCI-X devices on the bus segment sharing the interrupt signal lines. See Chapter 6 for more information on message interrupt request transaction packets.

Table 19.1 Interrupt Related Message

Name	TAG HDW Message Code Field [7::0]	TYPE HDW TC [2::0]	Data Field? In ATTRI. HDW Length [9::0]	Requester ID Field Entries
Assert_INTx	A = 0010 0000b B = 0010 0001b C = 0010 0010b D = 0010 0011b	000b	No Data Field Length = 00 0000 0000b = Reserved	BUS# = Actual DEVICE#= Reserved FUNC# = Reserved
Deassert_INTx	A = 0010 0100b B = 0010 0101b C = 0010 0110b D = 0010 0111b	000b	No Data Field Length = 00 0000 0000b = Reserved	BUS# = Actual DEVICE# = Reserved FUNC# = Reserved

Message Interrupt Requester Transaction Packet Flow

The operation of the INTx# signal lines in PCI and PCI-X is by the assertion and deassertion of a shared signal lines by open collector technology. PCI Express distinguishes two types of message interrupt requester transaction packets as assert or deassert by the encoding internal to the packet, as shown in Table 19.1. All message requester interrupt transaction packets flow upstream with the Root Complex as the eventual destination. If the legacy endpoints or bridges are directly connected to the Root Complex, the packet flows one-to-one directly upstream. However, the packet flow is not one-to-one if switches exist between the Root Complex and the legacy endpoints or bridges.

For emulation of a specific INTx# signal line (A, B, C, or D are each specific interrupts), the switches operate on behalf of the open collector technology. The message interrupt requester transaction packets are transmitted by the legacy endpoint or bridge on the downstream side of a link to the port on the upstream side of the link. If the port on the upstream side of the link is connected to the Root Complex, the message requester transaction packet has reached its final destination. If the port on the upstream side of the link is connected to a switch, the message interrupt requester transaction packet has reached its destination relative to the link but not the final destination. The switch must gather the received message interrupt requester transaction packets of a specific interrupt (A, B, C, or D) received on its downstream ports, and transmit from its upstream port representative message interrupt requester transaction packets for that specific interrupt as specified by the protocol discussed in the following section.

Legacy Interrupt Protocol Implementation by Legacy Endpoints, Bridges, and Switches

Two types of message interrupt requester transaction packets emulate the assertion and deassertion of a specific interrupt. These two types are the message Assert_INTx and the message Deassert_INTx requester transaction packet. The discussion of the LI protocol for the two ports on a link is illustrated by specific interrupts A and B (these represent the virtual interrupt signal lines A and B). In order to simplify the examples, the remapping of the specific interrupts is not included. See the remapping discussion later in this chapter.

Assertion Protocol Example of Interrupts A and B

Each upstream port of a legacy endpoint, bridge, and switch maintains the logical level of each specific virtual interrupt signal line: A and B. In the case of the upstream port of a switch, information about interrupts A and B are maintained for *each* downstream port. Assume the logical level of the virtual interrupt signal line for this example begins as deasserted for both interrupts A and B.

First Assertion of A and B When the upstream port determines that interrupt A has been asserted, it transmits upstream following these considerations:

- **Legacy endpoint:** If the upstream port is connected to a legacy endpoint, the assertion represents a request for interrupt service and the assertion of interrupt A is recorded.

- **Switch:** If the upstream port is connected to a switch, the assertion represents receipt of a message Assert_INTA requester transaction on a specific downstream port and the switch records the assertion of interrupt A including the specification of the downstream port.

- **Bridge:** If the upstream port is connected to a bridge, the assertion represents the assertion of the INTA# signal line on the downstream side of the bridge.

When the upstream port determines that interrupt B has been asserted, it transmits upstream a message Assert_INTB requester transaction. This message requester transaction does not affect the upstream port's records that interrupt A is presently asserted. The balance of the protocol is the same as outlined in the previous items listed except that it is applied to interrupt B.

Subsequent Assertion of A and B If the upstream port determines that interrupt A has been asserted again, it does not transmit a message Assert_INTA requester transaction upstream, because it has recorded that interrupt A has been asserted.

- **Legacy Endpoint:** If the upstream port is connected to a legacy endpoint, the assertion represents a request for a second interrupt service and is so recorded.

- **Switch:** If the upstream port is connected to a switch, the assertion represents receipt of another message Assert_INTA requester transaction. In the case of a switch, if the message Assert_INTA requester transaction received is on the same downstream port as the previous message Assert_INTA requester transaction, the upstream port records a second assertion for interrupt A for that specific downstream port. If the message Assert_INTA requester transaction received is on a different downstream port than the downstream port of the previous message Assert_INTA requester transaction packet, the upstream port records an assertion for interrupt A for that other specific downstream port. For this example assume that the second message Assert_INTA requester transaction packet is on the same downstream port as the previous message Assert_INTA requester transaction packet.

- **Bridge:** If the upstream port is connected to a bridge, it is not possible to determine a second assertion of INTA# signal line on the downstream side of the bridge because of open collector implementation.

> The upstream port of a legacy endpoint or switch has recorded two assertions of interrupt A and one assertion of interrupt B. In the case of a bridge, the number of assertions of the INTA# and INTB# signal lines are not recorded; the bridge only knows that INTA# and INTB# signal lines are both asserted. In all cases the upstream port of the legacy endpoint, switch, or bridge has transmitted upstream one message Assert_INTA requester transaction packet and one message Assert_INTB requester transaction packet. In the case of the single message Assert_INTA requester transaction packet, it mimics an open collector interrupt signal line where several assertions in parallel only appear as a single assertion.

Deassertion Protocol Example of Interrupts A and B

The deassertion of either interrupts A or B in this example results in the transmission of either message Dessert_INTA or message Dessert_INTA requester transaction packets, depending on the number of recorded assertions as follows.

- **Legacy Endpoint:** For a legacy endpoint, the deassertion of interrupt B results in the upstream port canceling the one recorded indication that interrupt B is asserted. In that there was only one recorded assertion, a message Deassert_INTB requester transaction packet is transmitted to reflect that no recorded assertions are outstanding. For a legacy endpoint, the deassertion of interrupt A results in the upstream port canceling one of the recorded indications that interrupt A is asserted. The one remaining recorded indication that interrupt A is asserted prevents the transmission of a message Deassert_INTA requester transaction packet. A subsequent deassertion of interrupt A results in the upstream port canceling the one remaining recorded indication that interrupt A is asserted and the transmission of a message Deassert_INTA requester transaction packet.

- **Switch:** For a switch, the receipt of a message Deassert_INTB requester transaction packet on the same downstream port as the previous message Assert_INTB requester transaction packet results in the upstream port canceling the one recorded indication that interrupt B is asserted. Subsequently, the upstream port transmits a message Deassert_INTB requester transaction packet.

If the receipt of a message Deassert_INTB requester transaction packet was on another downstream port with no recorded assertion of the interrupt B, the packet is ignored (not treated as an error) and no message Deassert_INTB requester transaction packet is transmitted by the upstream port. Switches must keep separate records of assertion of each specific interrupt received on each of its specific downstream ports. Similarly, the deassertion information received by switches on each of its downstream ports must be associated with a previous assertion relative to a specific interrupt and a specific downstream port. For a switch, the receipt of a message Deassert_INTA requester transaction packet on the same downstream port as the previous message Assert_INTA requester transaction packet results in the upstream port canceling one of the recorded indications that interrupt A is asserted. The upstream port subsequently does not transmit a message Deassert_INTA requester transaction packet. A subsequent receipt of a message Deassert_INTA requester packet on the same downstream ports results in the upstream port canceling the one remaining recorded indication that interrupt A is asserted. Subsequently the upstream port transmits a message Deassert_INTA requester transaction packet. If the receipt of a message Deassert_INTA requester transaction was on another downstream port with no recorded assertion of the interrupt A, the packet is ignored (not treated as an error) and no message Deassert_INTA requester transaction packet is transmitted by the upstream port. Switches must keep separate records of assertion of each specific interrupt received on each of its downstream ports. Similarly, the deassertion information received by switches on each of its downstream ports must be associated with a previous assertion relative to a specific interrupt and specific downstream ports.

- **Bridge:** For a bridge, the deassertion of the INTB# signal line by the same PCI or PCI-X device ("first") that asserted it results in the upstream port transmitting a message Dessert_INTB requester transaction packet, unless another PCI or PCI-X device ("second") asserted the INTB# signal line prior to the deassertion of the INTB# signal line by the "first" PCI or PCI-X device. Subsequently, the deassertion of the INTB# signal line by the "second" PCI or PCI-X device results in the upstream port transmitting a message Dessert_INTB requester transaction packet, unless another PCI or PCI-X device asserted the INTB# signal line, and so forth. For a bridge, the deassertion of the INTA# signal line by one of PCI or PCI-X devices that asserted is not seen by the bridge. The open collector technology permits the "other" PCI or PCI-X device asserting the INTA# signal line to mask it. The upstream port does not transmit a message Dessert_INTA requester transaction packet. If the "other" PCI or PCI-X device deasserts the INTA# signal line, the upstream port does sense the change and transmits a message Dessert_INTA requester transaction packet.

Legacy Interrupt Protocol Implementation by the Root Complex

The downstream ports of the Root Complex implement the LI protocol in a fashion similar to switches' downstream ports discussed in the previous section. Each individual downstream port is individually responsible for recording the numbers of times message Assert_ INTx requester transaction packets (relative assertions) are received for each specific interrupt x (A, B, C, and D). Each downstream port reduces (cancels) the recorded number of assertions for each message Deassert_INTx requester transaction packet (relative deassertion) received for each specific interrupt x (A, B, C, and D). The first recording of relative assertion and the cancellation of the last recorded relative assertion generates information internally to the Root Complex that is equivalent to a virtual assertion and deassertion of each specific interrupt x (A, B, C, and D) signal lines. If the receipt of a Deassert_INTx requester transaction packet is on downstream port with no recorded assertion of the interrupt x, the packet is ignored.

Boundary Conditions for Disabling INTx

From the previous discussion it is clear that the Root Complex's downstream ports and switches' upstream ports keep track of the message interrupt requestor transaction packets received. Under normal conditions, the number of virtual deassertions will eventually equal the number of virtual assertions as the interrupts are serviced. Atypical conditions occur when links enter the L_Inactive link activity state, which means pending virtual deassertion, information cannot flow upstream. The entrance into the L_Inactive link activity state when reset (Hot, Cold, or Warm), when the Physical Link = "0" as indicated by the Physical Layer, or when VC0 fails to achieve Flow Control Initialization. See Chapter 7 for more information on the link activity states.

Similarly, the software may disable message interrupt requester transaction packet transmission at different PCI Express devices via the configuration address space.

Under these two atypical conditions, the following protocols apply:

- Assume an upstream port (legacy endpoint, switch, or bridge) has transmitted a message Assert_INTx requester transaction and has not transmitted companion message Deassert_INTx requester transaction for the specific interrupt x. The upstream port is required to transmit a message Dessert_INTx requester transaction packet when its associated Interrupt Disable bit (Command Register of configuration register block) is set to "1." This applies to each specific interrupt x (A, B, C, and D) for each specific downstream port.

- Assume an upstream port of a switch has transmitted a message Assert_INTx requester transaction and has not transmitted companion message Deassert_INTx requester transaction for the specific interrupt x (A, B, C, and D) for a specific downstream port. The upstream port is required to transmit message Deassert_INTx requester transaction packet relative to when associated specific downstream port's link transitions to L_Inactive link activity state. This applies to each specific interrupt x (A, B, C, and D) for each specific downstream port.

- Assume a Root Complex downstream port reported an assertion of INTx and has not reported a deassertion of the INTx to the Root Complex for the specific interrupt x (A,B,C,D) for a specific downstream port. The downstream port is required to report a deassert_INTx to the Root Complex when the link of the downstream port transitions to L_Inactive link activity state. This applies to each specific interrupt x (A, B, C, and D) for each specific downstream port.

Unique Considerations for the Legacy Interrupt Protocol for Re-Mapping through Each Virtual PCI/PCI Bridge

As discussed earlier, the INTx# (INTA#, INTB#, INTC#, INTD#) signal lines on the downstream side of bridges (PCI Express/PCI, PCI Express/PCI-X Bridges) are the actual signal lines and not the virtual implementations on the upstream side of bridges. Historically, PCI and PCI-X have been slot-based platforms. There was a concern that add-in cards would select certain INTx# signal lines and cause unbalanced loading. Consequently, the interrupt protocol requires the remapping of the INTx# signal lines when the interrupt information traverses a PCI/PCI Bridge. To maintain software compatibility with PCI and PCI-X, the PCI Express specification also defines interrupt remapping for each virtual PCI/PCI Bridge.

As discussed in Chapter 3, three architectures can be applied to a PCI Express/PCI Bridge. In Figure 19.1, only one level of virtual PCI/PCI Bridge separates the message interrupt requester transaction from the PCI bus segment.

The remapping for Figure 19.1 is summarized in Table 19.2 and illustrated for a PCI platform. This particular bridge architecture has only one virtual PCI/PCI Bridge level. Once the interrupt information is mapped to the internal Local Bus (effectively a virtual PCI bus segment), no additional virtual PCI/PCI Bridge levels are necessary. Consequently, the specific INTx for the message interrupt requester transaction packet is listed in column 3 of Table 9.2.

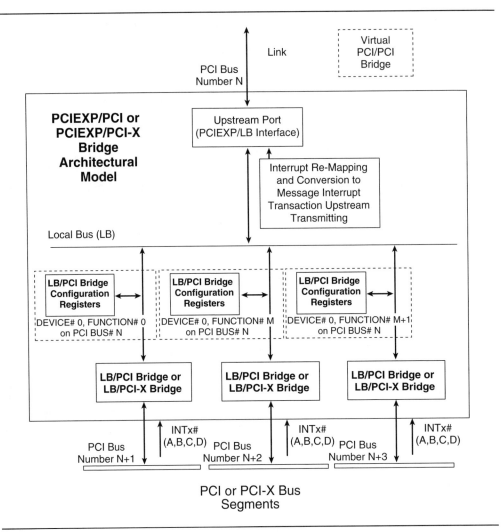

Figure 19.1 PCI Express/PCI Bridge: Single Level

Table 19.2 Interrupt Remapping

Column 1 Device # on PCI Bus Number N+1 to N+3 (Also Link Bus Number)	Column 2 For Device # in Column 1 Connected to the Following INTx#	Column 3 Maps to the INTx of the Following message ASSERT_INTx or DEASSERT_INTx Requester Transaction
0, 4, 8, 12, 16, 20, 24, 28	INTA	INTA
	INTB	INTB
	INTC	INTC
	INTD	INTD
1, 5, 9, 13, 17, 21, 25, 29	INTA	INTB
	INTB	INTC
	INTC	INTD
	INTD	INTA
2, 6, 10, 14, 18, 22, 26, 30	INTA	INTC
	INTB	INTD
	INTC	INTA
	INTD	INTB
3, 7, 11, 15, 19, 23, 27, 31	INTA	INTD
	INTB	INTA
	INTC	INTB
	INTD	INTC

Figures 19.2 and 19.3 implement one more virtual PCI/PCI Bridge levels between the internal virtual PCI bus segment and the upstream link. Consequently, the mapping summarized in Table 9.2 must be applied one more time with PCI Bus Number N+1 to N+3 in column 1 replaced by internal PCI Bus Number N.

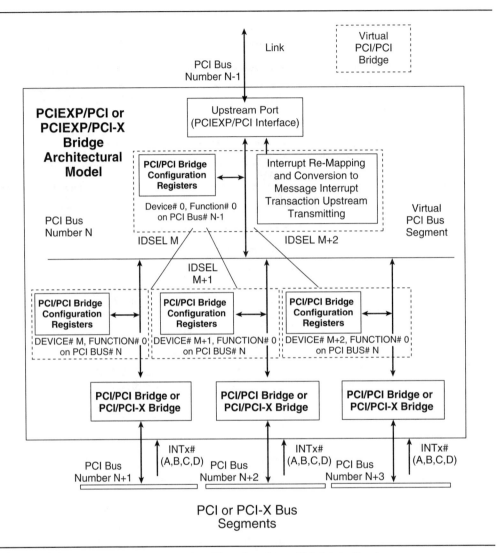

Figure 19.2 PCI Express/PCI Bridge: Two Levels and Multiple Devices

Chapter 19: Interrupts ■ **827**

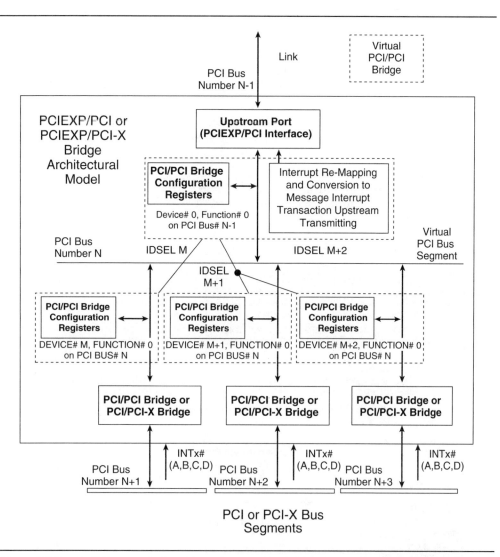

Figure 19.3 PCI Express/PCI Bridge: Two Levels, Multiple Devices and Functions

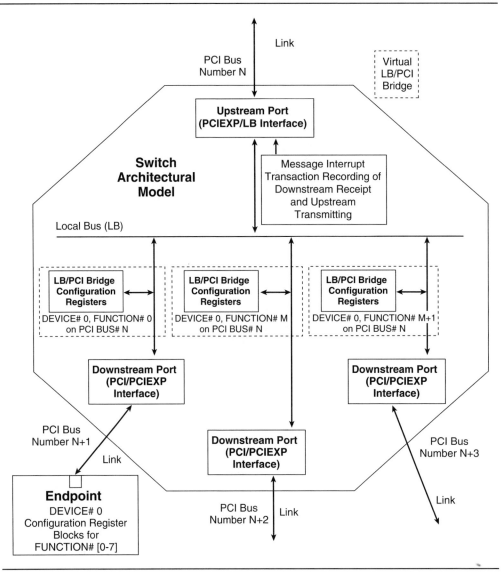

Figure 19.4 Switch: Single Level

Finally, as discussed in Chapter 3, the switches can be implemented with one or two virtual PCI/PCI Bridge levels. Figure 19.4 illustrates a switch with one virtual PCI/PCI Bridge level. By definition, the PCI Express link on the downstream port of a switch is only connected to DEVICE# 0. As shown in Table 9.4, the remapping for the interrupts for DEVICE# 0 is one-to-one. Consequently, a message interrupt requester transaction packet received on a downstream port of the switch for INTA maps to message interrupt requester transaction packet transmitted on the upstream port of the switch for INTA. A message interrupt requester transaction packet received on a downstream port of the switch for INTB maps to message interrupt requester transaction packet transmitted on the upstream port of the switch for INTB, and so forth.

Figures 19.5 and 19.6 illustrate switches with two virtual PCI/PCI Bridge levels. For the same reason discussed earlier (DEVICE# 0 on the downstream link), the message interrupt requester transaction packet received on a downstream port of the switch for INTA maps to virtual INT#A signal line. A message interrupt requester transaction packet received on a downstream port of the switch for INTB maps to virtual INT#B signal line. The subsequent message interrupt requester transaction packets transmitted from the switch's upstream port must be a remapping as summarized in Table 9.2, with PCI Bus Number N+1 to N+3 in column 1 replaced by internal PCI Bus Number N.

830 ■ The Complete PCI Express Reference

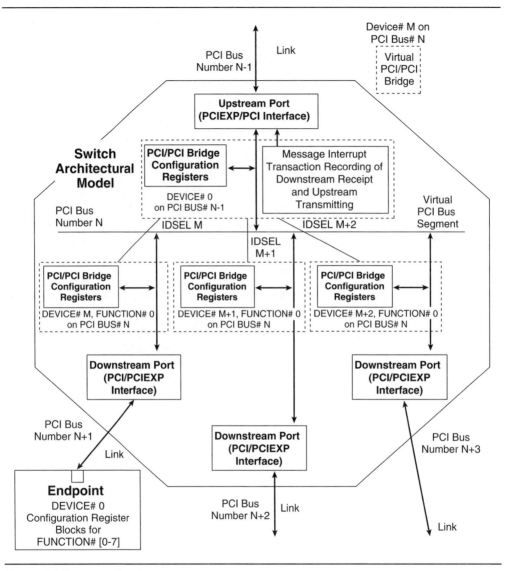

Figure 19.5 Switch: Two Levels and Multiple Devices

Chapter 19: Interrupts ■ **831**

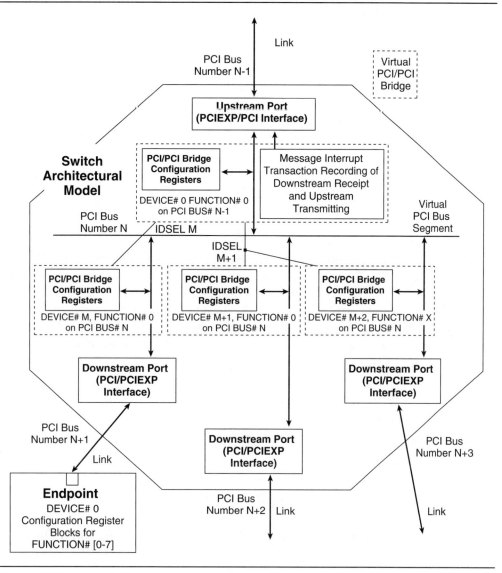

Figure: 19.6 Switch: Two Levels, Multiple Devices and Functions

Message Signaled Interrupt Protocol

The MSI protocol implementation and support are optional for legacy endpoints, bridges, and PCI and PCI-X devices below the bridge. MSI protocol implementation and support are required for PCI Express endpoints, switches, and Root Complex. MSI protocol implements memory write transactions between PCI or PCI-X devices below the bridge. If the MSI memory write transaction's destination is upstream of the bridge, the bridge must translate the MSI memory write transaction to a MSI memory write requester transaction packet. The MSI protocol can also be applied above the bridge between PCI Express devices. If the MSI memory write requester transaction packet's destination is downstream of the bridge, the bridge must translate the MSI memory write transaction packet to MSI memory write transaction.

The PCI specification Revision 2.3 defines the MSI protocol. The PCI Express specification does not provide any additional information or requirements relative to the MSI protocol. See the book and specification references in Chapter 1 for more detailed information.

The MSI protocol does implement the memory write requester transactions (covered in more detail in Chapter 6). Some items of note are:

- **ATTRI [1::0] field:** This field in the ATTRIBUTE HDW in the Header field of the memory write requester transactions must be set to 01b. This means that PCI strong ordering applies and no snooping is allowed.

- **TC [2::0] field:** This field in the TYPE HDW in the Header field of the memory write requester transactions can be any value. However, it is recommended that the TC value should be used that was used for the bus transaction transferring the data associated with the interrupt.
    - If the MSI is associated with legacy endpoints and bridges it is recommended that TC0 is used (TC [2::0] = 000b).
    - If the TC used in the MSI memory write requester bus transaction is not the same as used in the bus transaction transferring the data, the flow control can adversely affect the synchronization between the bus transactions associated with the MSI protocol. Implementation-specific methods may have to be employed to address the synchronization.

Chapter 20

Lock Protocol

This chapter discusses how the PCI Express specification supports a Lock protocol to provide backward compatibility to PCI and PCI-X software.

Introduction to the Lock Protocol

In a PCI or PCI-X platform the fundamental purpose of the Lock protocol is to ensure the HOST bus segment CPU exclusive access to PCI and PCI-X memory or I/O devices (targets) and platform memory. This is accomplished by Exclusive Hardware Access (EHA) or Exclusive Software Access (ESA); collectively called exclusive access. There have been many interpretations of the Lock protocol relative to the upstream and downstream accesses by PCI and PCI-X bus masters. The most recent interpretation permits PCI or PCI-X bus masters to implement the Lock protocol to establish EHA of any target (memory only) on the same or on a downstream bus segment. No Lock protocol is implemented for PCI or PCI-X bus masters to establish EHA of the platform memory.

PCI or PCI-X bus masters implement the Lock protocol to establish ESA of any target (memory and I/O) and platform memory. For both EHA and ESA, the targets to be locked may also be downstream of a PCI/LEGACY Bridge. The implementation of the LF protocol via ESA on a PCI Express platform is not defined. PCI and PCI-X compatible software is in no way prevented from executing an ESA to any PCI Express device. Indirectly, this implies that ESA is supported on a PCI Express platform.

A PCI platform has a unique consideration to maintain backward compatibility. The Lock protocol can be implemented by a LEGACY bus master on the LEGACY bus segment to establish EHA of the platform memory. Consequently, under this one unique situation the HOST/PCI and PCI/LEGACY Bridges must support upstream EHA to platform memory.

EHA Participants

The term Lock Function (LF) protocol simply implies that two PCI Express devices are the participants in EHA or ESA. The implementation of the LF via EHA on a PCI Express is more restrictive than PCI or PCI-X and is limited as follows:

- **HOST Bus Segment:** The HOST bus segment CPU via the Root Complex is the only PCI Express device that can initiate the LF protocol. Consequently, the Root Complex can only establish EHA to a PCI Express device downstream.

- **Legacy Endpoints and Bridges:** The only downstream PCI Express devices that can participate in the Lock protocol with EHA are legacy endpoints and bridges as targets. Downstream legacy endpoints and bridges cannot initiate the LF protocol in the upstream direction to establish EHA of the platform memory. PCI Express endpoints or switches cannot initiate or participate in the LF protocol with EHA. Switches only port the transaction packets associated with the LF protocol (Lock transactions) for the Root Complex achieving EHA to legacy endpoints and bridges downstream of the switch.

- **Platform Memory:** The PCI Express specification does not define if HOST bus segment CPU via the Root Complex can implement the LF protocol with EHA to the platform memory. An implementation of the Root Complex can optionally permit HOST bus segment CPU EHA to the platform memory. If the LF protocol with the EHA is implemented by one Root Complex, it may not be implemented by other Root Complexes that the software can be ported to. This results in a PCI or PCI-X software compatibility issue.

Lock Function (LF) Protocol

EHA within a PCI Express platform is implemented via the Lock Function (LF) protocol. The LF protocol defines a subset of Transaction Layer Packets to implement the EHA.

Lock Transaction Packets

The set of Transaction Layer packets (TLPs) defined by the LF Protocol are referred to as *lock TLPs*. All lock TLPs must be mapped to VC0 by assignment of TC0 to the packets. The LF does not apply and does not affect TLPs assigned to other VC numbers. The lock TLPs are

- **MRdLk:** Lock memory read requester transaction packets: Identical to the memory read requester transaction packets except the Lock bit (Type [0] field in TYPE HDW) is set to 1b and the TC [2::0] must equal 000b in the TYPE HDW of the Header field.

- **CplDLk and CplLk:** Lock completer transaction packets, identical to completer transaction packets with and without data except that the Lock bit (Type [0] field I TYPE HDW) is set to 1b and the TC [2::0] must equal 000b in the TYPE HDW in the Header field.

- **MWr:** Memory write transaction packets, identical to memory write requester transaction packets not used for the Lock protocol.

- **Message unlock requester transaction packets:** One of the message baseline requester transaction packets.

LF Protocol to Establish EHA

The LF protocol to establish the EHA between the Root Complex and a PCI Express device has several steps:

EHA Established

The HOST bus CPU via the Root Complex initiates the LF protocol to establish EHA by transmitting a lock memory read requester transaction (MRdLk) packet. The EHA is established at the legacy endpoint or bridge, and is defined as locked device with LF when it transmits a lock completer transaction packet with the successfully read data (CplDLK: STATUS [2::0] = 000b of STATUS HDW in the Header field). With the transmission of the lock completer transaction packet with the successfully read data, the PCI Express device is defined as a locked device. All of the functions internal to the locked device are part of the LF. The locked device cannot transmit any requester transactions on VC0. The EHA is established at the Root Complex, and is defined as the LF master when it receives a lock completer transaction packet with the successfully read data (CplDLK with STATUS [2::0] = 000b of STATUS HDW in Header field).

EHA not Established

The distinction between establishing or not establishing EHA is based on the response of the legacy endpoint or bridge. The EHA is not established at the legacy endpoint or bridge (no LF) if it transmits a lock completer transaction packet without data (not successfully read Cpl with STATUS [2::0] not = 000b) in response to the reception of the MRdLk. Subsequently, the EHA is not established at the Root Complex (no LF) if it receives a lock completer transaction without data (not successfully read Cpl with STATUS [2::0] not = 000b).

Other EHA Considerations

The Root Complex can have only one LF active at any time each hierarchy domain connects to a downstream port. To prevent a downstream deadlock, the Root Complex must not transmit a second lock memory read requester transaction packet while it is waiting for the receipt of a completer transaction packet associated with a previous lock memory read requester transaction packet. The PCI Express specification does permit LFs operating in parallel on the hierarchy domains of other downstream ports of Root Complex provide only one LF per hierarchy domain.

As will be discussed below, the unlock message requester transaction packet is transmitted by the Root Complex to terminate the LF. If an attempt has been made to established a LF but the attempt failed and EHA has not established, the Root Complex must transmit a unlock message requester transaction packet.

LF Protocol after EHA Established

Once the LF has been established between the Root Complex and a PCI Express device (locked device), only the lock TLPs can be transmitted between the Root Complex and the locked device. Only the Root Complex can access the locked device. If the locked device is connected directly to the Root Complex, no access issues occur. A locked device downstream of a switch or a bridge has other considerations that need to be reviewed.

Once LF has been established the Root Complex cannot access other PCIX devices connected via the downstream port involved with the LF. The PCI Express specification does not define if other downstream ports of Root Complex not involved with the LF can or cannot execute transactions unrelated to the lock protocol because the PCI and PCI-X specifications require that only one PCI or PCI-X device in the entire platform can be locked and the CPU on the HOST bus segment can only be accessing the locked device. This requirement assures other bus masters trying to access the locked device to receive the minimum number of Retry terminations. Given that the lock protocol is supported for PCI and PCI-X legacy reasons, this book assumes that the CPU on the HOST bus segment via the Root Complex is only accessing one locked device and is not accessing non-locked PCI Express devices on other downstream ports. However, no method forces the Root Complex to adhere to this assumption.

Restriction of Accesses to Locked Device

If the locked device is directly connected to the Root Complex, the only accesses to the locked device are from the Root Complex. The Root Complex can only have one LF active at any time on a specific downstream port. Also, once LF has been established, the Root Complex cannot access other PCIX devices connected via the downstream port involved with the LF.

Restrictions Specific to Switches

If the locked device is not directly connected to the Root Complex, the access restrictions are extended downstream to the intervening switches. If the locked device is downstream of a switch, the switch must block any downstream requester transaction packets transferring over VC0 of the switch's downstream port that transmitted the first lock memory read transaction packet when it recognizes that EHA has established at a downstream PCI Express device. The switch recognizes that EHA has been established at a downstream PCI Express device and it becomes the locked device when the switch ports upstream the first lock completer transaction with the successfully read data (CplDLK: STATUS [2::0] = 000b of STATUS HDW in Header field). Once the switch recognizes that a downstream PCI Express device has established EHA, the switch must *block* any upstream or downstream requester transaction packets transferring over VC0 of the switch's upstream port and the specific downstream port that transmitted the first lock memory read transaction packet. Transactions on other VC numbers of these two ports or any TLPs associated with other ports are not affected. Essentially, the switch's porting structure between these two ports becomes part of the LF for VC0. See "MWr Issue" at the end of chapter.

Restriction Specific to Bridges

If the locked device is downstream of a bridge, the protocol is the same as above except for the following consideration: Bridges only support VC0; no non-lock TLPs are possible once the upstream port of a bridge is supporting the LF.

LF Protocol to Terminate LF between Root Complex and Locked Device

The protocol to terminate the LF between the Root Complex and the downstream PCI Express device (locked device), switches, and bridges implements the message unlock requester transactions. Upon receipt of the unlock message requester transaction packet by the locked device or when it is ported through a switch or a bridge, the respective PCI Express devices cease support of the LF. These PCI Express devices are legacy endpoints, switches, bridges, and PCI or PCIX devices below the bridge. The source of the unlock message requester transaction packet is the Root Complex; consequently the Root Complex stops support of the LF at the time of transmission of the unlock message requester transaction packet.

> The Root Complex transmits an unlock message requester transaction packet to all downstream PCI Express devices (broadcast); consequently, all PCI Express devices that are not involved with the current LF or by definition cannot support the LF must ignore (discard without error) the unlock message requester transaction packet.

After the EHA has been established at the Root Complex, the subsequent receipt of a lock completer transaction packet without data (not successfully read CplLk with STATUS [2::0] not = 000b) by the Root Complex the Root Complex ceases the support of the LF. The source of the lock completer transaction packet without data (CplLk with STATUS [2::0] not = 000b) is the locked device; consequently it stops support of the LF function at the time of transmission of the CplLk or similar bus transaction for a PCI or PCI-X device below a bridge). However, the intervening switches and bridges maintain support of the LF. The Root Complex must subsequently transmit a message unlock requester transaction packet to inform the switches and bridges to cease support of the LF. The legacy endpoints or bridges downstream of the Root Complex must ignore and discard without error the unexpected unlock message requester transaction packet received.

MWr Issue

Once a switch or bridge has established the support of the LF, only lock TLPs are permitted on VC0; all others are blocked. One of the implied exceptions is the message unlock requester transition packet. The other implied exception is the memory write requester transaction packet (MWr). However, the PCI Express specification does not provide a special MWr for the LF; consequently the switches and bridges supporting the LF must implement the following, which is not defined in the PCI Express specification. When the LF is being established by the lock memory read transaction packet (MRdLk) the switch or bridge must copy the Requester ID. Through the use of the Requester ID, the switches and bridges can distinguish that a MWr is from the Root Complex (not blocked) versus from another PCI Express device (blocked) via peer-to-peer transactions. This prevents PCI Express devices like the legacy endpoints from accessing a locked device, which is the purpose of the LF protocol.

Also, the PCI Express specification does not define how to block a TLP per the LF protocol as discussed in the above section.

Chapter 21

Mechanical and Electrical Overview

This chapter provides a very brief summary of the mechanical and electrical elements, in order to tie this book in with the mechanical and electrical elements of the various electromechanical specifications.

Introduction

As discussed in Chapter 1, multiple specifications define PCI Express. Some of these specifications are also related to PCI and PCI-X. The four specifications directly related to PCI Express's mechanical and electrical are *PCI Express Base Specification revision 1.0a April 15, 2003, PCI Express Card Electromechanical Specification revision 1.0a April 15, 2003*, and *PCI Express Mini Card Electromechanical Specification revision 1.0 June 6, 2003*. The primary focus of this book has been the portions of the PCI Express Base specification that define the transaction protocols, platform architecture, and software elements compatible with PCI, not the specific mechanical or electrical elements defined in the above specifications.

Mechanical

There are two types of platforms and associated add-in cards defined by the PCI Express specification. One type is implemented by desktop PCs and servers and is defined as "non-mobile." The other type is implemented by any type of battery-operated PCI Express platform and is called "mobile." It is possible to design a non-mobile platform that has slots for mobile add-in cards, but to simplify the discussion this scenario is not addressed. The same comment applies for non-mobile add-in cards and mobile platforms.

Throughout this book the term add-in card is used and applies to both mobile and non-mobile add-in cards. Similarly, the term "PCI Express platform" applies to both mobile and non-mobile. The use of the adjectives "non-mobile" and "mobile" are used in this chapter due to the mechanical distinctions between the two. Otherwise, the PCI Express specification is the same for mobile and non-mobile implementations.

Non-Mobile Add-in Cards

There are several different non-mobile add-in card mechanical sizes and different link widths defined for each.

Link Widths, Connector Sizes, and Add-in Cards Sizes

In the same manner that PCI Express protects the software investment by using the PCI configuration mechanism, the mechanical definition of PCI Express protects the mechanical platform tooling investment made for PCI and PCI-X. Thus, even though the PCI and PCI-X bus segments have been replaced by the PCI Express links, the dimensions of the non-mobile PCI and PCI-X add-in cards for non-mobile platform have been retained. The three basic non-mobile add-on card sizes defined by PCI, PCI-X, and PCI Express are low profile, short length (variable height), and standard length. The PCI Express Card Electromechanical Specification provides all of the mechanical dimensions for non-mobile add-in cards.

As discussed throughout this book, to increase performance of PCI and PCI-X platforms, the bus segments evolved into essentially point-to-point interconnections. As previously discussed, PCI Express exploited this point-to-point interconnection by defining a minimal number of signal lines between PCI Express devices. The connector sizes implemented in each slot of non-mobile platforms can vary. Also, the size of the edge finger area of the non-mobile add-in cards can vary and this variation in the size of the edge finger area limits the size of edge fingers themselves. The location and number of lanes are predefined for these connectors and the edge fingers of the non-mobile add-in cards. The number of lanes that can be configured for the one link establishes the link width. Thus, the maximum link width that can be configured depends on the size of these connectors and the number of edge fingers of the non-mobile add-in in cards. By definition only one link can be configured per slot of a non-mobile platform.

The connectors defined for PCI Express non-mobile platforms have different pin counts from PCI or PCI-X, but are of the same connector design and type as PCI and PCI-X. Also, the non-mobile add-in card edges mate with the connector in the same fashion with edge fingers as PCI and PCI-X non-mobile add-on cards. The number of possible link widths of the add-in card slots for non-mobile platforms are x1, x2, x4, x8, x12, and x16. Consequently, x32 links cannot be implemented for non-mobile add-in cards according to the current PCI Express Card Electromechanical Specification revision 1.0a.

The different link widths are defined for the following connector sizes for non-mobile platforms: x1 = 36 pins, x4 = 64 pins, x8 = 98 pins, and x16 = 164 pins. The number of edge fingers on non-mobile add-in cards also comes in four sizes x1, x4, x8, and x16. The connector and non-mobile add-in card size notations of x1, x4, x8, and x16 simply define the maximum number of lanes that can be configured into a single configured link.

No connector or non-mobile add-in card sizes are defined as x2 or x12. A link width of x2 can be configured in a connector size and non-mobile add-in cards that define x4, x8, and x16. A link width of x12 can be configured in connectors and non-mobile add-in cards that define x16.

Table 21.1 summarizes the combination of connector and add-in card sizes for non-mobile implementations. Also in Table 21.1 are the possible lanes that can be configured for the combination of these connectors and add-in card sizes.

Table 21.1 Combinations of Non-Mobile Connector, Non-Mobile Add-in, and Link Width

Non-Mobile Connector Size	Non-Mobile Add-in Card Edge Finger Sizes that Can be Plugged In	Possible Link Widths that Can Be Configured
x1	x1	x1
x4	x1	x1
	x4	x1, x2, and x4
x8	x1	x1
	x4	x1, x2, and x4
	x8	x1, x2, x4, and x8
x16	x1	x1
	x4	x1, x2, and x4
	x8	x1, x2, x4, and x8
	x16	x1, x2, x4, x8, x12, and x16

As summarized in Table 21.1, the only edge fingers on the non-mobile add-in card that can be configured are those physically in a connector. Also, as discussed in Chapter 8, for all the lanes that can be configured into a configured link, some may not be configured. That is, even though mechanically the lanes on the connector of a non-mobile platform and the edge fingers on the non-mobile add-in card indicate a possible configured link width, the Link Training and Status State Machines of both PCI Express devices must sequence through the Detect, Polling, and Configuration Link States. The completion of this sequence establishes the actual configured lanes and thus the single configured link width.

To simplify the discussion, the focus is on the pin assignments listed in the Tables 21.2 through 21.5 for the connectors implemented on non-mobile platforms. Obviously, the edge fingers of the mating non-mobile add-in cards are defined in the same manner. Tables 21.2 through 21.5 list the pin assignments for the four connector sizes. A connector of size x1 only implements the pin assignments of Table 21.2. A connector of size x4 only implements the pin assignments of Tables 21.2 and 21.3. A connector of size x8 only implements the pin assignments of Tables 21.2 through 21.4. A connector of size of x16 implements the pin assignments of all Tables 21.2 through 21.5.

Table 21.2 Connector Pin Assignments for Non-Mobile

Pin# Side B	Signal Line Name	Pin# Side A	Signal Line Name
1	+12 V	1	PRSNT1#
2	+ 12 V	2	+12 V
3	Reserved	3	+ 12 V
4	GND	4	GND
5	SMCLK	5	TCK
6	SMDAT	6	TDI
7	GND	7	TDO
8	+3.3 V	8	TMS
9	TRST#	9	+3.3V
10	+3.3 Vaux	10	+3.3 V
11	WAKE#	11	PERST#
12	Reserved	12	GND
13	GND	13	REFCLK+
14	LANE# 0 Differential Pair Transmitter ... positive	14	REFCLK•
15	LANE# 0 Differential Pair Transmitter ... negative	15	GND
16	GND	16	LANE# 0 Differential Pair Receiver ... positive
17	PRSNT#2	17	LANE# 0 Differential Pair Receiver ... negative
18	GND	18	GND

Table 21.3 Connector Pin Assignments for Non-Mobile

Pin# Side B	Signal Line Name	Pin# Side A	Signal Line Name
19	LANE# 1 Differential Pair Transmitter ... positive	19	Reserved
20	LANE# 1 Differential Pair Transmitter ... negative	20	GND
21	GND	21	LANE# 1 Differential Pair Receiver ... positive
22	GND	22	LANE# 1 Differential Pair Receiver ... negative
23	LANE# 2 Differential Pair Transmitter ... positive	23	GND
24	LANE# 2 Differential Pair Transmitter ... negative	24	GND
25	GND	25	LANE# 2 Differential Pair Receiver ... positive
26	GND	26	LANE# 2 Differential Pair Receiver ... negative
27	LANE# 3 Differential Pair Transmitter ... positive	27	GND
28	LANE# 3 Differential Pair Transmitter ... negative	28	GND
29	GND	29	LANE# 3 Differential Pair Receiver ... positive
30	Reserved	30	LANE# 3 Differential Pair Receiver ... negative
31	PRSNT#2	31	GND
32	GND	32	Reserved

Table 21.4 Connector Pin Assignments for Non Mobile

Pin# Side B	Signal Line Name	Pin# Side A	Signal Line Name
33	LANE# 4 Differential Pair Transmitter ... positive	33	Reserved
34	LANE# 4 Differential Pair Transmitter ... negative	34	GND
35	GND	35	LANE# 4 Differential Pair Receiver ... positive
36	GND	36	LANE# 4 Differential Pair Receiver ... negative
37	LANE# 5 Differential Pair Transmitter ... positive	37	GND
38	LANE# 5 Differential Pair Transmitter ... negative	38	GND
39	GND	39	LANE# 5 Differential Pair Receiver ... positive
40	GND	40	LANE# 5 Differential Pair Receiver ... negative
41	LANE# 6 Differential Pair Transmitter ... positive	41	GND
42	LANE# 6 Differential Pair Transmitter ... negative	42	GND
43	GND	43	LANE# 6 Differential Pair Receiver ... positive
44	GND	44	LANE# 6 Differential Pair Receiver ... negative
45	LANE# 7 Differential Pair Transmitter ... positive	45	GND
46	LANE# 7 Differential Pair Transmitter ... negative	46	GND
47	GND	47	LANE# 7 Differential Pair Receiver ... positive
48	PRSNT#2	48	LANE# 7 Differential Pair Receiver ... negative
49	GND	49	GND

Table 21.5 Connector Pin Assignments for Non-Mobile

Pin# Side B	Signal Line Name	Pin# Side A	Signal Line Name
50	LANE# 8 Differential Pair Transmitter … positive	50	Reserved
51	LANE# 8 Differential Pair Transmitter … negative	51	GND
52	GND	52	LANE# 8 Differential Pair Receiver … positive
53	GND	53	LANE# 8 Differential Pair Receiver … negative
54	LANE# 9 Differential Pair Transmitter … positive	54	GND
55	LANE# 9 Differential Pair Transmitter … negative	55	GND
56	GND	56	LANE# 9 Differential Pair Receiver … positive
57	GND	57	LANE# 9 Differential Pair Receiver … negative
58	LANE# 10 Differential Pair Transmitter … positive	58	GND
59	LANE# 10 Differential Pair Transmitter … negative	59	GND
60	GND	60	LANE# 10 Differential Pair Receiver … positive
61	GND	61	LANE# 10 Differential Pair Receiver … negative
62	LANE# 11 Differential Pair Transmitter … positive	62	GND
63	LANE# 11 Differential Pair Transmitter … negative	63	GND
64	GND	64	LANE# 11 Differential Pair Receiver … positive
65	GND	65	LANE# 11 Differential Pair Receiver … negative

Table 21.5 Connector Pin Assignments for Non-Mobile *(continued)*

Pin# Side B	Signal Line Name	Pin# Side A	Signal Line Name
66	LANE# 12 Differential Pair Transmitter ... positive	66	GND
67	LANE# 12 Differential Pair Transmitter ... negative	67	GND
68	GND	68	LANE# 12 Differential Pair Receiver ... positive
69	GND	69	LANE# 12 Differential Pair Receiver ... negative
70	LANE# 13 Differential Pair Transmitter ... positive	70	GND
71	LANE# 13 Differential Pair Transmitter ... negative	71	GND
72	GND	72	LANE# 13 Differential Pair Receiver ... positive
73	GND	73	LANE# 13 Differential Pair Receiver ... negative
74	LANE# 14 Differential Pair Transmitter ... positive	74	GND
75	LANE# 14 Differential Pair Transmitter ... negative	75	GND
76	GND	76	LANE# 14 Differential Pair Receiver ... positive
77	GND	77	LANE# 14 Differential Pair Receiver ... negative
78	LANE# 15 Differential Pair Transmitter ... positive	78	GND
79	LANE# 15 Differential Pair Transmitter ... negative	79	GND
80	GND	80	LANE# 15 Differential Pair Receiver ... positive
81	PRST2#	81	LANE# 15 Differential Pair Receiver ... negative
82	Reserved	82	GND

An additional consideration of non-mobile add-in cards is that the PCI Express interconnection is not the only one currently defined. Non-mobile add-in cards also define the SMBus interconnection. Consequently, non-mobile add-in cards can be logically connected to the platform by the SMBus or PCI Express link or any combination.

Non-Mobile Add-in Card Size

The non-mobile add-in cards come in several board sizes: low profile, short length (variable height), and standard length. The possible edge finger sizes of the non-mobile add-in card are defined by the size of the board. Table 21.1 summarizes the combination of the non-mobile connector and add-in cards, and possible link widths. Table 21.6 correlates the board size with the possible non-mobile add-in card edge finger sizes and the possible link widths.

Table 21.6 Combinations of Non-Mobile Add-in Card Edge Finger Size, Non-Mobile Add-in Cards, and Link Width

Non-Mobile Add-in Card Edge Finger Sizes	Possible Link Widths that Can Be Configured	Non-Mobile Add-in Card Board Sizes
x1	x1	Short and Standard Length
x4	x1, x2, and x4	Standard Length and Low Profile
x8	x1, x2, x4, and x8	Standard Length and Low Profile
x16	x1, x2, x4, x8, x12, and x16	Standard Length and Low Profile

Mobile Add-in Cards: PCI Express Mini Cards

There is only one mobile add-in card mechanical size defined with one link width defined. There are provisions to define a wider link width at a later date.

Link Widths, Connector Sizes, and PCI Express Mini Cards

The traditional add-in cards for a mobile platform are PCMCIA and Cardbus. Neither of these mobile add-in cards implemented a specific "PCI-like" or "PCI-X-like" interconnection. In both cases a bridge between either a PCI or a PCI-X bus segment and the mobile add-in card is required. A mobile PCI Express platform can be designed with a bridge between PCI Express links and PCMCIA or Cardbus. However, the point-to-point attribute and the minimal signal lines make the PCI Express link a natural for a mobile add-in card interconnect. Consequently, the PCI-SIG defines a mobile add-in card that directly implements the PCI Express link as the interconnection. This mobile PCI Express add-in card is called the PCI Express Mini Card. The differences in the slots, connectors, and add-in cards of a non-mobile platform and that of a mobile platform for the PCI Express Mini Card are as the follows:

- A mobile connector is mechanically smaller than a non-mobile connector.

- Only one mobile connector size is defined with a single link width of x1 versus the four possible non-mobile connector sizes.

- Only one PCI Express Mini Card edge finger size is defined with a single link width of x1 versus the four possible non-mobile edge finger sizes.

- Non-mobile add-in cards come in several board sizes. PCI Express Mini Cards come in one board size.

- A non-mobile slot and add-in card define +12 volts, +3.3 volts, and +3.3 volts auxiliary (3.3 Vaux or just Vaux) power rails. The PCI Express Mini Cards and associated slots define +1.5 volts, +3.3 volts, and +3.3 volts auxiliary (3.3 Vaux or just Vaux) power rails. See Chapter 17 for more information.

- The pin assignments of the x1 non-mobile connector listed in Table 21.2 are different from those listed in Table 21.7 for the mobile connector for the PCI Express Mini Card.
    - As noted in the Table Notes for Table 21.7 some of the Reserved signal lines may be defined in the future for an additional lane, which would define a mobile connector and PCI Express Mini Card of x2 size.

Table 12-7 Connector Pin Assignments for Mobile

Pin#	Signal Line Name	Pin#	Signal Line Name
1	WAKE#	2	+3.3 V
3	Reserved (4)	4	GND
5	Reserved (4)	6	+1.5 V
7	Reserved	8	Reserved (2)
9	GND	10	Reserved (2)
11	REFCLK+	12	Reserved (2)
13	REFCLK-	14	Reserved (2)
15	GND	16	Reserved (2)
17	Reserved	18	GND
19	Reserved	20	Reserved (3)
21	GND	22	PERST#
23	LANE# 0 Differential Pair Receiver ... negative	24	+ 3.3 V
25	LANE# 0 Differential Pair Receiver ... positive	26	GND
27	GND	28	+ 1.5 V
29	GND	30	SMB_CLK
31	LANE# 0 Differential Pair Transmitter ... negative	32	SMB_DATA
33	LANE# 0 Differential Pair Transmitter ... positive	34	GND
35	GND	36	USB D-
37	Reserved (1)	38	USB D+

Table 12-7 Connector Pin Assignments for Mobile *(continued)*

Pin#	Signal Line Name	Pin#	Signal Line Name
39	Reserved (1)	40	GND
41	Reserved (1)	42	LED_WWAN#
43	Reserved (1)	44	LED_WLAN#
45	Reserved (1)	46	LED_WPAN#
47	Reserved (1)	48	+ 1.5 V
49	Reserved (1)	50	GND
51	Reserved (1)	52	+ 3.3 V

Table Notes:

1. Reserved for possible future implementation of LANE# 1.
2. Reserved for possible future implementation of SIM.
3. Reserved for possible future implementation of disable signal.
4. Reserved for possible future implementation of wireless coexistence control.

An additional consideration of the PCI Express Mini Card is that the PCI Express interconnection is not the only one currently defined. The PCI Express Mini Card also defines the SMBus and USB interconnections. Consequently, the PCI Express Mini Card can be logically connected to the platform by the SMBus, USB, PCI Express link, or any combination of these.

Mobile Add-in Card Size: PCI Express Mini Card

Unlike the non-mobile add-in card, only one board size is specified for the PCI Express Mini Card. Thus the current definition of a PCI Express Mini Card supporting only x1 link width applies to this single board size.

Cable Modules

Other modules that encapsulate add-in cards and connect to a PCI Express platform via a cable have been proposed. As of the print date of this book, the final specifications were not available.

Signal Line Definitions

The following are not all the signal lines defined for a slot. There are also power, ground, and all those associated with the lanes of the link. These have been listed in the tables earlier in this chapter and are not discussed below. The signal lines discussed below have unique definitions not discussed in detail in other chapters of this book.

Non-Mobile Add-in Card and Mobile Connector in Slot

The signal lines listed in Tables 21.2 through 21.5 are defined as follows. Unless otherwise noted, all of the following signal lines are required and all add-in cards are non-mobile.

- **REFCLK (Input to add-in card):** Reference Clock, similar to the CLK signal line of PCI and PCI-X except it is 100 megahertz ± 300 parts per million and used for the clock reference to transmit the bits of the symbol stream via 8/10b encoding (2.5 gigabits per second).
    - Consists of two differentially driven signal lines: REFCLK+ and REFCLK−.
    - The REFCLK is required to be provided on the connector of each add-in card slot, but the add-in card is not required to use it. The add-in card can implement another REFCLK provided it maintains a 600 parts per million data bit rate.
- **LANE# Differential Pair Transmitter (Output from add-in card):** The signal lines that implement the PCI Express lane of the link.
    - Consists of two differentially driven signal lines: positive and negative.
- **LANE# Differential Pair Receiver (Input to add-in card):** The signal lines that implement the PCI Express lane of the link.
    - Consists of two differentially driven signal lines: positive and negative.
- **PERST# (Input to add-in card):** Typically the PERST# signal line is used as a reference point for Cold Reset as an indicator for the power level of main power. It is also used as a reference point for a Warm Reset. See Chapter 13 for more information.

- **JTAG (Optional)**: Five connections used for platform or non-mobile add-in card production testing. JTAG is defined by IEEE Standard 1149.1.
    - TRST# (Input to add-in card): Test reset.
    - TCK (Input to add-in card): Test clock.
    - TDI (Input to add-in card): Test data in.
    - TDO (Output to add-in card): Test data out.
    - TMS (Input to add-in card): Test mode select.
- **SMBus (Optional)**: System Management bus (SMBus) can be implemented between PCI, PCI-X, and PCI Express components on the platform and the add-in cards. Defined by the SMBUS 2.0 specification, it is a low power and lower bandwidth serial bus. The bus consists of SMBCLK (input to the add-in card) and SMDAT (input/output). This general purpose bus is used for purposes that range from platform management to inputs for equipment.
- **PRSNT1# (Input to add-in card) and PRSNT2# (Output from add-in card)**: Originally defined by PCI and PCI-X as add-in card presence detect, these signal lines are defined by PCI Express for the Hot Plug protocol. See Chapter 16 for more information.
- **WAKE# (Output from add-in card)**: Required for add-in cards and platforms that implement the Wakeup protocol. This signal line is used in conjunction with Vaux as part of the Wakeup Protocol. See Chapter 13 for more information.
- **Power Pins (Input to add-in card)**: PCI Express non-mobile add-in card power rails are +12 volts, +3.3 volts and 3.3 Vaux volts. 3.3 Vaux is also known simply as Vaux. See Chapter 17 for more information.
- **Reserved**: These signal lines can only be redefined by the PCI-SIG. Neither platforms nor add-in cards can connect anything to these signal lines.
- **GND**: These signal lines are connected directly to the ground plane of the platform.

PCI Express Mini Card and Mobile Connector in Slot

The signal lines listed in Table 21.5 are defined as follows. Unless otherwise noted all of the following signal lines are required and all add-in cards are the PCI Express Mini Card.

- **REFCLK (Input to add-in card)**: Reference Clock, similar to the CLK signal line of PCI and PCI-X except it is 100 megahertz ± 300 parts per million and used for the clock reference to transmit the bits of the symbol stream per 8/10b encoding (2.5 gigabits per second).
  - Consists of two differentially driven signal lines: REFCLK+ and REFCLK−.
  - The REFCLK is required to be provided on the connector of each add-in card slot, but the add-in card is not required to use it. The add-in card can implement another REFCLK provided it maintains a 600 parts per million data bit rate.

- **LANE# Differential Pair Transmitter (Output from add-in card)**: The signal lines that implement the PCI Express lane of the link.
  - Consists of two differentially driven signal lines: positive and negative.

- **LANE# Differential Pair Receiver (Input to add-in card)**: The signal lines that implement the PCI Express lane of the link.
  - Consists of two differentially driven signal lines: positive and negative.

- **PERST# (Input to add-in card)**: Typically the PERST# signal line is used as a reference point for Cold Reset as an indicator for the power level of main power. It is also used as reference point for a Warm Reset. See Chapter 13 for more information.

- **USB (Optional)**: Universal Serial Bus. Defined by the USB Specification revision 2.0. The bus consists of differentially driven pair USB_D+ and USB_D− (input/output).

- **SMBus (Optional):** System Management bus (SMBus) can be implemented between PCI, PCI-X, and PCI Express components on the platform and the add-in cards. Defined by the SMBUS 2.0 specification, it is a low power and lower bandwidth serial bus. The bus consists of SMBCLK (input to the add-in card) and SMDAT (input/output). It is a general purpose bus used for purposes that range from platform management to inputs from equipment.

- **LED_WPAN, LED_WLAN#, LED_WWAN# (Output from add-in card):** Required when the associated function is implemented on the PCI Express Mini Card. These signal lines are output from the PCI Express Mini Card to drive LED indicators on the mobile platform.

- **WAKE# (Output from add-in card):** Required for add-in cards and platforms that implement the Wakeup protocol. It is used in conjunction with Vaux as part of the Wakeup Protocol. See Chapter 13 for more information.

- **Power Pins (Input to add-in card):** PCI Express Mini Card power rails are +1.5 volts, +3.3 volts and 3.3 Vaux volts. 3.3 Vaux is also known simply as Vaux. See Chapter 17 for more information.

- **Reserved:** These signal lines can only be redefined by the PCI-SIG. Neither platforms nor add-in cards can connect anything to these signal lines.

- **GND:** These signal lines are connected directly to the ground plane of the platform.

Electrical

This book's PCI Express electrical discussion includes the power requirements add-in cards, and an overview of the transmission and reception of the waveforms on the signal lines. Not covered in this book are extensive details of the electrical signaling requirements. The appropriate specifications provide sufficient information on these details. Also, the actual electrical signaling of a PCI Express component is partly dependent on the processing limitations of the component's vendor.

The electrical requirements for PCI Express and PCI Express Mini Card can be divided into two categories: power and waveform. The power category relates to the voltage and current defined for the platform and add-in cards, it is discussed in detail in Chapter 17. The signal category can be viewed in two parts: PHY and waveform. This is a very brief overview to introduce the reader to some key concepts that support topics discussed in this book. See the *PCI Express Card Electromechanical Specification revision 1.0a* and *Mini PCI Express Card Electromechanical Specification revision 1.0* for more detailed information.

PHY

The PHY defines a physical implementation for transmitting and receiving the Physical Packets between ports on a link. As discussed in Chapter 8 and summarized in Figure 21.1, Physical Packets that traverse the link contain LLTPs and DLLPs. At the transmitting ports, the Physical Packets are encoded into a serial bit stream. At the receiving ports, the Physical Packets are decoded from a serial bit stream. The serial bit stream is defined in 10-bit entities called sbytes. The 10-bit sbytes are defined as D and K symbols, thus it is actually a stream of symbols that traverse the link. See Chapter 8 for more information.

> The term PHY simply refers to the "physical" implementation of the Physical Layer. This term is also used in design papers from Intel referring to possible implementations of the Physical Layer. Other ASIC vendors may define a PHY implementation similar to or different from Intel's. The PHY discussed in this book may not be exactly the same as that defined by Intel and other ASIC vendors, but serves for purposes of explanation.

Transmitting Port

> The following discussion is an overview of the detailed information provided in Chapter 8, and is presented here to help with the PHY and waveform discussion.

The PHY transmitting protocol is illustrated in Figure 21.2. LLTPs and DLLPs from the Data Link Layer are converted from a parallel orientation to a serial byte stream. The addition of FRAMING bytes distinguishes LLTP from DLLP byte streams and demarcates the individual Physical

Packets. Each 8-bit byte of the Physical Packet is encoded to a 10-bit sbyte by 8/10b encoding. The *sbyte* is a term adopted by this book to identify this unique 10-bit byte. The sbytes are also called symbols; they ensure that sufficient transitions of the bits are in the symbol stream to allow the receiving port to maintain bit and symbol lock.

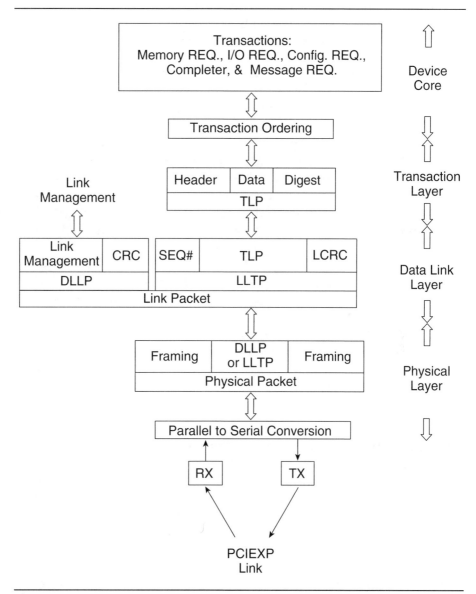

Figure 21.1 Physical Packet Transmission

As previously discussed, each link consists of one or more lanes. Each lane consists of two differentially driven pairs of signal lines, one pair for each direction. Once the 8/10b encoding has occurred, the resulting symbol stream representing the Physical Packet is parsed across the configured lanes of each configured link. The purpose of the parsing is to distribute the symbol stream across the link width to provide greater link bandwidth to transmit the Physical Packets.

If only one lane is available, no parsing occurs. As illustrated in Figure 21.2, two lanes are defined for the link. The symbol stream is parsed to the two lanes. Beginning with the FRAMING symbol, the first, third, fifth, seventh symbols (and so on) of the Physical Packet are transmitted onto LANE# 0. The second, forth, sixth symbols (and so on) of the Physical Packet are transmitted onto LANE# 1.

The last part of the PHY related to the transmitting port is to provide a bit period for each of the 10-bit sbytes for each symbol. That is, the differentially driven waveform representing the bit stream within each symbol must have transition points. These transition points are defined by a bit period. The bit period is generated from a 100 megahertz clock to provide a data bit rate of 2.5 gigabits per second. This is the only data bit rate supported by PCI Express specification revision 1.0a. Ten-bit periods define the common symbol period among the lanes of a specific link. That is, the lane to lane skew of the bit and symbol periods must be tightly controlled. Also, spread spectrum is applied to the REFCLK as discussed below.

As discussed in Chapter 8, Physical Packets may not be not available at all times to maintain a constant transmission of symbols across the link. The parser in Figure 21.2 integrates filler symbols with the symbol streams representing Physical Packets.

Receiving Port

The PHY receiving protocol is illustrated in Figure 21.3. The reception of the symbol stream requires the extraction of the reference clock. The 8/10b encoding provides a reference clock to which a PLL (phase lock loop) in the receiver can synchronize. This permits the receiver to determine the valid portion of each bit period. The resulting symbol streams from the two lanes are deparsed into a single symbol stream that is decoded from symbol stream of 10 bits into a stream of 8-bit bytes. The deparser also removes any filler symbols. The FRAMING demarcates the Physical Packets within the stream and distinguishes DLLPs from LLTPs.

Chapter 21: Mechanical and Electrical Overview ■ 861

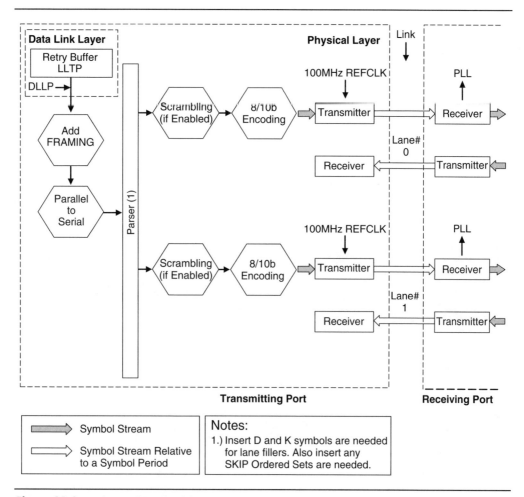

Figure 21.2 Physical Packet Transmission

The last part of the PHY related to the receiving port is to convert the serial stream of 8-bit bytes to a parallel orientation, and to identify direct DLLP and LLTP. The DLLP provides link management information to the Physical Layer and the LLTP is placed in an input buffer.

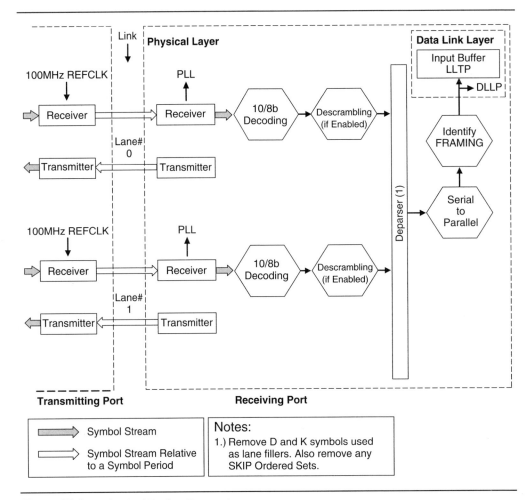

Figure 21.3   Physical Packet Reception

Scrambling

When a binary bit stream has a high frequency of bit transition points, measurable electromagnetic interference (EMI) may occur. If certain binary patterns are generated in a repetitive fashion, certain frequencies' EMI may need to be reinforced. The PCI Express specification has defined a scrambling protocol to address this issue by scrambling the exact binary pattern transmitted. The scrambling mechanism prevents the occurrence of repetitive binary patterns. Data scrambling can be enabled and disabled at different times. At the transmitting port, the scrambling is applied just prior to the 8/10b encoding, as shown in Figure 21.2. At the receiving port, the scrambling is applied just prior to the 8/10b encoding, as shown in Figure 21.3. See Chapter 8 for more information.

Waveform

As discussed in Chapters 1 and 8, one of the primary characteristics that differentiates PCI Express from PCI and PCI-X is the elimination of a CLK signal line to provide a reference sampling point for the valid period of other signal lines. Also, as discussed in Chapter 1, the PCI-X specification was revised to include source strobes to retain a reference sampling point to increase bus segment bandwidth. PCI Express eliminated the need for a CLK signal line or source strobes to provide a valid sampling point for the signal lines contain transaction information. PCI Express relies on 8/10b encoding at the transmitter to integrate a reference clock, and a phase lock loop at the receiver to extract the reference clock for sampling at the valid portion of the waveform.

Figure 21.4 illustrates the portion of the bit stream that comprises the symbol stream. Each symbol that traverses the link is constructed of a single sbyte of 10 bits. The data bit rate defined by PCI Express specification revision 1.0a is only 2.5 gigabits per second. This establishes the Typical Bit Period of each of the 10 bits within each symbol. The width of the Typical Bit Period will vary depending on the tolerance of the REFCLK ($\pm$ 300 parts per million) and the associated PHY circuitry. For this discussion, the variance in the Typical Bit Period is not be considered.

The Transmitter Valid Bit Period in Figure 21.4 is less than the Typical Bit Period. The Transmitter Valid Bit Period = Typical Bit Period − Transmitter Transition Range. The output of a port's transmitters for each lane of a link is a differentially driven pair of signal lines. The transition point at the boundaries of the each bit period is not exactly at the boundary of the Typical Bit Period due to the application of spread spectrum. For a fixed clock frequency, the electromagnetic interference (EMI) generated is very high for that specific frequency. If the frequency is varied slightly with time, the basic Typical Bit Period is maintained, but the value of the EMI of that specific frequency is reduced. The spread spectrum implementation is just another mechanism, along with scrambling, used to minimize the EMI of PCI Express platforms and add-in cards.

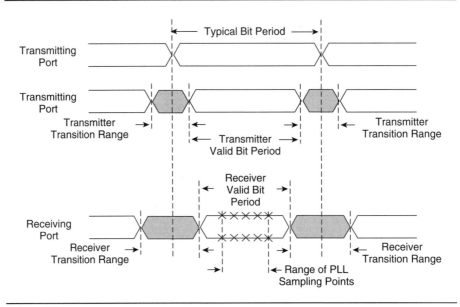

Figure 21.4 Waveform

As the waveform traverses the link, transmission line effects widen the transition range at the boundaries of Typical Bit Periods. Consequently, the waveform at the receiving port defines a Typical Bit Period established by the transmitting port and a Receiver Transition Range that is the Transmitter Transition Range plus the transmission line effects. As exemplified in Figure 21.4, the Receiver Valid Bit Period is less than the Typical Bit Period and Transmitter Valid Bit Period. The Receiver Valid Bit Period = Typical Bit Period – Receiver Transition Range.

The PHY mechanism at the receiving port contains a phase lock loop (PLL). As discussed in the Chapter 8, the symbol stream contains bit transition points for the 8/10b encoding that integrated a reference clock. The PLL synchronizes to the transition points provided in the bit and symbol lock on the frequency of the transition points to provide a reference clock at the receiving port. The purpose of the reference clock is to provide a sampling point within the Receiver Valid Bit Period. Due to the variance in the Receiver Transition Range and the nature of PLLs, the exact sampling point is within a certain range. Consequently, the limit of the data bit rate traversing the link is how small the Receiver Valid Bit Period becomes versus the exactness of the sampling points.

Chapter 22

Configuration Overview

This chapter gives a complete overview of PCI Express configuration space, including purpose, features, and usage. It explains the different types of configuration accesses, describes the configuration space hierarchy, and describes how the configuration space of individual PCI Express devices is initialized and addressed. The next chapter describes the function, architecture, and layout of PCI Express configuration registers.

Introduction

The PCI bus specification introduced *configuration address space* to go along with the common memory and I/O address spaces used with earlier buses. During runtime, an operating system may reallocate memory and I/O address resources as necessary, but configuration space is a fixed address window to each PCI device and does not change after bus enumeration, which typically occurs during system initialization. Each PCI device is always visible in configuration space, even before system software has allocated memory or I/O resources to any device.

The PCI Express specification is written to ensure system software compatibility between PCI and PCI Express. PCI Express is 100 percent binary compatible with configuration transactions as defined in the *PCI 2.3 Specification*. This means PCI configuration access mechanisms and registers map directly to PCI Express configuration access mechanisms and registers. The PCI Express specification is written to allow an upgrade of systems and devices from PCI to PCI Express with no change to the operating system.

But PCI Express also defines additional configuration registers and capabilities beyond PCI. When a PCI Express device uses only PCI configuration registers and capabilities, it is a *legacy* PCI Express device. When it uses additional features of PCI Express, it is a *native* PCI Express device.

This chapter introduces the PCI Express configuration environment. The next chapter describes PCI Express configuration registers. For additional details of legacy PCI configuration, see the books and specifications referred to in Chapter 1.

Features of Configuration Space

The fundamental PCI Express unit is a *device*. Examples of PCI Express devices are *endpoints* such as an embedded network controller chip or a plug-in card that operates a disk array, or *bridges* to remote PCI Express devices. Configuration space guarantees access to every PCI Express device in the system. Some features of configuration space include the ability to:

- Detect PCI Express devices.

- Identify the function(s) of each PCI Express device.

- Discover what system resources each PCI Express device needs. System resources include memory address space, I/O address space, and interrupts.

- Assign system resources to each PCI Express device.

- Enable or disable the ability of the PCI Express device to respond to memory or I/O accesses.

- Tell the PCI Express device how to respond to error conditions.

- Program the routing of PCI Express device interrupts.

Through configuration space, system software identifies every PCI Express device and assigns to each device a portion of system resources, changing that assignment during runtime if necessary as resource needs change. In this way, software optimizes the allocation of resources between devices, and ensures that no resource conflicts exist.

The Target of a Configuration Address

Historically, processors have been designed with two address spaces: memory and I/O. When a processor forms a memory address or an I/O address, the target of that address is a byte or word or DWORD in an I/O or memory device, such as a ROM chip holding the BIOS, a RAM chip, or a disk storage unit, for example. But when a processor forms a *configuration* address, the target of that address is always a byte, word, or DWORD in a specific PCI Express device.

A typical processor that can generate only memory addresses and I/O addresses needs a way to get to configuration address space. The PCI Express specification leaves this up to the implementation, but the most common method is described in the section "Creating a Configuration Address" later in this chapter.

The source of a PCI Express configuration read or write is always a processor, and the target is always a PCI Express device either in the Root Complex or somewhere in the PCI Express hierarchy below the Root Complex. As required in the PCI Express specification, PCI Express devices cannot initiate configuration reads or writes.

Configuration Space Size

Configuration space is strictly defined by the PCI Express specification. Each PCI Express *device* holds from one to eight *functions*. Every function is individually addressable using 16 configuration address bits, which means the PCI Express specification supports up to 65,536 configurable functions. This is the upper bound on the size of any PCI Express hierarchy.

> In practice, it is not possible to design a system with 65,536 unique functions using physical links, due to the point-to-point nature of the PCI Express link. A PCI Express bus within a switch component may have up to 32 devices (256 functions) attached, but a PCI Express link (a logical bus) between two components may have only one device (8 functions) at the downstream endpoint of the link.

Furthermore, each function decodes 12 configuration address bits for the bytes of configuration space within that function, which means each function holds up to 4096 bytes of configuration space. All of system configuration space is logically divided into functions, each of which has a 4096-byte block of configuration space.

Chapters 23 and 24 describe in detail the predefined format of this configuration block. Some byte locations in the block are required, others are optional. A PCI Express device may store as few as 13 bytes out of the 4096-byte block and still be fully compliant with the PCI Express specification.

Configuration Space Hierarchy

In a PCI Express system, the *bus* number, *device* number, and *function* number make up the 16-bit address for each unique function in a PCI Express hierarchy. You can think of the bus number as the street, the device number as the building on the street, and the function number as the apartment in the building.

At least one bus always exists: bus zero. A PCI Express hierarchy may have as many as 256 buses (numbered 0 to 255). Figure 22.1 shows a typical small PCI Express hierarchy in which bus 0 resides within the Root Complex and holds several devices. In a PCI Express system, bus 0 always exists within the Root Complex. Figure 22.2 shows a typical medium PCI Express hierarchy, with many subordinate buses. Figure 22.3 shows a typical large PCI Express hierarchy with a peer Root Complex and multiple subordinate buses. In every PCI Express hierarchy, a single Root Complex holds bus zero, and the total number of buses is never more than 256.

> The Root Complex may encompass two or more chips on a motherboard and bus 0 may pass between the chips, but how bus 0 is electrically routed within the Root Complex is implementation-specific and not defined in the PCI Express specification.

Buses hold devices, and each bus always has at least one device: device number zero. Some examples of devices are a network controller, a disk controller, or a bridge to subordinate PCI Express buses. Each bus holds up to 32 devices, numbered 0 through 31.

Each device performs one or more functions, represented by the function numbers in the configuration address. A configuration address always targets a specific function on a specific device. A disk controller device, for example, might support two disk interfaces and protocols, each a different function. Each function is allocated 4096 bytes, separate and unique from every other function. In a PCI Express hierarchy, each device on a bus holds up to eight functions. For each device, function number zero is required, and functions 1 through 7 are optional.

Chapter 22: Configuration Overview ■ 869

> Before bus enumeration, a device may increase the available configuration space available for one of its functions by claiming more functions, up to the maximum of eight.

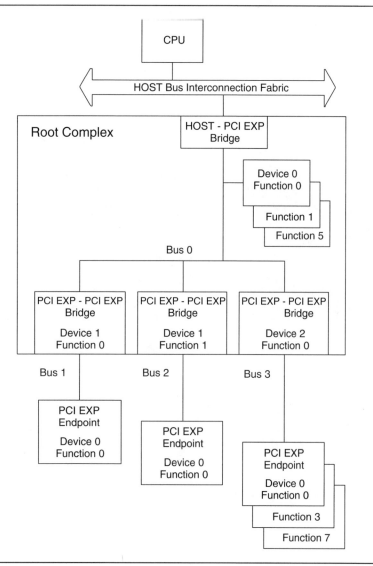

Figure 22.1 Small PCI Express Hierarchy

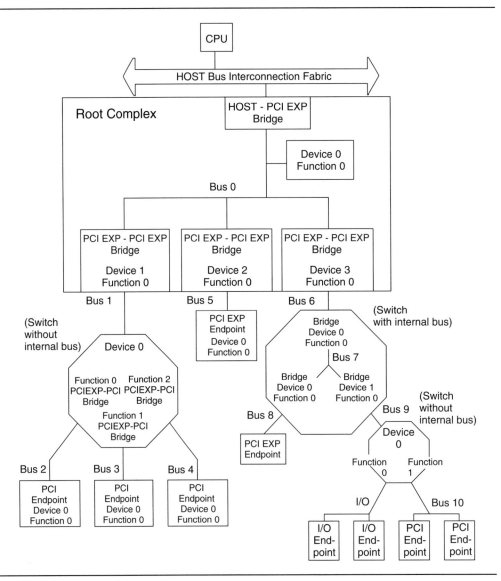

Figure 22.2 Medium PCI Express Hierarchy

Chapter 22: Configuration Overview   **871**

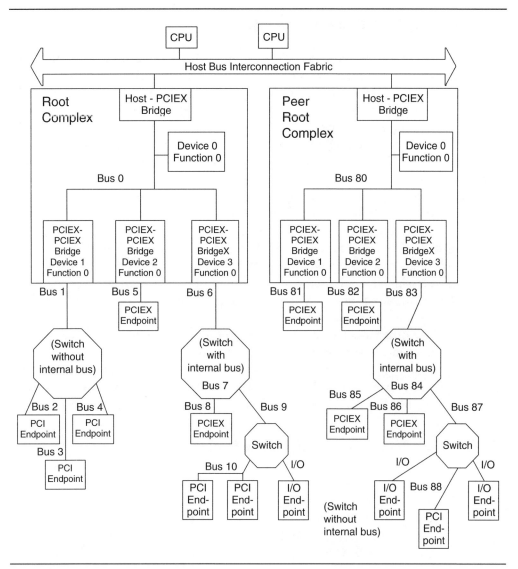

Figure 22.3 Large PCI Express Hierarchy

Creating a Configuration Address

PCI Express devices are configured by a processor reading and writing to addresses within that device's configuration space. The processor sends a configuration read or a configuration write to the Root Complex (or peer Root Complex), which then routes that access through its hierarchy to the correct bus, device, and function specified in the address.

Some processors do not themselves have a configuration address space and so must use memory or I/O space to initiate configuration accesses. The mechanism by which a processor instructs a Root Complex to read and write to configuration space in the PCI Express hierarchy is implementation-specific, but two methods are defined and commonly used.

One mechanism is defined in the PCI 2.3 specification. This mechanism converts *I/O* reads and writes from a processor into PCI configuration reads and writes in a host-to-PCI bridge. It is called PCI Configuration Mechanism #1. This mechanism works in PCI Express legacy-compatible Root Complexes and is defined in the following section, "Legacy PCI Express Configuration Mechanism."

The other mechanism is defined in the PCI Express specification. This mechanism converts memory reads and writes from a processor into PCI configuration reads and writes in a host-to-PCI bridge. It is called PCI Express Enhanced Configuration Mechanism and is defined in the section "Native PCI Express Configuration Mechanism" later in this chapter.

Legacy PCI Express Configuration Mechanism

The legacy PCI Express Configuration Mechanism uses I/O addresses 0CF8h and 0CFCh to translate processor accesses into PCI configuration accesses. The host-to-PCI bridge in the Root Complex captures I/O reads and I/O writes to these addresses and converts them to configuration reads and writes, which it passes through the PCI hierarchy to the target function. These I/O addresses are reserved for PCI configuration and do not conflict with other resources in a PCI-compatible system.

Config_Address Register

The Config_Address register is a 4-byte read/write register that resides in the PCI Express legacy Root Complex. The reset state of this register is zero; it must be written with a nonzero value to enable configuration accesses. The value written to Config_Address is stored within the Root

Complex; writing or reading Config_Address causes no activity on PCI Express. Sometime later when the processor writes or reads Config_Data, the Root Complex forwards that write or read on to the PCI Express bus as a configuration access using the value previously stored in the Config_Address register to create the configuration address for that access.

All writes and reads to Config_Address must be full DWORD accesses to address 0CF8h. A single-byte or single-word access to Address 0CF8h is treated as an I/O (not a configuration) access.

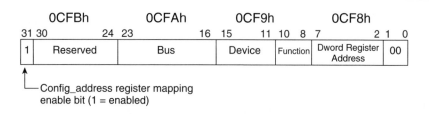

Figure 22.4 Config_Address Register Layout

Config_Data Register

The Config_Data register is a 4-byte read/write register that resides in the PCI Express legacy Root Complex. The reset state of this register is zero. The four bytes of this register are located at I/O addresses 0CFCh, 0CFDh, 0CFEh, and 0CFFh. Any byte, word, or DWORD value read or written to these locations is captured by the Root Complex and forwarded as a PCI Express legacy configuration access to the PCI Express hierarchy, using the address stored in the Config_Address register.

31	24 23	16 15	8 7	0
0CFFh	0CFEh	0CFDh	0CFCh	

Figure 22.5 Config_Data Register Layout

Config_Address and Config_Data Example

Suppose a processor writes to bus number 6, device number 2 on that bus, and function number 1 on that device. And within the block of memory reserved for that function, it writes 33h to address 52h.

This is a two-step process. First, the processor writes to the Config_Address register to set up the configuration address; then it writes the data to the Config_Data register. To set up Config_Address, the processor writes the hexadecimal DWORD value 80061150 to I/O address 0CF8h, as shown in Figure 22.6

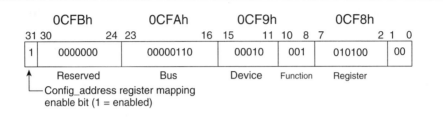

Figure 22.6 Config_Address Register Example

The enable bit set means that any subsequent reads and writes to Config_Data uses the value in this register to form the PCI Express legacy configuration address. Note that Config_Address stores not only the bus, device, and function number, but also the particular DWORD register within configuration space that is the target of the read or write. In this case, the targeted DWORD is at address 50h, which means one or more bytes at locations 50h, 51h, 52h, and 53h is the target of any subsequent Config_Data reads and writes.

The second step is to write the byte value 33h to the correct location in the Config_Data register. The processor performs a byte write to I/O address 0CFEh, as shown in Figure 22.7. The Root Complex captures this data, ignores the unwritten bytes, and packages the data with the address forming a PCI Express configuration write to the correct device and function within the PCI Express hierarchy. The format of PCI configuration accesses within the PCI Express hierarchy is defined in the section, "Native PCI Express Configuration Mechanism."

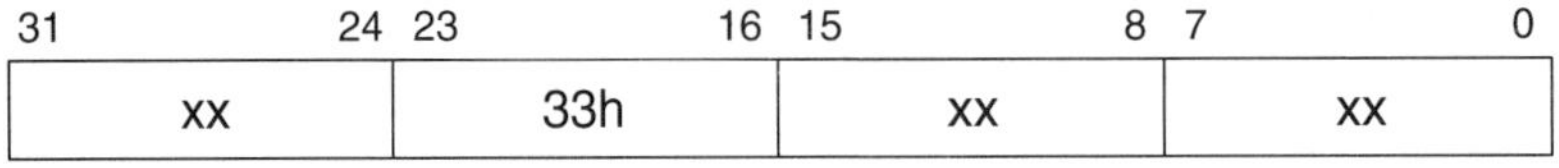

Figure 22.7 Config_Data Register Example

Limitation of Config_Address and Config_Data

PCI Configuration Mechanism #1 was defined to allow processors to access PCI (not PCI Express) configuration space. PCI configuration space defines 256 bytes of data per each unique function within a PCI hierarchy, not the 4,096 bytes defined for PCI Express native devices. PCI Express native devices in hierarchies configured using PCI Configuration Mechanism #1 must run in legacy mode with only 256 bytes per function available to the system.

Native PCI Express Configuration Mechanism

The "Enhanced PCI Express Configuration Mechanism" defined in the PCI Express specification maps all the PCI Express configuration registers into system memory space. The PCI Express hierarchy allows up to 65,536 separate functions, each of which may hold up to 4,096 bytes, for a total of 256 megabytes. This requires 28 address bits to uniquely decode every byte. These 28 bits are encoded onto address bits A27 down to A0, as shown in Figure 22.8. Address bits 29 and above are considered the "base address." During system initialization, system firmware determines the base address and communicates it to both the Root Complex and to the operating system. The method by which this is done is implementation-specific, and not defined in the PCI Express specification.

> The Enhanced PCI Express Configuration Mechanism implies that the Root Complex (and peer Root Complexes, if any) is mapped to system resources in a method other than using a memory-mapped base address, because it is not defined in the specification how the base address itself gets programmed into the Root Complex.

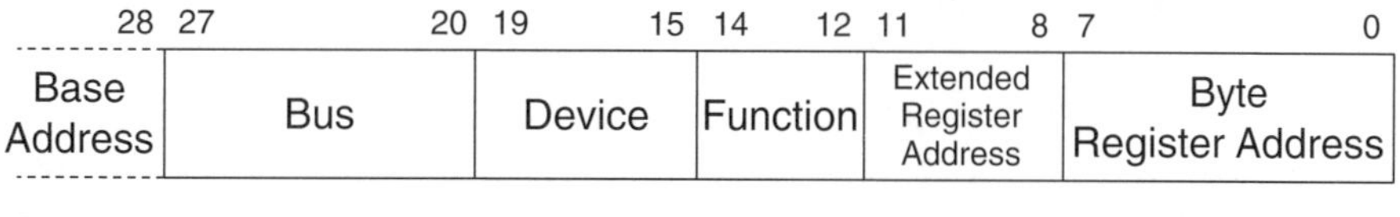

28 27	20 19	15 14	12 11	8 7	0
Base Address	Bus	Device	Function	Extended Register Address	Byte Register Address

Figure 22.8 Enhanced Configuration Mechanism Address Mapping

How Configuration Addresses Reach Their Destination

A configuration address consists of a bus number, a device number, a function number, and a register number:

- The bus number selects one of 256 possible buses.
- The device number selects one of 32 possible devices on the selected bus.
- The function number selects one of 8 possible functions on the selected device.
- The register number selects one of 64 possible DWORD registers (256 bytes total) within the selected function for PCI Express legacy configurations, or it selects one of 1024 possible DWORD registers (4096 bytes total) within the selected function for PCI Express native configurations.

Every PCI Express bridge knows the bus number to which it is attached, as well as the number for all subordinate buses attached below it in the PCI Express hierarchy. This information is configured during bus enumeration (see "How to Detect PCI Devices" later in this chapter). A configuration access routes through the hierarchy by having each bridge do one of two things: either claim that configuration access as belonging to a device within its subordinate bus hierarchy, or ignore the access. In this way a configuration access starts at the Root Complex and traverses through the hierarchy until the target function is reached.

Type 0 and Type 1 Configuration Accesses

All configuration accesses are encoded as "Type 0" or "Type 1." Type 1 means the access has not reached its destination bus, Type 0 means it has. A configuration access is passed from bridge to bridge as a Type 1 transaction until it finally reaches the bridge that holds its destination bus, and that bridge converts the access to a Type 0 configuration transaction. Only bridge devices claim type 1 transactions, any device may claim a Type 0 transaction.

The encoding of Type 0 or Type 1 is stored in the Type Field of the header (see Figure 23.3). When the Root Complex receives a configuration access from a processor (typically using an I/O-mapped or memory-mapped configuration mechanism) the Root Complex creates the appropriate Type field encoding and sends the configuration read or write to the destination bus.

Figure 22.8 shows the address encoding for both Type 0 and Type 1 configuration accesses. On PCI Express, the only difference between the two access types is the Type Field encoding in the transaction header. On legacy PCI buses, the Type 0 and Type 1 encoding is placed in the address itself, and the register field is only 8 bits, addressing 256 registers per function. A PCI Express-to-legacy-PCI bridge must convert the PCI Express encoding to the legacy PCI encoding. If the PCI Express encoding has a nonzero value in the Extended Register field, it cannot convert to legacy PCI, and the access is terminated as an Unsupported Request.

On legacy PCI buses, Type 0 accesses do not include the device number, so some technique is required to select the device. PCI buses use a select signal called "IDSEL" but this signal does not exist in PCI Express. PCI Express leaves device selection implementation-specific:

- A PCI Express bus between components is always point-to-point with just one device attached, so a select signal is not needed. Device 0 on the upstream port of a bridge always assumes it is selected when it sees a Type 0 configuration access.

- A PCI Express bus inside a switch or a Root Complex component can have multiple devices, so each device in the component must decode the device field in the configuration address, or else the component must drive an internal signal to select the target device during a Type 0 configuration access. In any case, since the entire bus is contained within the switch or Root Complex component, the PCI Express specification leaves the select mechanism implementation-specific.

Configuration Access Example 1

In this example, a processor writes to a configuration register in device 0, function 0, on bus 0. Refer to Figure 22.1.

Using a method understood by the Root Complex, the processor performs its configuration write. This method might be a pair of I/O writes to 0CF8h and 0CFCh for PCI Express legacy systems or a memory access for PCI Express native systems.

The Root Complex receives the configuration write and determines whether it needs to create a Type 0 or Type 1 configuration write. In this case, the answer might be neither. The Root Complex has two options:

- Route the configuration access to device 0 using any implementation-specific technique. Configuration accesses to devices within the Root Complex are a special case and can be handled in an implementation-specific manner.

- Route the configuration to device 0 using a Type 0 configuration access on bus 0, and select the device using an implementation-specific technique. When the device is selected it captures the function and register numbers from the Type 0 access and routes the data to the appropriate register.

Configuration Access Example 2

In this example a processor writes to a configuration register in device 0, function 0, on bus 3. Refer to Figure 22.1.

Using a method understood by the Root Complex, the processor performs its configuration write. This method might be a pair of I/O writes to 0CF8h and 0CFCh for PCI Express legacy systems or a memory write for PCI Express native systems.

The Root Complex receives the configuration write and routes it to the correct bridge. The Root Complex has two options:

- Route a Type 1 configuration access directly to device 2 within the Root Complex using any implementation-specific technique.

- Create a Type 1 configuration access on bus 0. The type 1 configuration access includes the bus number ("3") for the target. Device 2 on bus 0 notices that bus 3 belongs to its subordinate hierarchy, so device 2 claims the access.

After claiming the access, device 2 in the Root Complex determines that bus 3 is attached directly to its bridge and is not somewhere further down the hierarchy, so circuitry within device 2 converts the Type 1 configuration access to a Type 0 configuration access and sends it across bus 3.

Device 0 on bus 3 sees a Type 0 configuration access and immediately accepts it, knowing that any device receiving a Type 0 access is itself the target of that access. When it accepts the access, it routes the write data to the appropriate function and register.

Configuration Access Example 3

In this example a processor writes to a configuration register in device 0, function 1, on bus 8. Refer to Figure 22.2.

Using a method understood by the Root Complex, the processor performs its configuration write. This method might be a pair of I/O writes to 0CF8h and 0CFCh for PCI Express legacy systems or a memory access for PCI Express native systems.

The Root Complex receives the configuration write and routes it to the correct bridge. The Root Complex has two options:

- Route a Type 1 configuration access directly to device 3 within the Root Complex using any implementation-specific technique.

- Create a Type 1 configuration access on bus 0. The type 1 configuration access includes the bus number ("8") for the target. Device 3 on bus 0 notices that bus 8 belongs to its subordinate hierarchy, so device 3 claims the access.

After claiming the access, device 3 in the Root Complex determines that bus 8 is not attached directly to its bridge, so device 3 sends the Type 1 configuration access unchanged across bus 1.

Device 0 on bus 6 sees the Type 1 configuration access and determines that bus 8 belongs to its subordinate hierarchy, so device 0 accepts the access. After claiming the access, device 0 on bus 6 determines that bus 8 is not attached directly to its bridge, so device 0 sends the Type 1 configuration access unchanged across bus 7.

Device 0 on bus 7 sees the Type 1 configuration access and determines that bus 8 belongs to its subordinate hierarchy, so device 0 accepts the access. After claiming the access, device 0 on bus 7 determines that bus 8 is attached directly to its bridge, so device 0 converts the Type 1 configuration access to a Type 0 configuration access and sends it across bus 8.

Device 0 on bus 8 sees a Type 0 configuration access and immediately accepts it, routing the write data to the appropriate function and register. Device 0 on bus 8 knows it is the recipient of this Type 0 configuration access because it is the upstream port of a bridge device, and no other devices can exist on its bus.

Access Rules for Configuration Space

The objective of a PCI Express configuration access is to permit system software access to any specific device in the PCI Express hierarchy. Software can read or write any of the 256 register bytes in a PCI Express legacy device function, or any of the 4096 bytes in a PCI Express native device function. But some locations in the register map have reserved or fixed values, and other locations may not be implemented by a particular function.

Nevertheless, in all cases the following access rules apply.

Access Rules for Configuration Space Reads

- A PCI Express device is required to return data when its configuration space is read.

- A PCI Express component is required to return data when it receives a read to a bus, device, or function that does not exist. The data must be FFs with an "unsupported request completion status."

- A PCI Express device must reply with a normal completion when an unused or reserved register is read.

- A PCI Express device must reply with a zero for any bit that is unused or reserved, unless the specification explicitly says other-wise.

- A PCI Express device must return the data value that the device is actually using.

- Software must consider the value of reserved bits as undefined.

Access Rules for Configuration Space Writes

- A PCI Express device is required to ignore data during writes to reserved registers. The data is discarded, the write is changed to a no-op, and the write completes normally.

- Software must preserve the value of reserved bits when writing configuration registers. To do this, software must use a read-modify-write technique when writing to registers that hold reserved bits.

> Note: If two processors or two instruction threads are allowed to independently and simultaneously access PCI Express configuration registers, then a read-modify-write operation may be non-atomic. That is, two read-modify-write operations could overlap, thus destroying the data from one of them. One solution is to lock the read-modify-write sequence, but the PCI Express specification allows lock only for legacy PCI Express implementations, not for native implementations. Therefore, systems that allow multiple independent accesses to configuration space must ensure that only one configuration master is active at a time.

How to Detect PCI Express Devices

Software can detect all the devices in a PCI Express hierarchy by reading a single configuration space register at each possible bus and device combination. Software can follow this technique: for each bus number, read device 0, function 0, register Vendor ID (register offset 00, length 2 bytes). If a value other than FFFFh is read, the device is present. If the device is not present, move to the next bus number. If the device *is* present, move to the next device number on that bus until all possible devices have been tested.

Software can also detect all the functions on a given device with this simple technique: for function 0 on the given device, read the Header Type register (offset 0Eh, length 1 byte). If the most significant bit is set to 0, the device is a single-function device. If it is set to 1, it is a multi-function device. In either case, function 0 is always used. If it is a multi-function device, further discover which functions are used by reading the Vendor ID register for functions 1 through 7. If a value other than FFFFh is read, the function is present.

Note that historically some *single*-function PCI devices return a value other than FFFFh in the Vendor ID register for function numbers 1 through 7. Reading the Header Type register first will accurately tell whether the device is single-function or multi-function.

PCI Express Bus Enumeration

After reset, the host CPU can address bus number 0 in the Root Complex and the devices attached to it, but cannot address any other buses or devices. Every bus except for bus 0 needs its bus number assigned, which occurs after reset when system software performs PCI bus enumeration.

During enumeration, every bus is assigned a number. Configuration registers in each device that controls a bus contains the bus number. Refer to Figure 23.5 for the configuration registers involved. Following is a typical algorithm by which system software assigns bus numbers. Figure 22.9 illustrates this algorithm.

1. After reset, every bus is listed as zero. No device will respond to configuration accesses to any other bus number. The Root Complex is first to see configuration accesses to bus 0 and claims them.

> Note: A *peer* Root Complex cannot have the same bus number as the Root Complex, or both would respond to software accesses. Bus zero is always in the Root Complex. In systems with one or more peer Root Complexes, the primary bus in a peer Root Complex must reset to some number higher than the total number of buses in the Root Complex plus the total number of buses in any other peer Root Complexes enumerated before it.
>
> For example: In a system with a Root Complex and a single peer Root Complex, the peer Root Complex is typically hardwired to reset to bus number 128 (that is, 80h). After software enumerates each bus in the hierarchy of buses underneath the Root Complex with numbers from 1 up through the last bus (as high as 127 in this example), the software finds the peer Root Complex by testing for bus 80h (which it could then renumber), and proceeds to enumerate the hierarchy of buses underneath the peer Root Complex.

2. System configuration software identifies each device on bus 0 and discovers whether it includes any functions that are bridges to another bus. Bridge functions have a hardwired value of 06h in the Class Code Register. The devices can be searched in any order.

3. When a bridge function is found, the system configuration software writes to its Primary Bus Number and Secondary Bus Number registers. For the first bridge function below the host bridge (call it bridge 2) found during enumeration, the Primary Bus Number is 0 and the Secondary Bus Number is 1.

4. Before configuration software looks for another bridge on bus 0, it looks for subordinate bridges behind bus 1. It temporarily puts the value FFh in bridge 1's Subordinate Bus Number register, then checks for bridges on bus 1.

5. If configuration software finds a bridge on bus 1, it writes to that bridge's (bridge 6 in this example) Primary Bus Number and Secondary Bus Number registers. For bridge 6, the Primary Bus Number is 1 and the Secondary Bus Number is 2.

6. Before configuration software looks for another bridge on bus 1, it looks for subordinate bridges behind bus 2. It temporarily puts the value FFh in bridge 6's Subordinate Bus Number register, then checks for bridges on bus 2. For this example, no bridges are on bus 2, so configuration software writes the value 2 to bridge 6's Subordinate Bus Number register.

7. Now configuration software backtracks to bus 1 and looks for more bridges on that bus. It finds another bridge (bridge 7) and writes to its Primary Bus Number and Secondary Bus Number registers. For this example, the Primary Bus Number is 1 and the Secondary Bus Number is 3. Now configuration software puts a temporary value of FFh in bridge 7's Subordinate Bus Number register and looks for bridges on bus 3.

8. This recursive algorithm is followed until configuration software completely programs all the bridge functions with values in their Primary Bus Number, Secondary Bus Number, and Subordinate Bus Number registers.

Note: In this discussion the term *bridge* is used. To be precise, each bridge exists as a function on a device. A bridge could be the only function on a device (a single-function device) or there could be multiple bridges and other functions on a device (a multi-function device).

884 ■ The Complete PCI Express Reference

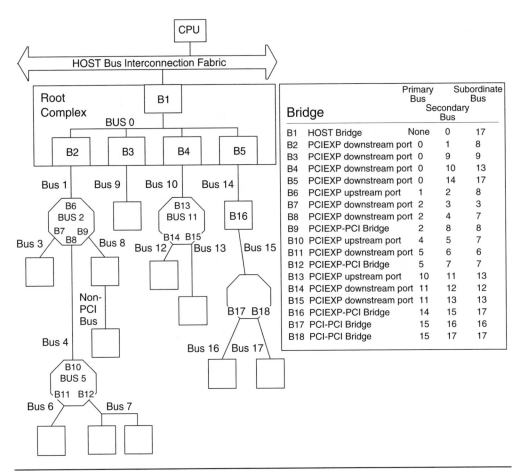

Figure 22.9 PCI Express Bus Enumeration Example

PCI Express Bridge Architecture

For designers familiar with legacy PCI bridges, PCI Express bridges may at first appear confusing. In fact, no single "PCI Express bridge" function exists. A PCI Express bridge is either a root port bridge, an upstream port bridge, or a downstream port bridge. While a legacy PCI bridge is identified as such in the Class Code register (see Figure 23.3 in Chapter 23), a PCI Express bridge (whichever type) is identified in the PCI Express Capabilities Register (see Figure 24.3 in Chapter 24).

Two Rules for Connecting PCI Express Bridges

The way in which PCI Express bridges connect to each other and to other devices has certain limitations. Specifically, the following two rules apply:

- All PCI Express devices that have functions other than root ports or downstream ports must connect to either a root port or a downstream port.

- Upstream ports connect on the south only to downstream ports; downstream ports connect on the north only to upstream ports (where north is closer to the Root Complex, and south is further from the Root Complex).

The following figures illustrate these rules.

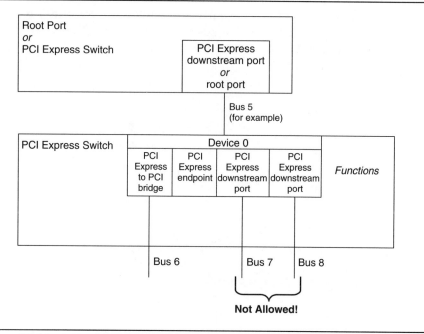

Figure 22.10 PCI Express Bridge Example #1

Figure 22.10 illustrates an illegal condition that violates the second rule. In this example, a PCI Express switch component includes four functions: one endpoint, one PCI Express to PCI bridge, and two PCI Express downstream ports acting as PCI Express bridges. This switch component does not have an internal bus. The PCI Express to PCI bridge and the endpoint are correctly placed as functions on device 0. But the PCI Express downstream ports cannot be placed as functions on device 0 because they connect on the north to a downstream or root port which is not allowed. PCI Express downstream ports must connect on the north to an upstream port.

Figure 22.11 illustrates an illegal condition that violates the first rule. In this example, a PCI Express switch component includes four functions: one endpoint, one PCI Express to PCI bridge, and two PCI Express upstream ports acting as PCI Express bridges. This switch component does not have an internal bus. All of these functions are correctly placed on device 0. But the PCI Express upstream ports cannot drive their south buses as external links because any device attached to that link violates the first rule. Every device must connect to either a root port or a downstream port.

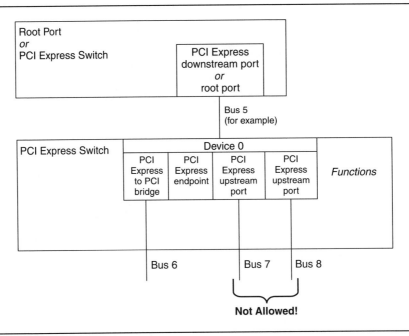

Figure 22.11 PCI Express Bridge Example #2

Figures 22.12, 22.13, and 22.14 give three solutions for designing the switch in the previous figures in a way that meets the rules for PCI Express bridges. All three of these solutions are equally valid. Figure 22.12 uses two internal buses, Figure 23.13 uses one internal bus and puts two devices on that bus, and Figure 23.14 uses one internal bus with one internal device that has two functions. Note that any PCI Express switch component that drives a south PCI Express link must have an internal bus.

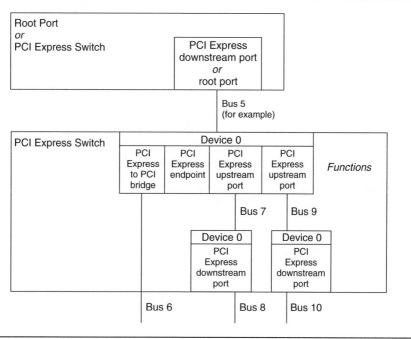

Figure 22.12 PCI Express Bridge Example #3

888 ■ The Complete PCI Express Reference

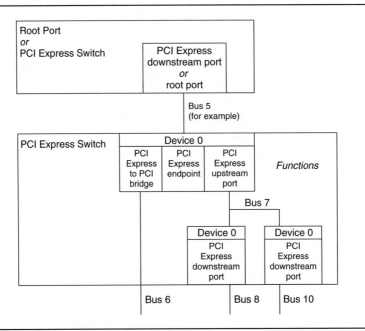

Figure 22.13 PCI Express Bridge Example #4

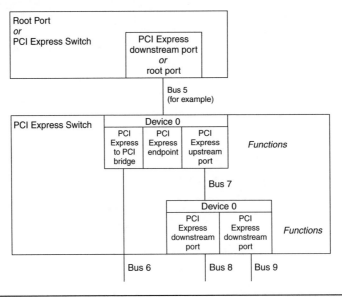

Figure 22.14 PCI Express Bridge Example #5

Figure 22.15 illustrates an illegal condition that violates both rules. In this example, a PCI Express switch component includes four functions: one endpoint, one PCI Express to PCI bridge, and two PCI Express downstream ports. This switch component has an internal bus and the four functions are in devices that reside on the internal bus. Both rules are violated because the upstream port connects to device 0 on bus 6, and that device has functions that are not downstream ports.

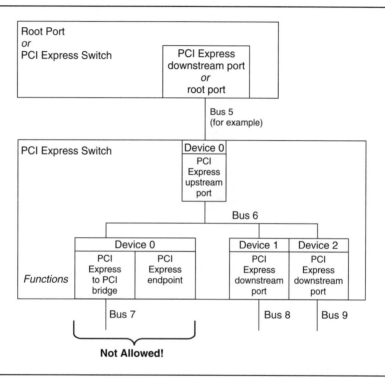

Figure 22.15 PCI Express Bridge Example #6

Figure 22.16 gives a solution for designing the switch in the previous figure in a way that meets the rules for PCI Express bridges. This solution is equally valid with the solutions in Figures 22.12, 22.13, and 22.14.

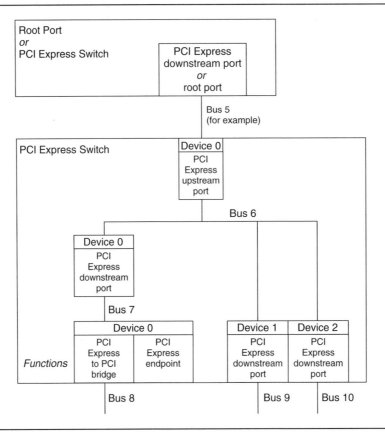

Figure 22.16 PCI Express Bridge Example #7

Other Rules for PCI Express Bridges

A PCI Express to PCI Express bridge in a switch consists of an upstream port function and a downstream port function. For these bridges, the following three rules apply:

- **Read requests must pass un-split through the bridge.**

    This rule means that a 4-kilobyte read request, for example, cannot be split into two or more read requests in the bridge. It is the responsibility of the destination of the read (for example, the root port) to split read requests if necessary. This rule also means that the *Max_Read_Request_Size* register (see Chapter 24) in the upstream and downstream ports should never be set to a value smaller than the value set in the *Max_Read_Request_Size* register of any devices that use the bridge.

- **Completions and writes must pass un-split through the bridge.**

    This rule means that a header with a data payload cannot be split into two or more headers with smaller payloads in the bridge. This rule also means that the *Max_Payload_Size* register (see Chapter 24) in the upstream and downstream ports should never be set to a value smaller than the value set in the *Max_Payload_Size* register of any devices that use the bridge.

- **ECRC must pass unchanged through the bridge.**

    There should be no need to change ECRC if the other rules are followed.

Chapter 23

Configuration Registers

The previous chapter presented an overview of PCI Express configuration space, including purpose, features, and usage. The different types of configuration access were explained, and the configuration space hierarchy described. This chapter describes the function, architecture, and layout of PCI Express configuration registers. The next chapter describes the architecture of PCI Express capabilities, and describes the capability functions that are required for PCI Express.

Introduction to Configuration Registers

A PCI Express bus is made up of PCI Express devices, each allotted a similar block of memory in configuration address space. Typically, each device interacts with the system during runtime using memory and I/O accesses, but devices are always accessible to the system using configuration address space. Every device, except for the Root Complex, is required to support configuration address space. Support for other address spaces is optional.

Out of the 4,096 bytes allocated to each device in configuration address space, some of those byte locations have fixed meanings as configuration registers, while others are left for the use of the device. In particular, the first 256 bytes of PCI Express configuration space have a fixed template, common with legacy PCI devices. The PCI Express configuration space is defined so that legacy PCI devices are accessible to a PCI Express system without any changes to the PCI device hardware or device driver software.

Basic Function of Configuration Space Registers

Some features of configuration space include the ability to:

- Detect PCI Express devices.
- Identify the function(s) of each PCI Express device.
- Discover what system resources each PCI Express device needs. System resources include memory address space, I/O address space, and interrupts.
- Assign system resources to each PCI Express device.
- Enable or disable the ability of the PCI Express device to respond to memory or I/O accesses.
- Tell the PCI Express device how to respond to error conditions.
- Program the routing of PCI Express device interrupts.
- Support runtime re-configuration of system resources to allow "Plug and Play" addition and "Hot Swapping" of PCI devices.

Many PCI Express configuration registers are optional, but every PCI Express function in a system must support this minimal set of registers:

- Vendor ID
- Device ID
- Revision ID
- Class Code
- Header Type

Configuration Space Architecture

Configuration space is allocated among the devices in a PCI Express hierarchy. Every function in every device gets its own equivalent block of configuration space. Six categories of devices may exist in a PCI Express hierarchy. The categories with their associated configuration space size and type are as follows:

- **Root Complex or Peer Root Complex:** Size is implementation-specific (includes Host to PCI Express bridge and Root Complex endpoint).

- **PCI Express to PCI Express bridge:** 4-kilobyte configuration space, Type 1 (includes PCI Express Root Port bridge, PCI Express Upstream Port bridge, and PCI Express Downstream Port bridge).

- **PCI Express to PCI bridge:** 4-kilobyte configuration space, Type 1.

- **PCI Express endpoint:** 4-kilobyte configuration space, Type 0.

- **PCI to PCI bridge:** 256-byte configuration space, Type 1 legacy.

- **PCI endpoint:** 256-byte configuration space, Type 0 legacy.

All of these devices have addresses allocated in PCI Express configuration space. Non-PCI Express (legacy) devices have smaller allocations and, in some cases, different register definitions not described in this book. For information on PCI device configuration, see *PCI Local Bus Specification, Revision 2.3*.

Figure 23.1 shows a large PCI Express hierarchy with all six categories of devices represented, along with a representation of where the configuration registers physically exist for each device.

896 ■ The Complete PCI Express Reference

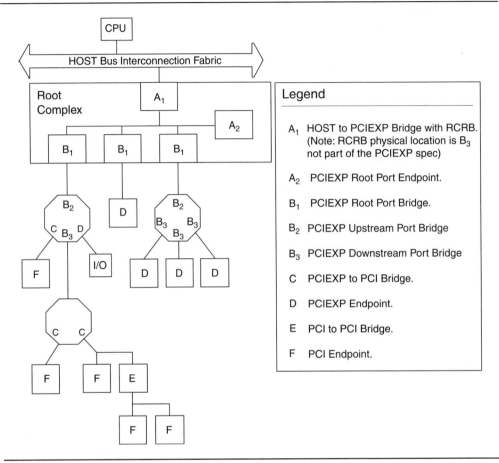

Figure 23.1 Large PCI Express Hierarchy

Configuration Space Layout

A basic rule of PCI configuration space is that every location is always readable. A typical PCI Express device does not need to support all of the predefined PCI Express registers, but it must respond to any access within its 4-kilobytes of configuration space.

Chapter 23: Configuration Registers ■ **897**

> Legacy PCI devices may exist in a PCI Express system that includes a PCI Express to PCI bridge. Legacy PCI devices contain functions that have exactly 256 bytes of configuration space. An attempt to read configuration space above 256 bytes on a PCI legacy function will cause the bridge to reply to the access with an Unsupported Request termination, equivalent to "Master Abort" in PCI legacy systems.

PCI Express Configuration space is logically divided into four regions, as shown in Figure 23.2.

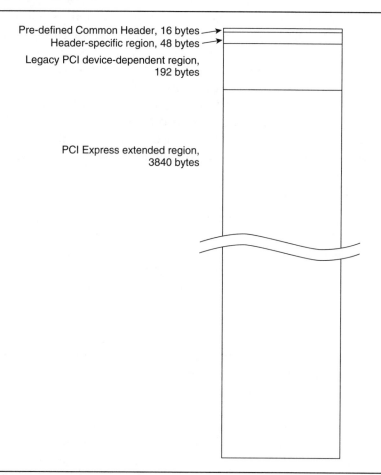

Figure 23.2 PCI Express Configuration Space Format

Root Complex Register Block

The Root Complex and any peer Root Complexes have a reserved block of registers called the Root Complex Register Block (RCRB). How these registers are made visible to the system, and what values they hold, is implementation-specific. In practice, these registers may be distributed across more than one component, and some of the registers may be mapped into system memory. In one example, the RCRB resides in configuration space as a function on bus zero, device zero, and matches closely the Type 1 header block in a PCI Express to PCI Express bridge device. The Root Complex can never be added or removed from a system, so some Type 1 header fields do not apply to the RCRB. Extending this example to a peer Root Complex, in this case the peer Root Complex's RCRB resides in the PCI Express configuration hierarchy as a function on device 0 of the bus enumerated for its root.

One important difference between the RCRB in the Root Complex and the RCRB in a peer Root Complex is that the peer Root Complex must support a Bus Number register and a Subordinate Bus Number register, while these fields are optional in the Root Complex.

RCRB Bus Number Register

The Root Complex always resides on bus 0 and this value does not change. If the Root Complex implements the Bus Number register, it is allowed to hardwire the value. All other Bus Number registers in all other devices in a PCI Express hierarchy must be writeable so that system software can enumerate the bus numbers after reset, and so that it can dynamically reconfigure the bus numbers during runtime following device add or drop events.

PCI Express buses are geographically addressable. That means every bus is unique and no two buses have the same number. In a typical system with a Root Complex and a peer Root Complex, the Root Complex has a range of buses numbered from 0 through N, and the peer Root Complex has a range of buses numbered from N+1 up to the last bus in the system. This typical system allows add-in PCI Express cards in the Root Complex hierarchy, so bus N is not determined until after reset and after bus enumeration of the Root Complex hierarchy. Therefore, the peer Root Complex in this typical system wires its Bus Number register to read 80h when reset. This value is higher than the highest bus number possible in the Root Complex hierarchy for this typical system. After the Root Complex hierarchy is enumerated, system software overwrites the 80h in the peer Root Complex Bus Number register with the value N+1, and proceeds to enumerate the peer Root Complex.

RCRB Subordinate Bus Number Register

As previously stated, the Root Complex always resides on bus 0. Typically, all the buses in the Root Complex hierarchy are enumerated before any buses in a peer Root Complex hierarchy. The highest bus number in the Root Complex hierarchy is typically placed in the Subordinate Bus Number register. However, the RCRB is not required to implement this register.

Knowing the last PCI bus number in the system is important for device drivers and for the operating system. However, if one or more peer Root Complexes exist, the Subordinate Bus Number register in the Root Complex or in any one of the peer Root Complexes might not be the number of the last bus in the system. To find this number, the best method is to use the system BIOS PCI routines. See *PCI and PCI-X Hardware and Software Architecture and Design* for more information about the PCI System BIOS Software Interface.

Configuration Header Register Block

Every PCI and PCI Express device supports the configuration space header registers. This block of registers uniquely identifies each device, provides basic control of the device, reports basic status of the device, and permits error handling for the device.

This 64-byte header region is strictly defined and every device must adhere to its organization. The structure has stayed the same from PCI to PCI Express. The following sections describe this region in more detail.

Common Header Region

The first 16 bytes of every 4,096 configuration block is the common header region, predefined for every device and unchanged from PCI through PCI Express. Figure 23.3 shows the configuration space Common Header Region.

```
  31                                              0
 ┌─────────────────────────┬─────────────────────┐
 │       Device ID         │      Vendor ID      │  00h
 ├─────────────────────────┼─────────────────────┤
 │        Status           │      Command        │  04h
 ├─────────────────────────┼─────────────────────┤
 │      Class Code                  │ Revision ID│  08h
 ├──────┬──────┬────────────────────┼────────────┤
 │ BIST │Header│     Reserved       │  Reserved  │  0Ch
 │      │ Type │                    │            │
 └──────┴──────┴────────────────────┴────────────┘
```

Figure 23.3 Configuration Space Common Header Region

The registers in the common header region are defined in detail below for PCI Express. Although this region is common between PCI and PCI Express, many of the registers are not used for PCI Express and are hardwired to an inactive value.

Vendor ID Register (Offset 00h)

Width	2 bytes
Valid Values	0000-FFFEh, Read Only
Description	This register identifies the manufacturer of the device. The PCI-SIG assigns vendor ID numbers. For example, a value of 8086h indicates the device is manufactured by Intel Corporation.

Device ID Register (Offset 02h)

Width	2 bytes
Valid Values	0000-FFFEh, Read Only **Note** 0000 indicates a reserved register; FFFF indicates a *no-response*.
Description	This register identifies this function, as designated by the manufacturer of the device. The manufacturer may choose any value for the device ID. Typically, software uses the device ID to identify the function and assign a device driver.

Command Register (Offset 04h)

Width	2 bytes
Valid Values	This field is a collection of bits, described below. Bits are Read-Only (RO) or Read/Write (RW).
Description	This register controls the ability of the function to respond to and initiate PCI Express transactions.

Bit	Register Description	Default	Type
0	**I/O Space Control** 0 Disable the function from responding to I/O accesses. 1 Enable the function to respond to I/O accesses. **Note** In a root hierarchy, any bridge that sets this bit to 0 stops I/O traffic going to subordinate functions. This bit does not affect configuration traffic.	0	RW
1	**Memory Space Control** 0 Disable the function from responding to memory accesses. 1 Enable the function to respond to memory accesses. **Note** In a root hierarchy, any bridge that sets this bit to 0 will stop memory traffic going to subordinate functions. This bit does not affect configuration traffic.	0	RW

Bit	Register Description	Default	Type
2	**Bus Master Enable**	0	RW
	0 Disable the function from generating reads or writes to memory or I/O spaces.		
	1 Enable the function to generate reads or writes to memory or I/O spaces.		
	Note If a function never initiates reads or writes then that function may hardwire this bit to 0. Functions that can initiate reads and writes set the bit to 1. When set to 1, software can write a 0, turning off bus mastering.		
	Note This bit affects requests only, not completions.		
	Note When set to zero, Message Signaled Interrupts (which are in-band memory writes) are disabled.		
	Note In a root hierarchy, any bridge that sets this bit to 0 (a *blocking bridge*) stops subordinate functions from reading or writing resources in the hierarchy above the bridge. However, subordinate functions that have this bit set to 1 may still initiate peer-to-peer accesses that do not go through the blocking bridge.		
3:5	**Hardwired**	000	RO
	PCI Express functions must hardwire these bits to 000.		
6	**Poison Reporting Control**	0	RW
	0 Disable poison reporting.		
	1 Enable poison reporting.		
	Note If poison reporting is enabled, then poisoned packets (received or generated) are reported in the Status Register (offset 06h), bit 8.		
7	**Hardwired**	0	RO
	PCI Express functions must hardwire this bit to 0.		

Bit	Register Description	Default	Type
8	**System Error Control** 0 Disable error reporting for fatal and nonfatal errors. 1 Enable error reporting for fatal and nonfatal errors. **Note** Two bits enable error reporting. If *either* bit is set, error reporting is enabled. This is one of the bits. The other bit is in the PCI Express Capability Structure, Device Control Register (offset 08h).	0	RW
9	**Hardwired** PCI Express functions must hardwire this bit to 0.	0	RO
10	**Interrupt Disable** 0 Enable generation of INTx interrupt messages. 1 Disable generation of INTx interrupt messages. **Note** When this bit is set, functions are prevented from generating INTx interrupt messages, and any INTx emulation interrupts already asserted must be deasserted.	0	RW
11:15	**Reserved**	00000	RO

Status Register (Offset 06h)

Width	2 bytes
Valid Values	This field is a collection of bits, described below. Bits are Read-Only (RO) or Read/Write-1-to-Clear (RW1C). The bits cannot be directly set by software.
Description	This register records events on the PCI Express bus attached to this function.

Note: Typically, the status register is not used by device drivers. The status register records catastrophic events that are not part of normal operation. System monitoring software might use the status register to determine the cause of a catastrophic event.

Bit	Register Description	Default	Type
7:0	**Hardwired** PCI Express functions must hardwire these bits to 00100000. **Note** Bit 5 must be set to 1, the other bits must be 0.	00100000	RO
8	**Poison Report** This bit is set by any function that initiates a request if either 1) the completion for its request returns poisoned, or 2) the request itself is marked poison. **Note** This bit is never set if the Poison Reporting Control bit in the Command Register (offset 04h) is clear. **Note** This bit is implemented only by bus masters.	0	RW1C
10:9	**Hardwired** PCI Express functions must hardwire these bits to 00.	00	RO
11	**Signaled Completer Abort** This bit is set by any function that responds to a request with a Completer Abort Completion Status.	0	RW1C
12	**Received Completer Abort** This bit is set by any function that initiates a request if the completion for its request has the Completer Abort Completion Status. **Note** All bus masters are required to implement this bit.	0	RW1C
13	**Received Unsupported Request** This bit is set by any function that initiates a request if the completion for its request has the Unsupported Request Completion Status. **Note** All bus masters are required to implement this bit.	0	RW1C

Bit	Register Description	Default	Type
14	**Signaled System Error** This bit is set when system errors are enabled and a device sends an ERR_FATAL or ERR_NONFATAL message. **Note** System errors are enabled by setting the System Error Control bit in the Command Register (offset 04h).	0	RW1C
15	**Detected Poison** This bit is set whenever a function receives a poisoned TLP. **Note** The Poison Reporting Control bit in the Command Register (offset 04h) has no effect on the Detected Poison status bit.	0	RW1C

Revision ID Register (Offset 08h)

Width	1 byte
Valid Values	00-FFh, Read Only
Description	This register identifies the revision level of this function, as designated by the manufacturer of the device. The manufacturer may choose any value for the revision ID.

Class Code Register (Offset 09h)

Width	3 bytes
Valid Values	Defined by PCI-SIG, Read Only
Description	This register identifies the generic class of devices to which this function belongs and (in some cases) its register level programming interface. Three bytes make up the Class Code Register: Byte 0Bh: Base Class Code. Byte 0Ah: Sub-Class Code. Byte 09h: Programming Interface. **Note** The PCI-SIG defines the allowed values for the Class Code Register. **Note** The purpose of the Class Code register is to allow generic device drivers to operate this function. For example, if a function is identified as a VGA device in the Class Code register, then it operates with a generic VGA device driver.

Reserved (Offset 0Ch)

Width	1 byte
Valid Values	00-FFh, Read/Write
Description	The value in this register has no impact on any PCI Express device functionality.

Note
In systems running legacy PCI code, the value in this register is typically equal to the number of DWORDs in a cache line. PCI Express devices use the payload size and request size registers in the PCI Express capability structure (described in the next chapter) to replace this register.

Reserved (Offset 0Dh)

Width	1 byte
Valid Values	00, Hardwired
Description	Not used by PCI Express.

Header Type (Offset 0Eh)

Width	1 byte
Valid Values	00h, 01h, 80h, 81h, Read Only
Description	The meaning of the valid values are as follows:

00h: This device has one function, using header type 0.

01h: This device has one function, using header type 1.

80h: This device has multiple functions, using header type 0.

81h: This device has multiple functions, using header type 1.

Note
Header type 0 refers to a PCI or PCI Express endpoint device. Header type 1 refers to a PCI or PCI Express bridge device.

Note
If software finds an undefined header type, it should disable the device by setting the bottom three bits in the Command Register to zero.

BIST (Built-In Self Test) (Offset 0Fh)

Width	1 byte
Valid Values	This field is a collection of bits, described in the following table. Bits are Read-Only (RO) or Read/Write (RW).
Description	This register provides control and status for a PCI Express function BIST. Functions that do not support BIST must return a value of 00h.
	Note The BIST could be invoked before the PCI hierarchy is fully enumerated. When a function is running its BIST, it should not generate any transactions on its PCI Express bus.

Bit	Register Description	Default	Type
3:0	**BIST Completion Status** 00h Function test passed. 01h-0Fh Function test failed. **Note** Device manufacturers define the failure codes for each device.	00	RO
5:4	**Reserved** These bits are hardwired.	00	RO
6	**BIST Start Control Bit** 0 BIST complete status. (The BIST clears this bit whenever its test completes.) 1 BIST start command. **Note** Software typically writes a 1 to this bit to start the BIST, then reads this bit to check for zero, indicating the BIST has completed. **Note** If the BIST does not complete its test and clear the bit to zero within two seconds, software should consider the BIST failed. **Note** This bit is hardwired to 0 if this function does not have a BIST.	0	RW

7	**BIST Support Status**
	   0 This function does not support a BIST.
	   1 This function supports a BIST.
		0	RO

In addition to the registers described in the previous section, three additional registers can be considered part of the Common Header Region. They are common to both header type 0 and header type 1. These registers are:

- Capabilities Pointer Register (Offset 34h)
- Interrupt Line Register (Offset 3Ch)
- Interrupt Pin Register (Offset 3Dh)

However, since these registers are placed within the region otherwise reserved for the type-specific headers, they will be described in the Type 0 Header Region and Type 1 Header Region sections.

Type 0 Header Region

After the Common Header Region, the next 48 bytes of every 4,096 configuration block is a pre-defined header region. For PCI Express systems, two header regions are predefined, Type 0 and Type 1. Type 0 is used by endpoint PCI Express devices. Figure 23.4 shows the configuration space Type 0 Header Region.

The registers in the type 0 header region are defined in detail here for PCI Express. Although this region is common between PCI and PCI Express, some of the registers are not used for PCI Express and are hardwired to an inactive value.

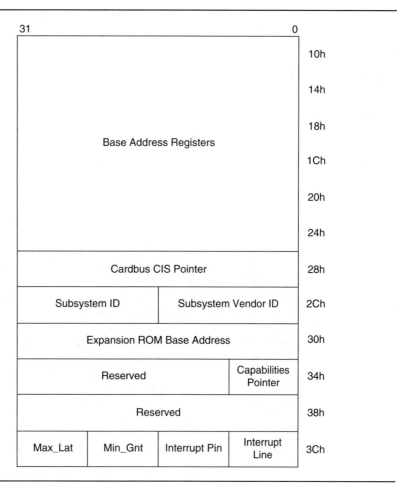

Figure 23.4 Configuration Space Type 0 Header Region

Base Address Registers (Offsets 10h, 14h, 18h, 1Ch, 20h, 24h)	
Width	4 bytes or 8 bytes
Valid Values	Device dependent
Description	These registers map the function's requested memory and I/O requirements into system memory and I/O space. The PCI Express function that owns these registers hardwires some of these bits to zero to tell system software its memory and I/O resource requirements. System software then writes the other bits to tell this function where in physical memory or I/O space those resources exist. This is described in detail at the end of this section.
	The PCI and PCI Express specifications do not allow a function to request a specific physical address or range of address in system memory. The function may only request a size, and system resources then assigns the base address for that size. For any size of memory or I/O requested, the system must assign a contiguous block of memory or I/O using a single base address to meet that request.
	I/O base addresses are always 32 bits, memory base addresses are either 32 or 64 bits. The function is allowed to mix I/O and memory base addresses in these registers. For example, base address register offset 10h could hold a 32-bit memory base address, offset 14h could hold a 64-bit memory base address (overlapping offset 18h), offset 1Ch could hold an I/O base address, offset 20h could be unused, and offset 24h could hold another 32 bit memory base address. In all cases, before the base address is assigned by system software, the function hardwires its memory or I/O size requests into the registers that will hold the base addresses.
	Following is a description of a base I/O address followed by a description of a base memory address. The fields are a collection of bits that are Read-Only (RO) or Read/Write (RW).

I/O Base Address Implementation

Bit	Register Description	Default	Type
1:0	**Reserved** These bits are hardwired. Bit 0 set to 1 tells system software that this register defines an I/O base address.	01	RO

31:2	**I/O Base Address**
	 These bits describe the base address of the requested I/O block. Before system software writes to this register, the register contains the function's encoded requested I/O block size.
	
	 How the requested I/O block size is encoded:
	
	     The function hardwires bits to zero starting at bit 2, describing the size of the requested I/O in DWORDs. The number of bits hardwired to zero describes the power-of-2 exponential number of requested DWORDs. For example, if the function requests 256 bytes (64 DWORDs) of contiguous I/O space, then given that $2^6 = 64$, six bits (bits 2 through 7) are hardwired to zero. Note that I/O size requests can be made only in powers-of-two.
	
	 How system software detects the size and encodes the base address:
	
	     When system software detects an I/O base address (bit 0 = 1), it then checks to find the requested size. To do that, system software writes FFFFh to this register. Then it reads the register. Any bits hardwired to zero describe the requested I/O block size. System software starts at bit location 2 and searches for the first bit set to 1. This bit is the binary weighted size of the requested I/O block. For example, if bit 8 is the first bit set to 1, then the requested I/O block size is 256 bytes ($2^8 = 256$). Finally, system software writes the base address of the block into the RW bits placed above the bits hardwired by the function.
	
	 Note
	 PCI Express functions are discouraged from using I/O address space. If a PCI Express function uses a base address to request a block of I/O, it is required to use another base address to request an equivalent block of memory space for the same resource. System configuration software then assigns memory address space to the function if the I/O space is not available.
		0000h	RO and RW

Memory Base Address Implementation

Bit	Register Description	Default	Type
0	**Reserved** This bit is hardwired to zero to tell system software that this register defines a memory base address.	0	RO
2:1	**Memory Type** 00h Base address is 32 bits, and may be set anywhere within the 32-bit range. 01h Base address is 32 bits, and must be set below 1MB in memory space. 10h Base address is 64 bits, and may be set anywhere within the 64-bit range. 11h Reserved.	00	RO

3	**Prefetchable**
	    0 Memory is not prefetchable.
	    1 Memory is prefetchable.
	 Note
	 PCI Express functions should set this bit to zero. PCI Express functions are not limited to the small read requests of PCI, but may request up to 4KB per read request. Requesting the exact amount of memory needed avoids the wasted overhead of speculative prefetching.
		0	RO

Bit	Register Description	Default	Type
31:4 or 63:4	**32 or 64 Bit Memory Base Address** Whether this field is 32 bits or 64 bits is defined in the Memory Type field. These bits describe the base address of the requested memory block. Before system software writes to this register, the register contains the function's encoded requested memory block size. How the requested memory block size is encoded: The function hardwires bits to zero starting at bit 4, describing the size of the requested memory in blocks or 16 bytes. The number of bits hardwired to zero describes the power-of-2 exponential number of requested blocks. For example, if the function requests 256 bytes (sixteen 16-byte blocks) of contiguous memory space, then given that $2^4 = 16$, four bits (bits 4 through 7) are hardwired to zero. Note that memory size requests can be made only in powers-of-two. How system software detects the size and encodes the base address: When system software detects a memory base address (bit 0 = 0, bits 2:1 do not equal 11b), it then checks to find the requested size. To do that, system software writes all 1s to this register. Then it reads the register. Any bits hardwired to zero describe the requested memory block size. System software starts at bit location 4 and searches for the first bit set to 1. This bit is the binary weighted size of the requested memory block. For example, if bit 8 is the first bit set to 1, then the requested memory block size is 256 bytes ($2^8 = 256$). Finally, system software writes the base address of the block into the RW bits placed above the bits hardwired by the function.	All 0s	RO and RW

Note: System software initializes all memory and I/O base address registers so that no system address conflicts occur. However, it is possible that memory or I/O resources may get depleted before all base address registers are initialized. In this case, system software initializes base address registers to zero when no system base address is available. System software also disables bits 2:1 in the Command Register (Offset 04h) when system resources are not available.

Note: Only system software resources should set base address values. If device specific resources (a device driver or device BIOS, for example) sets the base address values, system resource conflicts may occur.

Cardbus CIS Pointer Register (Offset 28h)

Width	4 bytes
Valid Values	Device dependent
Description	This register is optional and read-only. It is used by functions that share resources with a Cardbus card, using the Card Information Structure (CIS) standard. For more information on this register, see *PCI Local Bus Specification, Revision 2.3*.

Subsystem Vendor ID Register (Offset 2Ch)

Width	2 bytes
Valid Values	0000-FFFEh, Read Only
Description	This register identifies the manufacturer of the function. The PCI SIG assigns vendor ID numbers. For example, a value of 8086h indicates the function is manufactured by Intel Corporation.

Subsystem ID Register (Offset 2Eh)

Width	2 bytes
Valid Values	0000-FFFEh, Read Only **Note** 0000 indicates a reserved register; FFFF indicates a *no-response*.
Description	This register identifies this function, as designated by the manufacturer of the device. The manufacturer may choose any value for the subsystem ID.

Expansion ROM Base Address Register (Offset 30h)

Width	4 bytes
Valid Values	Device dependent, described below. Bits are Read-Only (RO) or Read/Write (RW).
Description	This register allows a PCI function's expansion ROM to be mapped into physical address space. **Note** This mapping can be dynamically turned on and off by system BIOS or system software such as the operating system or a device driver. This register is typically enabled only for copying the expansion ROM from the device into system memory.

Bit	Register Description	Default	Type
0	**Expansion ROM Decode Enable** 0 Disable decode of the Expansion ROM within system memory address space. 1 Enable decode of the Expansion ROM within system memory address space. **Note** The Command Register bit 1, Memory Space Control, must be set to enable this decode.	0	RO
10:1	**Reserved** These bits are hardwired.	000h	RO

Bit	Register Description	Default	Type

31:11	**Expansion ROM Base Address**
	
	These bits describe the base address of the function's expansion ROM mapped into system memory. Before system software writes to this register, the register contains the function's expansion ROM size.
	
	How the expansion ROM size is encoded:
	
	    The function hardwires bits to zero starting at bit 11, describing the size of the expansion ROM in 2KB blocks. The number of bits hardwired to zero describes the power-of-2 exponential number of requested blocks. For example, if the expansion ROM is 64K bytes (32 2K blocks), then given that $2^5 = 32$, five bits (bits 11 through 15) are hardwired to zero. Note that expansion ROM sizes can only be defined in powers of two.
	
	How system software detects the size of expansion ROM and sets its memory base address:
	
	    When system software detects an expansion ROM that can be mapped to memory (bit 0 is writable to one), it then checks to find the requested size. To do that, system software writes 1s to this field (bits 31:11). Then it reads the register. Any bits hardwired to zero describe the expansion ROM size. System software starts at bit location 11 and searches for the first bit set to 1. This bit is the binary weighted size of the expansion ROM. For example, if bit 16 is the first bit set to 1, then the expansion ROM size is 64KB (2^{16} = 64K). Finally, system software writes the base address of the block into the RW bits placed above the bits hardwired by the function.
		000h	RO and RW

Capabilities Pointer Register (Offset 34h)

Width	1 byte
Valid Values	40-FEh
Description	This register contains a pointer to a register in the function's device dependent region. The target register is the head of a PCI or PCI Express capability structure, which itself might point to another capability structure, in a linked list of structures. Capability structures are described in the following chapter.

Note
The value in this register has no alignment requirements. This register may point to a capability structure in the device dependent region starting on any byte address between 40h and FEh. However, it is recommended that all capability structures be DWORD-aligned. In the PCI Express extended configuration region (addresses 100h to FFFh), capabilities are required to be DWORD-aligned.

Interrupt Line Register (Offset 3Ch)

Width	1 byte
Valid Values	00-FFh. Bits are Read/Write (RW).
Description	This register maps a PCI Express interrupt to a system interrupt controller. Exactly how this mapping occurs is system-specific.

Note
This register must be implemented by any function that uses an interrupt pin (see Interrupt Pin Register). Values in this register are programmed by and used by system software. The function itself does not use this value.

Note
This register is supported for backward compatibility. Actual interrupt signaling on PCI Express uses in-band messages rather than physical pins.

Bit	Register Description	Default	Type
7:0	00h–FEh System interrupt controller line to which the function is connected. FFh The function is not connected to a system interrupt.	FFh	RW

Interrupt Pin Register (Offset 3Dh)

Width	1 byte
Valid Values	00–04h. Bits are Read Only (RO)
Description	For PCI, this register identifies which device interrupt pin the function uses. For PCI Express, this identifies the legacy interrupt message the function uses. **Note** This register is supported for backward compatibility. Actual interrupt signaling on PCI Express uses in-band messages rather than physical pins.

Bit	Register Description		Default	Type
7:0	00h	Interrupt pin or legacy interrupt message not used.	00h	RO
	01h	Interrupt pin or legacy interrupt message uses INTA.		
	02h	Interrupt pin or legacy interrupt message uses INTB.		
	03h	Interrupt pin or legacy interrupt message uses INTC.		
	04h	Interrupt pin or legacy interrupt message uses INTD.		
	05:FFh	Reserved.		

Min_Gnt Register (Offset 3Eh)

Width	1 byte
Valid Values	00h, Read Only (RO)
Description	This register is not used for PCI Express and is hardwired to 00h.

Max_Lat Register (Offset 3Fh)

Width	1 byte
Valid Values	00h, Read Only (RO)
Description	This register is not used for PCI Express and is hardwired to 00h.

Type 1 Header Region

The 48 bytes in configuration space that immediately follow the common header region are predefined as Type 0 or Type 1. Type 0 is used by endpoint PCI Express devices, and Type 1 is used by all other devices. Figure 23.5 shows the configuration space Type 1 header region.

31			0	
Base Address Register 0				10h
Base Address Register 1				14h
Secondary Latency Tmr	Subordinate Bus Number	Secondary Bus Number	Primary Bus Number	18h
Secondary Status		I/O Limit	I/O Base	1Ch
Memory Limit		Memory Base		20h
Prefetchable Memory Limit		Prefetchable Memory Base		24h
Prefetchable Base Upper 32 bits				28h
Prefetchable Limit Upper 32 bits				2Ch
I/O Limit Upper 16 bits		I/O Base Upper 16 bits		30h
Reserved			Capabilities Pointer	34h
Expansion ROM Base Address				38h
Bridge Control		Interrupt Pin	Interrupt Line	3Ch

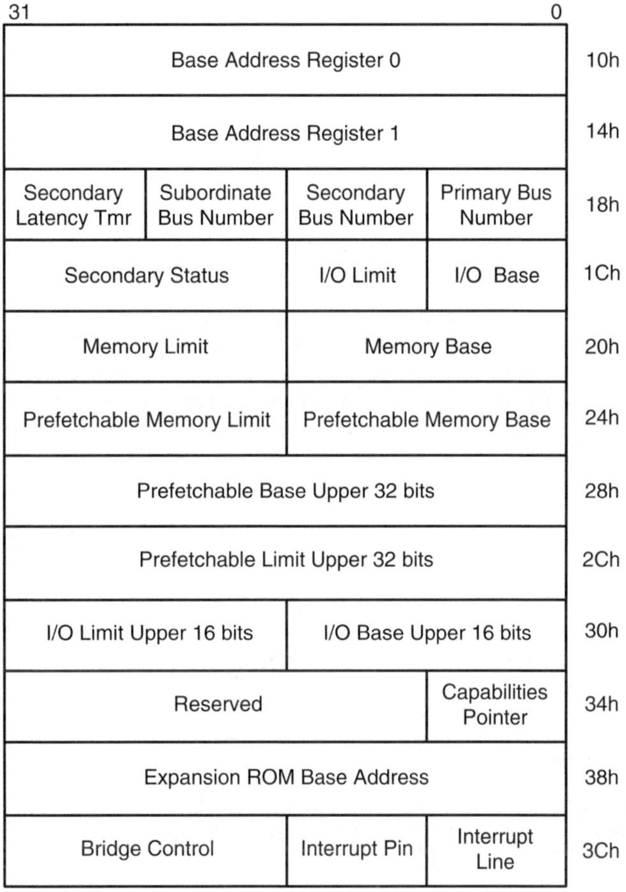

Figure 23-5 Configuration Space Type 1 Header Region

The Type 1 configuration space header region has the same definitions for PCI Express as for legacy PCI. However, many Type 1 registers are not used for PCI Express and are hardwired to 0, as defined in the next section.

Base Address Registers (Offsets 10h, 14h)	
Width	4 bytes or 8 bytes
Valid Values	Device dependent
Description	These registers map the function's requested memory and I/O requirements into system memory and I/O space. The PCI Express function that owns these registers hardwires some of these bits to zero to tell system software its memory and I/O resource requirements. System software then writes the other bits to tell this function where in physical memory or I/O space those resources exist. This is described in detail at the end of this section.
	The PCI and PCI Express specifications do not allow a function to request a specific physical address or range of address in system memory. The function may only request a size, and system resources then assign the base address for that size. For any size of memory or I/O requested, the system must assign a contiguous block of memory or I/O using a single base address to meet that request.
	I/O base addresses are always 32 bits, memory base addresses are either 32 or 64 bits. The function is allowed to mix I/O and memory base addresses in these registers. For one example, base address register offset 10h could be unused, and offset 14h could hold a 32-bit memory base address. In all cases, before the base address is assigned by system software, the function hardwires its memory or I/O size requests into the registers that will hold the base addresses.
	Following is a description of a base I/O address followed by a description of a base memory address. The fields are a collection of bits that are Read-Only (RO) or Read/Write (RW).

I/O Base Address Implementation

Bit	Register Description	Default	Type
1:0	**Reserved** These bits are hardwired. Bit 0 set to 1 tells system software that this register defines an I/O base address.	01	RO

Bit	Register Description	Default	Type
31:2	**I/O Base Address** These bits describe the base address of the requested I/O block. Before system software writes to this register, the register contains the function's encoded requested I/O block size. How the requested I/O block size is encoded: The function hardwires bits to zero starting at bit 2, describing the size of the requested I/O in DWORDs. The number of bits hardwired to zero describes the power-of-2 exponential number of requested DWORDs. For example, if the function requests 256 bytes (64 DWORDs) of contiguous I/O space, then given that $2^6 = 64$, six bits (bits 2 through 7) are hardwired to zero. Note that I/O size requests can be made only in powers of two. How system software detects the size and encodes the base address: When system software detects an I/O base address (bit 0 = 1), it then checks to find the requested size. To do that, system software writes FFFFh to this register. Then it reads the register. Any bits hardwired to zero describe the requested I/O block size. System software starts at bit location 2 and searches for the first bit set to 1. This bit is the binary weighted size of the requested I/O block. For example, if bit 8 is the first bit set to 1, then the requested I/O block size is 256 bytes ($2^8 = 256$). Finally, system software writes the base address of the block into the RW bits placed above the bits hardwired by the function. **Note** PCI Express functions are discouraged from using I/O address space. If a PCI Express function uses a base address to request a block of I/O, it is required to use another base address to request an equivalent block of memory space for the same resource. System configuration software then assigns memory address space to the function if the I/O space is not available.	0000h	RO and RW

Memory Base Address Implementation

Bit	Register Description	Default	Type
0	**Reserved** This bit is hardwired to zero to tell system software that this register defines a memory base address.	0	RO
2:1	**Memory Type** 00h Base address is 32 bits, and may be set anywhere within the 32-bit range. 01h Base address is 32 bits, and must be set below 1MB in memory space. 10h Base address is 64 bits, and may be set anywhere within the 64-bit range. 11h Reserved.	00	RO
3	**Prefetchable** 0 Memory is not prefetchable. 1 Memory is prefetchable. **Note** PCI Express functions should set this bit to zero. PCI Express functions are not limited to the small read requests of PCI, but may request up to 4KB per read request. Requesting the exact amount of memory needed avoids the wasted overhead of speculative prefetching.	0	RO

Bit	Register Description	Default	Type

31:4 or 63:4	**32 or 64 Bit Memory Base Address**
	
	Whether this field is 32 bits or 64 bits is defined in the Memory Type field.
	
	These bits describe the base address of the requested momory block. Before system software writes to this register, the register contains the function's encoded requested memory block size.
	
	How the requested memory block size is encoded:
	
	    The function hardwires bits to zero starting at bit 4, describing the size of the requested memory in blocks or 16 bytes. The number of bits hardwired to zero describes the power-of-2 exponential number of requested blocks. For example, if the function requests 256 bytes (sixteen 16-byte blocks) of contiguous memory space, then given that $2^4 = 16$, four bits (bits 4 through 7) are hardwired to zero. Note that memory size requests can be made only in powers of two.
	
	How system software detects the size and encodes the base address:
	
	    When system software detects a memory base address (bit 0 = 0, bits 2:1 do not equal 11b), it then checks to find the requested size. To do that, system software writes all 1s to this register. Then it reads the register. Any bits hardwired to zero describe the requested memory block size. System software starts at bit location 4 and searches for the first bit set to 1. This bit is the binary weighted size of the requested memory block. For example, if bit 8 is the first bit set to 1, then the requested memory block size is 256 bytes ($2^8 = 256$). Finally, system software writes the base address of the block into the RW bits placed above the bits hardwired by the function.
		All 0s	RO and RW

Note: System software initializes all memory and I/O base address registers so that no system address conflicts occur. However, it is possible that memory or I/O resources may get depleted before all base address registers are initialized. In this case, system software initializes base address registers to zero when no system base address is available. System software also disables bits 2:1 in the Command Register (Offset 04h) when system resources are not available.

Note: Only system software resources should set base address values. If device specific resources (a device driver or device BIOS, for example) sets the base address values, system resource conflicts may occur.

Primary Bus Number Register (Offset 18h)

Width	1 byte
Valid Values	00h to FFh
Description	This register gives the bus number on which this function resides. At reset, this value must be 00h. Configuration software writes this register when it enumerates the PCI Express hierarchy and assigns bus numbers. This register must be writeable, with one exception: The primary bus in the Root Complex is always bus zero, so functions permanently attached to that bus may hardwire this register to 00h.

Secondary Bus Number Register (Offset 19h)

Width	1 byte
Valid Values	01h to FFh
Description	This register gives the bus number immediately to the south of this bridge function. At reset, this value must be 00h. Configuration software writes this register when it enumerates the PCI Express hierarchy and assigns bus numbers. This register must be writeable. **Note** This register should not be changed by system software while any devices on the secondary bus have completion transactions pending anywhere in the system.

Subordinate Bus Number Register (Offset 1Ah)

Width	1 byte
Valid Values	01h to FFh
Description	This register gives the highest bus number in the bus hierarchy south of this bridge function. At reset, this value must be 00h. Configuration software writes this register when it enumerates the PCI Express hierarchy and assigns bus numbers. This register must be writeable.

Secondary Latency Timer Register (Offset 1Bh)

Width	1 byte
Valid Values	00h, Read Only (RO)
Description	This register is not used for PCI Express and is hardwired to 00h.

I/O Base Register (Offset 1Ch)

Width	1 byte
Valid Values	Device dependent, bits are Read Only (RO) or Read/Write (RW).
Description	This register is optional. If a bridge supports an I/O address range, then this register must be initialized by configuration software to define the base address of the I/O range used by this bridge and devices south of this bridge.

Bit	Register Description	Default	Type
7:0	If this register is *not used*, then the bits are hardwired to 00h. The following bit descriptions apply only in the case this register is used.	00h	RO

0	Bit 0 describes the I/O addressing capability of the bridge.
	
	    0 16-bit I/O addressing.
	    1 32-bit I/O addressing.
		0	RO

3:1	Reserved. These bits are hardwired to 000.	000	RO
7:4	These bits correspond to address bits 15:12 for I/O address decoding. Address bits 11:0 are assumed to be 000h. This means the I/O base address is always placed on a 4K address boundary. **Note** In systems that are limited to 64KB of I/O address space, PCI Express systems should avoid using this register and instead implement base address registers that support memory-mapped I/O.	000	RW

I/O Limit Register (Offset 1Dh)

Width	1 byte
Valid Values	Device dependent, bits are Read Only (RO) or Read/Write (RW).
Description	This register is optional. If a bridge supports an I/O address range, then this register must be initialized by configuration software to define the upper limit address of the I/O range used by this bridge and devices south of this bridge.

Bit	Register Description	Default	Type
7:0	If this register is *not used*, then the bits are hardwired to 00h. The following bit descriptions apply only in the case this register is used.	00h	RO

0	Bit 0 describes the I/O addressing capability of the bridge.
	  0 16-bit I/O addressing.
	  1 32-bit I/O addressing.
		0	RO

3:1	Reserved. These bits are hardwired to 000.	000	RO
7:4	These bits correspond to address bits 15:12 for I/O address decoding. Address bits 11:0 are assumed to be FFFh. This means the I/O limit address is always assumed to be placed at the top of an aligned 4K address boundary.	000	RW

Secondary Status Register (Offset 1Eh)

Width	2 bytes
Valid Values	This field is a collection of bits, described below. Bits are Read-Only (RO) or Read/Write-1-to-Clear (RW1C). The bits cannot be directly set by software.
Description	This register records events on the PCI Express bus attached to this function.

Bit	Register Description	Default	Type
7:0	**Reserved** PCI Express functions must hardwire these bits to 00h.	00h	RO

Bit	Register Description	Default	Type
8	**Master Data Parity Error**	0	RW1C
	This bit is controlled by the Parity Error Response Enable bit in the Bridge Control Register (offset 3EH, bit location 0). If the Parity Error Response Enable bit is clear, this bit is never set. If the Parity Error Response Enable bit is set, the following applies:		
	This bit is set by the bridge function if the function receives a completion with poisoned data, or if the function creates a write with poisoned data.		
10:9	**Hardwired**	00	RO
	PCI Express functions must hardwire these bits to 00.		
11	**Signaled Target Abort**	0	RW1C
	This bit is set when a Type 1 Configuration Space Header function completes a request using the Completer Abort Completion Status.		
12	**Received Target Abort**	0	RW1C
	This bit is set when a Type 1 Configuration Space Header function receives a completion with the Completer Abort Completion Status.		
13	**Received Master Abort**	0	RW1C
	This bit is set when a Type 1 Configuration Space Header function receives a completion with the Unsupported Request Completion Status.		
14	**Received System Error**	0	RW1C
	This bit is controlled by the SERR Enable bit in the Bridge Control Register (offset 3EH, bit location 1). If the SERR Enable bit is clear, this bit is never set. If the SERR Enable bit is set, the following applies:		
	This bit is set when the function sends an ERR_FATAL or ERR_NONFATAL message.		
15	**Detected Parity Error**	0	RW1C
	This bit is *not* controlled by the Parity Error Response Enable bit in the Bridge Control Register (offset 3EH, bit location 0). Regardless of the state of the Parity Error Response Enable bit, the following applies:		
	This bit is set whenever a Type 1 Configuration Space Header function receives a poisoned TLP.		

Memory Base Register (Offset 20h)

Width	2 bytes
Valid Values	Device dependent, bits are Read Only (RO) or Read/Write (RW).
Description	This register is initialized by configuration software to define the base address of the memory-mapped I/O range used by this bridge and devices south of this bridge.

Bit	Register Description	Default	Type
3:0	**Reserved** These bits are hardwired to 0h.	0h	RO
15:4	**Address bits 31:20** These bits correspond to address bits 31:20 for memory-mapped I/O address decoding. Address bits 19:0 are assumed to be 00000h. This means the memory mapped I/O base address is always placed on a 1M address boundary.	00000	RW

Memory Limit Register (Offset 22h)

Width	2 bytes
Valid Values	Device dependent, bits are Read Only (RO) or Read/Write (RW).
Description	This register is initialized by configuration software to define the upper limit address of the memory-mapped I/O range used by this bridge and devices south of this bridge.

Bit	Register Description	Default	Type
3:0	**Reserved** These bits are hardwired to 0h.	0h	RO
15:4	**Address bits 31:20** These bits correspond to address bits 31:20 for memory-mapped I/O address decoding. Address bits 19:0 are assumed to be FFFFh. This means the memory mapped I/O limit address is always assumed to be placed at the top of an aligned 1M address boundary.	00000	RW

Prefetchable Memory Base Register (Offset 24h)

Width 2 bytes

Valid Values Device dependent, bits are Read Only (RO) or Read/Write (RW).

Description This register is initialized by configuration software to define the base address of a prefetchable memory address range. However, if the function does not support prefetching, the function may hardwire this register to 0000h. See *PCI Local Bus Specification, Revision 2.3* for programming details.

Note
PCI Express devices should hardwire this register to 0000h and use the request length field in the TLP to indicate the largest possible request size.

Prefetchable Memory Limit Register (Offset 26h)

Width 2 bytes

Valid Values Device dependent, bits are Read Only (RO) or Read/Write (RW).

Description This register is initialized by configuration software to define the top address of a prefetchable memory address range. However, if the function does not support prefetching, the function may hardwire this register to 0000h. See *PCI Local Bus Specification, Revision. 2.3* for programming details.

Note
PCI Express devices should hardwire this register to 0000h and use the request length field in the TLP to indicate the largest possible request size.

Prefetchable Base Upper 32 Bits Register (Offset 28h)

Width 4 bytes

Valid Values Device dependent, bits are Read Only (RO) or Read/Write (RW).

Description This register is initialized by configuration software to define the upper 32 bits of the base of a prefetchable 64 bit memory address range. However, if the function does not support prefetching, the function may hardwire this register to 00000000h.

Note
PCI Express devices should hardwire this register to 00000000h and use the request length field in the TLP to indicate the largest possible request size.

Prefetchable Limit Upper 32 Bits Register (Offset 2Ch)

Width	4 bytes
Valid Values	Device dependent, bits are Read Only (RO) or Read/Write (RW).
Description	This register is initialized by configuration software to define the upper 32 bits of the top address of a prefetchable 64 bit memory address range. However, if the function does not support prefetching, the function may hardwire this register to 00000000h. **Note** PCI Express devices should hardwire this register to 00000000h and use the request length field in the TLP to indicate the largest possible request size.

I/O Base Upper 16 Bits Register (Offset 30h)

Width	2 bytes
Valid Values	Device dependent, bits are Read Only (RO) or Read/Write (RW).
Description	This register is initialized by configuration software to define the upper 16 bits of the 32-bit I/O base address used by this bridge and devices south of this bridge.

Bit	Register Description	Default	Type
15:0	If this register is *not used*, then the bits are hardwired to 0000h.	0000h	RO
15:0	If this register *is* used, then the bits specify the upper 16 bits (address bits 31:16) of a 32 bit I/O base address.	0000h	RW

I/O Limit Upper 16 Bits Register (Offset 32h)

Width	2 bytes
Valid Values	Device dependent, bits are Read Only (RO) or Read/Write (RW).
Description	This register is initialized by configuration software to define the upper 16 bits of the 32-bit I/O top address used by this bridge and devices south of this bridge.

Bit	Register Description	Default	Type
15:0	If this register is *not used*, then the bits are hardwired to 0000h.	0000h	RO
15:0	If this register *is* used, then the bits specify the upper 16 bits (address bits 31:16) of a 32-bit I/O top address.	0000h	RW

Capabilities Pointer Register (Offset 34h)

Width	1 byte
Valid Values	40-FEh
Description	This register contains a pointer to a register in the function's device dependent region. The target register is the head of a PCI or PCI Express capability structure, which itself might point to another capability structure, in a linked list of structures. Capability structures are described in the following chapter. **Note** The value in this register has no alignment requirements. This register may point to a capability structure in the device dependent region starting on any byte address between 40h and FEh. However, it is recommended that all capability structures be DWORD-aligned. In the PCI Express extended configuration region (addresses 100h to FFFh), capabilities are required to be DWORD-aligned.

Expansion ROM Base Address Register (Offset 38h)

Width	4 bytes
Valid Values	Device dependent, described below. Bits are Read-Only (RO) or Read/Write (RW).
Description	This register allows a PCI function's expansion ROM to be mapped into physical address space. **Note** This mapping can be dynamically turned on and off by system BIOS or system software such as the operating system or a device driver. This register is typically enabled only for copying the expansion ROM from the device into system memory.

Bit	Register Description	Default	Type
0	**Expansion ROM Decode Enable** 0 Disable decode of the Expansion ROM within system memory address space. 1 Enable decode of the Expansion ROM within system memory address space. **Note** The Command Register bit 1, Memory Space Control, must be set to enable this decode.	0	RO
10:1	**Reserved** These bits are hardwired.	000h	RO

Bit	Register Description	Default	Type
31:11	**Expansion ROM Base Address** These bits describe the base address of the function's expansion ROM mapped into system memory. Before system software writes to this register, the register contains the function's expansion ROM size. How the expansion ROM size is encoded: The function hardwires bits to zero starting at bit 11, describing the size of the expansion ROM in 2KB blocks. The number of bits hardwired to zero describes the power-of-2 exponential number of requested blocks. For example, if the expansion ROM is 64K bytes (32 2K blocks), then given that $2^5 = 32$, five bits (bits 11 through 15) are hardwired to zero. Note that expansion ROM sizes can only be defined in powers of two. How system software detects the size of expansion ROM and sets its memory base address: When system software detects an expansion ROM that can be mapped to memory (bit 0 is writable to one), it then checks to find the requested size. To do that, system software writes 1s to this field (bits 31:11). Then it reads the register. Any bits hardwired to zero describe the expansion ROM size. System software starts at bit location 11 and searches for the first bit set to 1. This bit is the binary weighted size of the expansion ROM. For example, if bit 16 is the first bit set to 1, then the expansion ROM size is 64KB ($2^{16} = 64K$). Finally, system software writes the base address of the block into the RW bits placed above the bits hardwired by the function.	000h	RO and RW

Interrupt Line Register (Offset 3Ch)

Width 1 byte

Valid Values 00-FFh. Bits are Read/Write (RW).

Description This register maps a PCI Express interrupt to a system interrupt controller. Exactly how this mapping occurs is system-specific.

Note
This register must be implemented by any function that uses an interrupt pin (see Interrupt Pin Register). Values in this register are programmed by and used by system software. The function itself does not use this value.

Note
This register is supported for backwards compatibility. Actual interrupt signaling on PCI Express uses in-band messages rather than physical pins.

Bit	Register Description	Default	Type
7:0	00h–FEh System interrupt controller line to which the function is connected. FFh The function is not connected to a system interrupt.	FFh	RW

Interrupt Pin Register (Offset 3Dh)

Width 1 byte

Valid Values 00-04h. Bits are Read Only (RO).

Description For PCI, this register identifies which device interrupt pin the function uses. For PCI Express, this identifies the legacy interrupt message the function uses.

Note
This register is supported for backward compatibility. Actual interrupt signaling on PCI Express uses in-band messages rather than physical pins.

Bit	Register Description	Default	Type
7:0	00h Interrupt pin or legacy interrupt message not used. 01h Interrupt pin or legacy interrupt message uses INTA. 02h Interrupt pin or legacy interrupt message uses INTB. 03h Interrupt pin or legacy interrupt message uses INTC. 04h Interrupt pin or legacy interrupt message uses INTD. 05:FFh Reserved.	00h	RO

Bridge Control Register (Offset 3Eh)

Width	2 bytes
Valid Values	This field is a collection of bits, described in the following table. Bits are Read-Only (RO) or Read/Write (RW).
Description	This register is an extension of the command register (offset 04h) with control for bridge functions. Some bits in this register control the operation of both the primary and secondary interfaces of the bridge. Other bits control the secondary interface where the command register controls the primary interface. The bits are described in the next table.

Bit	Register Description	Default	Type

0	Parity Error Response Enable
	
	This bit controls the response to poisoned TLPs, as logged in the secondary status register (offset 1Eh).
	
	    0 Ignore parity errors on the secondary interface.
	    1 Enable parity error detection and reporting on the secondary interface.
		0	RW

1	SERR Enable
	
	This bit controls forwarding of ERR_COR, ERR_NONFATAL and ERR_FATAL from secondary to primary, as recorded in the secondary status register (offset 1Eh).
	
	    0 Do not forward errors to primary interface.
	    1 Forward errors to the primary interface.
	
	Note
	Error reporting must be enabled in the Command Register (offset 04h) or in the PCI Express Capability Structure, Device Control Register (offset 08h) for errors to be captured on the primary interface so they can be forwarded.
		0	RW

2	ISA Enable
	
	    0 Forward all I/O addresses in the range defined by the I/O Base Register and the I/O Limit Register.
	    1 Do not forward the top 768 bytes if each 1K block of I/O.
		0	RW

Bit	Register Description	Default	Type
3	VGA Enable	0	RW
	0 Do not forward VGA-compatible addresses from the primary interface to the secondary interface.		
	1 Forward VGA-compatible addresses from the primary interface to the secondary interface.		
5:4	Reserved	00	RO
	PCI Express functions must hardwire these bits to 00.		
6	Secondary Bus Reset	0	RW
	Setting this bit triggers a warm reset on the corresponding PCI Express port and the PCI Express hierarchy domain subordinate to the port.		
15:7	Reserved	All 0s	RO
	PCI Express functions must hardwire these bits to 000h.		

Chapter 24

Configuration Capabilities

The previous chapter covered the function, architecture, and layout of PCI Express configuration registers. This chapter covers the architecture and function of PCI Express capabilities, and describes the capability functions that are required for PCI Express.

Introduction to Configuration Capabilities

The term *capabilities* has a precise meaning for PCI Express Configuration. It refers to a standard method for detecting and controlling advanced features of PCI and PCI Express devices.

Capabilities defined for PCI and PCI Express include:

- Power Management
- Hot Swap and Hot Plug
- Message Signaled Interrupts
- User-defined features

The capabilities configuration mechanism was introduced as an optional feature with the *PCI Local Bus Specification 2.2* and is a required feature of PCI Express. This chapter describes the capabilities architecture and summarizes the most widely used standard PCI Express capabilities.

Defined Capabilities

Appendix A lists the currently defined capabilities. Each capability has an ID number assigned by the PCI-SIG (**www.pcisig.com**). Note that each capability has an associated specification, which gives additional detail beyond the summary in this chapter.

Required Capabilities

All PCI Express functions must support the following capabilities:

- PCI Express Capability Structure
- PCI Power Management Capability Structure

All PCI Express functions that can generate interrupts must also support the following capability:

- Message Signaled Interrupts Capability Structure

These required capabilities may reside entirely in the lower 256 bytes of PCI Express register space. Each is described in more detail below.

Extended Capabilities

All PCI Express functions may optionally support *extended capabilities* in addition to the required capabilities.

PCI Express functions are allocated 4 kilobytes of register space. The first 256 bytes of that space are defined for PCI Express legacy compatibility. The remaining register space (4 kilobytes − 256 bytes) is reserved exclusively for PCI Express capability structures. This extended register space is not visible to legacy operating systems and is called the extended capabilities region. Certain capabilities are required to be placed in this region (not in the lower 256 bytes) if they are implemented, other capabilities can go in either location. All of the extended capabilities are optional. Some of these extended capabilities include:

- Advanced Error Reporting Capability
- Virtual Channel Capability
- Device Serial Number Capability
- Power Budgeting Capability

Configuration Capability Architecture

The 4-kilobyte register space allotted to each PCI Express function is divided into four regions: the device-independent region, the header type region, the device-dependent region, and the extended configuration region. Figure 24.1 illustrates these regions.

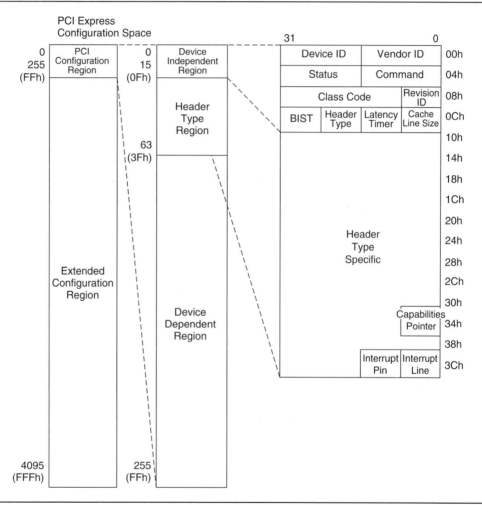

Figure 24.1 PCI Express Configuration Capability Registers

Device-independent Region

The device-independent region holds registers common to all PCI and PCI Express functions. All PCI and PCI Express functions are required to implement these registers. The first 16 bytes of configuration space, addresses 00h to 0Fh, make up this region.

Header Type Region

Byte address 0Eh in the device-independent region holds the *header type* register of the function. The *header type region* holds the registers predefined for that type, as determined by the PCI-SIG. The most common types are type 0 for most endpoint functions, and type 1 for bridge functions. The header type region is the 48 bytes (addresses 10h to 3Fh) following the device-independent region. In a minimal implementation, the header type region is optional for legacy PCI endpoint functions. PCI Express functions, however, are required to implement at least three bytes of the header type region common for all header types: addresses 34h, 3Ch, and 3Dh.

Device-Dependent Region

Word address 06h in the device-independent region holds the *status* register of the function. Bit 4 in this register tells whether a capabilities structure exists for that function. For PCI, the capabilities structure is optional and resides fully in this region. For PCI Express, the capabilities structure is required and resides in this region as well as in the extended configuration region. The device-dependent region is the 192 bytes (addresses 40h to FFh) following the header type region.

Extended Configuration Region

This region is visible only to PCI Express configuration software. Legacy PCI configuration software cannot access this configuration space. The extended configuration region is the 3,840 bytes (addresses 100h to FFFh) following the device-dependent region.

> If a PCI Express function places any of its capabilities structure registers in the extended configuration region, those registers are not visible to legacy PCI software. For backward compatibility, PCI Express functions should default to placing all their capabilities in the device-independent region, and use the extended region for enhanced functionality in a software environment aware of PCI Express.

Shared Registers

A PCI Express device (with up to eight functions) may implement certain capability registers one time and share those registers across all the functions in that device. A PCI Express device can save memory space with this technique. Registers that are guaranteed common across all functions (certain power budget or serial number registers, for example) can be shared.

Capabilities List Bit

PCI and PCI Express status register bit 4 enables the capabilities list. This bit is optional for PCI functions, but is hardwired set for all PCI Express functions. When set, this bit indicates that at least one capability is supported, but the number of capabilities that can be supported has no limit (except for the 4-kilobyte memory limitation or the PCI-SIG list).

How to Locate the Capabilities List

While PCI Express devices are required to support the capabilities list, legacy PCI devices may or may not support capabilities. In an environment that mixes PCI and PCI Express functions, software must detect whether each function supports a capabilities list.

Bit 4 of the status register tells whether a capabilities list is supported for that function. Software must check bit 4 of each function in each device to determine whether that function supports a capabilities list.

The RCRB structure may be different than this description. However, in a typical implementation, the RCRB resides in a function in the Root Complex (or peer Root Complex) that follows the organization of the PCI Express register block, and has a similar linked-list capabilities architecture. Root ports must implement the entire PCI Express capability structure.

> The PCI Express specification stipulates that the extended configuration region for the RCRB must "always begin at offset 0h" in the configuration address space for the RCRB. (For non-RCRB functions, the extended configuration region begins at offset 100h.) The specification continues, "Absence of any Extended Capabilities is required to be indicated by an Enhanced Capability Header with a Capability ID of FFFFh and a Next Capability Offset of 0h." (For non-RCRB functions, the absence of any extended capability structure requires a Capability ID of 0000h in the Capability ID register, at address 100h.) How the RCRB is configured is implementation-specific.

For functions that implement a capabilities list, the capabilities pointer in the header type region (address 34h) holds the configuration register address of the first capability. This first capability is always in the device-dependent region, because the 8-bit pointer field is not large enough to hold an address in the extended configuration region. All capabilities start on a DWORD boundary. Because of this, the last two bits of the capability pointer register are reserved and must read 00b.

> Every capability structure should link to a previous structure, in a chain going back to the first. This implies that the first capability structure in the extended configuration region is linked to a previous structure which resides in the device-dependent region. However, that particular link does not exist. The last capability structure in the device-dependent region ends its chain, and a new chain is started (if required) in the extended configuration region. In all PCI Express systems, software must check address 100h in each function's configuration address space for the existence of a Capability ID, starting the new chain.

How to Scan the Capabilities List

The first capability may reside anywhere in the device-dependent region (configuration addresses between 40h and FFh) on a DWORD boundary. Software may scan the capabilities list only if the first capability address meets these criteria. Each capability includes a pointer to the next capability in the linked list. The last capability in the list places 00h in the pointer field, automatically ending the list. PCI Express functions fix in hardware the values of the pointers for the capabilities linked list for each function, and the address/offset values of the linked list do not change during runtime. Capabilities are typically scanned during boot-up; the capabilities list for a PCI Express device does not change during operation except under special circumstances, such as a hot swap event.

Capability Structure Layout

Each capability in a linked list of capabilities has its own independent and dependent regions. The independent region is similar for all capabilities; the dependent region is unique to each capability and is defined by the PCI-SIG.

For all PCI and PCI Express capabilities, the independent region consists of two fields: the capability ID, and the pointer to the next capability. The size of these fields depends on whether the capability structure resides in the PCI-compatible device-dependent region (addresses 40h to FFh) or resides in the PCI Express extended configuration region (addresses 100h to FFFh).

Figure 24.2 illustrates the capability-independent region.

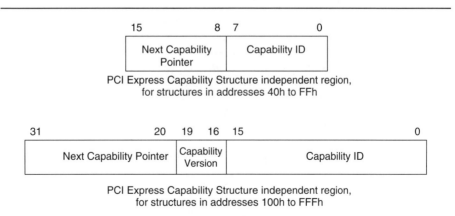

Figure 24.2 PCI Express Capability-Independent Region

Capability ID

Each capability has a unique ID assigned by the PCI-SIG. See Appendix A for a current list of capability IDs. The PCI-SIG requires that certain capability IDs are supported by all PCI Express functions, described in more detail later in this chapter.

Capability Version

This field is defined for capability structures that reside in the extended configuration region (addresses 100h to FFFh). This field should be considered an extension of the Capability ID field. Its value is assigned by the PCI-SIG.

Next Capability Pointer

Each capability is an absolute address into that function's configuration space. The Next Capability Pointer follows these rules:

1. Because each capability is DWORD-aligned, the lower two bits of the Next Capability Pointer field are always 00b.
2. The Next Capability Pointer describes a full address into that function's configuration space, not an offset from the current capability.
3. The Next Capability Pointer could hold a value that lands within the region of another capability. This may cause unpredictable behavior and should never happen. To check for this error condition, software must decode each capability ID while traversing the capabilities list, and compare it to a map of the current capabilities for that function.
4. The Next Capability Pointer could reference a region of memory above the top of configuration space for that function. Alternatively, the pointer could reference an address within configuration space, but too near the top of configuration space to hold all the registers defined for that capability. These may cause unpredictable behavior and should never happen. To check for these error conditions, software must track the configuration address ranges for each capability.

Capability Linked List

A PCI Express function holds several capabilities in a linked list. The list starts in the device-dependent region of configuration memory, and may optionally continue into the extended configuration region.

The Capability Linked List follows these rules:

1. Each capability structure is DWORD-aligned.
2. The Capabilities Pointer register in the Header Type region of configuration memory points to the address of the first capability in the linked list.
3. Each capability points to the next capability in the linked list, as an address into configuration memory.
4. No specific order or placement of capability structures is required. The capabilities may occur in any order, starting at any DWORD address in the device-dependent region, and may or may not be contiguous.
5. Each capability always has its Capability ID and its Next Capability Pointer as its first two bytes in the device-dependent region (addresses 40h to FFh) and as its first four bytes in extended configuration region (addresses 100h to FFFh). These fields make up the *capability-independent* region. The *capability-dependent* region immediately follows, and may be any size, as defined by the PCI-SIG.
6. No capability structure in the device-dependent region may extend past the top of the region (address FFh).
7. The last capability structure in the device-dependent region ends the chain without pointing to the first capability structure in the extended configuration region.
8. The first capability structure placed in the extended configuration region must start at address 100h.

> In all PCI Express systems, a capability structure must exist at address 100h in each function's configuration address space. If no capability structure is needed, a null capability structure is placed with the following encoding: Capability ID = 0000h, Capability Version = 0h, and Next Capability Pointer = 00h.

9. A value of 00h in the Next Capability Pointer register indicates that the linked list is complete; the current capability is the last one.

Required Capabilities Overview

This section summarizes the required capabilities for PCI Express. These capabilities have specific register assignments and definitions described in the PCI Express Specification.

PCI Express Capability Structure

Every PCI Express function includes this structure. This structure is not defined for PCI legacy functions. If this capability structure is missing, the function is by definition a legacy PCI function. With this capability structure, the function can operate in both PCI and PCI Express software environments, using the default register values described in the PCI Express Specification.

Following are some features of the PCI Express Capability Structure.

- The size of this structure depends on what features the function includes and where it exists. Extra registers are required for functions that include slots, and even more registers are required if the function is a root port.

- The entire structure must reside in the device-dependent region (the first 256 bytes) of the function's configuration space.

- The *Max_Payload_Size Supported* register in this capability structure describes the largest data payload that this function supports. Options include powers-of-2 between 128 bytes and 4096 bytes. This value is read-only to software. Software should never attempt to send a data payload to this function greater than the value hard-wired in this register. In fixing this value, the hardware design makes a tradeoff between bandwidth efficiency (a larger value) and reduced latency across multiple requests (a smaller value). By fixing a larger value, the hardware can more easily allow the software to make this tradeoff.

- The *Max_Payload_Size* register in this capability structure is where software makes the tradeoff between bandwidth and latency. Hardware cannot send a data payload larger than is indicated in this register, and software cannot program a size larger than is indicated in the *Max_Payload_Size Supported* register.

- The *Max_Read_Request_Size* register allows software to limit the size of read requests. If a function allows multiple agents to send read requests, a smaller value here could improve the read latency of each of the agents, sacrificing some bandwidth in the process. However, some PCI Express hardware implementations incorporate dynamic load balancing between read request streams, in which case software should always choose the largest possible value for this register to avoid limiting both bandwidth and latency.

- The *Maximum Link Speed* register describes the speed of the link. For PCI Express 1.0, only one speed, 2.5 gigatransfers per second per direction, is supported. The term "maximum" here does not mean slower speeds are allowed.

PCI Power Management Capability Structure

Every PCI Express function includes this structure. This capability structure is also defined for PCI legacy functions and for them it is optional. The definition and use of these bits is the same for PCI legacy and PCI Express systems. The *PCI Bus Power Management Interface Specification* describes this capability structure.

Following are some features of the PCI Power Management Capability Structure as used in PCI Express systems.

- PCI Express systems are required to support the D0 and D3 power management states, the D1 and D2 states are optional. D3 is a low-power or power-off condition, while D0 is a fully operational condition. D1 and D2 are low-power conditions.

- Bits 1:0 of the Power Management Status/Control register is the Power State Field. This field describes the current power management state of the function. Hardware sets these bits to the current state. Software can write to this field to change the state. If software attempts to write a state that is not supported, the write completes normally but the value in the register does not change.

Message Signaled Interrupts Capability Structure

Every PCI Express function that supports interrupts includes this structure. This capability structure is also defined for PCI legacy functions and for them is optional. The definition and use of these bits is the same for PCI legacy and PCI Express systems. The *PCI Local Bus Specification, Revision 2.3* describes this capability structure.

Following are some features of the Message Signaled Interrupts (MSI) Capability Structure as used in PCI Express systems.

- PCI Express functions that generate interrupts must be capable of generating them using MSI in-band write cycles rather than using legacy PCI INTx# sideband signals. When the MSI Enable bit is set, the PCI Express function is required to use MSI writes when it sends interrupts. Software must not assume that a PCI Express device supports INTx# sideband interrupts.

- A PCI Express function is allowed only one MSI capability structure, and each function in a device must have its own MSI capability structure.

PCI Express Capability Structure

The PCI Express Capability Structure is required for every PCI Express function. This 36-byte structure is strictly defined and every PCI Express function must follow its organization: the first 20 bytes are common to all PCI Express functions, the next 8 bytes are used only by root ports and functions with slots, and the last 8 bytes are used only by root ports. Figure 24.3 and the following register descriptions explain this capability structure in detail.

Chapter 24: Configuration Capabilities ■ 949

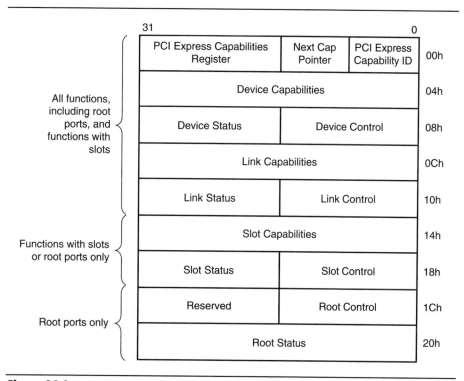

Figure 24.3 PCI Express Capability Structure

PCI Express Capability ID Register (Offset 00h)	
Width	1 byte
Valid Value	10h, Read Only
Description	The value 10h is defined by the PCI-SIG to describe the PCI Express Capability Structure. The PCI-SIG assigns capability ID numbers.

Next Capability Pointer Register (Offset 01h)

Width	1 byte
Valid Value	00h, 40h to FCh (must be DWORD-aligned), Read Only
Description	This value describes the address in this function's configuration space at which the next capability structure starts. The PCI Express Capability structure resides in the PCI device-dependent region (addresses 40h to FFh) so it can point only to other capabilities in that region. The value 00h indicates no further capability structures in this region — the PCI Express Capability structure is the last structure in this region (addresses 40h to FFh) for this function.

PCI Express Capabilities Register (Offset 02h)

Width	2 bytes
Valid Values	This field is a collection of bits, described in the next table. Bits are Read-Only (RO) or Hardware Initialized (HwInit). HwInit describes bits that are initialized at power-good reset and thereafter are Read-Only.
Description	This register identifies the PCI Express type and version of this function and describes some of its capabilities.

Bit	Register Description	Default	Type
3:0	**Capability Version** Indicates the PCI Express version number for this function. Valid version numbers are defined by the PCI-SIG. For PCI Express 1.0, this number must be 1h.	1h	RO

Bit	Register Description	Default	Type
7:4	**Function and Port Type** Indicates the type of PCI Express function. Valid encodings are defined by the PCI-SIG. Valid encodings are: 0000b PCI Express Endpoint function 0001b Legacy PCI Express Endpoint function 0100b Root Port of PCI Express Root Complex 0101b Upstream Port of PCI Express Switch 0110b Downstream Port of PCI Express Switch 0111b PCI Express-to-PCI/PCI-X Bridge 1000b PCI/PCI-X to PCI Express Bridge **Note** All encodings not defined by the PCI-SIG are reserved. **Note** All the non-endpoint functions require a Type 01h Configuration Space Header. **Note** PCI Express endpoint functions that require I/O resources during normal operation must respond with 0001b. If a PCI Express endpoint function does not require I/O resources, it can respond with 0000b.	Function-specific	RO
8	**Slot Implemented** 1 The PCI Express Link associated with this port is connected to a slot. 0 The PCI Express Link associated with this port is connected to an integrated component or is disabled.	0	Hw-Init
13:9	**Interrupt Message Number** If this function is assigned more than one MSI interrupt number, this register contains the offset between the base Message Data and the MSI Message that is generated when any of status bits in either the Slot Status register or the Root Port Status register of this capability structure are set. Hardware is required to update this field whenever the number of MSI Messages assigned to the device changes.	Function-specific	RO
15:14	**Reserved**	00b	RO

Device Capabilities Register (Offset 04h)	
Width	4 bytes
Valid Values	This field is a collection of bits, described in the following table. Bits are Read-Only (RO).
Description	This register describes specific capabilities of this function.

Bit	Register Description	Default	Type
2:0	**Max_Payload_Size Supported** This field describes the maximum payload size this function can support. Valid encodings are: 000b 128B max payload size 001b 256B max payload size 010b 512B max payload size 011b 1024B max payload size 100b 2048B max payload size 101b 4096B max payload size 11xb Reserved	Function-specific	RO

Bit	Register Description	Default	Type
4:3	**Phantom Functions Supported** **Note** This function is part of a device that may have up to 8 functions. If the device has less than 8 functions, the three-bit address field in the header describing the function number could possibly shrink to fewer bits, and the bit or bits saved may be used to expand the size of the outstanding transaction field (Tag Identifier field) in the header. Any bits removed from the function field are called *phantom function* bits. This field indicates the number of most significant bits of the function number field that are phantom bits and are logically combined with the Tag identifier. Valid encodings are; 00b No function number bits used for Phantom Functions; this device may implement all function numbers. 01b The most significant bit of the function number in the Requestor ID is used for Phantom Functions. The device may implement functions 0-3. Functions 0, 1, 2, and 3 may claim functions 4, 5, 6, and 7 as Phantom Functions respectively. 10b The two most significant bits of the function number in the Requestor ID are used for Phantom Functions. The device may implement functions 0-1. Function 0 may claim functions 2, 4, and 6 as Phantom Functions, function 1 may claim functions 3, 5, and 7 as Phantom Functions. 11b All three bits of function number in Requestor ID used for Phantom Functions; device must be a single function 0 device that may claim all other functions as Phantom Functions. **Note** Phantom Function support for this device must be enabled in the Device Control register.	00	RO
5	**Extended Tag Field Supported** This field indicates the maximum supported size of the Tag field for the device controlling this function. Valid encodings are: 0b 5-bit Tag field supported 1b 8-bit Tag field supported **Note** 8-bit Tag field support for this device must be enabled in the Device Control register.	0	RO

Bit	Register Description	Default	Type
8:6	**Endpoint L0s Acceptable Latency** This field describes how much latency an Endpoint can withstand when it transitions from the L0s state to the L0 state. (The greater the buffering in the endpoint function, the greater the latency it can withstand). **Note** This number is used by power management software to determine whether the transition to L0s can be accomplished with no drop in performance. Valid encodings are: 000b Less than 64 ns 001b 64 ns to less than 128 ns 010b 128 ns to less than 256 ns 011b 256 ns to less than 512 ns 100b 512 ns to less than 1 µs 101b 1 µs to less than 2 µs 110b 2 µs-4 µs 111b More than 4 µs	Function -specific	RO
11:9	**Endpoint L1 Acceptable Latency** This field describes how much latency an Endpoint can withstand when it transitions from the L1 state to the L0 state. (The greater the buffering in the endpoint function, the greater the latency it can withstand). **Note** This number is used by power management software to determine whether the transition to L0 can be accomplished with no drop in performance. Valid encodings are: 000b Less than 1µs 001b 1 µs to less than 2 µs 010b 2 µs to less than 4 µs 011b 4 µs to less than 8 µs 100b 8 µs to less than 16 µs 101b 16 µs to less than 32 µs 110b 32 µs-64 µs 111b More than 64 µs	Function -specific	RO

Bit	Register Description	Default	Type
12	**Attention Button Present** 1 An Attention Button is implemented on the card or module holding this function. 0 No Attention Button on the card or module. This bit is valid for the following PCI Express functions: • PCI Express Endpoint function • Legacy PCI Express Endpoint function • Upstream Port of PCI Express Switch • PCI Express-to-PCI/PCI-X bridge	0	RO
13	**Attention Indicator Present** 1 An Attention Indicator is implemented on the card or module holding this function. 0 No Attention Indicator on the card or module This bit is valid for the following PCI Express functions: • PCI Express Endpoint device • Legacy PCI Express Endpoint device • Upstream Port of PCI Express Switch • PCI Express-to-PCI/PCI-X bridge	0	RO
14	**Power Indicator Present** 1 A Power Indicator is implemented on the card or module holding this function. 0 No Power Indicator on the card or module This bit is valid for the following PCI Express functions: • PCI Express Endpoint device • Legacy PCI Express Endpoint device • Upstream Port of PCI Express Switch • PCI Express-to-PCI/PCI-X bridge	0	RO
17:15	**Reserved**	0h	RO

Bit	Register Description	Default	Type
25:18	**Captured Slot Power Limit Value** **Note** This field applies to upstream ports only. Combined with the Slot Power Limit Scale value, this field specifies the upper limit on power supplied by the slot. Power limit (in Watts) calculated by multiplying the value in this field by the value in the Slot Power Limit Scale field. This value is set by the Set_Slot_Power_Limit message or hardwired to 00h.	00h	RO
27:26	**Captured Slot Power Limit Scale** **Note** This field applies to upstream ports only. This field sets the scale used for the Slot Power Limit. Range of Values: 00b = 1.0x 01b = 0.1x 10b = 0.01x 11b = 0.001x This value is set by the Set_Slot_Power_Limit message or hardwired to 00b.	00b	RO
31:28	**Reserved**	0h	RO

Device Control Register (Offset 08h)

Width	2 bytes
Valid Values	This field is a collection of bits, described in the following table. Bits are Read-Write (RW) or Read-Write-Sticky (RWS). Sticky bits are not affected by reset, low-power states, or auxiliary power states.
Description	This register controls the operating parameters of this function.

Bit	Register Description	Default	Type
0	**Correctable Error Reporting Enable** 　0　Correctable errors are not reported. The ERR-COR message is never generated. 　1　Correctable errors are reported.	0	RW

Bit	Register Description	Default	Type
1	**Non-Fatal Error Reporting Enable** 0 Nonfatal errors are not reported. The ERR-NONFATAL message is never generated. 1 Nonfatal errors are reported.	0	RW
2	**Fatal Error Reporting Enable** 0 Fatal errors are not reported. The ERR-FATAL message is never generated. 1 Fatal errors are reported.	0	RW
3	**Unsupported Request Reporting Enable** 0 Unsupported Requests are not reported. 1 Unsupported Requests are reported.	0	RW
4	**Enable Relaxed Ordering** 0 This function may not set the Relaxed Ordering bit in the Attributes field of transactions it initiates. 1 This function may set the Relaxed Ordering bit in the Attributes field of transactions it initiates, as long as those transactions do not require strong write ordering. **Note** Relaxed ordering means that a completion is allowed to pass a posted request. **Note** This bit may be hardwired to 0 if a device never sets the Relaxed Ordering attribute in transactions that it initiates as a requester.	1	RW

Bit	Register Description	Default	Type
7:5	**Max_Payload_Size** This field sets the maximum payload size for this function. **Note** As a receiver, this function must handle TLPs up to and including the size defined here. As a transmitter, this function may not generate TLPs above the size defined here. **Note** This max payload size must not exceed that defined in the Device Capabilities register. Valid encodings for this field are: 000b 128B max payload size 001b 256B max payload size 010b 512B max payload size 011b 1024B max payload size 100b 2048B max payload size 101b 4096B max payload size 11xb Reserved	000b	RW
8	**Extended Tag Field Enable** 0 This function must use a 5-bit Tag field. 1 This function may use an 8-bit Tag field when sending request packets. **Note** Functions that cannot support 8-bit Tag fields should hardwire this bit to 0. **Note** See bit 5 in the Device Capabilities Register	0	RW
9	**Phantom Functions Enable** 0 This function may not use phantom functions. 1 This function is allowed to use phantom functions to extend its number of outstanding TIDs (transaction identifiers). **Note** Functions that cannot support phantom functions should hardwire this bit to 0. **Note** See bits 4:3 in the Device Capabilities Register	0	RW

Bit	Register Description	Default	Type
10	**Auxiliary (AUX) Power PM Enable** 0 This function may not draw AUX power. 1 This function is allowed to draw AUX power. **Note** Functions that cannot support AUX Power should hardwire this bit to 0.	0	RW
11	**Enable No Snoop** 0 This function may not set the No Snoop bit in the Requester Attributes of transactions it initiates. 1 This function is allowed to set the No Snoop bit in the Requester Attributes of transactions it initiates that do not require hardware enforced cache coherency. **Note** A function may set the No Snoop attribute only on a transaction when it can guarantee that the address of the transaction is not stored in any cache in the system. **Note** This bit may be hardwired to 0 if a device never sets the No Snoop attribute in transactions it initiates.	1	RW
14:12	**Max_Read_Request_Size** This field sets the maximum Read Request size for transactions initiated by this function. Valid encodings for this field are: 000b 128B max read request size 001b 256B max read request size 010b 512B max read request size 011b 1024B max read request size 100b 2048B max read request size 101b 4096B max read request size 11xb Reserved **Note** Functions that do not generate Read Requests larger than 128B may hardwire this field to 000b.	010b	RW
15	**Reserved**	0b	RO

Device Status Register (Offset 0Ah)

Width 2 bytes

Valid Values This field is a collection of bits, described in the following table. Bits are Read-Only (RO) or Read-Only-Write-1-to-Clear (RW1C).

Description This register gives the status of the operating parameters of this function.

Bit	Register Description	Default	Type

0	**Correctable Error Detected**
	   0 This function has not detected a correctable error.
	   1 This function has detected a correctable error.
	 Note
	 This register bit captures all correctable errors, irrespective of whether or not error reporting is enabled in the Device Control register.
		0	RW1C

1	**Nonfatal Error Detected**
	   0 This function has not detected a nonfatal error.
	   1 This function has detected a nonfatal error.
	 Note
	 This register bit captures all nonfatal errors, irrespective of whether or not error reporting is enabled in the Device Control register.
		0	RW1C

2	**Fatal Error Detected**
	   0 This function has not detected a fatal error.
	   1 This function has detected a fatal error.
	 Note
	 This register bit captures all fatal errors, irrespective of whether or not error reporting is enabled in the Device Control register.
		0	RW1C

3	**Unsupported Request Detected**
	   0 This function has not detected an unsupported request.
	   1 This function has detected an unsupported request.
	 Note
	 This register bit captures all detections of unsupported requests, irrespective of whether or not error reporting is enabled in the Device Control register.
		0	RW1C

Bit	Register Description	Default	Type
4	**AUX Power Detected**	0	RO
	0 The function has not detected AUX power.		
	1 The function has detected AUX power.		
	Note		
	Functions that do not require AUX power may hardwire this bit to 0.		
5	**Transactions Pending**	0	RO
	0 The function has received completions for all previous non-posted requests it initiated.		
	1 The function has not yet received completions for all outstanding requests.		
15:6	**Reserved**	000h	RO

Link Capabilities Register (Offset 0Ch)

Width	4 bytes
Valid Values	This field is a collection of bits, described in the following table. Bits are Read-Only (RO) or Hardware Initialized (HwInit). HwInit describes bits that are initialized at power-good reset and thereafter are Read-Only.
Description	This register describes specific capabilities of this link.

Bit	Register Description	Default	Type
3:0	**Maximum Link Speed**	0001b	RO
	This field describes the maximum Link speed of the given PCI Express Link. As of this writing, the only defined encoding is:		
	0001b 2.5 Gb/s Link		
	All other encodings are reserved.		

Bit	Register Description	Default	Type
9:4	**Maximum Link Width** This field describes the maximum width of the given PCI Express Link. Defined encodings are: 000000b Reserved 000001b x1 000010b x2 000100b x4 001000b x8 001100b x12 010000b x16 100000b x32 **Note** All links must operate at their maximum link width and at the x1 link width. Operating at other link widths is optional.	Function -specific	RO
11:10	**Active State Link PM Support** This field describes the level of active state power management supported on the given PCI Express Link. Defined encodings are: 00b Reserved 01b L0s Entry Supported 10b Reserved 11b L0s and L1 Supported	Function -specific	RO

Bit	Register Description	Default	Type
14:12	**L0s Exit Latency** This field describes the L0s exit latency for the given PCI Express Link. The value reported indicates the length of time this Port requires to complete its transition from L0s to L0. Defined encodings are: 000b Less than 64 ns 001b 64 ns to less than 128 ns 010b 128 ns to less than 256 ns 011b 256 ns to less than 512 ns 100b 512 ns to less than 1 µs 101b 1 µs to less than 2 µs 110b 2 µs-4 µs 111b Reserved	Function -specific	RO
17:15	**L1 Exit Latency** This field indicates the L1 exit latency for the given PCI Express Link. The value reported indicates the length of time this Port requires to complete its transition from L1 to L0. Defined encodings are: 000b Less than 1µs 001b 1 µs to less than 2 µs 010b 2 µs to less than 4 µs 011b 4 µs to less than 8 µs 100b 8 µs to less than 16 µs 101b 16 µs to less than 32 µs 110b 32 µs-64 µs 111b More than 64 µs	Function -specific	RO
23:18	**Reserved**	00h	RO
31:24	**Port Number** This field describes the PCI Express port number for the given PCI Express Link.	Function -specific	Hw-Init

Link Control Register (Offset 10h)

Width	2 bytes
Valid Values	This field is a collection of bits, described in the following table. Bits are Read-Write (RW) or Read-Only (RO).
Description	This register controls the operating parameters of this link.

Bit	Register Description	Default	Type
1:0	**Active State Link PM Control** This field controls the level of active state power management supported on the given PCI Express Link. Valid encodings are: 00b Disabled 01b L0s Entry Supported 10b Reserved 11b L0s and L1 Entry Supported	Function-specific	RW
3	**Read Completion Boundary (RCB)** **Note** The RCB parameter describes the naturally aligned address boundary on which a read request may be serviced with multiple completions. Valid encodings are: 0 64 byte 1 128 byte **Note** For a root port, this field is hardwired.	1	RW (RO for root ports)
4	**Link Disable** 0 The link is enabled 1 The link is disabled. **Note** This bit does not apply to endpoint functions or to upstream ports on a switch. For those functions, this bit is reserved. **Note** Writing to this bit immediately changes the bit value, and the new value can be read, even if the link has not completed its transition to enabled or disabled.	0	RW

Bit	Register Description	Default	Type
5	**Retrain Link**	0	RW
	0 No change.		
	1 Initiate link retraining.		
	Note This bit always returns 0 when read.		
	Note This bit does not apply to endpoint functions or to upstream ports on a switch. For those functions, this bit is reserved.		
6	**Common Clock Configuration**	0	RW
	0 This component and the component at the opposite end of this Link are operating with asynchronous reference clocks		
	1 This component and the component at the opposite end of this Link are operating with a distributed common reference clock.		
	Note Components utilize this common clock configuration information to report the correct L0s and L1 Exit Latencies.		
7	**Extended Sync**	0	RW
	0 Normal synchronization sequence.		
	1 Extended synchronization sequence which injects FTS ordered sets and extra TS2 at transition from L1 to L0.		
15:8	Reserved	00h	RO

Link Status Register (Offset 12h)	
Width	2 bytes
Valid Values	This field is a collection of bits, described in the following table. Bits are Read-Only (RO) or Hardware Initialized (HwInit). HwInit describes bits that are initialized at power-good reset and thereafter are Read-Only.
Description	This register gives the status of the operating parameters of this link.

Bit	Register Description	Default	Type
3:0	**Link Speed**	0001b	RO
	This field describes the Link speed of the given PCI Express Link. As of this writing, the only defined encoding is:		
	0001b 2.5 Gb/s Link		
	All other encodings are reserved.		
3:0	**Link Speed**	0001b	RO
	This field describes the Link speed of the given PCI Express Link. As of this writing, the only defined encoding is:		
	0001b 2.5 Gb/s Link		
	All other encodings are reserved.		
9:4	**Negotiated Link Width**	Function-specific	RO
	This field describes the negotiated width of the given PCI Express Link.		
	Defined encodings are:		
	000001b X1		
	000010b X2		
	000100b X4		
	001000b X8		
	001100b X12		
	010000b X16		
	100000b X32		
	All other encodings are reserved.		
10	**Training Error**	0	RO
	0 No link training error has occurred.		
	1 A link training error has occurred.		
	Note This is a read-only bit, writing it has no effect. The bit is cleared when the link trains successfully to the L0 Link state.		
	Note This bit does not apply to endpoint functions or to upstream ports on a switch. For those functions, this bit is reserved.		

Bit	Register Description	Default	Type

11	**Link Training**
	    0 Link training has completed.
	    1 Link training is in progress.
	 Note
	 This is a read-only bit, writing it has no effect. The bit is cleared when the link trains successfully.
	 Note
	 This bit does not apply to endpoint functions or to upstream ports on a switch. For those functions, this bit is reserved.
		0	RO

12	**Slot Clock Configuration**
	    0 The function uses an independent reference clock rather than a reference clock on the connector.
	    1 The function uses the same physical reference clock that the platform provides on the connector.
		Function -specific	RO

15:13	**Reserved**	0h	RO

Slot Capabilities Register (Offset 14Ch)

Width	4 bytes
Valid Values	This field is a collection of bits, described in the following table. Bits are Hardware Initialized (HwInit). HwInit describes bits that are initialized at power-good reset and thereafter are Read-Only.
Description	This register is used only for ports that have slots and for root ports. It describes specific capabilities of this slot.

Bit	Register Description	Default	Type

0	**Attention Button Present**
	    0 No Attention Button exists on the chassis for this slot.
	    1 An Attention Button is implemented on the chassis for this slot.
		Function -specific	Hw-Init

1	**Power Controller Present**
	    0 No Power Controller exists for this slot.
	    1 A Power Controller is implemented on the chassis for this slot.
		Function -specific	Hw-Init

Bit	Register Description	Default	Type
2	**MRL Sensor Present** 0 No MRL Sensor exists on the chassis for this slot. 1 An MRL Sensor is implemented on the chassis for this slot.	Function-specific	Hw-Init
3	**Attention Indicator Present** 0 No Attention Indicator exists on the chassis for this slot. 1 An Attention Indicator is implemented on the chassis for this slot.	Function-specific	Hw-Init
4	**Power Indicator Present** 0 No Power Indicator exists for this slot. 1 A Power Indicator is implemented on the chassis for this slot.	Function-specific	Hw-Init
5	**Hot-plug Surprise** 0 The card holding this function may not be removed from the system without prior notice. 1 The card holding this function may be removed from the system without prior notice.	Function-specific	Hw-Init
6	**Hot-plug Capable** 0 The card holding this function is not capable of supporting hot plug operations. 1 The card holding this function is capable of supporting hot plug operations.	Function-specific	Hw-Init
14:7	**Slot Power Limit Value** In combination with the Slot Power Limit Scale value, this specifies the upper limit on power supplied by the slot. **Note** The power limit (in Watts) is calculated by multiplying the value in this field with the value in the Slot Power Limit Scale field. **Note** This register must be implemented if the Slot Implemented bit is set. The default value prior to hardware/firmware initialization is 0000 0000b.	00h	Hw-Init

Bit	Register Description	Default	Type
16:15	**Slot Power Limit Scale** This register specifies the scale used for the Slot Power Limit Value. Valid range of Values: 00b = 1.0x 01b = 0.1x 10b = 0.01x 11b = 0.001x **Note** This register must be implemented if the Slot Implemented bit is set.	00b	Hw-Init
18:17	**Reserved**	00b	RO
31:19	**Physical Slot Number** This field indicates the physical slot number attached to this Port. This field must be hardware initialized to a value that assigns a slot number that is unique within the chassis. **Note** This field should be initialized to 0 for ports connected to devices that are integrated on the system board or integrated within the same silicon as the Switch device or Root Port.	0000h	Hw-Init

Slot Control Register (Offset 18h)

Width	2 bytes
Valid Values	This field is a collection of bits, described in the following table. Bits are Read-Write (RW).
Description	This register is used only for ports that have slots and for root ports. This register controls the operating parameters of this slot.

Bit	Register Description	Default	Type

0	**Attention Button Pressed Enable**
	  0 Disables the generation of a hot plug interrupt or wake message when an attention button is pressed.
	  1 Enables the generation of a hot plug interrupt or wake message when an attention button is pressed.
		0	RW

Bit	Register Description	Default	Type
1	**Power Fault Detected Enable**	0	RW
	0 Disables the generation of a hot plug interrupt or wake message on a power fault event.		
	1 Enables the generation of a hot plug interrupt or wake message on a power fault event.		
2	**MRL Sensor Changed Enable**	0	RW
	0 Disables the generation of a hot plug interrupt or wake message on a MRL sensor changed event.		
	1 Enables the generation of a hot plug interrupt or wake message on a MRL sensor changed event.		
3	**Presence Detect Changed Enable**	0	RW
	0 Disables the generation of a hot plug interrupt or wake message when a presence detect change is sensed.		
	1 Enables the generation of a hot plug interrupt or wake message when a presence detect change is sensed.		
4	**Command Completed Interrupt Enable**	0	RW
	0 Disables the generation of a hot plug interrupt when a command is completed by the Hot plug controller.		
	1 Enables the generation of a hot plug interrupt when a command is completed by the Hot plug controller.		
5	**Hot plug Interrupt Enable**	0	RW
	0 Disables generation of a hot plug interrupt on hot plug events.		
	1 Enables generation of a hot plug interrupt on hot plug events for which interrupts are enabled.		

Bit	Register Description	Default	Type
7:6	**Attention Indicator Control** Encoding of Attention Indicator: 00b Reserved 01b On 10b Blink 11b Off Reading this register returns the current state of the Attention Indicator. Writing this register controls the Attention Indicator as per the encodings given, and also causes the Port to send the appropriate ATTENTION_INDICATOR_* messages.	Function -specific	RW
9:8	**Power Indicator Control** Encoding of Power Indicator: 00b Reserved 01b On 10b Blink 11b Off Reading this register returns the current state of the Power Indicator. Writing this register controls the Power Indicator as per the encodings given, and also causes the Port to send the appropriate POWER _INDICATOR_* messages.	Function -specific	RW

10	**Power Controller Control**
	
	Encoding of Power Controller Control bit:
	
	    0 Power On
	
	    1 Power Off
	
	Reading this register returns the current state of the power applied to the slot.
	
	Writing this register controls the power going to the slot.
		Function -specific	RW

15:11	**Reserved**	00h	RO

Slot Status Register (Offset 1Ah)

Width	2 bytes
Valid Values	This field is a collection of bits, described in the following table. Bits are Read-Only (RO) or Read-Only-Write-1-to-Clear (RW1C).
Description	This register is used only for ports that have slots and for root ports. This register gives the status of the operating parameters of this slot.

Bit	Register Description	Default	Type
0	**Attention Button Pressed**	0	RW1C
	0 The attention button was not pressed.		
	1 The attention button was pressed.		
1	**Power Fault Detected**	0	RW1C
	0 The Power Controller did not detect a power fault at this slot.		
	1 The Power Controller detected a power fault at this slot.		
2	**MRL Sensor Changed**	0	RW1C
	0 An MRL Sensor state change was not detected.		
	1 An MRL Sensor state change was detected.		
3	**Presence Detect Changed**	0	RW1C
	0 A Presence Detect change was not detected.		
	1 A Presence Detect change was detected.		
4	**Command Completed**	0	RW1C
	0 The hot plug controller did not complete an issued command.		
	1 The hot plug controller completed an issued command.		
5	**MRL Sensor State**	0	RO
	0 Status of the MRL sensor is Closed.		
	1 Status of the MRL sensor is Open.		
	Note This status applies only when the MRL sensor is implemented.		

Bit	Register Description	Default	Type
6	**Presence Detect State** 0 Slot is empty. 1 A card is detected in the slot. **Note** This bit reflects the Presence Detect status determined through an in-band mechanism or through the Present Detect pins as defined in the *PCI Express Card Electromechanical Specification*. **Note** This register is required to be implemented on all Switch devices and Root Ports. The value of this field for Switch devices or Root Ports not connected to slots should be hardwired to 1. This register is required if a slot is implemented.	Function-specific	RO
15:7	**Reserved**	000h	RO

Root Control Register (Offset 1Ch)

Width	2 bytes
Valid Values	This field is a collection of bits, described in the following table. Bits are Read-Write (RW).
Description	This register is used only for root ports. This register controls the operating parameters of this root port.

Bit	Register Description	Default	Type
0	**System Error on Correctable Error Enable** 0 A System Error should not be generated if a correctable error is reported. 1 A System Error should be generated if a correctable error (ERR_COR) is reported by any of the devices in the hierarchy associated with this Root Port, or by the Root Port itself. **Note** The mechanism for signaling a System Error to the system is system-specific.	0	RW

Bit	Register Description	Default	Type
1	**System Error on Nonfatal Error Enable** 0 A System Error should not be generated if a nonfatal error is reported. 1 A System Error should be generated if a nonfatal error (ERR_NONFATAL) is reported by any of the devices in the hierarchy associated with this Root Port, or by the Root Port itself. **Note** The mechanism for signaling a System Error to the system is system-specific.	0	RW
2	**System Error on Fatal Error Enable** 0 A System Error should not be generated if a fatal error is reported. 1 A System Error should be generated if a fatal error (ERR_FATAL) is reported by any of the devices in the hierarchy associated with this Root Port, or by the Root Port itself. **Note** The mechanism for signaling a System Error to the system is system-specific.	0	RW
3	**PME Interrupt Enable** 0 Do not generate PME interrupts. 1 Generate a PME interrupt upon receipt of a PME message as reflected in the PME Status register bit. Alternatively, generate a PME interrupt when the PME Status register bit is set and this bit is set from a cleared state.	0	RW
15:4	**Reserved**	000h	RO

Root Status Register (Offset 20h)

Width	4 bytes
Valid Values	This field is a collection of bits, described in the following table. Bits are Read-Only (RO) or Read-Only-Write-1-to-Clear (RW1C).
Description	This register is used only for root ports. This register gives the status of the operating parameters of this root port.

Bit	Register Description	Default	Type
15:0	**PME Requestor ID** This field holds the PCI requestor ID of the last PME requestor.	0000h	RO
16	**PME Status** 0 The most recent PME does not necessarlly have the same requestor ID indicated in the PME Requestor ID field. 1 The most recent PME has the requestor ID indicated in the PME Requestor ID field. **Note** When this bit is set, subsequent PMEs are kept pending until the status register is cleared by software by writing a 1.	0	RW1C
17	**PME Pending** 0 No PME is pending. 1 The PME Status bit is set and another PME is pending. **Note** When PME Pending is set, it is cleared only when the following sequence occurs: 1) the PME Status bit is cleared by software, 2) hardware updates the PME Requestor ID field, and 3) hardware sets the PME Status field. This three-step sequence repeats until no more PMEs are pending.	0	RO
31:18	**Reserved**	0000h	RO

Chapter 25

Real-Time Analysis

This chapter provides an illustrated review of many aspects important to PCI Express. The test setup uses Computer Access Technology Corporation's (CATC) protocol analyzers and a test bench to capture the illustrations shown in this chapter. CATC develops advanced protocol analysis expert systems used by semiconductor, device, system, and software companies to monitor communications traffic, diagnose design and operational problems, and confirm interoperability and standards compliance. For more information, visit www.catc.com.

Configurations

PCI Express illustrations captured for this chapter were collected in x1, x4 and x8 configurations. The following describe the test bench setup for x1, x4 and x8 configurations.

X1 and X4 Configuration Test Setup

In the x1 or x4 configurations, the protocol analyzer is connected to a host PC via a USB 2.0 cable. The analyzer (CATC PETracer[†] for the x1 and a CATC PETracer[†] ML for the x4 configuration) is cabled to a slot interposer, which sits between the root complex PCI Express connector and the Device Under Test (DUT) for data capture. Figure 25.1 illustrates a typical test bench configuration for x1 or x4.

Figure 25.1   Typical Test Bench Configuration for x1, x2, and x4

X8 Configuration Test Setup

The x8 test setup utilizes two CATC PETracer[†] ML analyzers in a stacked configuration that are connected to the x8 interposer card. Figure 25.2 illustrates a x8 configuration.

Figure 25.2   Typical Test Bench Configuration for x8

Mid-Bus Probe:

Interposer cards are used to debug the DUT when PCI Express slots are available. Mid-bus probe is an alternative mechanism intended for debug purposes when PCI Express slots are not available. Figure 25.3 illustrates a mid-bus probe application.

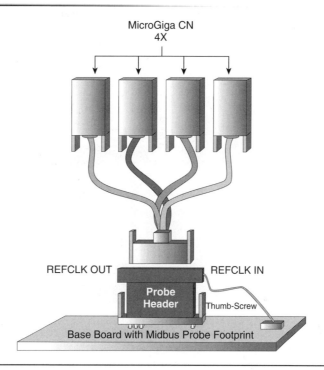

Figure 25.3 Mid-bus Probe

PCI Express Transaction Flow Overview

This chapter includes a detailed review of the Transaction Layer, showing different types of Transaction Layer packets and different scenarios for communication between PCI Express components; the Data Link Layer, showing the mechanism for acknowledgments and retries on a PCI Express link; and the Physical Layer, showing different types of PCI Express packet framing, applying packet symbols to lanes, scrambling and 8/10-bit encoding. PCI Express Configuration Space accesses and the Virtual Channel mechanism are also discussed. Figure 25.4 illustrates PCI Express traffic flow.

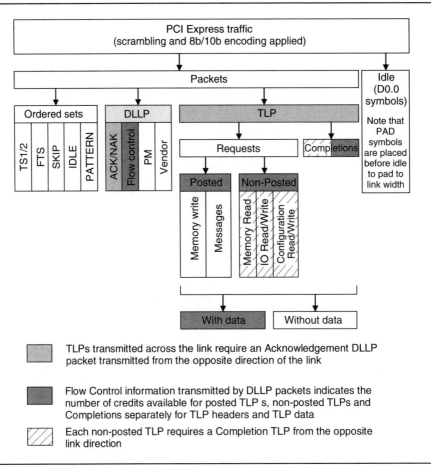

Figure 25.4 PCI Express Traffic Flow

Transaction Layer

The Transaction Layer facilitates the exchange of data between PCI Express components by means of TLPs that are transmitted across the PCI Express fabric.

TLP Fields

TLPs consist of three distinct segments:
1. Header
2. Data (optional)
3. ECRC (optional)

Figure 25.5 shows the possible configurations for these segments.

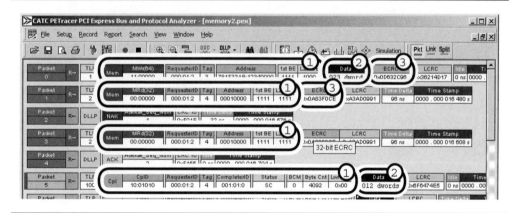

Figure 25.5 TLP Segments

TLP header layouts differ between TLP types, but the header fields illustrated in Figure 25.6 are common for all TLP types.

Figure 25.6 TLP Header

Specifically, the TLP header layout includes:

- Header type
- TC: traffic class
- TD: indicates that ECRC is present at the end of the TLP
- EP: indicates that the TLP is poisoned
- Attributes: snoop and ordering
- Requester ID
- Tag (valid only for non-posted TLPs—TLPs that require completion)

Additionally, all TLP requests have a Byte Enable field in the TLP header that indicates which bytes are valid in the data payload.

Figure 25.7 shows an example of the TLP header layout for a 32-bit Memory Read request.

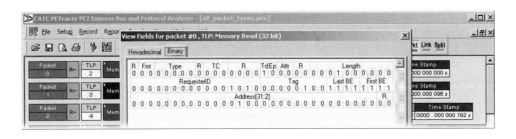

Figure 25.7 TLP Header Layout

TLP Routing

Figures 25.8 through 25.10 show the three methods used for routing TLPs across the PCI Express fabric:

- Address (Memory Read/Write, I/O Read/Write TLPs)
- ID (Configuration Read/Write, Completion TLPs)
- Implicit (Message TLPs)

Figure 25.8 TLP Address Routing

Figure 25.9 TLP ID Routing

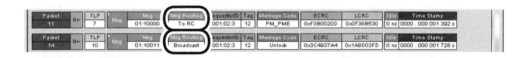

Figure 25.10 Implicit TLP Routing

TLP Request/Completion

TLP Requests are separated into two groups:

- Non-Posted: TLPs that require one or more Completion TLPs from the opposite link direction (Memory Read, I/O Read/Write, Configuration Read/Write)

- Posted: TLPs that do not require a Completion TLP (Memory Write, Messages)

TLP Requests along with one or more Completion TLPs that correspond to this request are called a *split transaction*. (Multiple completions can only be issued for Memory Read requests.) The Requester ID and Tag fields that are present in the TLP Request and TLP Completion(s) identify split transactions, as shown in Figure 25.11.

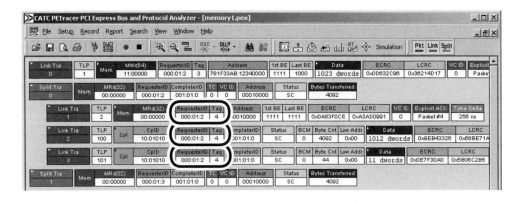

Figure 25.11 Split Transaction

By default only 32 outstanding requests are allowed for a device (only five bits of the Tag field are used).

It is possible to use all eight bits in the Tag field by setting the Extended Tag Field Enable bit in the Device Control Register of the PCI Express Capability Structure in the device's configuration space. Each Completion TLP has Status field in TLP header, as illustrated in Figure 25.12.

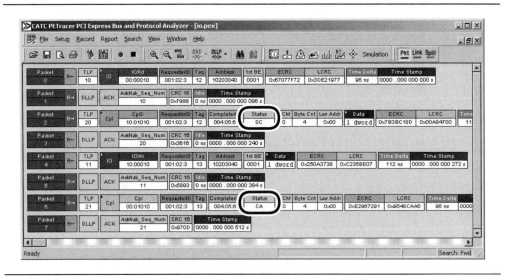

Figure 25.12 Completion Status

There are four different completion status values:

- Successful Completion.

- Unsupported Request: indicates some action or access to some space that is not supported by the device.

- Configuration Request Retry Status: indicates that the device is temporarily unable to process a Configuration request, probably due to device initialization.

- Completer Abort: indicates that a request violates the programming model of the device, or that a device-specific error condition has occurred.

There may be a number of Completion TLPs corresponding to a single Memory Read request.

The Byte Count field in the Completion TLP header is used to determine the remaining number of bytes to complete the request, illustrated in Figure 25.13.

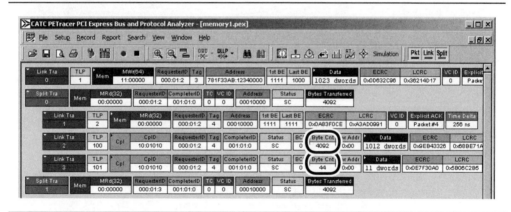

Figure 25.13 Completion Byte Count

This field indicates the number of bytes required to complete the request, including the number of bytes returned by this Completion TLP.

Data Link Layer

The Data Link Layer handles the acknowledgment and retry sequences for all TLPs. To do this, the Data Link Layer applies a Sequence Number for each TLP. Acknowledgement DLLPs reference the Sequence Number.

There are two Acknowledgement DLLP types:

- ACK: acknowledges all TLPs with a Sequence Number less than or equal to AckNak_Seq_Num

- NAK: acknowledges all TLPs with a Sequence Number less than or equal to AckNak_Seq_Num and initiates retries of TLPs with a Sequence Number greater than AckNak_Seq_Num

In Figure 25.14, a DLLP with AckNak_Seq_Num 101 acknowledges all TLPs with Sequence Numbers 101 and below.

Figure 25.14 ACK DLLP

TLPs are stored in a retry buffer until the Acknowledgement DLLP for the Sequence Number assigned to the TLP is received. A retry is initiated if:

- A NAK DLLP is received or

- No ACK/NAK DLLPs are received for a specified timeout period (Note: this timeout period is determined by the link width and the Max_Payload_Size field of the Device Control register in the PCI Express Capability Structure of the device's configuration space.)

In Figure 25.15, DLLP 2 with AckNak_Seq_Num equal to 1:
1. acknowledges all TLPs with Sequence Numbers 1 and below.
2. causes TLP 2 to be retransmitted, as can be seen directly below the DLLP NAK in the illustration.

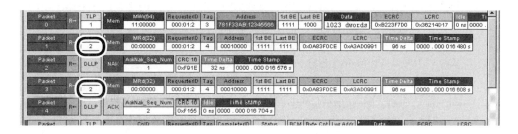

Figure 25.15 NAK DLLP

Figure 25.16 shows how the Data Link Layer facilitates data integrity by applying a 32-bit CRC to each TLP and a 16-bit CRC to each DLLP.

Figure 25.16 Data Integrity: CRC

Physical Layer

There are three different packet types from the point of view of the Physical Layer:

- Transaction Layer Packets (TLPs)
- Data Link Layer Packets (DLLPs)
- Ordered sets

PCI Express packets have the following algorithm of transmission on the PCI Express link.

1. Framing and lane application

   A. TLPs and DLLPs

   In the case of TLPs and DLLPs, framing symbols are applied to the beginning and ending of the packet buffer. The resulting buffer is then distributed across all link lanes for transmission purposes and is reassembled on the receiving side.

   The example in Figure 25.17 shows how TLP data (0x00 0x65 0x4A 0x00 0x80 0x08 0x01 0x08 0x00 0x2c 0x00 0x0a ... 0x58 0x06 0xC2 0xB5) would be distributed and transmitted across all lanes of an eight-lane link after it is framed by an STP (K27.7) and END (K29.7).

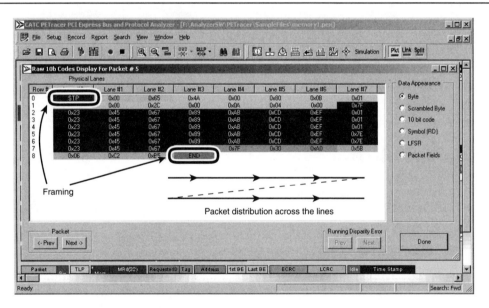

Note:

1. TLP data could also be framed with an EDB (K30.7) as an end symbol to indicate a nullified TLP.
2. DLLP packet framing symbols are SDP (K28.2) and END (K29.7).

Figure 25.17 TLP/DLLP Framing and Line Application

B. Ordered sets

Ordered set data differs from TLPs and DLLPs in that the data is applied to all lanes simultaneously starting with a COM (K28.5) symbol as shown in Figure 25.18.

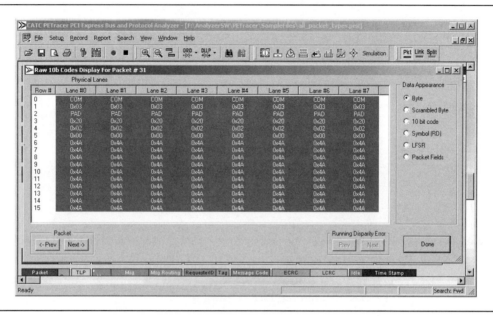

Figure 25.18 Ordered Set Framing and Line Application

2. Scrambling

Data scrambling is performed on a per-lane basis using the X16 +X5 +X4 +X3 +1 polynomial for TLP and DLLP data. No data scrambling is performed on ordered sets.

The example in Figure 25.19 shows how the TLP data, considered in example 1(a), would look with scrambling applied.

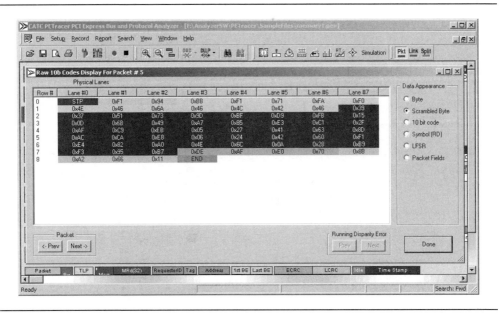

Figure 25.19 Scrambling

Figure 25.20 shows how the Linear Feedback Shift Register (LFSR) is advanced every symbol time.

Chapter 25: Real-Time Analysis ■ **991**

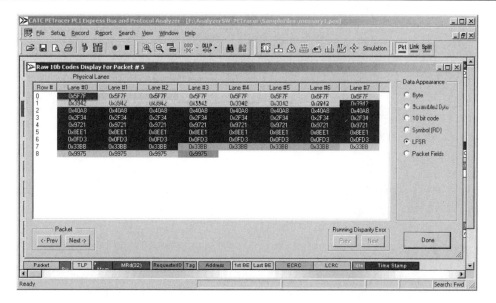

Note:

1. No scrambling is performed for ordered sets.
2. LFSR advancing would not take place for a SKP (K28.0) symbol.

Figure 25.20 Advancing LFSR

3. 8/10-bit encoding

Figure 25.21 shows how each scrambled byte is encoded using 8b/10b as the encoding scheme.

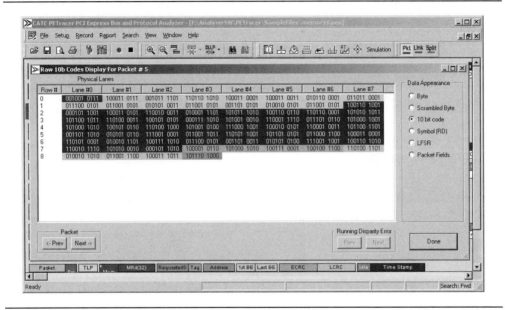

Figure 25.21   10-Bit Codes

Configuration Space

Each logical device on the PCI Express bus utilizes configuration space compatible with PCI 2.3 for each of its logical functions. The first 256 bytes of the PCI Express 4096 byte configuration space are in the region compatible with PCI 2.3.

Access to the device configuration space is facilitated by means of Configuration Requests, where the Register field is the address in the configuration space, as shown in Figure 25.22.

Figure 25.22   Configuration Space Addressing

Figure 25.23 illustrates an example of the Configuration Request TLP header layout.

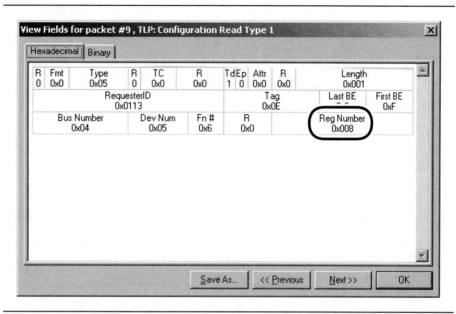

Figure 25.23 Configuration TLP Header Layout

The data of Configuration Write requests and the data of Completions for Configuration Read requests relates to the configuration space register with the address specified in the request, as shown in Figure 25.24.

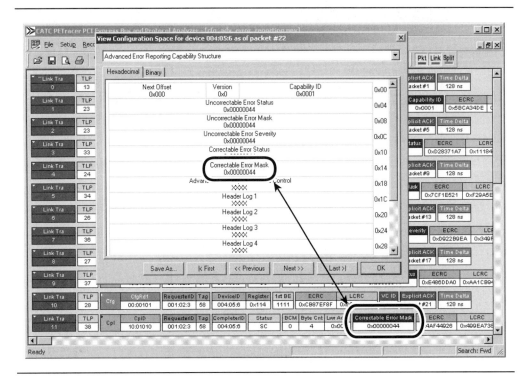

Figure 25.24 Configuration Request Data

Type 0 and Type 1 Configuration Requests are always sent from the Root Complex.

Type 0 Configuration Requests are handled by endpoint devices, root ports, switches, and bridges.

Type 1 Configuration Requests are forwarded by root ports, switches, and bridges to downstream ports, which modify the request type if necessary. Type 1 Configuration Requests are unsupported by endpoint devices. Figure 25.25 is an example of Configuration Request routing.

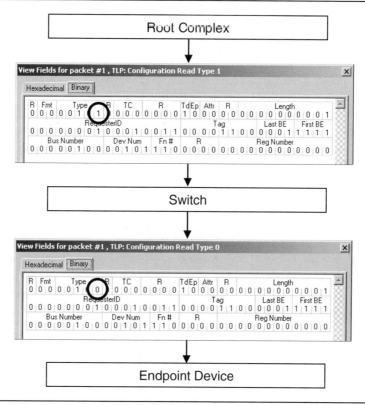

Figure 25.25 Configuration Request Routing

The first 64 bytes of the PCI Express Configuration Space represent the Configuration Space Header. Endpoint devices have a Type 0 Configuration Space Header while switches have a Type 1 Configuration Space Header.

The first 256 bytes of the PCI Express Configuration Space contain a linked list of Capability Structures compatible with PCI 2.3. The address of the first Capability Structure is located in the Configuration Space Header, as shown in Figure 25.26.

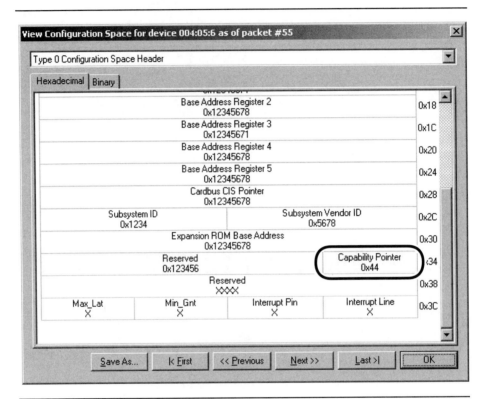

Figure 25.26 PCI Capability Pointer

As illustrated in Figure 25.27, each Capability Structure contains its own ID (shown as 1), which is assigned by the PCI-SIG, and pointer (shown as 2) to the next Capability Structure.

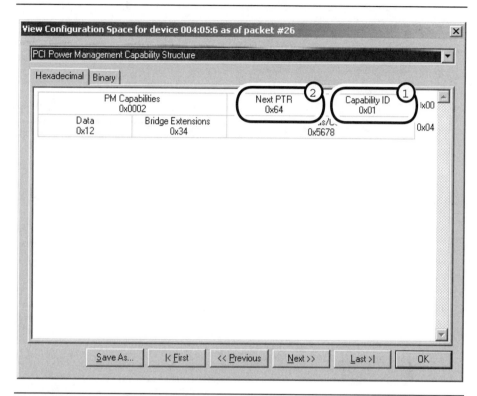

Figure 25.27 PCI Capability Structure

The last Capability Structure in the linked list is identified by a zero value in Next PTR field.

The following Capability Structures are required for all PCI Express devices:

- PCI Power Management Capability Structure (ID 0x01)
- MSI Capability Structure (ID 0x05)
- PCI Express Capability Structure (ID 0x10)

Bytes 256 through 4095 of the PCI Express Configuration Space contain a linked list of PCI Express Extended Capability Structures. The address of the first Extended Capability Structure is byte 256. Each Extended Capability Structure starts with a PCI Express Enhanced Capability

pability Structure starts with a PCI Express Enhanced Capability Header, which contains the ID of Extended Capability Structure (assigned for each type of Capability Structure by the PCI-SIG), shown as 1 in Figure 25.28, and an offset to the next Extended Capability Structure, shown as 2.

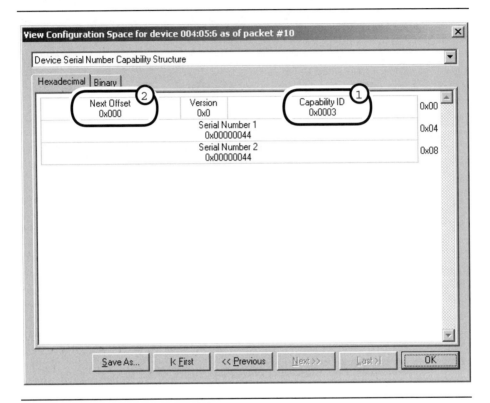

Figure 25.28 PCI Express Capability Structure

The last Capability Structure in the linked list is identified by a zero value in the Next Offset field.

PCI Express devices may optionally support the following Extended Capability Structures:

- Advanced Error Reporting Capability (ID 0x0001)
- Virtual Channel Capability (ID 0x0002)
- Device Serial Number Capability (ID 0x0003)
- Power Budgeting Capability (ID 0x0004)

Traffic Class and Virtual Channels

The Virtual Channel (VC) mechanism is used to support multiple independent traffic flows across a single physical link. Each port, switch, or device can support from one to eight Virtual Channels with each Virtual Channel containing its own independent fabric resources (such as queues and buffers).

Each TLP Header contains a TC (Traffic Class) field, as shown in Figure 25.29, which is mapped to the Virtual Channel on the PCI Express link that this TLP is to use.

Figure 25.29 TC Field of TLP Header

The Traffic Class of the TLP is invariant when the TLP is transferred across PCI Express fabric.

The Virtual Channel is determined for each link as this TLP is about to be transferred.

The system software defines the mapping between the Traffic Class and the Virtual Channels. This mapping is stored in the device configuration space within Virtual Channel Capability and accessed by the Configuration Requests, as illustrated in Figure 25.30.

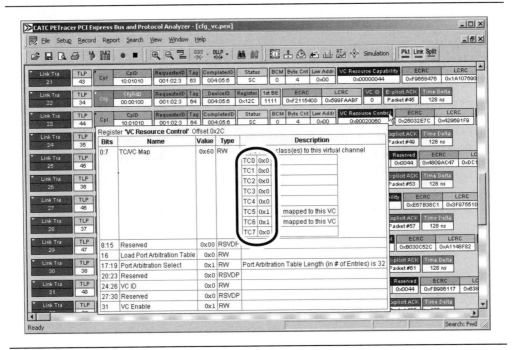

Figure 25.30 TC/VC Mapping

Note: Mapping of TC0 to VC0 is fixed and by default is supported by all PCI Express devices.

Through the use of DLLP Flow Control, each Virtual Channel is able to maintain independent flow control on the PCI Express link. In the example shown in Figure 25.31, you can see that flow control credits are maintained separately for headers and data of posted requests, non-posted requests, and completions.

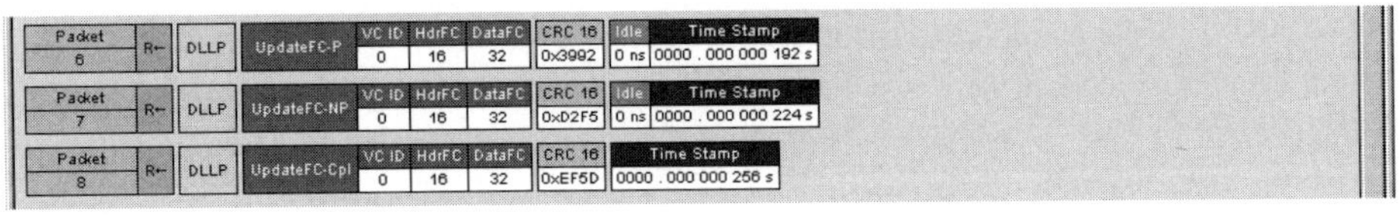

Figure 25.31 Flow Control DLLP

Appendix A

Capability IDs

Table A.1 Capability IDs for Conventional PCI Configuration Space

ID	Capability
00h	Reserved.
01h	PCI Power Management Interface: This capability structure provides a standard interface to control power management features in a PCI device. It is fully documented in the *PCI Power Management Interface Specification,* and in the *PCI Express Base Specification,* available from the PCI-SIG.
02h	AGP: This capability structure identifies a controller that is capable of using Accelerated Graphics Port features. Full documentation can be found in the *Accelerated Graphics Port Interface Specification.* This is available at `www.agpforum.org`.
03h	VPD: This capability structure identifies a device that supports Vital Product Data. It is fully documented in the *PCI Local Bus Specification, Revision 2.3,* available from the PCI-SIG.
04h	Slot Identification: This capability structure identifies a bridge that provides external expansion capabilities. Full documentation of this feature can be found in the *PCI to PCI Bridge Architecture Specification,* available from the PCI-SIG.
05h	Message Signaled Interrupts: This capability structure identifies a PCI function that can do message signaled interrupt delivery. It is fully documented in the *PCI Local Bus Specification, Revision 2.3,* available from the PCI-SIG.

Table A.1 Capability IDs for Conventional PCI Configuration Space *(continued)*

ID	Capability
06h	CompactPCI Hot Swap: This capability structure provides a standard interface to control and sense status within a device that supports Hot Swap insertion and extraction in a CompactPCI system. This capability is documented in the *CompactPCI Hot Swap Specification PICMG 2.1, R1.0* available at `www.picmg.org`.
07h	PCI-X: Refer to the *PCI-X Addendum to the PCI Local Bus Specification* for details, available from the PCI-SIG.
08h	Reserved for AMD.
09h	Vendor-Specific: This ID allows device vendors to use the capability mechanism for vendor-specific information. The layout of the information is vendor-specific, except that the byte immediately following the Next pointer in the capability structure is defined to be a Length field. This Length field provides the number of bytes in the capability structure (including the ID and Next pointer bytes).
0Ah	Debug port.
0Bh	CompactPCI central resource control: This capability is defined in the *PICMG 2.13 Specification,* available at `www.picmg.com`.
0Ch	PCI Hot-Plug: This ID indicates that the associated device conforms to the Standard Hot-Plug Controller model.
0Dh	Bridge subsystem vendor ID.
0Eh	AGP 8x.
0Fh	Secure Device.
10h	PCI Express Capability: This capability identifies a PCI Express device and indicates support for PCI Express features. The PCI Express Capability structure is required for PCI Express devices. It is fully documented in the *PCI Express Base Specification,* available from the PCI-SIG.
11h (1)	MSI-X.
12h-FFh	Reserved.

Table Note:

1. As this book goes to press, Capability IDs 0Dh, 0Eh, 0Fh, and 11h are reserved. However, the PCI-SIG expects to assign those IDs the capabilities listed in this table.

Table A.2 Capability IDs for Extended PCI Express Configuration Space

ID	Capability
0000h	Reserved.
0001h	Advanced Error Reporting: This optional extended capability may be implemented by PCI Express devices supporting advanced error control and reporting. It is fully documented in the *PCI Express Base Specification*, available from the PCI-SIG.
0002h	Virtual Channel Enhancement: This capability structure is required for PCI Express functions that use a traffic class other than TC0 or a virtual channel other than VC0. It is fully documented in the *PCI Express Base Specification*, available from the PCI-SIG.
0003h	Device Serial Number: This optional extended capability uniquely identifies each device with its own 64-bit serial number. This capability is fully documented in the *PCI Express Base Specification*, available from the PCI-SIG.
0004h	Power budgeting: This capability allows the system to properly allocate power to devices that are added to the system at runtime. It is fully documented in the *PCI Express Base Specification*, available from the PCI-SIG.
0005h–FFFFh	Reserved.

Glossary

Abbreviations

"0" Logically low voltage on signal line.

"1" Logically high voltage on signal line.

0b Logical low as a binary bit.

1b Logical high as a binary bit.

h At the end of a series of numerals 0 through 9 and /or a–f, this signifies hexadecimal notation.

d At the end of a series of numerals 0 through 9, this signifies decimal notation.

Gb/sec Gigabits per second.

GB/sec Gigabytes per second.

KHz Kilohertz.

MB Megabyte(s).

MHz Megahertz

μsec. Microsecond(s).

μF. Microfarad(s).

nsec Nanosecond(s).

nF Nanofarad(s).

psec Picosecond(s).

pF Picofarad(s).

Terms

A

add-in card Collectively refers to either the non-mobile type or mobile type cards that can be inserted and removed from a slot on a PCI Express platform. The non-mobile type is used in a non-mobile PCI Express platform. The mobile type is used in mobile PCI Express platforms and is also called PCI Express Mini Card.

address space Entity that contains addressable DWORDS. There are four types of address space in PCI Express:

configuration 4096 bytes per function (configuration address block).

I/O 32 bits in size.

memory 32 or 64 bits in size.

message Not a traditional address space. Addressing is defined by routing encoding between PCI Express devices. *See* routing.

advanced peer-to-peer communication *See* communication.

advanced peer-to-peer link *See* link.

B

bridge *See* PCI Express device.

bus The generic name for the overall interconnection in a PCI or PCI-X platform. *See* bus segment and link.

bus segment Part of the bus definition for interconnects in the PCI and PCI-X portion of the platform.

HOST bus segment Interconnects platform CPUs, cache, and Root Complex.

Virtual PCI bus segment Implemented for software and configuration compatibility with PCI, within the Root Complex, bridges, and switches. From the software and configuration viewpoint virtual PCI bus segments are "actual" PCI bus segments. From the implementation viewpoint, they may or may not be actual bus segments.

C

configuration address space *See* address space.

configuration register block The 1024 (for PCI and PCI-X) and 4096 bytes for PCI Express configuration address space assigned per function.

Root Complex Register Block (RCRB) Register Block defined for the Root Complex and located in memory or I/O address space. It is the configuration register block that is associated with the virtual HOST/PCI Bridge.

CRC Cyclic redundancy check.

CRC CRC version used to provide error detection over the link management packet of a version of a link packet.

ECRC CRC version used to provide error detection over the Header and Data fields of a transaction packet.

LCRC CRC version used to provide error detection over the SEQ# and transaction packet of a version of a link packet.

communication

advanced peer-to-peer communication Message advanced switching requester transactions used for PCI Express devices to communicate between two PCI Express fabrics.

in-band communication Message baseline and message vendor-defined requester transactions used for PCI Express devices to communicate within a PCI Express fabric.

completer *See* participant.

completer destination *See* participant.

completer source *See* participant.

cross-link *See* link.

D

D symbol *See* symbol.

Data field Field within the transaction packet that contains data.

Data Link Layer *See* layer.

decoding *See* encoding.

device Historically, a PCI-compliant component or add-in card. In PCI Express, a physical or virtual component containing at least one and up to eight functions (each function contains a configuration register block). At least one device and as many as 32 devices may reside on each virtual PCI bus segment within the Root Complex or a switch.

Digest field Field within the transaction packet that contains ECRC. *See also* CRC.

DLLP Data Link Layer Packet. *See also* packet

DLCMSM Data Link Control and Management State Machine implemented in the Data Link Layer of PCI Express devices.

DL_Active *See* link activity state.

DL_Down *See* Link_DOWN.

DL_Inactive *See* link activity state.

DL_Init *See* link activity state.

DL_Up *See* Link_UP.

downstream PCI Express devices attached below the Root Complex on the side opposite the HOST bus segment. Downstream was defined as *south* in earlier PCI specifications.

DWORD A naturally-aligned four-byte block of memory, I/O, or configuration address spaces. The least significant two bits of a DWORD address are 00b.

E

ECRC *See* CRC.

encoding There is 8/10b encoding and an associated 10/8b decoding implemented in the Physical Layer. The purpose of the 8/10b encoding is to integrate a reference clock within the Physical Packet with the conversion of 8 bits to 10 bits. The decoding does the reverse.

endpoint *See* PCI Express device.

F

FRAMING *See* packet.

function Entity within a device with a configuration register block that is accessed with PCI Express configuration requester transactions. One to eight functions can be contained in a single device. *See* PCI device and PCI Express device.

H

HDW Header Word used in the Header field of the transaction packet.

Header field Field within the link management packet that contains HDWs.

hierarchy The collection of all PCI Express devices connected to a specific Root Complex and the Root Complex itself.

hierarchy domain The collection of all PCI Express devices downstream of a specific Root Complex downstream port.

I

I/O address space *See* address space.

ID Routing *See* routing.

Implied Routing *See* routing.

In-band communication *See* communication.

isochronous transfer A type of data transfer that relies on a known latency and bandwidth to ensure proper operation.

K

K symbols *See* symbol.

L

lane A set of two differentially driven pairs of signal lines. Within the set one pair drives an 8/10-bit encoded bit stream one direction while the other pair of the set drives the 8/10-bit bit stream the other direction. There are multiple lanes within each link.

layer The divisions in the PCI Express protocol to transfer transactions over links through the use of packets. The layers create and process the different types of packets.

Data Link Layer Executes and manages the transfer on the link of the link packets containing transactions packets. The Data Link Layer also coordinates link states (active, inactive, and so on).

Physical Layer Transfers Physical Packets that contain link packets. The Physical Layer initializes the link and encodes a clock.

Transaction Layer Processes Transaction Layer Packets and supports transaction ordering.

LCRC *See* CRC.

link The interconnect between PCI Express devices associated with the same Root Complex. A link contains from 1 to 32 lanes.

advanced peer-to-peer link A unique interconnect between independent PCI Express fabrics (also called a *cross-link*).

link activity state A logic state of the Data Link Control and Management State Machine (DLCMSM) in the Data Link Layer of a PCI Express device. The link activity states are DL_Inactive, DL_Init, and DL_Active. Link activity states are used as part of Flow Control Initialization to establish available buffer space to receive Transaction Layer Packets.

Link_DOWN Status indicator of the DLCMSM in the Data Link Layer to the Transaction Link Layer.

link management packet *See* packet.

link packet *See* packet.

link state A logic state of the Link Training and Status State Machine (LTSSM) in the Physical Link Layer or logic state of the link.

Link Training The protocol that includes the Detect, Polling, and Configuration link states to establish a configured link.

Link_UP Status indicator of the DLCMSM in the Data Link Layer to the Transaction Link Layer.

legacy function Element in PCI Express derived from PCI or PCI-X. This is not PC legacy as previously defined in other books and specifications. In this book PCI legacy is defined as PC-compatible functions (keyboard, mouse, interrupt controller, and so on).

legacy PCI Express Element of PCI Express using the configuration and software models of PCI for compatibility.

LLTP *See* packet.

LTSSM Link Training and Status State Machine: Logic implemented in the Physical Layer of PCI Express devices.

M

memory address space *See* address space.

message address space *See* address space.

message advance switching transaction A type of requester transaction used to support communication between PCI Express fabrics over advance peer-to-peer links.

message baseline transaction A type of requester transaction used to support interrupt, hot plug /removal, error, power management, slot power, and lock.

message vendor-defined transaction A type of requester transaction defined by the vendor of the PCI Express device.

N

native PCI Express *See* legacy PCI Express.

north *See* upstream.

O

Ordered Set A series of K and D symbols transmitted across the link to implement transitions between states of the Link Training and Status State Machines (LTSSM) and the link states of the associated link. One type (Skip) is also used to compensate for the differences between reference clocks at the two ends of the link.

P

packet General term for one of the three possible packets transferred in the Root Complex and PCI Express fabric.

link packet One of two types: One type contains Sequence number, transaction packet, and LCRC field, also known as LLTP (Link Layer Transaction Packet. The other type contains the link management and CRC, also known as DLLP (Data Link Layer packet).

physical packet contains FRAMING K symbols and link packets

transaction packet Also known as TLP (Transaction Layer Packet) contains Header field, Data field, and Digest field.

participant A PCI Express device that transmits or receives packets containing a specific transaction. Participants are the source or destination of the transaction. PCI Express devices that are switches and that are porting the packet from one port to another port are not participants. The requester transaction and completer transaction are defined as independent transactions even though the completer transaction is associated (in response) with the requester transaction.

completer Name used by the PCI Express specification for the PCI Express device executing a completer transaction.

completer destination Receives completer transaction packets. The completer destination is also the PCI Express requester source of the associated requester transaction.

completer source Transmits requester transaction packets. The completer source is also the PCI Express requester destination of the associated requester transaction.

requester Name used by the PCI Express specification for the PCI Express device executing a requester transaction.

requester destination Receives requester transaction packets. The requester destination subsequently becomes the PCI Express completer source if the completer transaction is defined for the requester transaction.

requester source Transmits requester transaction packets. The requester source subsequently becomes the PCI Express completer destination if completer transaction is defined for the requester transaction.

PAW Packet Word in the Packet field of the Data Link Layer Packet used for link management information

Paw field Field within the link management packet that contains PAW.

PC-compatible function *See* legacy function.

PCI Peripheral Component Interconnect. PCI refers to the original PCI up through Revision 2.3 (local bus specification), exclusive of any PCI-X enhancements.

PCI bus master PCI resource that becomes owner of the associated bus segment.

PCI device The PCI devices are Host/PCI Bridge, PCI bus master, target, and PCI/PCI Bridge. The primary use of PCI devices in PCI Express are as virtual PCI-compatible elements with a 256 bytes configuration register block. Also, any resource downstream of a PCI Express/PCI Bridge.

PCI resource Any component on a PCI bus segment.

PCI-X Enhancement to PCI done prior to PCI Express. Refers to PCI-X up through Revision 2.0 (addendum bus specification).

PCI-X bus master PCI resource that becomes owner of associated bus segment.

PCI-X device PCI-like device compatible with one of the PCI-X specifications.

PCI-X resource Any component on a PCI-X bus segment.

PCI Express Replacement of PCI and PCI-X bus specification to provide platforms with greater performance. Retention of the configuration and software models for compatibility to PCI software.

PCI Express device Collective name for the following items
    bridge A component that interconnects a PCI Express link with a PCI or PCI-X bus segment or bus segments. PCI Express/PCI Bridges and PCI Express/PCI-X Bridges are collectively referred to as bridges. That is, the PCI Express link is upstream of the bridge with PCI or PCI-X bus segments on the downstream side of the bridge.
    Virtual PCI/PCI Bridge The entity compatible with PCI configuration space in the Root Complex, switches, and bridges.
    Reverse bridge A PCI/PCI Express or PCI-X/PCI Express Bridge. That is, the PCI Express link is downstream of the reverse bridge with a PCI or PCI-X bus segment on the upstream side of the reverse bridge.

endpoint collective name for legacy endpoint and PCI Express endpoint. An endpoint is a component that is downstream of the Root Complex or a switch and contains one device with one to eight functions.

Root Complex The component in a PCI Express hierarchy that connects to the HOST bus segment on the upstream side with one or more PCI Express links on the downstream side. The component also contains the interface to the platform memory.

switch A component that interconnects an upstream link to a downstream link or links.

PCI Express fabric Collection of all links, switches, bridges, and endpoints associated with a specific Root Complex.

PCI Express Mini Card *See* add-in card.

peer-to-peer transaction *See* transaction.

Physical Layer *See* layer.

Physical LinkUp Status indicator of the LTSSM in the Physical Layer to the Data Link Layer.

Physical Packet *See* packet.

platform memory The portion of memory address space upstream of the Root Complex.

powered-off The PCI Express device on the platform or a portion of the hierarchy domain is provided NO power from main power and NO power from Vaux.

powered-on The PCI Express device on the platform or a portion of the hierarchy domain is provided power from main power and optionally may or may not be provided power from Vaux.

Q

Quality of Service The ability to define the latency, effective bandwidth, bandwidth consistency, and total latency.

R

Read Completion Boundary (RCB) A memory address space boundary naturally aligned 64 byte address. Completer transactions can provide data requested by a memory read requester transaction in one completer transaction packet or set of completer transaction packets. For the set of completer transaction packets each packet must return a length of data that does not end between RCBs.

requester *See* participant.

requester destination *See* participant.

requester source *See* participant.

resets Fundamental resets are defined as Cold Reset and Warm Reset. A Cold Reset is related to the PERST# signal line deassertion (hardware reset) and main power is applied after platform powered-off. A Warm Reset is related to the PERST# signal line deassertion (hardware reset) and may not include main power applied after platform powered-off. Warm reset may only be a PERST# signal line deassertion with main power retained without the platform being powered off. Hot reset is defined for each specific link and is one of the link states.

reverse bridge *See* PCI Express device.

RID Routing *See* routing.

Root Complex *See* PCI Express device.

Root Complex Register Block (RCRB) *See* configuration register block.

routing A method to direct to the transaction's destination without the use of the typical address bits.

ID (Identification) Routing Used by configuration requester transactions, completer transactions and message vendor-defined requester transactions. ID routing implements BUS#, DEVICE#, and FUNCTION# to define which PCI Express device is the destination.

Implied Routing Used by message baseline requester transactions and message vendor-defined requester transaction; uses the r field in the message transaction to define which PCI Express device is the destination.

RID (Route Identifier) Routing Used by message advance switching requester transactions to define the PCI Express resource in another PCI Express device. *See* advanced peer-to-peer link.

S

SEQ# Sequence number. Attached by the link layer to a link packet containing a transaction packet to ensure strict order of packets traversing a link. *See* packet.

sleep The PCI Express device on the platform or a portion of the hierarchy domain is provided NO power from main power but is provided power from Vaux.

slot The entity on the platform that contains a connector into which an add-in card can be inserted.

source *See* participant.

south *See* downstream.

switch *See* PCI Express device.

symbol The 10-bit encoded version of a byte used in physical packets. There are two types: K symbols are special symbols used for FRAMING and Ordered Sets. D symbols are used for the bytes of the LLTP or DLLP contained in the packet and for filler. D symbols are also used for lane filler.

T

target For PCI and PCI-X, the resource that participates with the bus master in transactions, not defined for PCI Express.

TLP *See* packet.

transaction Collectively refers to all memory, I/O, configuration, message transactions; includes both requester transactions and completer transactions. The term also includes peer-to-peer transactions and advance peer-to-peer transactions

    memory transaction Collectively refers to memory requester and associated completer transactions.

    I/O transaction Collectively refers to I/O requester and associated completer transactions.

    configuration transaction Collectively refers to configuration requester and associated completer transactions.

    message transaction Collectively refers to message baseline, message vendor-defined, and message advance switching requester transactions.

    peer-to-peer transaction Collectively any transaction flowing upstream and then downstream (inflection in a switch) without traversing the Root Complex.

Transaction Layer *See* layer.

transaction packet *See* packet.

U

upstream PCI Express devices attached above another PCI Express device in the direction of the Root Complex; defined as *north* in earlier PCI specifications.

V

virtual PCI bus segment *See* bus segment.

virtual PCI/PCI Bridge *See* PCI Express device.

Index

A

Ack and Nak DLLPs, 473, 600, 607, 649, 753, 756, 759, 986
Ack and Nak DLLPs packets, 356
Ack and Nak packets, 72, 73
active-state PM protocol, 774
active-state power management, 85
active-state power management (PM) protocol, 648, 774, 759, 768
add-in card present, 776, 781
add-in cards, 86, 87, 93, 638, 657, 775, 776, 797, 798, 803, 841
    defined, 1006
address elements, 218
address mapping, 540, 576
address protocols, 121, 150, 161
address spaces, 83, 1006
addressing, 284, 295

advance error reporting
    capabilities, 998, 1003
    protocol, 176, 202, 485, 486, 500
    protocol registers, 494
advanced peer-to-peer communications, 62, 122, 1007
advanced peer-to-peer links, 35, 47, 79, 418, 419, 675, 811, 1010
advanced peer-to-peer transactions, 808
advanced switching
    about, 807
    protocol, 807
aliasing, 150
application for quality of service, 526, 581
arbitration, 43
    generally, 4, 12, 14, 624
    priority schemes, 624
    TLPs, 607
arbitration tables, 621

ASIC circuitry, 505
assertion protocol, 817
assertions, 821
attention buttons, 776, 779
attention indicators, 776, 779
attribute ordering, 506

B

backward compatibility, 833
BAD DLLP errors, 471
BAD LLTP errors, 473
bandwidth
    actual, 625
    link, 628
    maximizing, 502
beacons, 645, 655, 661, 690
BIOS software interface, 899
bit and symbol lock, 639
bit lock, 407
bit streams, 359
board sizes for non-mobile add-in cards, 850
bridge architecture
    examples, 890
    generally, 891
bridge connection rules, 891
bridges, 42, 44, 56, 93, 101, 217, 519, 629, 642, 659, 866, 1013
    and power fence protocol, 752
    architectural model, *135*, *166*
    architecture, 119
    specifications, 112
    TLP porting, 62
    types, 112
buffer management, 526, 553
buffer overflow, 442
buffer space, 320, 524, 525, 553
buffer space and flow control, 340
buffers, 54, 72, 74, 78, 91, 321, 325, 331, 501, 503, 504, 505, 507, *511*, 553, 555, 608, 617, 861
buffers, infinite, 572
bus enumeration, 869, 883
bus segments, 14, 1006
bus transactions, 119, 501
buses, 868
    defined, 1006
byte streams, 359, 381

C

cable modules, 853
cache coherency, 193, 197, 201, 209, 212, 218
cacheable memory, 124
cachelines, 517
capabilities
    defined, 937
    extended, 938
    required, 938, 946
capabilities list
    locating, 941
    scanning, 943
capabilities pointer register, 908, 917, 931
capability
    defined, 1002
capability architecture, 937
capability functions of PCI Express, 937
capability ID registers, 942
capability IDs, 1001, 1002, 944
capability linked lists, 945
capability structure, 943, 946
    PCI Express, 975
capability structures, 996
capability-dependent regions, 945
capability-independent regions, 945

CATC, 977
chassis, 782
    defined, 776
class codes, 894, 905
cold resets, 662, 663, 664, 671
common header regions, 908
communication between components, 979
compatibility
    software, 17
completer abort, 432
completer abort errors, 428, 430, 431, 436, 437, 459
completer destinations, 57, 60, 69, 425, 512, 1012
completer IDs, 175, 176, 198
completer sources, 57, 59, 60, 61, 69, 425, 512, 524, 1012
completer transaction packets, 182, 198, 200
    examples, 188
completer transactions, 152, 225
completers, 1012
completion timeout errors, 444
completion timeout protocol, 444
completion timeout timers, 444
completion TLPs, 985
compliance
    electrical specifications, 724
    patterns, 729
    testing, 696
    timing specifications, 724
Computer Access Technology Corporation (CATC), 977
Config_Address Register, 875
configuration, 865
configuration access examples, 879
configuration access types, 877

configuration address space registers, 781, 791
configuration address space registers for slot power, 801
configuration address spaces, 79, 80, 82, 93, 114, 121, 133, 813, 865, 875, 893, 1006
configuration addresses, 867, 875, 876
configuration capability architecture, 941
Configuration Header Register Block, 935
configuration link states, 408, 478, 639, 665, 672, 676, 693, 713, 844
configuration mechanism, 875
configuration read and write execution, 200
configuration read requests, 994
configuration read transactions, 197
configuration register blocks, 601, 642, 774, 1007
configuration registers, 882, 935, 894
    about, 893
configuration space, 865, 867, 992, 995
configuration space access rules, 881
configuration space accesses, 980
configuration space addressing, 992
configuration space architecture, 896
configuration space headers, 899, 995
configuration space layout, 897
configuration transaction, 164

configuration transaction packets
  treatment of other fields, 200
configuration transaction
  protocol, 201
configuration transactions, 152,
  218, 1017
configuration write requests, 994
configuration write transaction, 197
configured lanes, 412
control elements, 218
cooling issues, 798
correctable errors
  defined, 421
CRC errors, 471, 472, 478, 601
CRCs, 65, 369, 480, 987, 1007
cross-links, 809. See advanced peer-to-peer links
cut-through, 601
cut-through protocol, 569
cyclic redundancy check, 215

D

"D" states, 648, 656, 744, 741, 758, 794
D symbols, 372, 374, 382, 401, 402, 407, 475, 729
D0 state, 653, 742
D1 state, 742
D2 state, 742
D3 state, 653, 661, 743
D3cold state, 744
D3hot state, 743
data bit rates, 696, 864
data corruption, 442
data elements of transaction
  packet, 220
data field formats, 312
data fields, 172, 173, 215, 218, 365, 1007
data integrity, 987

Data Link Control and Management
  State Machine (DLCMSM), See
  DLCMSMs
data link layer, 986, 1010
data link layer errors
  BAD LLTP (link packet
    containing TLP), 470
data link layer errors (DLLPE),
  See DLLPEs
data link layer errors (DLLPEs)
  BAD DLLP (link management
    packet), 470
  generally, 424, 469, 470
  Retry_Num# Rollover, 470
  Retry_Timer Timeout, 470
data link layer packets (DLLPs),
  See DLLPs
Data Link Layer Packets (DLLPs),
  See DLLPs
data link layer protocol errors
  (DLLPEs), 473
data link layers, 65, 66, 68, 69, 72, 74, 171, 315, 643, 979
data poisoning, 449, 462, 469
data poisoning errors, 442
data prefetching, 125
data rates
  defined, 378
data scrambling, 862, 989
data widths, 116
deadlocks, 501, 503, 505, 618
deassertions, 821
debugging, 979
decoding protocols, 381
delayed transaction protocol, 14, 17, 24
deparsing, 379, 384
detect link states, 408, 420, 477, 639, 653, 655, 658, 662, 665, 672, 696, 844

detected parity bits, 423
detecting devices, 881
device
    defined, 1008
device control registers, 984
device failure, 442
Device IDs, 894, 900
device serial number capabilities, 998
device under test (DUT), 977
devices, 51, 54, 55, 58
    interconnections for, 41, 47
digest field formats, 313
digest fields, 115, 172, 173, 215, 218, 224, 365, 1008
disable bits, 676
Disabled link state, 672, 722
disabling INTx
    boundary conditions, 821
disparity, 407
DL_Active, 1008
DL_Active state, 321, 324
DL_Down, 1008
DL_Inactive, 1008
DL_Inactive state, 320, 322
DL_Init, 1008
DL_Init state, 320, 323
DL_Up, 1008
DLCMSMs, 319, 322, 640, 663, 664, 672, 1008
DLLP conversion, 369
DLLPEs, 331, 473, 587
DLLPs, 66, 68, 73, 215, 315, 317, 342, 359, 365, 369, 381, 553, 582, 677, 741, 858, 860, 987, 1008
    PCI-PM Software protocol, 753

DLLPs and LLTPs
    conversion to serial protocol, 362
downstream
    defined, 1008
DUT, 979
DWORDs, 71, 621, 805, 867, 942, 945, 1008

E

ECRC, 65, 215, 484, 891, 1007
ECRC errors, 432, 434, 440, 479
ECRC, CRC, and LCRC errors
    generally, 424
edge finger area of non-mobile add-in cards, 842
EHA, 834, 835, 836, 839
electrical Idle operating system (OS) protocol, 399
electrical IDLE operating system (OS)
    generally, 400
electrical requirements, 857
electrical signaling, 857
electrical specifications, 724, 841
electrical specifications of
    PCI Express, 88
electromagnetic interference (EMI), 862
electromechanical specifications, 841
EMI, 862
encoding
    defined, 1008
encoding protocols, 381
endpoints, 42, 54, 59, 61, 98, 132, 502, 505, 519, 543, 642, 659, 866, 1008, 1014
    and power fence protocol, 750
ERR_COR, 424, 486
ERR_FATAL, 424, 486

ERR_NONFATAL, 424, 486

error checks, 74

error forwarding, 444, 449, 462, 469

error logging, 484

error messages, 205

error pollution, 485

error protocol and PCI, 422

error related messages, 289, 487

error reporting, 424, 484

errors, *See* physical layer errors. *See* physical layer errors. *See* data link layer errors (DLLPEs). *See* transaction layer errors

exclusive hardware access (EHA), 833

exclusive software access (ESA), 833

execution order, 218

execution protocol, 193

F

fast training sequence (FTS) protocol, 397

fatal errors
  defined, 421

FCCs, 555, 559

FCPEs, 446

FIFO (first-in and first-out) transmissions, 503

FIFO buffers, 504, 508, 523, 555, 608, 616

FIFO ordering, 510

first error pointers, 500

flow control and buffer space, 340

flow control credits (FCC), 445, 553

flow control DLLPs, 317, 323, 328, 350, 1000

flow control initialization, 321, 322, 341, 639, 646, 655

flow control initialization protocol example, 339

flow control protocol, 75, 219, 321, 581, 628, 662
  elements, 520, 526
  execution, 559
  generally, 519

flow control protocol errors (FCPEs), 446, 447

FRAMING, 361, 362, 372, 381, 601, 858, 860, 1009

FRAMING K symbols, 342, 355, 387, 404, 471

frequency support, 116

functions, 867, 1009

fundamental reset protocol, 664

fundamental resets, 662, 663, 664, 671, 675

G

gate circuitry, 566, 578

granularity and transaction ordering, 516

H

HDWs, 173, 174, 193, 197, 202, 222, 1009

header field formats, 224
  configuration read transactions, 273
  configuration write transactions, 282
  I/O read transactions, 254
  I/O write transactions, 263
  memory read transactions, 238
  memory write transactions, 245
  message advanced switching transactions, 305
  message baseline transactions, 293

message vendor-defined transactions, 299
header fields, 172, 173, 175, 193, 197, 215, 218, 224, 365, 369, 1009
   about, 175
header log registers, 500
header type, 894
header words, *See* HDWs
hierarchies, 11, 44, 869, 870, 871, 1009
hierarchy domains, 44, 645, 655, 658, 749, 836, 1009
higher layer direction, 675
HOST bus segment CPUs, 813
HOST bus segments, 42, 48, 80, 95, 97, 217, 484, 813, 833, 834, 1006
HOST/PCI bridges, 95
hot plug (HP) capabilities, 86, 775, 937
hot plug (HP) protocol, 115, 158, 638, 646, 746, 775, 796
hot plug (HP) software, 731, 797
hot plug messages, 205
hot plug related configuration registers, 790
hot plug related messages, 290
hot plugs, 1002
Hot Reset link state, 672, 676, 724
hot resets, 662, 666, 668, 671
hot swap capabilities, 937

I

I/O address space, 79, 113, 121, 126, 865
I/O read and write execution, 196
I/O read transactions, 193
I/O transaction execution protocol, 197
I/O transaction packets, 197
   treatment of other fields, 197
I/O transaction protocol, 193
I/O transactions, 151, 218, 809, 1017
I/O write transactions, 193
ID (identification) routing, 1016
ID routing, 121, 202, 211, 218
IDLE D symbols, 387, 404
implementations
   variant, 111
implied routing, 84, 123, 154, 156, 202, 211, 218, 284, 295, 1016
in-band communications, 62, 122, 809, 1007
integrity elements, 220
interposer cards, 979
interrupt line register, 908, 917, 933
interrupt pin register, 908, 918, 933
interrupt protocol, 813
interrupt related messages, 204, 289
interrupt signal lines, 813
interrupts, 84, 115, 485, 776, 793, 832, 948
INTx# signal lines, 813, 814
isochronous transfers, 523, 581, 636, 1009

K

K symbols, 361, 372, 374, 381, 386, 396, 398, 401, 402, 407, 475, 729

L

L0 link state, 639, 644, 646, 662, 671, 678, 681, 682, 684, 734, 735, 762, 786
L1 link state, 648, 671, 678, 688, 736, 763

L1 link state transitions, 769

L2 link state, 638, 644, 645, 646, 649, 654, 658, 671, 692, 739, 786

L2/3 link state, 648, 651

L2/3 Ready link state, 672, 679, 689, 738

L3 link state, 638, 644, 649, 671, 692, 740, 786

lane parsing
    example, 383
    generally, 383

lane polarity, 407

lane polarity inversion, 639

lane reversal, 419, 420

lanes, 1009

layers, 1010

LCRC, 66, 72, 368, 369, 482, 582, 1007

LCRC errors, 471, 478

LCRC mapping, 481

legacy (LI) protocol
    implementation, 821

legacy bridges, 817

legacy endpoints, 817, 834, 839

legacy function, 1010

legacy interrupt (LI) protocol, 822, 817, 823

legacy PCI Express, 1011

legacy switches, 817

LF protocol, 833, 834, 835, 839

Link _DOWN status, 323

link activity states, 324, 1010

link activity states and link states differentiated, 674

link capabilities registers, 601, 774

link configuration protocol, 420
    DSD iterative process, 416
    generally, 413
    USD iterative process, 414

link control registers, 676

link flow control, 78

link layer transaction packets, *See* LLTPs

link management, 861

link management packet, 1010

link packets, 1012

link retraining, 409, 639

link state transitions, 675

link states, 318, 359, 408, 673, 734
    defined, 1010

link topology, 411

link training, 409, 639, 644, 645, 646, 655, 662, 665, 692, 1010

Link Training and Status State Machine (LTSSM), *See* LTSSMs

link widths
    mobile add-in cards, 851
    non-mobile add-in cards, 842

Link_DOWN, 319, 640, 1010

Link_DOWN status, 322

Link_UP, 319, 640, 1010

Link_UP status, 324

linked lists, 945

links, 359, 641, 642, 1010

livelock, 618

livelocks, 501, 503, 505, 618

LLTP conversion, 365

LLTP format, 358

LLTP tracking, 526, 581, 604

LLTPs, 64, 69, 72, 171, 215, 315, 317, 318, 323, 342, 355, 359, 365, 381, 503, 526, 555, 677, 758, 858, 860

lock functions, 58, 85

lock protocol, 839

lock protocol messages, 208

lock protocol transaction related messages, 293

lock TLPs, 835

locked devices, 836
loopback link states, 672, 676, 728
loopback protocol, 724, 725
Lower Power Sleep State, 739
Lower Power Standby State, 737
lower power states, 794
lowered power, 646
LTSSMs, 319, 322, 361, 408, 638, 662, 664, 671, 672, 686, 692, 844, 1011
LTSSMs and transmitters, 685

M

main power, 650, 651, 653, 654, 657, 688, 690, 733, 748
malformed errors, 427, 428, 429, 431, 434, 438, 439, 441, 442, 443, 459, 449, 450, 451, 452, 463, 469
malformed packets, 546
mapping
    asymmetrical, 536
    peer-to-peer, 539
    symmetrical, 535
master data parity error bits, 423
May Pass entries, 513, 516
mechanical specifications, 841
mechanical specifications of PCI Express, 86
memory address space, 79, 113, 121, 126, 865, 1006
memory read transaction execution, 188
memory read transaction packets, 179
    treatment of other fields, 192
memory read transactions, 125, 175, 176, 224
memory write transaction execution, 192
memory transaction packets, 193
memory transactions, 151, 218, 809, 1017
memory write requester transaction packet (MWr), 839
memory write transaction packets
    treatment of other fields, 192
memory write transactions, 121, 175, 176
message address space, 115, 140, 1006
message advanced switching, 35, 79, 123, 141, 142, 160
message advanced switching transaction protocol, 213
message advanced switching transactions, 213, 303, 811, 1011
message baseline requester transactions, 758, 776, 781, 786, 811
message baseline transaction execution, 208
message baseline transaction packets, 202, 203
    treatment of other fields, 208
message baseline transactions, 141, 149, 154, 156, 284, 295, 1011
message elements, 221
message error requester transaction packet, 382
Message Error Requester Transaction Requester and Functions, 487
message error requester transactions, 485
    severity, 487
message interrupt requester transaction packets, 829
message interrupt requester transactions, 813
message interrupt requests, 84

message PM_PME completer transactions, 746
message PM_PME requester transactions, 745, 746
Message PM_Turn_Off and PME_TO_Ack requester transactions, 747
message signal interrupt (MSI), 814
message signaled interrupt (MSI), 1001, 937, 948
message signaled interrupt (MSI) protocol, 832
message transaction protocol, 213
message transactions, 84, 218, 1017
 types, 141
message vendor-defined requester transactions, 811
message vendor-defined transaction execution packets, 211
message vendor-defined transaction packets, 202, 211, 293
 treatment of other fields, 212
message vendor-defined transactions, 141, 142, 154, 159, 1011
messaging, 217
mid-bus probes, 979
mini cards, 858
minimal error reporting protocol, 485, 486, 491, 492
mobile add-in cards, 853
MRL (manually operated retention latch), 776, 778
MRL sensors, 776, 778
MSI, 814
MSI (memory signaled interrupt) transactions, 84, 948
MSI capability structures, 997
MSI protocol, 832
MWrs, 839

N

native PCI Express
 defined, 1011
nonfatal errors
 defined, 421
non-mobile add-in cards, 853
north, 1011
nullified TLP packet errors, 471
nullified TLPs, 569, 601

O

Off State, 740
ordered sets, 361, 382, 395, 404, 989, 1011
ordering of packets, 193, 197

P

packet flow, 530
packet flow rates, 528
packet flow, optimizing, 581
packet framing, 980
packet merging, 530
packet ordering, 201, 209, 212
packet sources, 539
packet transmission priorities, 530
packets, 1012
PAD K symbols, 382, 393, 402, 404, 409
parallel bus implementations, 24
parallel to serial conversion of DLLPs and LLTPs, 362
parsing
 examples, 387
parsing protocol, 389, 391
participants, 57, 1012
participants of transactions, 150
PAW, 1013
Paw field, 1013
PC-compatible function

defined, 1013
PCI, 1013
PCI bus architecture, 4
    generally, 18
PCI bus masters, 502, 833, 1013
PCI bus power management
    interface, 753
PCI bus protocol, 813
PCI bus segments, 113, 116, 432,
    441, 842
PCI devices, 217, 1013
PCI Express
    defined, 1013
    generally, 23
    implementation, 29
    improvements to platform
        bandwidth, 23
    key terms, 30
    messages, 34
    specifications, 3
PCI Express device
    defined, 1013
PCI Express fabric, 1014
PCI Express layers, *172, 216, 542*
PCI Express mini cards, 851,
    856, 1014
PCI Express platform, *94*
    generic architecture, 43
    illustrated, *30*
PCI Express Specific performance, 39
PCI Express specifications, 93
PCI Express Virtual Channel
    Capability Structure, *See* PVCCS
PCI Express/PCI bridges, 492, 512,
    824, 827
PCI Express/PCI-X bridges, 492, 512

PCI platforms, 505
    transaction ordering, 501
PCI resources, 1013
PCI software, 26
PCI Special Interest Group, 3
PCI-PM software protocol, 676,
    679, 733, 758
PCI-SIG, 3, 938, 945
PCI-X, 1013
PCI-X bus architecture
    features, 21
    generally, 1
PCI-X bus masters, 502, 833, 1013
PCI-X bus protocol, 813
PCI-X bus segments, 113, 116, 432,
    441, 842
PCI-X devices, 217, 1013
PCI-X modes, 116
PCI-X platform architecture, 22
PCI-X platforms, 505
    transaction ordering, 501
PCI-X resources, 1013
peer root complexes, 882
peer-to-peer packet flow, 529
peer-to-peer port arbitration, 605
peer-to-peer TLPs, 616
peer-to-peer transaction packets, 513
peer-to-peer transactions, 46, 50,
    808, 809, 839, 1017
pending transmissions, 762, 763
phase lock loop (PLL), 863, 864
PHY, 862, 864
physical layer, 359
physical layer errors
    generally, 424, 474
    Receiver Errors, 474
    Training Errors, 474

physical layers, 66, 69, 72, 73, 318, 979, 987, 1010, 1014
physical link layers, 671, 672
Physical LinkUp, 1014
physical packet reception, 476
physical packet transmission, *89*
physical packets, 63, 125, 131, 171, 319, 359, 404, 475, 503, 526, 858, 1012, 1014
    transmission, 382
    transmission and reception, 862
platform memory, 48, 124, 833, 1014
platform sleep, 648, *650*
PLLs, 372, 373, 379, 734
PM_Request_Ack DLLPs, 755
PME Service Timeout protocol, 746
PME software, 651, 731
PMEs, 775
point-to-point interconnections, 24
poisoned TLP errors, 431, 434, 442, 443
polling compliance, 696
polling configuration, 699
polling link states, 408, 477, 639, 665, 672, 693, 700, 844
polling speed, 700
port arbitration, 512, 526, 581, 605, 611
port arbitration protocol, 504, 577, 612, 631
ports
    receiving portions, 764
    slot placement, 800
power budgeting capabilities, 998
power consumption, 731
power controllers (fault), 776, 780
power cycling, 638, 650

power fence protocol, 750
power indicators, 776, 777
power limits, 798, 801, 804
power management, 85, 158, 206, 797, 937
    powered off, 648
power management (PM) protocol, *647*, 731
power management (PM) software, 637, 731
power management capabilities register (PMCR), 733
power management capability structures, 947, 997
power management control/status register (PMCSR), 747
Power Management DLLPs, 352
power management event (PME) software, 638, 797
power management events (PMEs), 637, 731, 775
power management interface, 1001
power management link states, 736
power management protocol, 350, 637
power management related message, 292
Power Removal Pending State, 738
power states
    powered off, 638, 650, 653
    powered on, 651, 653, 654
    sleep, 648, 654
power, main, *See* main power
powered-off, 644, 692, 731, 733, 1014
powered-off preparation, 648, *650*
powered-on, 731, 733, 1014

priority schemes
  arbitration, 624
protocol analysis, 977
PVCCS, 543, 544, 545, 547, 552

Q

quality of service (QoS), 33, 75, 523, 581, 627, 1014
quantities and transfer equations
  example, 564, 568

R

RCBs, 180
RCRBs, 898, 942, 1015
Read Completion Boundary (RCB), 1015. See RCBs
read transactions, 218
real-time analysis
  bench configuration, 978
  methodology, 977
Receiver errors, 382, 395, 404, 407, 475, 476
Receiver Overflow errors, 445, 446, 574
receivers and LTSSMs, 686
receiving ports, 316, 555, 582, 583, 666
Recovery link state, 672, 678, 720
REFCLK, 378, 652, 734, 773, 774, 854, 863
reference clocks, 359, 379, 854, 863
relaxed ordering of attributes, 506, 515
repeater/completer protocol, 219, 225, 245, 255, 264, 274
requester destinations, 57, 425, 512, 1012
requester IDs, 121, 177, 193, 197, 202, 217

requester sources, 57, 425, 512, 524, 1012
requester/completer protocol, 26, 39, 65, 121, 130, 171, 217, 501, 503, 520
requesters, 1012
resets, 1015
restricting access to locked devices, 839
retrain bits, 675
retry buffers, 589, 603
retry protocol, 588
Retry_Num# Rollover errors, 473
Retry_Timer Timeout errors, 473
Reverse bridges, 1013
Revision IDs, 894, 905
RID (route identifier) routing, 202, 1016
root complex, 502, 503, 605, 994, 1014
root complex error command and status configuration block registers, 497
root complex register block (RCRB), 49, 138, 898, 1007
root complexes, 51, 57, 93, 98, 124, 140, 164, 519, 543, 643, 868, 898
round robin arbitration, 611
round robin priority, 578
round robin priority schemes
  programmable weighted round robin, 620
  round robin, 621
  Time-Based weighted round robin (TBWRR), 631
  weighted round robin, 621
routing, 1016
routing protocols, 121, 150, 154
running disparity, 407

S

sbyte streams, 360
sbytes, 859
Scheduling LLTP Transmission, 598
scrambling, 406, 862, 980, 989
scrambling, disabling of, 406
secondary status register, 423
SEQ#, 1016
serial protocols for transferring DLLPs and LLTPs, 362
SERR# Enable bits, 441
settling time, 360
signal lines, 854
Skip OS, 396
sleep, 637, 644, 645, 731, 794, 1016
    defined, 638
Sleep to Powered-on transactions, 745
slot, 1016
    defined, 776
slot capabilities register, 803
slot identification, 1001
slot power (SP) protocol, 797, 798
slot power (SP) software, 731, 799
Slot Power (SP) software, 638, 797
slot power items
    additional, 805
slot power limit messages, 293
slot power limit protocol, 803
slot power messages, 207
slot power protocol, 638
slot power registers, 800
slot status registers, 793
slots, 782
    placement with ports, 800
SMBus interconnections, 850
software compatibility, 865
south
    defined, 1016
split transaction protocol, 36
split transactions, 983
standby power, 805
standby power state, 682
status registers, 423
sticky bits, 641, 644, 654, 667
strong ordering of attributes, 506, 512, 514
strong ordering or attributes, 513
supported requester errors, 466
switches, 42, 53, 58, 93, 99, 132, 507, 510, 519, 543, 641, 1014
    and power fence protocol, 751
    architectural model, *134*, *165*
switching
    about, 807
symbols, 1016
synchronous bus protocol, 8

T

TAG, 217
Tag#, 175
Tag# fields, 168, 193, 197, 202
targets, 1017
TC numbers, *See* traffic class numbers
TC to VC mapping, 576, *1000*
testing components, 724
timing specifications, 724
TLP address routing, *983*
TLP arbitration, 607
TLP fields, 981
TLP header, *981*
TLP header layout, *982*
TLP ID routing, *983*
TLP request/completion, 983

Index ■ **1033**

TLP routing, 982
TLP segments, *981*
TLPs, 58, 64, 65, 69, 115, 150, 171, 172, 215, 217, 221, 315, 322, 359, 503, 526, 541, 555, 741, 758, 987
    downstream flow to one endpoint, *522*
    flow, 530
    for PCI-PM software protocol, 744
    nullified, 569
    upstream flow from two endpoints, *524*
    with switches, *525*
traffic class (TC) fields, 999
traffic class (TC) numbers, 192, 196, 201, 208, 212, 217, 506, 533, 539, 541, 581, 628
    and errors, 441
traffic classes, 76, 526, 528, 553, 628
traffic flow, *980*
training errors, 409, 477, 478
training sequence 1 and 2 OSs, 402, 407, 409, 414, 415, 417, 420
Training Sequence and Fast Training OSs, 361
transaction IDs, 169, 177, 198
transaction integrity, 217
transaction layer errors
    basic, 423, 444
    Completer Abort, 426
    Completion Timeout, 426
    completion timeout errors, 423
    ECRC, 426
    FCPE (Flow Control Protocol Error), 426
    flow control protocol errors (FCPEs), 424
    generally, 423
    Malformed, 426
    Poisoned TLP (Error Forwarding), 426
    Receiver Overflow, 426
    receiver overflow errors, 424
    TLP fields checked for errors, 424
    Unexpected Completer, 426
    Unsupported Requester, 426
transaction layer implementation, 221
transaction layer packets, *See* TLPs
transaction layers, 65, 68, 71, 74, 171, 215, 221, 359, 979, 981, 1010
transaction ordering, 501, 507, *508*, *509*, 523, 524, 605
    example, 513
transaction ordering protocol, 504, 507
transaction packets, 1012
transaction posting, 125, 127
transactions, 62, 171, 1017
transmission order, 381, 575
transmitters and LTSSMs, 685
transmitting and receiving packets example, 69
transmitting and receiving physical packets, *364*
    steps, 362, *363*
transmitting ports, 316, 555, 582, 583, 666
TS ordered sets, 475
TS1 and TS2 OSs, *See* training sequence 1 and 2 OSs
type 0 configuration requests, 994
type 0 header region, 918
type 1 configuration requests, 995
type 1 header region, 908
type supported errors, 427

U

undefined errors, 744

unexpected completer errors, 434, 440, 460, 466

unlock message requester transaction packet, 838

unsupported errors, 428

unsupported requester errors, 429, 430, 431, 436, 437, 440, 442, 443, 448, 459, 461, 463, 465, 468, 742, 743

unsupported requesters, 432, 452

upstream
   defined, 1017

user-defined features, 937

V

Vaux, 638, 645, 651, 652, 653, 654, 657, 661, 662, 688, 689, 733, 748, 805, 851

VC arbitration, 613

VC arbitration protocol, 504, 614

VCs, 582. *See* virtual channels

Vendor IDs, 894, 900

vendor-defined messages, 299

Vendor-Specific DLLPs, 354

virtual channel (VC) arbitration, 512, 526, 540, 578, 581, 605

virtual channel (VC) capabilities, 998

virtual channel (VC) numbers, 556, 581
   and errors, 441

virtual channels (VCs), 76, 78, 526, 528, 553, 582, 628, 980, 999

virtual channels, PCI Express platform model with, *521*

virtual HOST/PCI bridges, 49, 50, 129, 138, 605

virtual interrupt signal lines, 817

Virtual PCI bus segments, 49, 1006

virtual PCI/PCI bridges, 49, 50, 95, 98, 99, 129, 138, 422, 605, 676, 825, 829, 1013

visual channel (VC) arbitration, 613

voltage tolerances, 799

W

wake events, 655, 656

WAKE# signal line, 659, *660*

wakeup events, 655, 746, 792, 793

wakeup protocol, 638, 645, 646, 654, 659, 732

warm resets, 646, 662, 663, 664

wattages, 799

waveforms, 864

WORDs to BYTEs
   converting, *367*

write transactions, 218

www.pcisig.com, 3, 938

"As the pace of technology introduction increases it's difficult to keep up. Intel Press has established an impressive portfolio. The breadth of topics is a reflection of both Intel's diversity as well as our commitment to serve a broad technical community.

I hope you will take advantage of these products to further your technical education."

Patrick Gelsinger
Senior Vice President and Chief Technology Officer
Intel Corporation

Turn the page to learn about titles from Intel Press for system developers

ESSENTIAL BOOKS FOR SYSTEM DEVELOPERS

Break Through Performance Limits with PCI Express

Introduction to PCI Express†
A Hardware and Software Developer's Guide
By Adam Wilen, Justin Schade, and Ron Thornburg
ISBN 0-9702846-9-1

Written by key Intel insiders who have worked to implement Intel's first generation of PCI Express chipsets and who work directly with customers who want to take advantage of PCI Express, this introduction to the new I/O technology explains how PCI Express is designed to increase computer system performance. The book explains in technical detail how designers can use PCI Express technology to overcome the practical performance limits of existing multi-drop, parallel bus technology. The authors draw from years of leading-edge experience to explain how to apply these new capabilities to a broad range of computing and communications platforms.

Introduction to PCI Express explains critical technical considerations that both hardware and software developers must understand to take full advantage of PCI Express technology in next generation systems.

From these key Intel technologists, you learn about:

- Metrics and criteria for developers and product planners to consider in adopting PCI Express

- Applications for desktop, mobile, server, and communications platforms that will benefit significantly from PCI Express technology

- Implications for hardware and software developers of the layered architecture of PCI Express

- Comparison of features of PCI Express, PCI-X, and PCI

❝ *This book helps software and hardware developers get a jump start on their development cycle that can decrease their time to market.* ❞

Ajay Kwatra, Engineer Strategist, Dell Computer Corporation

● **Serial ATA Storage Architecture and Applications**
Designing High-Performance, Low-Cost I/O Solutions
By Knut Grimsrud and Hubbert Smith
ISBN 0-9717861-8-6

Serial ATA, a new hard disk interconnect standard for PCs, laptops, and more is fast becoming a serious contender to Parallel ATA and SCSI. Computer engineers and architects worldwide must answer important questions for their companies: "Why make the change to Serial ATA? What problems does Serial ATA solve for me? How do I transition from parallel ATA to Serial ATA and from SCSI to Serial ATA?" The authors of this essential book, both Intel Serial ATA specialists, have the combined expertise to help you answer these questions. Systems engineers, product architects, and product line managers who want to affect the right decisions for their products undoubtedly will benefit from the straight talk offered by these authors. The book delivers reliable information with sufficient technical depth on issues such as Phy signaling and interface states, protocol encoding, programming model, flow control, performance, compatibility with legacy systems, enclosure management, signal routing, hot-plug, presence detection, activity indication, power management, and cable/connector standards.

"This book provides explanations and insights into the underlying technology to help ease design and implementation."

Rhonda Gass, Vice President, Storage Systems Development, Dell Computer Corporation

● **USB Design by Example**
A Practical Guide to Designing I/O Devices
By John Hyde
ISBN 0-9702846-5-9

John Hyde, a twenty-three year veteran of Intel Corporation and recognized industry expert, goes well beyond all the Universal Serial Bus specification overviews in this unique book, offering the reader the golden opportunity to build unparalleled expertise, knowledge, and skills to design and implement USB I/O Devices quickly and reliably. Through a series of fully documented, real-world examples, the author uses his practical customer training experience to take you step by step through the process of creating specific devices. As a complete reference to USB, this book contains design examples to cover most USB classes and provides insights into high-speed USB 2.0 devices.

"We could implement a USB design with this book alone."

Chris Gadke,
Design Engineer,
Tektronix, Inc.

● Building the Power-Efficient PC
A Developer's Guide to ACPI Power Management
By Jerzy Kolinski, Ram Chary, Andrew Henroid, and Barry Press
ISBN 0-9702846-8-3

An expert author team shows developers and integrators how to address the increasing demand for energy conservation by building power-managed PCs. Learn from key engineers responsible for the development of ACPI Power Management the practical knowledge and design techniques needed to implement this critical technology. The companion CD includes sample code, complete power management documentation, Intel® power management tools, and links to references.

Learn how to build power-efficient PCs from the experts

● InfiniBand† Architecture Development and Deployment
A Strategic Guide to Server I/O Solutions
By William T. Futral
ISBN 0-9702846-6-7

InfiniBand, a contemporary switched fabric I/O architecture for system I/O and inter-process communication, offers new and exciting benefits to architects, designers, and engineers. Intel I/O Architect William Futral was a major contributor to InfiniBand architecture from its inception. Currently, he serves as Co-Chair of the InfiniBand Application Working Group. His comprehensive guide details the InfiniBand architecture, and offers sound, practical expert tips to fully implement and deploy InfiniBand-based products, including deployment strategies, InfiniBand-based applications, and management.

Develop leading edge server I/O solutions

- **IXP1200 Programming**
  The Microengine Coding Guide for the Intel® IXP1200
  Network Processor Family
  By Erik J. Johnson and Aaron Kunze
  ISBN 0-9712887-8-X

As very deep submicron ASIC design gets both more costly and time-consuming, the communications industry seeks alternatives providing rich services with higher capability. The key to increased flexibility and performance is the innovation incorporated in the IXP1200 family of network processors. From engineers who were there at the beginning, you can learn how to program the microengines of Intel's IXP12xx network processors through a series of expanding examples, covering such key topics as receiving, processing, and transmitting packets; synchronizing between hardware threads; debugging; optimizing; and tuning your program for the highest performance.

Increase performance with this hands-on coding guide

- **The Virtual Interface Architecture**
  A Guide to Designing Applications for
  Systems Using VI Architecture
  By Don Cameron and Greg Regnier
  ISBN 0-9712887-0-4

The VI architecture addresses the long-standing problem for systems that need an efficient interface between general-purpose computers and high-speed switched networks. In this book, Intel architects outline the motivation, benefits, and history of the Virtual Interface Architecture. Code examples guide you through the syntax and semantics of the VI Provider Library API. With this reference, hardware and software engineers can apply the VI Architecture to development of scalable, high-performance, and fault-tolerant systems.

Design scalable, high-performance, fault-tolerant systems

Please go to this Web site

www.intel.com/intelpress/bookbundles.htm

for complete information about our popular book bundles. Each bundle is designed to ensure that you read important complementary topics together, while enjoying a total purchase price that is far less than the combined prices of the individual books.